2^5 Problems for STEM Education

2^5 Problems for STEM Education

Valery F. Ochkov
Moscow Power Engineering Institute

CRC Press is an imprint of the
Taylor & Francis Group, an **informa** business
A CHAPMAN & HALL BOOK

CRC Press
Taylor & Francis Group
6000 Broken Sound Parkway NW, Suite 300
Boca Raton, FL 33487-2742

First issued in paperback 2021

ISBN-13: 978-0-367-34525-9 (hbk)
ISBN-13: 978-1-03-217462-4 (pbk)
DOI: 10.1201/9780429344251

Publisher's Note
The publisher has gone to great lengths to ensure the quality of this reprint but points out that some imperfections in the original copies may be apparent

Visit the Taylor & Francis Web site at
http://www.taylorandfrancis.com

and the CRC Press Web site at
http://www.crcpress.com

In memory of Alexey Ochkov

Contents

Acknowledgment

The author expresses deep gratitude to colleagues and friends for their help in writing this book: Natalia Slivina, Elena Bogomolova, Jana Kalova, Julia Chudova, Natalia Ochkova, Katarina Pisačić, Gizem Karaali, Inna Vasileva, Irina Kirichenko (ladies first), Evgeny Nikulchev, Massimiliano Nori, Fred Kohlhepp, Alan Stevens, Werner Exinger, Michael Kirsanov, Alexey Sokolov, Vladimir Chudov, Alvaro Dias, Luc Mekes, Alvaro Falconi, Mati Hainloo, Sergey Fedorovich, Alexander Eliseev, Alexander Bobriakov, Sergey Khorkov, Volker Lehner, Sasha Gurke, Andreas Look, Viktor Korobov, Alexander Solodov, Vitaly Samsonov, Aung Thu Ya Tun and Jury Fedorov.

The author is immensely grateful to Massimiliano Nori, Fred Kohlhepp Alan Stevens, and Dmitry Petrov for their great help in translating this book into English and for valuable comments.

Author

Valery F. Ochkov is a Professor at Moscow Power Engineering Institute (Technical University, MPEI, www.mpei.ru) in the Department of Theoretical Basics of Thermal Engineering (TOT), founder and general manager of Trieru (www.trie.ru) an engineering consulting firm that develops simulators and analytical software for the power industry. He also works at Joint Institute for High Temperatures (www.jiht.ru) of the Russian Academy of Sciences and at Moscow High School at MPEI. He is a member of the working group on thermodynamics of the International Association for the Properties of Water and Steam (IAPWS) and PTC Mathcad Community (https://community.ptc.com).

He has authored more than 25 books in Russian and English and numerous journal articles on the use of math software for solving the problems in thermodynamics, heat and mass transfer, and fluid dynamics. Professor Ochkov is one of the creators of WaterSteamPro™, an application for simulating thermophysical properties of water and steam and also a co-creator of web calculations for the Elsevier/Knovel website.

Teaching of Informatics, Mathematics, Physics, Chemistry with Mathematical Packages Or

Between Scylla and Charybdis

The history of the development of methods for solving problems in mathematics, physics and chemistry in school and at university is, among other things, the history of struggling *with computational tools.* First (in the oral calculations classes, for example), the use of pen and paper was forbidden; then (when studying calculation with a "stick" pen on paper), the calculator was forbidden; and, finally, when solving more complex problems, using a computer with modern mathematical programs, such as Mathematica, Maple, MATLAB®, Mathcad, SMath, etc. (or, rather, computer programs), though not explicitly prohibited, was alleged to be the equivalent of solving simple problems (such as 65 times 9, for example) with a calculator. This prohibition was not applied to the computational tools themselves when specially selected examples in computer science and programming classes were being taught, but only to their use to solve problems in arithmetic, trigonometry, linear algebra and mathematical analysis classes. However, modern schoolchildren and students do not understand this; for them the computer is ubiquitous in both work and play.

With mental calculation, everything is more or less clear. These exercises are wonderful gymnastics *for the mind* (see Study 3). The use of calculators in these classes is equivalent to the addition of sports simulator hydraulic boosters. Counting "column" can also be considered as a gym for the mind and good motor skills for the hands. But there is another reason. We must be able to do mental calculations in the absence of pencil and paper; further, we must be able to calculate with pencil and paper, in the absence of a calculator, etc. However—to get fire by friction if there are no matches at hand; to be able to determine cardinal points through the trees in the forest, if there is no compass or navigator at hand; to be able etc. Unfortunately, or fortunately, with the development of civilization, we have forgotten how to do mental calculation, to obtain fire by friction, to orient ourselves on the terrain by natural signs, etc. The schoolteacher of the author of this book did not tire of repeating in the lessons on arithmetic, that if we, his students did not learn to count quickly and accurately in the mind or, at least, on a sheet of paper, we would be weighed and cheated in stores. Now this motivation to develop an oral calculation does not work anymore, because in modern supermarkets, modern methods of "hanging and counting" are used.

Opponents of the use of modern computer solutions for school and university tasks in mathematics, physics, chemistry, etc., also rely on a number of other arguments, which they, however, do not openly talk about.

First, and very unfortunately, many schoolteachers and university teachers simply cannot work with modern computer mathematical programs and/or do not know about their capabilities. These teachers have mastered the computer, but at the level of office programs (text editor, table processor, email, work on the Internet) and the basics of operating (file system, etc.), but they do not want to go further, or cannot, justifying it by saying that such programs are harmful to learners.

Second, the introduction of these programs in the educational process requires a thorough revision of the content and methods of teaching, as well as the rewriting of textbooks and problem books in mathematics, physics and chemistry or, at least, their substantial reprocessing. The examples in the problem books, of course, are rewritten. Before, there were, for example, "funts and arshins" (Russian obsolete units equivalent to pound and yard, respectively, see Study 6), and now "meters and kilograms". Previously, the problem was: "The digger has dug so many meters of ditch", and now "the computer has such volume of memory", but the essence of the tasks and the method of solving them, alas, do not change.

Third, the previously mentioned computer programs are quite expensive. Our schools and universities were not able to afford many of them. However, this problem is now solved. Firms that develop mathematical programs provide substantial discounts to educational institutions and in some cases supply programs free of charge. This has a commercial goal, of course. After graduation, students who mastered the free program will hopefully buy it themselves or ask their employer to buy it. Teachers should not complain about the high cost of programs or the inability to work with pirated copies for various reasons, but look for ways to solve this problem, by contacting, for example, software developers and their dealers. In addition, you need to remember that there are free versions of the programs. For example, PTC, the developer of the Mathcad program, allows a month to work with the full version of Mathcad Prime, and then with its truncated version Mathcad Express, which nevertheless allows solving rather complicated mathematical problems.

And one more very important point.

At present, powerful software tools for the computer modeling of physical processes are being intensively introduced into the educational process of technical universities. If we talk about universities in the energy field, in which the author of this book works, then there are programs such as ANSYS or COMSOL (modeling of heat and mass transfer processes), ThermoFlow (calculations of thermodynamic cycles), FlowVision (modeling of hydrodynamics). New laboratory work is created on the basis of similar programs, seminars are held and lecture courses are taught in many advanced universities of energy and power engineering.

Unfortunately, these programs are often used as "black boxes"; initial data is entered, the "box cover" is closed (press the Calculate button), then the user opens the "cover" and takes out the results: tables with figures, graphs, three-dimensional virtual objects (which can even be printed on a 3D printer), animation, etc. This means students have very vague (if any) ideas about what is happening in these "black boxes". After all, these programs were created not for educational, but for commercial purposes, and the design code in them is hidden from extraneous views (protected from industrial espionage). Also, there are often cases when this code is hidden because it is very old and primitive, and the "modernity" of the program is visible only on its interface (all the steam went off in a whistle). The benefit of this educational process is often close to zero. However, there are educational technologies in which a real experiment (for example, heating or cooling a physical body with temperature being measured on its surface using a thermal imager) is carried out in the laboratory, followed by the modeling and mapping of this process on the computer. The comparison of images with the thermal imager and with the display screen stimulates discussion of the issues related to the essence of a physical phenomenon, to the methods and the errors of the tools of the instrumental control, as well as to the limitations and the assumptions of computer programs: this process is modeling.

We have now examined a kind of Scylla, meaning by this "sea monster" the way in which packages like ANSYS or COMSOL are *implemented* in the learning process rather than meaning the packages themselves. Now let's talk about the second "monster"—about a kind of Charybdis.

Currently, a large number of programs have been accumulated in the so-called high-level programming languages (Fortran, Pascal, BASIC, C, Python, etc.) for calculating the processes that are modeled in the previously described "black boxes". Teachers of older generations are trying to "re-teach" modern students these programs (codes) in order to explain the essence of the calculations. But for most students, the languages of these programs are incomprehensible. And, most importantly, the motivation of the majority of the students, who will never seriously code (especially in these languages), is not clear, since they will use ready-made software products (the "black boxes") or *mathematical packages* for solving their particular problems: calculations with complex formulas, solving equations and systems (algebraic and differential), constructing graphs, creating animations, processing statistical data, etc.

But many teachers still try to teach students to work with "their" programs. There are many possible explanations for this. First, do not "dismiss an equivalent product"—created with much effort and thoroughly tested for decades; for example, in "good old Fortran". Second, many teachers do not know about the possibilities of new software or consider that they (the teachers) are incapable of studying something new. Third, the appeal of the "good old horse", sorry, Fortran or BASIC, "which does not spoil furrows". This is explained by the fact that in many organizations where students end up working after university, Fortran is that "horse", which is still used to "plow", guided among other things by the principle that the best is the enemy of the good.

We can swim between Scylla and Charybdis, if we use mathematical packages, such as Mathcad, for modeling physical phenomena with details of their essence and methods of implementation on the computer. That is the main idea of this book.

But why Mathcad? What good is it for these kinds of calculations? First, of course, the fact that the author is well aware of this package and has written several books on it.

However, there are, of course, objective indicators of the convenience of working with Mathcad.* We list them.

1. Good documentation of calculations. Calculations made in Mathcad can be printed and submitted for review or shown to a person (teacher) who has never worked on a computer.† Mathcad reflects the appearance of calculations made on paper, thanks to the WYSIWYG mode—what you see is what you get—"what you see (on the display screen), then you will get (on printed paper)". The printouts of the solutions received in Mathcad can be left in an archive so that someone reading them 50–100 years later will understand what is written there and be able to reproduce it in the new software environments that will have appeared by that time. And this is very important.

In the area of computer information technologies there is currently a crisis, which we have already touched on above, and which is conventionally called the "nightmare of legacy software". What is it? Imagine a firm, corporation or university that after 30–40 years of intensive use of computers has developed and accumulated a large number of programs of varying complexity for calculating and modeling various processes, devices and technologies in various fields of science and technology, and also for the creation and management of databases (knowledge).

In place of morally and physically obsolete computers come new computers with new hardware and new operating systems. Computers are integrated into local networks, which participate in the process of "globalization" and are integrated into the Internet—wired and wireless. The development of wireless communication between computers and servers has led to the emergence of "cloud" information technology. Modernization of the computer park often leads to the fact that old applications stop running on new or updated computers—workstations and servers. Sometimes it's just impossible to read a program from old media (punched cards, punched tape, a diskette of various diameters, obsolete "flash drives", etc.), since new computers do not have appropriate reading devices. We either have to completely abandon such programs and create new ones from scratch, or spend time and money on developing

* The book's calculations are slightly diluted with solutions in alternative mathematical environments: SMath (Figure 1.8), Maple (Figure 17.3, Figure 24.4, Figure 25.3 and Figure 26.34) and Mathematica (Internet version, see Figure 7.5, Figure 26.32, Figure 26.33 and Figure 28.8).

† At the end of Study 9, we show one of the differences between Mathcad and manual formulas. It will also be said there that these differences are gradually disappearing.

or purchasing certain utilities (emulators) to run old programs on new or updated computers.

But that's not so bad. The real trouble begins when specialists, who created and supported these programs, leave the "firms, corporations, universities" and the young people who replace them can neither modernize the programs nor adapt them to new requirements. The reason for this would be an insufficient documentation of codes, the lack of appropriate tools and the simple impossibility of correctly reading a program written in "dead" programming languages. If, however, it is difficult to recreate or modernize the old program, using old or even new programming languages, the "nightmare of legacy software" will reappear after a while.

Good documentation and detailed calculations are very important in the educational sphere. It's no secret that many teachers of higher educational institutions are opposed to the use of computers for calculating educational tasks. We have already looked at with this problem. These teachers quite reasonably believe that such tasks need to be addressed solely by use of the "pen" and not the computer, or, at least, by the means that they themselves used as students. But these teachers are referring to the use of computers as "black boxes" (see above), in which the "initial" data are "input" and from which ready-made answers are "taken out". Mathcad is by no means a "black box", which is contraindicated for students.

2. Working with units of measurements. Electronic tables and programming languages played a cruel joke on us—they made us accustomed to working without physical quantities. More accurately, we were taught to work with quantities devoid of dimension (reduced to basic SI), and to keep in mind the units of their measurement (SI units—meters, seconds, pascal, kelvin, joules, watts, etc.), which is very inconvenient and fraught with errors in the calculations. Also, the basic SI units themselves are inconvenient: the basic unit of pressure (pascal) is very small and always requires multipliers of kilo or mega, the temperature in kelvins often requires conversion to Celsius or Fahrenheit degrees, etc. This problem will be considered in more detail in Study 1. This feature allows us to call Mathcad not just mathematical, but a physical and mathematical package. It was created, incidentally, by a nuclear physicist by education—Allan Razdow.

3. A flexible system of variable names. Variables and functions in Mathcad, with rare exceptions, have the same names that were assigned to them in certain scientific and technical disciplines long before the advent of computers. For example, the Greek letter η with various indices in thermodynamics is the efficiency (thermal, internal relative, etc.), and the Greek letter λ is the thermal conductivity. This, along with the use of traditional writing of mathematical constants, operators and functions, as well as upper and lower indexes, makes the "language" of Mathcad available to the uninitiated (see paragraph 1 above) without much additional comment. The notation in Mathcad practically coincides with generally accepted mathematical notation: integral, derivative, modulus, degree, etc.

Another four features, that are also available in other mathematical packages (MATLAB, Maple, Mathematica, SMath, Derive, etc.), are listed here:

4. Numerical and symbolic mathematics, which make it possible to use a rich library of numerical methods for solving problems, anticipating or supplementing them with attempts (successful or less successful) to solve the problem analytically. Mathcad was originally created as a numerical mathematics package, to which later (in the fifth version) the core of symbolic mathematics was attached from Maple. This core was then (in the thirteenth version) replaced by the core of symbolic mathematics from MuPAD. Similarly, the "numerical" package MATLAB, initially also had a symbolic engine from Maple that was later replaced by MuPAD. Conversely the Maple and Mathematica packages were originally packages of symbolic mathematics with elements of numerical calculations.

5. Mathcad includes powerful and flexible tools for creating 2D and 3D plots, as well as animations. This allows you to quickly and easily visualize the source, intermediate and summary data without calling external procedures or writing auxiliary

programs, which contributes to a better understanding of the essence of the calculation, identifying possible errors and false ways to solve a problem within it.

6. Coding. In Mathcad, mathematical actions on the worksheet are executed on a regular sheet of paper from left to right and from top to bottom. But sometimes it is necessary to change this order of calculation. This opportunity (programming) is provided in Mathcad, and is successfully used not only by advanced users, but also by those who once mistakenly believed that they would never program. Mathcad's programming tools allow us to solve rather complicated tasks that do not fit in the narrow framework of the sequential algorithm (from left to right and from top to bottom).

7. Mathcad provides the ability to expand the list of available functions in three ways:

   - The first method is to link Mathcad by a DLL (Dynamic Link Library) mechanism with functions written, for example, in C programming language.

   - The second is a link from the working document to other Mathcad-sheets. Because of this link, the user variables and functions stored in the linked document (file) will become available (visible, as the programmers say) also in the working document. These and other Mathcad-sheets can be downloaded to your computer or to a local network and used as templates. However, you could do this if your computer has a reliable, high-speed connection to the Internet.

   - The third method: Mathcad tools allow the user to write and debug functions quickly that return, for example, material properties (example in Figure 2.3 in Study 3), based on formulas, tables or graphs taken from an external source—from a paper or electronic book, as well as from the Internet.

These three possibilities could and should be used in order to make available in the calculation document, for example, functions that have as arguments the parameters of a particular point of the thermodynamic cycle and return the necessary thermophysical properties of the working body at this point: specific enthalpy, specific entropy, density, specific heat capacity (isobaric or isochoric), thermal conductivity, viscosity, etc. Without this, it is impossible to calculate heat engineering processes (a specialization of the author).

And lastly. Calculations created in Mathcad can be placed on the server (on the Internet or on the local network) without much trouble to work without Mathcad itself. So, for example, with many calculations described in this book, you can work through the Internet using Mathcad Calculation Server technology, without installing any additional programs on your computer.

In this book, all these possibilities will be shown using concrete physical and mathematical examples.

And more.

Almost all problems of the book were solved in Mathcad. PTC Academic is committed to providing access to Mathcad for free.

PTC Mathcad Express, a lighter version of PTC Mathcad is available for student and educators' download at https://www.ptc.com/en/products/mathcad-express-free-download.

If you are interested in purchasing full-featured PTC Mathcad for university or school - please contact PTC academic team at https://www.ptc.com/en/academic-program/contact-us.

Formulas for Calculations Or

Chekhov's Gun

ABSTRACT

Mathematical expressions can be added or edited in electronic media as embedded calculations. Formulas in an electronic document are not just expressions written in a formal language, they are also an object that could be studied, helping to comprehend a text and avoid many typos and mistakes typical of mathematical formulas found in print. This Study describes some features of formulas embedded in online scientific literature and educational web apps. The Study is based on the experience of teaching engineering subjects and math assisted by the visualization of calculations. The author examines key features and practical application of calculations and formulas embedded into the scientific and educational texts.

Mathematics: Formulas.

Physics: Units, empirical and pseudo-empirical formulas, molarity, molality.

IT: Cloud calculation, modern publishing process.

Art: Anton Chekhov.

Study website: https://community.ptc.com/t5/PTC-Mathcad-Questions/PhysMathStudy1/m-p/85566.

A lot of changes are taking place in the educational and scientific content publishing arena. Traditional print is enhanced and often supplanted by electronic materials with embedded features such as web links, bookmarks, comments, references to forums, and video clips or animations [1].

Embedded calculations play an important role among these features, enhancing comprehension of the math while enabling live computation [2].

There is a wide variety of software capable of supporting embedded calculations. For example, the symbolic math engine Maple has found wide use in academia for differentiation and integration. The popular programming language Python is used for realization of algorithms in TeX documents [3]. Math formulas on HTML pages can be supported by Mathcad [4]. Cloud computing applications are also becoming more widespread [5].

Regardless of the technology used, a modern electronic textbook is not just a web-enabled text. It is also a self-contained learning tool [6]. It helps us to comprehend math better, allows for calculation and visualization of solutions in the form of graphs, and can be used for in-text exercises.

If there is an equation in a book, the reader should be able to use it for calculation right there. Additionally, readers should be able to change the variables in the equation. This also applies to tables and graphs containing functional dependencies [7].

The following are some observations from our pedagogical practice that illustrate the usefulness of embedded equations in education.

1. In a physics department, students can conduct lab experiments and then use a computer to access a web page with live calculations, to enter data points and to fit a curve using the least-squares method. This method used to be the normal approach. Now, an increasing number of students use smartphones and tablets. When asked why they do not use the lab computer, students often reply: "I hardly know how to work with a standard keyboard, and I can input data faster on the tablet."

2. During math, physics and chemistry classes, students often use their mobile devices (smartphones, tablets, notebooks) to go to a math website and to key in an algebraic or differential equation (see Figure 1.2) in order to find an analytic or numeric solution that validates calculations written by the lecturer on the blackboard.

3. Finally, during a physics class, a student might derive an equation describing string vibration and then create an animation for its visualization.

There is a new trend in the content and in the production of scientific papers, handbooks, monographs and textbooks containing calculations. In the old days, authors submitted their work to publishers in the form of typed manuscripts with handwritten formulas and special characters. Now, publishers require electronic files with the text in MS Word and formulas created in a math editor or native software; the formulas are scanned and inserted as images into the text.

Formulas in scientific papers and textbooks are there for the purposes of calculation. In electronic formats, they can be used for computing right on the screen. Printed formulas can be easily validated by using an interactive electronic version. This is a revolutionary change!

The production of electronic handbooks with embedded interactive equations is described in [8]. Making equations interactive reveals many typographical and other errors in the formulas. For example, multiple errors were found during the development of a series of web-based handbooks called "Heat Technology and Power Engineering" and corrected for the subsequent edition.

During peer review of scientific papers, we are often asked by editors to validate numeric solutions. While the damage caused by typos in regular text is negligible, those in formulas can lead to serious consequences. For example, during the review of an article on the thermal conductivity of an aqueous solution of NaCl, calculations were validated using the Mathcad Calculation Server. This revealed that some equations and coefficients in the article had typos. When these were corrected the article was published free of error and with a reference to "live" computing.

In this chapter of the book, we will examine various aspects of computing with embedded formulas and discuss ways of improving the quality and precision of solutions by using validated formulas.

Modern calculation apps such as Mathcad can handle units of measurement. This is very convenient due to the automatic conversion of input and output values with different units, and the control over their dimensional compatibility. However, there are some issues.

There are two types of scientific equations—physical and empirical. The former deal with physical values and can be used with any appropriate units of measurement. Good examples of these formulas are $E = m \cdot c^2$ and $F = m \cdot a$. Calculation apps will solve these equations correctly regardless of the units as long as they have the right dimension, and they will indicate an error when the dimensions are not compatible. Empirical equations, created as a rule by statistical treatment of experimental (empirical) data, require numeric input in the stated units of measurement. Probably the simplest and best known empirical formula is that linking the height of an average man with his weight [4]: *height of a person in centimeters is his weight in kilograms plus 100.* This formula, written in Mathcad, and the units of measurement are shown in Figure 1.1. The results in Figure 1.1 are rounded, although variables in equations contain 15 digits in the mantissa. This detail is an important aspect of the creation of "live" equations. In paper handbooks, monographs and textbooks, calculation of the systems of equations is done by rounding intermediate results and then using these results in other equations. In computer calculations, variables retain maximum precision. The results of these calculations would thus be slightly different from those in paper publications.

In handbooks, the basic units of measurement are shown in the legends for physical formulas. However,

What should be the weight of Hercule Poirot - Agatha Christie's hero?

$Height := 5\ ft + 5\ in \qquad Weight := \left(\frac{Height}{cm} - 100\right) \cdot kg = 144\ lb$

FIGURE 1.1 Input of a simple empirical equation.

$\eta := 34\%$

$b := \frac{123}{\eta} = 361.765$ Correct answer, but without units

$Q_{ef} := 7000 \frac{kcal}{kg}$ The calorific value of equiv. fuel

$\frac{1}{Q_{ef}} = 122.835 \frac{gm}{kW \cdot hr}$ Why the result is not 123?

$b := \frac{1}{Q_{ef} \cdot \eta} = 361.28 \frac{gm}{kW \cdot hr}$ Correct answer with units

FIGURE 1.2 Working with pseudo-empirical formulas.

machine computing of these formulas is done without using the units. Here is a simple example: $F = m{\cdot}a$, where F is force in N (newton), a is acceleration in m/s^2, and m is mass in kg. For modern calculation applications input of the units is redundant because all appropriate conversions for the derived units are done automatically.

We should not forget about physical problems with solutions that contain numbers beyond the range of computer calculations. They could have both very large and very small parameters, resulting in incorrect rounding due to simultaneous processing of large and small numbers. Also, solution algorithms for math models could become unsuitable if model equations contain variables that differ by several orders of magnitude. In this case, the errors for the large parameters could be on the same scale as the parameters themselves and could significantly distort the values for the small parameters. Before solving these equations, it is advisable to modify them to reduce the spread between the magnitudes of parameters and to reduce the number of arithmetic operations leading to an accumulation of errors.

A solution is to replace the variables with dimensionless quantities prior to calculation, making all variables in a math model practically the same magnitude. As a rule, the equations themselves become simpler due to the elimination of "extraneous" constants and quantities, and the real parameters of the process being modeled are disclosed. The equations are made unit-independent by converting absolute values of some properties to relative values, specific to a particular physical model. Usually, these are length, speed, time and energy-related or mechanical properties.

Many physical constants are dimensionless by definition. These include, for example, the Mach number (compressibility factor), Reynolds number (viscosity factor), Poisson's ratio (a parameter of material elasticity) and the Prandtl number (reflecting the relationship between viscosity and thermal conductivity). An angle measured in radians is a dimensionless quantity as well, and it is calculated as a ratio between the length of an arc and a radius of a circle.

Let's use oscillation of a weight attached to a spring to illustrate how an equation can be made dimensionless. Suppose at time t_0 the initial extension of the spring is x_0 (m), with parameters k (N/m), v_0 (Hz) and m (kg). The characteristic time would have the dimension of $\sqrt{m/k}$ and, if the spring is stretched instantaneously, the interval of calculations would be so small it would be comparable to computational error, and representative values of x could be so large as to lead to a loss in precision. Let's make the quantities dimensionless by introducing new variables $\tilde{x} = x/x_0$, $\tilde{t} = t\sqrt{k/m}$ and $\tilde{v} = v\sqrt{m/k}$. The system of equations with new variables can be expressed as $\begin{cases} \dot{\tilde{x}} = \tilde{v}, \\ \dot{\tilde{v}} = -\tilde{x}. \end{cases}$.

Here, representative values of all the variables are comparable and, therefore, the system of equations can be solved numerically with the required precision. In addition, by solving this system once, we can obtain a solution for the whole range of problems for different values of x_0, k and m. To get back the units, we need to reverse substitute the variables.

Getting rid of the units of measurement involves transitioning from the absolute physical quantities to the relative quantities specific to a given physical process. This transition is governed by the theory of similarity. For example, the transition from laminar flow to turbulent occurs at a certain value of Reynolds number, depending on the flow velocity, geometry and fluid viscosity.

Making calculated quantities dimensionless opens the possibility of adding Reynolds and Prandtl numbers without getting a program error. Adding viscosity to thermal conductivity is not possible or, more precisely, it is not possible using software that can control units. This and other reasons lead to the return to dimensional physical quantities and rejection of their "shadows"—dimensionless quantities—the obvious drawbacks of which are felt by those who use pseudo-units such as degrees. Modern software has the means to remedy this situation.

Another obvious case of dimensionless quantities is the use of decibels. Bel is a common logarithm of a ratio of two unidimensional physical quantities, and a decibel is, of course, one-tenth of a bel. By measuring a property in decibels we are actually converting a physical quantity into a dimensionless value on a logarithmic scale. Usually, decibels are used for measuring the volume of sound by comparing the intensity of sound waves I with the lowest level of sound perceivable by a human ear I_0 (typically $I_0 = 0.01$ W/m^2). We can calculate the ratio I/I_0 directly, but the range of this ratio is very wide, making its use inconvenient. This is resolved by using logarithmic scale and the prefix deci, $-0.1 \cdot \log(I/I_0)$. A very loud noise of a sledgehammer has a volume of 80 dB, whereas a conversation in a room has a volume of 60 dB and a barely audible sound such as the rustling of leaves is approximately 10 dB. Decibels, as well as other dimensionless quantities, make calculations easier but, again, not if we use a computer.

Some time ago, during transition to the SI system of units, STM publishers and journal editors stopped accepting manuscripts that used other, still common, units of measurement. This transition was painful for engineers and scientists who used foot, pound and other non-metric units. In spite of the prevalence of SI units, there are some customary, non-metric units that are still being used in many countries. For example, we (in Russia) will never learn to measure atmospheric pressure in millibars (hectopascals), preferring millimeters of mercury (torrs). Now, publishers are beginning to deviate from the rules requiring the use of SI units in articles and books. One of the reasons for that is unit-enabled calculation software.

Some equations in journals and books could be called pseudo-empirical. These are basically physical equations that in reality are empirical because they require units of measurement for computation (see Figures 1.3 and 1.4). The following is an example from power engineering. At a given efficiency (η) of a power plant, the specific fuel consumption (b) can be calculated as $b = 123/\eta$. This means that a power plant with efficiency 34% consumes 362 (123/0.34) grams of fuel to generate one kilowatt-hour of electricity. This simple equation would result in an incomplete solution in Mathcad, i.e. a solution without the units (see the first line in calculations shown in Figure 1.2). So, what is the problem? The reason is that, in the past, this equation was adopted for manual calculation and for calculation with software that could not handle the units (such as spreadsheets). However, if we remember the equivalence between fuel and its calorific value (Qef), we can restore the original equation $b\ 1/(Qef \cdot \eta)$ and use it without any problems in Mathcad (see the last line in Figure 1.2). The result will have units and, in addition, will be more accurate. A power engineer using the original equation would find out that his/her plant consumes almost half a gram less fuel. On a large scale it represents a huge energy saving.

The case shown in Figure 1.2 is quite common. The lack of units in the result shown in the first line in Figure 1.4 points to the pseudo-empirical nature of this equation.

In the next example, the pseudo-empirical nature of the formula is not so apparent. This formula is used to convert concentration from one unit to another. When taken from a handbook and input into Mathcad, it produces an answer with the correct units of measurement and, incredibly, with plausible results. The goal is to convert the molarity (the ratio of the solute to the volume of the solution) of an aqueous NaCl solution to its molality (ratio of the solute to the mass of solution). The formula for this conversion can be easily found in a number of chemical handbooks and on the web (see Figure 1.3).

We can copy an equation for calculation of molality (L) based on molarity (M) from the table shown in Figure 1.3. To calculate, we need to input additional values: molecular mass of NaCl (mp) and solution density (q). Copy, input and get an incorrect answer (see the 2nd line of the calculations in Figure 1.4).

In the table shown in Figure 1.3, the variable K is defined as a mass percent without disclosing whether it is a mass of the solution or a mass of the solvent. An analysis of this table reveals that it is the mass of solution and not of the solvent. However, other cases are not so apparent, leading to calculation errors. For example, in many analytical chemistry handbooks, solubility in water is given as a ratio of the mass of solute to the mass of solvent by default, without an explanation. On http://twt.mpei.ac.ru/MCS/Worksheets/Thermal/T90-T68.xmcd this quantity is provided in two units of measurement (ratio to the mass of solvent and to the mass of solution) to prevent calculation errors. There are lots of these "defaults". Take temperature for example. Here is a typical problem: the temperature at the inlet

http://erichware.info/sposob/ximia/ximikonc.htm

	0 < K < 100 % mass ratio	T titer	L molality	M molarity
M =	$\frac{10 * q * K}{wp}$	$\frac{1000 * T}{wp}$	$\frac{1000 * q * L}{1000 + wp * L}$	$\frac{1000 * yp}{v}$
L =	$\frac{1000 * K}{wp * (100 - K)}$	$\frac{1000 * T}{wp * (q - T)}$	$\frac{1000 * yp}{mb}$	$\frac{1000 * M}{1000 * q - wp * M}$
T =	$\frac{q * K}{100}$	$\frac{mp}{v}$	$\frac{q * wp * L}{1000 + wp * L}$	$\frac{wp * M}{1000}$
K =	$\frac{mp * 100\,\%}{mb + mp}$	$\frac{100 * T}{q}$	$\frac{100 * wp * L}{1000 + wp * L}$	$\frac{wp * M}{10 * q}$

wp – molar mass of solute (g/mole) mb – mass of solvent (g)
mp – mass of solute (g) yp – moles of solute (mole)
v – volume of solution (cm^3) q – density of solution (g/cm^3)

FIGURE 1.3 Equations for calculation of different types of concentration.

$$M := 2\,\frac{mol}{L} \quad q := 1.076\,\frac{gm}{cm^3} \quad wp := 58.44\,\frac{gm}{mol}$$ Input data

$$L := \frac{1000 \cdot M}{1000\,q - wp \cdot M} = 1.859\,\frac{mol}{kg}$$ Similar, but the wrong answer

$$L := \frac{1000\,\frac{M}{mol \cdot L^{-1}}}{1000\,\frac{q}{gm \cdot cm^{-3}} - \frac{wp}{gm \cdot mol^{-1}} \cdot \frac{M}{mol \cdot L^{-1}}} \cdot \frac{mol}{kg} = 2.085\,\frac{mol}{kg}$$ Empirical formula

$$L := \frac{M}{q - wp \cdot M} = 2.085\,\frac{mol}{kg}$$ Correct answer - the physical formula

FIGURE 1.4 Working with a pseudo-empirical formula for conversion of different units of concentration.

of a heat exchanger is given as 25°C, find the temperature at the outlet if the temperature of the fluid in the heat exchanger increases by 5°C. The answer is 30°C. However, if this problem is input into Mathcad as it is, the answer will be paradoxically 25°C + 5°C = 303.15°C. The answer can be explained by remembering that 5°C equal 5K but 25°C is equal 298.15K. Most users understand this default (Celsius scale and centigrade degree) as it is found in many handbooks, but computers do not. Also, in these calculations, users should account for the year of the temperature scale, 1968 or 1990, in which the temperature is given.

The problem here is that the equations in Figure 1.3 have been adopted for the convenience of manual computation by using non-basic ("chemical") units of measurement: concentration, density (g/cm^3 instead of kg/m^3), molar mass (g/mol vs. kg/mol), mass (g vs. kg), volume (cm^3 vs. m^3), etc. The formula for computation of molality from molarity can, of course, be used as an empirical one (see above) by adding the required units of measurement (see 3rd line in Figure 1.6) to obtain the correct result. However, it is better to go back to the initial physical formula by removing the coefficients "1000"—(see the last line in Figure 1.4), realizing that 1000 is the number of grams in a kilogram and the number of cubic centimeters in a liter, etc.

As a result, most practicing chemists, when they need to convert concentration from one unit to another, try to avoid the existing formulas like those shown in the table in Figure 1.3. Instead, they calculate concentrations using ratios. These ratios are not required, however, if they have access to a computer with Mathcad or to a website with an appropriate computational application. There he/she can input and solve an algebraic equation linking, for example, the moles and the mass of the solute to different units of concentration. Mathcad has a symbolic math engine that allows computing with the symbols of variables instead of their numerical values. Mathcad calculations with formulas for conversion of solution concentration from one unit to other using algebraic equations are shown in Figure 1.5. This is a true improvement with respect to handbook formulas. One can see both the formulas used for the calculation and the corresponding physical law-based equations from which the formulas are derived!

In Figure 1.6 an example solution is provided for the conversion of molality to molarity using a formula, modified as shown in Figure 1.5 on the basis of the physical law of the conservation of mass. The operator **clear**$_{\mathbf{sym}}$ that is in the collapsible area is used to remove numeric values of variables during symbolic transformations.

Figure 1.7 shows a website for the conversion of concentrations using the above formula. This site was created some time ago and should be updated to show not just the formula itself but also the equation from which it was derived (see Figure 1.6).

from molarity to molality

$$\mathrm{M} := \mathrm{M}\cdot v = \mathrm{L}\cdot \left(\underbrace{v\cdot q}_{\text{"Solution mass"}} - \underbrace{\mathrm{M}\cdot v\cdot wp}_{\text{"Solute mass"}} \right)_{\text{"Solvent mass"}} \xrightarrow{solve,\,L} \frac{M}{q - M\cdot wp}$$

from molality to molarity

$$\mathrm{M} := \mathrm{M}\cdot v = \mathrm{L}\cdot \left(\underbrace{v\cdot q}_{\text{"Solution mass"}} - \underbrace{\mathrm{M}\cdot v\cdot wp}_{\text{"Solute mass"}} \right)_{\text{"Solvent mass"}} \xrightarrow{solve,\,\mathrm{M}} \frac{L\cdot q}{L\cdot wp + 1}$$

from titre to molality

$$\mathrm{L} := \frac{\mathrm{T}\cdot v}{wp} = \mathrm{L}\cdot \left(\underbrace{v\cdot q}_{\text{"Solution mass"}} - \underbrace{\mathrm{T}\cdot v}_{\text{"Solute mass"}} \right)_{\text{"Solvent mass"}} \xrightarrow{solve,\,\mathrm{L}} \frac{\mathrm{T}}{wp\cdot(q - \mathrm{T})}$$

from molality to titre

$$\mathrm{T} := \frac{\mathrm{T}\cdot v}{wp} = \mathrm{L}\cdot \left(\underbrace{v\cdot q}_{\text{"Solution mass"}} - \underbrace{\mathrm{T}\cdot v}_{\text{"Solute mass"}} \right)_{\text{"Solvent mass"}} \xrightarrow{solve,\,\mathrm{T}} \frac{L\cdot q\cdot wp}{L\cdot wp + 1}$$

FIGURE 1.5 Generation of formulas for conversion of different types of concentration units.

Calculation of molality of aqueous NaCl solution from its molarity

$$\mathrm{M} := 2\,\frac{mol}{L} \qquad \mathrm{q} := 1.076\,\frac{gm}{cm^3} \qquad \mathrm{wp} := 58.44\,\frac{gm}{mol} \qquad \text{Input data}$$

$$\mathrm{clear_{sym}}(\mathrm{M}, \mathrm{q}, \mathrm{wp})$$

$$\mathrm{L} := \mathrm{M}\cdot v = \mathrm{L}\cdot(v\cdot \mathrm{q} - \mathrm{M}\cdot v\cdot \mathrm{wp}) \xrightarrow{solve,\,\mathrm{L}} \frac{M}{q - M\cdot wp} = 2.085\,\frac{mol}{kg} \qquad \text{Answer}$$

FIGURE 1.6 Computation of molality using modified formula.

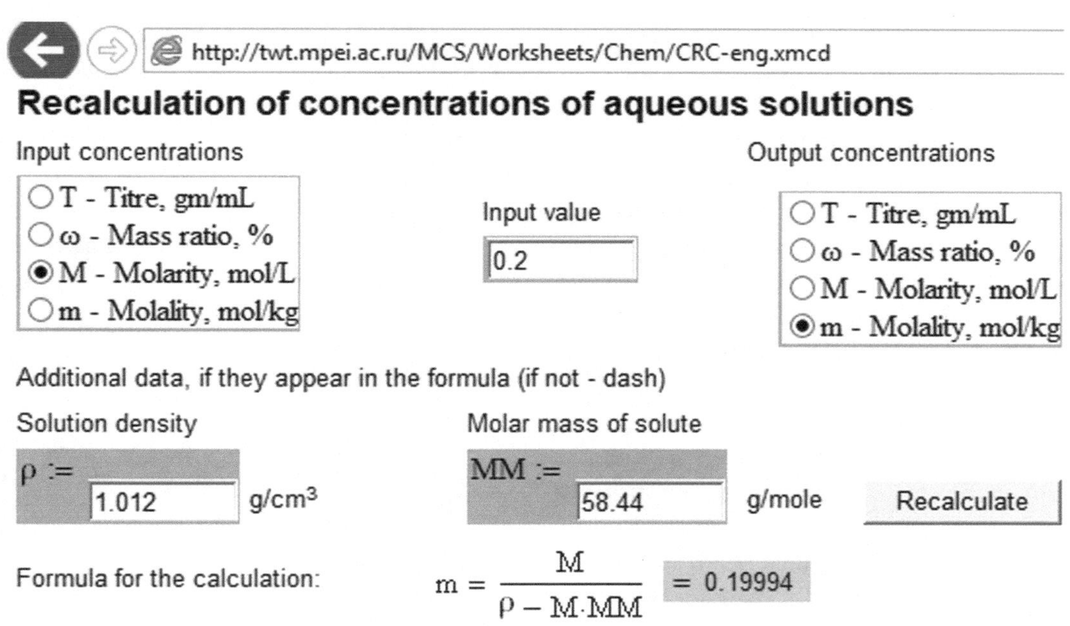

FIGURE 1.7 Website for the conversion of different types of concentrations (http://twt.mpei.ru/MCS/Worksheets/Chem/CRC-eng.xmcd).

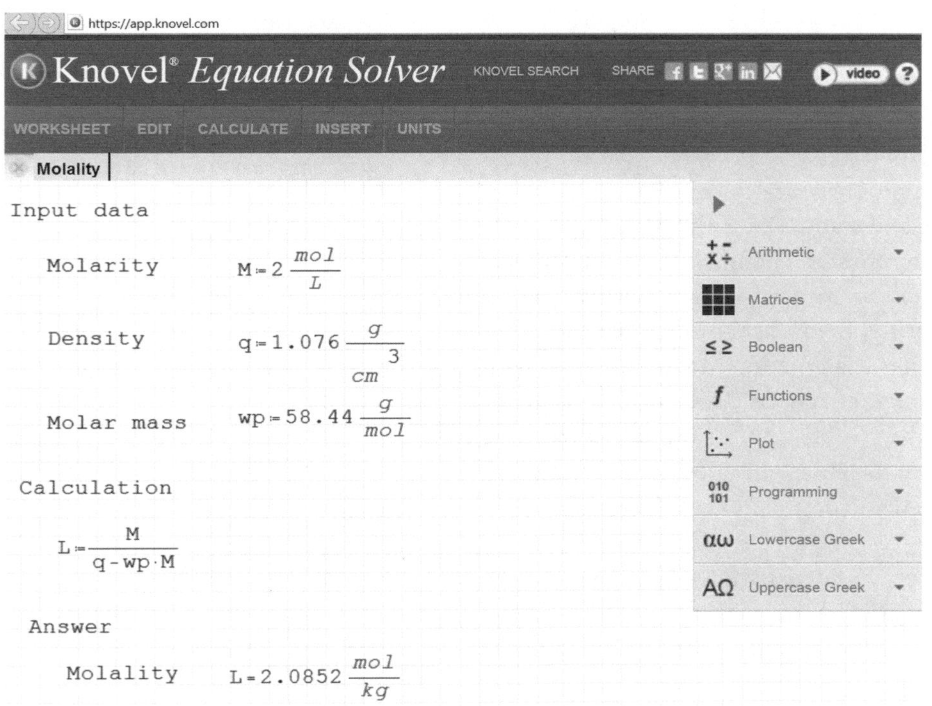

FIGURE 1.8 Working with Knovel Interactive Equations (https://app.knovel.com).

Similar results (Figure 1.8) can be obtained by entering relevant formulas online using Elsevier's Knovel Interactive Equations (https://app.knovel.com/ie/#welcome).

Knovel Interactive Equations is an Elsevier application originally developed in 2012–14 by a team led by Knovel co-founder Sasha Gurke. It is based on a proprietary web-enabled math engine that supports a growing collection of several hundred validated equations and working examples in several subject areas, including Chemistry and Chemical Engineering, Electronics and Semiconductors, General Engineering, Mechanics and

Mechanical Engineering, Metals and Metallurgy and Oil and Gas. Users can browse or search the collection, use the built-in Equation Solver to calculate and export calculations for reports or sharing knowledge. They can also create a worksheet from scratch by combining text, math, images and plots. The application has easy one-click access to a toolbox containing math functions, engineering units, programming structures and math symbols used in engineering formulas.

Reference literature contains a large number of pseudo-empirical formulas similar to those shown in the figures above. Using these formulas without proper adjustment can lead to computational errors, but the correct simplification of these formulas, the exclusion of the coefficients 1000, 100, 3600 from them, and so on, leads to the fact that these formulas begin to produce the correct result (see Figure 1.2). Their modification, so that more convenient units can be used, is no longer required in the computer age.

The website www.trie.ru, with interactive calculations, has been developed and has been in operation for almost 10 years. During this time, >20,000 interactive calculations have been posted on this site, mostly related to power engineering, energy conservation, energy efficiency, structural materials and properties of heat-transfer fluids. In addition, there are calculations in the areas of general science and engineering and educational subjects such as math, physics, chemistry (see above), heat exchange, thermodynamics, gas and hydrodynamics, electrical engineering, theoretical mechanics and other academic disciplines. Calculations are augmented with functions that can be referenced. These functions, when referenced, become visible in documents containing calculations. Many formulas are duplicated as text that can be easily copied and pasted in spreadsheets and programming code.

The technique of "live calculations" developed by the author is currently used by the International Association for the Properties of Water and Steam (IAPWS) on its website www.iapws.org. This site has official formulas for water and steam in static PDF and, now, in "live" calculation formats.

We currently collaborate with Elsevier on Knovel Interactive Equations.

A calculation server that supports www.trie.ru is being developed at the Moscow Power Engineering Institute within the framework of the national research program.

REFERENCES

1. Zeidmane, A. Development of mathematics competences in higher education institutions. *International Journal of Engineering Pedagogy (iJEP)*, 2013, Volume 3, pp. 11–17.
2. Booth, S. Learning and teaching for understanding mathematics. In *Proceedings of the 12th SEFI MWG Seminar*; M. Demlova, D. Lawson (Eds.), Vienna, 2004, pp. 12–25.
3. Poore, G.M. Reproducible documents with PythonTeX. In *Proceedings of the 12th Python in Science Conference*; Stéfan van der Walt (Ed.), 2013.
4. Ochkov, V.F.; Orlov, K.A.; Ko, C.K. "Cloud" functions and templates of engineering calculations for nuclear power plants. *Thermal Engineering*, 2014, Volume 61, pp. 760–764.
5. Pluzhnik, E.V.; Nikulchev, E.V. Virtual laboratories in cloud infrastructure of educational institutions. In *2014 International Conference on Computer Technologies in Physical and Engineering Applications ICCTPEA*, 2014, p. 139.
6. Flegg, J.; Mallet, D.; Lupton, M. Students' perceptions of the relevance of mathematics in engineering. *International Journal of Mathematical Education in Science and Technology*, 2012, Volume 43, no. 6, pp. 717–732.
7. Ochkov, V.F.; Ustyuzhanin, E.E.; Ko, C.K., Shishakov, V.V. Thermophysical databases: From tables to interactive internet resources and cloud templates. *High Temperature*, 2015, Volume 53, no. 4, pp. 515–520.
8. Ochkov, V.F.; Orlov, K.A.; Frenkel, M.L.; Ochkov, A.V.; Znamenskiy, V.E. "Cloud" service on the properties of working media for thermal engineering calculations. *Thermal Engineering*, 2012, Volume 59, no. 7, pp. 566–572.0960120000

Swan, Crayfish and Pike Or

Scalar and Vector in Computer Calculations

ABSTRACT

The Study considers two fundamental types of data—a scalar and a vector (array). It is necessary to have the ability to work with these, otherwise it is impossible to solve school and university problems in mathematics, physics, chemistry and other technical educational disciplines on a computer. Some fundamental principles of teaching computer science at school and university are touched on.

Mathematics: Scalar, vector, matrix, column vector, row vector, two-dimensional, three-dimensional and *n*-dimensional space, absolute value, length of vector, vector norm, linear function, equation of a line in a plane, linear interpolation, systems of linear algebraic equations, ordinary differential equations, initial conditions.

Physics: Scalar and vector physical quantity, density of saline solution, air resistance.

IT: One-dimensional, two-dimensional and nested array, numerical solution of differential equations with physical quantities, work with planar computer graphics, structured programming.

Art: Poem, written by Ivan Krylov.

Study website: https://community.ptc.com/t5/PTC-Mathcad-Questions/PhysMathStudy2/m-p/51200.

Once Crawfish, Swan and Pike
Set out to pull a loaded cart,
And all together settled in the traces;
They pulled with all their might, but still the cart refused to budge!
The load it seemed was not too much for them:
Yet Crawfish scrambled backwards,
Swan strained up skywards, Pike pulled toward the sea.
Who's guilty here and who is right is not for us to say-
But anyway the cart's still there today.

I. A. Krylov

In order to investigate what students have been taught at school, and what they are studying at university in parallel courses of physics, thermodynamics, theoretical mechanics and other disciplines, lecturers in computer science and mathematics from Moscow Power Engineering Institute (MPEI) sometimes do surveys. To the question, whether they know the meaning of scalar and vector physical quantity, all students always respond in unison: "We do!" In students' responses one can detect the continuation of the answer. All pupils and students know clearly that mass is a scalar quantity, that force is a vector quantity and that problems of physics should be solved taking into account these fundamental concepts.

Programming languages, spreadsheets and math packages, currently widely used in teaching computer science and other disciplines at schools and universities, are based on scalars and vectors.

In programming languages, the concept of an array is one-dimensional, two-dimensional or multidimensional. Spreadsheets (SuperCalc, VisiCalc, Excel, etc.) were created in order to store and to process tables of various data.

MATLAB®, a programming language for technical computing, which is often compared with Mathcad, consists of two parts—MAT stands for matrix and LAB stands for laboratory.

In mathematical packages, especially in Mathcad, the concept of an array is included explicitly. Calculations can be done by users in array (matrix) with one column, which is called a vector in Mathcad. However, it should be clearly understood that a Mathcad vector carries at least two meanings. We will try to explain them by solving two simple tasks.

Figure 2.1 shows the calculation of the acceleration of a body with three known forces pointing in the three directions of the axes x, y and z (our task is solved in three-dimensions). Nevertheless, it is possible to transform the problem into two dimensions or even in a single dimension. Furthermore, a person could transform it into the space of four or more dimensions, which is often described in science fiction, but is actually used in mathematics.

In the calculation shown in Figure 2.1, two initial conditions were introduced. The force as a vector with three components (1000 N, 2000 N and –3000 N) and the mass—a scalar quantity with the value 20 kg. Then we calculated the length of the vector ||Force||—we determined the absolute value of the force acting upon the body, and the acceleration vector (three of its components along x, y, z axes). Many users of Mathcad mistakenly believe that the length of a vector is the number of its components, instead of the square root of the sum of squares of the values of its components. This confusion is due to the fact that, in Mathcad, the number of vector components is calculated using a built-in function called length (Figure 2.2). The operator for the "real" length of the vector (the vector norm) is in Mathcad 15 $|v|$ and in Mathcad Prime $||v||$.

REMARK

We utilized one of the useful tools of Mathcad Prime, the table. We filled in a part of a rectangular table—three rows and two columns (Figure 2.1), entering three force values (vector) and a mass value (scalar).

The problem of the acceleration of the body, solved in the previous version of Mathcad, is shown in Figure 2.2. In that version the force is not entered as a vector with three components (as in Figure 2.1), but as *Force* ^{x}, *Force* ^{y} *and Force* ^{z}. First, in the upper section of the calculation

Force (*N*)	*Mass* (*kg*)
1000	20
2000	
−3000	

$$\|Force\| = 3741.7\ N$$

$$Acceleration := \frac{Force}{Mass} = \begin{bmatrix} 50 \\ 100 \\ -150 \end{bmatrix} \frac{m}{s^2}$$

$$\|Acceleration\| = 187.08\ \frac{m}{s^2}$$

$$Force := Mass \cdot Acceleration = \begin{bmatrix} 1000 \\ 2000 \\ -3000 \end{bmatrix} N$$

FIGURE 2.1 Problem of a body motion in the space (Mathcad Prime).

$$ORIGIN := 1 \quad x := 1 \quad y := 2 \quad z := 3$$

$$Length(v) := \sqrt{\sum_{i=1}^{length(v)} (v_i)^2}$$

$$Force_x := 1000N \quad Force_y := 2000N \quad Force_z := -3000N \quad Mass := 20kg$$

$$length(Force) = 3 \qquad Length(Force) = 3741.7N$$

$$Accelration := \frac{Force}{Mass} = \begin{pmatrix} 50 \\ 100 \\ -150 \end{pmatrix} \frac{m}{s^2} \qquad Length(Accelration) = 187.08\frac{m}{s^2}$$

$$\text{Сила} := Mass \cdot Accelration = \begin{pmatrix} 1000 \\ 2000 \\ -3000 \end{pmatrix} N \qquad Length(Mass \cdot Accelration) = 3741.7N$$

FIGURE 2.2 Problem of a body motion in the space (Mathcad 15).

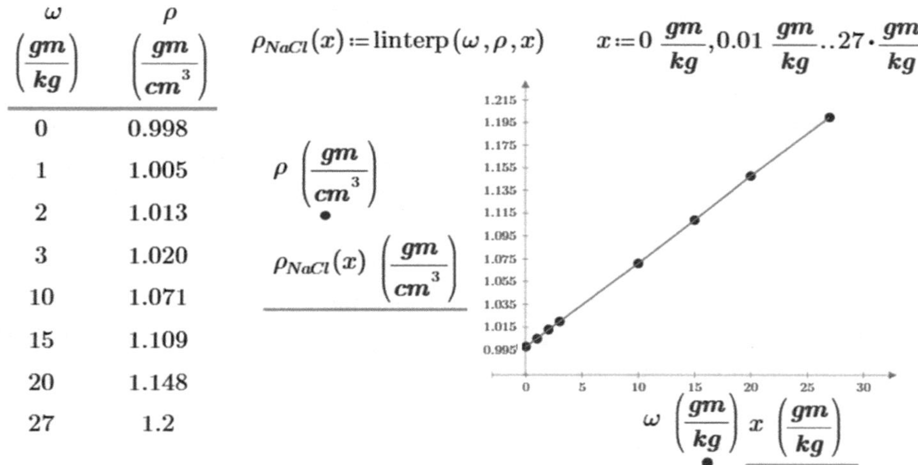

FIGURE 2.3 Data interpolation in Mathcad.

(top of Figure 2.2 up), the system variable ORIGIN* is set to 1. Second, variables x, y and z are set to 1, 2 and 3, respectively.

In the figure "acceleration" is misspelled. Also when teaching Mathcad I advise students never to use "m" for mass. I reserve "m" for the unit meter. One example: *mass* := 20 kg, not m := 20 kg

The components of the vector of force are denoted as $Force_x$, $Force_y$ and $Force_z$, instead of $Force_1$, $Force_2$ and $Force_3$, as this is more appropriate for the problem at hand and for decomposition of the vector into the three coordinates x, y and z.

Looking at Figures 2.1 and 2.2, one could imagine a real force (arrow-vector) of 3741.7 N attached to the body that is decomposed to three coordinates x, y and z with values of 1000 N, 2000 N and –3000 N, respectively. Such vectors (arrows) are often drawn in textbooks and physical books, illustrating the forces (Figure 7.7), e.g. direction, electric and magnetic fields and other vector physical quantities. By the way, in several textbooks and popular physical publications there are forces similar to those in the famous poem written by Krylov,† in which a swan, a crayfish and a pike are connected to a force vector. We will consider forces impacting a planet or a satellite due to the action of the Newton's law of universal gravitation. These forces have effects along two or three axes of the space.

The second type of data that also has the form of a vector in Mathcad, but is not so physical, is an array of data in a column. Figure 2.3 shows the solution to a problem of the linear interpolation of points.

The core of the problem is the following. Discrete values of the mass fraction, ω (the ratio of mass of a substance to the mass of the total mixture), of an aqueous solution of sodium chloride and its density ρ are given. It is necessary to create a continuous function that returns discrete values of the density of the salt solution. This function is needed to convert one type of concentration to another (Figures 1.6, 1.7). More information on this statistical operation will be provided in Study 14.

The coordinates of the interpolated data, which are displayed in Figure 2.3, are stored in two arrays (vectors ω and ρ). These arrays are vectors. However, they are not physical vectors, like the ones presented in Figures 2.1 and 2.2. Of course, it is possible to calculate the length of the vectors ω and ρ (the square root of the sum of the squares of the components), nevertheless, these values will not have any physical meaning (not the same meaning as in the problem displayed in Figures 2.1 and 2.2). In Figures 2.3 what is actually shown is a database of two arrays (ω and ρ) with eight records that are kept in the form of two vectors of the same size. It is possible to display these data graphically, to process them statistically (as shown in Figure 2.3), to sort and to order them. But by no means do these "vectors" have any relationship to n-dimensional space.

* The system variable ORIGIN is equal neither to zero nor to one, but to two (2) in only in one of the calculations described in this book (see Figure 16.2 in Study 16).

† Ivan Andreyevich Krylov, Russian fabulist, 1769–1844 (The swan, the pike and the crawfish, all of them pulling the cart in a different direction. As a result, the cart does not move, see poem at the beginning of this Study). We could also mention another fable related to "animal difficult direction", due to the Italian writer Gianni Rodari https://lasardegnainblog.wordpress.com/2015/03/13/gianni-rodari-the-young-prawn-il-giovane-gambero/.

We can draw a straight line through points as shown in Figure 2.3, and describe the dependence of density on the concentration of the saline solution by a simple formula $\rho = a + b \cdot \omega$. (A description of this procedure is introduced in Study 14.) However, the physical nature of the formula, which is based on a table of measured density and concentration values, will be lost in this case. So it is better to keep the formula as displayed in Figure 2.3, by inserting the two vectors (ω and ρ) into the function (Figure 2.4). This function will contain discrete data in this case. The function will also show the range of applicability of the formula for the mass fraction (from 0 to 27%). This helps to eliminate an extrapolation error.

The second difference between the vectors, shown in Figures 2.1 and 2.2 on the one hand, and the array shown in Figure 2.3 on the other, is in the numbering of the components. Generally, the numbering of a vector of components starts from 1 while the numbering of array elements starts from 0. In Figure 2.3 the setting, ORIGIN = 0, is the default. A third difference is the following. The components of the vectors showed in Figures 2.1 and 2.2 are very often called components (vector coordinates), while the components of the array in Figure 2.3 are commonly called elements (elements of a column matrix). Nevertheless, this difference is not so important.

As the reader has probably already noticed, the problems displayed in Figures 2.1–2.3 are solved with physical units: meters (m), seconds (s), kilograms (kg), etc. This is a very strong tool of Mathcad, which allows us effectively to check the correctness of the calculations (not to mix meters with kilograms, etc., roughly speaking) and to arrange a convenient input and output for the numerical values of the variables. An interesting semantic and technological question is whether an array/vector of Mathcad could have elements with differing dimensions. For example, whether one element has the dimension of length, the second of time with the third being dimensionless.

```
ρNaCl(ω) := "Density of NaCl water solution as function of the mass ratio"
            ω ← ω / (gm/kg)
            if ω<0
              return error ("ω<0")
            if ω>27
              return error ("ω>27 gm/kg")
            M ← [ 0      1      2      3     10     15     20     27 ]
                [ 0.998  1.005  1.013  1.02  1.071  1.109  1.148  1.2 ]
            ωt ← (M^⟨ORIGIN⟩)^T
            ρt ← (M^⟨ORIGIN+1⟩)^T
            linterp(ωt, ρt, ω)·gm/cm^3

ρNaCl(−1·gm/kg) = ?     ρNaCl(30·gm/kg) = ?  [ω>27 gm/kg]     ρNaCl(17·gm/kg) = 1.125 gm/cm^3
```

FIGURE 2.4 Calculation of the density as a function of the mass ratio.

In traditional Mathcad (e.g. Mathcad 15), such a possibility is not allowed. However, in Mathcad Prime it is permitted. The technological aspect of the issue is solved here, but what about the semantic one? Are there any real mathematical problems that require using vectors and arrays with different dimensions for their solutions?

Figure 2.5 illustrates such a problem: two points are given in the plane, with coordinates $[x_1, y_1]$ and $[x_2, y_2]$. We need to find the values of coefficients a and b of the equation of the line $y(x) = a + b$ passing through these two points. The solution of this problem in Mathcad Prime is shown in Figure 2.5.

The problem of the equation of a straight line passing through two points in the plane* could be reduced to the solution of a system of two linear algebraic equations (LAE) in the matrix form $M \cdot x = v$. Here M is the square matrix of the coefficients of the unknown x (x is the column vector), v is the column vector of zero-degree terms. To solve the system, we need to find the values of vector x, which consists of two components. The first value is quantity with the dimension of length, the second is dimensionless. The matrix M also contains values of different dimensions, the first column contains dimensionless values, while the values in the second column have the dimension of length. The question

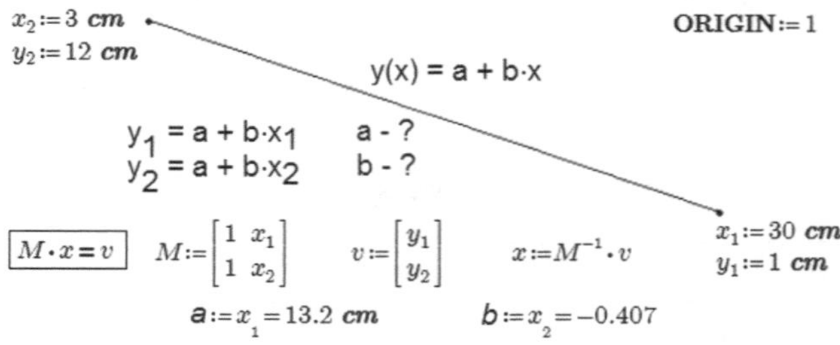

FIGURE 2.5 The problem of setting the equation of a straight line.

* We will solve this problem in Study 8 (Figure 8.6), when simulating the motion of car windscreen wipers, fixed at two points.

arises whether the vectors *x* and *v* (Figure 2.5) can be called vectors in a physical sense, as are the ones shown in Figures 2.1 and 2.2. Of course not! However, in the course of linear algebra, in the section dealing with the solution of systems of linear algebraic equations, the structures *x* and *v* are called vectors and *M* is called a matrix.

According to the criteria of classical programming languages, a vector is just a one-dimensional array, i.e. an array with only one index, while a matrix is a two-dimensional array, i.e. an array with two indexes. In programming, multidimensional vectors (three-, four-, …) are often used. If a scalar can be likened to a point, a vector to a line segment, a matrix to a rectangle, then a three-dimensional array could be likened to a parallelepiped.

In Mathcad, these three-dimensional structures can be created as a nested array. An example is illustrated in Figure 2.6, where the storing of discrete values of the function $f(x,y,z)$ is shown. Here $f(0.5, 15, 0.2) = 0.81$, $f(1, 10, 0.24) = 0.64$, etc.

A nested array is a vector or a matrix (a simple array) with one, two or more non-scalar elements, and has one or more vectors or matrices as elements. Talking about problems of physics that are solved in Mathcad, the term "tensor" should be mentioned. A tensor corresponds to an array of three-dimensions or more. In physics and engineering applications tensors of so-called "rank" 2 are very common and they look like a matrix (e.g. https://en.wikipedia.org/wiki/Cauchy_stress_tensor). A typical nested array in Mathcad is a vector storing three matrices—the intensity of the three colors of an image (Figure 5.7).

AFTERWORD

Going back to the beginning of Study 2, where we mentioned the author's experience in communication with students of computer science and mathematics at schools and universities, it is possible to observe a remarkable phenomenon. The author of this book has been giving lectures to students of computer science for nearly a quarter of a century (http://twt.mpei.ac.ru/ochkov/Potoki.htm). Twenty years ago, students were strongly interested in learning programming, they were almost hooked on it. Today, for many students some kind of allergy to learning how to code arises. There could be two reasons for this sad phenomenon. The first relates to a school side effect—students often reject what is taught at schools. For example, many people have an aversion to the compulsory books that are taught at schools. Twenty years ago, coding at schools was a new subject for students, and they were interested in it. Now they are not. The second reason of the students' loss of interest in learning coding seems related to the fact that teachers often assign very boring problems. These problems might have been interesting and challenging in the last century, nevertheless, new alternative methods may be available to solve these problems. Pupils and students

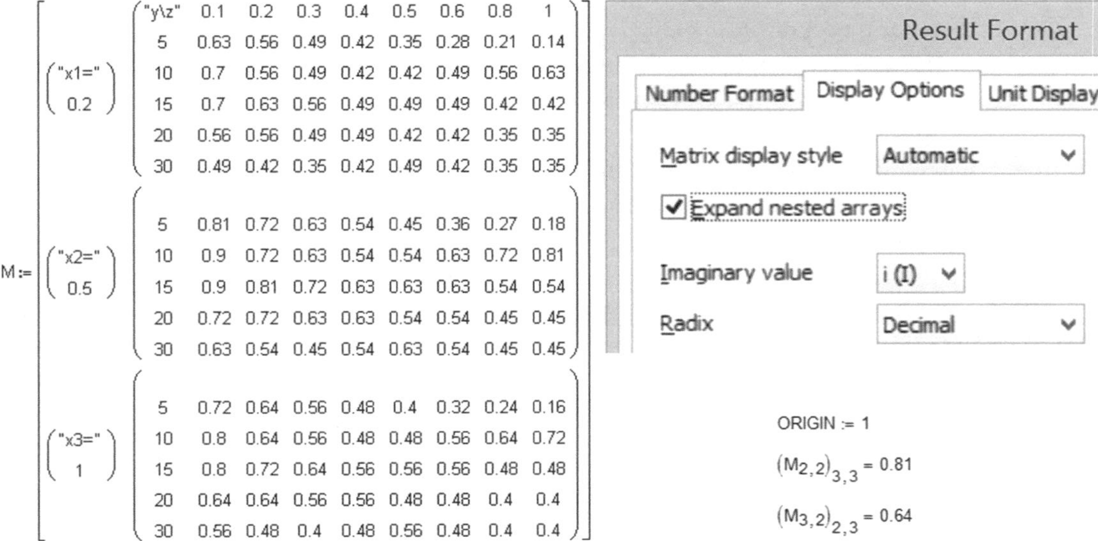

FIGURE 2.6 An example of a nested array in Mathcad.

perceive this situation and they know how to solve the problem presented in new and different ways. Let us give an example. Figure 2.2 shows how the function "length", which returns the length of a vector, is created using a summation operator in the current Mathcad. But at schools and universities some teachers and lecturers of informatics still require pupils and students to create this or other similarly simple functions using classical programming (Basic, Pascal, etc.). Teachers ask this, not knowing, or ignoring, the fact that in modern software built-in functions and operators are implemented. By the way, by means of coding in Mathcad (Figure 2.7), the user-defined function called the length can be created by using a FOR loop (loop with a parameter).

For example, in Mathcad there is a very interesting function called *match*(z, A), that returns the coordinates of the element with a value z in the matrix A. If there are more elements with the same value z, then the *match* function returns a nested array (Figure 2.6)—vector, whose elements (new vectors with two components) will give the row index and the column index of the position of element z. (We will use this function in programming the traveling salesman problem in Study 15.) Of course, the teacher could ask students to create an appropriate program with loops and statements that realizes the *match* function. However, it seems better to assign to students problems whose solution is not so well known or is not prepared in advance. To assign students to code the *match* function is the same as to ask them to write a program for the calculation of the *sinus* function using the Taylor series and to conceal that the *sinus* function is built into any calculator and programming language.

This issue could be more widely applied and transferred to other school subjects and university courses, such as physics, mathematics or chemistry.

Actually, many school disciplines are still taught using methods used before computers or even calculators were available. We mentioned it in the remark of Study 1. Let us mention physics, for instance, with which we started Study 1. When a pupil or student is working on a physics assignment, he/she must remember or

```
ORIGIN ≡ 1
Length(v) := | Sum ← 0
             | for i ∈ 1.. length(v)
             |    Sum ← Sum + (v_i)^2
             | return √Sum

Force := ( 1000N  )
         ( 2000N  )
         ( -3000N )

length(Force) = 3
Length(Force) = 3741.66N
```

FIGURE 2.7 Program in Mathcad 15.

find an appropriate formula or a set of formulas in the textbook. These formulas are only special cases of the solutions of equation systems (for example algebraic, differential, integral, etc.) that reflect, for example, the fundamental laws of conservation of mass or energy, transfer of energy from one form to another. Many years ago, an emphasis on the use of ready-made solutions, rather than on finding the problem solution in its general form, was proposed for a simple and obvious reason. At that time, when educational methods of problem-solving were created, there weren't many effective and affordable ways to solve systems of equations that could reflect the fundamental laws of physics, chemistry and other scientific and academic disciplines. Now, however, these tools have been developed and this requires a fundamental revision of problem-solving techniques. Textbooks should be rewritten as well. We introduced this topic in more detail in Study 1.

Let us give a specific example. We need to calculate how high a stone will travel if it is thrown straight up. This problem could be solved using the formula written in a textbook, where the height is calculated from initial velocity and acceleration of free fall. If the problem is more complicated, for instance, if the stone is thrown not strictly vertically, but at a certain angle to the vertical, we will need to find new formulas. Nevertheless, it is possible to approach the problem in a different way. We can take it as a system of differential equations and solve it analytically (to get the same formulas with particular solutions) or numerically.

In this case, a person does not need to look for any formula. What is needed is knowledge of the fundamental laws of nature, to be able to translate them into the language of mathematics (to set systems of the same differential equations and others) and to solve them using modern computer technologies. Another advantage of this approach is that many restrictions that reduce the accuracy of the results can be removed. In the case of the stone we could also take into account air resistance, changing its density with height, changing the value of free-fall acceleration with height, the curvature of the earth's surface and more.

All these phenomena can be included if we shift from creating sets of formulas to the solution of the fundamental equations. The example in Figure 2.8 shows how to solve smartly the problem of an accelerating car with mass m and cross-sectional area s, which moves through the air of density ρ where the friction coefficient is k.

N	m	S	ρ	k	v_0	t_{end}
(kgf)	(kg)	(m^2)	$\left(\frac{kg}{m^3}\right)$	$\left(\frac{N}{N}\right)$	(kph)	(s)
300	1200	2.5	1.2	2	0	30

Solve

$$v(0\ s) = v_0$$

Inertia + Air resistance = Driving force

$$m \cdot v'(t) + k \cdot \rho \cdot S \cdot v(t)^2 = N$$

$$v := \mathbf{odesolve}\left(v(t), t_{end}\right)$$

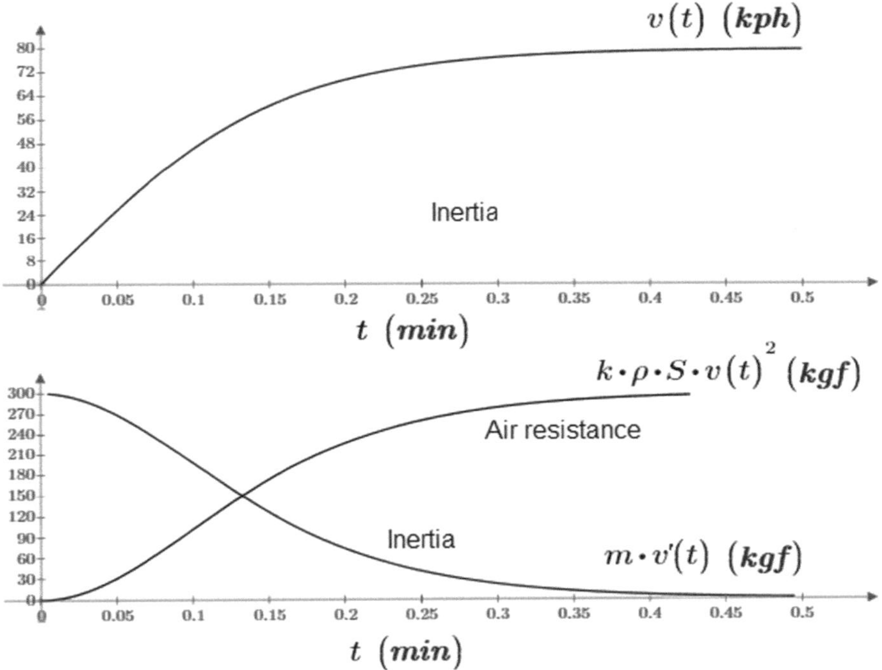

FIGURE 2.8 Problem about an accelerating car.

If required, the change of the driving force N according to time, the rolling resistance of wheels on the road and other factors can also be taken into account. In Study 4 a similar problem will be solved for a motorboat (Figure 4.21) and in Study 9 for an Earth satellite.

Finally, the author once spoke to teachers from a very good Moscow school about his experience with the development of using of mathematical packages in teaching mathematics, physics and chemistry.

The teachers told him that it is very good and attractive, but their school is now renamed; it is no longer "a secondary general education school" but "a preparatory school for the common state exam in mathematics, physics, chemistry, computer science, …". Exam questions are made in such a specific way that to answer them it is not necessary to use a computer or even a simple calculator. This in spite of the fact that students, engineers, scientists or office workers currently cannot work without these technologies. (This issue relates to the isolation of school from real life.) Earlier, before the era of the state exams, teachers used to focus on students' strengths in math, and would make lessons more creative and interesting, preparing creative and original problems beyond those in the textbooks. The challenge of the problems would have been tailored according to a particular student's abilities, thus making them appropriate to the student's level. Current trends and demands are to prepare all students for the state exam, without exception. Based on the results of these exams, students, teachers and schools are evaluated. There is another obstacle. Of course, one can agree that the state exam system helped poor students and below-average teachers to reach better results, but unfortunately, this system also lowered the standard for many above-average students and teachers. This educational model may be optimal for some people. However, if we want to maintain a scientific standard and level, the educational model, very often bureaucratic rather than professional, must be changed.

We suggest the future of state exams in mathematics, physics or chemistry might look like the following. A tested student uses a computer to solve the exam assignment. To pass the exam it is possible to use all available and permitted ways. The student is allowed to work with a sheet of paper, any software of his/her own choice, to search the Internet, to use any user forum and to ask for information, or even to organize a discussion about the problem. Answers given by students could be evaluated automatically using a computer system. Of course, questions and problems stated in such an exam would need to be sophisticated and the preparation of the assignments would be very difficult (but wouldn't pose an insurmountable obstacle).

CONCLUSIONS

Without perfect understanding of basic concepts, such as scalar, vector, matrix, table, array (one-, two- and multidimensional), database and knowledge base, learning informatics will be very difficult. At high schools it is possible and necessary to organize some preparation courses for the state exams. However, we recommend the use of computers in such courses. Moreover, using computers is strongly preferable in lessons focused on mathematics, physics, chemistry or computer science. Problems similar to those which are presented in this book are appropriate for the exam. Incidentally, in Study 14 we will discuss the correlation of the results in the state exams of mathematics with the real mathematical knowledge of university freshmen.

The Gold-Bug Or

Roman-Arab Cryptography

ABSTRACT

This Study explains numeral systems used in mathematics, how to work with variables, how to store character strings (texts) and simple and complex methods of information encryption. The Study discusses the problems of translating the numbers of the Roman system into Arabic and back. The work of an ATM to translate a numerical record of amounts into a text record is discussed.

Mathematics: Numeral systems: positional and non-positional, probability, system of linear algebraic equations (SLAE), fractions, greatest common divisor (GCD), simple multipliers, derivative.

Physics: Heating of a parchment.

IT: Binary, octal, decimal and hexadecimal number format, pre-test loop, string variables and functions for working with them.

Art: Edgar Allan Poe, *The Gold-Bug.*

Study website: https://community.ptc.com/t5/PTC-Mathcad-Questions/PhysMathStudy-3/m-p/38576.

A **numeral system** is a method of writing numbers and corresponding rules of operation over them. People have come up with several numeral systems. It is rather difficult to classify them, because they have to be compared for different properties.

The first property is notation, i.e. a set of symbols is used to write numbers. The notation can be digital (0, 1, 2, 3, 4, 5, 6, 7, 8, 9) alphabetic (I, V, X, L, C, D, M), icon-based (hieroglyphs, cuneiform and other icons) or mixed (numbers and letters or letters with icons). For example, in the Old Russian system, letters were used to write numbers, supplemented by different icons—tildes, arrows and solid and discontinuous circles. At the beginning of the eighteenth century Russia used a mixed system of writing numbers, consisting of Cyrillic and Arabic numerals. For example, on some copper half-coins (coins of denomination 1/4 kopecks) there are dates that appear as 17K (1720) and 17K1 (1721).

The second property is the base of the number system, i.e. the basic set of numbers (in different notations) that is sufficient for the expression of all the numerical values that are meaningful for practical use for the life of a person of the corresponding civilization and culture. Such systems include unary (a stick), binary (0 and 1), quinary, octal, decimal, duodecimal, vigesimal and sexagesimal. For example, in hexadecimal, the base is the number 16, which is indicated by "10", and the series of natural numbers from 1 to 16 is written as 1, 2, 3, 4, 5, 6, 7, 8, 9, A, B, C, D, E, F, 10.

The third property is a set of operations with the characters of the number entry. There are additive and multiplicative systems.

The fourth property is the value of the symbol of the number entry. There are non-positional systems in which the value of the symbol does not depend on its position in the number entry. For example, in the

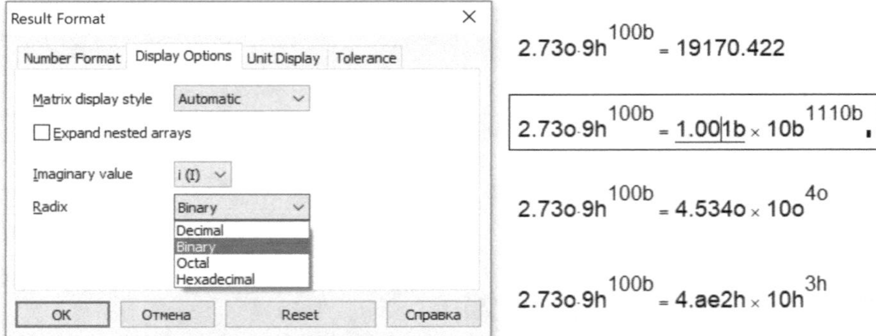

FIGURE 3.1 Numeral systems in Mathcad 15.

Roman number XX (twenty) the symbol X occurs twice and, in each case, means the same value—the number 10, with twice 10 in the total giving 20. However, there are also positional systems. In them, the value of a symbol uniquely depends both on its position and on the rule of reading the number (from left to right or vice versa). In this case, the "weight" of each symbol varies depending on its position in a sequence of symbols representing the number. For example, in the Arabic decimal number 353.3 the first three means three hundred, the second three means three units, and the third three means three-tenths of a unit. In some positional systems there are special rules for reading and writing numbers. Thus, in the Old Russian Glagolitic numeral system, numbers, with the exception of those from 11 to 19, were pronounced from right to left (for example, 27 would be seven-by-twenty). Some "failures" in decimal can also be observed at the present time. Modern Germans, for example, say not "twenty-one", but "one and twenty", not "seventy-four", but "four and seventy". The same German will never say "one thousand two hundred", he will say "twelve hundred". The French number 75 is pronounced as 60 and 15. The subject of counting using countable sticks is non-positional, single and additive. Whichever stick from the set we take, it always means "one". But this system can be complicated by coloring the sticks in different colors and giving different colors a different "weight".

In the third millennium BC or thereabouts, the ancient Egyptians came up with their particular numeral system in which they used special icons—hieroglyphics—to designate key numbers 1, 10, 100, etc. All other numbers were composed of these base numbers using the addition operation. The numeral system of Ancient Egypt was decimal, non-positional and additive.

An important topic of computer science lessons at school is in fact the numeral system: *binary, octal, decimal, hexadecimal.* Students should be able, at least, to convert the decimal number to binary and vice versa. Mathcad 15 has built-in tools for working with these numbers. In Figure 3.1, you can see a number composed of three digits in different forms (2.73o is an octal number, 9h is a hexadecimal number and 100b is a binary number). They are "printed" in four built-in formats in Mathcad—binary, octal, decimal and hexadecimal. If the number in Figure 3.1 does not end with the suffix b (**b**inary), o (**o**ctal) or h (**h**exadecimal), then it is the decimal number—see the first answer in Figure 3.1.

The interesting error with the hexadecimal form of the number record is $h = 3$ $2h = 2$. What's the matter, why is 2 times 3 (times the variable h), equal to two, and not to six. Many users are perplexed and suggest that either Mathcad has broken down or has a bug, or that they themselves have gone crazy. Similar incidents are collected at the author's website "Mathcad Mad House—Crazy House Mathcad", at the PTC community forum.* The fact is that the expression $2h$ is not only a duplication of the value in the variable h (and Mathcad is so efficient that it adapts to the general mathematical rule that allows implicit multiplication), but also simply a numerical constant that is 2 in hexadecimal format, that is, 2 also in the decimal format. Hence the conclusion is that the multiplication symbol

* Another similar and very common incident with the multiplication operator is the following. If we enter two operators f:= 2 f (1 + 2) and press the "=" key after the second statement, we get a non-numerical answer (6), and an error message. Mathcad will calculate that we are trying to call a user function with name f, and so not multiply the value of the variable named f by what is written in parentheses. Mathcad often adds a multiplier to the user, but this option is sometimes cumbersome.

FIGURE 3.2 Website for converting numbers written in one numeral system to another numeral system.

should also be introduced where mathematicians usually do not introduce it—between a numerical constant and a variable.

In Mathcad 15, it's easy to create a user-defined function for converting numbers from one numeral system to another. Figure 3.2 shows such a calculation, accessible through the Internet, which allows the representation of numbers in various numeral systems: from binary to hexadecimal. Thus, the ternary number 020202 is equivalent to the sevenfold number 350, as it is evident from Figure 3.2. This calculation website is created in line with the idea set out in Study 1—that we give out not only the result of the calculation but also all the formulas through which they are executed.

With figures in the binary, octal and other representations in everyday life, we almost do not have a case.* When solving a problem on a computer or simply by using a calculator, we input the usual decimal numbers. The computer translates them into a binary format, performs the necessary actions with them, receives a binary response, which is converted to a decimal format and then displayed on the screen.

Let us now talk about notations for numbers. More familiar to us is the Roman representation of numbers—Roman numerals. We deal with them almost every day, and it does not seem strange. And not only when we look at a clock with Roman numbers on the dial, but also when we withdraw money with card in an ATM (**automated teller machine**), for example. At the ATM we enter the Arabic number ($389, for example), and the ATM gives us the Roman number (Figure 3.3): three hundred eighty-nine and dollars. We conduct a "Roman" account of money also when we are asked to write down in financial documents a monetary sum not by numbers, but in words: "Three hundred and eighty-nine dollars".

Figure 3.3 shows a program function written in Mathcad Prime, which converts the amount of money

* Therefore, probably in Mathcad Prime there are no more binary, octal and hexadecimal input and output formats.

```
f(n) := | M ← [ 100      50      20      10      5       2      1     ]
        |      [ "100$ " "50$ " "20$ " "10$ " "5$$ " "2$ " "1$ " ]
        | R ← ""
        | while n>0
        |   | while n≥M_{0,i}
        |   |   | R ← concat(R, M_{1,i})
        |   |   | n ← n − M_{0,i}
        |   | i ← i+1
        | R

f(389) = "100$ 100$ 100$ 50$ 20$ 10$ 5$$ 2$ 2$ "
```

FIGURE 3.3 Program function for decomposing an amount of money into banknotes.

into a set of banknotes. If in this program the matrix of money is replaced by a matrix of Roman numbers (Figure 3.4), then it will convert Arabic numerals into Roman numerals.

The algorithm for converting an Arabic number to a Roman one is pretty simple: the Arabic numeral n, "Roman banknotes and coins" (the second row of the M matrix) is removed until the "Arab" money runs out (while $n > 0$). In this case, the algorithm checks whether there is a corresponding ith banknote or coin for this operation (while $n \geq M_{0,i}$). The concat function built into Mathcad concatenates a string of characters—the denomination of banknotes and coins.

It is more difficult to write an inverse function—a function for converting a Roman number into Arabic. We carry out this operation in our minds when we see an old building with an "encrypted" year of construction in the form of a Roman number on a pediment or a monument with "Roman" birth and death dates. The task can be simplified by using the direct (initial) function of converting an Arabic number to a Roman one (Figures 3.3 and 3.4): all Arabic numerals starting from number one, are translated into a "Roman" form, which is then compared with the original Roman number. If there is coincidence, the found Arabic number is displayed "for printing" (see Figure 3.5a).

The author saw in his dacha[*] an unusual use of Roman numbers. A bathhouse was built on a nearby plot—a log house was assembled from logs, which were marked by notches: pairs of Roman numbers: I, II, III, IV, etc. The first notch signified the wall number of the bath from the first (I) to the fourth (IV), and the second—the number of the log in this wall from the lowest to the highest. This frame was made somewhere far away and transported to a new location. Before they left, the carpenters marked the walls with Roman numerals because Arabic numerals would be difficult to write with an ax or a grinder. They were quite competent carpenters. After all, it was possible to make just the notches I, II, III, IIII,[†] IIIII, IIIIII, etc.

To the scholastic form of writing numbers we can add simple fractions, which we will examine below (see Figure 3.7).

Roman numbers could be considered as some kind of encrypted Arabic numbers. In this regard, during lessons of computer science it is possible to offer students the following game: to create a matrix similar to the one shown in Figure 3.4, and to encrypt Arabic numbers on the matrix for other students to decipher.

Informatics is the science of working with information. In particular, its protection from external corruption. The section of computer science applied to protect written texts is called *cryptography*. The decoding of this cryptographic text is colorfully described in the story of Edgar Allan Poe "The Gold-Bug". This American writer is rightfully called the father of the detective genre in literature.

The plot of the story is uncomplicated. A person finds a parchment on the beach, brings it home and accidentally leaves it near the fire.

After heating, mysterious symbols appear on the parchment: 53 ‡‡ † 305)) 6 *; 4826) 4 ‡.) 4 ‡); 806 *; 48 † 8¶60)) 85; 1 ‡ (;: * 8 † 83 (88) 5 * †; 46 (88 * 96 *?; 8) * ‡ (485); 5 * † 2: * ‡ (; 4956 * 2 (5 * –4) 8¶8 *; 4069285);) 6 † 8) 4 ‡ ‡; 1 (9; 48081; 8: 8; 1; 48; 85; 4) 485; 528,806 * 81 (‡ 9, 48, (88; 4 (‡? 34; 48) 4 ‡; 161;: 188; ‡ ?;

We can sometimes see such chains of "mysterious" characters on a computer's display when an incorrect encoding of the text is selected.

The person who finds the parchment, deciphers the text and finds a treasure hidden by pirates.

This cipher was based on replacing the letters of the source text with different letters or symbols. This is one of the oldest and easiest ways to encrypt texts. To date, a large number of encryption systems of varying complexity have been developed. A new impetus to this

[*] A dacha is a seasonal or year-round second home, often located in the exurbs of Russian-speaking and other post-Soviet countries. The author's dacha is one of the sources of the themes of this book: see Study 7 about a brick suspended on electric wires, or Study 12 about a fire bucket.

[†] Roman numerals sometimes write four as IIII, but never write forty as XXXX.

```
Arab2Rim(n) := | M ← [ 1000  900   500  400   100  90    50   40    10   9     5    4     1  ]
               |     [ "M"   "CM"  "D"  "CD"  "C"  "XC"  "L"  "XL"  "X"  "IX"  "V"  "IV"  "I" ]
               | R ← ""
               | while n > 0
               |   | while n ≥ M_{0,i}
               |   |   | R ← concat(R, M_{1,i})
               |   |   | n ← n − M_{0,i}
               |   | i ← i + 1
               | R
```

Arab2Rim(1828) = "MDCCCXXVIII"   *Arab2Rim*(2019) = "MMXIX"

FIGURE 3.4 Conversion of Arabic numbers into Roman.

(a)

```
Rim2Arab(R) := | n ← 1
               | while Arab2Rim(n) ≠ R
               |   | n ← n + 1
               | n
```

Rim2Arab("MDCCCXXVIII") = 1828

Rim2Arab("MMXIX") = 2019

(b)

```
RA(n) := | AV ← (1000 900 500 400 100 90 50 40 10 9 5 4 1)^T
         | RV ← ("M" "CM" "D" "CD" "C" "XC" "L" "XL" "X" "IX" "V" "IV" "I")^T
         | (A ← 0  i ← 0)
         | while strlen(n) > 0
         |   | while RV_i = substr(n, 0, strlen(RV_i))
         |   |   (A ← A + AV_i  n ← substr(n, strlen(RV_i), strlen(n) − strlen(RV_i)))
         |   | i ← i + 1
         | return A

RA("MCMXLVIII") = 1948     RA("MCMXCVIII") = 1998     RA("MCMXXVIII") = 1923 ???
```

FIGURE 3.5 Translation of a Roman number into Arabic: (a) using the previously defined direct function; and (b) without using the previously specified direct function (Why upside down—see the first assignment for the readers at the end of this Study).

process was the development of computer information technologies. Even when sending information within a computer, we "encrypt" it—by translating the usual letters and numbers into zeros and ones: into chains of "mysterious symbols".

Computers, when sending information to each other over local networks or over the Internet, can use special encryption algorithms to ensure that the transmitted information cannot be read by someone who is not supposed to.

Deciphering the text based on the replacement of characters consists of two stages—routine and creative respectively. The routine stage is the calculation of the frequency of use of certain symbols in the source text. The hero of the story of Edgar Allan Poe did it manually. We can count it on a computer by writing a simple program in the Mathcad language—see Figure 3.6.

A string of mysterious characters from the story of Edgar Allan Poe is inserted into the Text variable (see Figure 3.6) (copied from the Internet). Only the beginning of this chain with 203 characters is shown. The Mathcad built-in function strlen (str = string and len = length) counts the number of characters in the variable storing the string. We need one more Mathcad string

```
text := "53‡‡†305))6*;4826)4‡.)4‡);806*;48†8¶60))85;1‡(;:‡*8†83(88)5*†;46(;88*96*?;8)*‡(;48

v := | [i ← 0  v[0,0] ← "5"  v[0,1] ← 1]
     | for j ∈ 1..strlen(text) − 1
     |   | [New_Symbol ← "Yes"  Symbol ← substr(text, j, 1)]
     |   | for k ∈ 0..i
     |   |   | if v[k,0] = Symbol
     |   |   |   | [New_Symbol ← "Нет"  v[k,1] ← v[k,1] + 1]
     |   | if New_Symbol = "Yes"
     |   |   | [i ← i + 1  v[i,0] ← Symbol  v[i,1] ← 1]
     | csort(v, 1)^T
```

$$v = \begin{bmatrix} \text{“\_”} & \text{“.”} & \text{“¶”} & \text{“?”} & \text{“:”} & \text{“3”} & \text{“9”} & \text{“2”} & \text{“0”} & \text{“†”} & \text{“1”} & \text{“(”} & \text{“6”} & \text{“5”} & \text{“*”} & \text{“‡”} & \text{“)”} & \text{“4”} & \text{“;”} & \text{“8”} \\ 1 & 1 & 2 & 3 & 4 & 4 & 5 & 5 & 6 & 8 & 8 & 10 & 11 & 12 & 13 & 16 & 16 & 19 & 26 & 33 \end{bmatrix}$$

FIGURE 3.6 Program for counting the frequency of symbols in a text.

function—the function substr that returns a part of the string starting from the right place (the second argument of the substr function), and the required length (the third argument). These three built-in functions were also used in the translation of the Roman number into Arabic—see the inverted Figure 3.5b.

In the program shown in Figure 3.6, two columns will be formed in the matrix *v*—the symbol itself and the frequency of its occurrence in the text. The first line of the program records the first row of this matrix—in our case numbers 5 and 1. Further in the program, a cycle with the parameter *j* passes through the string variable *Text* from its second element (the first one is already fixed in the matrix *v*) to the last. It is assumed that the next character of the *text*, fixed in the *Symbol* variable, was not previously encountered: the value "Yes" is entered in the *NewSymbol* variable. In the second (nested) cycle with the parameter *k*, this assumption is checked by a search of previously formed rows of the matrix *v*.

If it turns out that the next character already occurred in the text ($Symbol = v_{k,0}$) and our assumption becomes erroneous (the *NewSymbol* variable takes the value "No"), then the value of the element of the matrix $v_{k,1}$ is increased by one. If the next character from the *Text* variable in the matrix *v* is not fixed yet (the variable *NewSymbol* has kept the value "Yes" to exit the loop with the parameter *k*), then a new row is created in the matrix v ($i \leftarrow i + 1$), where the new symbol itself is entered ($v_{i,0}$ ← Symbol) and one ($v_{i,1} \leftarrow 1$). The program returns the sorted and transposed matrix *v*—see the bottom of the figure.

Our encrypted text contains 203 characters. There are only 20 distinct symbols in the text, of which 8 occur most often (33 times). This short text could, of course, be analyzed without a computer, as the hero of the story of Edgar Allan Poe did (the story was published for the first time in 1848).* But through our program you could analyze more extensive texts. This will allow you to find out which letters in a particular language are used more often, and which are less frequent. This same work can be done taking as a basis not letters but individual words. In the English language on which the story "The Gold-Bug" was written, the most "frequent" letter is the letter e, and the most frequent word is the definite article. This was the key to deciphering the text in the story of Edgar Allan Poe (the creative part of the decipherment). How?! Read or reread the story again (https://www.eapoe.org/works/tales/goldbga2.htm).

Edgar Allan Poe's story "The Gold-Bug" aroused great public interest in cryptography in his time (mid-nineteenth century). There is also the famous story of Arthur Conan Doyle "The Adventure of the Dancing Men",† whose plot is based on the decoding of chains of symbols as well.

---

* This Edgar Allan Poe short story is available for instance at: https://poestories.com/read/goldbug.

† "The Adventure of the Dancing Men" is available at: https://sherlock-holm.es/stories/pdf/a4/1-sided/danc.pdf.

However, if in the past only the military and diplomats were engaged in cryptography,* nowadays everyone faces this important section of computer science directly or indirectly, when talking on a cell phone or taking money from an ATM.

We hope that our simple program that helps decrypting a simple code will become a starting point for learning more complex algorithms of cryptography. In the meantime, you can play this game. Two people develop a code for correspondence based on replacing one character with another, and the third "intercepting" and analyzing these messages reveals "encryption". In such messages, spaces and punctuation marks are usually omitted and capital letters are ignored. This was done by encrypting the story of Edgar Allan Poe. In the story of Arthur Conan Doyle, a dancing man with a flag in his hand marked the beginning of the word. How much time is needed to decrypt these intercepted messages is a separate matter. Rather, we focus on the task of creating a program for the disclosure of the cipher key. Part of this program is shown in Figure 3.6.

The cipher from Edgar Allan Poe's story is based on replacing the letters of the source text with different letters or symbols. This is one of the oldest and easiest ways to encrypt texts. This code is the **replacement cipher**. However, in the **permutation cipher**, the symbols are reversed, observing a certain regularity.

By now, a great number of encryption systems of various complexity have been developed. Claude Shannon summarized the experience of developing ciphers in his paper "Communication Theory of Secrecy Systems" (1949, http://pages.cs.wisc.edu/~rist/642-spring-2014/shannon-secrecy.pdf). He found out that in any complex ciphers, replacement ciphers, as well as shift ciphers or their combinations, can be identified as typical components.

When using digital computers, it is convenient to represent information in the form of sequences of zeros and ones. On the other hand, words in any alphabet can easily be translated into binary words. Consider, for example, the telegraph code—the old technical application of the binary number system. It consists of 32 symbols: binary words of length 5. We associate with each letter a binary word of length 5 as follows: space → 00000, A → 00001, B → 00010, B → 00011, G → 00100, D → 00101, …, → 11110, Y → 11111.

It is easy to compile a program for telegraphic encryption and decryption. Let's instruct the reader how to do this.

The question of what is the smallest length of a word in the telegraphic coding is solved very simply for each alphabet. Each letter in the alphabet has its own serial number. In the Latin alphabet (basic version) there are only 26 letters. In the binary notation, we have $26 = 16 + 8 + 2 = 2^4 + 2^3 + 2^1 = 11010$. Again we have a five-character binary word. And how many letters should be in the alphabet so that a telegraphic code contains only four-letter words? It's easy to count: $1111 = 1 + 2 + 4 + 8 = 15$.

In real devices, when transmitting signals consisting of zeros and ones, due to various forms of interference, the number of digits may change. Then the correct interpretation can be carried out only with a certain probability.

Let the same symbol be received with the probability *p1* when the symbol "0" is transmitted, and when the symbol "1" is transmitted, the same symbol will be obtained with the probability *p2*. Let m be the number of zeros and n the number of ones in a transmitted message. Then the probability that the message will be received correctly, is given by the formula

$$P = p_1^m \cdot p_2^n.$$

Since any probability is equal to a number from 0 to 1, and the natural powers of such numbers (not equal to 1 and 0) are less than the number itself, it becomes clear that as the number of letters in the message increases, the probability of obtaining an absolutely accurate message decreases.

We suggest the reader writes a program, considering for example, *p1* = 0.98 and *p1* = 0.97, assuming m and n equal to 10, from 10 to 100, more than 100 (different).

In real life, a probability above 0.95 is considered sufficient. It is possible to find out at what maximum message length and for what variants of values of m and n the required probability is not less than 0.95. For this it is sufficient to solve the inequality by a numerical search method. Obviously, it's very difficult to do this manually, but with the help of a computer it's easy enough. Reader! To work!

If a logically connected message is transmitted, for example, an everyday or literary text in some language, then a small number of defects in the received message

* On this subject, for instance, the 2001 movie *Enigma* was partly inspired by the life of the great mathematician Alan Turing and his contribution to the Allied forces in the Second World War https://en.wikipedia.org/wiki/Enigma_(2001_film).

is not an obstacle to understanding. For example, when talking on a mobile phone or on a walkie-talkie, if there are small jams on the line, we can understand the interlocutor even from fragments of phrases.* But if logically unrelated symbols or sets of numbers are transmitted, the incorrectly received message will not only be useless, but can also cause serious problems. To insure against such cases, the messages are duplicated, i.e. They are transmitted two or three (or more) times. Then, when a message is sent k times, the probability that at least one of the transmission attempts will lead to an absolutely accurate result is calculated by the formula

$$P = 1 - \left(1 - p_1^m \cdot p_2^n\right)^k .$$

Obviously, for a large number of attempts, the probability will be close to one. However, there is no sense in making a large number of attempts, it is better to look for the best option. For example, to get $P \geq 0.99$ implying $\left(1 - p_1^m \cdot p_2^n\right)^k \leq 0.01$, the reader could write the corresponding program after selecting an interesting case for the message.

Now we will consider another type of cipher. A permutation code is based on the permutation of letters in the source text. Usually this text is divided into segments of equal length and each segment is encrypted (i.e. the letters are interchanged) independently.

The simplest permutation cipher is the following. We take a rectangular table of appropriate size (matrix), we enter the original text into it "by lines" without spaces, while the encrypted text is read "by columns". Obviously, you should choose the message size so that there are several variants of the matrix.

Since all the lines must be filled, at the end of the true message you can assign arbitrarily the requested number of different letters. For example, let's consider a message consisting of 114 letters. Since 114 = 6 • 19, we break the text into 6 or 14 lines and encrypt it accordingly. It is possible to add 6 arbitrary letters at the end, and we will then have more options, because 120 = 2 • 60 = 3 • 40 = 4 • 30 = 5 • 24 = 6 • 20 = 8 • 15 = 10 • 12, P. Any of the options suits. The reader can write this simple program independently.

There are further clever ways of writing and decoding.

If it is required to encrypt n natural numbers $x_1, \ldots, x_n$, then we can proceed as follows. Instead of these numbers, you can pass other natural numbers $b_{11}, \ldots, b_{nn}$, $a_1, \ldots, a_n b$, agreeing in advance that $x_1 b_{11} + \ldots + x_n b_{1n} = a_1$, etc. In this case, the original numbers are uniquely located as a solution of a system of n linear equations with n unknowns. This is another problem for the reader!

ASSIGNMENTS FOR READERS

1. Create a program that translates a Roman number into Arabic, not through using the reverse program (see Figure 3.5), but directly. If there are difficulties, we can look at Figure 3.5b, just by turning the book upside down.

2. Write a program to determine the frequency of encountering a word in a text. The text is not stored in a variable as a string (see Figure 3.6), but in a text file on a disk, reading it using the built-in Mathcad tools provided for this purpose.

3. Create a program for the encryption of numbers based on Fibonacci numbers

* The reason for Maximilian Andreevich's coming to Moscow was a telegram received late in the evening two days before with the following content: Have just been run over by tram-car at Patriarch's Ponds funeral Friday three pm come. Berlioz.Maximilian Andreevich was considered one of the most intelligent men in Kiev, and deservedly so. But even the most intelligent man might have been nonplussed by such a telegram. If someone sends a telegram saying he has been run over, it is clear that he has not died of it. But then, what was this about a funeral? Or was he in a bad way and foreseeing death? That was possible, but such precision was strange in the highest degree: how could he know he would be buried on Friday at three pm? An astonishing telegram!However, intelligence is granted to intelligent people so they can sort out entangled affairs. It was very simple. A mistake had been made, and the message had been distorted. The word 'have' had undoubtedly been taken from some other telegram in place of the word "Berlioz", which got moved and wound up at the end of the telegram. With such an emendation, the meaning of the telegram became clear, though, of course, tragic.Mikhail Bulgakov, *Master and Margarita*, Richard Pevear and Larissa Volokhonsky, English translation and notes

Solvers Or

The Mathcad Magnificent Seven

ABSTRACT

In this Study it is shown which methods are available in Mathcad for the solution of equations and systems of equations, which equations could have been solved before the computer era and which ones can be solved only now. A new original method for assessing the quality of methods for solving equations based on the analysis of mosaic patterns is discussed.

Mathematics: Equation roots, zero of a function, maximum of a function, linear system of algebraic equations, polynomial, section area, types of equations, analytical methods for the solution of an equation, optimal chain line (catenary).

Physics: Path, time, velocity, equation of motion, uniform motion, alternating motion, sliding friction.

IT: Analytical and numerical solution of an equation, computer analytical transformations, Newton method (tangents), bisection method, tracing tools for numerical solutions, tracing and enlarging graphics, accuracy of solving systems of equations, color in programs.

Art: Vincent van Gogh, *Landscape in the Over after the Rain*; Leo Tolstoy novel, *Anna Karenina*; knitting art

Study website: https://community.ptc.com/t5/PTC-Mathcad/Chapter-4/m-p/606046.

INTRODUCTION

Many problems in mathematics, physics, chemistry, mechanics, thermodynamics and other school and university disciplines reduce to solving equations and systems of equations. Therefore, it will be useful to find out which tools for solving such problems are available in Mathcad—which is very popular among schoolchildren and students of engineering and science. These tools are grouped into the "Solving Equations" group of the built-in Mathcad functions that use various *numerical methods.* In Mathcad 15, you can find the names of these methods if you right-click on the names of some of the functions that implement them. Through the same click you can change some of the settings of these numerical methods.

In the group "Solving equations" there are traditionally seven functions (see the second title of the Study). There is a classic Western *The Magnificent Seven* of this name, the Hollywood adaptation of the philosophical drama Akira Kurosawa's *The Seven Samurai.* In the American film, the "main" character of the seven cowboys who defended peasants from bandits, was Chris Adams, played by Yul Brynner. Seven is a kind of sacred number* not only in culture and history (seven ancient sages, seven wonders of the world, seven days of the week, seven notes in the scale, etc.), but in natural science—seven colors of the rainbow, seven basic units of measurement of the international SI system, etc.

* It is said that also among Pythagoreans seven (7) was a sacred number for several reasons: "The number seven was also sacred because it was the number of planets and the number of strings on a lyre, and because Apollo's birthday was celebrated on the seventh day of each month." For more about Apollo see Study 19.

Furthermore, in Mathcad there is the **solve** operator for the symbolic (analytical) solution of equations and their systems. The description of these tools will be made on some simple school "water" examples.

PROBLEM 1

A motorboat passes along a river in one direction ($L = 10$ km), and then returns to the starting point, spending 1 hour 45 minutes (t) on this "cruise". The question is, what is the speed of the water flow in the river (unknown x), if the boat's own speed (v is the speed in the stagnant water) is 12 kph.

Previously, similar "school" problems were solved in several actions. But not every task can be solved step-by-step. That's why people invented algebra. In ancient times, for example, until the formula for the roots of the quadratic equation was derived, not every such equation could be solved step-by-step, and the solutions were very ingenious. By the way, our task for this motorboat cannot be solved step-by-step either. Reader, find, if you can, a step-by-step solution of this problem and compare the solution found with that given below. The first step of such a solution can be as follows: 2 • 10 km/12 kph = 1 hr + 40 min: movement in flowing, not in stagnant water, increased the journey time by 5 minutes. Many students, adapting to the somewhat obsolete requirements of teachers, conduct step-by-step calculations on a computer in the environment of the same Mathcad package, and rewrite the answer in a settlement note, imitating a manual calculation.

Now, in connection with the extensive use of computers in the educational sphere, it is customary to compose and then solve the equations, choosing their suitable roots, and we discussed this in Study 1. Let's proceed along this path, however, by making an equation, let's try to solve it not on paper, but on the computer in Mathcad.

In our problem of the motor boat, the time in the path t is the total time spent on a one-way trip $L/(v+x)$ (conditionally we assume that this is along the river), and in the opposite direction (against the current) $L/(V - x)$. Therefore, our equation will have the form:

$$(L/(v+x)) + (L/(v-x)) = t$$

Solve

Let's start with the solution of the resulting equation by means of Mathcad symbolic mathematics. The formal, more correct and longer name of symbolic mathematics is *computer analytical transformations*, but we have adopted a loan translation from English—*symbolic math*. We will use this name in the following.

If *numerical mathematics* (which, we repeat, also exists in Mathcad and forms its basis) operates with numbers stored in variables, then the *symbolic math* works first with the symbol variables themselves, and then with the numbers that are stored in them.

Figure 4.1 shows the solution of the equation of the forward and backward motion of the motorboat along the river with the help of the **solve** operator of Mathcad *symbolic mathematics* (this and some other figures will show menu items and toolbars of Mathcad Prime and Mathcad 15 for solving the described problems).

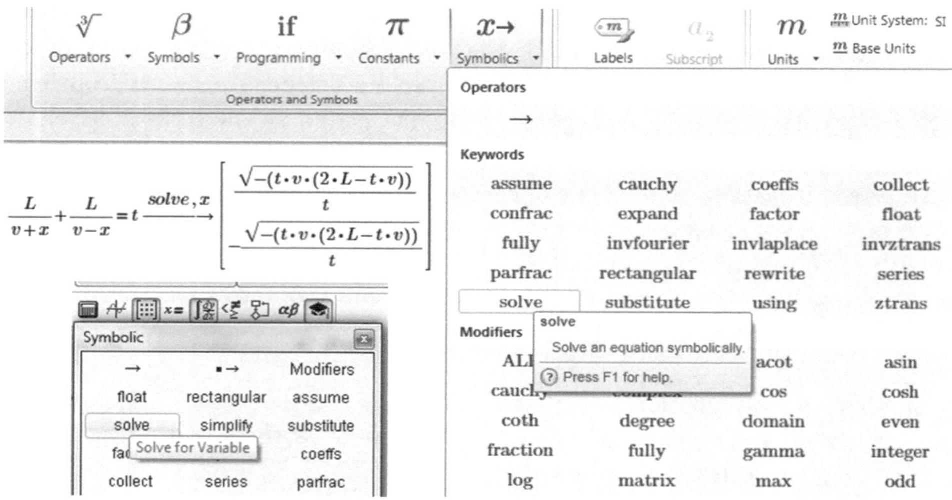

FIGURE 4.1 Analytical solution of problem about the motion of a motorboat: Mathcad Prime and Mathcad 15.

$$L := 10\ km \qquad t := 1\ hr + 45\ min \qquad v := 12\ kph$$

$$\frac{\sqrt{t \cdot v \cdot (t \cdot v - 2\ L)}}{t} = 2.619\ kph$$

FIGURE 4.2 The solution of the problem of the motorboat from the formula found in Figure 4.1.

From the resulting general analytical solution (from a vector with two formula elements—see Figure 4.1), one element can be copied, the original values of the variables L, t and v can be substituted into it (see Figure 4.2) and a numerical answer obtained for the flow velocity of water in the river. The answer will be given in meters divided by a second (Mathcad by default uses SI, the international system of calculus), and is corrected by the user for more relevant kilometers in the hour (kph). Mathcad variables store not numbers, but physical quantities (length, time, force, mass, etc.), which is very useful for calculations in problems with a physical meaning. This greatly speeds up and simplifies calculations, avoids mistakes in them and also automates the corresponding conversion of units of measurement.

The question is, why, then, is numerical mathematics also in Mathcad, if the problem can be solved simply and elegantly with the help of symbolic mathematics? The fact is that symbolic mathematics, aimed at issuing all solutions in the form of formulas (absolute accuracy!), often does not cope with a more complicated problem, and this is shown in Figures 4.3 and 4.4.

In Figure 4.3 in the equation of motion of a motorboat, one x was squared. Some of the physical justification for the equation is gone (the velocity is added with the square of the velocity, which can't be undone).

Note

All modern textbooks and problem books on mathematics are filled with such "non-physical" formulas.

$$a := \frac{L}{v + x} + \frac{L}{v - x^2} = t \xrightarrow{solve,\, x} ?$$

The symbolic result returned is too large to display, but it can be used in subsequent calculations if assigned to a function or variable.

FIGURE 4.3 Finding the root of the equation: a very voluminous hidden answer.

$$a := \frac{L}{v + x^2} + \frac{L}{v - x^3} = t \xrightarrow{solve,\, x} ?$$

FIGURE 4.4 Find equation root: solution not found (see Figure 4.19 as well).

And this is not very good, or rather, it is very bad. It's good when the formula hides some physical reality. This direction of mathematics is conventionally called "natural mathematics". The problem of dimensions disappears immediately, as soon as we remember that even at the time of François Vieta, the principle of homogeneity dominated, which required in such cases to multiply X by one unit, getting everywhere the square of speed. But, as usual, in modern mathematics, multiplication by one is not performed, and dimensions are ignored. A typical problem, often offered to schoolchildren and students for solving within the framework of the course of mathematical analysis, which we already mentioned in Study 3, is a comparison of the function itself and its derivative. But a comparison of a function with its first derivative is equivalent to a physically incorrect comparison of the path (length) and speed—of different physical quantities.

But now for the main non-physical meaning of the equation shown in Figure 4.3. What is important is that the Mathcad package, having solved this slightly complicated equation, could not display the answer because it turned out to be cumbersome. But this is not so bad, since it could be overcome by reading the comment in Figure 4.3. Figure 8.13 of Study 8 shows how such a volumetric response of symbolic mathematics is pulled out. The real "trouble" is shown in Figure 4.4 for an even more complicated equation. If, for example, one X is squared and the other is raised to the cube, the symbolic mathematics of Mathcad "lifts its hands up" and says "I give up!".

If the constants of this "non-physical" equation are replaced by dimensionless numerical values of known quantities, then at least one real root of this equation can be determined—see Figure 4.5, where this problem is solved graphically.

root+root

Figure 4.5 shows that our equation, transformed into a user function by transferring the variable `t` to the left side, has at least one real root close to the number 0.9.

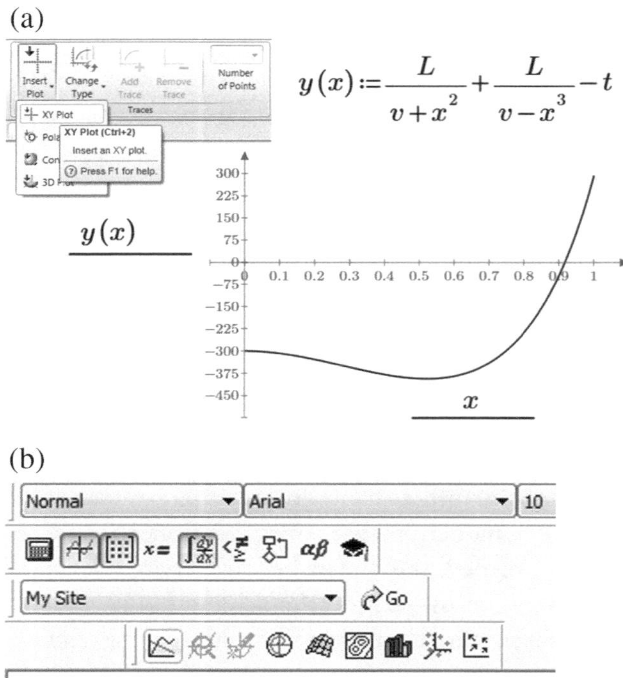

FIGURE 4.5 Graphical zero search: (a) Mathcad Prime, (b) Mathcad 15.

The Mathcad built function *root* can help to refine the numerical value of this root—see Figures 4.6 and 4.7.

Figure 4.6 shows the workings of the *root* function with four arguments and Figure 4.7 with two. In both cases, the answer is displayed with three characters

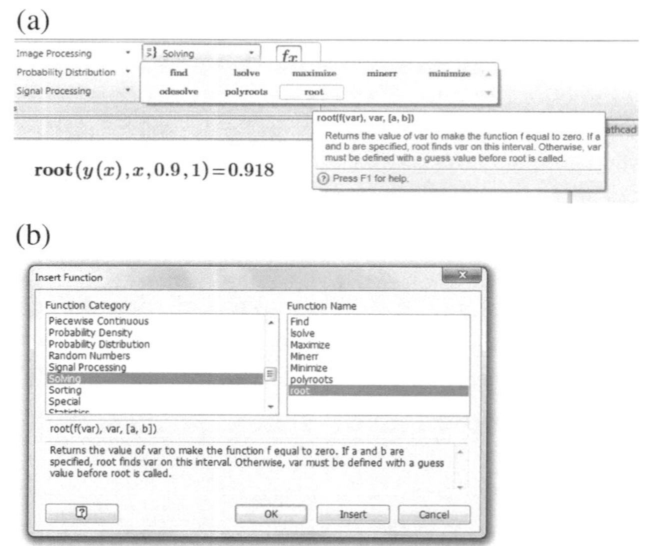

FIGURE 4.6 Working in Mathcad with the built-in function root with four arguments.

$$x := 1 \quad \mathbf{root}(y(x), x) = 0.918$$

FIGURE 4.7 Working in Mathcad with the built-in function root with two arguments.

after the decimal point. But you can also output a larger number of characters, up to 15. In the first case (Figure 4.6), the zero of the function *y(x)* is found by dividing it in half on the interval specified by the third and fourth arguments of the *root* function. In the second case (Figure 4.7), the zero of the function is calculated by the method of secants based on the first assumption $x = 1$. In Mathcad, to calculate the zero of the user function, in fact, there are two built-in functions called *root*.

Figure 4.8 shows the operation of the *root* function on a more complex example—with the user function *sin(x)/x*, which has an infinite number of zeros. On segment 2 to 7, the function *y(x)* has two zeros (π and 2π), but the four-argument *root* function did not return a response, since the function *y(x)* has the same signs at the ends of this segment and the *root* function assumes that there is not root of the equation. There are already five zeros on the segment from 1 to 17 zeros, one of which (9.425) was found by the four-argument function *root*. At the ends of the segment from 1 to 17, the function *y(x)* has different signs. At the first approximation, equal to 0.01, the two-argument function *root* did not give the nearest zero (3.14), but "very far": 298.451. It is possible to understand these peculiarities of the root function only after a detailed consideration of the numerical methods

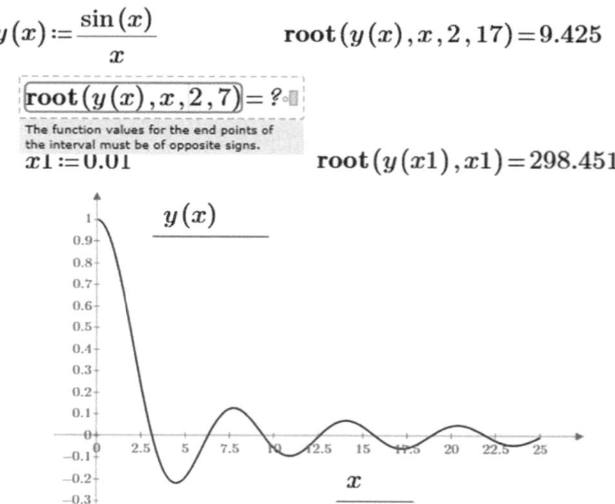

FIGURE 4.8 Features of the two-argument root function.

$$\frac{L}{v+x^2}+\frac{L}{v-x^3}=t \xrightarrow{solve,x} \begin{bmatrix} 0.91753907432064754652 \\ -0.10072745491075931273-0.5044730777236 81i \\ -0.10072745491075931273+0.50447307772368065275i \\ -0.35804208224956446053-1.43206766221807i \\ -0.35804208224956446053+1.4320676622180746717i \end{bmatrix}$$

FIGURE 4.9 The numerical answer of the symbolic operator.

implied in this function—the method of dividing a segment in half and the method of secants.

Earlier, we noted that the symbolic mathematics Mathcad does not operate with numbers, but with symbols—by the variables themselves, storing or not storing numbers. But it is not so.

If any variable stores a numeric value, the symbolic math will work not with the variable itself (with the symbol), but with the number stored in this variable. Figure 4.3 showed a mistake of Mathcad's symbolic mathematics in solving a fairly simple equation. But if all the variables of this equation except the variable *x* are given numerical values, then the symbolic operator *solve* successfully copes with the task—see Figure 4.9.

Figure 4.9 shows that the "symbolic" *solve* operator, unlike the "numeric" *root* function, issued all five roots of the equation (one real and four complex with an imaginary part) without setting the interval (Figure 4.6) or the first assumption (Figure 4.7). In addition, while the numerical mathematics, as we have already noted, is by default limited to three characters after the decimal point for the output of the "print" answer, in this case the symbolic operator *solve* returned numerical solutions with twenty decimal places. With a "numerical" answer, the number of significant digits can be increased up to 15, while with a symbolic number up to 250.

Note To deprive a variable of its numerical value for subsequent analytic transformations, it is possible to use operators: $clear_{sym}(a)$ (Mathcad Prime) or $a := a$ (Mathcad 15).

If our equation with numerical constants (see Figure 4.9) is further complicated, then at some stage the *solve* operator cannot find the roots. The *root* function will still produce the root, although only one of many, based on the specified search interval (Figure 4.6) or on the first approximation (Figure 4.7). In this case, you will have to set the search interval, proceeding from the certainty that there is a root on this interval. The method of secants, with the wrong first approximation, does not give the desired result at all. This is such a kind of compromise. Hence the general rule: it is necessary to first try to solve the mathematical problem posed analytically in a general way, without attaching concrete numerical values to the variables (Figure 4.1) or giving numerical values to all variables (Figure 4.9). If the answer cannot be found, then we will have to proceed to the search for particular solutions by numerical methods, complementing them with the analysis of graphs.

Figure 4.10 shows the use of the graph and the *root* function in its two variants to solve our motorboat problem. Interesting fact. The two-argument function *root* with the first approximation equal to zero, did not produce the expected positive root but an unexpected negative root. This nuance can be understood if we again take into account the peculiarities of the secant method when searching for the zeros of the function and after plotting the plot not in the interval from −3 to +3 km/hr, but on the interval −13 to +13 kph, covering the points of discontinuity, which we will do below. The fact is that the *root* function with two arguments works like this. The user sets one search reference point

$t := 1\ hr + 45\ min \qquad v := 12\ kph \qquad L := 10\ km$

$y(x) := \frac{L}{v+x} + \frac{L}{v-x} - t \qquad x := -3\ kph, -2.99\ kph..3\ kph$

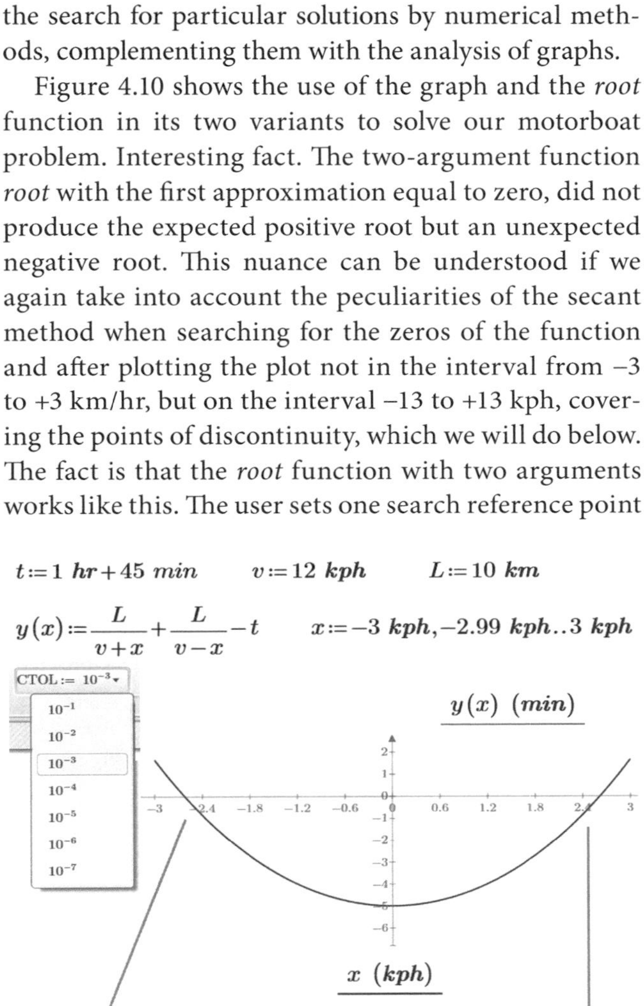

CTOL := 10^{-3}

10^{-1}, 10^{-2}, 10^{-3}, 10^{-4}, 10^{-5}, 10^{-6}, 10^{-7}

$y(x)\ (min)$

$x\ (kph)$

$\text{root}(y(x), x, 0\ kph, 10\ kph) = 2.619\ kph$

$\text{root}(y(x), x, 0\ kph, -10\ kph) = -2.619\ kph$

$x := 0\ kph \qquad \text{root}(y(x), x) = -2.619\ kph$

FIGURE 4.10 Graphical and numerical (via the root function) solution of the problem of the motorboat.

(the first *guess*, but this is by no means always the first *approximation*—see the example in Figure 4.8). Next, Mathcad fixes the second point to the right of this point at a distance *CTOL* (by default 0.001, in our case meters, i.e. 1 m) and passes through these two points a secant that will be close to the tangent). This secant crosses the X-axis somewhere—this will be the third point, the next point of the iterative root search. Next, the classical secant method is realized. The right point of discontinuity of our analyzed function "throws" the search in the region of negative values. We can say that the *root* function with two arguments contains a hybrid of Newton's method (tangents) and the method of secants.

Note To increase the accuracy of finding the function zero with the Mathcad built-in *root* function, you cannot decrease the value of the CTOL system variable (see the list to the left of the graph in Figure 4.10), but instead need to multiply the analyzed function by 10^3, 10^4, etc. The choice of accuracy of calculations is a separate task. On the one hand, increased accuracy will never be superfluous, and, on the other hand, it leads to a slowdown in calculations and a breakdown in a number of cases. All in all a measure is needed!

The equation of motion of the motorboat in Figure 4.1 can be converted to a quadratic one. This method is usually resorted to in schools. Schoolchildren, as a rule, are taught to solve quadratic equations analytically only—see the text in Figure 5.13 in Study 5. How such a transformation can be done in Mathcad is shown in Figure 4.11.

The *simplify* operator adjusts the left-hand side of the original expression to a common denominator, multiplies both sides of the equation by the resulting denominator and transfers all the terms to the left-hand side of the equation (Figure 4.11). In this way, a function is selected that is equated to zero. The operator *coeffs* finds the coefficients of this polynomial function (in this case square). This quadratic equation can be solved by the *solve* operator, but see below.

(a)

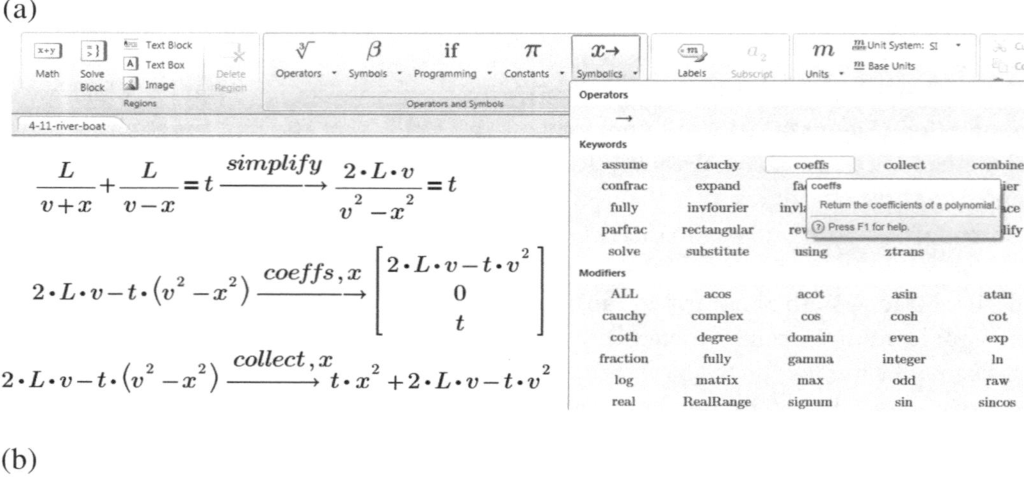

(b)

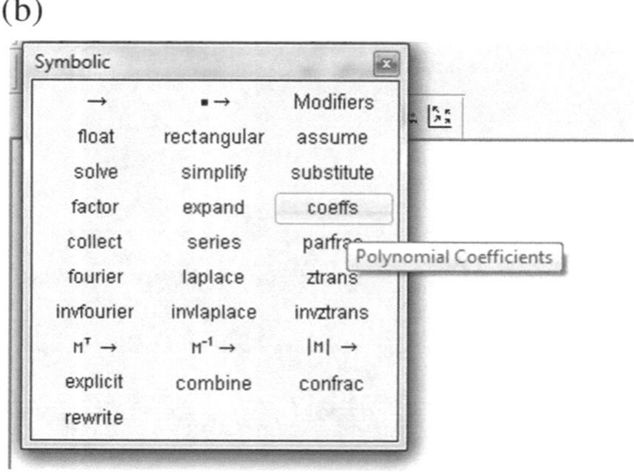

FIGURE 4.11 Determining the coefficients of a polynomial: Mathcad Prime and Mathcad 15.

Note. A quadratic function obtained after the transformation of the initial equation of motion of the motorboat is not equal to the original function, but only has two common roots with it. This can be seen by looking at the graphs shown in Figure 4.12. Since by the condition of the problem $v = 12$ kph, and the fact that the equation of motion does not make sense at $x = \pm v$, the function on the left side of the equation has discontinuities at x equal to 12 and minus 12 kph. Of course, there is no quadratic function of such discontinuities.

Polyroots

If the expression is a polynomial (square, for example, see above), then you can find all its zeros by using one more function from the "Mathcad Magnificent Seven"—the *polyroots* function. It has as its argument the vector of the coefficients of the polynomial and returns its zeros (a vector that is one element shorter than the vector argument), i.e. the solution to our problem—see Figure 4.13.

In our problem of the motion of the motorboat, the polynomial turned out to be quadratic and its roots, we repeat, could be found through the operator of symbolic mathematics *solve* (see Figure 4.1). But in the case of high-degree polynomials, the *solve* operator does not work. This is where the numerical built-in *polyroots* function comes in handy.

Find

Show the work of another function from the "Mathcad Magnificent Seven"—*Find* functions—will help us with an additional motorboat.

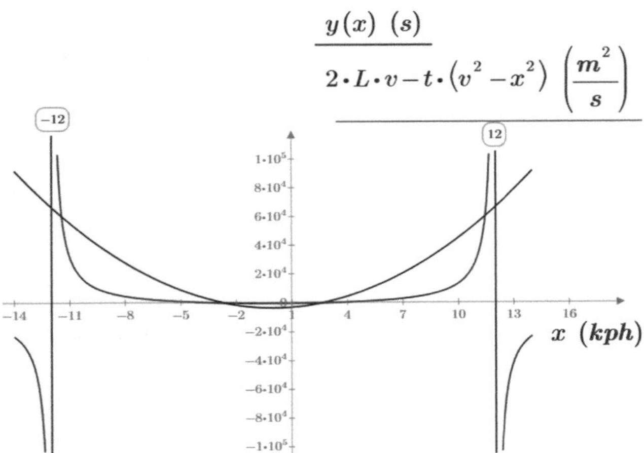

FIGURE 4.12 The initial and quadratic equation of motion of a motorboat.

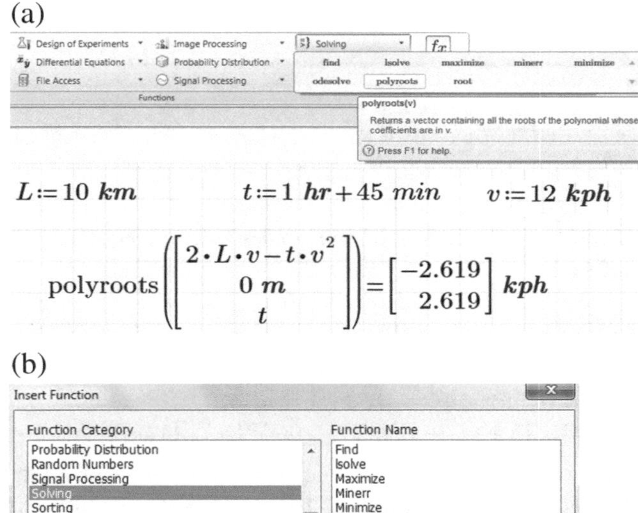

FIGURE 4.13 Search for zeros of a polynomial in Mathcad: Mathcad Prime and Mathcad 15.

PROBLEM 2

From two piers on a rectilinear section of the river, two motorboats leave simultaneously and move toward each other. They meet at the point dividing this section of the river in the ratio of the golden section. Find the speed of the second boat v_2 and the speed of the water flow in the river v if the speed of the first boat v_{21} is known, the distance between the piers L and the time t the boats take to the meeting point. (A ridiculous version of the problem: two trains came from two stations on a single track to meet each other, and yet did not come across each other! Why? *Answer*: not fate!).

In this problem we use the famous "Golden Section" method, i.e. the division of a segment into two unequal parts, so that the ratio of the length of the smaller part of the segment to the length of the greater part is the same as the ratio of the length of the greater part to the length of the whole segment (see Figure 4.14). This property of the golden section is remembered by many who do not know anything about the formula of the golden section.

The golden section is inserted in the task not accidentally. You can search in your memory or in the references (paper or Internet) for the golden section formula. But we can do otherwise. Remembering the basic idea of Study 1, we can write in Mathcad the equation of

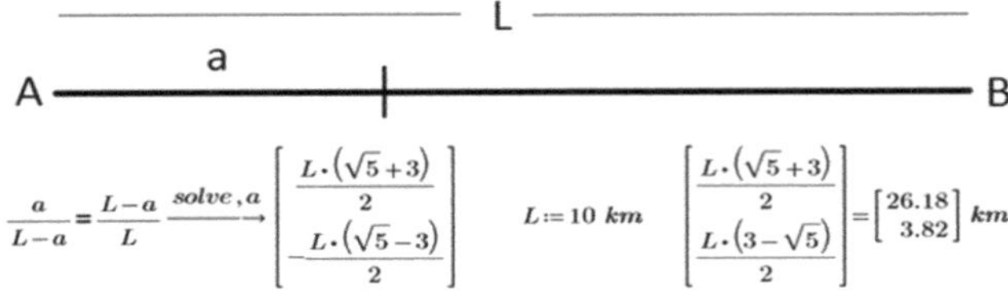

FIGURE 4.14 Solution of the golden section equation in Mathcad.

the golden section with reference to our problem of the motorboats and solve it analytically, having obtained the desired formula—see Figure 4.14.

In Figure 4.14, the *solve* operator issued two solutions, of which only the second solution is suitable—3.82 km. The first solution (26.18 km) lies outside the considered segment. Symbolic math, we repeat, gives out all the answers, from which we still need to be able to choose the right one. Or to be able to make the operator *solve* only for one answer.

Figure 4.15 shows the solution of the problem of two motorboats in Mathcad Prime and Mathcad 15, the problem reduces to solving a system of two equations with two unknowns. The solution was found using the *Find* function, which requires an initial guess to the solution.

The built-in *Find* function changes the value of its arguments, starting from the initial approximation/assumption, so that the equations of the system become identities. Rather, almost identities. The fact is that both the *root* functions (Figures 4.7 and 4.9) and the *Find* function (Figure 4.7) return values that differ from the exact solutions by an amount not exceeding the modulus of the value stored in the CTOL system variable. After all, what is the root of the equation?! The root is the value of the variable under which the equation becomes an identity. But with a numerical (approximate!) solution, it is not always possible to find the exact root. And time to spend on it is not advisable. Substitution of the approximate value of the root into the equation leads to the fact that the right and left sides of the equation differ from each other by the value stored in the variable CTOL, which by default is 0.001. This value can be changed by solving a specific problem. We can see the features of the search for the four roots of the system of two non-linear equations: the ellipses equations and the Bernoulli lemniscate equations (see also Figure 13.14 from Study 13).

Note. In Mathcad Prime, in comparison with Mathcad 15, the technology of solving equations with

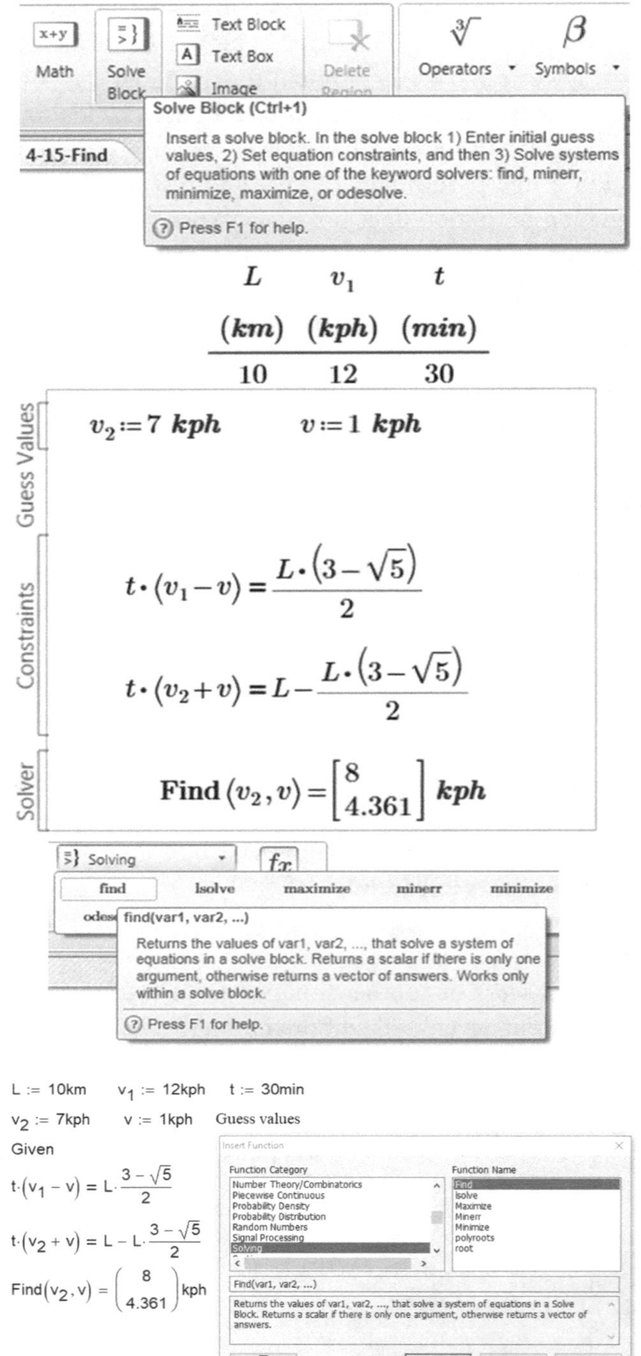

FIGURE 4.15 Solution of the system of equations with the help of the function Find: Mathcad Prime and Mathcad 15.

the help of the *Find* function has significantly changed. In Mathcad Prime, you do not need to enter the Given keyword anymore. It is enough to enter the *Solve* area with three sub-areas. The keyword Given was often refused, because many users pressed the space bar after this word, thereby turning the keyword into a comment and they did not understand why Mathcad would then refuse to solve the system of equations. In Figure 4.15 the solutions in both versions of Mathcad are shown for comparison.

Note Variable *L*, which in Figure 4.15 is assigned an initial value of 10 km (L := 10 km), underlined by a wavy line indicating some abnormal design situation in Mathcad 15. The *L* default variable in Mathcad 15 stores the value of one liter (unit of capacity) and this value is overridden by the user. In Mathcad Prime this incompleteness (inconvenience) is fixed—there you can have two independent variables *L*: a unit of capacity and a separate user variable, storing, as in our case, the distance.

lsolve

It can be understood that the system of two algebraic equations of motion of the two motorboats facing each other, shown in Figure 4.15, is *linear*, and it is possible to apply to it one more function from the "Mathcad Magnificent Seven"—the *lsolve* function, intended to solve exactly the systems of linear (l) algebraic equations (SLAEs)—see Figure 4.16.

In Figure 4.16, the system of equations identical to the system in Figure 4.15, with the help of identical algebraic transformations, is reduced to the form of a classical linear system: on the left are the unknowns v_2 and v with their coefficients and on the right are the free terms. The *lsolve* function has two arguments: the coefficient matrix for unknown SLAE (we have this *M*) and the vector of the free terms of *V*. The *lsolve* function returns the found values of the vector as unknown. When solving SLAU using the *lsolve* function (see Figure 4.16), it is not necessary to enter the initial approximations (see Figure 4.15). This system has an unique solution, or there are no solutions at all, or there are infinitely many solutions.

Note The second argument of the *lsolve* function can be not only a square (classical case of SLAE), but also a rectangular matrix representing an underdetermined or overdetermined system.

Minimize and Maximize

For the next function of the "Magnificent Seven", the function *Minimize*, we use an example of an *optimization problem*, also related to "water transport".

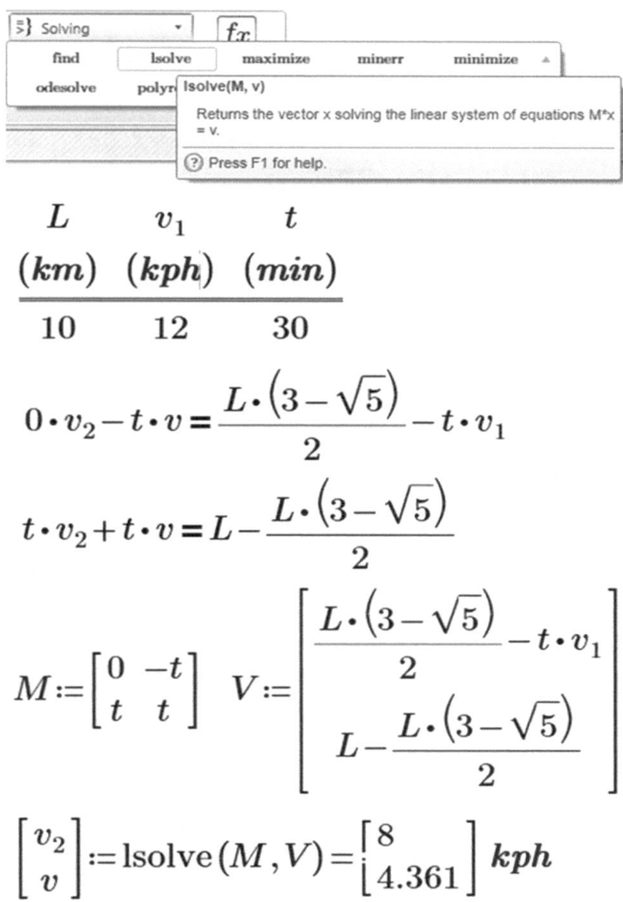

FIGURE 4.16 SLAE solution in Mathcad Prime.

PROBLEM 3

Determine the cruising speed of a ship—the speed at which the cost of its operation will be minimal.

The problem is extremely simplified—the cost of operating the ship consists of two parts: the crew's hourly wage, that is proportional to the time of the ship's travel (inversely proportional to the vessel's speed), and the fuel costs proportional to the square of the ship's speed (proportionality factors—*a* and *b*). Increasing the speed of the vessel, we save on the crew's salary, but we have to spend more money on fuel. Let's try to find the optimal solution here!

Figure 4.17 shows the solution to this typical optimization problem using the built-in *Minimize* function with a graphical illustration of the solution.

The *Minimize* function changes the value of its second argument, starting from the given assumed value (we have 10 kph) so that the value of the first argument

€	a $(€\cdot hr^{-1})$	b $(€\cdot km^{-3}\cdot hr^{2})$	$Cost(v)$
¤	100	0.03	$\frac{a}{v}+b\cdot v^{2}$

$v:=10\ knot \qquad Cost(v)=25.249\ \frac{€}{mi}$

$v_{opt}:=\mathbf{Minimize}(Cost, v)=6.402\ knot$

$Cost(v_{opt})=20.361\ \frac{€}{mi} \qquad \frac{a}{v_{opt}}=13.574\ \frac{€}{mi} \qquad b\cdot v_{opt}^{2}=6.787\ \frac{€}{mi}$

$v:=3\ knot, 3.01\ knot..10\ knot$

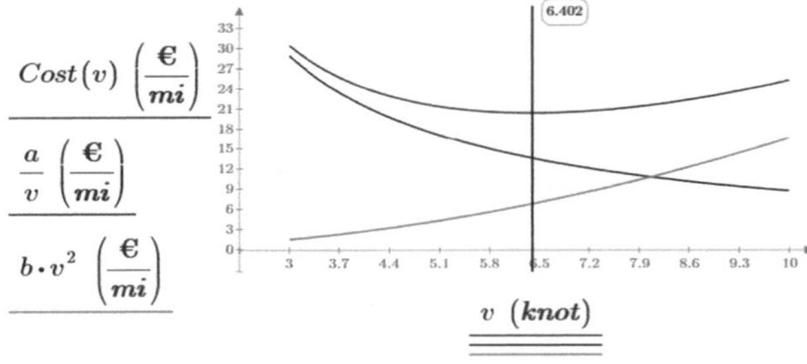

FIGURE 4.17 Finding the cruising speed of the ship by Mathcad numerical mathematics.

(the target function *Cost*) takes a minimum value. If we did not minimize costs, but maximized, for example, the profit of the owner of the ship, then it would be necessary to replace the *Minimize* function with the *Maximize* function. In optimization problems, there are often restrictions—the speed of the vessel, for example, cannot exceed the maximum permissible. In this case, the *Minimize* or *Maximize* functions will need to be placed in the *Constraints* area of the Solve block, shown in Figure 4.15.

Finding the minimum of our target function *Cost* can also be achieved using Mathcad symbolic mathematics, as shown in Figure 4.18.

$$\frac{d}{dv}\left(\frac{a}{v}+b\cdot v^{2}\right)\xrightarrow{solve,\,v}\begin{bmatrix}\left(\frac{a}{2\cdot b}\right)^{\frac{1}{3}}\\ -\frac{\left(\frac{a}{2\cdot b}\right)^{\frac{1}{3}}}{2}+\frac{\sqrt{3}\cdot\left(\frac{a}{2\cdot b}\right)^{\frac{1}{3}}\cdot 1i}{2}\\ -\frac{\left(\frac{a}{2\cdot b}\right)^{\frac{1}{3}}}{2}-\frac{\sqrt{3}\cdot\left(\frac{a}{2\cdot b}\right)^{\frac{1}{3}}\cdot 1i}{2}\end{bmatrix}$$

$€:=¤ \qquad a:=100\ \frac{€}{hr} \qquad b:=0.03\cdot €\cdot\frac{hr^{2}}{km^{3}} \qquad v_{opt}:=\left(\frac{a}{2\cdot b}\right)^{\frac{1}{3}}=6.402\ knot$

FIGURE 4.18 The cruising speed of the ship by Mathcad symbolic mathematics.

$$\frac{d}{dv}\left(\frac{a}{v}+b\cdot v^{n}\right)\xrightarrow{solve,\,v}\ ?$$

No symbolic rezult was found!

FIGURE 4.19 Misfire when working with Mathcad symbolic math.

Figure 4.18 searches for the zeros of the first derivative of the function in terms of unit costs per kilometer of the vessel's path. But if the fuel costs depend on the speed of the vessel, taken not to the second power, but to the power of n (this coefficient close to two is refined experimentally) then the symbolic mathematics will not cope with such a complicated task (Figure 4.19) and we will have to return to numerical methods for solving the problems (Figure 4.17).

Minerr

The last function* of the "Mathcad Magnificent Seven" is the *Minerr* function (minimal error). If the *Find* function (see Figure 4.15) does not find a solution to the equation system, it returns an error message. In this situation the function *Minerr* will not return an error message, but the values of its arguments (residual system), under which the system of equations will be as close as possible to the system of identities—the point of the last approach to the solution. In older versions of Mathcad, there were no *Minimize* and *Maximize* functions, and optimization tasks had to be handled through the *Minerr* function. Let's try it ourselves. Figure 4.20 shows how this function solves the problem of the vessel cruising speed: with optimal traffic, the cost of operating the vessel will be maximally close to zero (the dream of all ship owners).

The *Minerr* function will help us in Study 8 to make the animation of the Chebyshev rule (see Figures 8.8 and 8.9) with a truncated single rod.

The function Minerr can be considered the main one in the "Mathcad Magnificent Seven", as. It can replace the *Find* function, the *root* function (in its two variants), the *polyroots* function, the *lsolve* function and, in some cases, the *Minimize* and *Maximize* functions. When using the *Minerr* function, it is necessary

* The last one on the list, but not the last one in importance—we look further.

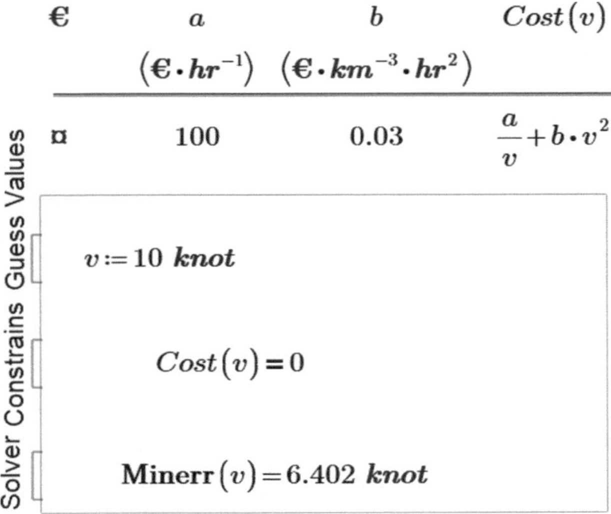

FIGURE 4.20 Solution of the optimization problem using the Minerr.

to provide a check for solutions. It is not uncommon for results to be erroneous, most often due to the fact that of several roots there is an unreal (or non-interesting) root. The fact is that the *Minerr* function tries to find the maximum approximation to the required number by minimizing the root mean square error of the solution. It is necessary to make sure beforehand that the solution exists and to indicate the initial approximation to the solution as accurately as possible.

On the other hand, the main (irreplaceable) function of the "Mathcad Magnificent Seven" can be considered the *root* function, since it alone remains activated when Mathcad Prime finishes its monthly trial using Mathcad Express and migrates to Mathcad Express.

Computer mathematics with universal and hidden methods of analytical and numerical solutions makes us forget about the types of equations. Let's remember them! See "Types of equations" on the website for this Study.

CONCLUSIONS

Each of the "Mathcad Magnificent Seven" functions considered has its own peculiarities and limitations. Before proceeding with a task, one should consider which of these functions will lead to the goal in the best possible way.

A schoolboy, a student, an engineer or a scientist all need to master the "Mathcad Magnificent Seven"—numerical, graphical and analytical methods for solving problems in order to successfully solve their educational or professional problems on a computer.

Readers can find a sub-chapter "Van Gogh in Mathcad" on the website for this chapter.

ASSIGNMENTS FOR READERS

1. Choose a physics problem to which you can apply most of the functions described in this Study.
2. Create programs for the methods for the numerical solution of algebraic equations.
3. A swimmer, whose speed v_1 is half the speed of the river v2, wants to swim the river so that he will drift as little as possible. At what angle across the river should he head?
4. A fragmentation grenade is tested at the center of the bottom of a cylindrical pit 20 m deep. The fragments formed during the explosion, whose speed does not exceed 300 m/s, should not even reach the edge of the pit. What should be the minimum diameter of the pit?

Guess the Image Or

The Matrix Game

ABSTRACT

The Study explains how to get a matrix with numbers from an image (graphic file) and what you can do with this matrix. A new interesting computer game and the methods of processing graphic information on a computer are discussed.

Mathematics: Matrix, columns, rows, matrix transposition, determinant, inverse matrix, integer, ellipsoid, quadratic equation with three unknowns, Bernoulli lemniscate, Bernoulli "volumetric" lemniscate, triangle, triangle area, cosine theorem, quadratic equation, Pythagorean theorem, similar triangles.

Physics: Color.

IT: Searching for information on the Internet, storing color and black-and-white images in a computer, converting an image into a matrix and into an embedded data array, a cycle with a parameter, a nested loop with a parameter, computer animation, image file formats, image manipulation functions.

Art: Alberto Korda, *Che Guevara.*

Study website: https://community.ptc.com/t5/PTC-Mathcad/Chapter-5/m-p/606052.

There is a television game called "Guess the Melody". The first notes of a piece of music are played, but the winner is the one who names the work first.

Such a game can be transferred from audio to video—to the visual plane—and called "Guess the image" or, to be more precise, "Guess a person or something else from their incomplete image". The person's portrait in this game needs to be shown on the screen, but not all at once, rather as a melody—by separate consecutive notes, sorry, separate random dots (raster elements) gradually filling the screen. The more points (dots) that are shown, the more likely it is the image will be identified. Below is a description of a program for Mathcad 15 that creates this "portrait melody".

A portrait of a man displayed on a computer monitor* is, as a rule, a lot of multicolored points or dots. Such an image, in mathematical terms, is called a *matrix*, each element of which stores an integer from 0 to 256, if we talk about an image with only shades of gray. A color image is comprised of three matrices that store the intensity values of three colors (RGB: Red, Green, Blue).

Recently, information technology has been rapidly developing in the direction of the *recognition of images.* Here are two typical examples of the practical use of this:

A surveillance camera takes a picture of a car that has exceeded the speed limit on the road or committed another violation, and the computer processes this picture, determines the registration number of the car and issues a fine to the offender.

At passport control at the border, a digital camera captures the face and/or the iris of the eye of the monitored person, and the computer compares this data with

* Or on the TV screen. But at the present time (the era of digital television), these two devices (computer and television) are increasingly in a single "box".

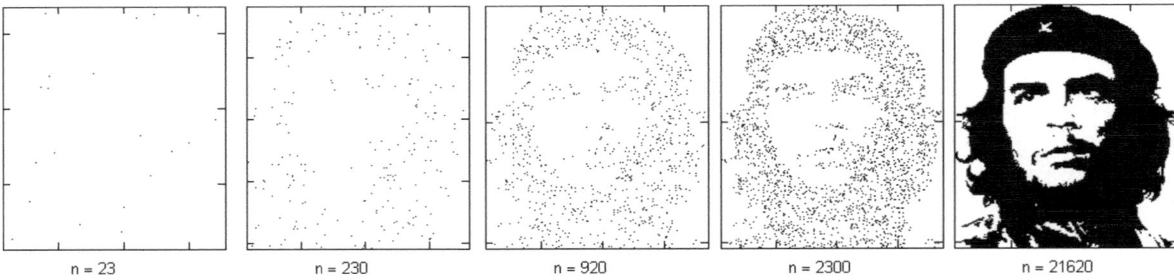

FIGURE 5.1 Point drawing of the portrait of Che Guevara by Alberto Korda.

FIGURE 5.2 Search for a portrait on the Internet and save it to disk.

those that are "sewn" into the chip of the presented passport.

Not far away is the time when we will be able to approach the ATM and say: "Give me a hundred rubles!". And this device will fulfill our request, recognizing us in appearance and/or voice.

The game "Guess the image" described in this Study not only helps us entertain friends at a party with an interesting contest, but also reveals some secrets about the modern technologies of storage and recognition of graphic objects.

In Figure 5.1, you can see "playing" one of the most recognizable of graphic objects—a portrait of Che Guevara,* looking at us from numerous wall graffiti, posters and T-shirts.

The portrait of Che Guevara, shown in the right-hand image of Figure 5.1, consists of 21,620 black dots on a white background.† In Figure 5.1 it is drawn with different numbers of points (n): for $n = 230$ our hero is not yet recognizable, but by $n = 920$ you can raise your hand and shout: "I recognize it!"

A computer portrait is drawn like this on the Internet. The image of Che Guevara‡ is saved to disk in a black-and-white format of a bmp file—see Figure 5.2.

Figure 5.2 shows a search using the Google portrait of Che Guevara on the Internet. It is saved to disk using the menu command that appears when you right-click. Portraits cannot only be taken from the Internet, but also created by yourself—take pictures of friends and relatives on a digital camera and save files on a disk for further processing.

In Mathcad 15 there is a command on the insertion menu in the calculation of the figure (Figure 5.3) with which you can draw the selected image in the Mathcad sheet by specifying the file name. You can just paste the picture from the clipboard into the worksheet with Ctrl + v, but in this case the drawing will not be tied to the selected Mathcad variable. In Figure 5.3, it has the name *Che*.

Inserting a picture into a Mathcad-sheet is in itself a very useful tool. Using the if function (if), you can change the name of the graphic file and load this or that picture depending on the results of a calculation. (For

* Che Guevara—a very controversial figure. Some consider him a hero, a fighter for freedom. Others are see him as almost a terrorist. In a word, everyone draws him in black-and-white colors. We will also do this, but in a literal, not figurative sense—we will draw his portrait only with black-and-white dots.

† There is such a direction in the art of pointillism—the manner of writing separate smears of the correct point (round) or rectangular shape.

‡ In the same place on the Internet you can find portraits and other recognizable people (architectural silhouettes, industrial objects, etc.) for our game "Guess the image."

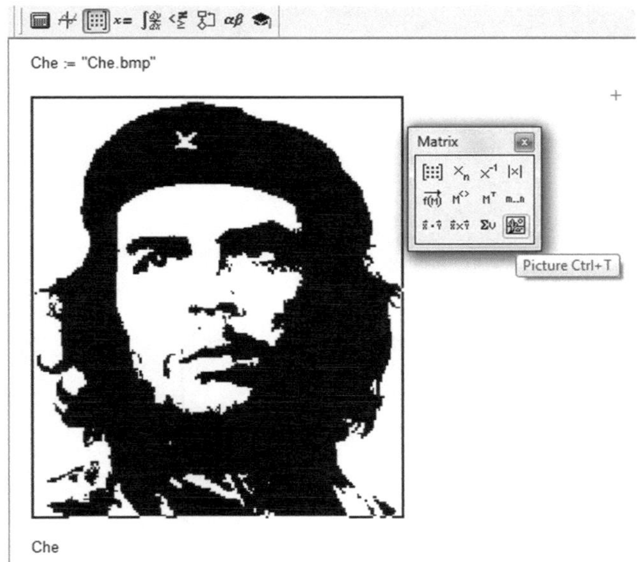

FIGURE 5.3 Inserting a picture into a Mathcad-sheet.

example, this was done by showing the formulas for recalculating the types of concentration as in Figure 1.9 in Study 1.)

The black-and-white picture stored on the disk as a bmp file can be turned into *a matrix*, the elements of which will fix the color of the image point: 0, black to 256, white. For this, there is a function *READBMP* in Mathcad, see Figure 5.4.

In Figure 5.4 we have "printed out" the lower right corner of the matrix M to show how the fragment of the left shoulder of Che Guevara is stored in the matrix. In the matrix M, 207 rows (from zero to 206) and 189 columns (from zero to 188), store information about 39,123 points of the portrait, of which 21,620 are black. Figure 5.4 shows how, using the rows and columns built-in functions in Mathcad, the number of rows and columns in the matrix *M* was counted. These numbers are needed for further image processing.

If we have a matrix (a collection of numbers grouped by rows and columns), then we can do a lot with it: put it on its side (in mathematics this is called the transposition of the matrix), flip it, sort it and so on. If the matrix is square, then the determinant of the matrix can be computed. A matrix with a non-zero determinant can be inverted (calculate the inverse* matrix). All these tools are in Mathcad.

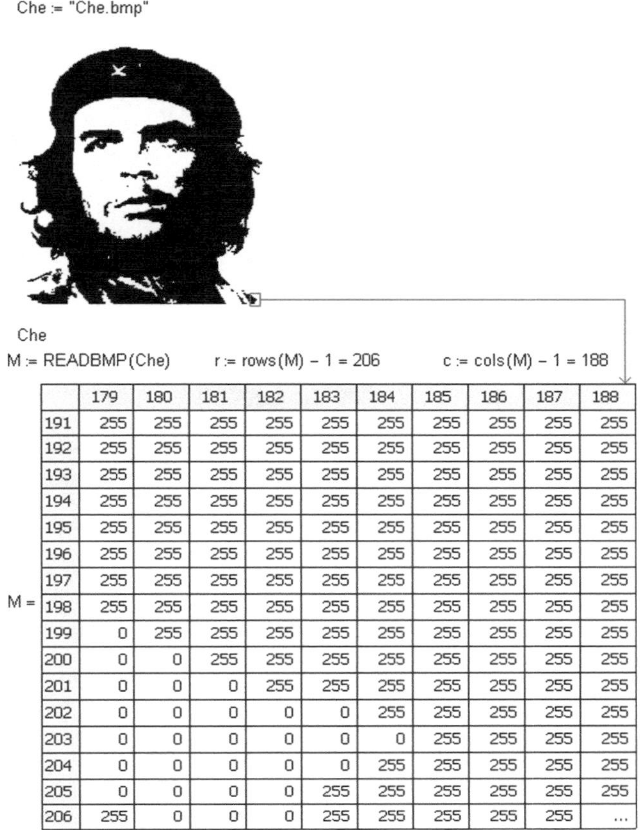

	179	180	181	182	183	184	185	186	187	188
191	255	255	255	255	255	255	255	255	255	255
192	255	255	255	255	255	255	255	255	255	255
193	255	255	255	255	255	255	255	255	255	255
194	255	255	255	255	255	255	255	255	255	255
195	255	255	255	255	255	255	255	255	255	255
196	255	255	255	255	255	255	255	255	255	255
197	255	255	255	255	255	255	255	255	255	255
198	255	255	255	255	255	255	255	255	255	255
199	0	255	255	255	255	255	255	255	255	255
200	0	0	255	255	255	255	255	255	255	255
201	0	0	0	255	255	255	255	255	255	255
202	0	0	0	0	0	255	255	255	255	255
203	0	0	0	0	0	0	255	255	255	255
204	0	0	0	0	0	255	255	255	255	255
205	0	0	0	0	255	255	255	255	255	255
206	255	0	0	0	255	255	255	255	255	...

FIGURE 5.4 Convert an image to a matrix.

Moreover, Mathcad is equipped with a special image processing library, see Figure 5.5.

We will decompose our matrix into two vectors with the names *Ion* and *Iat*, which will store the coordinates of the black points of our black-and-white portrait of Che Guevara. This is done in the first program shown in Figure 5.6, with two loops that sort through the matrix in columns and rows. If the next element of the matrix *M* is zero (the black point of the portrait), then the number of the given column is written in the vector *X*, and the number of the row in the vector *Y*.

The second program, shown in Figure 5.6, the core of which is the function *rnd* (which will be the protagonist of Study 17), returns random numbers and mixes the elements of the vectors *Ion* and *Iat*. After that, using the Mathcad animation tools it's easy to draw our hero point-by-point—see his incomplete portrait in Figure 5.6 and the individual stages of this "drawing" in Figure 5.1. If the vectors *Ion* and *Iat* do not mix, then the picture will be displayed on the computer screen line-by-line,

---

* Multiplication of the inverse matrix by the original matrix yields a unit matrix—a matrix with units on the main diagonal and with zeros in the remaining places.

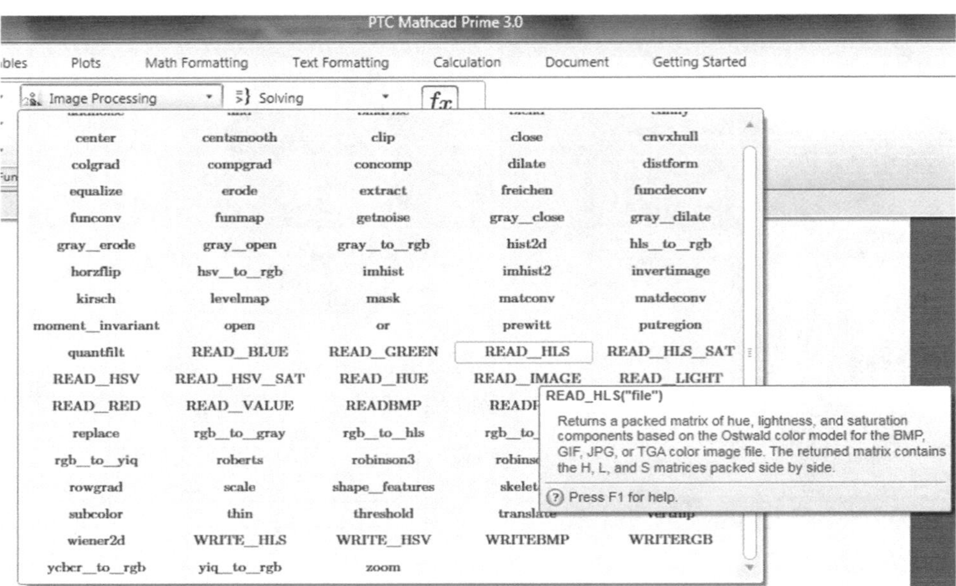

FIGURE 5.5 Mathcad library functions work with images.

and you can clearly see the image rasters. So, you too, can play the game “Guess the image”.

The programs and animations described here are on the website for this Study. In particular the details of the “Guess Image” game with color images are discussed using the example of the famous portrait of Andy Warhol’s “Liz”. In the upper left corner of this figure is a printout of the matrix *M*, which holds the portrait in three vectors, each of which stores an intensity matrix of one of three colors. With these matrices we perform the operations shown in Figure 5.6 and create an animation, the frames of which are displayed in Figure 5.1, but in color.

The childhood, adolescence and youth of the author of this book took place during the birth and flowering of not only cosmonautics (see Study 9 and 24) but also of analog television. He remembers how his parents saved money for the first KVN-49 TV, then a lens was fitted to it (the zoom function—see the last line in Figure 5.5), the announcer explaining how to switch to the second program by switching the lever on the back panel of the TV. The Polytechnic Museum in Moscow holds the first Soviet color television with three beams in a tube for three colors appeared, and later a flat and a digital TV, etc. Many memories are inspired by this Study.

The most recognizable female image is, of course, that of the Mona Lisa. The author of this book has another book with a co-author (see http://twt.mpei.ac.ru/ochkov/MC_PC/index.htm). On the cover of that book was a portrait of Mona Lisa, printed with noughts and crosses. By then such printouts were no longer a wonder but were looked at with interest. The co-author (a soldier, a signal officer) brought the newly printed book home and showed it to his wife. She asked: “Who is this!?” He replied: “She is Mona Lisa or Gioconda”. His wife threateningly exclaimed: “What is this Gioconda!? This Raika is your front-line radio operator, with whom you proclaimed love…” Scandal almost with assault! The co-author told all this and showed one Raika photo—a complete resemblance. Well, not with a portrait from the Louvre, but with a picture on the cover of the book. See also Figure 15.7 in Study 7.

GUESSING NOT THE IMAGE BUT MELODY

The author of this book once proposed to sound the beginnings and ends of all functional program blocks in the system of the European musical scale:

- functional block—do (C)
- the “until”—re (D)
- alternatives—mi (E)
- the cycle “before”—fa (F)
- cycle with parameter—so (G)
- Multiple branching—la (A)
- call procedure—ti (B)

```
(lon) := | n ← 0
(lat)    | for  i ∈ 0.. c
         |   for  j ∈ 0.. r
         |     if  M_{j,i} = 0
         |         | X_n ← i
         |         | Y_n ← -j
         |         | n ← n + 1
         | (X)
         | (Y)

(lon) := | n ← last(lon)
(lat)    | for  i ∈ 0.. n - 1
         |     | RN ← floor(rnd(n))
         |     | while  lon_RN = "-"
         |     |    RN ← floor(rnd(n))
         |     | (lon1_i)    (lon_RN)
         |     | (lat1_i) ←  (lat_RN)
         |     | lon_RN ← "-"
         | (lon1)
         | (lat1)
```

$$FRAME := 500 \qquad n := ceil\left(\frac{last(lon)}{940}\right)\cdot FRAME \qquad i := 0..n$$

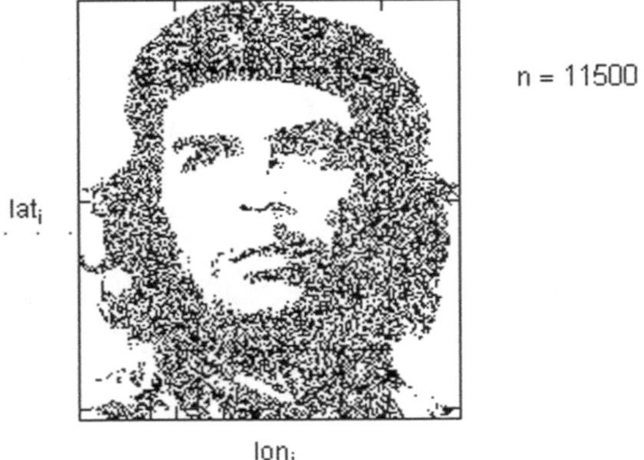

FIGURE 5.6 Converting a matrix into two vectors.

In this case, the depth of the occurrence of a particular structure is fixed by the duration of the sound. Running the program on a computer with a sound generator and speakers generates a melody, listening to which, you can follow the progress of the solution of the task, fix the loops, etc. You can also play a new game "Guess the program"—the melody created according to the above scheme is played, and the players must guess what it is: sorting an array, solving a system of equations, numerically integrating, etc. See http://twt.mpei.ac.ru/ochkov/128.

ASSIGNMENTS FOR READERS

Create a program and/or find a built-in function that calculates the number of points of a different color in the graphic file.

Create a program and/or find the built-in function to create a negative or other alternative image of a graphic file.

Calculate the frequency of meeting points of a color in a graphic file and build a histogram from these data (see Figures 14.19, 14.20 and 14.21 in Study 14).

For the sake of interest and pleasure, let's replace the points in the image matrix with chocolate sweets in confectionery boxes. Solve this problem. At the confectionery factory there is a machine for packing chocolate sweets. It is a mechanical manipulator, capable of moving along one straight line perpendicular to the conveyor, along which empty boxes with cells for sweets move. Candy instantly enters the cell as soon as the manipulator is above it. Let us consider the motion of the automaton and boxes in the XOY plane. The coordinates of the cells (x, y) are natural numbers, and, in the initial position, (all coordinate values are given in inches). The belt of the conveyor starts to move in the direction opposite to the axis OX, with a velocity $v = 1$ inch/s. Simultaneously, the manipulator moves from the origin along the OY axis at a constant speed without stopping. What is the maximum number of sweets that can be placed by the manipulator during a single crossing of the conveyor and at what speed should it move?

Solve problem 4, provided that the manipulator must place the candies in a "gift box" of non-rectangular shape, i.e. the number of chocolates in each row is subject to some dependence. How in this case to define the speed function of the manipulator?

How Much Or

The Old Problem in a New Way

ABSTRACT

The Study describes how to use a computer to solve mathematical problems taken from literature. How to solve ancient problems on a computer and how to make these tasks more modern are shown. Methods for symbolically solving equations on a computer are discussed.

Mathematics: Finite-search method, cubic equation, imaginary unit, complex number, system of linear algebraic equations (SLAE), integer equation, factorial, cylinder volume, cone volume, underdetermined system of equations.

Physics: Ancient and modern systems of physical quantities and their units of measurement, surface area, volume, displacement.

IT: Computer tools for analytical transformations, computer tools for working with physical quantities, cycle with a parameter, nested loop, conditional operator, variable range, pre-test loop, graphical solution of a system of equations.

Art: Anton Chekhov *The Tutor*; Fyodor Dostoyevsky *The Gambler*; Jean-Baptiste Poquelin (Molière) *Le Bourgeois gentilhomme.*

Study website: https://www.ptcusercommunity.com/thread/127824.

1001 Problems for Mental Calculation was, and still is, a very interesting book by Sergei Alexandrovich Rachinsky (2017, the Bely Gorog Publishing House - https://www.ozon.ru/context/detail/id/6443736/).*

It was published, in print, in the nineteenth century. Nowadays it has been revived at www.1001problem.ru with a convenient means of searching for and commenting on problems. On the same website there is a link to a very interesting discussion that unfolded in the journal *Science and Life* on the problem of mental calculation in the modern educational process—http://www.nkj.ru/archive/articles/6347.

Owners of Apple devices, such as the iPhone and the iPad, can download Rachinsky's book at no charge from the online store AppStore and place it in their iBooks library.

The website 1001problem.ru emphasizes that all the problems of the book must be solved purely in the mind without involving any calculation devices and tools: pen and paper, calculator and so on. With a paper version of the book, this requirement is simple—to be left with the problems book in a forest and to solve in the mind for life. However, with the electronic versions of

* Sergei Alexandrovich Rachinsky (May 2 (14), 1833, p. Tatevo, Smolensk Province–May 2 (15), 1902 was a Russian scientist, teacher, educator, professor at Moscow University, botanist and mathematician. Corresponding Member of the Imperial St. Petersburg Academy of Sciences. Court counselor (translated from http://ru.wikipedia.org/wiki/Рачинский,_Сергей_Александрович).

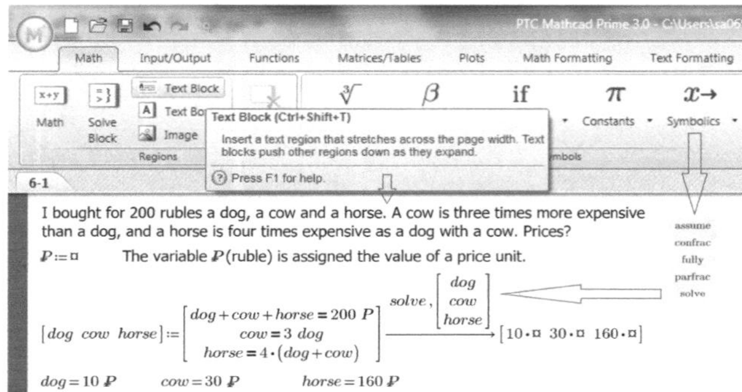

FIGURE 6.1 The problem of a dog, a cow and a horse.

the book it is very difficult to fulfill this requirement. A second, obsolete name of the computer is electronic computer. And the very word computer means a calculator. On a computer (even if it is not a traditional "calculator", but a smartphone or "tablet"), there are calculators at hand. There is not only a trivial calculator connected to the operating system, but also programs specially designed for solving complex mathematical problems: Mathematica, Maple, MATLAB®, Mathcad, SMath, etc. Here are some examples of solving problems from Rachinsky's book performed in Mathcad—see Figures 6.1 through 6.5.

It's a sin, of course, to look down on the author of the problems book and on his readers. They say that a person could go mad, coming up with the problems[*] or solving them in his/her mind, while now we are literally completing them in two calculations—one or two and the answer is ready.

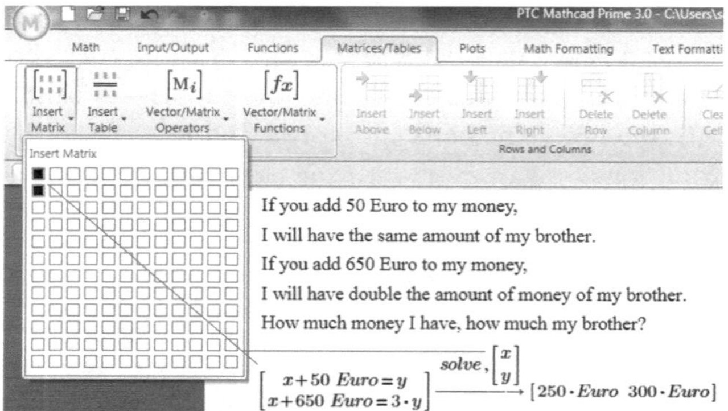

FIGURE 6.2 The problem of the money of two brothers.

[*] I wonder if Rachinsky made his problem "purely in the mind" or he still used a pen, paper or bills—a computer of the nineteenth century.

What could we reply to this!? Why do we turn again to some old problems in our computer age!?

First, they are not just "old", but "old, good problems".

Second, mental calculation is wonderful exercise for the mind.[†] And we have noted this in many of the Studies in this book. The use of a computer or a calculator is equivalent to the addition of a sports simulator (a modern gymnastic apparatus for the body[‡]) with hydraulic amplifiers.

The solution of Rachinsky's problems basically reduces to finding the roots of systems of algebraic equations (see Study 4). In Mathcad, a ***solve*** operator is provided for solving such problems. In some problems it needs to be supplemented with a ***substitute*** operator that replaces one variable with another or with an expression.[§] With the help of this operator, we told the computer that there are three yards (*Rus.* Arshins) (Figure 6.3) in a fathom (*Rus.* Sazhens), 32 lots in a pound, and three golden_ones in a lot (Figure 6.5[¶]), i.e. we taught the computer to handle the old Russian measures of weight and length that were in use at the time of writing the problem book. The problem in Figure 6.4 teaches us to work directly with old non-decimal units of length.[**] Such a problem, incidentally, is also relevant today. Who can mentally calculate immediately, how many years, months, weeks, days, hours, minutes and seconds there

[†] In our time *programming* is the best gymnastics for the mind. But far from all these, "mental exercises" are available to all. Some forums (forum of mathematical packages http://forum.exponenta.ru, for example) require four numbers to be added to be admitted. Many, sitting at a computer, cannot do this in the mind and resort to the help of a calculator. Mental calculations also remind us of accounts in "math history" about people who have incredible calculation skills: https://en.wikipedia.org/wiki/Mental_calculator, though here our book refers to "standard" calculation mental skills.

[‡] The author stumbled upon such a photograph on the Internet. An empty staircase. On the right and left of it are escalators that raise and lower quite obese people. At the top, at the end of the stairs is a building with the signboard "Fitness club". See http://www.goodexperience.com/broken/i/04/02/america-fitness-s.jpg.

[§] Figures 6.1–6.3 show additional tools for solving problems and processing the solutions: the operators of symbolic mathematics Mathcad (Figure 6.3), the command of the comment insertion menu (Figure 6.1) and the menu command for inserting a vector with two elements (2-by-1 matrix) and unknown equations (Figure 6.2).

[¶] Modern units of measurement are built into Mathcad, but there are no old Russian units in it. The old Russian units of measurement (as arshin) were redefined relative to the English system by Peter the Great in the eighteenth century. Also Cyrillic numerals were replaced with Hindu-Arabic as part of his civil script reform initiative.

[**] The transition from the non-decimal system of calculus (in a verst 500 fathoms, in a fathom 3 yards (arshins), in a yard (arshin) 16 versts) to the decimal (in a kilometer there are 1000 meters, in a meter 100 cm, in a centimeter 10 millimeters, etc., to all the nanoscience quantities)—this is the first blow to mental calculation. A calculator with a computer just finishes it.

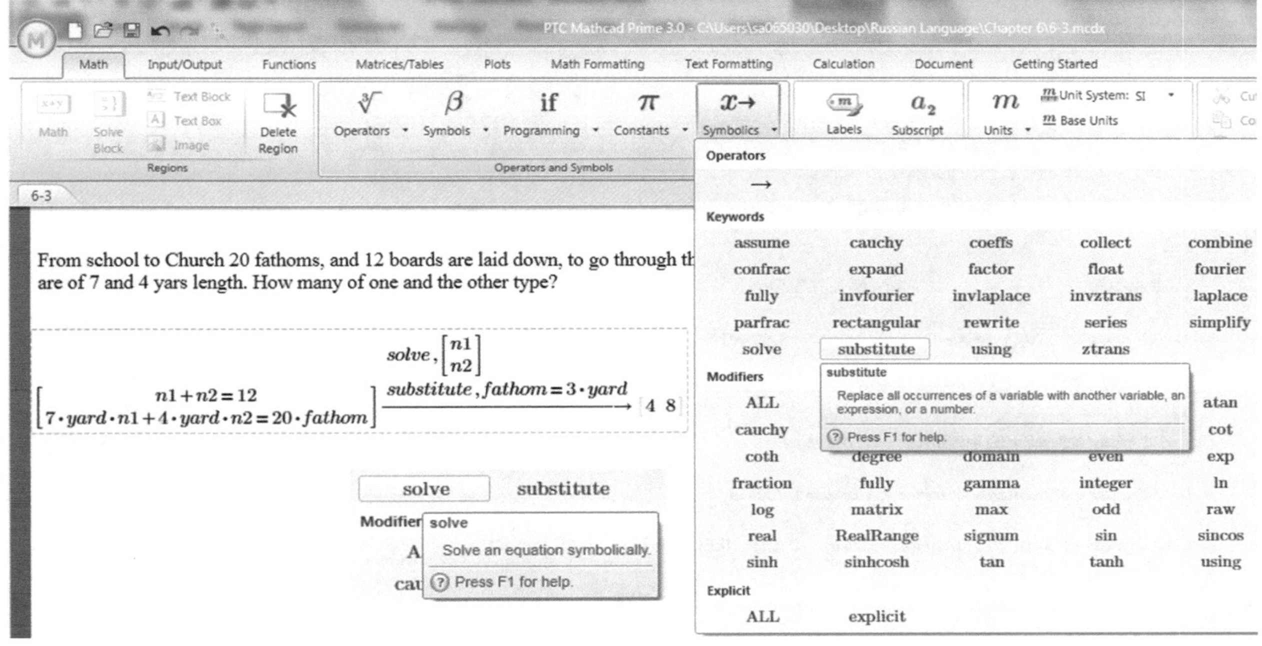

FIGURE 6.3 The problem of the road to the church.

1000 tip,top - how many fathoms, yards and tip,top ?

$$\frac{1000\ tiptop}{fathom} \xrightarrow[substitute,\ yard=16\ tiptop]{substitute,\ fathom=3\ yard} \frac{125}{6} = 20.833$$

$$\frac{1000\ tiptop - 20 \cdot fathom}{yard} \xrightarrow[substitute,\ yard=16\ tiptop]{substitute,\ fathom=3\ yard} \frac{5}{2} = 2.5$$

$$\frac{1000\ tiptop - 20 \cdot fathom - 2\ yard}{tiptop} \xrightarrow[substitute,\ yard=16\ tiptop]{substitute,\ fathom=3\ yard} 8 = 8$$

Answer 1000 tip,top = 20 fathom 2 yards 8 tip,top

FIGURE 6.4 The decomposition of 1000 apices into fathoms, arshins (yards) and tiptop.

The copperman had 8 pieces of copper each weighing 1 pound 8 lots. From this copper he made kettles, each weighing 1 pound 21 lots 1 golden one. How many kettles?

$$8 \cdot (pound + 8 \cdot lot) = n \cdot (pound + 21 \cdot lot + golden\_one) \xrightarrow{\substack{substitute,\ pound=32 \cdot lot \\ substitute,\ lot=3 \cdot golden\_one \\ solve,\ n}} 6$$

FIGURE 6.5 The problem of the copper kettles.

are in a billion seconds?!* Figure 6.6 shows the solution of this problem in Mathcad.†

So, what new slant can be put on the old arithmetical problems!?

First, the texts of problems from Rachinsky's book end with numerical answers with or without measurement units.‡ Our computer solutions to problems (see Figures 6.1 through 6.5) can be considered as answers to problems. Moreover, in such answers, a modern computer solution of problems is also shown. A problem book with these answers turns into a manual for schoolchildren, i.e. a problem book with a detailed description of solutions.

Second, in modern mathematics the authors of math problems designed for mental calculations for solving with pen and paper and/or a calculator are able to use not only "a pen with paper and a calculator", but they can also use the above mathematical programs for preparing the problems and, most importantly, to verify the correctness and uniqueness of the solutions—see footnote 1. The teacher here must go ahead of the disciples. To us it often turns out that schoolchildren know perfectly well how to use a computer while the teacher is afraid to approach it. At the same time, this "computerphobia" is masked by statements such as "the computer in the learning process only brings harm—preventing,

* "Decimal" did not fully cover the measurement system. In addition to time, we can mention angle units with degrees, minutes and seconds.

† Comparing Figure 6.4 with Figure 6.6, where the same problems are solved, it can be noted that in the first case, the *symbolic mathematics* of Mathcad is used, and in the second case the *numerical* mathematics is used.

‡ In modern problem books, answers are usually placed at the end of the book.

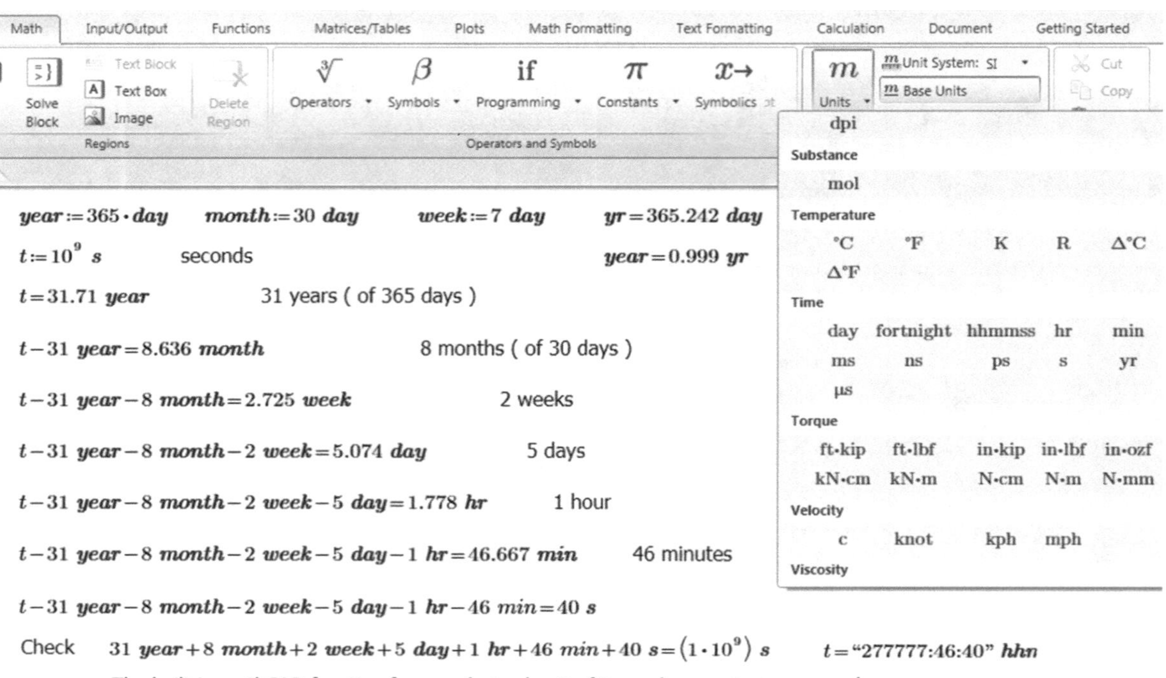

FIGURE 6.6 The decomposition of a billion seconds.

for example, the development of mental calculation techniques".

Third, the problems of Rachinsky, and other similar ones, can be modernized so that it would be rather difficult to solve them without a computer. See the example in Figure 6.7.

Both the schoolchild and the illiterate Russian peasant of the nineteenth century knew perfectly well how many pounds in a pood, or how many tiptops in a yard or yards in a fathom. A modern schoolchild will look in a reference book or on the Internet for this information. And if there is no reference book at hand, and the network is temporarily unavailable!?* No problem! The ratio of the pound, the lot and the golden_one is sewn into the problem of the copperman (Figure 6.5) with the answer (6 teapots). Solving the new problem (Figure 6.7), the student will have to excel not in mental calculation, but in the ability to code a program for a computer. Figure 6.7 shows one of the possible solutions of the problem, based on the search of all possible options (see also Study 17 for more about searching): in a pound there could be from 1 to 100 lots, and in a lot from 1 to 100 golden_ones. Incidentally, almost all the problems of Rachinsky are chosen so that they have integer answers. When solving these problems, it is possible to successfully use the brute-force method that has blossomed in the era of computers with their incredible speed. What is the cube root of 1728? I simply look through the numbers in a reasonable range (from 1 to 100, for example) and remember the number whose cube is 1728. Mentally of course, you will not do it, but on a computer—please.† Let the reader try to solve the problem in Figure 6.1 without indicating that 12 boards were laid from school to church: "From school to church there are 20 fathoms, and boards are laid to pass through the mud.‡ There are boards of 7 and 4 yards in length. How many we have of those and others ones? [*Answer*: 4 and 8 or 8 and 1]", see Figure 6.1a. And if the short board is one yard long, then there will be eight options for laying the boards! Although the correct answer will be eight plus one board—we

* Many cannot work on a computer with the Internet turned off, even if the Internet is not needed and, moreover, harmful (i.e. a distraction from performing urgent work). It is akin to a feeling that you are locked in a room for a while. There is no need to go out, but it's still unpleasant.

† The cube root of 1728 is not one number, but three: twelve and two more complex numbers. This can be seen if we solve the equation $x^3 = 1728$ in Mathcad. A search here can get only one answer. The same incomplete answer is given also by calculators with computers, see Figure 6.9.

‡ This Russian problem has not been solved so far. Let's remember about our two troubles: "about fools and roads!" The author hopes that this book will somehow correct the first trouble.

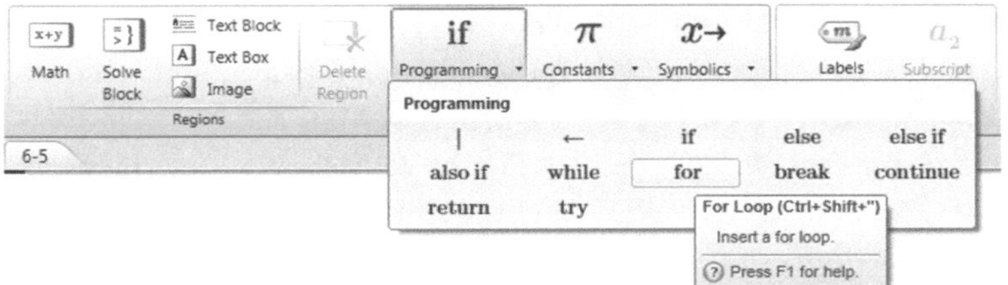

The copperman had 8 pieces of copper each weighing 1 pound 8 lots. From this copper he made 6 kettles,each weighing 1 pound 21 lots 1 golden one. How many lots in a pound (Nlot) and how many golden one in a lot (Ngolden_one) ?

$Answer :=$ pound ← 1
  i ← 0
  for $Nlot \in 1..100$
    for $Ngolden\_one \in 1..100$
      if $8 \cdot \left(pound + 8 \cdot \frac{pound}{Nlot}\right) = 6 \cdot \left(pound + 21 \cdot \frac{pound}{Nlot} + \frac{pound}{Nlot \cdot Ngolden\_one}\right)$
        $M^{\langle i \rangle} \leftarrow \begin{bmatrix} Nlot \\ Ngolden\_one \end{bmatrix}$
        $i \leftarrow i + 1$
  M

$Answer = \begin{bmatrix} 32 & 34 \\ 3 & 1 \end{bmatrix}$ In a pound 32 lots, in a lot 3 golden_one

FIGURE 6.7 The definition of the old Russian units of mass (weight).

put first long boards on the ground, and then when the next long one does not fit on the road, then we put short boards on the road. Other scientific disciplines are mixed with mathematics, for example: the price per meter of a long board (rub/m) is usually less than that of a short one: the second answer in the problem of boards (eight boards of seven yards and one board per one yard) will also be cheaper due to the fact that nine boards are more easily laid than twelve (see Figure 6.8).

The problems of Rachinsky are also interesting because it is possible to find out how people lived in the nineteenth century, what they were doing, what they were selling and what were the prices. In literature (and Rachinsky's problems, among other things, are also distinguished by a good literary style) you can find mathematical calculations dealing, mainly, with financial problems. Here are three of them.

PROBLEM 1

[1] Anton Chekhov's short story "*The Tutor*"* begins with this dialog (https://www.ibiblio.org/eldritch/ac/tutor.htm):

"Now on arithmetic ... Take a board. What is the next problem? Next we'll do some arithmetic. Fetch your slate! I want you to do this problem."

Pete spits on his slate and rubs it dry with his sleeve. His tutor picks up the arithmetic and dictates the following problem to him.

"If a merchant buys 138 yards of cloth, some of which is black and some blue, for 540 roubles, how many yards of each did he buy if the blue cloth cost 5 roubles a yard and the black cloth 3?' Repeat what I have just said."

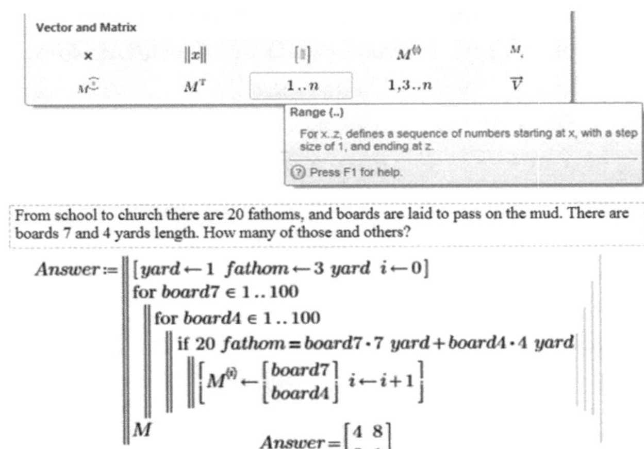

FIGURE 6.8 The problem of a road paved with boards.

* It is possible to read it at: https://www.ibiblio.org/eldritch/ac/tutor.htm. To the interested readers, Other Chekhov short stories are available here: http://www.free-short-stories.org.uk/anton-chekhov-free-short-stories.htm.

Peter repeats the problem and instantly and silently begins to divide 540 by 138.

"What are you doing? Wait a moment! No, no, go ahead! Is there a remainder? There ought not to be. Here, let me do it!"

Ziboroff divides 540 by 138 and finds that it goes three times and something over. He quickly rubs out the sum.

"How queer!" he thinks, ruffling his hair and flushing. "How should it be done? H'm—this is an indeterminate equation and not a sum in arithmetic at all—"

The tutor looks in the back of the book and finds that the answer is 75 and 63.

"H'm—that's queer. Ought I to add 5 and 3 and divide 540 by 8? Is that right? No that's not it. Come, do the sum!" he says to Pete.

"What's the matter with you? That's an easy problem!" cries Udodoff to Peter. "What a goose you are, sonny! Do it for him, Mr. Ziboroff!"

Gregory takes the pencil and begins figuring. He hiccoughs and flushes and pales.

"The fact is, this is an algebraical problem," he says. "It ought to be solved with x and y. But it can be done in this way, too. Very well, I divide this by this, do you understand? Now then, I subtract it from this, see? Or, no, let me tell you, suppose you do this sum yourself for to-morrow. Think it out alone!"

Pete smiles maliciously. Udodoff smiles, too. Both realize the tutor's perplexity. The high-school boy becomes still more violently embarrassed, rises, and begins to walk up and down.

"That sum can be done without the help of algebra," says Udodoff, sighing and reaching for the counting board. "Look here!"

He rattles the counting board for a moment, and produces the answer 75 and 63, which is correct.

"That's how we ignorant folks do it."

This system of two linear algebraic linear equations (SLAE) with two unknowns (namely, it can be seen in the above dialog: $x + y = 138$ and $5\ x + 3\ y = 540$) can be solved without a computer and even without a calculator, by performing simple arithmetic operations on paper or just in the mind. Now of course, very few people will do it on a counting board (as Udodoff the elder, Petya's father did). We will solve this problem on the computer (on a modern "counting board"), using the program Mathcad—see Figure 6.10.

Figure 6.10 shows how, in Mathcad, the merchant and the cloth problem is solved with the help of the ***solve*** operator, designed to search for the roots of single algebraic equations and their systems, both linear and nonlinear: "That's in our way, *computer way*"!

Chekhov himself does not provide the "ignorant" solution in the story, but it can easily be reconstructed, since the problem has a standard arithmetic solution based on logic and requiring six arithmetic operations:

"Suppose that all the purchased cloth was blue, then a lot of 138 yards (arshins) would cost 5 * 138 = 690 rubles, but this is 690 – 540 = 150 rubles more than what was paid in reality. The "overrun" of 150 rubles indicates that in the lot there was a cheaper, black, cloth—3 rubles per yards (arshin). We have this cloth so much that from the two ruble difference (5 – 3 = 2 rubles), 150 "extra" rubles are obtained. That is, 150/2 = 75 yards (arshin) of black cloth, hence 138 – 75 = 63 yards (arshin) cloth of blue."

"The rattling of the counting board",* performed by Udodoff, looked like this:

1. First of all, Udodoff Sr. "composes" the number 138: a bone (abacus bead) on the first wire, three on the second, eight on the third.
2. Then he must multiply 138 by 5. To simplify the calculation, instead, he first multiplies 138 by 10 without doing any manipulation, just mentally transferring all the bones one row higher, then dividing by 2: on each wire, starting from the bottom, he flips half a stone. On the third wire, where eight beads are deposited, he folds four; on the middle wire of three bones, he folds two, while for one of them he mentally replaces the ten lower ones and divides them in half, that is, adds five bones to those on the next wire; on the top wire he removes one bone, adding five to the bones on the second wire. As a result, there are no bones on the upper wire, there are six on the second, and nine on the third. Total: 690.
3. Further Udodoff, the elder, needs to subtract 540 from 690: from the second wire, five seeds

* The abacus: in Western Europe it was reintroduced (since it disappeared around the sixteenth century) from Russia by the French mathematician Jean Viktor Poncelet (who was a prisoner in Russia while serving in Napoleon's army), mostly for teaching and didactic purposes.

are removed, from the third, four. Leaving 150 remaining.

4. Now 150 should be divided in half (see above)—it turns out to be 75.
5. It remains to subtract 75 from 138. Again he "composes" 138, on the second wire you need to drop seven, but there are only three. There are not enough for four, so Udodov the elder leaves six bones on the wire (if he is too lazy to subtract four out of ten in his mind, he can throw all ten bones on the second wire to the left and discard the "undervalued" four bones), and from the first wire he removes one bone. It remains to drop five bones from eight on the third wire. It turns out to be 63.

The mother of the author of this book could calculate square roots on the counting board. However, she could not calculate cubics—see Figure 6.9.

In the problem of the merchant and the cloth, the number of equations is *equal* to the number of unknowns. But in life and, accordingly, in fiction, this is not always the case. In addition, the problems themselves are not always formulated clearly, and can be "blurred" throughout the text—see the second and third problems.

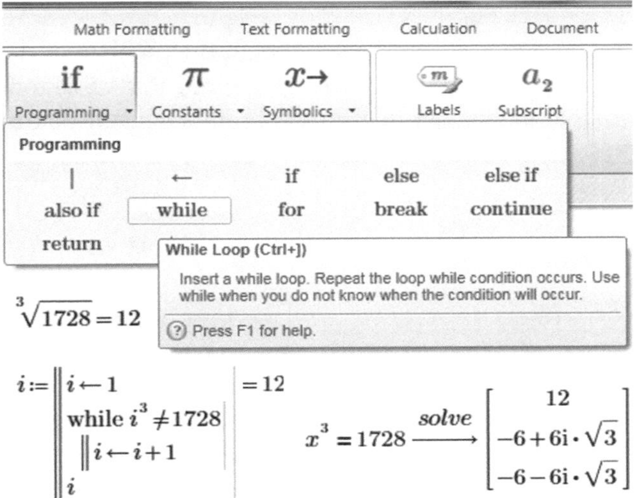

FIGURE 6.9 Solution of the problem of the merchant and the cloth in Mathcad Prime.

$$\begin{bmatrix} x+y=138\ yard \\ 5\,\frac{Rubles}{yard}\cdot x+3\cdot\frac{Rubles}{yard}\cdot y=540\ Rubles \end{bmatrix} \xrightarrow{solve,\begin{bmatrix} x \\ y \end{bmatrix}} [63\cdot yard\ \ 75\cdot yard]$$

FIGURE 6.10 Search for a cubic root.

PROBLEM 2

In a story from another Russian writer Fyodor Dostoyevsky, "The Gambler", seven quotations can be found* in which the exchange rates of European currencies during the writing of the story (the second half of the eighteenth century) are encrypted:

Quotation 1

"You will receive this money immediately," the general replied, blushing a little, rummaged in his bureau, consulted in a notebook, and it turned out that he had about one hundred and twenty rubles for my money. "How do we count," he said, "must be translated into thalers." Yes, take a hundred thalers, round-robin - the rest, of course, will not be lost.

and also

You should get me these four ***Friedrich d'or and three florins*** *for this calculation.*

Quotation 2

Polina was just angry when I gave her only ***seven hundred gulden****.*

and also

Listen and remember: take these ***seven hundred florins*** *and go play, win as much as you can on roulette; I need money, no matter what.*

Quotation 3

I started by taking out ***five Friedrich d'or*** *that is,* ***fifty gulden****, and put them on a rosary.*

Quotation 4

"Yes, I tried and I won twelve thousand florins!" What twelve, and about the gold? With the gold, almost ***thirteen*** *thousand will come out. This is how much in our currency? Six thousand, whatever that, will be?*

* And in the era of the Internet and digital libraries, it's much easier to do this: I've set the search keyword and calculated how many times it appears in the text. https://en.wikisource.org/wiki/The_Gambler.

I answered that at the current currency exchange they are about seven, perhaps, up to ***eight thousand*** *rubles.*

Quotation 5

"Oui, madame," the croupier politely confirmed, "just like any single rate should not exceed ***four thousand florins*** *at once, according to the statute," he added in explanation.*

And also

I put the biggest allowed bet, in ***four thousand gulden****, and lost.*

Quotation 6

She had to get exactly ***four hundred and twenty Friedrich d'or*** *that is,* ***four thousand florins and twenty Friedrichs d'or****.*

Quotation 7

"Pauline, ***twenty-five thousand florins*** *are* ***fifty thousand francs****, even more."*

In the story of A.P. Chekhov the equations are spelled out explicitly; whereas in Dostoevsky's "*The Gambler*" they are ciphered in quotations. Having decided upon the corresponding system of equations, you can find the exchange rates (thaler, Friedrichs d'or, florin, gulden and franc) in relation to the ruble:

120 rubles = 100 talers + 4 Friedrichs d'or + 3 florins (quotation 1—see above)

700 gulden = 700 florins (quotation 2)

5 Friedrichs d'or = 50 gulden (quotation 3)

13 000 florins = 8000 rubles (quotation 4)

4,000 florins = 4000 gulden (quotation 5)

420 Friedrichs d'or = 4000 florins + 20 Friedrichs d'or (quotation 6)

25,000 florins = 50,000 francs (quote 7)

If the number of equations (we have seven) exceeds the number of unknowns (we have five), then such system of equations is called *overdetermined*.

The problem of "*The Gambler*" can also be solved without the computer, by consistently calculating the rates of individual currencies (the gulden is equal to the florin, the Friedrichs d'or is equal to ten gulden, etc.), and determining their value relative to the ruble. However, as in the case with the first problem from Chekhov's "*The Tutor*" https://www.ibiblio.org/eldritch/ac/tutor.htm), we ask the computer to do this. Let the computer substitute and simplify what it considers necessary—see Figure 6.11.

Figure 6.11 shows the solution of the problem of the currency rates using the ***solve*** operator in Mathcad. From the solution shown in Figure 6.11, it is evident that in the second half of the nineteenth century the thaler was

$$\begin{bmatrix} 120\ Rubles = 100\ Taler + 4 \cdot Friedrichsdor + 3 \cdot Florin \\ 700\ Gulden = 700\ Florin \\ 5 \cdot Friedrichsdor = 50\ Gulden \\ 13000\ Florin = 8000\ Rubles \\ 4000\ Florin = 4000\ Gulden \\ 420 \cdot Friedrichsdor = 4000\ Florin + 20 \cdot Friedrichsdor \\ 25000\ Florin = 50000 \cdot Frank \end{bmatrix} \xrightarrow{solve, \begin{bmatrix} Taler \\ Friedrichsdor \\ Florin \\ Gulden \\ Frank \end{bmatrix}, float, 2} [0.94 \cdot Rubles\ \ 6.2 \cdot Rubles\ \ 0.62 \cdot Rubles\ \ 0.62 \cdot Rubles\ \ 0.31 \cdot Rubles]$$

FIGURE 6.11 Solving the problem of the exchange rate in Mathcad Prime.

$$\begin{bmatrix} \begin{pmatrix} 200 \cdot louis \\ + 120\ louis \\ + 140\ louis \\ + 1832 livres\ ... \\ + 2780 livres\ ... \\ + 4379\ livres + 12\ sols + 8\ deniers\ ... \\ + 1748\ livres + 7\ sols + 4\ deniers \end{pmatrix} = 15800 \cdot livres \\ (200 + 120 + 140) \cdot louis = 5060 \cdot livres \\ 20\ sols = livre \end{bmatrix} \xrightarrow{solve, \begin{pmatrix} louis \\ livres \\ sols \end{pmatrix}} (2640 deniers\ \ 240 deniers\ \ 12 deniers)$$

sol = 12 deniers livres = 20 sols louis = 11 livres

FIGURE 6.12 The first problem of Molière.

worth 94 kopecks, the Friedrich d'or 6 rubles 20 kopecks, the florin and gulden each 62 kopecks and the franc 31 kopecks. In this situation, the general paid the teacher (Alexei Ivanovich) 117 rubles, 73 kopecks ("about 120 rubles"—see quotation 1 above), and the "granny" won 7930 rubles at roulette ("up to eight thousand rubles"—see quotation 4).

Note. The fact that florins and gulden are equal in value, the Friedrich d'or is ten times higher in value than a florin or a gulden and the franc is half the value of a florin or a gulden, is also clear without a computer. But we specifically introduced all seven quotations into the computer to get an overdetermined system of equations.

PROBLEM 3

A good example of monetary calculations can be found in Molière's *Le Bourgeois gentilhomme* (http://www.gutenberg.org/files/2992/2992-h/2992-h.htm).

DORANTE: I want to settle this matter with you, and I came here to make up our accounts together. <...> Do you remember well all the money you have lent me?

MONSIEUR JOURDAIN: I believe so. I made a little note of it. Here it is. Once you were given two hundred Louis d'or.

DORANTE: That's true.

MONSIEUR JOURDAIN: Another time, six score.

DORANTE: Yes.

MONSIEUR JOURDAIN: And another time, a hundred and forty.

DORANTE: You're right.

MONSIEUR JOURDAIN: These three items make four hundred and sixty Louis d'or, which comes to five thousand sixty livres.

DORANTE: The account is quite right. Five thousand sixty livres.

MONSIEUR JOURDAIN: One thousand eight hundred thirty-two livres to your plume-maker.

DORANTE: Exactly.

MONSIEUR JOURDAIN: Two thousand seven hundred eighty livres to your tailor.

DORANTE: It's true.

MONSIEUR JOURDAIN: Four thousand three hundred seventy-nine livres twelve sols eight deniers to your tradesman.

DORANTE: Quite right. Twelve sols eight deniers. The account is exact.

MONSIEUR JOURDAIN: And one thousand seven hundred forty-eight livres seven sols four deniers to your saddler.

DORANTE: All that is true. What does that come to?

MONSIEUR JOURDAIN: Sum total, fifteen thousand eight hundred livres.

DORANTE: The sum total is exact: fifteen thousand eight hundred livres. To which add two hundred pistoles that you are going to give me, which will make exactly eighteen thousand francs, which I shall pay you at the first opportunity.

In this passage there are the following units of value: Louis d'or, livre, sol, denier, pistol and franc. The ratio between Louis d'or and livre is explicit in the text: 460 Louis = 5060 livres, hence, Louis = 11 livres. The search for the remaining ratios requires some mental effort and Mathcad.

It is easy to see that 8 denier and 4 denier sum to one sol, i.e. 12 denier. Similar calculations, connected with the fact that the total amount is a round number, can be applied to the sum (12 sol + 8 sol = ...), but stop! Let's move the solution to Mathcad (Figure 6.12).

From the solution (Figure 6.12) it is clear that 1 sol = 12 denier, 1 livre = 20 sols (240 denier), and 1 Louis d'or = 11 livres (12 sols + 8 sols = 1 livre—we repeat, this problem could be solved without a computer.

It is more difficult with pistols and francs. This problem has not been fully determined.

Figure 6.13 shows a solution to this problem under the assumption that the livres were equal to the franc. Here we can talk about two different names for the same coins.* In a hundred years, historians will calculate the

$$\begin{pmatrix} 15800\ \text{livres} + 200\ \text{pistoles} = 18000\ \text{francs} \\ \text{pistoles} = 11\ \text{livres} \end{pmatrix} \xrightarrow{\text{solve},\begin{pmatrix}\text{livres}\\ \text{pistoles}\end{pmatrix}} (\text{francs}\quad 11\text{francs})$$

livres = francs   pistoles = louis

FIGURE 6.13 The second problem of Molière.

* *Le Bourgeois gentilhomme* (*The Bourgeois Gentleman* or *The Middle-Class Aristocrat* or *The Would-Be Noble*) is a five-act *comédie-ballet*—a play intermingled with music, dancing and singing—written by Molière, first presented on 14 October 1670 before the court of Louis XIV at the Château of Chambord by Molière's troupe of actors. Between 1360 and 1641, coins worth 1 livre were minted, known as francs. This name persisted in common parlance for 1 livre but was not used on coins or paper money (from Wiki).

ratio between the dollar and the "buck," forgetting that it's the same thing.

If the number of equations of an algebraic system is *less* than the number of unknowns, then such a system is called *underdetermined*. The tutor Ziboroff from the Chekhov story tried to solve the problem of the merchant and the cloth erroneously. However, it was clearly determined—there are two algebraic equations with two unknowns in it; the system itself has one single solution. An underdetermined problem can be seen in the description of the submarine *Nautilus*—"hero" of the Jules Verne novels *Twenty Thousand Leagues under the Sea** and *The Mysterious Island*—see Study 32.

REFERENCE

Ochkov Valery, Look Andreas. Math Lessons in Classical Literature. *Journal of Humanistic Mathematics*, 2015, Volume 5, no. 2, pp. 121–132 (http://twt.mpei.ac.ru/ochkov/MathLit.pdf).

* *Twenty Thousand Leagues under the Sea* by Jules Verne is available at: http://www.gutenberg.org/files/164/164-h/164-h.htm.

Catenary Or

The Fifth Element

ABSTRACT

The Study explains how a computer can calculate the parameters of a power cable suspended over a road and many other interesting things related to the catenary function. Furthermore, electrical circuits, spotlights and lenses are be examined. The Study introduces and discusses a new type of equation—a software equation, which is formed by means of programming.

Mathematics: Exponent, hyperbolic sine and cosine, catenary function, derivative, parabola, catenoid, curve length, arc length, Taylor series expansion, Fourier series expansion, linear algebraic equations system (SLAE), matrix rank, Kronecker-Capelli theorem, the limit, the sinusoid, the base of the natural logarithm.

Physics: Arch of maximum strength, friction of a rope with a bollard, a rope suspended over a bridge, aerial ropeway, balance of forces and of moments of forces, center of mass, square wheel, kinematics of the bicycle, electrical circuit, Ohm's law, electromotive force, law of light reflection, Snell's law.

IT: Numerical and analytical solution of algebraic and differential equations and their systems, flat and three-dimensional computer graphics, programming equation, computer animation, programming, cycle with parameter, debugging of computer calculations.

Art: The Arch in St. Louis.

Study website: https://community.ptc.com/t5/PTC-Mathcad/Catenary-Physic-Math-IT/m-p/494899.

As noted earlier, in the educational process today there is a tendency to study *computer science* and *mathematics* simultaneously.

This is the subject of this Study, and of a series of others Studies in this book, involving math, physics (theoretical mechanics and theoretical fundamentals of electrical engineering) and computer science.

At their lectures students of electricity are told this "terrible" and instructive story. In winter, an overhead power transmission line (OTL) was thrown across a Siberian river. Everything seemed to be correctly envisaged, designed and done, however, so that the high voltage wires should not have interfered with shipping navigation.* In the springtime, when ice was flowing down the river and navigation was beginning again, a ship touched a wire under tension with her metal mast. It is highly dangerous just to swim near high-voltage wires even without touching them, since an electrical discharge could occur! What was the reason for this engineering error? In spring, because of the warmth, the steel wires across the river lengthened due to the thermal expansion of the metal, and the gap between the lower point of the wire and the river became shorter

* As engineers, we repeat, we have to solve not just engineering problems, but problems with elements of economy. In the case of power lines, it was possible to increase the height of two coastal masts or tighten the wires for insurance, but this would lead to a rise in the cost of building and operating the power transmission line.

than it was during winter.[*] In addition, the water level in the river was abnormally high. What was the cause of the accident—the thermal expansion of the wires (they are also heated a little by the electric current), high water levels in the river and/or something else? Nobody could answer this question without referring to the calculations shown below.

If somebody takes an absolutely flexible wire (chain) and hang it on two supports, then the shape of the curve that the chain will form will be given by a function, which is called—*catenary*.[†] The chain could be broken by pulling it at the two ends, but it cannot be broken by bending it.[‡] On the links of the sagging chain it is only tensile forces that are acting and there are no bending forces, such as those that take place, for example, in a rigid beam. If the sagging chain is, let's say, "frozen" and turned "upside down", then you will get *an arch* that also has this unique property—only compression forces act on the elements of this arch and there are no bending forces there. This property of the arch has been a priori and intuitively used since ancient times: the value of the resistance to compression of many construction materials is much higher than the value of the bending resistance: for example, a brick can be broken into two halves with the hands, but it is almost impossible to crush it by hand.

In monographs and handbooks, as well as on the Internet,[§] it is possible to find the formulas for the catenary by solving the differential equation of the balance of forces acting on an elementary element (sorry for the tautology—on the reference) of the chain. In all these formulas the most "persistent" elementary function is probably the *exponential function*,[¶] e^x, the main property of which consists in the fact that it can be differentiated an arbitrary number of times (take the derivative with respect to x), but it is "like water off a duck's back" since it remains itself and it does not change its appearance. This is unlike the parabola, whose graph is similar to the catenary (we will mention this below). The parabola, in contrast to the catenary, turns to zero after only taking the derivative three times. With a catenary this does not happen: $ax^2+bx+c \rightarrow 2ax+b \rightarrow 2a \rightarrow 0$.

The formula of the catenary can be found not only in books or on the Internet, but also on architectural objects that exploit the beauty and the unique strength property of the inverted chain—see in Figure 7.1 the famous archway (Gateway Arch) erected in the American city of St. Louis on the west bank of the Mississippi river.[**]

At the base of the arch it is written its formula in feet:

$$y = 757.7' - 127.7' \cdot \mathrm{ch}(x / 127.7')$$

In the formula of the catenary, the *hyperbolic cosine ch*, consisting of the average of two exponentials, appears:

FIGURE 7.1 The St. Louis arch has the form of an inverted catenary (Daniel Schwen, Attribution-ShareAlike 2.5 Generic [CC BY-SA 2.5]).

[*] There is this problem: how much the Trans-Siberian trains tracks are shortened in winter. In this case it is possible to fix two temperatures, to establish the coefficient of linear thermal expansion of the rail steel and to solve the problem by simple multiplication. It is also possible to simulate the profile of the temperature field of the rails in winter and in summer and to solve the problem through programming (finite element method) or by integration.

[†] This function is denominated catenary from Latin for chain (*catena*). The word was introduced in English by Thomas Jefferson, the American Founding Father who was the principal author of the Declaration of Independence and later served as the third president of the United States. The word was introduced in a letter to Thomas Paine on the construction of an arch for a bridge. Was it perhaps a bridge over the Mississippi river?

[‡] This property is called absolute flexibility

[§] Now, we repeat, at the occurrence of a question the arm more and more often stretches for an answer not to a shelf with books but to the mouse of the computer.

[¶] It is possible to find a rigorous mathematical derivation of the equation of the catenary on various websites—see for instance http://mathforum.org/calcpow/images/catenarytextexact.htm.

[**] If somebody looks at this arch from the other bank, e.g. from the eastern bank of the river, then on the water surface you can see the sagging chain as the reflection of the arch. The city itself, by the way, is in the state of Missouri.

$ch(x) = (e^x + e^{-x})/2$. In Mathcad notation, it is indicated as *cosh*. If we replace the plus with a minus in this formula, we get the *hyperbolic sine* (*sinh* or *sh*—see Figure 7.9 below), which appears in the formula for the derivative of the catenary. In the second derivative, the hyperbolic cosine reappears. This (*cos-sin-cos-sin*) and other features relate ordinary trigonometric functions with their hyperbolic equivalent. Another example is the hyperbolic identity $ch(x)^2 - sh(x)^2 = 1$, which is similar to the Pythagorean trigonometric identity and could be used to find an analytical solution of the integral in the problem of the closed chain on the hangers (see Study 31: Mad Chain).

Figure 7.2 shows the calculation of the dimensions of the arch of the city of St. Louis in meters, with the help of the formula in feet presented above.

In Figure 7.2 is shown not only the graph of the function written on the St. Louis arch, but also its height (the value of the function $y(x)$ *at* $x=0$) and the "pitch" of the arch, i.e. the distance L between its supports on the ground. The arch fits into a square with a side of exactly 630 feet (210 yards or about 192 meters).

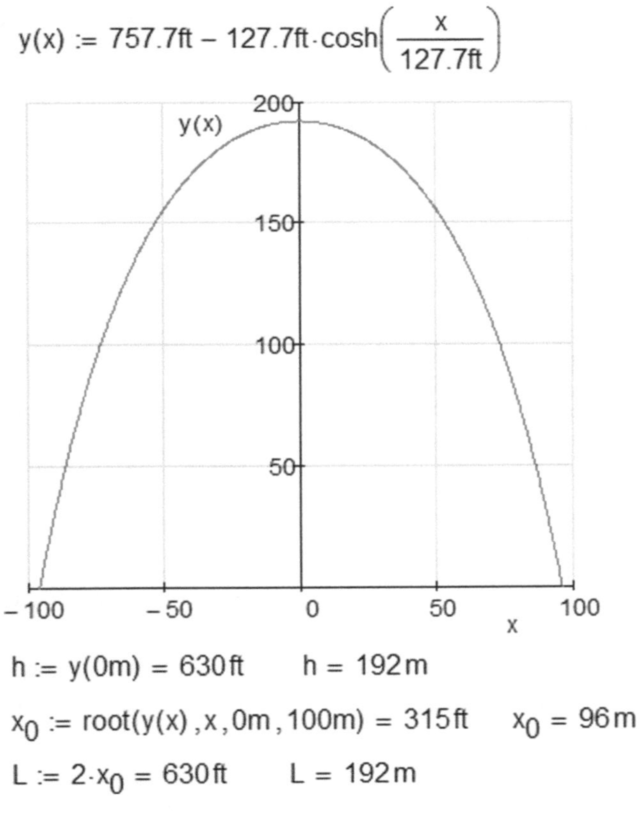

$$y(x) := 757.7\text{ft} - 127.7\text{ft}\cdot\cosh\left(\frac{x}{127.7\text{ft}}\right)$$

y(x)

200, 150, 100, 50

−100, −50, 0, 50, 100

x

$h := y(0\text{m}) = 630\text{ft} \qquad h = 192\text{m}$

$x_0 := \text{root}(y(x), x, 0\text{m}, 100\text{m}) = 315\text{ft} \qquad x_0 = 96\text{m}$

$L := 2\cdot x_0 = 630\text{ft} \qquad L = 192\text{m}$

FIGURE 7.2 Calculation of the geometrical parameters of the St. Louis arch in Mathcad.

The traditional (canonical*) form of the catenary equation, given in all the reference books, is $y = a \cdot ch(x/a)$. It is concise, but it is not entirely convenient in practical calculations, such as those that will be considered below. Therefore, we will work with another formula for the catenary that has one argument x and three parameters that specify not only its shape, but also its position relative to the chosen coordinate system. We could not wait and in fact we already worked with this formula in Study 4—see Figure 4.49.

At the author's website at twt.mpei.ac.ru/MCS/Worksheets/chain-eng.xmcd there is an online interactive open calculation of the sagging chain, as shown in Figure 7.3. When this calculation was created, we used the catenary in the form $y = h + a \cdot ch((x - x_0)/a) - 1$, in which x_0 is the value of the argument for which the catenary has either a minimum (sagging chain—see Figure 7.3) or a maximum (arch†—see Figures 7.1 and 7.2); h is the value of the catenary at x_0 (the minimum or the maximum value, the arch height, for example). The parameter a also characterizes the form of the catenary: for $a > 0$ it is a sagging chain, for $a < 0$ it is an arch. The chain will approach a straight line, but it will never become straight if the chain is tightened from both ends: recall the guitar string. Because of its absolute flexibility, the cord will always sag a bit and no force will counteract the action of the gravity force no matter how small. And we can also consider, for example, brickwork, the horizontal inter-brick lines (gaps with mortar) of which are, strictly speaking, not straight lines, but segments of a catenary. When laying a brick wall, bricklayers stretch a string line horizontally to control the flatness and horizontality of the laying bricks. But this line will still slacken. It can be stretched by putting it either through a block or through a smooth cylindrical rod with a

* Canonical in the sense that a coordinate system is chosen for which only a parameter (a) is shown in the formula. The other two parameters become equal to zero. In the formula of the St. Louis arch, two parameters are shown, presented in the form of the two constants: 757.7 and 127.7 feet. In Study 9 we tried to convert the hyperbole into the canonical form: we will discuss the techniques for the transformation of a plane curve of the second order having five parameters to a canonical form having two parameters. The canonical parabolic notation also has one parameter p: $y = 2p\cdot x^2$. A quadratic equation has three parameters.

† On the St. Louis arch it would actually be better to write down our form of the catenary equation, from which the height of the arch and its width (half-width) at the base would be clearly seen. At the left beam at the base, it would be possible to fix the origin of the coordinates with two arrows (horizontal and vertical). Now the formula, written at the base of the arch, implies that the origin is on the ground below the point of maximum arch height (see Figure 7.2).

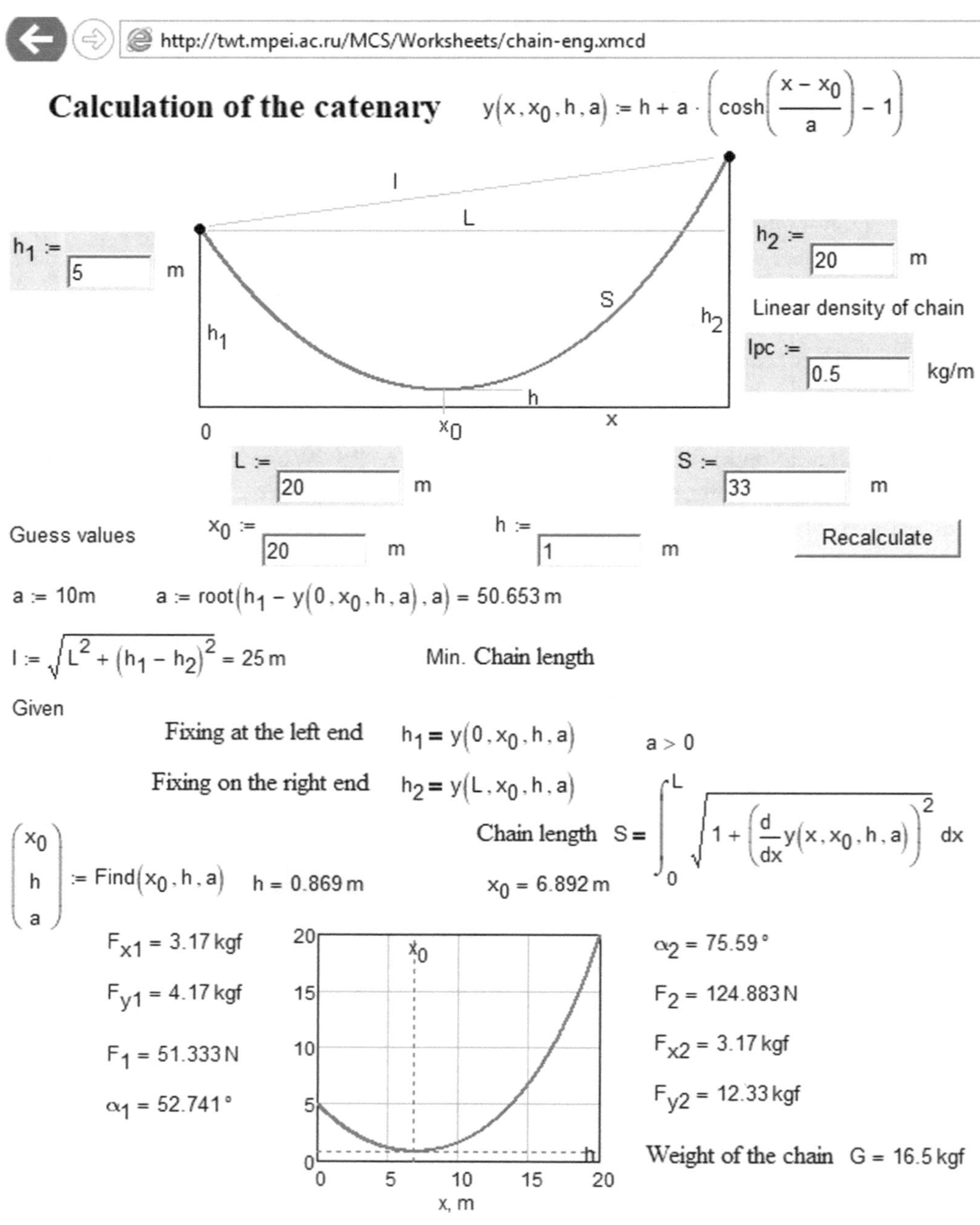

FIGURE 7.3 Website for the calculation of the sagging chain (twt.mpei.ac.ru/MCS/Worksheets/chain-eng.xmcd).

suspended load. This vertical end of the rope also serves to control the verticality of the masonry of the brick wall. Here, by the way, another exponent "floats up"—the friction force of the rope around the cylinder, on which it is attached or wound up, is proportional to the exponent of the winding angle. This method was used in ancient times for mooring ships to the pier. The formula with the exponent was derived by Euler in the eighteenth century and it is known as the capstan equation.

The website shown in Figure 7.3, requires the visitor to enter the values of the heights of the chain suspension on two posts (the masts of a power line—see the beginning of this Study and the "terrible electrical story"): height h_1 and height h_2, the horizontal distance L between supports and the value of the chain length S. From these data, the geometry of the chain is determined, in particular the value h of the minimum gap between the chain and the zero mark (ground or water in the river). In addition, the angles α_1 and α_2 at which our chains will pull the supports are calculated. These angles are useful for solving the problem of a chain with a suspended load (see below). The problem reduces to solving a system of three equations with three unknowns—the parameters x_0, h and a, forming a catenary. The first equation is the abscissa of the point of attachment of the chain at the left support, the second equation is the abscissa of the point of attachment of the chain at the right support and the third equation is the formula for the length of the curve.*

In the calculation shown in Figure 7.3, the Given–Find block was used (see Study 4: Solvers). The solution is then checked by comparing the curve on the graph drawn with a real metal chain at the right scale. Figure 7.4 shows the check of our calculation—the printout of the calculation shown in Figure 7.3 is inserted in a wooden frame, to which the chain for the check is nailed. This "work of art" hangs in the office of the author and attracts the attention of many mathematicians and non-mathematicians. This website could be used to analyze the "terrible story" given at the beginning of the Study. For this purpose, it is necessary to enter into the calculation the "winter" and "summer" values of the wire length and to calculate the "winter" and "summer" gaps over the river. However, it is also necessary to take into account the linear extension of the wire due to its own weight, which in turn also depends on the temperature of the wire. So it is not so straightforward.

A second "work of art" of the author, also associated with the catenary, is shown in Figure 7.5. This chain (suspension) bridge was drawn in the following way. When the version of Mathcad with the built-in function *Odesolve* was released (see Study 9: Celestial Mechanics), the author immediately tried to test it on this simple boundary-value problem: given four points for fastening a bridge circuit (0-0, ground; 2-5, first pillar; 9-5, second pillar; and 11-0, land on the

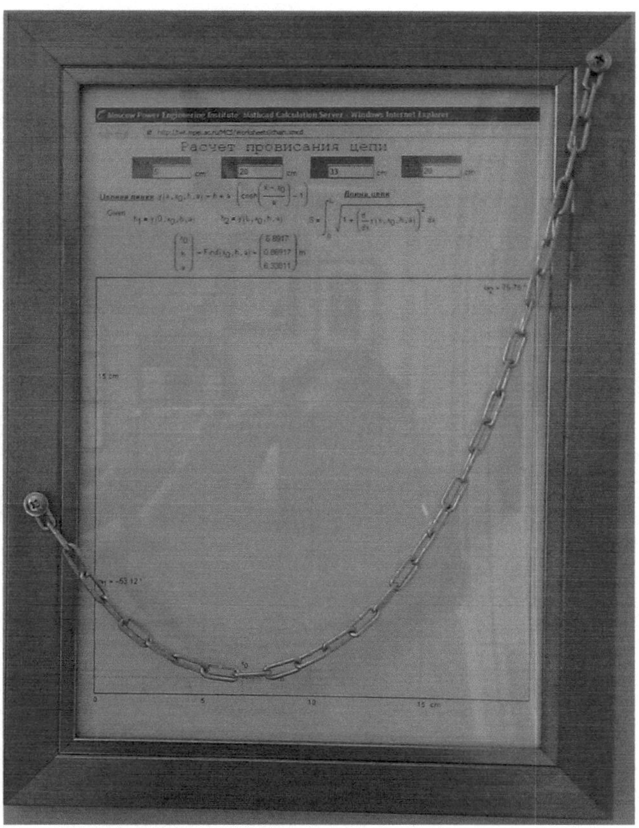

FIGURE 7.4 Real and virtual chains (with author's image).

other bank of the river), it is necessary to determine the shape of a chain corresponding to the property: the second derivative is equal to the function itself (the second derivative of the hyperbolic cosine is equal to the hyperbolic cosine). The solution of this problem (a, numerical, and b, analytical) is shown in Figure 7.5 with a graphical representation on the plane, while Figure 7.6 is in 3D. Figure 7.6 shows "two chains" with four fastening points each with the addition of fog,† perspectives, road markings with a dividing line, knots for securing vertical ropes and a flying eagle. The vertical ropes themselves are not drawn. The chain (rope) of the suspension bridge did not have the shape of a catenary but of a parabola as we have already mentioned. In this case, the loads from the roadway to the vertical ropes will be constant

* This equation and its derivation are also easily found in reference books or on the Internet. In the integrand, the Pythagorean theorem is clearly recognizable. In case of the catenary, it can be simplified by eliminating the definite integral on the segment from 0 to L. But we will not do this, leaving the "original" form of the expression. This is better for educational purposes, in accordance with the ideas expressed in the first Study of this book.

† Fog, perspective and lighting are the main tools for increasing the volume of three-dimensional objects on a flat computer display or on a sheet of paper. About perspective, as is known, Renaissance artists were very concerned, see for instance *Leonardo da Vinci Notebooks, IV Perspective of Disappearance*, https://en.wikiquote.org/wiki/Leonardo_da_Vinci#IV_Perspective_of_Disappearance

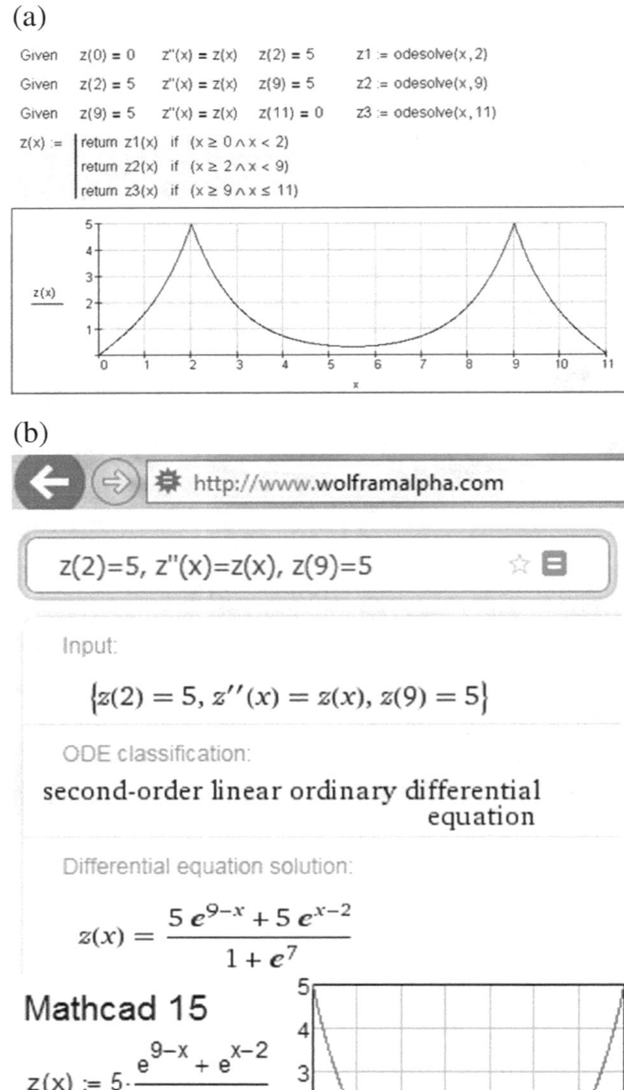

FIGURE 7.5 Solution to the virtual suspension bridge problem: (a) numerical solution in Mathcad 15 (three spans), (b) analytical solution on the Wolfram website with a plot obtained with Mathcad 15 (central span).

along the bridge because of their uniform arrangement. Consequently in Figure 7.6 we see only the bridge while under construction, on whose pontoons the canvas of the carriageway has been brought and to which the vertical ropes of a certain length are still going to be fastened, so that the chain will change its shape from the catenary to the parabola.* Strictly speaking, catenary and parabola are two different curves whose graphs will never coincide. However, as is known, all infinitely differentiable functions (and the catenary is of this type) can be represented by an infinite sum of power functions. This representation is called Taylor series. Thus, if in the expansion $a \cdot ch(x/a) = a \cdot [1 + (1/2!) \cdot (x/a)^2 + (1/4!) \cdot (x/a)^4 + (1/6!) \cdot (x/a)^6 + \ldots]$ we keep only the first two terms, then it is evident that, with a certain error, the catenary approaches the parabola. Obviously, the error is less, the closer the point x is to zero and the larger is the modulus of the parameter a. These differences are clearly seen in Figure 7.29. Figure 7.30 shows how Mathcad works with the Taylor series using the catenary example.

There are not only aesthetic but also "engineering and mathematical" feelings that arise when you admire these "works of fine art", shown in Figures 7.1, 7.4 and 7.6.

The sagging chain has another interesting property—its form does not depend on the mass of the chain. So, both the thin cord (the brick mason) and the massive chain (the monument fence) will sag equally if they have the same lengths and coordinates of the fastening points. In this regard, another famous physical law with similar

* "Another way to draw the desired parabola on the prism is as follows. Let's put two nails into the wall at the same height above the horizon and at such a distance from each other, so that it is equal to the double width of the rectangle on which it is desirable to build a parabola; between one and the other nail we hang a thin chain that would hang down and be so long that the lowest point of it would be from the level of the nail at a distance equal to the length of the prism. This chain, hanging down, will be arranged in the form of a parabola, so that, marking its trace on the wall with a dotted line, we get a complete parabola, cut in half by a perpendicular drawn through the middle of the line connecting both nails. Galileo Galilei, *Discourses and Mathematical Demonstrations Relating to Two New Sciences* (1638).

However, the master was mistaken. Between the parabola and the curve of the sagging chain there will be a slight difference. Only half a century later, Johann Bernoulli, Gottfried Leibniz and Christian Huygens derived the equation of the catenary. It involves a parameter whose changes determine various sagging curves of the chain. We owe the emergence of the name "catenary" to Huygens. By the way, if students are shown a real sagging chain in math, physics or computer science classes (the author always does this before reading this lecture study) and asked what formula describes it, then students, echoing Galileo, will say in chorus that it is a parabola. Another way to show the catenary is to pull a soap film inside a wire ring and place it horizontally. The film will sag and assume the shape of the surface formed by the rotation of the catenary around the vertical axis. However, after in the *Discourses*, on the fourth day, Galileo corrected himself (in Italian: "Quarta giornata del dialogo Discorsi e dimostrazioni intorno a due nuove scienze", from https://flore.unifi.it/retrieve/handle/2158/1108601/288291/CatenRivista.pdf) when Salviati (the author's alter ego) explains that the parabola is only an approximation of the catenary "le quali assai si avvicinano alle paraboliche" as per original in Latin language.

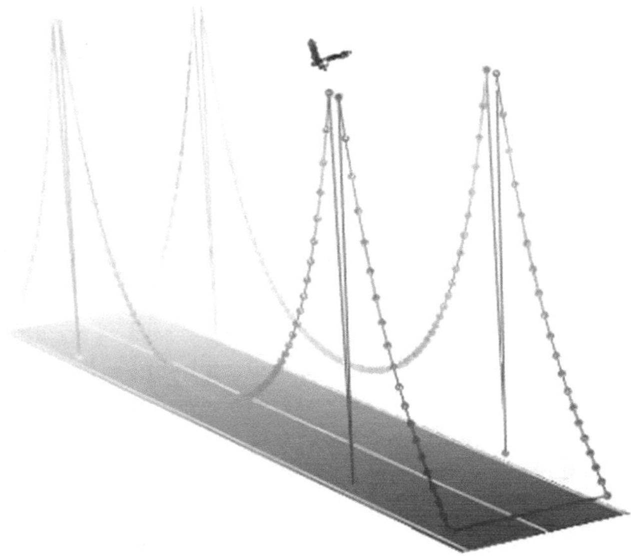

FIGURE 7.6 Virtual suspension bridge.

paradoxical elements comes to mind—the rate of fall of a body in a vacuum is the same for a feather and for a brick (see Study 9, where this issue will be described in more detail).

The problem of the sagging chain, which we have just described, is rather trivial. However, a more complex and "fresher" problem came to the mind of the author during a walk in a suburban area, where not only fire buckets hang (see Study 12), but also electric wires stretch between pillars. A brick was hung on a wire so that the wind did not swing the bare electrical wire strongly and it did not touch the neighboring one.

The author's thoughts immediately switched from brick to a suspension cableway, and immediately the following idea arose for calculating all this and creating an animation: on two pillars of height H_1 and H_2, an absolutely flexible rope (cable, chain) of length S is suspended and a specific (linear) mass g_c, along which a load of weight G slides. How will the shape of this rope change and what forces will act on the rope at the points of support and at the point of fastening of the load? In Study 4 we have already solved a simplified problem: we determined the length of a chain without load, which will pull with minimum force its support points located on the same level.

The assignment of the cable suspended over the road was posted at the PTC Community forum by a Finnish student whose thesis was associated with it. This is a common modern practice—the engineer, having received a new assignment, immediately looks for similar solutions, learning not from his own, but from other people's mistakes. The student, in this situation, is looking for what is known in Russian as "fish"—a similar student work done before.* In the era of the Internet, this has become much easier, asking for help from the audience of social and professional networks.

We will deal here with two catenaries that are "suspended" between each of the pillars and the point of suspension of the load. If the weight of the load increases, then these curves will tend to straight lines and the problem will be greatly simplified. If the weight of the load is reduced to zero, then these two catenaries must merge into one, which we have calculated earlier (see Figure 7.3 in this Study). The problem of the cable over the road with two segments of a catenary and a load can be reduced to solving a system of nine equations, four of which take into account the geometry, and five, the balance of forces and moments of forces (see the diagram in Figure 7.7).

These are the equations.

Geometry

1. The sum of the lengths of the two branches (parts) of the chain is equal to a given value S.
2. The left branch of the chain is attached to the left pillar and it ends at the point where the load is suspended.
3. The right branch of the chain begins at the point of suspension of the load and it is attached to the right pillar.
4. Two branches of the chain intersect at the point of suspension of the load.

Balance of Forces

5. The sum of the vertical components of the reaction of the supports ($F_{y1} + F_{y2}$) is equal to the sum of the weight of the chain and the weight of the suspended load. We note at once that the two horizontal components of the chain reaction in

* The text "fish" (also a filler text or dummy text), in the jargon of text designers, is a text inserted into the layout and not carrying a semantic load (Wikipedia). On the web, you can find other interpretations of this term. So the composer composes the music for a song, for which there is still no text, on the "fish"—according to the text the name of the fish is made: "cod, herring, perch, catfish...". The main thing here is for the rhythm to be respected. By the way, among Kiev students, "goat" seems used instead of "fish": "do you have a course goat?"

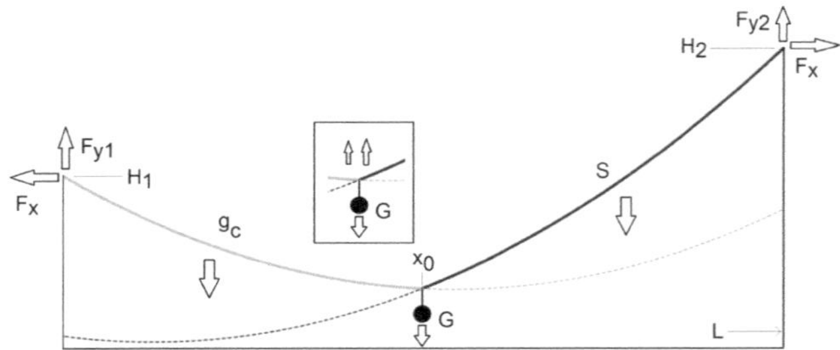

FIGURE 7.7 Forces acting on a sagging chain with a load.

the supports are equal ($F_{x1}=F_{x2}=F_x$), since there are no other horizontal forces in the system.

6 and 7. If $F_{x1}=F_{x2}=F_x$ and the values of the derivatives of the two catenaries at the support points are known, then the values of F_{y1} and F_{y2} can be calculated, taking into account that the value of the derivative is the slope of the tangent at this point. Here we make also use of the fact that the resultant force is along the chain, and there is not a normal component to the chain as imposed by the condition of mechanical equilibrium for a completely flexible chain.

8. At the point of intersection of the chain branches, the weight of the load is balanced by the two vertical forces from the left and right branches of the chain. Also in this equation the values of the vertical components of the forces are expressed by the horizontal components by using the derivatives of the catenaries.

9. The equality of the moments of forces acting on the system at the point of intersection of the branches of the chain: the three forces (F_{y1}, F_x of the right support and the weight of the right branch of the chain) rotate the system clockwise, and the other three forces (F_x of the left support, F_{y2} and the weight of the left branch of the chain) rotate the system counterclockwise. One assumption is made here, the validity of which we will discuss at the end of the Study: the ordinates of the centers of mass of the two branches of the chain are in the middle of the segments 0-x_0 (the left branch of the chain) and x_0-L (the right branch).

Figure 7.8 shows the beginning of the calculation document (Mathcad Prime 3) with a table of input data and two user functions with the catenary itself (it has an

H_1 (m)	H_2 (m)	L (m)	S (m)	G (kg)	g_c $\left(\frac{kg}{m}\right)$	x_0 (m)
7	12	120	121	50	1	60

$$y(x,x0,h,a):=h+a\cdot\left(\cosh\left(\frac{x-x0}{a}\right)-1\right) \qquad y'(x,x0,a):=\sinh\left(\frac{x-x0}{a}\right)$$

FIGURE 7.8 Input data on the problem of the sagging chain with a load.

argument x and three parameters, h, a and x_0, see also Figure 7.3) and its derivative (an argument x and two parameters a and x_0). Parameters in the equations when called will have indices 1 and 2, indicating respectively the left or the right branch of the chain.

Figure 7.9 shows the nine equations written in the constraints area of the solve block of Mathcad Prime (see Study 4).

Figure 7.10 shows three graphs showing the sagging chain at three load positions: at $x_0=5\ m$, $x_0=60\ m$ and $x_0=115\ m$. The dashes indicate the continuations of the two branches of the catenary.

Figure 7.11 shows a frame of the animation of the load movement along the ordinate axis.

The created mathematical model will not work at zero mass of the chain ($g_c=0$), because in this case, the equations of the catenary (see Figure 7.8) will need to be changed to the equations of the straight line. Yes, and this weightless chain is unreal—it must have zero resistance to rupture. However, the model should work at zero load weight ($G=0$) and produce a result similar to that shown in Figure 7.3. Figure 7.12 shows the animation frame of the zero load movement along a chain consisting of two merged chain branches.

But if the load is near the left or right pillar (Figure 7.13), then at zero load weight, the two branches of the chain diverge, indicating a defect in the model, which is due to the assumption in the ninth equation (the balance of the moments of forces relative to the point of suspension of the load) noted above.

Figure 7.13 did not bother the author. Suddenly, someone will design a rope suspension with our assumption for the fifth element,* sorry, for the ninth equation of the

* "The fifth element"—a fantastic action movie with elements of comedy. At the heart of the story it is the philosophical parable of love. It is the only "fifth element" capable of uniting the four primary materials that make up the universe. Our fifth equation of balance of moments of forces "combined" four disparate primary equations of balance of forces into a single correct mathematical model.

Constrains

$$S=\int_{0\,m}^{x_0}\sqrt{1+\left(y'\left(x,x0_1,a_1\right)\right)^2}\,\mathrm{d}x+\int_{x_0}^{L}\sqrt{1+\left(y'\left(x,x0_2,a_2\right)\right)^2}\,\mathrm{d}x \quad \text{length of two parts of chain}$$

$$H_1=y\left(0\,m,x0_1,h_1,a_1\right) \quad \text{Left support} \qquad H_2=y\left(L,x0_2,h_2,a_2\right) \quad \text{Right support}$$

$$y\left(x_0,x0_1,h_1,a_1\right)=y\left(x_0,x0_2,h_2,a_2\right) \quad \text{hanging point}$$

$$F_{y1}+F_{y2}=g\cdot\left(G+g_c\cdot S\right) \quad \text{chain weight with load}$$

$$F_{y1}=-F_x\cdot y'\left(0\,m,x0_1,a_1\right) \quad \text{vertical reaction of the left support}$$

$$F_{y2}=F_x\cdot y'\left(L,x0_2,a_2\right) \quad \text{vertical reaction of the right support}$$

$$G\cdot g=F_x\cdot\left(y'\left(x_0,x0_2,a_2\right)-y'\left(x_0,x0_1,a_1\right)\right) \quad \text{Balance of forces at the point of suspension}$$

Balance of moments of forces relative to the suspension point of the load

$$F_{y1}\cdot x_0+F_x\cdot\left(H_2-y\left(x_0,x0_1,h_1,a_1\right)\right)+g_c\cdot g\cdot\int_{x_0}^{L}\sqrt{1+\left(y'\left(x,x0_2,a_2\right)\right)^2}\,\mathrm{d}x\cdot\frac{L-x_0}{2}=$$

$$=F_{y2}\cdot\left(L-x_0\right)+F_x\cdot\left(H_1-y\left(x_0,x0_1,h_1,a_1\right)\right)+g_c\cdot g\cdot\int_{0\,m}^{x_0}\sqrt{1+\left(y'\left(x,x0_1,a_1\right)\right)^2}\,\mathrm{d}x\cdot\frac{x_0}{2}$$

FIGURE 7.9 The system of equations of the problem of the sagging chain with a load.

balance of moments of forces, and the rope will break. The idea that the maximum load on the rope occurs when the load is in the middle of the rope, and not at its edges, was not reassuring—see the values of F_l and F_2 in Figure 7.10. It is reminiscent of a story that is told about a famous Russian mechanical engineer. In his student years, he used to work part-time in an engineering office. As a test, he was given the design of a bridge and was asked to check it. The student sat down for work, found a mistake in the project and ran to the department, where he shouted: "This bridge cannot be built—it collapses!" He was told that the bridge had already been built and it had collapsed. The student was recruited. It is also a reminder that an engineer designing bridges had to stand under them during the first test of the bridge with maximum load or to go first with this load on the bridge. When working on the English version of this book, the tragedy of the collapse of the bridge (Ponte Morandi) in Genoa (Italy) occurred. This was a cables-stayed bridge with cables in prestressed concrete. In this case cables do not behave as parabolas, nor as catenaries, but as straight lines.

Figure 7.14 shows the refined equation of the balance of the moments of forces, into which was inserted the determination of the moment of forces from the weight of the sagging chains, set by means of a cycle with a parameter. The chain is divided into 1000 elementary pieces (elements) with their own moment of forces, and then the summation is performed.

The first operator in Figure 7.14 suggests how an expression with trigonometric functions can be replaced by a simpler expression with a square and a square root.

As is known, equations may be classified as algebraic and non-algebraic (transcendental—see Study 4), which in turn are divided into rational, irrational, logarithmic, exponential and others. Equations are also integral or differential (see Study 9: Celestial Mechanics). In our case it turned out to be a kind of *programming* equation. In this type of equations, at least one member is a program code.

A person could even think to try to replace the summation in a cycle by a definite integral*—see Figure 7.15, where a calculation check of the moment of forces acting on the right branch of the chain is made.

However, the summation in the cycle (see Figure 7.14 and the first operator in Figure 7.15) is preferable to the integral (see the second operator in Figure 7.15), if we follow the basic idea of Study 1, which states that a formula should make clearly visible its physical and mathematical essence. In the sum, it is immediately visible (the

* In some cases, you can get rid of the integral by taking it — by finding the antiderivative of the integrand: in some cases the programmer equation can be reduced to the integral, and the integral to the usual algebraic or transcendental (non-algebraic) one.

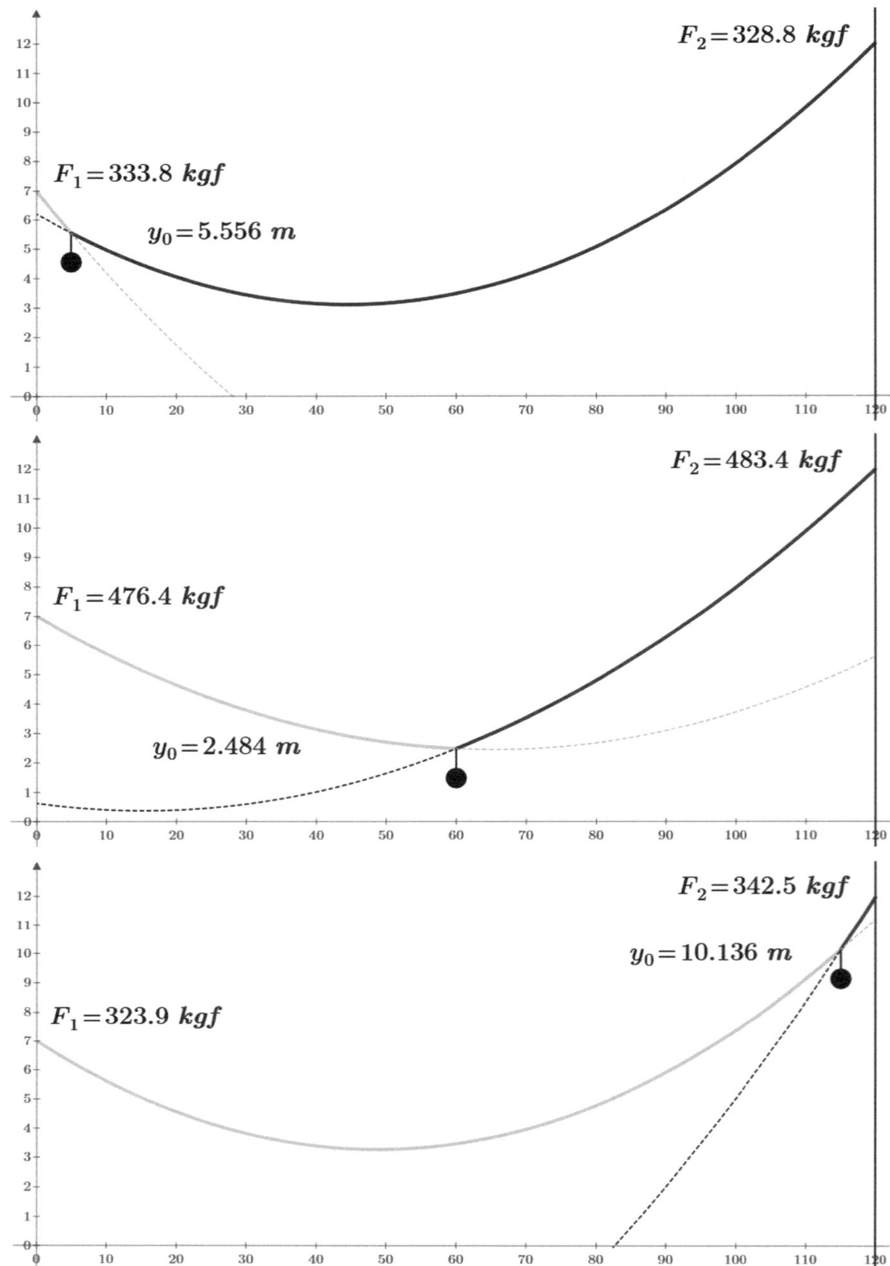

FIGURE 7.10 The shape of the sagging chain with three load positions (the axes are not in scale).

summation of the moments of forces from the elementary sections of the chain $M = M + g \cdot \Delta m \cdot l$), while in the integral it is not.

Testing with the help of the tweaked mathematical model of our aerial cable-way showed that the two curves of the catenary merge into one, even when the point mass is close to zero—compare Figure 7.13 with Figures 7.16 and 7.17.

A small, but important digression from the main theme of this Study.

Figures 7.14 and 7.15 show the *codes* created with Mathcad tools. It's no secret that in many schools and colleges, computer science classes are actually exercises in coding basics, although computer science and coding are, as mathematicians say, intersecting sets, but not coinciding. Fuzzy sets—see Study 12.

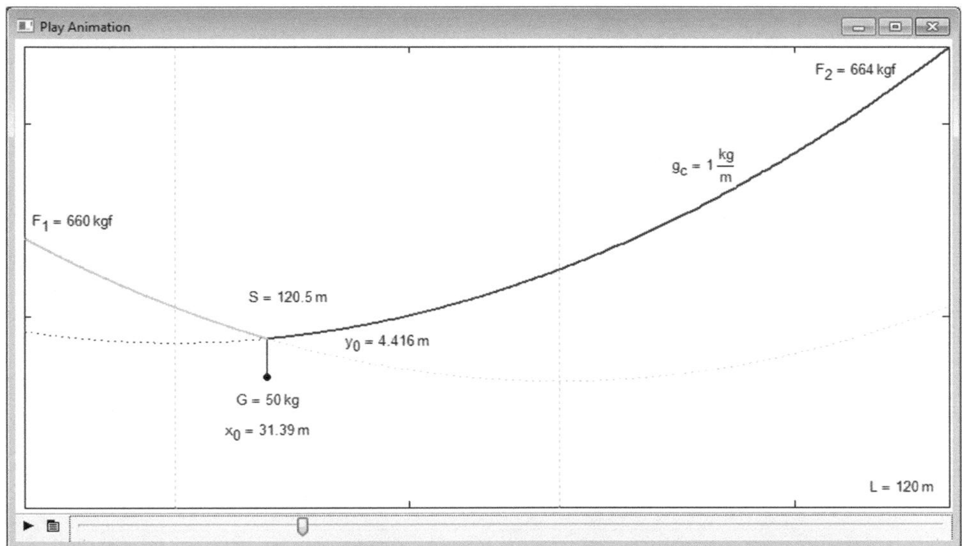

FIGURE 7.11 Frame animation of the movement of the load suspended on the chain.

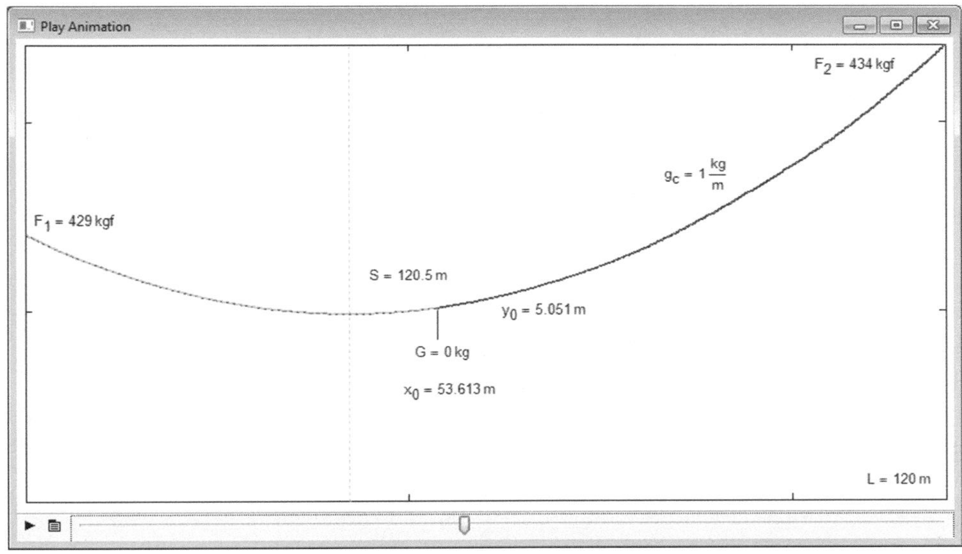

FIGURE 7.12 Testing the model of a ropeway with zero load—everything is okay.

In Mathcad, it is possible solve fairly complex problems without having to resort to programming and executing operators in such (natural*) order from left to right and/or from top to bottom. However, there are procedures to follow when this order needs to be changed—for example, to execute a particular block of operators and not execute another one. Sometimes it is required to execute some groups of operators several times, for example, the implementation of the successive approximation method, which is often used in engineering practice.† Or, as in our case with a loaded chain, in order to calculate the moment of

* This order (from left to right) can be called natural, if you do not consider some languages (Hebrew, Arabic), where they write from right to left.

† The method of successive approximation can be applied if, for example, it is necessary to calculate the thickness of a cable with a known specific tensile strength in the problem of the cable-way. The thickness of the cable is given and its cross-section is then calculated. According to this, the force at which the cable breaks is determined. Then the maximum force that will act on the cable when it is suspended with a load over the road is also calculated. If this second force is greater than the first, then the thickness of the cable is increased, and the calculation is repeated. If the second force is significantly lower than the first, then the thickness of the cable is reduced. Such successive approximations are made until an acceptable result is obtained. This can be done by repeating the given calculation fragment several times, or by inserting a cycle.

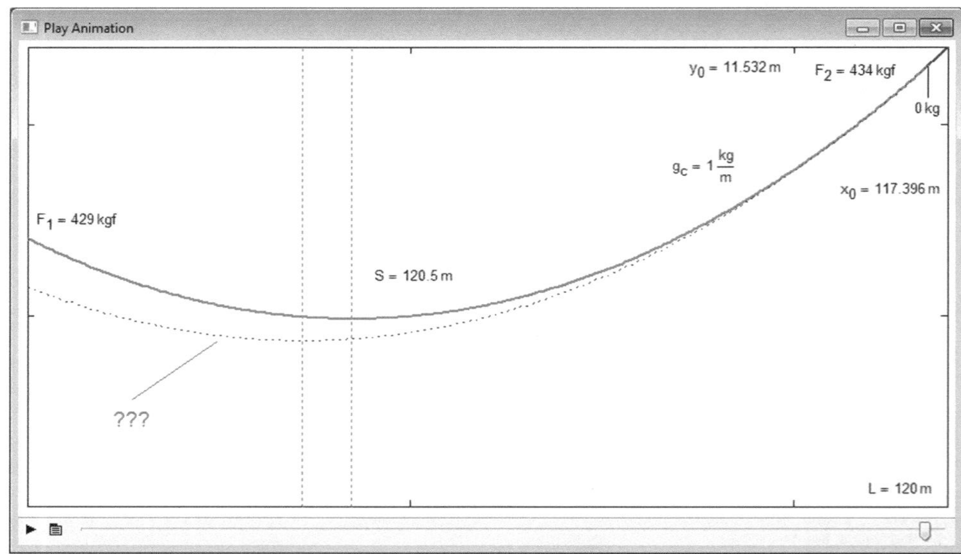

FIGURE 7.13 Testing the model of a ropeway with zero load—an error at the right edge.

$$\frac{\Delta L}{\cos(\operatorname{atan}(x))} \rightarrow \Delta L \cdot \sqrt{x^2+1}$$

$$F_{y1} \cdot x_0 + F_x \cdot (H_2 - y(x_0, x0_1, h_1, a_1)) + \left| \begin{array}{l} M \leftarrow 0 \cdot \boldsymbol{kgf} \cdot \boldsymbol{m} \\ n \leftarrow 1000 \\ \Delta L \leftarrow \dfrac{L - x_0}{n} \\ \text{for } i \in 1..n \\ \quad \left| \begin{array}{l} l \leftarrow \dfrac{\Delta L \cdot i + \Delta L \cdot (i-1)}{2} \\ \Delta m \leftarrow g_c \cdot \Delta L \cdot \sqrt{(y'((x_0 + l), x0_2, a_2))^2 + 1} \\ M \leftarrow M + g \cdot \Delta m \cdot l \end{array} \right. \\ M \end{array} \right. =$$

$$= F_{y2} \cdot (L - x_0) + F_x \cdot (H_1 - y(x_0, x0_1, h_1, a_1)) + \left| \begin{array}{l} M \leftarrow 0 \cdot \boldsymbol{kgf} \cdot \boldsymbol{m} \\ n \leftarrow 1000 \\ \Delta L \leftarrow \dfrac{x_0}{n} \\ \text{for } i \in 1..n \\ \quad \left| \begin{array}{l} l \leftarrow \dfrac{\Delta L \cdot i + \Delta L \cdot (i-1)}{2} \\ \Delta m \leftarrow g_c \cdot \Delta L \cdot \sqrt{(y'((x_0 - l), x0_1, a_1))^2 + 1} \\ M \leftarrow M + g \cdot \Delta m \cdot l \end{array} \right. \\ M \end{array} \right.$$

FIGURE 7.14 Refined ninth equation.

force applied to an irregular body.* The first but not only attribute (an integral feature) of a program in Mathcad is *the change in the natural order of execution of operators.*

At the beginning of the programming era, this order was changed as follows: by inserting labels together with conditional and unconditional operators for transition to the labels in the program code. In Mathcad, as well as in other contemporary programming languages, this is

* In Mathcad, there is a handbook with formulas for the moment of inertia of some standard bodies: cylinder, cone, plate, disk, etc.

$$M := \left| \begin{array}{l} M \leftarrow 0\ \boldsymbol{kgf \cdot m} \\ n \leftarrow 100000 \\ \Delta L \leftarrow \dfrac{L - x_0}{n} \\ \text{for } i \in 1..n \\ \quad \left| \begin{array}{l} l \leftarrow \dfrac{\Delta L \cdot i + \Delta L \cdot (i-1)}{2} \\ \Delta m \leftarrow g_c \cdot \Delta L \cdot \sqrt{y'(x_0 + l, x0_2, a_2)^2 + 1} \\ M \leftarrow M + g \cdot \Delta m \cdot l \end{array} \right. \\ M \end{array} \right. = 1817.49257527\ \boldsymbol{kgf \cdot m}$$

$$M := g_c \cdot g \cdot \int_{x_0}^{L} (x - x_0) \cdot \sqrt{1 + (y'(x, x0_2, a_2))^2}\, \mathrm{d}x = 1817.49257527\ \boldsymbol{kgf \cdot m}$$

FIGURE 7.15 The replacement of the sum (programming) by the integral in the ninth equation.

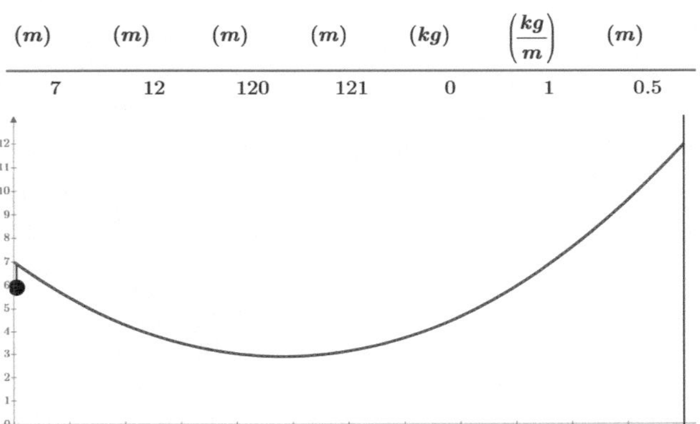

FIGURE 7.16 Load with zero mass at the left support of the suspended power cable over a road.

basically done* not through labels, but with the help of *control structures blocks* — basic and auxiliary.† In their implementation, these *control structures* are combined into program blocks—the second attribute of a Mathcad program. These blocks, which, as a rule, have one input and one output, are executed as a single operator. In Mathcad, program blocks are visually marked with a vertical bar—solid single (Mathcad 15) or double dotted (Mathcad Prime‡). These lines appear automatically when a new operator is added to a block (by clicking

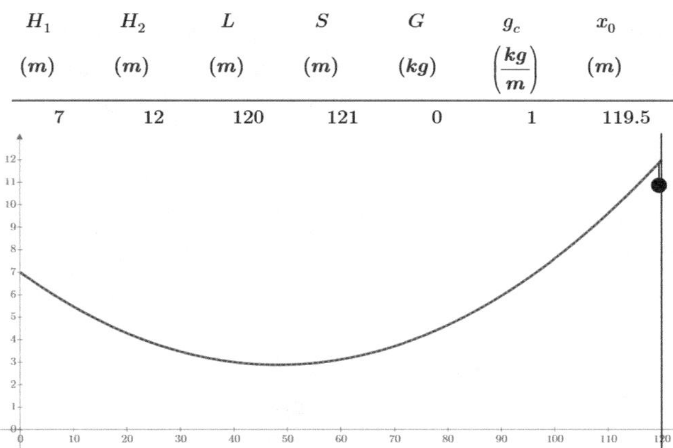

FIGURE 7.17 A cargo with zero mass at the right support of the suspended power cable over a road (see Figure 7.13).

the **Add line** button or **|** in the programming panel). In Mathcad Prime, a gray dashed line is added to the right of the block operators, in order to simplify the program editing procedure—to enter a new block of operators (the cursor is to the left of the gray vertical line), or to enter a new operator in the block (the cursor is to the right of the gray vertical line).

In the environment of the programming language Pascal, which many studied intensively at school and which is largely no longer used in a practical way,§ the beginning and the end of the program block were fixed by the keywords `begin` and `end`, respectively. In Mathcad, an operator block could also be formed by inserting operators into a matrix with a single line — see, for example, the second line in the program in Figure 18.2 in Study 18. However, this is a not referenced technique, which, nevertheless, it is sometimes used to make the program (or its individual program blocks) fit entirely on the display screen, so it is easier to analyze and edit it.

In Mathcad Prime, programs began to resort to such "vertical compression" more often because it became

* In Mathcad, not in general, but always.

† In Mathcad, the main structural control construct is the while loop (pretest loop). The other programming tools (if and for statements) could be implemented by means of the while statement. The operators return, break and continue were introduced in Mathcad relatively recently. They allow early termination of the execution of the statements in a program block.

‡ When preparing some of the figures for this book, this dotted line was replaced by a solid line. In addition, the second line was sometimes removed.

§ This is in support of the thesis about the separation of school from life. History repeats itself. In Old Russia, in the Gymnasium, students studied Latin and Ancient Greek. But no one was speaking in these languages and even scientists from other countries had stopped corresponding in Latin, having switched to German or French. In contemporary Russia they continue to teach the Pascal language, although nobody uses it for practical purposes. In Italy there are supporters and opposers to studying the ancient language studies (Greek and Latin) at the "Liceo Classico". The first group insisting on the formative characters of the classical education, the other on practical purposes of the educational process. This is very old debate, even in the time of Seneca, AD 65 it was said, *non scholae, sed vitae discendum est*; and also in the novel *Anna Karenina* there are some references to this question.

possible to store in the matrix operators with different dimensions (length and weight, for example), which was impossible in the traditional Mathcad 15 environment.

The third attribute of a program is *local variables*—variables that are visible only in the program itself. In all the versions of Mathcad, starting with Mathcad 5, when programming tools appeared, the local variable is specified by a ← (left arrow) operator, whose form (but not essence) in Mathcad 15 can be changed to "=" or ":=". In Mathcad Prime, this option is not yet available.

From the above three attributes of programming, in truth, only the first can be considered an attribute (an inalienable attribute), a change in the natural order of execution of operators. The remaining two (operator blocks and local variables) may or may not be present in a program.

One of the reasons for the wide popularity of Mathcad is that users, solving their routine problems with its help, could flawlessly and unassumingly enter a wonderful world of programming. Many people erroneously believe that they could never write a computer program, whether complicated or not, and they are afraid to deal with programming languages, limiting themselves to simple calculations using, for example, spreadsheets. Programming languages pushed away many ordinary users of computers because it was necessary to do some additional, often incomprehensible, operations in their environment, before performing the necessary calculations. In the environment of the Pascal language, for example, which we repeat is still widely used in schools and universities to teach programming, you must first define beforehand variables and their type: logical (boolean), integer, real, real with double precision, etc. This operation was understandable and justified at a time when computer memory was one of the limiting parameters in solving problems. Now it's easier to enlarge the memory of a computer than to set the type of variables.

Working in Mathcad, an ordinary user can solve rather complex problems (for example, solve algebraic and differential equations (see Study 9), build graphics, create animation (see Study 8), conduct complex statistical analysis (see Study 14)) without resorting to programming. And we have already noted this. However, the same user can quietly begin to use some elements of programming in his calculations, enriching his arsenal of tools. Some users, not knowing about the presence in Mathcad of certain built-in functions, begin to create them using the Mathcad programming tools. So, for example, if you want to find the minimum or maximum element of a vector or a matrix, then you can use the *for* loop with the *if* statement inside to sort through all the elements of the array and select the desired one. Also, a user could remember (or find out) that in Mathcad there are the corresponding built-in functions *min* and *max* for this purpose.

Important note

Programming is necessary to solve complex problems for which there are no built-in tools. Programming is also a very powerful tool for the development of "intelligence and ingenuity". This is a very effective exercise for the mind. Usually this proposition refers to mathematics—elementary (school) and university (higher). After studying mathematics, a person will perhaps never again solve equations and systems of equations, take integrals, calculate limits, invert matrices or determine their ranks. However, having developed his intellect through mathematics, a person can successfully solve his everyday and work problems. Of course, it is also necessary to remember that programming is a very popular profession, especially in those cases when a person has acquired another specialty—heat engineering, for example.

Many people like to solve all sorts of puzzles in their leisure time: charades, rebuses, crossword puzzles, sudoku, etc. So, compiling and debugging a program, even if it does not represent any practical value, is a good puzzle in leisure hours when you want to occupy your mind and hands with an interesting problem.

But back to the aerial cable.

Our calculations show that a bridge or an aerial cable suspended over a road must be tested not only at maximum, but also at a minimum (zero) load. And even with a negative load—see Figure 7.18.

We have already stressed specifically that an engineer should solve not only technical problems but also technical and economic problems. If, for example, an abnormally high object should swim under our power line just once, then it is possible not to build tall masts or tighten wires unnecessarily, but only lift them once with the help of, for example, a helicopter.* By the way,

* Another solution occurs in the case where a steamboat with a high stack is sailing along a river, then the stack can be temporarily lowered to pass underneath the bridge or the power line.

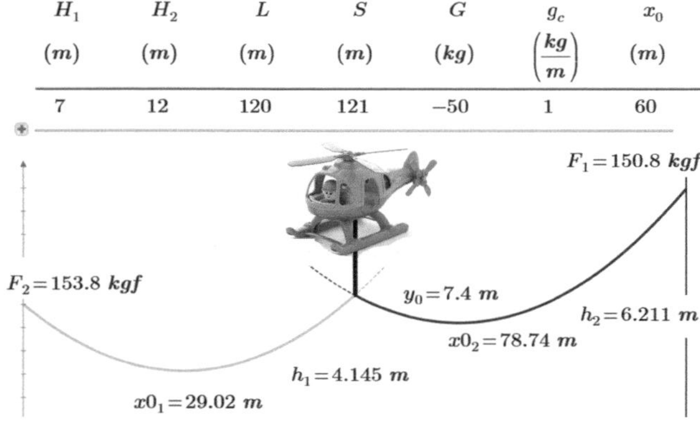

H_1 (m)	H_2 (m)	L (m)	S (m)	G (kg)	g_c $\left(\frac{kg}{m}\right)$	x_0 (m)
7	12	120	121	−50	1	60

FIGURE 7.18 An aerial cable with a negative mass load.

you can see that by the external suspension of a flying helicopter, the cable bends not in a straight line, but also along a catenary. And this can also be simulated using the approaches described in this Study.

This fifth element, sorry, the ninth equation, had another story. First, to solve this problem, equations were built taking into account the geometry of the suspended cable and the balance of the forces only (see above). When the *Find* function began to produce reasonably plausible answers, based on these results it was decided to stop. However, when the model was animated, it was noticed that the virtual rope somehow twitches strangely, and for some positions of the load turned into an arch. This caused alarm and made it necessary to analyze the mathematical model more thoroughly. It became clear that there were nine unknowns and eight equations in it—the system was underdetermined (see Study 6) and had an infinite number of solutions that made the animation twitch. In this situation, the *Find* function, in theory, should return a text message about an infinite number of solutions, rather than a vector of one of the possible solutions, depending on the first approximation. I had to recall once again the university course in theoretical mechanics and to introduce the "fifth element" into the system—the ninth equation, who takes into account the balance of moments of forces, with which we had to adjust our model a bit.

Our calculation could be complicated by hanging a second or a third load on the chain/rope and expecting that the real chain (or the *Find* function, that at some stage will not be able to solve the resulting complex system of equations) would be likely to break. Complications can also be made in other directions. It is possible, for example, to include the elasticity of the rope and the change in its length under the action of a tensile force, which, by the way, has a different value at different points of the rope. It is possible to solve the differential equation of the oscillation of the cable with the load, remembering where our calculation began—from the suburban electric station with a suspended brick.

SQUARE WHEEL

The arch shown in Figure 7.1 is installed, as already noted, on the banks of the Mississippi river.* Usually on a city's embankments a place is created for walking—walking and cycling. So, it is possible to recommend the building of a bike lane at the foot of this arch for riding a bicycle, not with ordinary, but with square wheels. And in order to avoid shaking and the bicycle moving along a purely straight line, the bicycle path should resemble an old washing board whose profile consists of segments of a catenary.

The idea of the square wheels is discussed on the PTC Community website https://community.ptc.com/t5/PTC-Mathcad/Bike-and-Catenary/m-p/614226#M186150.

And another good idea. Bicycles with square wheels for riding on the special lane at the foot of the arch in St. Louis can be stored in a circular (cylindrical) room with the dome roof made in the form of a catenoid—the surface obtained rotating a catenary around the vertical axis passing through the maximum point. According to this, it is possible to optimize the surface of the roof so that it will stand on the support ring with minimum force.

CONTINUING STUDY 7

About the chain. More precisely, about the chain for the bicycle, an excessive sagging could ruin a bike ride at the foot of the arch in St. Louis (see Figures 7.1 and 7.19). Switching the speed of the bicycle, the chain shifts from gear to gear. Let's try to animate this kinematic model.

The input data for the calculation of the animation shown in Figure 7.19 are the constant angular velocity of the bicycle pedals (one revolution per second) and the geometry of the bicycle and of its gears. The linear speed of the bike is calculated (17.925, 22.407 and 29.876 kph)

* The very same city of St. Louis is in the state of Missouri. This is done to ensure that schoolchildren do not have easy geography. The city is just south of the confluence of the Missouri river and the Mississippi river that marks Missouri's border with the state of Illinois.

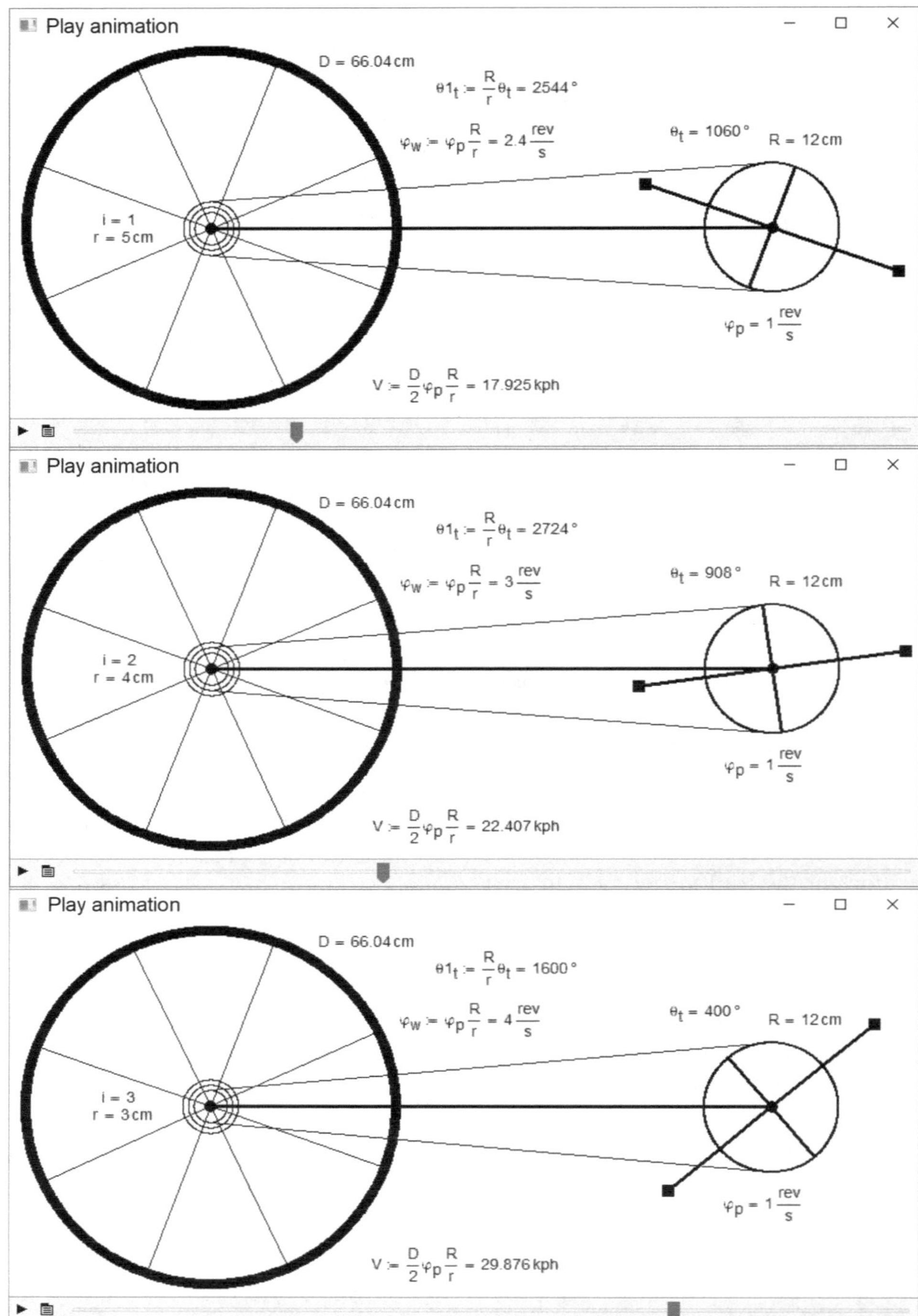

FIGURE 7.19 Three frames of animation of three speeds of a bicycle.

at three speeds: $i = 1$, 2 and 3—three gears on the rear wheel hub. And there can be up to nine gears. And up to five gears are on the pedaling hub. Calculating their sizes and combinations for shifting the chain is an interesting engineering problem.

Also about a Chain (Circles), but Electrical

Students at technical universities usually study two sections of higher mathematics in the first year—mathematical analysis and linear algebra. Elements of mathematical analysis (analysis of functional dependencies and differential and integral calculus) are discussed above. But if we describe what happens to our electric current, flowing in the catenary wires of the power line across the river, then we could also present elements of linear algebra.*

And what happens to the current "flowing in the catenary wires" across the river? Voltage is lowered and it is distributed over electrical networks, which are also sometimes called chains.

Here is a typical problem of electrical engineering—the determination of the current values in specific branches of an electrical circuit, depending on a set of values of local resistances and electromotive force (EMF). It reduces to a system of algebraic equations, see Figure 7.20.

Figure 7.20 shows the solution of an electrical problem by using the Solve block: initial approximations to the solution are introduced: one ampere is assigned to all six required current values. However, the reader could try to guess other values. The choice of these values is an important step in the solution of a problem.

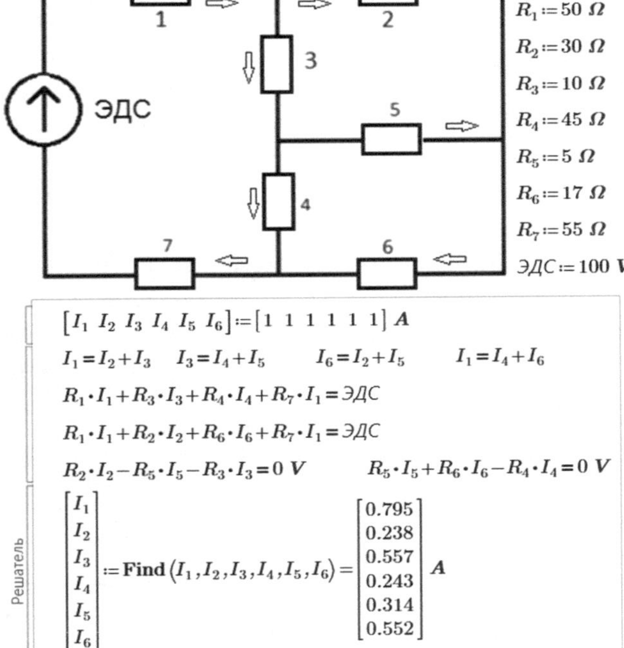

FIGURE 7.20 An example of electrical engineering calculation in Mathcad Prime.

Mathcad sometimes cannot solve a system and asks to change guess values. An engineer who solves his problem on a computer should know approximately what solution he needs and specify a first approximation close to this point. Furthermore, based on Figure 7.20 the equations of the current balance (Kirchhoff's first law) and the balance of "energies" over separate closed branches of the electrical network (Kirchhoff's second law) are set. The Find function returns the value of six currents, giving an approximate solution to the original system. The problem has eight equations and six unknowns. The problem, therefore, is *overdetermined*,† since the number of equations exceeds the number of unknowns. In principle, here two equations are superfluous. However, we could add a couple of equations more,‡ applying Kirchhoff's second law to the closed branches $EMF\text{-}R_1\text{-}R_3\text{-}R_5\text{-}R_6\text{-}R_7$ and $EMF\text{-}R_1\text{-}R_2\text{-}R_5\text{-}R_4\text{-}R_7$ of the electrical circuit. The problem will be even more overdetermined, but the solution will remain the same. This situation only indicates that some equations are linearly expressed through others and they could be removed from the system. However, to analyze linear dependence and to simplify the system of equations—this is a complicated problem. Therefore, in the case of existence of a unique solution of the system (and we have a specific current in the circuit), we can shift the solution of the overdetermined system to the shoulders of the computer.

The solution shown in Figure 7.20 may well suit a user if he is sure of the uniqueness of the solution, i.e. in the ability to search for this solution by any available means. However, even here it is required to clarify the uniqueness of which solution we are talking about: solving the electrotechnical problem or solving a particular system of linear algebraic equations, which in general could

---

* By analogy, if we would have discussed quantum mechanics instead of chain-like electrical wires, we could have referred to Heisenberg's formulation for linear algebra and Schroedinger's formulation for analysis.

† Not every problem in which the number of equations is greater than the number of unknowns is overdetermined. For example, if the equations are linearly dependent, then the described system can be either determined (in fact, this is shown in the example of the calculation) or underdetermined. The ratio of the number of equations and the number of unknowns is not a criterion for overdetermination. Perhaps, in general, one should not pay so much attention to this subtle question. Maybe it's better to say that there are all kinds of systems, but the package itself can (or cannot!) figure out a solution.

‡ If the problem is overdetermined, then why add equations? Or why can they not be added, if they are physically necessary? Something should be justified. Probably, refer to the theory of circuit design. In any case, before considering the problem of the numerical solution of the system, it is necessary to justify the adequacy of the chosen mathematical model (having the form of a system of linear algebraic equations) of the physical model of the examined electrical circuit.

poorly describe a physical model. In addition, it could be confusing that we have to specify initial approximations to the solution.* Usually this is done when a system has several solutions, while in this case it is only necessary to find one. The user should at least know approximately how many solutions will satisfy a given real problem, and what tactics he should use when solving a specific problem. Several solutions are characteristic of non-linear systems, and our system, we repeat, is linear. A system modeling a real physical process usually has a unique solution, but there are cases when there are no solutions or, vice versa, an infinite number of solutions. Now we will show how it is possible to determine these three cases in a linear system of algebraic equations.

Figure 7.21 gives a solution to our problem with one of the Mathcad tools, designed to solve systems of *linear* algebraic equations—the function lsolve (l = linear).

In the upper part of Figure 7.21 our equations according to the first and second Kirchhoff's laws are added, supplemented by the "zero" terms and the coefficients so that a system of linear equations is obtained in its classical written form: the coefficients for unknowns, the unknowns and the right-hand sides of the equations. After this, we can form a matrix M of coefficients for unknowns and a vector of free terms v. The lsolve function returns a solution that coincides with the solution in Figure 7.20, but without any "initial approximation", which was a source of confusion.

$$1\cdot I_1-1\cdot I_2-1\cdot I_3+0\cdot I_4+0\cdot I_5+0\cdot I_6=0\ A$$
$$0\cdot I_1+0\cdot I_2+1\cdot I_3-1\cdot I_4-1\cdot I_5+0\cdot I_6=0\ A$$
$$0\cdot I_1-1\cdot I_2+0\cdot I_3+0\cdot I_4-1\cdot I_5+1\cdot I_6=0\ A$$
$$1\cdot I_1+0\cdot I_2+0\cdot I_3-1\cdot I_4+0\cdot I_5-1\cdot I_6=0\ A$$
$$(R_1+R_7)\cdot I_1+0\ \Omega\cdot I_2+R_3\cdot I_3+R_4\cdot I_4+0\ \Omega\cdot I_5+0\ \Omega\cdot I_6=\text{ЭДС}$$
$$(R_1+R_7)\cdot I_1+R_2\cdot I_2+0\ \Omega\cdot I_3+0\ \Omega\cdot I_4+0\ \Omega\cdot I_5+R_6\cdot I_6=\text{ЭДС}$$
$$0\ \Omega\cdot I_1+R_2\cdot I_2-R_3\cdot I_3+0\ \Omega\cdot I_4-R_5\cdot I_5+0\ \Omega\cdot I_6=0\ V$$
$$0\ \Omega\cdot I_1+0\ \Omega\cdot I_2+0\ \Omega\cdot I_3-R_4\cdot I_4+R_5\cdot I_5+R_6\cdot I_6=0\ V$$

$$M:=\begin{bmatrix}1&-1&-1&0&0&0\\0&0&1&-1&-1&0\\0&-1&0&0&-1&1\\1&0&0&-1&0&-1\\R_1+R_7&0\ \Omega&R_3&R_4&0\ \Omega&0\ \Omega\\R_1+R_7&R_2&0\ \Omega&0\ \Omega&0\ \Omega&R_6\\0\ \Omega&R_2&-R_3&0\ \Omega&-R_5&0\ \Omega\\0\ \Omega&0\ \Omega&0\ \Omega&-R_4&R_5&R_6\end{bmatrix}\quad v:=\begin{bmatrix}0\ A\\0\ A\\0\ A\\0\ A\\\text{ЭДС}\\\text{ЭДС}\\0\ V\\0\ V\end{bmatrix}\quad \text{lsolve}(M,v)=\begin{bmatrix}0.795\\0.238\\0.557\\0.243\\0.314\\0.552\end{bmatrix}A$$

FIGURE 7.21 Solution of an overdetermined system of linear algebraic equations in Mathcad Prime.

From our system, we can remove two extra equations,† as we have already noted earlier. After this, the matrix M becomes square, and it will be possible to calculate the determinant and find the inverse matrix. The inverse matrix could be multiplied by the vector of free terms and obtain the solution of the problem. This is where a knowledge of linear algebra, which was acquired in University, will come in handy! In Figure 7.22 a "classical" system of six linear algebraic equations with six unknown "classical" methods is solved—vectorial multiplication of the inverted matrix of unknown coefficients by the vector of free terms. It is possible to ensure that we have a unique solution with the help of the rank function, which returns the rank of the square matrix M and of the extended matrix $M1$. The Rouché–Capelli theorem‡ says that "a system of linear algebraic equations is consistent if and only if the rank of its basic matrix is equal to the rank of its extended matrix, also the system has a unique solution if the rank is equal to the number of unknowns while an infinite set of solutions if the rank is less than the number unknowns".

In our system of six linear algebraic equations with six unknowns (see Figure 7.22), three equations are Kirchhoff 's first laws and three are Kirchhoff's second laws. If this proportion is violated, writing, for example, two Kirchhoff's first law for two "tees" and four second Kirchhoff's laws for four closed loops, then it is possible to obtain a wrong answer.§ And a most difficult aspect it is that this solution will be very similar to the true solution.¶ If our model is changed (by setting two Kirchhoff's first laws, and not three, but four Kirchhoff's second rules), then the ranks of the new main and new extended matrices will be equal to 5 (see Figure 7.23), not to 6 see Figure 7.22), which is an indication of an error in

* Why is that? Is it always? For which systems? Who or what would make the initial approximations required in this case? It is necessary to know the specifications of the numerical method implemented in the package.

† Why are they superfluous, and why are they the superfluous ones, and not others? From the point of view of the theory of the systems of linear algebraic equations, superfluous equations are found, and not uniquely, only after finding the rank of the extended matrix. Here, it is probably more appropriate, without going into mathematical details, to give a physical or technical justification for the allocation of extra equations.

‡ The theorem is known also as the Kronecker-Capelli theorem (in Poland, Romania and Russia), the Rouché-Fontené theorem (in France) or the Rouché-Frobenius theorem (in Spain and many countries in Latin America).

§ Also you cannot obtain it? How is this determined?

¶ What does it mean "similar"? If the solution is unique and approximately (numerically) coincides with the one sought, then for technical calculations this is not a nuisance, but a norm. If, however, the mathematical theory proves the absence of a solution, while it is present (that is, computed), or the existence of an infinite number of solutions (the number of unknowns greater than the rank of the system), and all are calculated, then an incorrect model is used. And this is a very big problem.

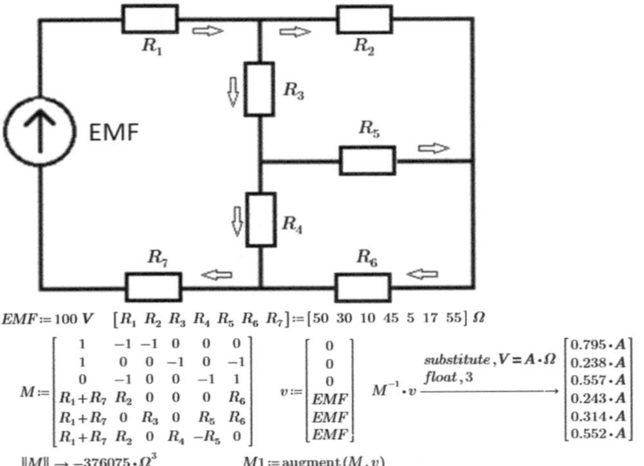

$$EMF := 100\,V \quad [R_1\ R_2\ R_3\ R_4\ R_5\ R_6\ R_7] := [50\ 30\ 10\ 45\ 5\ 17\ 55]\,\Omega$$

$$M := \begin{bmatrix} 1 & -1 & -1 & 0 & 0 & 0 \\ 1 & 0 & 0 & -1 & 0 & -1 \\ 0 & -1 & 0 & 0 & -1 & 1 \\ R_1+R_7 & R_2 & 0 & 0 & 0 & R_6 \\ R_1+R_7 & 0 & R_3 & 0 & R_5 & R_6 \\ R_1+R_7 & R_2 & 0 & R_4 & -R_5 & 0 \end{bmatrix} \quad v := \begin{bmatrix} 0 \\ 0 \\ 0 \\ EMF \\ EMF \\ EMF \end{bmatrix} \quad M^{-1} \cdot v \xrightarrow[\ ]{substitute,\, V=A\cdot\Omega \atop float,\,3} \begin{bmatrix} 0.795\cdot A \\ 0.238\cdot A \\ 0.557\cdot A \\ 0.243\cdot A \\ 0.314\cdot A \\ 0.552\cdot A \end{bmatrix}$$

$$\|M\| \rightarrow -376075\cdot\Omega^3 \qquad M1 := \text{augment}(M, v)$$

$$\text{rank}(M) \rightarrow 6 \quad \text{rank}(M1) \rightarrow 6 \quad \text{cols}(M) \rightarrow 6$$

Only one solution!

FIGURE 7.22 Analysis and solution of the "classical" system of linear algebraic equations in Mathcad Prime with the use of the principles of linear algebra.

the solution, since our system has 6 unknowns (see the Rouché–Capelli theorem above). The conclusion is this: for the problem to be solved, the mathematical model that is adequate to the physical one must contain exactly six linearly independent equations that can be properly extracted from a larger (in excess) number of consistent equations, using either a mathematical approach (calculating the rank of the matrix) or a physical approach (establishing the physical significance of each of the equations). The first approach requires a good knowledge of the general theory of the solution methods for systems of linear algebraic equations, the second, in addition to knowledge of the theory of circuits, requires well-developed engineering intuition and skill in solving similar problems. However, if the reader does not aim to reduce the number of equations, then the solution of the overdetermined algebraic system could be easily left to the computer.

In a meeting before an exam in computer science students of the author of this book confessed that they had

$$M := \begin{bmatrix} 0 & -1 & 0 & 0 & -1 & 1 \\ 0 & 1 & 1 & -1 & 0 & -1 \\ 0 & R_1+R_7 & R_1+R_7 & R_3+R_4 & R_3 & 0 \\ 0 & R_1+R_2+R_7 & R_1+R_7 & 0 & 0 & R_6 \\ 0 & R_2 & 0 & -R_3 & -(R_3+R_5) & 0 \\ 0 & 0 & 0 & -R_4 & R_5 & R_6 \end{bmatrix} \quad v := \begin{bmatrix} 0 \\ 0 \\ \text{ЭДС} \\ \text{ЭДС} \\ 0 \\ 0 \end{bmatrix}$$

$$M1 := \text{augment}(M, v)$$

$$\text{rank}(M) \rightarrow 5 \qquad \text{rank}(M1) \rightarrow 5 \qquad \text{cols}(M) = 6$$

FIGURE 7.23 Analysis of an incorrectly written system of linear algebraic equations.

passed the last exam very badly—the test for linear algebra, which, according to the curriculum, students follow in parallel to the courses of mathematical analysis and computer science. The core of the course of mathematical analysis is the study of functional dependencies. While linear algebra with its vectors and matrices is aimed at solving *systems of linear algebraic equations* (SLAE). Analyzing the failures in the exam (difficulty of understanding) of students in linear algebra, this author, while illustrating the tools of Mathcad, investigated a system of three linear equations and presented the geometric interpretation for each case of the solution—see Figures 7.24 through 7.26.

Note The expression $rank(A) = rank(A1) = 1$ does not mean that the ranks of the matrices A and $A1$ are equal to one, but that the ranks of these matrices are equal. In Mathcad, a unit (not zero) is true in Boolean terms, while zero is false (see Study 11).

In Figures 7.24 through 7.26 Mathcad shows the three cases that arise when solving systems of three linear algebraic equations: an infinite number of solutions (three planes intersect not at a point but on a straight line—Figure 7.24), the absence of a solution (three planes intersect in pairs on different straight lines—Figure 7.25) and a single solution (three planes intersect at one point in Figure 7.26). After this analysis of the problem with its graphical interpretation, the students told the author that if they had been shown all this in the classes on linear algebra, then they would have performed much better in the exam: there would not have been a simple memorization of the theorems, but a clear understanding as well, if not of the essence of the problem, but at least of its formulation.

Here, too, it seems desirable and practical to solve problems of linear algebra in seminars not with pen and paper, but using a display on a computer or even on a large screen hanging in the classroom where the course of linear algebra is taught. There are laser devices that allow a teacher to draw the planes shown in Figures 7.24 through 7.26 above the heads of the students in the audience and even rotate them in space.

ELECTRICAL COMMENTARY

In classical electrical engineering, the basic model of an electrical circuit is represented in the form of a system of equations for the currents using Kirchhoff's laws as follows. First, a scheme is substituted for the circuit. For this, topological concepts are used:

- Branch is a section of a circuit with a single current value
- Node is a junction of branches
- Contour is a closed loop that passes through the nodes and includes branches in sequential order.

Figure 7.27 shows an example.

In this scheme:

- Elements of the chain (E, $R_1, \dots, R_7$)
- Branches ($I_1, \dots, I_6$) are numbered and the arrows of tentatively positive current directions are placed
- Nodes are numbered **(1, ..., 4)**
- Contours **(I, II and III)**

Next, the number of sought currents $NX = nB$, the number of nodes $NV = 3$ are determined.

Then part of the system of equations is written according to the first Kirchhoff's law (NN node equations). For this it is necessary and sufficient to choose $NN = NV - 1$ nodes. Linearly independent equations

A system of linear equations with n variables has a solution if and only if the rank of its coefficient matrix A is equal to the rank of its augmented matrix A1. if n = rank(A), the solution is unique, otherwise there are infinitely many solutions.

Case 1:

$$A := \begin{bmatrix} 1 & 2 & 3 \\ 4 & 5 & 6 \\ 7 & 8 & 9 \end{bmatrix} \quad B := \begin{bmatrix} 1 \\ 2 \\ 3 \end{bmatrix} \quad A1 := \text{augment}(A, B) = \begin{bmatrix} 1 & 2 & 3 & 1 \\ 4 & 5 & 6 & 2 \\ 7 & 8 & 9 & 3 \end{bmatrix}$$

$$A \cdot x = B$$

rank of basic matrix	$\text{rank}(A) = 2$	Infinite set of solutions
rank of augmented matrix	$\text{rank}(A1) = 2$	$\text{rank}(A) = \text{rank}(A1) = 1$
number of unknowns	$n := \text{cols}(A) = 3$	$\text{rank}(A) < n = 1$

Graphical illustration - three planes intersect in a line

$$z1(x, y) := A_{1,1} \cdot x + A_{1,2} \cdot y + A_{1,3} \cdot z = B_1 \xrightarrow{solve, z} \frac{1}{3} - \frac{2 \cdot y}{3} - \frac{x}{3}$$

$$z2(x, y) := A_{2,1} \cdot x + A_{2,2} \cdot y + A_{2,3} \cdot z = B_2 \xrightarrow{solve, z} \frac{1}{3} - \frac{5 \cdot y}{6} - \frac{2 \cdot x}{3}$$

$$z3(x, y) := A_{3,1} \cdot x + A_{3,2} \cdot y + A_{3,3} \cdot z = B_3 \xrightarrow{solve, z} \frac{1}{3} - \frac{8 \cdot y}{9} - \frac{7 \cdot x}{9}$$

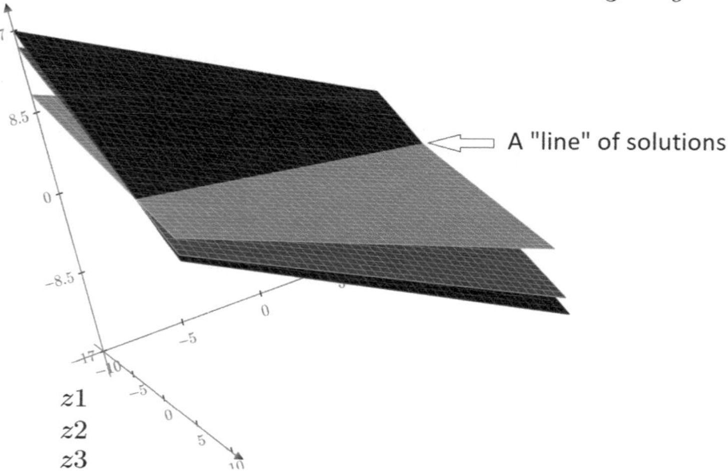

FIGURE 7.24 The SLAE solution in Mathcad is an infinite set of solutions.

Case 2:

$$A:=\begin{bmatrix}1&2&3\\4&5&6\\7&8&9\end{bmatrix}\quad B:=\begin{bmatrix}1\\2\\1\end{bmatrix}\quad A1:=\text{augment}(A,B)=\begin{bmatrix}1&2&3&1\\4&5&6&2\\7&8&9&1\end{bmatrix}$$

rank of basic matrix	$\text{rank}(A)=2$	No solution
rank of augmented matrix	$\text{rank}(A1)=3$	$\text{rank}(A)=\text{rank}(A1)=0$
number of unknowns	$n:=\text{cols}(A)=3$	$\text{rank}(A)<n=1$

$$z1(x,y):=A_{1,1}\cdot x+A_{1,2}\cdot y+A_{1,3}\cdot z=B_1\xrightarrow{solve,z}\frac{1}{3}-\frac{2\cdot y}{3}-\frac{x}{3}$$

$$z2(x,y):=A_{2,1}\cdot x+A_{2,2}\cdot y+A_{2,3}\cdot z=B_2\xrightarrow{solve,z}\frac{1}{3}-\frac{5\cdot y}{6}-\frac{2\cdot x}{3}$$

$$z3(x,y):=A_{3,1}\cdot x+A_{3,2}\cdot y+A_{3,3}\cdot z=B_3\xrightarrow{solve,z}\frac{1}{9}-\frac{8\cdot y}{9}-\frac{7\cdot x}{9}$$

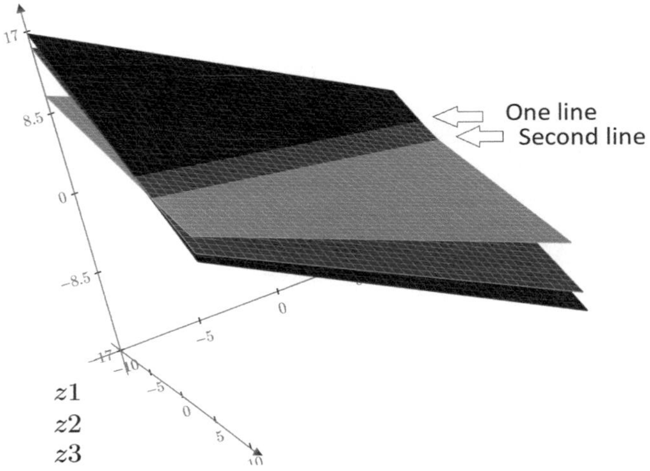

FIGURE 7.25 The SLAE solution in Mathcad—there is no solution.

are always obtained. For example, $NN=3$. For nodes 1, …, 3, it follows that:

$$-I1+I2+I3=0,$$

$$-I3+I5+I4=0,$$

$$I1-I4-I6=0.$$

When writing equations, it is possible to use the sign rule as in the Ostrogradsky-Gauss theorem (with the sign "–" currents directed toward the node), since the first Kirchhoff's law is a direct consequence of div $J=0$.* If after calculations we get $I<0$, then the current has a different direction.

For the remaining contour equations $NK=NX-NV$, linear independence can be provided if each contour contains at least one element that does not enter into other contours. In the example, we selected contours containing fewer elements (I, II, III):

$$(R1+R7)I1+R3I3+R4I4=E,$$

$$R2I2-R5I5-R3I3=0,$$

$$R5I5+R6I6-R4I4=0.$$

Here, for the electromotive force E and the currents I, the rule of signs determines + if the direction of the circuit contour and of the EMF or the current coincide.

Thus, the problem of forming a system of equations for electrical circuits is solved step by step with a choice of nodes and contours ensuring the completeness of the system of equations and the linear independence of the equations in the system. The solution to the above approach using the Mathcad Prime 3 solve block is shown in Figure 7.28.

Note In the scheme representing the circuit, the matrix method is not used to solve the problem of the analysis of circuits. The inhomogeneity of the coefficients

* This is the divergence that, together with the rotor, is seen by negligent students in nightmares before the mathematics exam. It's a pity that in this we book did not pay proper attention to these important concepts of mathematics.

Case 3:

$$A:=\begin{bmatrix}1&2&3\\4&5&6\\7&8&1\end{bmatrix}\quad B:=\begin{bmatrix}1\\2\\1\end{bmatrix}\quad A1:=\text{augment}(A,B)=\begin{bmatrix}1&2&3&1\\4&5&6&2\\7&8&1&1\end{bmatrix}$$

rank of basic matrix $\text{rank}(A)=3$ One solution

rank of augmented matrix $\text{rank}(A1)=3$ $\text{rank}(A)=\text{rank}(A1)=1$

number of unknowns $n:=\text{cols}(A)=3$ $\text{rank}(A)=n=1$

$$z1(x,y):=A_{1,1}\cdot x+A_{1,2}\cdot y+A_{1,3}\cdot z=B_1\xrightarrow{solve,z}\frac{1}{3}-\frac{2\cdot y}{3}-\frac{x}{3}$$

$$z2(x,y):=A_{2,1}\cdot x+A_{2,2}\cdot y+A_{2,3}\cdot z=B_2\xrightarrow{solve,z}\frac{1}{3}-\frac{5\cdot y}{6}-\frac{2\cdot x}{3}$$

$$z3(x,y):=A_{3,1}\cdot x+A_{3,2}\cdot y+A_{3,3}\cdot z=B_3\xrightarrow{solve,z}1-8\cdot y-7\cdot x$$

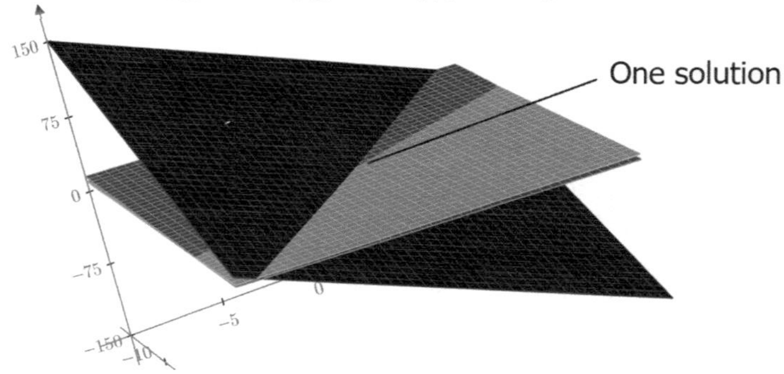

First Method

$$x:=\begin{bmatrix}0\\0\\0\end{bmatrix}$$

$$A\cdot x=B$$

Verify

$$x:=\text{find}(x)=\begin{bmatrix}0.250\\0.000\\0.250\end{bmatrix}\quad A\cdot x=\begin{bmatrix}1.000\\2.000\\1.000\end{bmatrix}$$

Second Method $x:=A^{-1}\cdot B=\begin{bmatrix}0.250\\0.000\\0.250\end{bmatrix}$

Third Method $x:=\text{lsolve}(A,B)=\begin{bmatrix}0.250\\0.000\\0.250\end{bmatrix}$

FIGURE 7.26 The SLAE solution in Mathcad is a unique solution.

makes difficult to form the matrix of the system from the scheme of the circuit. The "Node Potential" method is applied instead. In this method, the unknowns are the potentials of the nodes of the circuit. To a node in the scheme is assigned a value of zero (reference node). For this node, the equation is not written. The matrix of the coefficients contains only conductance. At the same time, the dimensionality of the system of equations is significantly reduced. Consequently, in the example above only three equations remain. The branch currents are calculated in a second stage by using the already found node potentials.

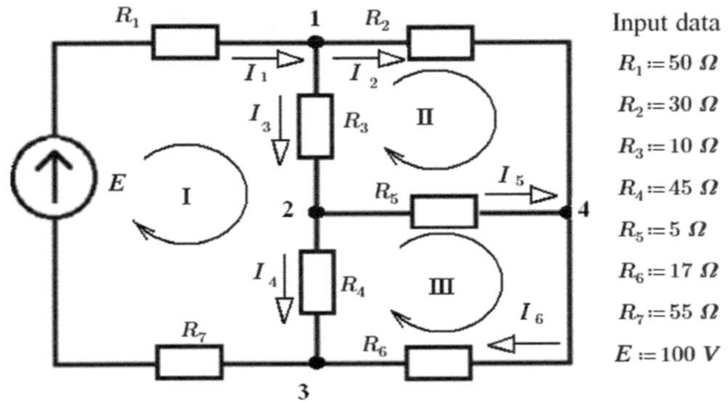

FIGURE 7.27 Beginning of the classical solution of a classical electrotechnical problem.

AND THE LAST

With the formula written for the arch in St. Louis (Figure 7.1), it's not so simple. First, this arch is not a one-dimensional line, but a curved hollow beam with real, rather large, sizes in section (inside the arch people can climb to its upper point). It can be assumed that this formula refers to the upper or lower generatrix of the arch or to some middle line. Second, it follows from the course of construction mechanics that the arch structure has the greatest strength if it is made in the form of a parabola, and not in the form of a catenary. If the arch has a zero thickness (inverted thin chain), then the catenary is more suitable. If the thickness of the arch is increased, i.e. to build a real, rather than a virtual arch, the catenary should gradually move into a parabola. If, on this arch, someone pours in some soil and makes a straight road along the construction, having built an arch bridge, for example, the arch formula should return to the catenary. We can give the owners of the arch in St. Louis another good idea: draw a parabola on the arch using, for example, a laser, emphasizing the complexity and beauty of the corresponding calculations. Only the thickness of the arch would suffice for this show! Figure 7.29 shows the beginning of this calculation—the 630-foot square line (*y(x)* is the upper curve) and the parabola (*p(x, a)* is the average curve). In addition, another catenary (*y1(x)*—lower curve) is drawn, in which the parameters are manually chosen so that it touches the parabola at two points, and it mimics the lower generatrix of the arch in St. Louis.

A parabola with same height and base span (a square with a side of 630 feet—see the two upper curves in Figure 7.29), incidentally, will be shorter than the catenary by almost 10 meters. The construction of the parabolic arch will require less building material, but will it be more beautiful than the catenary arch? This question

Initial Values $I_6] := [0\ 0\ 0\ 0\ 0\ 0]\ A$

Ограничения Начальные прибли

Constraints Node 1

Node 2

$I_1 = I_4 + I_6$ Node 3

$R_1 \cdot I_1 + R_3 \cdot I_3 + R_4 \cdot I_4 + R_7 \cdot I_1 = E$ | Contour I

$R_2 \cdot I_2 - R_5 \cdot I_5 - R_3 \cdot I_3 = 0$ | Contour II

$R_5 \cdot I_5 + R_6 \cdot I_6 - R_4 \cdot I_4 = 0$ | Contour III

Решатель

Solution

$$\mathbf{Find}(I_1, I_2, I_3, I_4, I_5, I_6) = \begin{bmatrix} 0.795 \\ 0.238 \\ 0.557 \\ 0.243 \\ 0.314 \\ 0.552 \end{bmatrix} A$$

FIGURE 7.28 "Electrotechnical" calculation of the electrical circuit.

$y(x) := 757.7' - 127.7' \cdot ch\left(\frac{x}{127.7'}\right)$ $y_1(x) := 715' - 108' \cdot ch\left(\frac{x}{107.7'}\right)$

$h := y(0m) = 630 \cdot ft$ $h = 192m$

$x_0 := root(y(x), x, 0m, 100m) = 315 \cdot ft$ $x_0 = 96m$

$L := 2 \cdot x_0 = 630 \cdot ft$ $L = 192m$

$p(x, a) := h + a \cdot x^2$

$a := 0m = p(x_0, a)$ solve, $a \rightarrow -\frac{0.020833194002156725894}{m}$

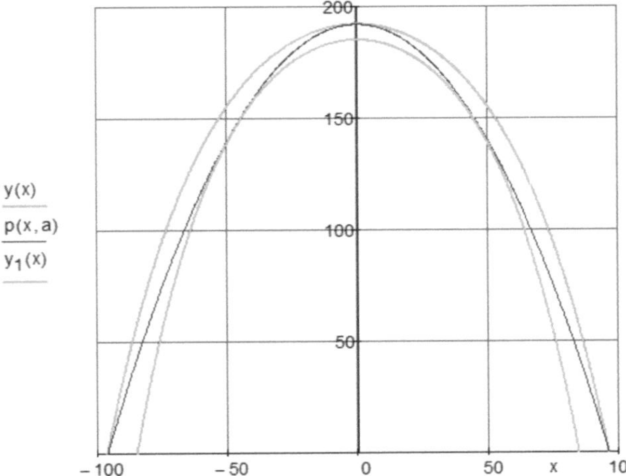

FIGURE 7.29 Parabola sandwiched between two catenaries.

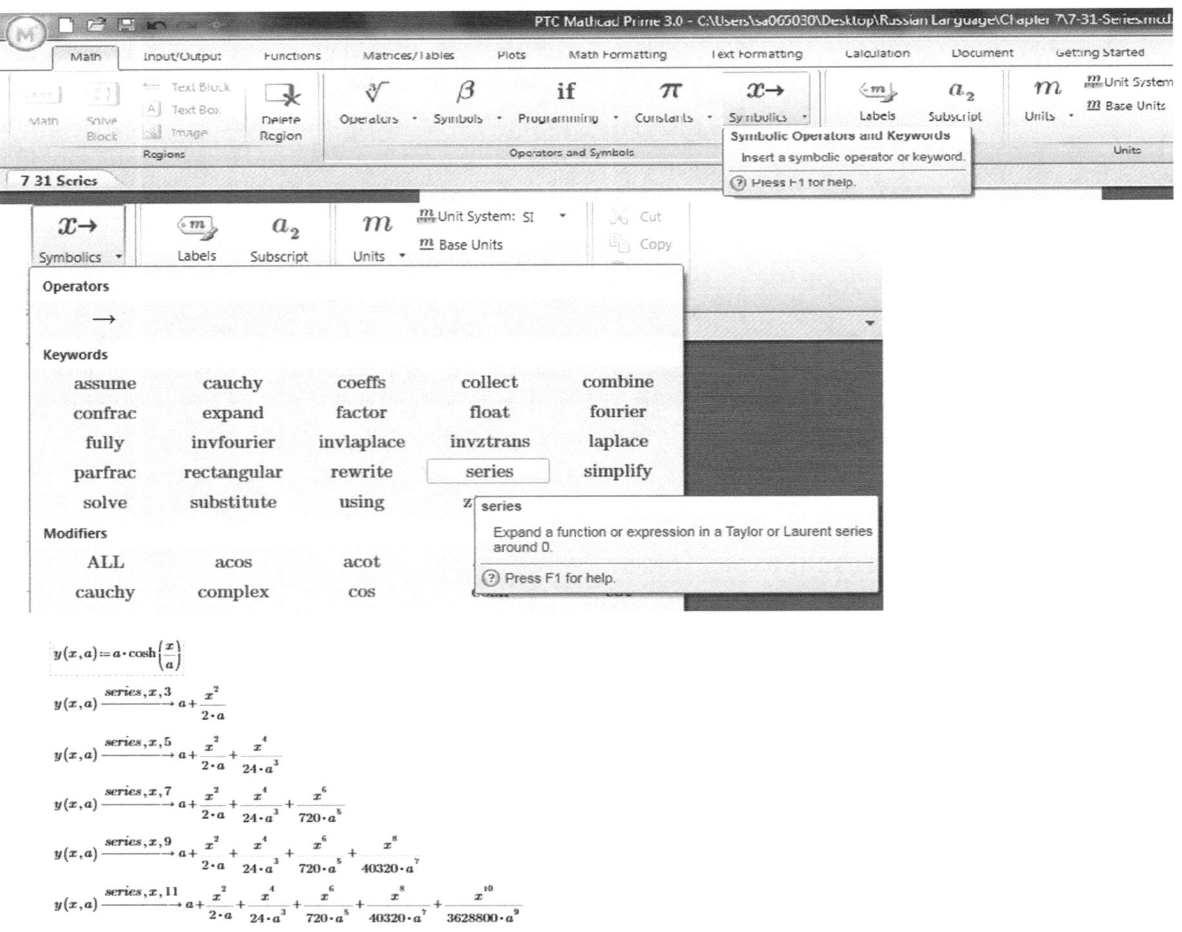

FIGURE 7.30 Expansion of the catenary in a Taylor series in Mathcad.

could not be answered by simple calculations. Here we need the theory of fuzzy sets, operating with linguistic variables. The elements of this theory will be considered in Study 12. In addition, one must take into account that in ancient times people did not know either the exponential or the catenary, but they were well acquainted with the power functions, in particular with the parabola. Figure 7.30 shows how in Mathcad it is possible to "help the ancient people"—by expanding the catenary in a Taylor series (in a power series), which we mentioned earlier and which in our case will consist of even powers of 2, 4, 6, 8, 10, etc.

It is easy to spot the factorials of even numbers 2, 4, 6, 8, 10 in numbers 2, 24, 720, 40320, 362880 and to derive a general expansion formula, by using the summation operator. We invite readers to derive this formula. By the way, in almost all the reference tables for the Taylor series, the first place is assigned to the exponential function that forms the catenary. One of the definitions of the number e is based on the summation of an infinite series. In Figure 7.31 it is shown how to calculate this number in Mathcad.

We also invite readers to create this animation: where the animation frames are the exponent of a power function that approaches a given catenary

$$\sum_{n=0}^{\infty} \frac{1}{n!} \rightarrow e$$

$$\lim_{x \rightarrow \infty} \left(1 + \frac{1}{x}\right)^x \rightarrow e$$

$$\int_1^a \frac{1}{t}\, dt = 1 \xrightarrow{solve, a} e$$

$$\frac{d}{dt} a^t = a^t \xrightarrow{solve, a} \begin{cases} \text{if } t \le 0 \le \mathrm{Re}(t) \vee t \le 0 \wedge \mathrm{Re}(t) < 0 \vee 0 \le \mathrm{Re}(t) \wedge 0 < t \vee \mathrm{Re}(t) < 0 < t \\ \quad e \\ \text{else} \\ \quad undefined \end{cases}$$

FIGURE 7.31 Four ways of calculating the number e in Mathcad.

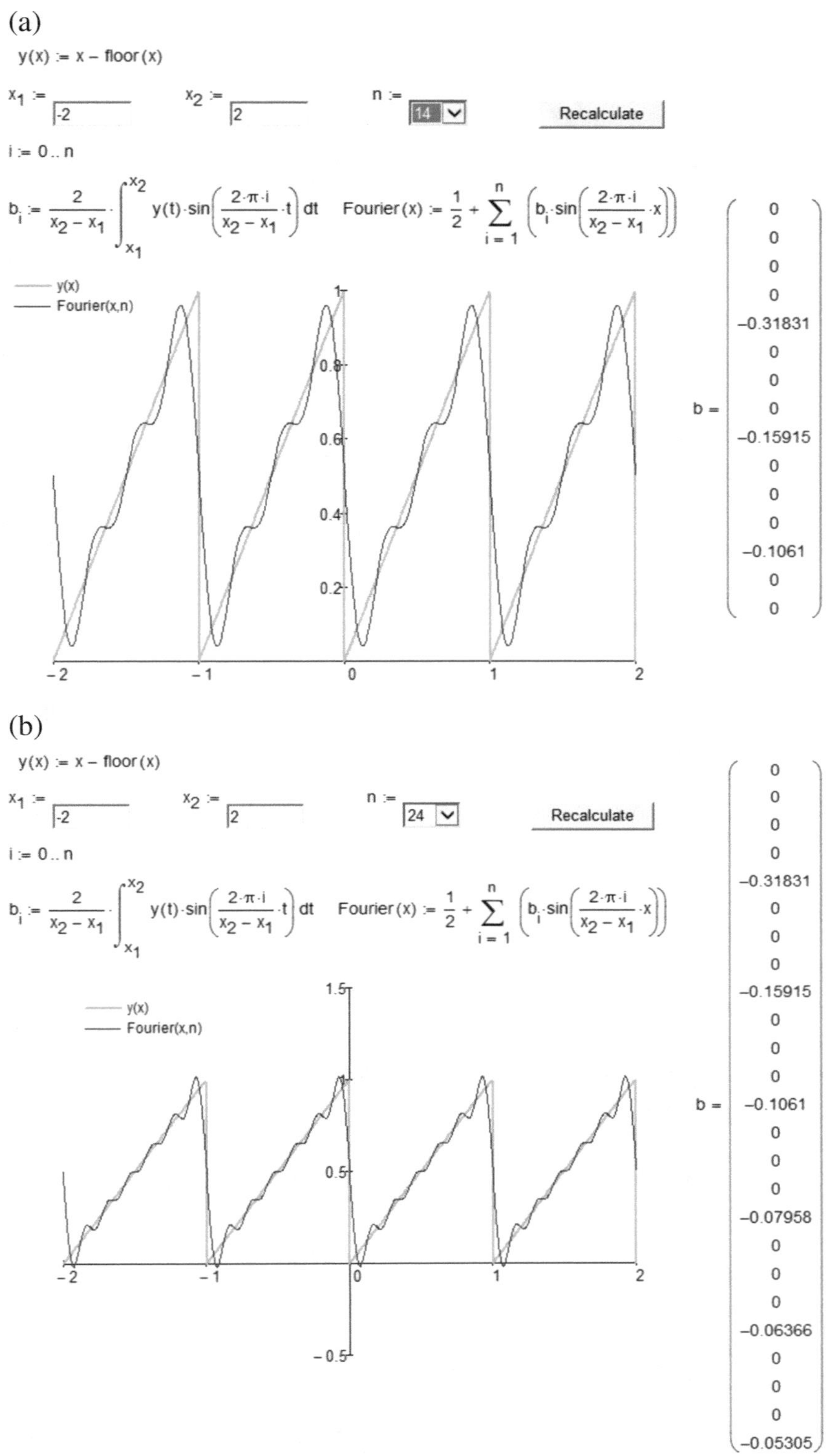

FIGURE 7.32 Expansion in a Fourier series in Mathcad.

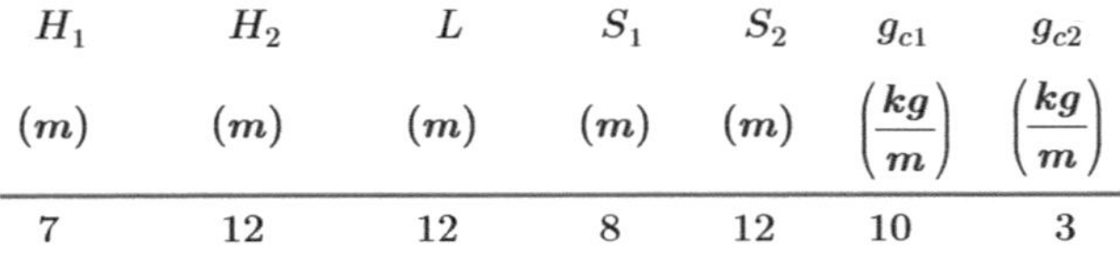

H_1 (m)	H_2 (m)	L (m)	S_1 (m)	S_2 (m)	g_{c1} $\left(\frac{kg}{m}\right)$	g_{c2} $\left(\frac{kg}{m}\right)$
7	12	12	8	12	10	3

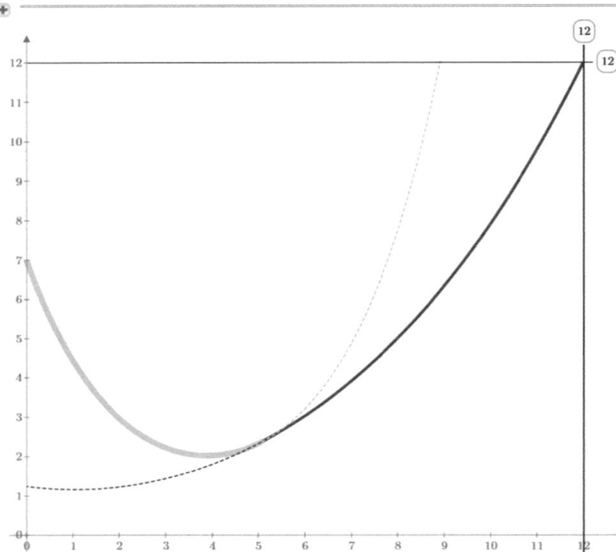

FIGURE 7.33 Sagging of the compound chain.

function as this exponent degree increases. This animation (the curve itself and the power series) can be projected at the arch in St. Louis, promoting mathematics to the visitors. Another problem: to find the parameters of a quadratic equation that best approaches the equation of the catenary. This can be done by tabulating two functions and using the least squares method described in Study 14.

Since we mentioned the Taylor series (say, A), we must also recall the Fourier series (say, B)—what kind of mathematical analysis is possible without these two series! Mathcad can also decompose functions into trigonometric Fourier series whose terms contain sines and/or cosines. Our calculation of the electrical circuit (Figures 7.20 through 7.23, 7.27 and 7.28) implied that the current in the circuit is constant. However, in real household and industrial electrical circuits, the current is variable and does not always correspond in shape exactly to the sinusoid. Alternating current is produced in generators by a rotating rotor. Several generators with different values of rotational speed, phase shift and amplitude could be placed in parallel and generate alternating currents of almost any shape. This is the main objective of the Fourier series, which has been proven to many generations of electrical students. Figure 7.32 shows how by setting three (Figure 7.32(a)) or six generators in the network (Figure 7.32(b)) with their own sinusoids, it is possible to generate an electrical current whose shape will approach a sawtooth shape as the number of sinusoidal current generators increases (from 14 to 24).

ASSIGNMENTS FOR READERS

1. Find with Mathcad or with Internet the formula for the catenary obtained as a result of solving the differential equation of the balance of forces acting on the elementary element (link) of the chain.
2. Solve the problem of the shape of a chain suspended on two supports. *Two* or even *three* weights are attached to chain.
3. Solve the problem of the shape of a chain suspended on two supports and which is supported by a movable or fixed cylinder (roller) from below. The chain rolls along this system to a certain equilibrium state. When the roller is stationary, the frictional resistance of the rope against the cylinder must be taken into account according to the Euler formula, in which the exponent is present.
4. Determine the sagging of a chain composed of two chains with different linear masses. An example of such calculation can be seen in Figure 7.33. What should I write in the hidden area?
5. A jewelry chain is connected in a closed chain and stretched in a horizontal direction, grabbing it from two points—try to put it around your neck, for example. Determine the capture points (the angle of the sector of the circle of the chain laid out on a table) under which ... Formulate the conditions yourselves in this problem. One possible condition is indicated at the end of Study 4 (minimum effort to break the chain at the fastening points).
6. Calculate the sagging of a horizontal string (rope), which bricklayers pull when they are laying a brick wall in order to control the alignment of the laying of bricks (another version of Problem 3).
7. Introduce in the above calculations the chain elongation, assuming that the elongation of the

elementary section is proportional to the applied force (Hooke's law). The proportionality factor can be quite large—you can hang on two poles, for example, an elastic band and a spring.

8. Find the shape of a chain (cable) suspended from one end and blown (washed) by a stream of air (water): for example, a helicopter flies and hangs a rope for suspending the load with it; or a ship sails with a scrap of anchor chain dangling under water. The frictional forces of air or water on the rope or chain (forces deflecting the chain or the rope from the vertical) can be modeled by adopting the assumptions and limitations described in Study 9 (skydiving and diver jumping). In addition, it will be necessary to take into account the effect of the arrangement of the chain or of the rope in the streamlined environment.
9. Draw a room in Mathcad (for the storage of bicycles with square wheels in the city of St. Louis), with a cylinder for walls and a catenoid for the roof. The roof parameters should be optimal in terms of its load on the walls.
10. The mathematical pendulum, as you know, is a thin, inextensible and weightless thread, to which a point-like load hangs. Let's make also this thread weighty, remove the suspended load and, anticipating Study 9, simulate the oscillation of this pendulum—the fragment of a chain (a series of connected pendulums).
11. Try to solve the problem of the optimal length of the chain of the aerial cable (see Figure 7.7), at which the maximum possible tensile strength of the chain will be minimal.
12. A chain is suspended on two very high towers, located far apart from each other. At such height it is necessary to take into account the dependence of the acceleration of gravity on height and the non-parallelism of the vectors of gravity. Derive the formula of the catenary for this general case.
13. There is a smooth rigid cone with a surface area $S = \pi\ 1^2 / 3$, where 1 is the generator of the cone. From a flexible inextensible thread of length $l_0 = 34$ cm $< l$ make a ring and put it on the cone. Then a force F directed along the generator of the cone to the side opposite to the vertex is applied to one of the points of the ring. At what minimum distance x from the top of the cone will the thread pass, and when will it become fully stretched?

Live Kinematic Schemes Or

*Arm of Hella**

ABSTRACT

The Study explains how it is possible to create animated clips based on mathematical models of various mechanisms and how to use the Mathcad website for educational work. The issues of optimizing the operation of mechanisms are considered. A new type of visual art based on mathematical models and animations of kinematic mechanisms is discussed.

Mathematics: Equation of the circle, equation of the straight line, system of linear algebraic equations.

Physics: Steam engine, car wipers, lambda-mechanism, plantigrade machine, shock absorber.

IT: Tools of the computer analytical transformations, variable range, animation, approximate solution of a system of equations, combination of symbolic and numerical solutions, Mathcad user forum.

Art: Mikhail Bulgakov, *Master and Margarita*; Fedor Dostoevsky, "The Gambler"; new form of fine art

Study website: https://community.ptc.com/t5/PTC-Mathcad/Chapter-8/m-p/607677.

In 1705, the English ironmonger Thomas Newcomen† built a steam engine, that was used for pumping out water from mines.‡ It was an atmospheric-type steam engine. Saturated steam vapor was supplied from the boiler to the vertical cylinder. When the piston reached the upper limit point, the steam supply valve was closed and another valve was opened, supplying cold water to the cylinder. The steam condensed and the piston reversed. Then the water from the piston was poured out, fresh steam was fed from the boiler and everything was repeated.

The Study website contains the author's animation of this steam engine, whose two frames are shown in Figure 8.1.

There is a legend connected with this machine. In 1712 the boy Humphrey Potter was appointed as operator of this type of steam pump. To carry out the working cycle, he had to open the steam supply valve when the piston was in its lower position, and he had to open the water supply valve when the piston was fully raised. However, this clever boy figured out how to get rid of this monotonous work.§ With the help of cords and catches the boy connected the valve handles to the beam of the steam engine. It turned out to be a kind of wedge mechanism. Now the valve was controlled by the piston

* Hella is a fictional character from the novel *Master and Margarita* by the Russian writer Mikhail Bulgakov. The novel was written between 1928 and 1940, but unpublished in book form until 1967. Hella is a beautiful, redheaded succubus. She serves as maid to Woland and his retinue. Described as being "perfect, were it not for a purple scar on her neck"—the scar suggesting that she is also a vampire. Reference to her arm will be clear later in this Study.

† As Wikipedia mentions, the inventor Thomas Newcomen was an ironmonger by trade and a Baptist lay preacher by calling.

‡ Water will also accumulate in the rectilinear tunnel of the gravity train, which is described in Study 9, and there will be the need to pump the water out.

§ In the book of Mark Twain, *Tom Sawyer* (Chapter Two) there is a story in which Tom Sawyer was able to get rid of the monotonous whitewashing fence job: http://www.pbs.org/marktwain/learnmore/writings_tom.htm l. Although Tom made use of political (soft power) skills instead of the technical acumen shown by Humphrey.

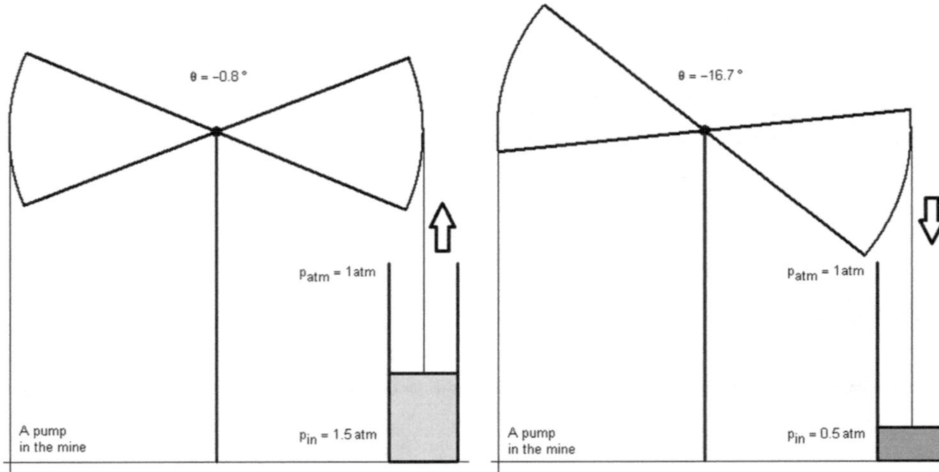

FIGURE 8.1 Two frames of animation of the atmospheric steam engine.

itself, and the boy could play with his comrades in the yard. Thus, the *slide valve* was invented—an automatic steam distribution device for the steam engine.*

A steam engine "in iron" can now be seen only in museums. On the Internet, there are many animations of the steam engine itself or of it when it has been installed in a locomotive or in a steamboat. With the help of the mathematical program Mathcad it is quite possible to create an animation of a steam engine. This feature is discussed in this Study.

Figure 8.2 shows a Mathcad-sheet, a drawing and then an animated horizontal steam engine, or rather its central part consisting (see left to right) of the flywheel of radius *R*, the connecting rod (rod 1–2) of length l_1, the shaft (rod 2–3) of length l_2, the cylinder and the piston.

At the beginning of the calculation, the input data are set with the help of the assignment operator (: =)—the radius R of the flywheel, the radius r of the circle, along which the left end of the connecting rod (point 1) rotates, the length of the connecting rod l_1, the length of the rod l_2 and the diameter of the piston *D*. After this, the angle of rotation θ of the flywheel is also input. In Figure 8.2 this value is specified by two operators θ:= *FRAME*° and θ:= 40°. For debugging the program, the second operator is used that allows the user to manually change the angle of rotation of the flywheel in the range from 0° to 360° and to see how the relative positions of the parts of the steam engine change. When creating the animation (as will be discussed in detail below), the operator θ:= 40° is disabled or erased altogether, so that the first operator θ:= *FRAME*° works. The operator is disconnected using a special switch that appears after clicking the right mouse button when the cursor is on the given operator. When the operator is switched off from calculation there is a label—a black square on the right side. The system variable *FRAME* is zero by default, therefore, the value of the angle θ is zero. However, when creating the animation, Mathcad itself changes the value of the *FRAME* variable (see below), setting the flywheel to a different angle: 0°, 1°, 2°, etc. up to 360°.

The kinematic scheme of our steam engine has four joints, which we numbered as 0, 1, 2 and 3. The first joint is fixed. The vertical coordinates of the joints 2 and 3 will remain unchanged. At the zero joint, we place the origin of the coordinates: y_0 := 0m x_0 := 0m. (The variables y_0 and x_0 are not zero values of the *y* and *x* vectors, but variables with a lower index named *y0* and *x0*, where the zeros (indices) are only lowered slightly down. In Mathcad, lower indexes of variables are entered using a point in the variable name *y.0* and *x.0*. When the user moves the cursor away from this variable name, the point disappears, and the index moves down.) The coordinate system is Cartesian. However, some other mechanical schemes (e.g. the pendulum swing model) are better "placed" in the polar coordinate system, which is also available in Mathcad. The correct choice of the origin and of the coordinate system can greatly simplify the solution of a problem. It is also very important to determine the joints of a kinematic scheme and

* A "zolotnik" (*orig. Russian*, a slide valve meaning literally "a golden one") is also an old Russian unit of mass—see Figure 6.7 in Study 6.

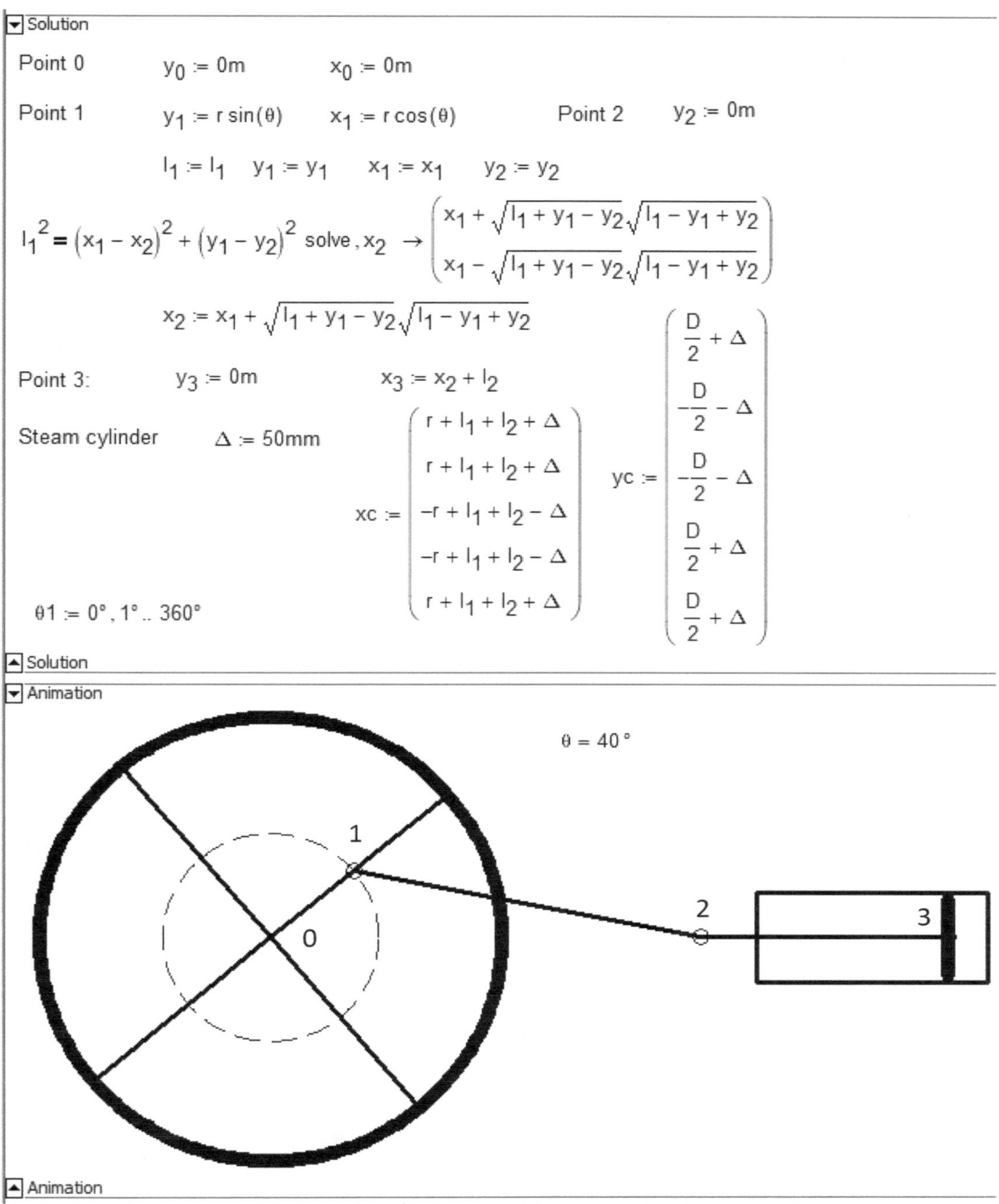

FIGURE 8.2 Mathcad-sheet, drawing of a steam engine.

their numbering. Our problem is planar, but in Mathcad there are also the three-dimensional graphics, which allow the realization of 3D kinematic schemes with a third coordinate z.

To obtain the scheme of the steam engine, it is necessary to calculate the coordinates of the points 1, 2 and 3, depending on the value of the angle θ. This operation is done in the Mathcad-sheet area (Figure 8.2) denominated *Solution*. The operators defining the coordinates of the points are enclosed in an *area* that could be hidden in order to see the scheme of the steam engine right after the operators for specifying the angle θ of rotation of the flywheel. This method simplifies the process of debugging the program: it is not necessary to change the value of

θ, and then scroll the display screen vertically to see the picture. An area collapses with the command *Collapse* from the list that appears when the right mouse button is pressed. While an area is inserted into the calculation with the command *Insert* from the same menu. The area can also be protected from editing (with or without a password).

A more (or less) complicated calculation is required for determining the abscissa of point 2—the values of the variable x_2: analytically (symbolically) the equation of the length of the connecting rod is solved. This non-linear equation, following from the Pythagorean theorem, has two solutions (*roots*)—the connecting rod with the specified length l_1 and the coordinates of the fixing point on the flywheel (x_1 and y_1) can be either on the left or on the right of the flywheel. The operator of symbolic mathematics *solve*, returns two x_2 solutions in form of a vector. We take the first element of this vector as the solution of our problem—the coordinate x_2. Operators $l_1 := l_1$, $y_1 := y_1$, $x_1 := x_1$ and $y_2 := y_2$, written below before the *solve* operator, "suppress" the numerical values of the variables l_1, y_1, x_1 and y_2 so that the *solve* operator returns expressions for x_2, rather than the numerical values of these expressions. Symbolic mathematics operators (in Figure 8.2 they are kept on a gray background) can be removed from the calculation after their use. However, we left them there in order to explain this step. In truth, the equation of the rod position can also be solved mentally without involving computer mathematics of analytical (symbolic) transformations. Nonetheless we used these tools to show how symbolic mathematics helps to solve problems in Mathcad.

The steam cylinder is a rectangle that we draw through its four vertices. Vectors *xc* and *yc* store the coordinates of these vertices. The variable Δ is the value of the "clearance" of the cylinder: the cylinder length is greater than the extreme points of the piston stroke by Δ, and the diameter of the cylinder is larger than the piston diameter by 2Δ.

Figure 8.3 shows the process of inserting, filling and formatting the Cartesian graph, to draw the steam engine. The graph has seven curves (in all there may be up to sixteen of them), whose description (seven pairs of expressions) is represented on the chart axes and listed in the chart formatting window:

<u>Curve 1</u>. The circumference of the flywheel of the steam engine is drawn as follows. The variable $\theta 1$ is set, storing not a scalar value but discrete values of the angle from 0° to 360° in steps of 1° (see the operator at the bottom of the solution in Figure 8.2). A parametric curve is constructed on the graph (Figure 8.3) with the expressions *R·sin(θ1)* on the Y-axis and *R·cos(θ1)* on the X-axis, which allows a circle to be drawn (in the figure this is still an ellipse) of radius *R*. If the variable $\theta 1$ stores

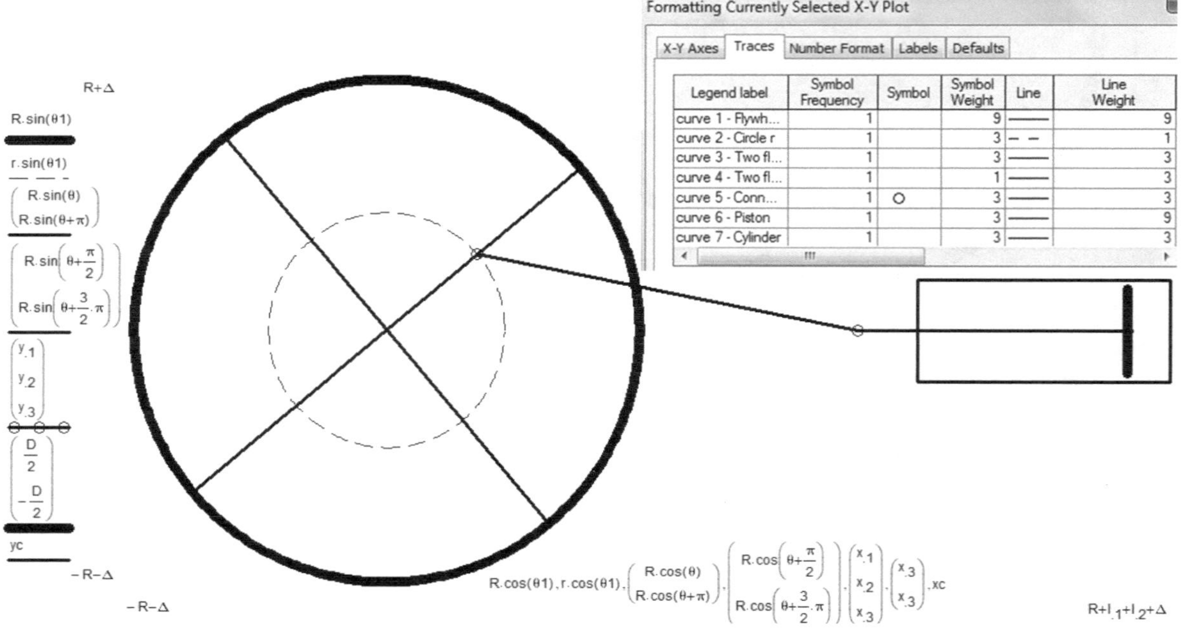

FIGURE 8.3 Formatting the chart—steam engine scheme.

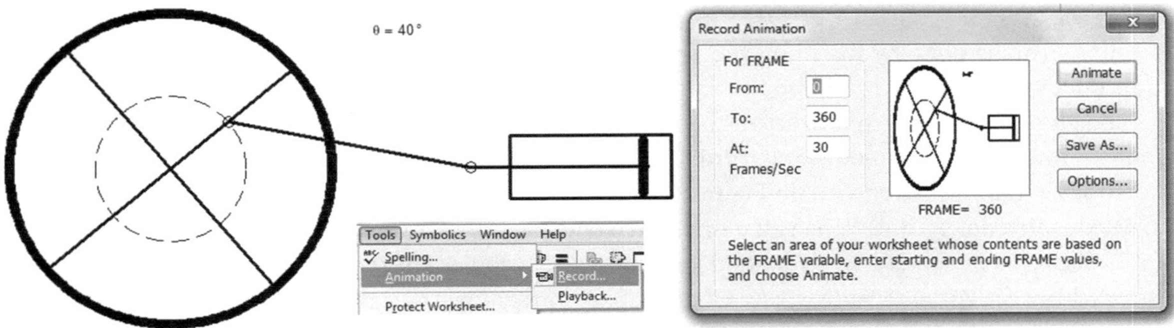

FIGURE 8.4 Creating the steam engine animation.

discrete data not along the entire circumference, but only on a part ($\theta 1$:= 40°, 41°, …, 90°, for example), then only an arc, not the whole circle, will be drawn. Such dotted arcs can be seen in Figure 8.6 below.

Curve 2. The circumference of the motion of the joint (point 1) of the connecting rod on the flywheel is constructed in the same way as the circle of the flywheel, except that this circle is smaller and it is drawn with a thin dashed line instead of a continuous thick line. The centers of our two circles with radii R and r are at the origin. However, if we add constants to the expressions $R\cdot sin(\theta 1)$ and $R\cdot cos(\theta 1)$, then the circle will be drawn with a displacement relative to the origin. If we do not add constants to the expressions $R\cdot sin(\theta 1)$ and $R\cdot cos(\theta 1)$, but quantities that depend on the variable θ (animation variable), then the circle will move. For example, it is possible to animate a rolling wheel.

Curves 3–7. The spokes of the flywheel of the steam engine are drawn in order to "fix" the connecting rod on the flywheel and show the rotation of the flywheel during the animation. We limited ourselves to four spokes (two crossed diameters), but more spokes can be drawn. In Study 7 the animated wheel of the bicycle has eight spokes. The spokes, like other straight segments, as well as broken lines consisting of straight lines, are drawn by assigning a pair of vectors that store the coordinates of the points: two points for the spokes and the piston, three points for the connecting rod and the piston rod (points 1, 2 and 3) and five points for the steam cylinder. We placed the setting of the pair of vectors for the cylinder in the *Solution* area in order not to clutter the graph area.

The size of the graph should be defined so that the flywheel looks round (see Figure 8.2; in Figure 8.3 the flywheel looks oval due to the fact that the proportions of the graph are somewhat distorted when it is formatted). Changing the size of the chart is carried out by stretching its frame. In addition, when creating an animation, it is necessary to fix the ranges of the graph along its axes. By default, Mathcad itself sets the ranges, according to the maximum and minimum values for the abscissa and ordinate axes, which can vary. However, in the process of creating an animation, these ranges must be fixed: at the end of the graph we specified $-R-\Delta$ and $R+\Delta$ (X-axis) and $-R-\Delta$ and $R+l_1+l_2+\Delta$ (Y-axis), where Δ is a certain gap not only for the cylinder of the steam engine (see above) but also for the graph. If the ranges of the chart are not fixed, then the animation will be "curved": the circle will deform into an ellipse, etc.

Before the era of cinema and television, there was a popular toy consisting of a notebook with a man drawn on several pages in different poses. If the notebook was simply scrolled, then the movement of the little man (animation) would not be seen. But if the pages of the notebook were bent and quickly released, then there would be an animation—the drawn man would apparently start moving.* The same situation could be observed with our calculation of the geometric parameters of the steam engine. If somebody manually changes the value of the variable θ and collapses the *Solution* area, he may see how the position of the piston, rod, connecting rod and flywheel changes (this is done when debugging the program), but the animation does not work. Animation occurs when the frame rate is at least ten frames per second. For "flipping through" our "notebook" in Mathcad with images of the steam engine in different phases of the motion *special tools for creating* animations are called for—see Figure 8.4.

To create the animation, first, it is necessary to transfer the control of the independent animation variable (in this case, the variable θ—the angle of rotation of

* We will say more about this toy in Study 13.

the flywheel) to the system variable FRAME (animation frame counter). This is done by disabling the operator θ := 40°. Second, it is necessary to call up the animation creation dialog box (Figure 8.4) and to set the necessary animation parameters in this window, and third, to drag the mouse to select the animation area. In Figure 8.4 this area is marked with a dotted line. When an animation is created, it is displayed in the *Record* animation window. Now, after clicking the *Animate* button (see Figure 8.4), Mathcad itself changes the value of the angle of rotation of the flywheel of the steam engine θ by the system variable *FRAME* and stores the animation frames. When the animation is complete, a window with the animation appears—see Figure 8.5. This animation could be run in Mathcad, and it is possible to save it as an avi-file on the disk and to run the file without Mathcad in the environment of any player of Windows animation. That's all! The steam engine was created, and we felt ourselves to be some kind of "creator"—who can breathe life into immovable virtual objects.

The reader can complicate the steam engine—add to it the already mentioned legendary slide device. But if this seems too complicated, it is possible to first practice on another kinematic problem—the problem of the motion of the wipers of a car. While the steam engine could be seen only in a museum or on the Internet, we encounter car wipers almost every day.

Car wipers are also surrounded by legends.* It is said that after a concert a conductor came back home in the rain. The driver often had to stop the car to wipe the windshield. And the conductor at that time still "lived" his concert, the musical beat on his mind and saw before him a "metronome"—a hand that wipes raindrops from the glass. The next day, it was installed on the hood of the car not an imaginary but a real metronome with a brush attached to it. In this way, they say, car windshield wipers were invented.

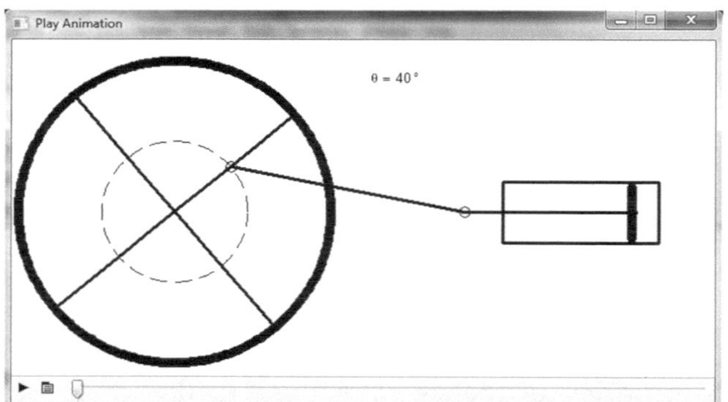

FIGURE 8.5 Animation of the steam engine.

In the era of the steam engines, the problem was to convert the translational motion into rotational motion. When solving the problem of the steam engine (Figures 8.2–8.5), we slightly cheated. By changing the angle of rotation of the flywheel, it was not the translational movement of the piston that was converted into the rotational movement of the flywheel, but the rotational movement of the flywheel was converted into the translational movement of the piston. We actually implemented not a steam engine but a piston compressor driven by an electric motor or an internal combustion engine. Nevertheless, we have got the animation, and to find out "who is driving whom" is not so important.

At present, in connection with the widespread distribution of electric motors with a rotor and a stator, the problem of converting rotational motion to the translational one has become more urgent. Car wipers are a typical example of such a conversion.

Figure 8.6 shows the calculation using Mathcad 15 of the joints of the kinematic scheme of the car wipers (a parallelogram, in which one of the vertices is moved by the crank), depending on the angle of rotation of the electric motor. The core of the calculation is the determination of the coordinates of joint 2. Here it will be necessary to solve not an equation, as in the case of the steam engine (Figure 8.2), but a *system* of two non-linear algebraic equations. And we will not solve this system analytically (symbolically, since we can try to solve it analytically, but the solution would be very cumbersome), but numerically by the Given-Find block built into Mathcad (see Figure 8.6). The numerical solution of the system of equations requires a first approximation to the solution, which is provided by the operators x_2:= b and y_2:= h. It is very important to set the correct first approximation so that the Given-Find block produces the "correct" solution in the entire range of the angle θ change from 0° to 360°. After all, again, the wiper "rod" (line 1–2) with an erroneous first approximation may be on the "erroneous" side of point 5. In this case, the motion of the car wipers may be accompanied by "convulsions"—the connecting rod will appear to the right,

* An interesting movie about windshield wipers: https://en.wikipedia.org/wiki/Flash_of_Genius_(film), where the college engineering professor Robert Kearns invented a wiper blade mechanism modelled on the human eye, which blinks every few seconds rather than continuously, i.e. the intermittent windshield wiper systems used on most automobiles from 1969 to the present.

Point 0 $x_0 := 0mm$ $y_0 := 0mm$ Point 1 $x_1 := r\cos(\theta) = 60\,mm$ $y_1 := r\sin(\theta) = 0\,mm$

Point 5 $x_5 := b = 150\,mm$ $y_5 := h = 240\,mm$

Point 2: We solve the system of two algebraic equations

Given

Initial values $x_2 := b + 30mr\ y_2 := h$

$$L^2 = (x_1 - x_2)^2 + (y_1 - y_2)^2 \qquad r_1^{\,2} = (x_5 - x_2)^2 + (y_5 - y_2)^2$$

$$\begin{pmatrix} x_2 \\ y_2 \end{pmatrix} := \text{Find}(x_2, y_2) = \begin{pmatrix} 225.44 \\ 213.377 \end{pmatrix} mm$$

Point 3 $x_3 := x_2 - a - b$ $y_3 := y_2$

Point 4 $x_4 := x_5 - a - b$ $y_4 := y_5$

Point 6 on a straight line with points 3 and 4: $y = A + B\,x$

$$\begin{pmatrix} A \\ B \end{pmatrix} := \text{lsolve}\left[\begin{pmatrix} 1 & x_4 \\ 1 & x_3 \end{pmatrix}, \begin{pmatrix} y_4 \\ y_3 \end{pmatrix}\right]$$

Given

Initial values $x_6 := x_4$ $y_6 := y_4 + L_1$

$$L_1^{\,2} = (x_4 - x_6)^2 + (y_4 - y_6)^2 \qquad y_6 = A + B x_6$$

$$\begin{pmatrix} x_6 \\ y_6 \end{pmatrix} := \text{Find}(x_6, y_6) = \begin{pmatrix} -1254.4 \\ 506.233 \end{pmatrix} mm$$

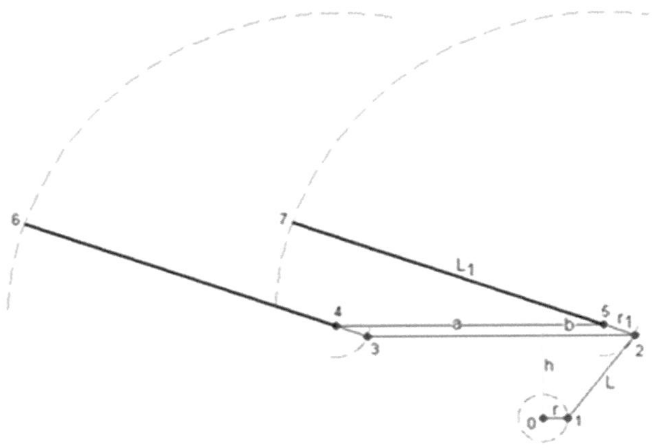

FIGURE 8.6 The problem of car windshield wipers.

then to the left of the attachment point. In principle, the first approximation to the solution should not be a constant, but a variable that depends on the animation variable θ.

Point 6 lies on a straight line with points 3 and 4. Here we first have to solve the system of two linear algebraic equations using the *lsolve* function:

$$y_3 = A + B x_3$$

$$y_4 = A + B x_4$$

and to determine the values of the coefficients A and B,* and then to solve another system of two equations, one of which is non-linear, and to find the coordinates of point 6. When the coordinates of all points are defined, it is easy to create an animation of the car wipers motion according to the technique described above for the steam engine—see Figures 8.3 and 8.4.

At the end of the nineteenth century, at the World Exhibition in Paris, the plantigrade machine of the Russian mathematician Pafnuty Tchebyshev received a gold medal.† In this machine, the rotational motion turned into a motion similar to the one of the foot when walking on an even surface.

Figure 8.7 shows the kinematic diagram of a "leg" of the Tchebyshev's plantigrade machine, as well as the trajectory of point 5 and the printout of the vectors x and y—coordinates of six mechanism points (joints), two of them (1 and 2) are fixed with respect to the origin located

* We solved this problem in Study 2 (see Figure 2.4) in Mathcad Prime, where the dimensionality mechanism works completely and arrays can store different dimensional quantities, for instance a dimensionless value and a length. In the problem of the wipers of Figure 8.6 it is necessary to disconnect the tools of dimensional values. This nuance is examined in an example about solving differential equations in Study 9.

† In Mathcad, by the way, there is a built-in function bearing the name of this famous mathematician—Tcheb. From Wiki (https://en.wikipedia.org/wiki/Pafnuty_Chebyshev): "Tchebyshev is considered to be a founding father of Russian mathematics. Among his well-known students were the mathematicians Dmitry Grave, Aleksandr Korkin, Aleksandr Lyapunov, and Andrei Markov. According to the Mathematics Genealogy Project, Tchebyshev has 12,422 mathematical "descendants" as of July 2018."

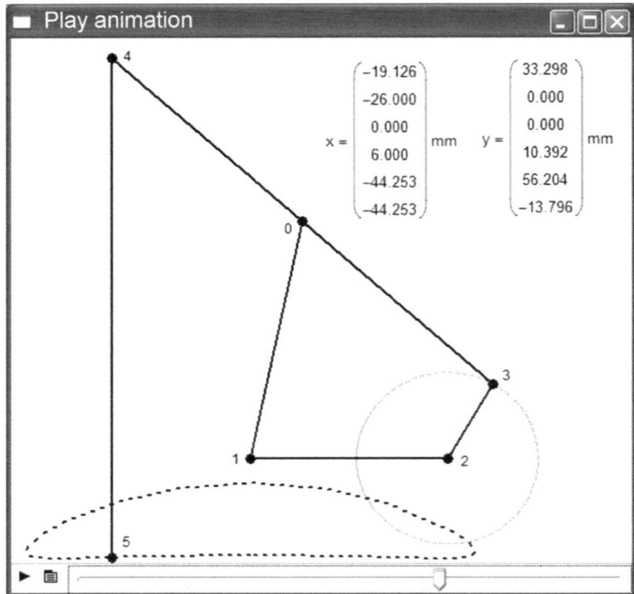

FIGURE 8.7 The Tchebyshev plantigrade machine.

at the second point. The numbering of points, again, starts from zero and goes in such a way that the "foot" of Tchebyshev's plantigrade machine can be drawn without taking the pencil from the paper and not drawing a line twice. This allows us to represent a closed curve on the graph using a pair of vectors x and y. The segment of the straight line 2–3 performs a circular motion around point 2. The coordinates of the point 0 can be found by solving a system of two algebraic equations, as we did in the problem of the car wipers (Figure 8.6). We invite the reader to write the Mathcad-sheet for creating the animation of the "Tchebyshev's leg" by the methods described above. Distances between the points are not indicated—they can simply be measured by applying a ruler to Figure 8.7. It is very difficult to force Mathcad to draw the trajectory of the motion of point 5 (or point 4—line 4–5 is strictly vertical). There are different ways to realize this closed dotted curve. It is possible, for example, to manually ("worker-peasant approach") tabulate the "closed" function by changing the angle θ values from 0° to 360° in 10° increments and recording the obtained coordinates of the point in two vectors with 37 elements, and then displaying these two vectors in the plot, connecting them with a dotted line.* Point 4 moves synchronously with point 5, point 0 describes an arc of a circle, point 3 makes a circular motion, and points 1 and 2 are fixed. In this problem, the main thing is to choose the lengths of the levers so that the trajectory of the points 4 and 5 on a certain section is closest to the horizontal line, and the entire plantigrade machine (and it has two pairs of such legs moving with a shift of 180°) moves without tangible vertical fluctuations.

Tchebyshev had many followers in terms of creating plantigrade machines, translating the rotational motion into a smooth horizontal motion. The most famous of them is the Danish Theo Jansen, who designed and built a machine, whose scheme is shown in Figure 8.8. In this machine, it is possible also to change the lengths between the points (the joints of the "foot") so that the motion of the "stop" reference point is as horizontal as possible. In a Jansen plantigrade machine, only points 0 and 1 are fixed, the rest are either in a circular motion or in a more complex movement (see the dotted lines in Figure 8.8). Tchebyshev also had many predecessors in terms of creating plantigrade machines, but P. L. Tchebyshev

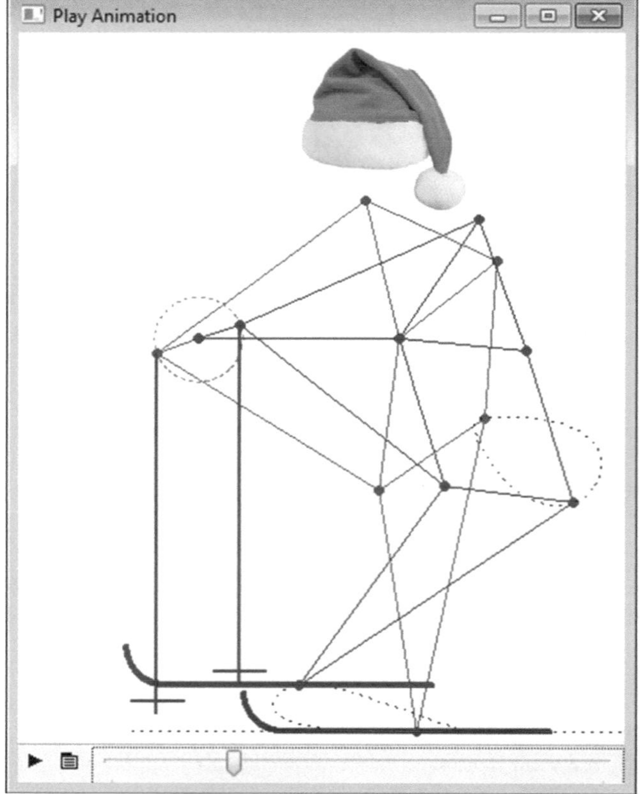

FIGURE 8.8 The Jansen plantigrade machine.

* The "intelligent approach" will be to form a user function of the problem solution that constructs a closed dotted curve of the trajectory of the point motion at each step. This approach is implemented in the author's animation of an oscillating cylinder steam engine – see PTC Community forum.

was the first who, when constructing such mechanisms, began to use a complex mathematical apparatus. Prior to Tchebyshev, such devices were made empirically—adjusting the length of the rods. Figure 8.7 shows a device that "draws" a certain oval, flattened from one side. However, mathematicians have proved that it is possible to create a more complex, though similar, mechanical device, which, for example, could reproduce the signature of any person.

Figure 8.8 shows the two "legs" of Jansen, one of them moves with a lag of 180°. Theo Jansen built a machine out of wood with a dozen legs, which are driven by a windmill and, with a screech, move along the beach. This video can be found on YouTube. We have attached a hat, skis and ski poles to this animation. It turned into a good animated Christmas card.

A few words about this website. This website is created and maintained by PTC (www.ptc.com)—the owner of Mathcad. Part of this website is given to the community of Mathcad users so that they can communicate with each other, ask questions, keep blogs, share useful information, etc. On this website, you can create groups for specific topics. One of these "groups" has a name that coincides with the first title of this Study. From the same website, it is possible to download a free trial version of Mathcad.

The reader himself can use the words "steam machine", "Tchebyshev's plantigrade machine", "Theo Jansen" and others to find animations on the Internet, described in the Study. Though it's one thing to look for and to launch Internet animations, and it's another thing to create them in virtual or real form!

When creating animations in Mathcad with its rich mathematical apparatus, it is possible to solve not only algebraic equations (this was discussed in Study 4), but also *differential equations and systems* (Study 9). In this case, it is permissible to take into account not only the *kinematics* of models, but also their *dynamics*. The mechanism points of such models will move not only due to the rigid links shown in the Studies, but also under the influence of gravity, inertia and other real physical forces. Thus, in Study 9, it is described how it is possible to simulate and to animate in Mathcad the motion of planets and their satellites under the influence of inertia and of the force of universal gravitation, and the motion of a pendulum under the action of the same inertia force, the force of gravity, the tension of the thread (rigid or elastic) and the resistance force of the viscous medium, where the pendulum swings.

One more useful application of animation is the creation of clips explaining the essence of solved problems—see Figure 10.7 (Animation frames for making boxes from a workpiece) in Study 10.

THE ARM OF HELLA

> She began to hurry, stuck her red-haired head through the vent, reached her arm down as far as she could, her nails clawing at the lower latch and shaking the frame. Her arm began to lengthen, rubber-like, and became covered with a putrid green. Finally the dead woman's green fingers got hold of the latch knob, turned it, and the frame began to open.*
>
> **M.A. Bulgakov** *Master and Margarita*

Figure 8.9 shows an animation frame of the rule of Tchebyshev—a device with rigid connections (links) that converts the circular motion of the point 1 into the almost rectilinear oscillatory motion of the point 4. When implementing this model in Mathcad 15, the built-in function *Find* is used to solve systems of nonlinear algebraic equations describing the relationships in the mechanism. The lengths of the rods must be selected or calculated so that the maximum straightness of the motion of point 4 is observed. If at some angle of the lever 0–1 the length of the rod 2–3 is not enough, then the real mechanism is stalled, while in the mathematical model, the function *Find* returns an error message indicating that there is no solution. However, if the function *Find* is replaced by the *MinErr* function, then there is no stalling in the mechanism and no errors in the calculation, but there will be ... the extension of the rod 2–3 (from 30 to 35.8 cm; see Figure 8.10), like Hella's arm from Bulgakov's novel (Figure epigraph to this part of the Study).

Elongation or shortening of the rod in a kinematic scheme (using a "soft" rather than rigid connection) can substantially smooth out the course of the same plantigrade mechanisms, which do not always have to move along a purely rectilinear surface. This approach,

* From Chapter 14, Glory to the Cock, in *Master and Margarita*, Penguin Books, 1997, translated By Richard Pevear and Larissa Volokhonsky (English Translation and Notes).

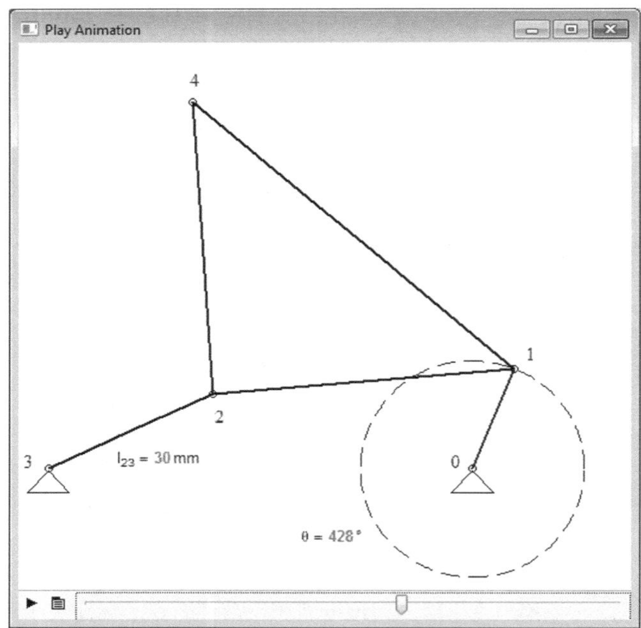

FIGURE 8.9 Normal length of the rod 2–3 of the Tchebyshev rule.

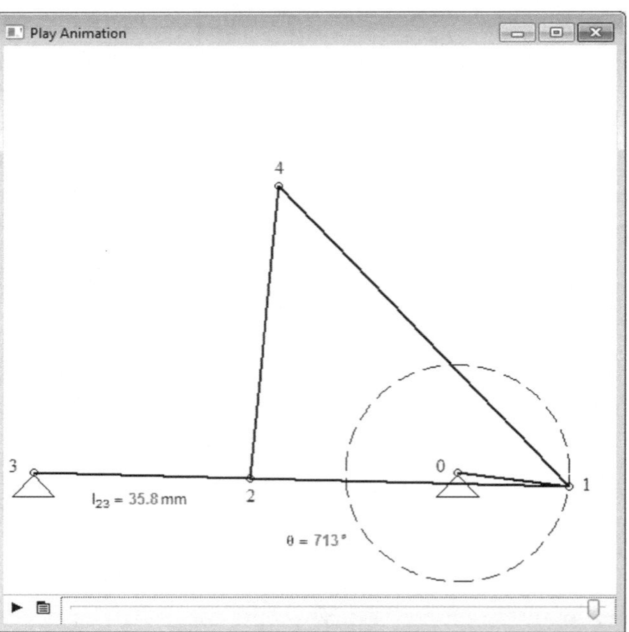

FIGURE 8.10 Extension of the rod 2–3 of the Tchebyshev rule.

by the way, was introduced long ago for "self-wheeled carriages"—cars and motorcycles, equipped with leaf springs, coil springs and shock absorbers.

JOURNEY INTO THE WORLD OF SCIENCE AND ART ON THE PLANTIGRADE MACHINE

The website of the divertissement https://community.ptc.com/t5/PTC-Mathcad-Questions/Tchebyshev-Math-Journey/td-p/500239.

> I sat down and went. In the railway carriage it is quiet, there are no pushes.
>
> **F. M. Dostoyevsky,** *The Gambler*

A century-and-a-half ago only very rich and physically healthy people could afford to travel the world. But with the advent of modern vehicles, such a pleasure became available to the "broad masses of working people", and not just to the elite ("very rich and physically healthy people"): to go on a plane, car or train and, in a short time, comfortably to reach every corner of the Earth.*

Something similar can be said about mathematics. Previously, only selected people could get into its jungle—people with special mathematical abilities (with a special "mathematical ear") and having the appropriate mathematical education. But at the present time this circle has significantly expanded due to the emergence of computer mathematical packages that facilitate the journey into the world of mathematics. Conditionally it could be said that a type of mass mathematical tourism has appeared.

And one more important introductory remark.

Why do we study mathematics at schools and universities?

First, in order to be able to master other academic disciplines: physics, chemistry, theoretical mechanics, fluid dynamics, material resistance, engineering graphics, economics, finance, etc. That's why a course of mathematics is read at the very beginning of study at universities!

Second, one must always remember that mathematics is the best gymnastics (fitness) for the mind. Studying mathematics, we develop our mental abilities, which will be useful for us in solving not only purely mathematical, but also everyday problems.

And third, the study of mathematics (a journey into its "jungle") is in itself an interesting and fascinating

* The words in the epigraph are pronounced by the paralyzed "granny" who "lost her legs" and who is transferred from place to place in a chair. She quietly lived in Moscow (in the second half of the nineteenth century), but dismissed the doctors, boarded the train and "fell down like a snow on her head" to her relative—the general who was in Roulettenberg (the German city of Baden-Baden), waiting for the death of the "granny" in order to get her inheritance – see http://ilibrary.ru/text/67/p.1/index.html.

activity that can be practiced as a hobby. But without mathematical computer packages, it was almost impossible for ordinary people to do this.

Let's illustrate these propositions by simple and entertaining mechanical-mathematical examples.

The Polytechnic Museum of Moscow keeps a device that has a direct relation to travel and "around the world" and "in mathematics". This is the Tchebyshev plantigrade machine, which converts rotational motion into translational motion—see Figure 8.11.

Figure 8.11 is taken from the website of mathematical studies http://www.Studies.ru/ru/Studies/Tchebyshev-plantigrade-machine, where it is possible to see the animation of the motion of this machine, as well as a detailed description of the device. The central link of this exotic vehicle is the mechanism schematically shown in Figure 8.12 (it is also called a λ-mechanism). Points 1 and 2 are fixed, point 3 rotates around point 2, point 0 moves along an arc of a circle, and point 4 on a closed curve ("mushroom's head"), the lower part of which is close to a straight horizontal line. The movement of this point is transmitted to the "feet" of the machine shown in Figure 8.11.

These devices were made even before Tchebyshev, choosing *empirically* the length of the segments L_{01}, L_{03}, L_{12} and L_{04}, as well as the height h so that the plantigrade machine went without vertical pushes. But it was only this great Russian who began to apply *higher mathematics* for the design of such devices. We can say that Tchebyshev saddled this "wooden horse" and made a trip into the World of Mathematics. Let's do it. For us there is no need of special mathematical talents and corresponding mathematical education and knowledge because we have at hand a computer with the mathematical program Mathcad. We will solve the problem in a *hybrid* way (see Study 25)—using both analytical transformations (computer *symbolic mathematics*) and *numerical* (approximate) methods of solution.

A little about numerical mathematics.

At present, numerical mathematics replaces symbolic mathematics, the mathematics of analytical evaluations, in solving not only purely practical problems, but also in educational purposes. The "culprits" of this process are computers. A vivid concrete example. If we want to embody our plantigrade machine in metal or wood (see Figure 8.11), then we will have to calculate not only its geometry (see Figure 8.12 and other figures below), but also the shape and the dimensions (profiles) of its parts. They should be strong enough that the machine does not break when walking, but

FIGURE 8.11 Tchebyshev's plantigrade machine.

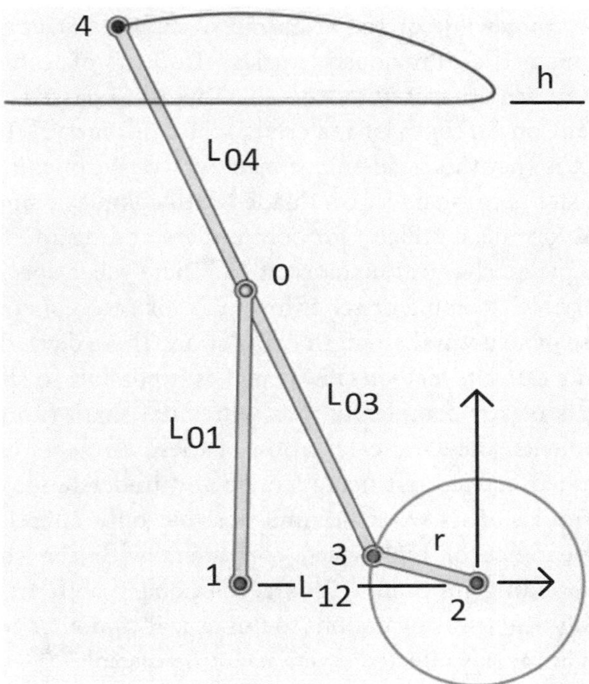

FIGURE 8.12 Scheme of the central mechanism of the plantigrade machine.

$$r := 15m \quad L_{12} := 35m \quad L_{03} := 35m \quad L_{01} := 35m \quad L_{04} := 35m \quad h := 55m \quad \alpha := FRAME\ deg$$

$$x_1(L_{12}) := -L_{12} \quad y_1 := 0m \quad x_2 := 0m \quad y_2 := 0m \quad x_3(\alpha, r) := r \cdot \cos(\alpha) \quad y_3(\alpha, r) := r \cdot \sin(\alpha)$$

$$L_{03}^2 = (x_0(\alpha, r, L_{12}, L_{03}, L_{01}) - x_3(\alpha, r))^2 + (y_0(\alpha, r, L_{12}, L_{03}, L_{01}) - y_3(\alpha, r))^2$$

$$L_{01}^2 = (x_0(\alpha, r, L_{12}, L_{03}, L_{01}) - x_1(L_{12}))^2 + y_0(\alpha, r, L_{12}, L_{03}, L_{01})^2$$

$$x_0(\alpha, r, L_{12}, L_{03}, L_{01}) := \frac{x_3(\alpha, r)^2 + L_{01}^2 - L_{03}^2 + y_3(\alpha, r)^2 - x_1(L_{12})^2 - \frac{y_3(\alpha, r) \cdot \left[y_3(\alpha, r)^3 - x_1(L_{12}) \cdot \sqrt{}\right.}{}}{}$$

$$y_0(\alpha, r, L_{12}, L_{03}, L_{01}) := \frac{y_3(\alpha, r)^3 - x_1(L_{12}) \cdot \sqrt{\left(L_{01}^2 + 2 \cdot L_{01} \cdot L_{03} + L_{03}^2 - x_1(L_{12})^2 + 2 \cdot x_1(L_{12}) \cdot x_3(\alpha\right.}}{}$$

$$a(\alpha, r, L_{12}, L_{03}, L_{01}) := \frac{x_3(\alpha, r) \cdot y_0(\alpha, r, L_{12}, L_{03}, L_{01}) - y_3(\alpha, r) \cdot x_0(\alpha, r, L_{12}, L_{03}, L_{01})}{x_3(\alpha, r) - x_0(\alpha, r, L_{12}, L_{03}, L_{01})}$$

$$b(\alpha, r, L_{12}, L_{03}, L_{01}) := \frac{y_3(\alpha, r) - y_0(\alpha, r, L_{12}, L_{03}, L_{01})}{x_3(\alpha, r) - x_0(\alpha, r, L_{12}, L_{03}, L_{01})}$$

$$x_4(\alpha, r, L_{12}, L_{03}, L_{01}, L_{04}) := x_0(\alpha, r, L_{12}, L_{03}, L_{01}) + \frac{L_{04}}{L_{03}} \cdot (x_0(\alpha, r, L_{12}, L_{03}, L_{01}) - x_3(\alpha, r))$$

$$y_4(\alpha, r, L_{12}, L_{03}, L_{01}, L_{04}) := a(\alpha, r, L_{12}, L_{03}, L_{01}) + b(\alpha, r, L_{12}, L_{03}, L_{01}) \cdot x_4(\alpha, r, L_{12}, L_{03}, L_{01}, L_{04})$$

FIGURE 8.13 Beginning of the calculation of a plantigrade machine.

not too massive, so that the machine can move easily. Calculations of this kind are conducted within the framework of the scientific discipline "Strength of materials". Previously, Russian students of technical universities used to have this saying: "I passed the exam on Strength of materials so I can marry!" The fact is that this academic subject was very difficult to master because it was overloaded with complex empirical formulas, difficult for both understanding and for simple mechanical memorization. There was a special "terrible" formula or set of formulas for each concrete case of structural strength calculation. These days, the *finite element method* (FEM) makes it possible to split parts of any complexity into individual small (finite) elements, then the calculation of these elements can applied carried out using simple and understandable formulas. This work became possible only after the appearance of high-speed computers with the corresponding programs. This revolutionary shift from analytical transformations to numerical computer calculations has affected other scientific disciplines—for example, hydrodynamics and heat and mass transfer. However, within engineering practitioners, there are still "partisans" of the "rules of thumbs" and "back of envelope calculations".

The core of the calculation of the plantigrade machine is the following: it is necessary to calculate the lengths of the segments L_{01}, L_{03}, L_{12} and L_{04}, as well as the height h for a given value of the radius r so that point 4 on the lower part of its trajectory moves along an almost straight horizontal line, consequently a plantigrade machine with such a central mechanism would go as far as possible without vertical pushes.

Figure 18.13 shows the beginning of the Mathcad-sheet with the calculation of the geometry of the mechanism shown in Figure 8.12. In the first line of the sheet we see the input of the initial data—estimated values of the lengths of the links of the central mechanism of the plantigrade machine.* The value of the angle α of turning the link 2–3 is tied to the system variable *FRAME*,

* They are entered with meters as the unit of length. However, this is a convention. There may be other meanings, and other units of length (centimeters, feet, inches or arshins with vershoks, which were in use in Russia at the time of Tchebyshev: 3 arshins = 7 feet, 16 vershok = 1 arshin). The main thing is that this method allows you to control the dimensions and enter and output lengths with the necessary units of length.

which will change from 0 to 360 and it will determine the frames of the animation, which we will create later. When debugging the calculation associated to a graph (a separate animation frame) with the image of our mechanism (see below), we will assign (global assignment) a value of 30° to *FRAME* and see the geometry of the mechanisms when turning the lever 2–3 to 30 degrees. When creating the animation, it is necessary to disable this operator so that Mathcad can create the animation. Mathcad itself will change the value of the variable *FRAME* and store the animation frames in the avi-container.*

Points 1 and 2 are fixed, while points 0, 3 and 4 are in motion. Their coordinates are given in the form of functions with names x_0, y_0, x_3, y_3, x_4 and y_4 with arguments α, r, L_{12}, L_{03}, L_{01} and L_{04}. Working with functions, rather than with the correspondent scalar values, allows us, first, to create graphs and animations with the drawing of the trajectories of all points† and, second, to carry out *optimization*—to find the optimal lengths L_{12}, L_{03}, L_{01} and L_{04}, as well as the height h, which ensure the maximum rectilinear motion of the plantigrade machine—movement without "pushes".

The position of point 0 (the "central" point of our mechanism) is determined through the analytical (symbolic) solution of a system of two algebraic nonlinear equations written at the beginning of Figure 8.13. These equations were solved symbolically in a separate Mathcad sheet (it is also available on the Study website). The system of equations has two roots, one of them was transferred to the sheet in Figure 8.13 and formed the functions named x_0 and y_0. The second extra root captures the geometry of the mechanism shown in Figure 8.12, if point 0 is not higher, but lower than points 1 and 3. These roots (algebraic expressions) are very large Figure 8.13 shows only their beginning.‡

Point 4 is on a straight line passing through points 0 and 3. To determine the coordinates of point 4, use the equation of the straight line $y = a + b \cdot x$, whose coefficients a and b are also symbolically determined in a separate document through the solution of a system of two algebraic equations and carried over in the document in Figure 18.13. This Mathcad sheet is also posted on the Study's website.

After the coordinates of all five points of our kinematic scheme have been specified or calculated, it is easy to display it on the graph—see Figure 8.14.

We can see in the plot of Figure 8.14,§ that the given values of the variables L_{12}, L_{03}, L_{01} and L_{04} and h (see the beginning of Figure 8.20) are far from optimal and our step-plantigrade machine during the motion will strongly shake the rider. How this could be avoided? Pafnuty Tchebyshev (https://en.wikipedia.org/wiki/Pafnuty_Tchebyshev) took pen and paper, strained his entire mathematical talent, made a bunch of analytical evaluations and eventually found the answer: a set of complex formulas by which it is possible to calculate the optimal configuration of the central mechanism of the plantigrade machine. We do not have such a mathematical talent, but there are optimization tools built into Mathcad.¶ And now, problems of this kind, we repeat, are being solved more and more often not analytically, but numerically, using a computer.

On the lower part of the "mushroom cap" (Figure 8.14), points are marked that indicate the rotation of the link 2–3 along −100, −90, −80, ..., 0, 10, 20, ..., 90 and 100 angular degrees. These angles are stored in a program-created vector named A. It is necessary to make sure that these points are not in an arc, but in a straight line. In order to obtain this, it is simple, for example, to minimize the sum of the squares of the deviations of these points from the horizontal line!

Figure 8.15 shows how the objective function with the name S and with five arguments, *optimization parameters*, is specified. The function S returns the

* Audio Video Interleave: a container or wrapper format is a metafile format whose specification describes how different elements of data and metadata coexist in a computer file.

† It is possible to find many drawings and animations of various kinematic devices on the Internet. However, all of them, as a rule, are given without drawing the trajectories of the motion of the linkage points (joints).

‡ In the calculation, it could be simpler to use not symbolic mathematics and not user functions, but numerical mathematics (the Given-Find block) and scalar user values. However, in this case, we repeat, it would be impossible to draw the trajectories of the motion of the points, to create an animation with the display of the trajectories of the points movement and their optimization.

§ It is slightly stretched horizontally. This happens when the graph is edited—click on it with the mouse.

¶ By the way, the Tcheb function is a built-in function in Mathcad and other popular mathematical packages. The Tcheb function returns the Tchebyshev polynomial. Because of this and, of course, for other reasons (his contributions, for instance, in prime number theory: the Bertrand–Tchebyshev theorem, and statistics: the Tchebyshev inequality), Pafnuty Tchebyshev became the most famous Russian mathematician abroad.

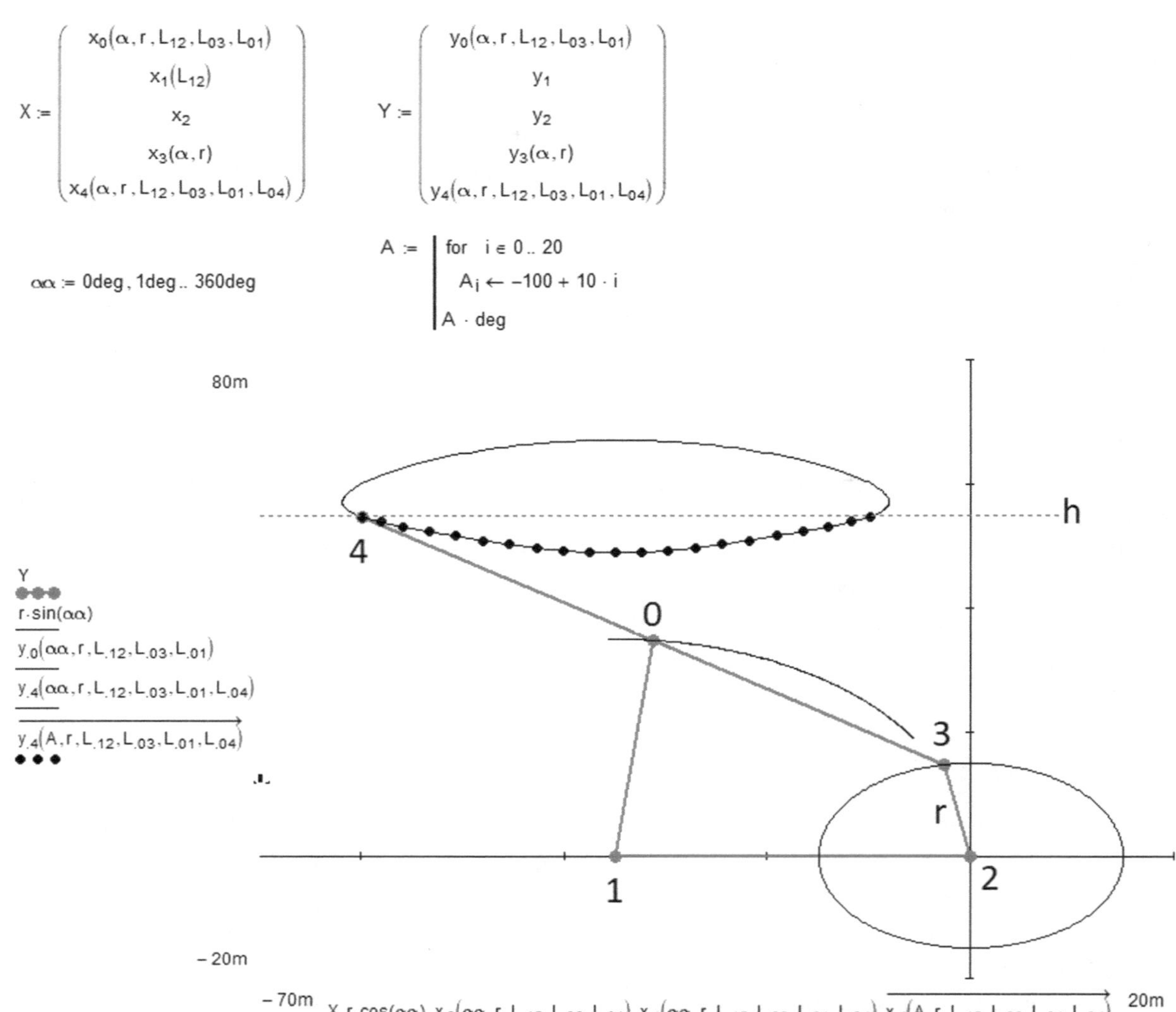

FIGURE 8.14 The non-optimized central unit of Tchebyshev's plantigrade machine.

$$S(L_{12}, L_{03}, L_{01}, L_{04}, h) := \sum_{i=0}^{20} \left(y_4(A_i, r, L_{12}, L_{03}, L_{01}, L_{04}) - h\right)^2$$

$$S(L_{12}, L_{03}, L_{01}, L_{04}, h) = 341.443\,\mathrm{m}^2$$

$$\begin{pmatrix} L_{12} \\ L_{03} \\ L_{01} \\ L_{04} \\ h \end{pmatrix} := \mathrm{Minimize}(S, L_{12}, L_{03}, L_{01}, L_{04}, h) = \begin{pmatrix} 31.233 \\ 39.39 \\ 39.384 \\ 39.362 \\ 63.804 \end{pmatrix} \mathrm{m}$$

$$S(L_{12}, L_{03}, L_{01}, L_{04}, h) = 0.035\,\mathrm{m}^2$$

FIGURE 8.15 Optimization of the plantigrade machine.

mentioned sum of squares of deviations of points from the horizontal line. With the optimization parameters specified by the first line in Figure 8.13, this amount is equal to 341.5 m^2. It is possible to request Mathcad to change the optimization parameters L_{12}, L_{03}, L_{01} and L_{04} and h so that this amount becomes minimal. This "request" is performed by the Mathcad built-in function *Minimize*: the sum of the squares of deviations of points from the line became 0.035 m^2, and our plantigrade machine began to go almost without pushes—see Figure 8.16.

Figure 8.16 shows that after optimization the values of variables L_{03}, L_{01} and L_{04} became almost equal to each

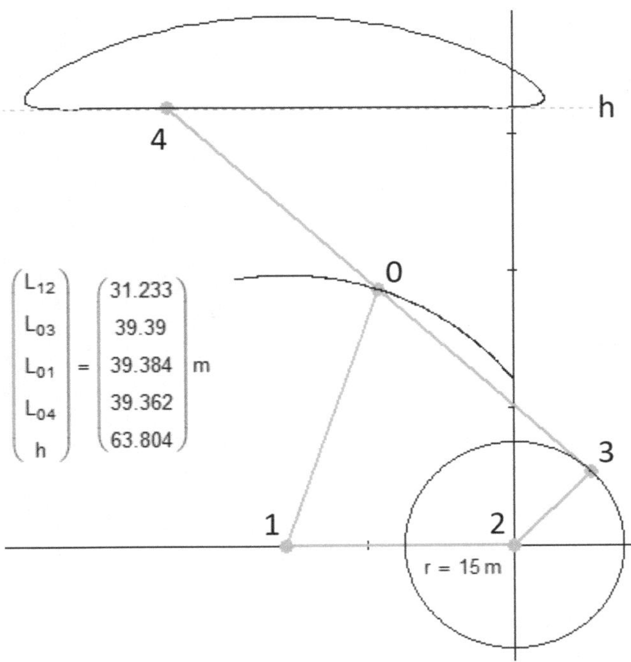

FIGURE 8.16 Optimized central mechanism of Tchebyshev's plantigrade machine.

other. The conclusion is that they are in fact equal, and that Tchebyshev had proved this analytically long ago. So, the number of independent parameters of our optimization problem can be reduced from five to three. This reduction is important for manual analytical calculations. Today, when we increasingly solve similar problems on the computer, this *optimized optimization* becomes less important.

Figure 8.17 shows four frames of the animation for different values of the angle of rotation α. The same animation, we repeat, could be seen on the website https://community.ptc.com/t5/PTC-Mathcad-Questions/Tchebyshev-Math-Journey/td-p/500239. There the reader can find the corresponding Mathcad sheets in two versions—Mathcad 15 and Mathcad Prime.

LET'S MOVE AHEAD

On the plantigrade machine a person can travel not only to the world of science (mathematics and mechanics—see above), but also to the world of art, without having the appropriate talents. And with the advent of computers, in particular computer graphics, new forms of visual art have emerged, accessible to the "broad masses of working people". We will propose one more form of fine art connected with our linkages!

Figure 8.18 shows how the two links pivoted at one end at hinges 1 and 2 (fulcra of the levers in a more dynamics language), and rotating around these points, move three other links (one of them slides in support 3) and draw a funny face. It is possible to change the parameters of this linkage and make the face look cuter or more amusing.

Figure 8.19 shows how we can draw a bird with the help of a kinematic scheme. Points 1, 3 and 7 are fixed. Point 2 moves right-left, up-down, left-right and up-down, drawing a cross. Point 4 moves in a circular orbit. Points 5 and 6 write out intricate trajectories that can be seen on the site's website.* The final point 8 draws our blue bird.

Figure 8.20 shows a more complex hinge mechanism that depicts a certain lace: with four links that rotate around four hinges, other articulated links are attached to them, the joints of which draw closed curves. To this design, it is possible to add new hinges and to weave a new "lace"—see the website https://community.ptc.com/t5/PTC-Mathcad-Questions/New-art-form-with-Mathcad/mp/537923 with the corresponding drawings, animations and Mathcad files.

The number of levers that rotate around points can be increased, the radii of rotation, the speed of rotation, the lengths of the connected links, and so on can be changed. However, it is possible to limit oneself to only two initial levers, and to rotate them not only in a circle, but also in a "square".

The levers shown in Figure 8.21 draw a contour of a lemon. Instead of a square, it is possible to use a triangle or other polygons and build entertaining, and in some ways beautiful, closed curves. These attempts by the author and his colleagues are shown on the forum with the name "Circle + square = lemon" (https://community.ptc.com/t5/PTC-Mathcad-Questions/Square-circle-lemon/m-p/538487).

* https://community.ptc.com/t5/PTC-Mathcad-Questions/Square-circle-lemon/m-p/538487.

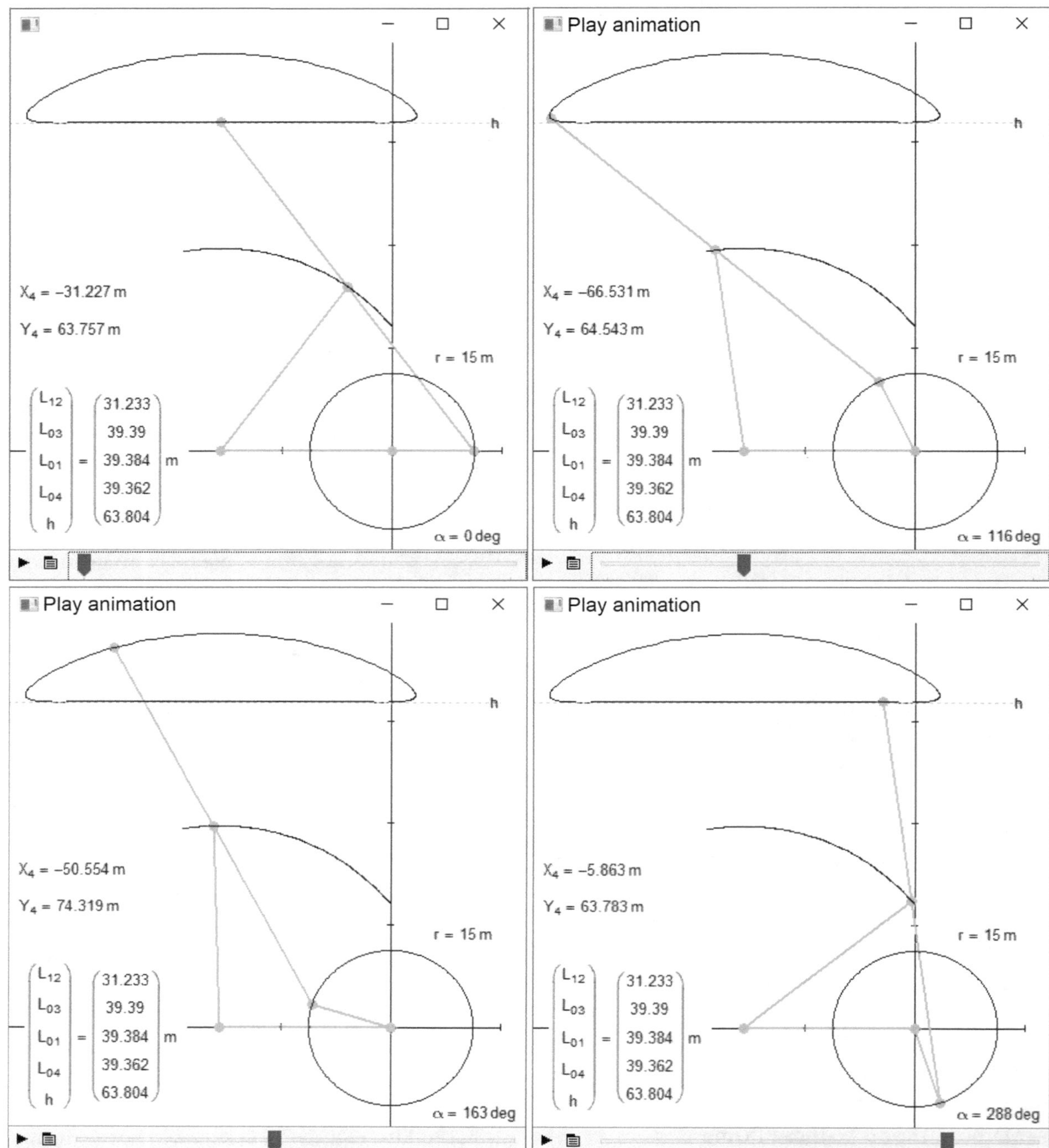

FIGURE 8.17 Frames of animation of the central mechanism of Tchebyshev's plantigrade machine.

AND FINALLY

In Mathcad Prime, alas, there are no tools for creating animations. However, this problem can be solved as follows—by creating several picture frames, which then can be used to create animation with tools that are available on the Internet. With this approach, for example, the author's animation of the chain sagging (Study 7 theme) was created, consisting of two sections with different specific (linear) mass and pulled up at one end—see the PTC Community forum.

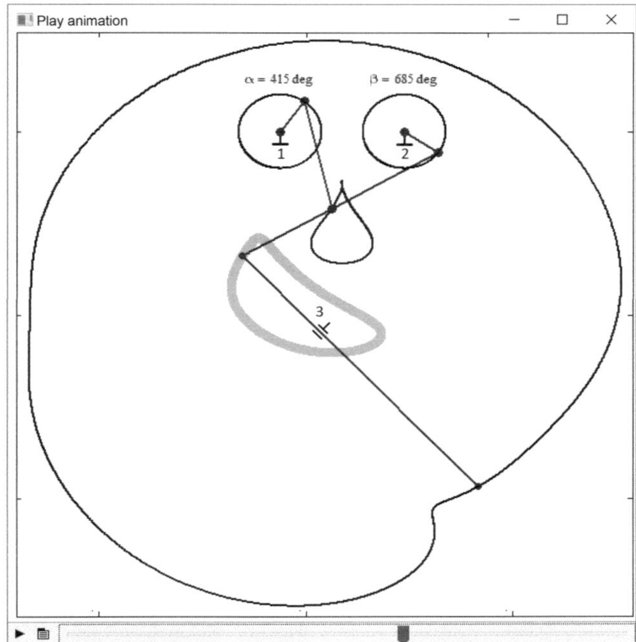

FIGURE 8.18 Drawing a funny face with a five-link kinematic device.

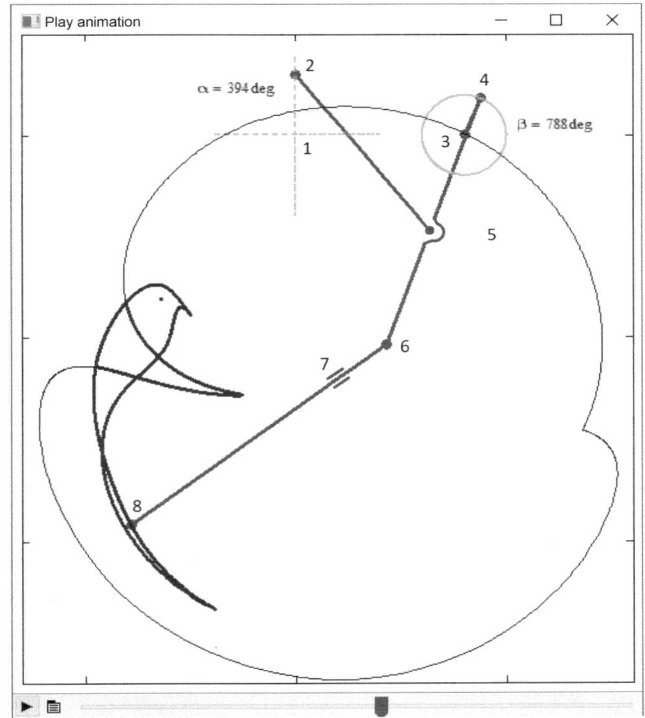

FIGURE 8.19 Drawing a bird with the help of a kinematic scheme.

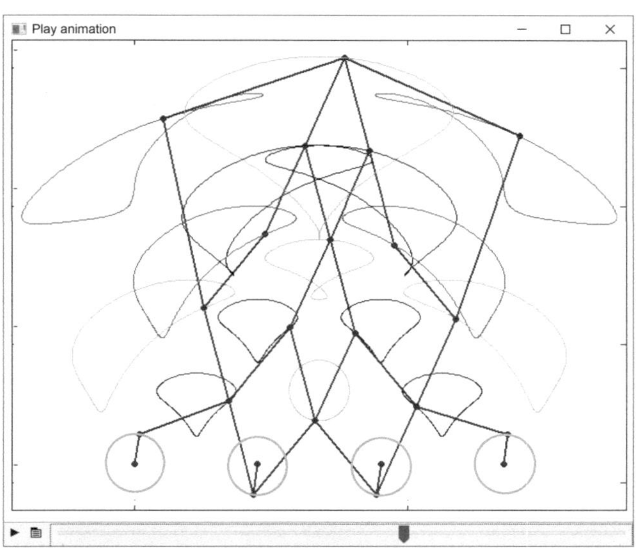

FIGURE 8.20 Lace of the trajectories of the joints of a linkage.

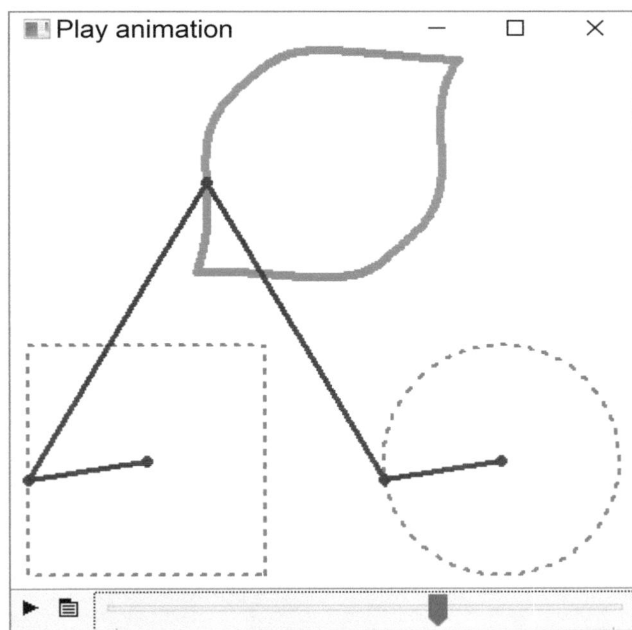

FIGURE 8.21 Drawing the contour of a lemon with the help of a linkage.

ASSIGNMENTS FOR READERS

1. Create an animation of the steam engine shown in Figure 8.1, and of the steam engine mentioned in footnote 10.

2. Change the length of the levers of the Tchebyshev rule, shown in Figure 8.9, so that point 4 moves in as straight a line as possible. One of the solutions is to put several points on the assumed straight line and reduce the solution to the search of the root of a system of equations. We will use this approach for the determining the coefficients of the second-order curve equation describing the motion of a cosmic body near the Earth (see Figures 9.16–9.18 in Study 9).

Celestial Mechanics Or

Kepler's Watch

ABSTRACT

The Study explains how it is possible to calculate and to animate the motion of material points (planets and satellites) under the influence of inertia, gravity and medium resistance forces. Various interesting configurations of planets and their satellites are considered. The motion of a comet is studied on the basis of solving systems of differential and algebraic equations.

Mathematics: Ordinary differential equations (ODEs), system of ordinary differential equations, initial conditions, boundary value problem, integral curves, definite integral, derivative, spherical volume, circle area, second-order curve, hyperbola, ellipse, parabola, Euler's method for solving systems of differential equations, Runge-Kutta method, Gaussian curve (normal distribution), canonical transformation, attractor, system of linear homogeneous algebraic equations, set of proportional solutions.

Physics: Celestial mechanics, the law of universal gravitation, Newton's second law, Kepler's second law, Kepler's clock, gravitational constant, acceleration of gravity, medium resistance, planets, orbit correction, satellite interception, satellite exchange, parachute jump, rocket motion, barometric formula, satellite of the Earth, meteorite near the Earth, landing of a spacecraft, mathematical and physical pendulum, gravitational train, damped oscillations, profile of tram rails, roller coaster (Russian mountains).

IT: Numerical and symbolic solution of differential equations and their systems, methods for solving differential equations and their systems, computer animation, numerical error.

Art: A.S. Pushkin *The Portrait*, *The Stone Guest*, the movie *Armageddon*.

Study website: https://community.ptc.com/t5/PTC-Mathcad/Celestial-Mechanics/m-p/562213.

Dialogue during a lesson of computer modeling.

Student: "Please look, how the two bodies on my screen rotate precisely in orbits!"

The teacher, frowning at the display: "Here you do not have two bodies rotating, but three, as Isaac Newton, too, turns over in his coffin…"

Let's start with the "upset" Newton.

One often hears this rhetorical question: "How would differential calculus look now if Newton had had a computer?"

History does not have the subjunctive mood, but somebody believes that if Newton had had a computer, then there would not be any differential calculus. However, there would be a great amount of solved and apparently solved problems, a lot of tables and curves on the computer displays, from which it would be extremely difficult to obtain any conclusions or more general laws—"necessity [people without a computer] is the mother of invention." Others argue that the "computerized Newton" would have helped us avoid many mistakes and errors, saved us from looking for analytical solutions where there are none, and ultimately led to a more rapid development of science and technology in general and of mathematics in particular. However, the correct answer, probably, is this: "If Newton had had a computer, it would mean that differential calculus as a science would already have existed for three hundred years before Newton, and Newton himself (one of the creators of the differential calculus) would have had another name."

All these arguments about differential calculus can be also applied to the law of universal gravitation—the second great child of the great and "uncomputerized" Newton, in fact according to the legend this law "entered his head thanks to a falling apple". Let's see how modern computers can help or, on the contrary, can prevent us from understanding a given physical problem.

The following is a story behind this Study. The Study was sent in form of an article to a journal, on whose editorial board is the author of a program for calculating the motion of the planets. He developed it for 30 years, using the Pascal programming language. Now the mathematical packages that have replaced programming languages allow us to perform these calculations more easily and more quickly, with units of measurement, excellent plots and animations, as will be shown in this Study.

The article received a response with comments "on the case" and some of them sounded like a hidden insult. I did it half-jokingly as I have done in most of my life, being a non-specialist in this field (celestial mechanics). Comments "on the case" were incorporated in the Study. "Insulting" remarks were simply taken into account.

This episode reveals an important problem. Often people create programs and write books wearing two hats. They are specialists in a certain field of science and technology (power engineering, for example), though fate forces them to engage in mathematics and computer science. As a consequence, they often receive reproaches from both sides.

So, let's start this Study with celestial mechanics.

The essence of the matter is that not only specialized software packages, but also universal mathematical programs allow us to solve problems about the motion of material points (planets, satellites, comets, etc.) of a given mass according to Newton's second law $m \cdot a = \Sigma F$ and to the law of universal gravitation $F = G \cdot m_1 \cdot m_2 / r^2$.

For example, in Mathcad there is the *Odesolve* function for numerically solving systems of ordinary differential equations (ODE), to which the problem of the motion of material points is reduced. If the use of this function is supplemented with animation tools (Study 8), then it is possible to get some very interesting and instructive solutions related to the theme of this Study.

Let's look at a few examples.

Example 1 (Test). The Magnificent Eight

The motion of three material points along an orbit in the shape of the infinity sign (or the "Magnificent Eight").

We can find analytical solutions of this particular case of motion of three material points* in books. We now solve this problem in Mathcad numerically and compare the results of the numerical and analytical solutions.

In order to solve numerically an ODE system in Mathcad 15, it is necessary to enter the word `Given`, to write the system of equations, using the apostrophe (it is entered by the keyboard shortcut *Ctrl + F7*) in order to denote the first derivative, or the double apostrophe for the second derivative, and also to introduce the initial conditions—the position of the material points and their velocities at the initial time (at the start, $t = 0$). In our case, the differential equations of motion of the material points are written in pairs—along the horizontal (x) and vertical (y) directions.

Figure 9.1 shows a system of six ODEs entered for three material points (black, red and blue†): the product of the mass by the acceleration (the left side of the equation) is equal to the sum of the two gravitational forces acting on the considered material point from its two "neighbors". Or rather, not the accelerations and the forces themselves, but their projections in the horizontal‡ (x)

* For all cases the analytical solutions for the motion of two points are reducible to ellipse (circle), parabola, and hyperbola. For three or more points, the analytical solutions are possible for a limited number of cases.

† Mathcad 15 has the ability to edit the variables used in the calculation in various colors, instead of assigning indices: 1, 2, 3, etc., to the variables. But this book was published in a black and white version, where the colors of the variables will be lost. Readers can see color variables on the Study website. Furthermore if a colored edition fell into the hands of a color-blind person, it also would not be very good. Once upon a time, the author sent an article about color in programs to a magazine. The article was not published—the reviewer said that color displays and printers would not appear soon if at all. However he himself (the reviewer) was color blind. Five years after the attempt to publish that article, the Visual Pascal language was introduced, where color was first used to highlight programmer constructs: built-in, user operators, remarks, error messages, etc.

‡ In outer space, of course, there are no horizontal and vertical directions in physical sense. They will appear only when we later consider similar problems near the Earth.

Given

$$x1(0)=x1_0 \quad x1'(0)=vx1_0 \quad m1\cdot x1''(t)=\frac{G\cdot m1\cdot m2\cdot(x2(t)-x1(t))}{\left[\sqrt{(x1(t)-x2(t))^2+(y1(t)-y2(t))^2}\right]^3}+\frac{G\cdot m1\cdot m3\cdot(x3(t)-x1(t))}{\left[\sqrt{(x1(t)-x3(t))^2+(y1(t)-y3(t))^2}\right]^3}$$

$$y1(0)=y1_0 \quad y1'(0)=vy1_0 \quad m1\cdot y1''(t)=\frac{G\cdot m1\cdot m2\cdot(y2(t)-y1(t))}{\left[\sqrt{(x1(t)-x2(t))^2+(y1(t)-y2(t))^2}\right]^3}+\frac{G\cdot m1\cdot m3\cdot(y3(t)-y1(t))}{\left[\sqrt{(x1(t)-x3(t))^2+(y1(t)-y3(t))^2}\right]^3}$$

$$x2(0)=x2_0 \quad x2'(0)=vx2_0 \quad m2\cdot x2''(t)=\frac{G\cdot m2\cdot m1\cdot(x1(t)-x2(t))}{\left[\sqrt{(x2(t)-x1(t))^2+(y2(t)-y1(t))^2}\right]^3}+\frac{G\cdot m2\cdot m3\cdot(x3(t)-x2(t))}{\left[\sqrt{(x2(t)-x3(t))^2+(y2(t)-y3(t))^2}\right]^3}$$

$$y2(0)=y2_0 \quad y2'(0)=vy2_0 \quad m2\cdot y2''(t)=\frac{G\cdot m2\cdot m1\cdot(y1(t)-y2(t))}{\left[\sqrt{(x2(t)-x1(t))^2+(y2(t)-y1(t))^2}\right]^3}+\frac{G\cdot m2\cdot m3\cdot(y3(t)-y2(t))}{\left[\sqrt{(x2(t)-x3(t))^2+(y2(t)-y3(t))^2}\right]^3}$$

$$x3(0)=x3_0 \quad x3'(0)=vx3_0 \quad m3\cdot x3''(t)=\frac{G\cdot m3\cdot m1\cdot(x1(t)-x3(t))}{\left[\sqrt{(x3(t)-x1(t))^2+(y3(t)-y1(t))^2}\right]^3}+\frac{G\cdot m3\cdot m2\cdot(x2(t)-x3(t))}{\left[\sqrt{(x3(t)-x2(t))^2+(y3(t)-y2(t))^2}\right]^3}$$

$$y3(0)=y3_0 \quad y3'(0)=vy3_0 \quad m3\cdot y3''(t)=\frac{G\cdot m3\cdot m1\cdot(y1(t)-y3(t))}{\left[\sqrt{(x3(t)-x1(t))^2+(y3(t)-y1(t))^2}\right]^3}+\frac{G\cdot m3\cdot m2\cdot(y2(t)-y3(t))}{\left[\sqrt{(x3(t)-x2(t))^2+(y3(t)-y2(t))^2}\right]^3}$$

FIGURE 9.1 Entering an ODE system in Mathcad 15.

and vertical (y) directions.* Each equation can be divided by the mass of the corresponding planet, leaving only the second derivative on the left. In addition, you could remove the variable G (gravitational constant), because in our calculations it is adopted (for here and below) as the unit. This simplifies and speeds up the calculation. But we intentionally leave these variables in the equations in order to preserve at the maximum extent the physics of the problem: on the left the product of the mass of the point times the acceleration, on the right the force acting on the point—a planet or a satellite.

The solve block of an ODE system in Mathcad 15 ends with *Odesolve*† (Figure 9.2), that has the following arguments:

- the vector of the names of the functions that are a solution of the ODE system (in Mathcad Prime, we need to specify the function names in parentheses in its argument—see below);
- the independent variable, which is numerically integrated (argument of the sought functions; in Mathcad Prime this variable is not specified);
- the final (right) value of the integration interval (the "left" value is zero);

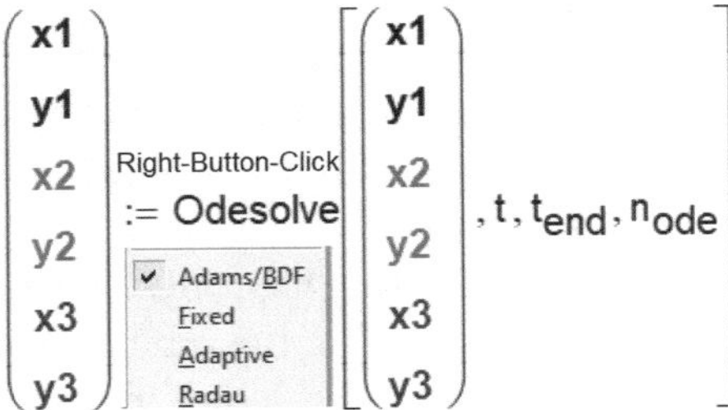

FIGURE 9.2 Setting Odesolve to the desired method for solving systems of ordinary differential equations (on the color web site of this Study, we have color variables x and y (black), x and y (red) and x and y (blue); but on the b/w book we have black variables x1 and y1, x2 and y2, and x3 and y3).

* Some readers, seeing a cube, and not a square in the denominator of the fractions in Figure 9.1 and not noticing at the same time the numerator, could be reminded of an old anecdote: "If instead of an apple a larger object (a coconut, for example) fell on Newton's head, then in the law of universal gravitation the separation distance would be raised not to the power of two but to the power of three."

† In Mathcad, there are other tools for the numerical solution of differential equations (ordinary and partial derivatives, boundary problem), but they are not considered in this Study.

$$\begin{pmatrix} m1 & x1_0 & y1_0 & vx1_0 & vy1_0 \\ m2 & x2_0 & y2_0 & vx2_0 & vy2_0 \\ m3 & x3_0 & y3_0 & vx3_0 & vy3_0 \end{pmatrix} := \begin{pmatrix} 1 & 0.97000436 & -0.24308753 & \frac{0.93240737}{2} & \frac{0.86473146}{2} \\ 1 & -0.97000436 & 0.24308753 & \frac{0.93240737}{2} & \frac{0.86473146}{2} \\ 1 & 0 & 0 & -0.93240737 & -0.86473146 \end{pmatrix}$$

FIGURE 9.3 Initial conditions for the "Magnificent Eight" problem with the motion of three planets.

- the number of points on which the sought functions must be tabulated (optional parameter—by default it is set to 1000 points).

In addition, by pressing the right mouse button, the user could select for the *Odesolve* function one of four possible methods for solving numerically the problem:

- Adams/BDF—mixed Adams algorithm and inverse differentiation formula;
- Fixed—integration by the Runge-Kutta method with fixed step;
- Adaptive—integration by the Runge-Kutta method with variable step;
- Radau—RADAUS algorithm for stiff ODE systems.

In Mathcad Prime, there is no possibility of changing the method of solving an ODE. There it is available only with the Adams/BDF method.

If three initial points are taken from one book, also shown in Figure 9.3, then the solution of the ODE system (Figure 9.1) is the functions whose graphical mapping (trace-to-track) is shown in Figure 9.4(a).

Comparison of "telemetry data taken from the orbit" and shown in Figure 9.4(a), with the data from one book, gives a complete match of the results, which may indicate the adequacy of our model set in the equations of Figure 9.1. However, to be precise, the following happens. At about the first one hundred "loops" of the orbit, there is no deviation of the numerical solution from the analytical one. However, with an increase in the value of t, a certain "blurring" of the orbit occurs as a result of the accumulated error in the numerical solution of the problem (Figure 9.4(b)). This feature can be clearly seen in a second example, where new equations and new terms are added to the model (see Figure 9.1) for the two additional points—brown and green.

Example 2 (Test Continuation). Round Dance of the Planets

On the tips of a five-pointed star, five material points (planets) with equal masses are placed. The initial velocities have equal absolute value and are directed clockwise along the tangent to the circle into which the star is inscribed (see Figure 9.5(a)). The motion of the planets from these initial positions and velocities, called a round dance, is shown in Figure 9.5(b).

A "cosmic round dance", like any round dance, cannot last indefinitely—at some point it disintegrates, and its participants scatter in different directions. Something similar can be observed in our dance of planets as the value of the variable of numerical integration *t* is increased—see Figure 9.5(c)*

In the problem of the motion of material points, a third dimension can also be introduced, transferring it from the plane to the volume (into outer space†). For this it is necessary and sufficient to write another triad of equations in the system shown in Figure 9.1 and to add the initial data. Figure 9.6 shows the "round dance" of three planets in the three-dimensions. The third axis is directed upward (it gives the same initial uniform motion of the planets upward along the *z*-axis), and the curves are the plots of the three solutions. This plot (Figure 6.1) can also be considered as the change with time of the location of the points on

* Such "works of art" are often created by children: at first everything goes reasonably well, and then the child gets tired of drawing correctly and painting on the almost finished picture a "daub". It can be said that Mathcad is "tired of drawing for us a beautiful cosmic flower" (Figure 9.5(b)) and finished all with a "daub" (Figure 9.5(c)).

† We used to say "flights in interplanetary space". But it is more correct to say "flights on the interplanetary plane". Most calculations of flights of natural and artificial cosmic bodies are conducted in two dimensions after choosing an appropriate plane.

(a)

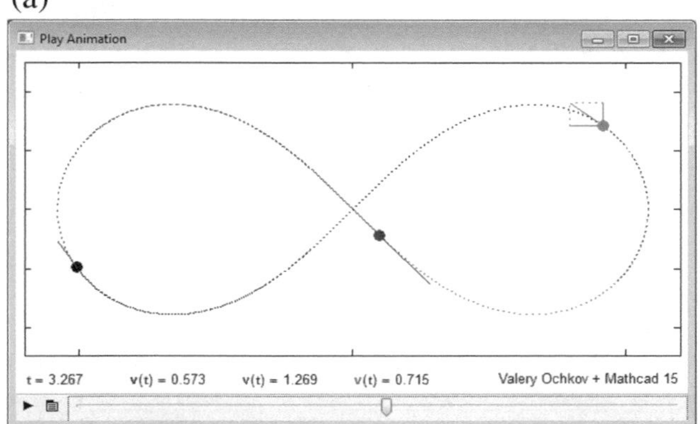

(b)

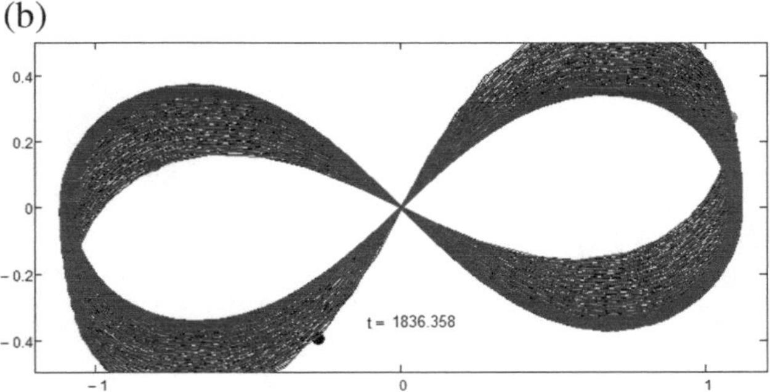

FIGURE 9.4 (a) A frame of the animation of the motion of three material points (planets) in an orbit of an infinity symbol shape; planet velocity vectors are shown, for the red planet the velocity vector is additionally decomposed along the x and y axes. (b) Graphical representation of the loss of accuracy of the numerical solution of the ODE system.

the plane (see Figure 9.4(a)) (these plots are named integral curves).

Figure 9.4(a) actually shows one of the three projections of the motion of three points (Figure 9.6), when the z-axis is directed toward the observer. The trajectories shown in Figure 9.6,—can be interpreted as weaving three strands of hair in a pigtail, which is clearly visible during the animation of this process. If the initial velocities for the three planets are different, one can get very tangled trajectories of motion, and from them it will be difficult to isolate the components associated with the physics of the problem and its numerical solution will lose accuracy. An analytical (absolutely exact) solution is impossible in this case.

Note on 3D plots

Volumetric plots—this is more a kind of advertising thing than a tool for visualizing the results of "rigorous" calculations, calculations that have a practical, rather than demonstrational (presentation, advertising, etc.) purpose, especially if this result is transferred to printed paper, where it is not possible to rotate the volumetric plot and to examine it from another direction in the same way as when it is displayed on a screen.

Therefore, when creating a Mathcad-sheet, a user should immediately focus on a simple, "undistorted" plot, remembering that the display screen and printer paper are flat objects. Analogously, real rigorous problems can be based on a set of functions of three, four or more arguments whose plots must be shown in the corresponding four-, five- or more-dimensional space. However, we have it in only three-dimensions: the line (1D), the plane (2D) and the volume (3D). It should also be remembered that unnecessarily distorted plots very often hide a trivial calculation, masking it with complex volumetric surfaces including perspective, backlight, "fog" and other "bloat". A complicated calculation that has a clear practical value, as a rule, is illustrated by simple "flat" plots showing the basic regularities of calculation. These calculations, for example, include families of curves, etc., and not drawings of surfaces that are pleasant to admire, but difficult to use.

If it is necessary to use 3D plots for showing some abstract or real bodies (motionless or moving) rather than for displaying functions of two arguments, then, of course, it's better to use specialized tools rather than Mathcad.

But let us return to the "planar" cosmic cases, bearing in mind, first, that the planets of the solar system, for example, move in orbits lying approximately in one plane, and, second, that the numerical methods built into Mathcad fortunately cope with planar problems.

Two of our previous examples ("The Magnificent Eight" and "The Round Dance") are tests for comparing numerical and analytical solutions when the error of numerical methods has not yet reached a critical level. The adequacy of the solution for values of t that are not too large* allows us to apply this solution method (see Figures 9.1 and 9.2) to cases not yet investigated, for which analytical solutions are unknown or impossible. To achieve this goal, it is necessary (sufficiency is a separate question) to add new equations with new

* We have still to determine what "not too large" is in quantitative terms.

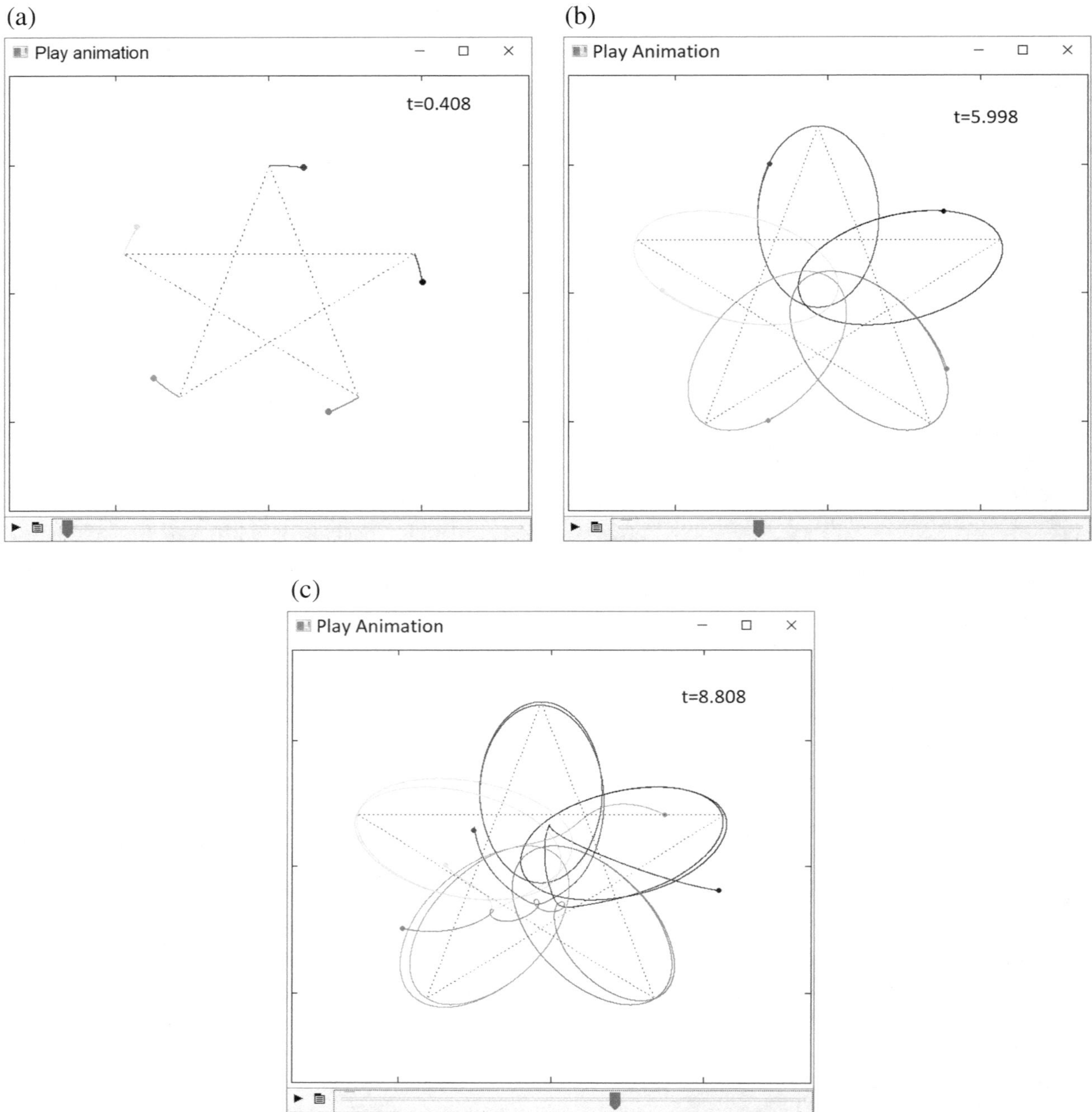

FIGURE 9.5 Frames of animation of the five-planet dances: (a) start. (b) The complete coincidence of the numerical solution with the analytical. (c) The loss of accuracy of the numerical solution.

terms on the right-hand side, to change the initial conditions and to choose an appropriate method for the numerical solution.

Example 3. Interception of a Satellite

A red planet, paired with a blue satellite, is moving toward a black planet. Each one has its own mass, respectively: 2, 0.01 and 20 units of mass (later we will consider kilograms and tons). What will happen to these celestial bodies when they become close to each other? To investigate this situation, we will perform some numerical experiments and solve the problem by using different methods of solution. In Figure 9.7 is an animation frame, which shows that at the moment of the planets'

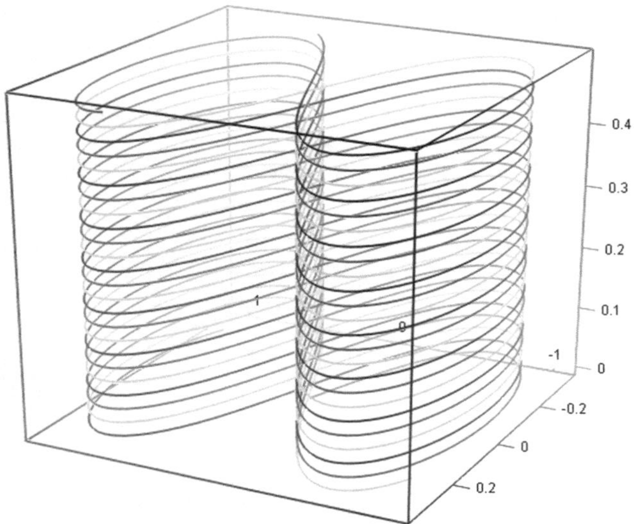

FIGURE 9.6 The three-dimensional problem of the motion of three material points.

approaching, the blue satellite changes its "patron" and starts to rotate around the arc of the black planet. The red planet continues its celestial path in solitude.

While creating the animation shown in Figure 9.7,* the interception of the satellite was modeled using the Radau method (see Figure 9.2) for the *Odesolve* function. If the solution method is changed from Radau to Adams/BDF, we will see a qualitatively different picture: the interception of the satellite will not take place, and only some deviation of its orbit around the red planet will be

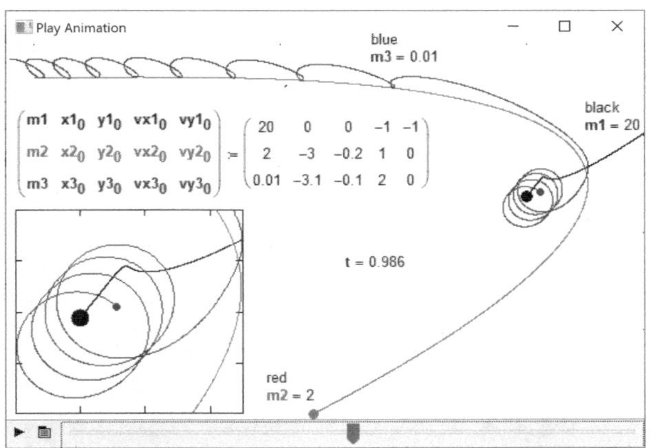

FIGURE 9.7 Animation frame for satellite interception.

* The figure on the left shows an enlargement of the interception area of the satellite.

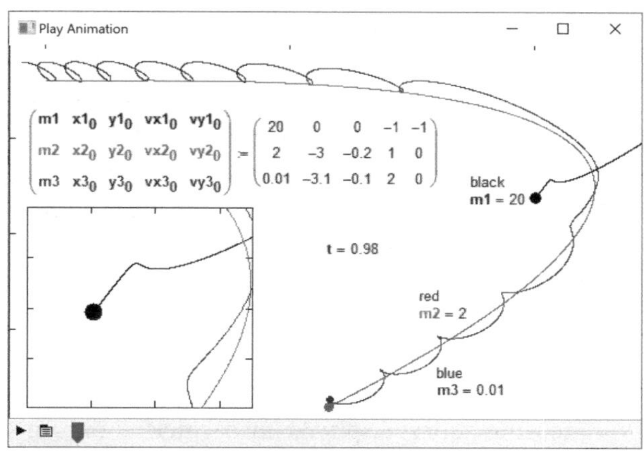

FIGURE 9.8 The animation results using not the Radau but the Adams/BDF numerical methods of the ODE solution (no satellite interception).

observed due to the approach of the black planet to the system "red planet/blue satellite", see Figure 9.8.

Figure 9.8 shows that two different methods for solving the ODE system (Radau and Adams/BDF) give the same results first (before the encounter with the black planet), and then (after meeting with the black planet) the results diverge: with the Radau method we observe interception of the satellite by the black planet near the red planet, while the Adams/BDF method shows only a deviation of its orbit near the old "patron", the red planet. Fixed and Adaptive methods (integration with a constant and variable step), as you can see from the corresponding animation by visiting the above-mentioned website, are of no use in this case—failures of the solution are evident from the very beginning of the process of the numerical solution of the problem (integration). So which is correct and why is the other wrong? It is an equation!

So, we have a real physical process, its physical model, the mathematical model of the physical model (equations and initial conditions), and several computational (numerical) models of the same process: calculation modules and formulas for the Radau, Adams/BDF, Fixed and Adaptive methods (see Figure 9.3 above). Numerical models give some approximations to a mathematical model that approximates a physical model, which in turn approximates a real physical process. At each stage of the simulation, for various objective reasons, we make mistakes, and therefore it is necessary to separate insignificant mistakes from the principal ones. Errors of these numerical models are fundamental: four calculation schemes yielded four different results, which strongly depended on the method of solution. And not only quantitatively (see Figures 9.4 and 9.5), but also qualitatively (see Figure 9.8). This means that the problem is not solved in general. Although, the first phase of the motion is the same for the two methods (see Figure 9.8), and therefore, are most likely correct, the initial results can be used in the numerical modeling of the problem being solved.

The animations displayed in a single frame in Figures 9.5, 9.6, 9.7 and 9.8 once again focus our attention on the "brilliance and misery" of numerical methods for solving ODEs: they allow us to solve problems that are "beyond the teeth" of the analytical methods, and to create quite entertaining and instructive animations though they can also "lie in your eyes". However, we will return to this point below.

Here an important question arises—is it possible to use an animation to evaluate the quality of various methods of solving ODE systems? The perturbations of the trajectories, the "daubs" indicate that the method of solution used is not suitable, at least for the chosen range of integration. On the other hand, the correctness of the trajectories of planets and satellites, their similarity to real processes, can serve as a kind of signal of the good quality of the applied methods of solution. However not always—see Figure 9.8, where both the trajectories of the blue satellite are "beautiful" and similar to real trajectories.

Example 4. Exchange of Satellites

Figure 9.9 shows another non-trivial and more complex case—an exchange of satellites. The cases shown earlier in Figures 9.7 and 9.8 have analogies with everyday life: there is a certain married couple (the "planet-satellite" system), that is approached by a certain extraneous subject. He can either "beat off" the satellite (Figure 9.7) or bring some "perturbations" into the life of the couple (Figure 9.8).

Figures 9.7, 9.8 and 9.9 have this somewhat frivolous interpretation. Because it is not possible to take these "everyday" trajectories seriously there are doubts about the adequacy of the numerical methods used to solve these problems. However,

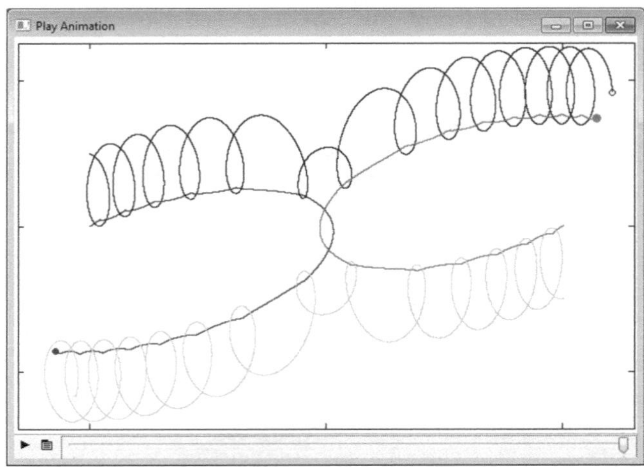

FIGURE 9.9 Animation frame for the satellite exchange.

the cases are interesting and based on "appropriate" physics. And we need to agree with this. Or better: you can choose to agree or disagree.

Example 5 (Hybrid: Checking the Old and Showing the New One). Round Dance (pas de deux) of Four Planets with Four Satellites

Figure 9.10(a) shows the start frame of the animation of a system of eight material points: four planets (each of mass 0.5) and four satellites (each of mass 0.0001). The starting conditions are as follows: the planets are placed in the corners of a square with sides equal to unity and their speeds are set equal to 0.5 directed respectively upward (blue planet), right (black), down (red) and left (brown). Five, not four, planets are spinning, without satellites, in the dance shown in Figure 9.5 (performing a certain pas de deux). However, in this example satellites with zero initial speed are additionally placed in the middle of the square sides.

Animation of the motion of these eight celestial bodies is almost impossible to describe in words. Each satellite will try to alternately "attach" to each of the four planets, making all kinds of "pas" (Figure 9.10(b)). At the end of the "dance" these satellites will fly out of the "round dance"—see Figure 9.10(c). On the above-mentioned website it is possible to see the animation of the motion of one of the satellites, while the remaining three satellites are not visible (Figure 9.10(d)). It clearly shows how the satellite changes partners in this "dance". At the same time, other satellites follow the same "motions", with the difference that their

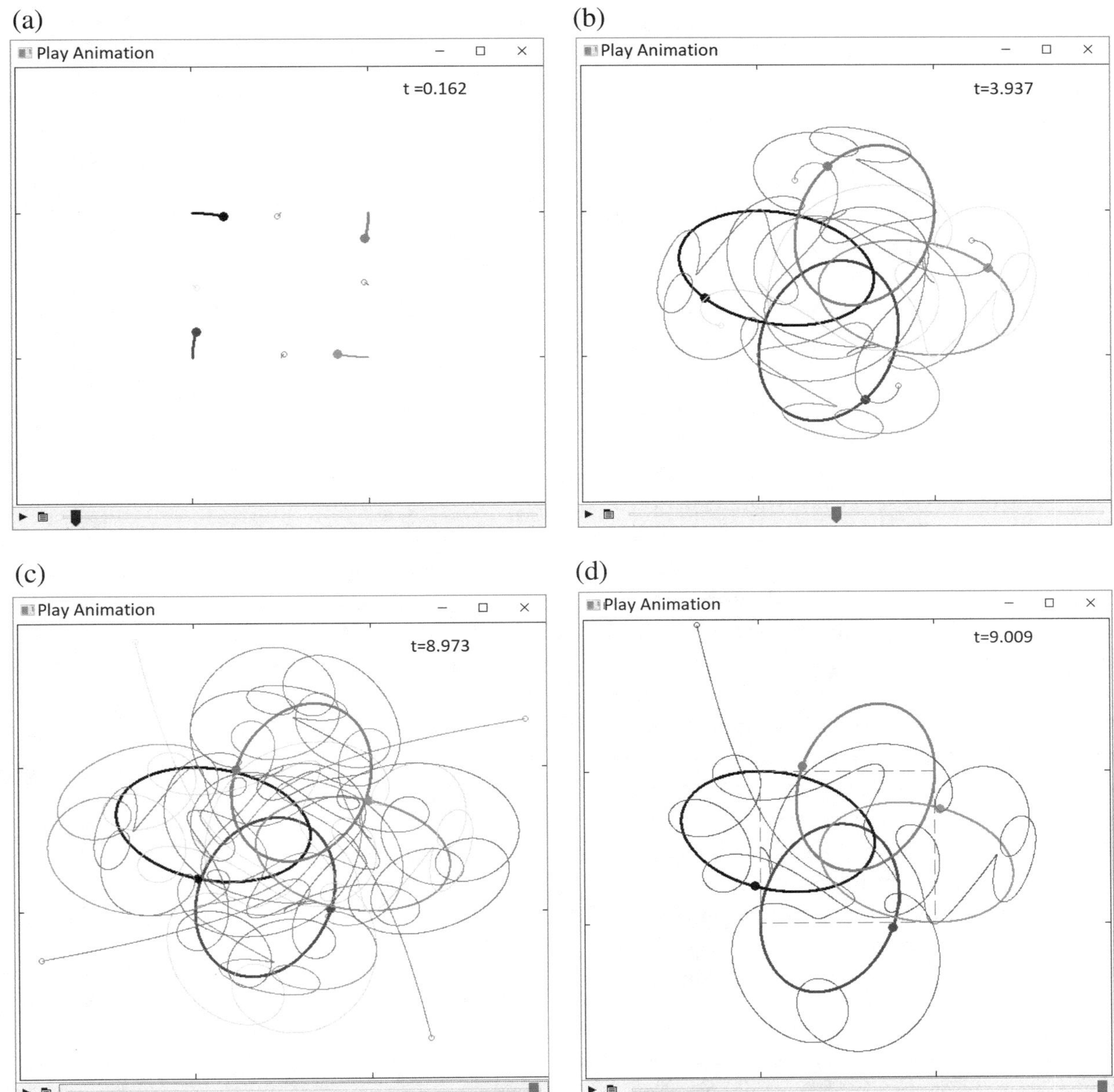

FIGURE 9.10 Motion animation frames of four planets and four satellites. (a) Start. (b) The flight of satellites around the planets. (c) The departure of the satellites of their "round dance". (d) The trajectory of one satellite (the other three trajectories are invisible).

flight trajectories are rotated by 90° relative to the central point (center of mass) (Figure 9.10(c)). It remains an open question as to whether the satellites will return back to the "round dance". It is difficult to answer this question by a numerical experiment—here the whole picture will be spoiled by the build-up of numerical errors. (see, for example, Figure 9.5(b)). An attempt to answer (or, rather, a way of solving) this particular problem will be given below. The symmetry of the flight of planets and satellites, recorded in Figure 9.10(c), supports the conclusion that the numerical solution of the problem is sufficiently accurate.

Example 6. Earth Satellite or Kepler Clock

The new version of Mathcad, Mathcad Prime, does not have animation yet, but it can solve ODEs with units of measurements of physical quantities. In the previous versions (e.g. Mathcad 15), these units were ignored and dimensionless units of length, speed, mass, etc., were used, something that is fraught with errors. Somewhere in the system of equations, it is possible that a three might mistakenly be replaced by a two, for example. Nevertheless, the system might still be solved, but gives the wrong answer. In Mathcad Prime, an error of this type will be immediately noticed. In this case, the units of measurement are recalculated, which is very useful as is shown in the figures below.

Figure 9.11 shows the orbit of rotation of an Earth satellite with real values for distances, velocities and, most importantly, for the gravitational constant *G*. These initial data were selected so that the period of rotation of the satellite was 12 hours. In addition, sectors of the elliptical orbit that illustrate Kepler's second law are added on the plot: the areas of these sectors at equal (hourly) intervals are equal (with accuracy of hundredths of a percent—see Figure 9.11(b)) in the numerical solution of the problem, which is a further confirmation of the correctness of our model.* The animation itself can be considered as another unusual mathematical clock, in which the hands move along an elliptical dial with a variable speed.† These watches will look good in Kepler's museums, planetariums and observatories.

Figure 9.11 shows that at about "eight o'clock" the satellite is closest to the Earth (perigee) and at this point it will have the maximum speed

(a)

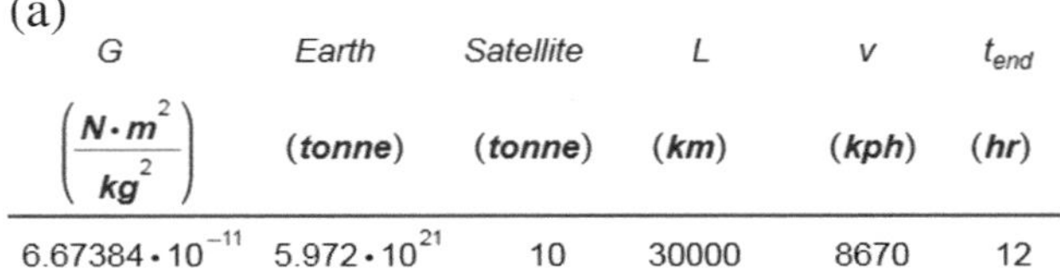

G $\left(\frac{N\cdot m^2}{kg^2}\right)$	Earth (tonne)	Satellite (tonne)	L (km)	V (kph)	t_{end} (hr)
$6.67384\cdot10^{-11}$	$5.972\cdot10^{21}$	10	30000	8670	12

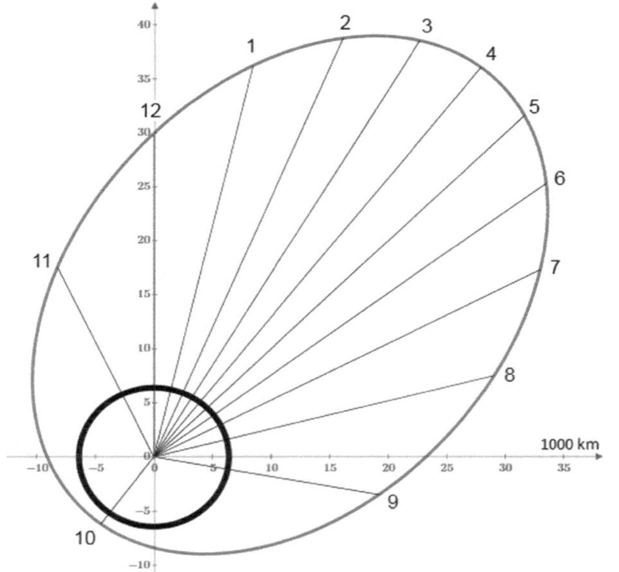

(b)

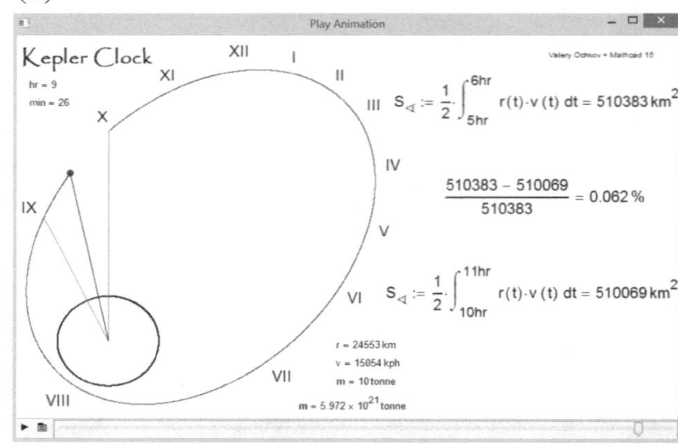

FIGURE 9.11 Kepler's clock: (a) Mathcad calculation, (b) animation frame.

according to Kepler's second law. In the region of two o'clock (apogee), the speed of the satellite will be minimal.

Near the Earth it is possible to abandon the law of universal gravitation, for the potential gravity field near our planet, characterized by free-fall acceleration, and to take into account the effect of the Earth's atmosphere. This problem is also easy to solve with the help of Mathcad—see Figure 9.12.

* Our ellipse can be reduced to the canonical form and the values of its parameters—the coefficients of the second-order curve equation, the axes, the foci, and so on could be calculated.

† Let's try to make of this formula an invention. The watch of Kepler is a traditional clock with hands and dial. The difference is that the dial of Kepler's clock is not a circle, but an ellipse with numbers 1, 2, 3, ..., 12 (I, II, III, ..., XII). The hands of such clocks (hour, minute and second) move in the circle unevenly. The lengths of the hands also change as they move. The ends of the hands describe elliptical trajectories. Additionally on the dial there is a sector covering two adjacent hours, between the two hours there is another hand at a given time. This sector changes its location and shape every five seconds, but its area remains constant according to Kepler's second law. Around the place where the hands are fastened, a circle depicting the Earth is drawn, and at the end of the second hand there is a circle depicting the Earth's satellite. The Kepler clock could be also made with only one hand that shows the minutes of the hour covered by the ellipse sector (see Figure 9.11(b)). The Kepler clock will be very appropriate in the schoolrooms of mathematics and physics, in planetariums, observatories, museums, attracting people's attention, forcing them to think about the laws of celestial mechanics, in particular, the second law of Kepler.

G	$Earth$	R	x_0	y_0	v_{x0}	v_{y0}	t_{end}
$\left(\frac{N\cdot m^2}{kg^2}\right)$	$(tonne)$	(km)			$\left(\frac{km}{s}\right)$	$\left(\frac{km}{s}\right)$	(hr)
$6.67384\cdot10^{-11}$	$5.972\cdot10^{21}$	6370	$2R$	$2R$	−0.5	−7	12

$$x(0\ s)=x_0 \qquad x'(0\ s)=v_{x0} \qquad x''(t)=-G\cdot\frac{Earth\cdot x(t)}{\left(\sqrt{x(t)^2+y(t)^2}\right)^3}$$

$$y(0\ s)=y_0 \qquad y'(0\ s)=v_{y0} \qquad y''(t)=-G\cdot\frac{Earth\cdot y(t)}{\left(\sqrt{x(t)^2+y(t)^2}\right)^3}$$

$$\begin{bmatrix}x\\y\end{bmatrix}:=\mathbf{Odesolve}\left(\begin{bmatrix}x(t)\\y(t)\end{bmatrix},t_{end}\right)$$

FIGURE 9.12 Blocks of the initial data and of the numerical solution of the ODE system.

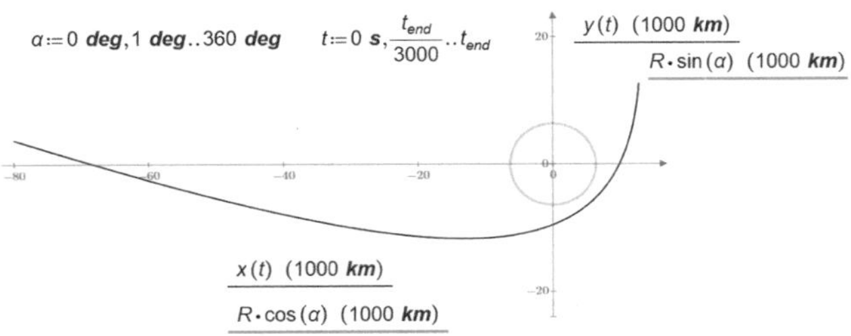

FIGURE 9.13 Graphical representation of the numerical solution of the ODE system.

LAWLESS COMET

"Just like a lawless comet, Amidst the planets, calculated ..."

A.S. Pushkin "*The Portrait*"**

Figure 9.11 shows the motion of a satellite in an elliptical orbit. But there are celestial bodies ("lawless comets", for example) that move along different trajectories, and this is reflected also in the epigraph to this part of this Study.

There are trajectories in the shape of ellipses (a special case is the circle), hyperbolas and parabolas (a rare transitional case from hyperbola to ellipse).

From our calculations based on the numerical solution of an ODE with initial conditions, one can often see only some arc of the trajectory, according to which it is difficult to determine whether it is an *ellipse*, a *hyperbola* or a *parabola*. However, this can be understood if we recall that the curves under consideration are plane curves of the second order. In the equations of these curves (see the upper part of Figure 9.14) there are six coefficients. We can calculate them (graphically or through the calculation of invariants) to determine the type of curve.

Here comes to mind another quotation from A.S. Pushkin, though from another of his works, *The Stone Guest*:†

Don Juan.
There's nothing visible
Of her beneath her somber widow's veil;
I just but glimpsed a trim and narrow heel.

Leporello:
That's quite enough for you. Imagination
will in a jiffy study you out the rest;
Your fancy's quicker than the painter's brush.
The starting point is all the same to you—
The forehead, or the foot.

However, we have not only "imagination", but also powerful computing tools for solving systems of algebraic equations (see Figure 9.14). The "narrow heel" is our points of the numerical solution of the ODE system (see Figures 9.12 and 9.13), and the "rest" is the entire hyperbola, or rather the coefficients of its equation. And the fact that this is a hyperbola is proved by the value of the invariant D, which is less than zero (for an ellipse it is greater than zero, and for a parabola it is zero), and by the corresponding graphs in Figure 9.15.

Figure 9.12 shows the block of input data and the block of the numerical solution of the ODE system that displays the motion of a material point (satellite) near another material point (Earth) with a mass much greater than the mass of the first point so that the displacement of the second point can be neglected.‡

Figure 9.13 shows a graphical representation of the solution of the problem of the motion of a celestial body of small mass (an asteroid, for example) near the Earth in 12 hours.

It is rather difficult to understand from Figure 9.13, whose partial arc we have obtained—ellipse, hyperbolas or parabolas. It is possible to increase the integration limit§ (by increasing the value of the variable t_{end}) and to see if the curve will curl (fold) into an ellipse or will go to infinity, remembering at the same time about the accuracy of the solution (see the figures above). However, it is possible to do otherwise—to restore the whole curve of

* A translation of this short poem of Alexander Pushkin is available at https://www.poetryloverspage.com/yevgeny/pushkin/portrait.html.

† *The Stone Guest* (Russian: «Каменный гость», *translit.* Kamennyy gost') is a poetic drama by Alexander Pushkin based on the Spanish legend of Don Juan. Pushkin wrote *The Stone Guest* in 1830 as part of a collection of four short plays known as Little Tragedies. A translation is available at (at least in the USA) https://archive.org/stream/littletragedies00alek#page/66.

‡ In the problem shown in Figure 9.11, we did not make this simplification.

§ But, we repeat, it is necessary to remember that this often results in the accumulation of error—see Figures 9.4(b) and 9.5(c) above.

(a)

Solve

$$a_{11}\cdot x^2+2\,a_{12}\cdot x\cdot y+a_{22}\cdot y^2+2\,a_1\cdot x+2\,a_2\cdot y+a_0=0$$

Guess Values

$$\begin{bmatrix}a_{11} & a_{12} & a_{22} & a_1 & a_2 & a_0\end{bmatrix}:=\begin{bmatrix}1\,\frac{1}{m^2} & 1\,\frac{1}{m^2} & 1\,\frac{1}{m^2} & 1\,\frac{1}{m} & 1\,\frac{1}{m} & 1\end{bmatrix}$$

Constraints

$$a_{11}\cdot x(0\,hr)^2+2\,a_{12}\cdot x(0\,hr)\cdot y(0\,hr)+a_{22}\cdot y(0\,hr)^2+2\,a_1\cdot x(0\,hr)+2\,a_2\cdot y(0\,hr)+a_0=0$$
$$a_{11}\cdot x(1\,hr)^2+2\,a_{12}\cdot x(1\,hr)\cdot y(1\,hr)+a_{22}\cdot y(1\,hr)^2+2\,a_1\cdot x(1\,hr)+2\,a_2\cdot y(1\,hr)+a_0=0$$
$$a_{11}\cdot x(2\,hr)^2+2\,a_{12}\cdot x(2\,hr)\cdot y(2\,hr)+a_{22}\cdot y(2\,hr)^2+2\,a_1\cdot x(2\,hr)+2\,a_2\cdot y(2\,hr)+a_0=0$$
$$a_{11}\cdot x(3\,hr)^2+2\,a_{12}\cdot x(3\,hr)\cdot y(3\,hr)+a_{22}\cdot y(3\,hr)^2+2\,a_1\cdot x(3\,hr)+2\,a_2\cdot y(3\,hr)+a_0=0$$
$$a_{11}\cdot x(4\,hr)^2+2\,a_{12}\cdot x(4\,hr)\cdot y(4\,hr)+a_{22}\cdot y(4\,hr)^2+2\,a_1\cdot x(4\,hr)+2\,a_2\cdot y(4\,hr)+a_0=0$$
$$a_{11}\cdot x(5\,hr)^2+2\,a_{12}\cdot x(5\,hr)\cdot y(5\,hr)+a_{22}\cdot y(5\,hr)^2+2\,a_1\cdot x(5\,hr)+2\,a_2\cdot y(5\,hr)+a_0=0$$
$$a_{11}\cdot x(6\,hr)^2+2\,a_{12}\cdot x(6\,hr)\cdot y(6\,hr)+a_{22}\cdot y(6\,hr)^2+2\,a_1\cdot x(6\,hr)+2\,a_2\cdot y(6\,hr)+a_0=0$$
$$a_{11}\cdot x(7\,hr)^2+2\,a_{12}\cdot x(7\,hr)\cdot y(7\,hr)+a_{22}\cdot y(7\,hr)^2+2\,a_1\cdot x(7\,hr)+2\,a_2\cdot y(7\,hr)+a_0=0$$
$$a_{11}\cdot x(8\,hr)^2+2\,a_{12}\cdot x(8\,hr)\cdot y(8\,hr)+a_{22}\cdot y(8\,hr)^2+2\,a_1\cdot x(8\,hr)+2\,a_2\cdot y(8\,hr)+a_0=0$$
$$a_{11}\cdot x(9\,hr)^2+2\,a_{12}\cdot x(9\,hr)\cdot y(9\,hr)+a_{22}\cdot y(9\,hr)^2+2\,a_1\cdot x(9\,hr)+2\,a_2\cdot y(9\,hr)+a_0=0$$
$$a_{11}\cdot x(10\,hr)^2+2\,a_{12}\cdot x(10\,hr)\cdot y(10\,hr)+a_{22}\cdot y(10\,hr)^2+2\,a_1\cdot x(10\,hr)+2\,a_2\cdot y(10\,hr)+a_0=0$$
$$a_{11}\cdot x(11\,hr)^2+2\,a_{12}\cdot x(11\,hr)\cdot y(11\,hr)+a_{22}\cdot y(11\,hr)^2+2\,a_1\cdot x(11\,hr)+2\,a_2\cdot y(11\,hr)+a_0=0$$

Solver

$$\begin{bmatrix}a_{11}\\a_{12}\\a_{22}\\a_1\\a_2\\a_0\end{bmatrix}:=\mathrm{Find}\,(a_{11},a_{12},a_{22},a_1,a_2,a_0)=\begin{bmatrix}(108.22724800507\cdot10^{-15})\,\frac{1}{m^2}\\(148.549565390801\cdot10^{-15})\,\frac{1}{m^2}\\(83.8987903667927\cdot10^{-15})\,\frac{1}{m^2}\\(3.1515511012128\cdot10^{-6})\,\frac{1}{m}\\-3.4201697225529\cdot10^{-6}\,\frac{1}{m}\\-72.5605605935609\end{bmatrix}$$

(b)

Guess Values

$$\begin{bmatrix}a_{11} & a_{12} & a_{22} & a_1 & a_2\end{bmatrix}:=\begin{bmatrix}\frac{1}{m^2} & \frac{1}{m^2} & \frac{1}{m^2} & \frac{1}{m} & \frac{1}{m}\end{bmatrix}$$

Constraints

$$a_{11}\cdot x(8\,hr)^2+2\,a_{12}\cdot x(8\,hr)\cdot y(8\,hr)+a_{22}\cdot y(8\,hr)^2+2\,a_1\cdot x(8\,hr)+2\,a_2\cdot y(8\,hr)=-a_0$$
$$a_{11}\cdot x(9\,hr)^2+2\,a_{12}\cdot x(9\,hr)\cdot y(9\,hr)+a_{22}\cdot y(9\,hr)^2+2\,a_1\cdot x(9\,hr)+2\,a_2\cdot y(9\,hr)=-a_0$$
$$a_{11}\cdot x(10\,hr)^2+2\,a_{12}\cdot x(10\,hr)\cdot y(10\,hr)+a_{22}\cdot y(10\,hr)^2+2\,a_1\cdot x(10\,hr)+2\,a_2\cdot y(10\,hr)=-a_0$$
$$a_{11}\cdot x(11\,hr)^2+2\,a_{12}\cdot x(11\,hr)\cdot y(11\,hr)+a_{22}\cdot y(11\,hr)^2+2\,a_1\cdot x(11\,hr)+2\,a_2\cdot y(11\,hr)=-a_0$$
$$a_{11}\cdot x(12\,hr)^2+2\,a_{12}\cdot x(12\,hr)\cdot y(12\,hr)+a_{22}\cdot y(12\,hr)^2+2\,a_1\cdot x(12\,hr)+2\,a_2\cdot y(12\,hr)=-a_0$$

Solver

$$\begin{bmatrix}a_{11}\\a_{12}\\a_{22}\\a_1\\a_2\end{bmatrix}:=\mathrm{Find}\,(a_{11},a_{12},a_{22},a_1,a_2)=\begin{bmatrix}(1.4915435067719\cdot10^{-15})\,\frac{1}{m^2}\\(2.0472490947877\cdot10^{-15})\,\frac{1}{m^2}\\(1.1562584951361\cdot10^{-15})\,\frac{1}{m^2}\\(43.4333826097209\cdot10^{-9})\,\frac{1}{m}\\-47.1353810518701\cdot10^{-9}\,\frac{1}{m}\end{bmatrix}\qquad a_0=-1$$

FIGURE 9.14 Numerical solution of the system of algebraic equations. (a) Solution of the homogeneous system of 13 equations with a set of proportional solutions. (b) Solution of the system of five equations with five unknowns and a unique solution. (c) Solution of a linear system of five equations with five unknowns and a unique solution.

(c)

$$M := \begin{bmatrix} x(7\,hr)^2 & 2\,x(7\,hr)\cdot y(7\,hr) & y(7\,hr)^2 & 2\,x(7\,hr) & 2\,y(7\,hr) \\ x(8\,hr)^2 & 2\,x(8\,hr)\cdot y(8\,hr) & y(8\,hr)^2 & 2\,x(8\,hr) & 2\,y(8\,hr) \\ x(9\,hr)^2 & 2\,x(9\,hr)\cdot y(9\,hr) & y(9\,hr)^2 & 2\,x(9\,hr) & 2\,y(9\,hr) \\ x(10\,hr)^2 & 2\,x(10\,hr)\cdot y(10\,hr) & y(10\,hr)^2 & 2\,x(10\,hr) & 2\,y(10\,hr) \\ x(11\,hr)^2 & 2\,x(11\,hr)\cdot y(11\,hr) & y(11\,hr)^2 & 2\,x(11\,hr) & 2\,y(11\,hr) \end{bmatrix}$$

$$v := \begin{bmatrix} 1 \\ 1 \\ 1 \\ 1 \\ 1 \end{bmatrix} \quad M^{-1}\cdot v = \begin{bmatrix} \left(1.49154295494464\cdot 10^{-15}\right)\dfrac{1}{m^2} \\ \left(2.04724825269486\cdot 10^{-15}\right)\dfrac{1}{m^2} \\ \left(1.15625764036381\cdot 10^{-15}\right)\dfrac{1}{m^2} \\ \left(43.4333638807231\cdot 10^{-9}\right)\dfrac{1}{m} \\ -47.1353781403196\cdot 10^{-9}\dfrac{1}{m} \end{bmatrix}$$

FIGURE 9.14 (CONTINUED) Numerical solution of the system of algebraic equations. (a) Solution of the homogeneous system of 13 equations with a set of proportional solutions. (b) Solution of the system of five equations with five unknowns and a unique solution. (c) Solution of a linear system of five equations with five unknowns and a unique solution.

the second order by the coefficients of its equation. How this could be done it is shown in Figure 9.14.

Figure 9.14 shows three solutions. The first one (Figure 9.14(a)) can be called a "frontal attack": initial approximations (assumptions) to the solution are given, together with thirteen equations fixing the position of the satellite at the initial instant of time ($t = 0$), after an hour of flight, after 2 hours of flight and so on up to 12 hours of flight. Then the *Find* function, which determines the values of its arguments that convert the equations into identities, is called. If other initial estimated values for the solution are provided, then another answer will be obtained, which will also be correct. In this case, the values of six unknown coefficients for the first set of initial approximations will be proportional to the values of these same unknowns for the second set of initial approximations. And this is understandable, since it is possible to multiply all the coefficients in the equation of the curve shown in Figure 9.14(a) by the same number to get a new equation for x and y that is equivalent to the original one. This property allows us to fix the value of one of the six coefficients and then determine the remaining ones. Thus, by specifying different initial approximations for all six unknown coefficients, we obtain different sets of coefficients that are proportional to each other and that correspond to an equation of the same second-order curve.

In Figures 9.14(b) and 9.14(c), only five equations are left in the problem. A curve of second order can be reconstructed from five points. The coefficient a_0 is assigned the value minus one. In this case the solution will be unique, and the initial approximations for the remaining five coefficients will no longer affect the response (see Figure 9.14(b)), so the problem can be reduced to solving a system of *linear* algebraic equations that does not require initial approximations (see Figure 9.14(c)). The linear system in Mathcad can also be solved using the *Find* function (see Figure 9.14(b)). However for this problem, Mathcad has specialized tools: for example, it is possible to multiply the inverted coefficient matrix of the unknowns by the vector of the free terms (see Figure 9.14(c)).

In Figure 9.15, first, it is shown that the principal invariant D of the equation of the plane curve of the second order is less than zero and, then, two halves of the hyperbola are constructed—the top (blue) and the lower (up, black). Analytical expressions for these halves are obtained by solving the equation of the curve with respect to the variable y and by generating two functions named y_{top} and y_{up}. On the "up" branch of the hyperbola (the bottom, black half) our red points are threaded.

$$D := \left\| \begin{bmatrix} a_{11} & a_{12} \\ a_{12} & a_{22} \end{bmatrix} \right\| = -0.000002467 \frac{1}{(1000\ km)^4} \qquad D<0 \qquad \textit{Hyperbola}$$

$$y_{up}(x) := -\frac{a_2 + x \cdot a_{12} + \sqrt{x^2 \cdot a_{12}^2 - a_{11} \cdot a_{22} \cdot x^2 + 2 \cdot x \cdot a_2 \cdot a_{12} - 2 \cdot a_1 \cdot a_{22} \cdot x + a_2^2 - a_0 \cdot a_{22}}}{a_{22}}$$

$$y_{top}(x) := -\frac{a_2 + x \cdot a_{12} - \sqrt{x^2 \cdot a_{12}^2 - a_{11} \cdot a_{22} \cdot x^2 + 2 \cdot x \cdot a_2 \cdot a_{12} - 2 \cdot a_1 \cdot a_{22} \cdot x + a_2^2 - a_0 \cdot a_{22}}}{a_{22}}$$

$$x1 := -200 \cdot 1000\ \mathbf{km}, -199.99 \cdot 1000\ \mathbf{km} .. 200 \cdot 1000\ \mathbf{km} \qquad t1 := 0\ \mathbf{hr}, 1\ \mathbf{hr} .. 12\ \mathbf{hr}$$

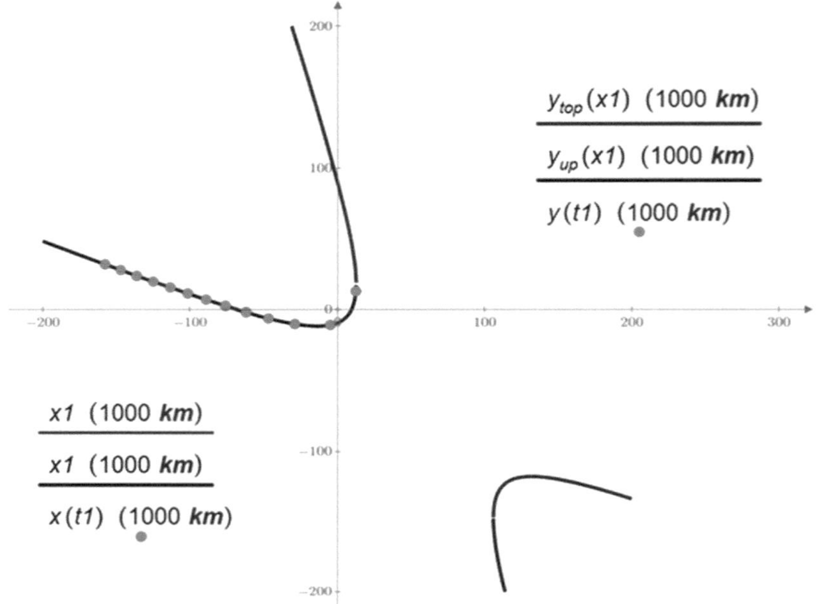

FIGURE 9.15 The construction of a hyperbola by its equation.

This calculation can be continued by determining, for example, the values of the coefficients a and b of the canonical equation of the hyperbola:

$$\frac{x^2}{a^2} - \frac{y^2}{b^2} = 1$$

The first step of this transformation will be to find the angle to which the axes of the Cartesian graph should be rotated*—see Figure 9.19.

We suggest readers complete this transformation on their own, having at hand all six coefficients a of the equation of the second-order plane curve. We have already found the angle of rotation of the two branches of the hyperbola (see Figure 9.16—search for one of the two roots of the quadratic equation in the interval from 0 to 90 angular degrees). It remains only to determine the value of the shift of the branches of the hyperbola, so that it takes the canonical form—symmetric with respect to the axes and the origin.

The left five red dots shown in Figure 9.15 can be considered as telemetry data of the motion of an asteroid near the Earth. Having these data, scientists can calculate how the asteroid will continue: whether it will fly past the Earth (namely it is shown in Figures 9.13 and 9.15), or collide with the Earth. To prevent a global catastrophe, a team can be sent to the outer space, for example, to explode the asteroid—see the movie *Armageddon*.

$$\phi := \mathbf{root}\left(\tan(\phi)^2 - \frac{a_{11} - a_{22}}{a_{12}} \tan(\phi) - 1, \phi, 0\ \mathbf{deg}, 90\ \mathbf{deg}\right) = 47.341\ \mathbf{deg}$$

FIGURE 9.16. Angle of rotation of the axes.

* This formula is obtained by introducing a generic set of coordinates (x_1, y_1) rotated clockwise by an angle phi and setting the "new" $x_1 y_1$ coefficient (the new a_{12}) equal to zero.

G	m_e	R_e	m_s	ρ_s	h_0	k	t_{end}
$\left(\frac{N\cdot m^2}{kg^2}\right)$	$(tonne)$	(km)	$(tonne)$	$\left(\frac{kg}{m^3}\right)$	(km)		(hr)
$6.67384\cdot10^{-11}$	$5.972\cdot10^{21}$	6370	10	800	40	0.00003025	4.25897

$$\rho(h):=\left\| \begin{array}{l} L\leftarrow 0.0065\,\frac{K}{m} \\ T_0\leftarrow 288.15\,K \\ T\leftarrow T_0-L\cdot h \\ M\leftarrow 28.9644\,\frac{gm}{mole} \\ p\leftarrow 101325\,Pa\cdot\left(1-\frac{L\cdot h}{T_0}\right)^{\frac{g\cdot M}{R\cdot L}} \\ \rho_{возд}\leftarrow \frac{p\cdot M}{R\cdot T} \end{array}\right.$$

Air density

$$\rho(0\,m)=1.225\,\frac{kg}{m^3} \qquad \rho(h_0)=0.000061545\,\frac{kg}{m^3}$$

$$v_0:=\sqrt{\frac{G\cdot m_e}{R_e+h_0}}=7.885\,\frac{km}{s}$$ Start satelitte velosity

Volume, diameter and section of the satelitte

$$V_s:=\frac{m_s}{\rho_s}=12.5\,m^3 \qquad d_s:=\sqrt[3]{\frac{6\cdot V_s}{\pi}}=2.879\,m \qquad S_s:=\pi\cdot\left(\frac{d_s}{2}\right)^2=6.512\,m^2$$

FIGURE 9.17 Modeling the drag of a satellite in the dense layers of the atmosphere: beginning of the calculation.

Solve

$$x(0\,s)=0\,m \qquad v_x(t)=x'(t) \qquad m_s\cdot v_x'(t)=-G\cdot\frac{m_s\cdot m_e\cdot x(t)}{\left(\sqrt{x(t)^2+y(t)^2}\right)^3}-k\cdot v_x(t)\cdot|v_x(t)|\cdot S_s\cdot\rho\left(\sqrt{x(t)^2+y(t)^2}-R_e\right)$$

$$v_x(0\,s)=v_0$$

$$y(0\,s)=R_e+h_0 \qquad v_y(t)=y'(t) \qquad m_s\cdot v_y'(t)=-G\cdot\frac{m_s\cdot m_e\cdot y(t)}{\left(\sqrt{x(t)^2+y(t)^2}\right)^3}-k\cdot v_y(t)\cdot|v_y(t)|\cdot S_s\cdot\rho\left(\sqrt{x(t)^2+y(t)^2}-R_e\right)$$

$$v_y(0\,s)=0\,\frac{m}{s}$$

$$\begin{bmatrix} x \\ y \\ v_x \\ v_y \end{bmatrix}:=\textbf{Odesolve}\left(\begin{bmatrix} x(t) \\ y(t) \\ v_x(t) \\ v_y(t) \end{bmatrix}, t_{end}, 10^5\right)$$

FIGURE 9.18 Modeling of the satellite drag in the dense layers of the atmosphere: continuation of the calculation.

When this Study was written (May 2015), news circulated* that the *Progress* spacecraft that launched the International Space Station did not enter the calculated orbit and should have shortly fallen toward the Earth. This event prompted the author to try to simulate the behavior of an uncontrolled artificial Earth satellite—see Figures 9.17, 9.18 and 9.19.

In Figure 9.17, the initial data are set in Mathcad Prime: the gravitational constant G, the parameters of the Earth (its mass m_e and radius R_e), the parameters of the artificial satellite (its mass m_s, density ρ_e and height above ground at the initial moment h_0), the air friction coefficient of the satellite k and the flight time t_{end}. Further in the calculation shown in Figure 9.17, the functional dependence of the air density with height $\rho(h)$ is introduced (see also Figure 9.12), the speed of rotation v_0 of the satellite along the circular orbit at height h_0 above the Earth's level (the first cosmic velocity) and the parameters of the artificial satellite: its volume, diameter and the cross-sectional area under the assumption that it is a sphere of radius R_e and density ρ_e.

Figure 9.18 shows the solution operators for a system of four ordinary differential equations simulating the motion and fall of an artificial satellite in the Earth's orbit. To the force of attraction, determined by the law of universal gravitation, was added the drag force arising from air friction. This last force is proportional to (drag coefficient k) the cross-sectional area of satellite S_s, the air density ρ, and the square of the velocity (see also Figures 9.12, 9.13(a) and 9.13(b)). In this case, the square of the velocity is not inserted in the calculation, but the product of the velocity by the absolute value of the velocity is used in order to determine the correct direction of the force—it acts in the direction opposite to the direction of the velocity vector.

Figure 9.19 shows the graphical representation of the solution of the problem of the reduction of the satellite's orbit up to the moment of contacting the Earth's surface: the time dependence of the height $h(t)$

* https://www.theguardian.com/science/2015/apr/29/russia-tries-to-make-radio-contact-with-progress-spacecraft-on-iss-mission.

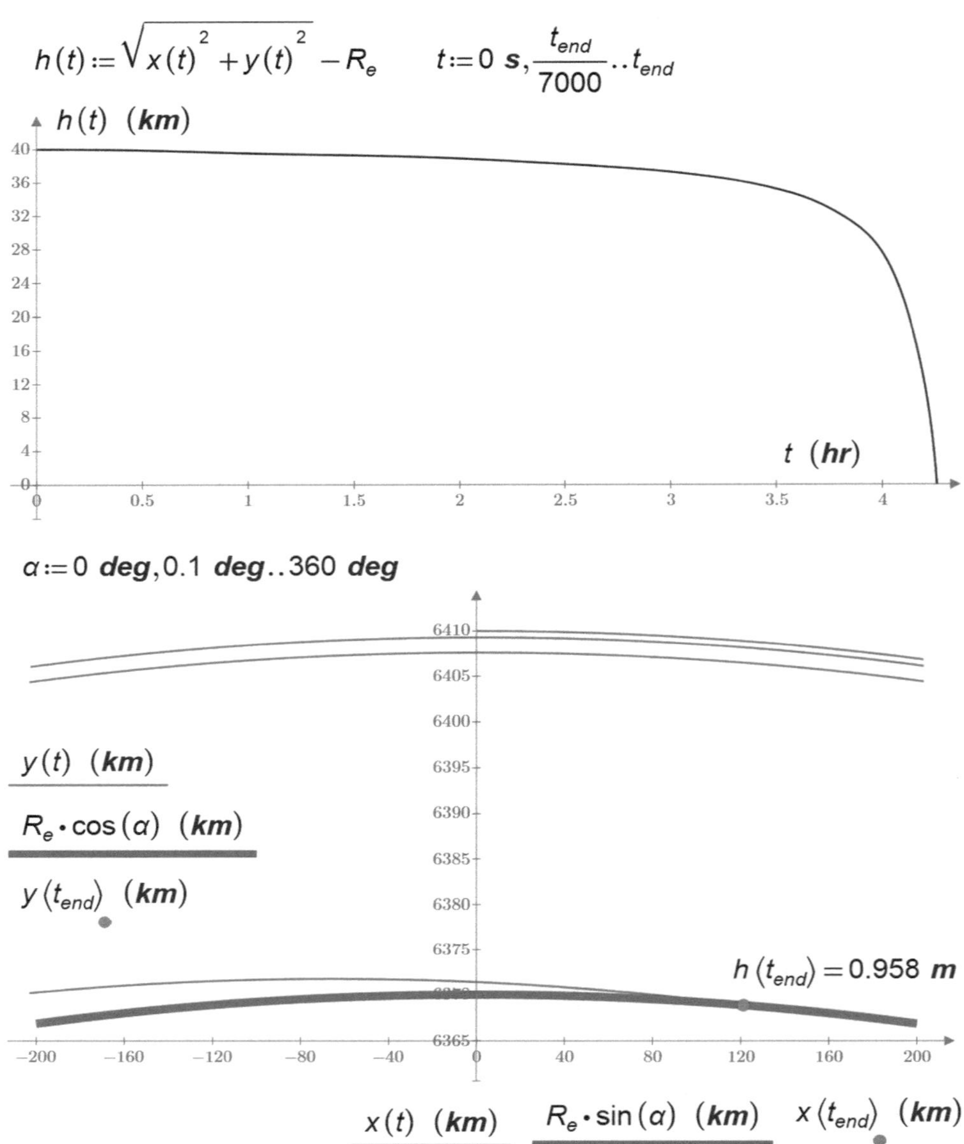

FIGURE 9.19 Modeling the drag of a satellite in the dense layers of the atmosphere: end of the calculation.

(upper graph) and some fragments of the orbit trajectory (in the lower graph it is shown also the Earth surface (blue) arc as function of the x, where the satellite arrives from a height of 6410 km, after making three loops).

On the Study website there is a model and an animation of the reverse case where, due to the inclusion of an engine, the satellite moves to a new, higher orbit.

All the calculations files presented in this Study (Mathcad 15 and Mathcad Prime 3) can be downloaded from the Study website. There the corresponding animations are also available. Their individual web addresses and frames were listed above.

And Last but One for the Most Curious and Meticulous

What algorithms are embedded in the numerical methods for solving ordinary differential equations?

The simplest method for solving these problems was proposed by the great Euler. If we talk about a problem with initial conditions* (e.g. the Cauchy problem), then Euler's method is applied as follows.

If the initial (first) value of the unknown function is known, the segment of integration is divided into n short

* However there is also the boundary value problem, where the conditions are defined at the ends of the integration segment. Below, one of the ways to solve this problem will be briefly described.

segments of length Δ and the value of the desired function at the second point is calculated from the value of the derivative at the first point—and this is the right side of the differential equation:* $y_2 = y_1 + y'_1 \cdot \Delta$. This operation is performed in a cycle ($y_i = y_{i-1} + y'_{i1} \cdot \Delta$)† in which two vectors are formed, storing the discrete values of the argument and the desired function. If the number of points n is large enough and the desired function is not too "steep",‡ the Euler method provides quite acceptable solutions, although it is not built into Mathcad (see the list of built-in methods in Figure 9.2). We have just described the application of the Euler method to a first-order differential equation. In our problems of celestial mechanics, systems of second-order equations are relevant: since, in the mathematical models, the acceleration, the second time derivative of the displacement, occurs. However, Euler's method can be applied not only to a single equation, but also to a system, and a single second-order equation can be transformed into two first-order equations. This solution approach is shown in Figure 9.20 with reference to another classical problem related to the law of universal gravitation and to gravity: the problem of the oscillation of the pendulum.§

The mathematical model of the oscillation of the pendulum takes into account the fact that its angular acceleration is proportional to the product of the acceleration of gravity times the sine of the angle of deviation from the vertical of the pendulum cord and it is inversely proportional to the length of the pendulum cord. If the sine of the angle is replaced by the angle itself (and this can be done for values of the modulus less than about 7°), then this equation (the equation of the *mathematical* pendulum) can be solved analytically—see Figure 9.21(a), where the Mathematica website was used for this purpose.

If, in the equation of motion of the pendulum, the sine is introduced, then on the above-mentioned website the solution will still be obtained, though it can be called analytical with an effort, because there will be a *special* function in it—the Jacobi elliptic function (*cd*—see Figure 9.21(b)). Looking ahead to a similar problem—the problem of the gravitational train, it can be noted that there is no *sin* in the equation initially (see Figure 9.27). However, in the equation of motion of such a semi-fantastic train it is possible to introduce the force of air resistance, proportional to the square of its speed. In this case, in the solution in Figure 9.27 there will be one more special function—the Bessel function, which, by the way, is built into Mathcad.¶ An analytical solution is usually understood as a solution made only of elementary functions. However, the term "analytical" has too many different interpretations. Many believe that a solution obtained with the help of any known special functions and even of any series (and even integrals) could be considered analytical. Computers only added misunderstandings on this issue.

However, let's return to numerical methods for the solutions.

It can be seen from Figure 9.20 that for $n = 100$ (the number of partitions of the integration interval) the solution of the system of the two ODEs by the Euler method turned out to be rather rough, deviating significantly from the true solution. The "true" (analytical) solution is drawn with a line on the graph, while the numerical (approximate) solution is marked with dots to emphasize once again that the Euler method did not return the desired function, but a set of points that somehow need to be transformed into a function using, for example, the interpolation methods, which we will consider in Study 14.

In case of a single first-order differential equation our *Euler* function (Figure 9.20) returns the desired points in form of a matrix with two columns—discrete values of the argument and the corresponding discrete values of the function. If, on the other hand, we have not only an equation, but a system of equations, additional columns will appear in the matrix. In Figure 9.20, two

* Because of this relation, the method is also known as the forward Euler or explicit method, to distinguish it from the backward Euler or implicit method, where $y_2 = y_1 + y'_2 \cdot \Delta$.

† In the programs in Figures 9.19 and 9.22, the lower index of the variable (the vector element) is replaced by the "superscript" (column of the matrix with one row). This was done in order to display the final matrix of solution. These nuances were described in the forum https://www.ptcusercommunity.com/thread/85205.

‡ The word "steep" could mean many things... it's like in youth slang. In our problems, the behavior of the satellites and planets was "steep" and consequently required rather complicated methods for solving differential equations.

§ On the PTC Community website it is possible to find the author's animation of the single pendulum in a viscous environment, on an elastic thread, and also the animation of two, three and four connected pendulums on a rigid unbending thread. Pendulums in this case describe intricate curves, similar to those shown in Figure 9.10. And there it is also possible to see clearly the failure of the numerical solution.

¶ Bessel functions are especially important for many problems of wave propagation in cylindrical and spherical coordinates (see for instance: https://www.researchgate.net/publication/317785584_Sound_propagation_in_porous_materials_with_annular_pores).

$$\text{Euler}(y_b, x_b, x_e, n, f) := \left\| \begin{array}{l} \text{"ODE Solution by Euler method"} \\ \left[X^{\langle 0 \rangle} \leftarrow x_b \;\; Y^{\langle 0 \rangle} \leftarrow y_b \;\; \Delta \leftarrow \dfrac{x_e - x_b}{n} \right] \\ \textbf{for } i \in 1..n \\ \quad \left\| \begin{array}{l} Y^{\langle i \rangle} \leftarrow Y^{\langle i-1 \rangle} + f\left(X^{\langle i-1 \rangle}, Y^{\langle i-1 \rangle}\right) \cdot \Delta \\ X^{\langle i \rangle} \leftarrow X^{\langle i-1 \rangle} + \Delta \end{array} \right. \\ \text{augment}\left(X^T, Y^T\right) \end{array} \right.$$

$L := 1\ m$ $\quad \alpha_0 := 7°$ $\quad n := 100$ $\quad t_e := 3\ s$

$$f(t, x) := \begin{bmatrix} x_1 \\ -\sin\left(x_0\right) \cdot \dfrac{g}{L} \end{bmatrix} \qquad \begin{array}{l} \alpha'(t) = v(t) \\ v_\alpha'(t) = -\sin(\alpha(t)) \cdot \dfrac{g}{L} \end{array}$$

$$M := \text{Euler}\left(\begin{bmatrix} \alpha_0 \\ 0\ \dfrac{deg}{s} \end{bmatrix}, 0\ s, t_e, n, f \right) \qquad t := 0\ s, \frac{t_e}{300} .. t_e$$

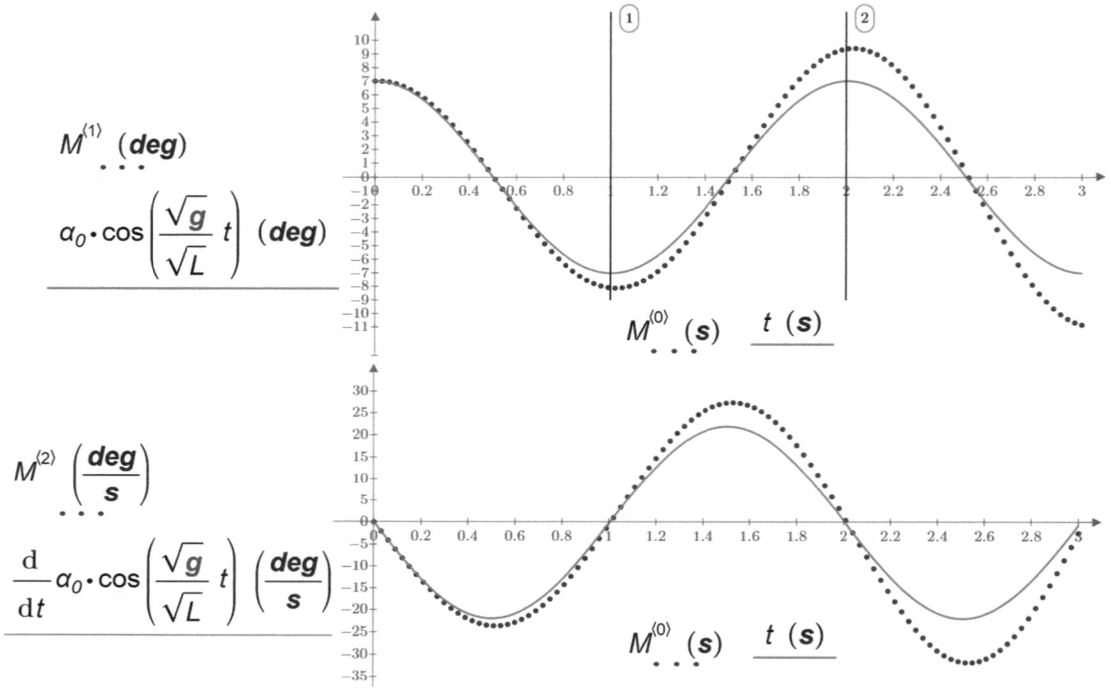

$M^{\langle 1 \rangle}$ (deg)

$\alpha_0 \cdot \cos\left(\dfrac{\sqrt{g}}{\sqrt{L}}\, t\right)$ (deg)

$M^{\langle 0 \rangle}$ (s) $\quad t$ (s)

$M^{\langle 2 \rangle}$ $\left(\dfrac{deg}{s}\right)$

$\dfrac{d}{dt}\alpha_0 \cdot \cos\left(\dfrac{\sqrt{g}}{\sqrt{L}}\, t\right)$ $\left(\dfrac{deg}{s}\right)$

$M^{\langle 0 \rangle}$ (s) $\quad t$ (s)

FIGURE 9.20 The Euler method, applied to solve a system of two ordinary differential equations.

graphs are specifically shown (graph of the deviation of the pendulum's chord from the vertical and the angular velocity of the pendulum) obtained after "extracting" the desired columns of the matrix M: $M^{<0>}$ and $M^{<1>}$.

If, in the solution shown in Figure 9.20, the value of n is increased from 100 to 1000 or to an even greater value, then the points on the graph merge into a line (giving the illusion of an analytical solution with the function in the answer) and lie close to the curve, the solution will be more or less accurate. However, this will increase the computational time* accordingly. Because of this, and also for other reasons (because of the "steepness", i.e. the very rapid change of the unknown functions in some ranges), "trickier" methods for the numerical solution of the ODE have been developed.

* Now thanks to the use of powerful computers this change in the computational time is almost not felt, but in those days when the described methods were invented and there were no computers at all, this difference in computational time was significant and even determining when choosing one or another method of solution.

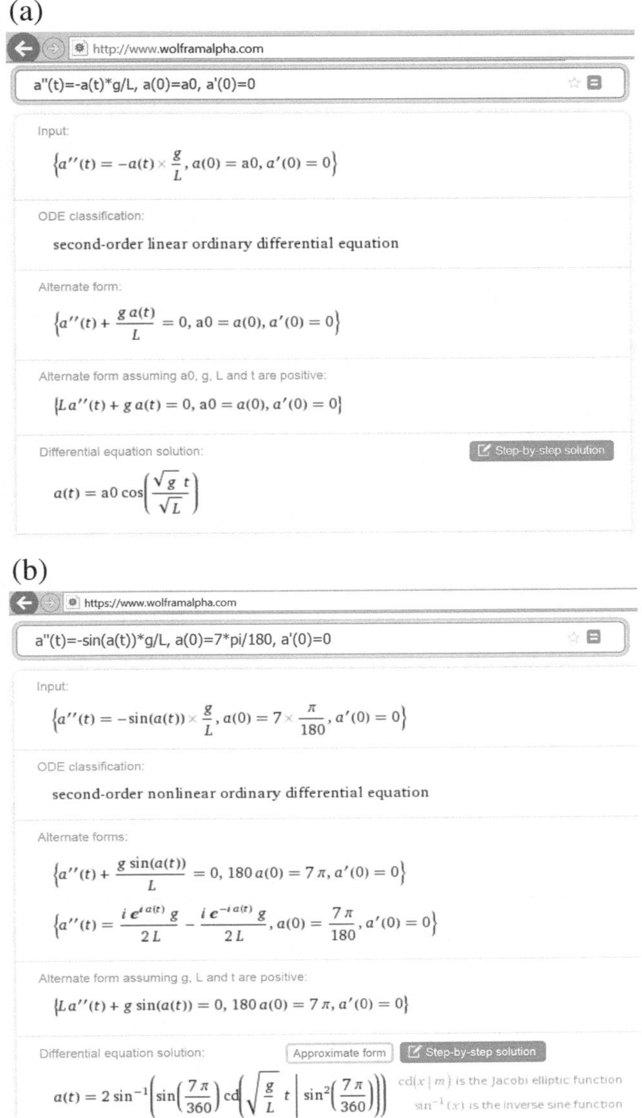

FIGURE 9.21 The analytical solution of the pendulum differential equation: (a) mathematical, (b) physical.

The most famous of them was developed around 1900 by the German mathematicians K. Runge and M. Kutta. The Runge-Kutta method (Figure 9.22) calculates the value of the next point of the sought function with a certain "adjustment" at half the value of Δ. It is built into almost all computer computational programs, including Mathcad—see Figure 9.2 above. Later in the twentieth century even more complex algorithms for the numerical solution of an ODE were developed—the Runge-Kutta method with variable step Δ, the Adams method, the BDF method, the Radau method, etc. We "played" with these methods by solving the problem of the interception of the satellite—see Figure 9.9. The choice of the numerical method of solving an ODE for a given problem is a subtle science and a high art. In the *Odesolve* function, which we used for the calculations, an interpolation algorithm is added in order to produce a function (namely a set of functions), instead of a matrix of discrete values. Because of this, sometimes there is a misunderstanding—some believe that if this built-in function doesn't return a table, then an analytical solution of the problem is obtained. But is it worth discussing the non-solve block methods that return tables of values?

If, however, it is necessary to solve a boundary-value problem, rather than an initial conditions problem (a Cauchy problem), then the shooting method is used. The boundary-value problem applied to the oscillation of the pendulum can be formulated as follows: the values of the angle of deviation of the pendulum cord from the vertical at two points at a given time are known, it is necessary to determine the position of the cord at other instants of time. The shooting method works as follows. The missing initial condition (the angular velocity of the cord for instance, having in mind the problem of the pendulum) is provided. The problem then reduces* to the known problem with the initial condition, and by solving it one could find the sought value at the other end of the integration segment. If it does not coincide with the value originally specified ("miss"), then it leads to a "shoot adjustment"—changing the desired parameter value at the starting point: the speed of the pendulum cord or the angle of the gun's inclination. This operation is repeated until the problem is solved with the specified accuracy. It is possible to measure the position of the satellite at two points in time and calculate the trajectory (orbit-boundary-value problem) by the previously described method. It is possible to do this also by measuring not only the position of the satellite, but also its speed at some point in time (the problem with the initial conditions). If there is no influence of third cosmic bodies, then it is possible to measure the position of the satellite at five points of the motion (at different points in time) and to solve the system of non-linear but

* How to boil water in a kettle? Answer. You need to pour water into it and put it on the fire. A new challenge: the kettle already has water. The answer is mathematics. Water should be poured. After this, the problem is reduced to the previous one. Jokes are jokes, but many people pour water out of the kettle in the morning, believing that it has deteriorated overnight, and pour it into the kettle with a new tap or filter.

$$rkfixed(y_b, x_b, x_e, n, f) := \left\| \begin{array}{l} \text{"ODE Solution method Runge-Kutta"} \\ \left[X_1 \leftarrow x_b \;\; Y_1 \leftarrow y_b \;\; \Delta \leftarrow \frac{x_e - x_b}{n} \right] \\ \text{for } i \in 2..n \\ \quad \left\| \begin{array}{l} k_1 \leftarrow f\left(X_{i-1}, Y_{i-1}\right) \\ k_2 \leftarrow f\left(X_{i-1} + \frac{\Delta}{2}, Y_{i-1} + \frac{k_1}{2} \cdot \Delta\right) \\ k_3 \leftarrow f\left(X_{i-1} + \frac{\Delta}{2}, Y_{i-1} + \frac{k_2}{2} \cdot \Delta\right) \\ k_4 \leftarrow f\left(X_{i-1} + \Delta, Y_{i-1} + k_3 \cdot \Delta\right) \\ Y_i \leftarrow Y_{i-1} + \frac{k_1 + 2 \cdot k_2 + 2 \cdot k_3 + k_4}{6} \cdot \Delta \\ X_i \leftarrow X_{i-1} + \Delta \end{array} \right. \\ \text{augment}(X, Y) \end{array} \right.$$

$L := 1\ \mathbf{m}$ $\quad \alpha_0 := 7°$ $\quad n := 100$ $\quad t_e := 3\ \mathbf{s}$

$i := 1..n \qquad t := 0\ \mathbf{s}, \frac{t_e}{300}..t_e \qquad n = 100$

$$f(t,x) := \begin{bmatrix} x_2 \\ -\sin\left(x_1\right) \cdot \frac{g}{L} \end{bmatrix} \qquad \begin{array}{l} \alpha'(t) = v(t) \\ v'(t) = -\sin(\alpha(t)) \cdot \frac{g}{L} \end{array} \qquad M := rkfixed\left(\begin{bmatrix} \alpha_0 \\ 0 \;\; \frac{\text{deg}}{\mathbf{s}} \end{bmatrix}, 0\ \mathbf{s}, t_e, n, f \right)$$

$\left(\left(M^{\langle 2 \rangle}\right)_{i,1}\right)_1$ **(deg)**

$\alpha_0 \cdot \cos\left(\frac{\sqrt{g}}{\sqrt{L}}\ t\right)$ **(deg)**

$\left(M^{\langle 1 \rangle}\right)_i$ **(s)** $\quad t$ **(s)**

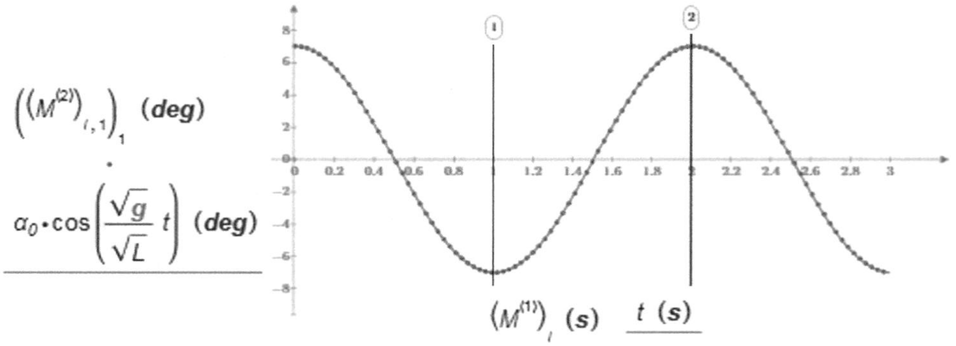

FIGURE 9.22 The Runge-Kutta method, applied to solve a system of two ordinary differential equations.

linear algebraic equations (see Figure 9.14(c)), meaning that the satellite moves strictly on an elliptical (circular), hyperbolic or parabolic (a rare case) trajectory. If there are more than three cosmic bodies, then there are no analytical solutions here.*

Mathematical modeling of more or less complex (real) physical objects often reduces to solving several thousand differential equations. And often it turns out that no developed and tested tools built into the mathematical packages Mathcad, Maple, Mathematica, ANSYS, COMSOL, etc., are suitable for their solution. Here it is necessary to make an individual design scheme with a lot of assumptions, simplifications, linearizations and other "tricks." And this can only be done by a specialist in this area of modeling (space flights, energy, radio engineering, etc.), who knows what is important and what is secondary. Coupled with a good mathematician, of course.

* Mathematicians and astronomers searched them for centuries, until they were convinced that they were not. Although—how to be sure. Some problems of heat conduction, which reduce to the solution of differential equations, did not have analytic solutions. However then the mathematicians came up with new functions (Bessel functions) and began to believe that there were analytical solutions, though they contained these very functions of Bessel. They even created special tables, similar to the sine or logarithm tables. In principle, for our problem of the motion of a satellite and two planets (Figure 9.7), we can also come up with and tabulate the special functions that form the analytical solution of this problem.

In Mathcad (unlike Maple and Mathematica), there is no analytical (symbolic) solution for differential equations. Because of this, we used third-party tools—see, for example, Figure 9.21.

In the impossibility of a direct analytical solution, Mathcad, or rather, its developers, is often blamed for the ODE function. However, here it is necessary to understand that a complicated differential equation is not solved symbolically, and if it is solved, the solution turns out to be very puzzling and inconvenient for further use. This can be seen for the original equation written in the window in Figure 9.21a, with the sine term replaced (see Figure 9.21b), increasing the swing of the pendulum.

The animation (see Figure 9.23) shows that "Euler slightly deviates on the corners," while Runge paired with Kutta "clearly moves his step along the line of the analytical solution". Similar behavior is shown in Figures 9.20 and 9.22.

The PTC Community forum discusses the process of creating the programs shown in Figures 9.19 and 9.22. There it is possible to find a program and an example of animation of the numerical solution of an ODE system by the Runge-Kutta method with a variable step (Adaptive—see Figure 9.2), where the step changes

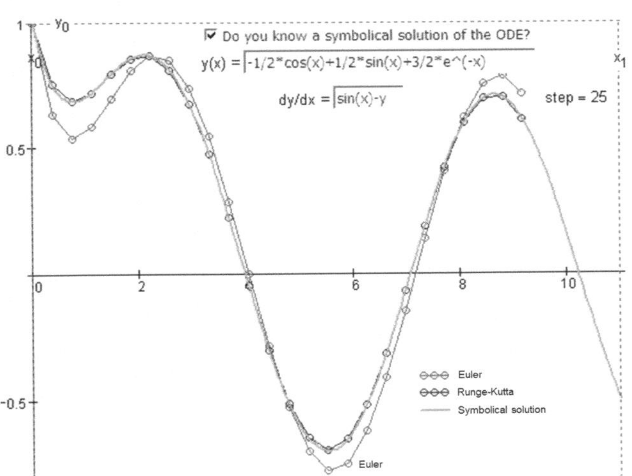

FIGURE 9.23 A frame of animation comparing the Euler method and the Runge-Kutta method.

depending on the magnitude of the derivatives—the right parts of the ODE.

Concluding this Study, we use one more quote from A.S. Pushkin: "About how many wonderful discoveries the enlightenment prepares for us ...",* replacing the word "enlightenment" with the word IT in it.

ASSIGNMENTS FOR READERS

1. To invent and implement on the computer new interesting "dances" of stars (celestial bodies).
2. Create an animation that includes the following phases of the flight to the moon: starting from the Earth's surface, moving to Earth's orbit, starting from the Earth's orbit to the Moon, moving to the Moon with the engines turned off, moving to the Moon's orbit, landing and then returning to Earth.
3. After setting the geometry of the roof-like tunnel, calculate the point at which a person sitting in the car or walking "will see the light at the end of the tunnel".
4. Create an animation of a train along bends, similar to the one shown in Figure 9.32, however not along a curve of the normal distribution, but along segments of curves whose curvature varies smoothly from 0 to `1/R` (the first curve), then with a constant curvature `1/R` (the second curve) and then the rails whose curvature changes smoothly from `1/R` to 0 (the third curve).
5. Usually, calculations similar to the one shown in Figure 9.34, do not take into account the geographical latitude of rocket launch sites. However, it is known that cosmodromes need to be placed closer to the equator. Show it by calculation!

* This is an extract (October–November 1829) from Alexander Pushkin, *Lyric Poetry*, 1826–1836.

Recursion Or

A Knight's Tour

Hermann went out of his mind and is now confined in room Number 17 of the Obukhov Hospital. He never answers any questions, but he constantly mutters with unusual rapidity: "Three, seven, ace!" "Three, seven, queen!"

A.S. PUSHKIN, "*THE QUEEN OF SPADES*"*

ABSTRACT

The Study explains what recursion is and how it is possible and should be applied to solve various mathematical problems on a computer. Fractals, simple and optimized, are also presented. The algorithm of the game Hanoi Tower and the construction and animation of fractals are considered.

Mathematics: Factorial, parallelepiped volume, local extremum of a function of one variable, differential equation, level lines, saddle point, series, local extremum of a function of two variables, fractal, Fibonacci numbers, Ackermann numbers.

Physics: Chess

IT: Recursive algorithm, computer analysis of functional dependencies.

Art: Alexander Pushkin, *The Queen of Spades.*

Study website: https://community.ptc.com/t5/PTC-Mathcad/Chapter-10/m-p/607926.

Almost everyone knows what the factorial of the number n is. It is the product of all natural numbers from 1 to n included. For the factorial, the notation $n!$ is applied, which is also used in Mathcad. So, for example, $5! = 1 \bullet 2 \bullet 3 \bullet 4 \bullet 5 = 120$. Such a classical definition of the factorial does not encompass the zero. For the convenience of writing general formulas in which 0! usually is a factor, mathematicians agreed to count $0! = 1$.

Remarkably, it is easy to count the factorial of zero (not to determine, but to count) by recursion (from Latin *recursio*, return). Figure 10.1 shows a program function in Mathcad Prime to calculate factorials using recursion.

A recursive function is a function that calls itself when it is evaluated. It usually depends on a parameter, whose change produces a recursive descent or ascent according a so-called recurrence relation.†

This is how a humorous guide to informatics defines recursion: "Recursion—see recursion." The easiest way to illustrate recursion geometrically is by repeating some geometric elements (possibly with a change in scale) many times (even an infinite number of times) in order to get self-similar geometric objects. A classic example of infinite recursion is a pair of images in two mirrors facing

* http://www.online-literature.com/alexander-pushkin/2800/, translated by H. Twitchell.

† Recurrence relation: a mathematical relationship expressing f_n as some combination of f_i with $i<n$. When formulated as an equation to be solved, recurrence relations are known as recurrence equations, or sometimes difference equations. http://mathworld.wolfram.com/RecurrenceRelation.html.

$$
Factorial(n) := \left\| \begin{array}{l} \text{if } n=5 \\ \quad \| \text{return } 120 \\ \text{if } n>5 \\ \quad \| \text{return } (Factorial(n-1) \cdot n) \\ \text{if } n<5 \\ \quad \left\| \text{return } \dfrac{Factorial(n+1)}{n+1} \right. \end{array} \right.
$$

$$Factorial(4)=24 \qquad Factorial(6)=720$$

$$Factorial(5)=120 \qquad Factorial(0)=1$$

FIGURE 10.1 The recursive function "Factorial".

each other: they form two corridors of diminishing and fading mirror reflections. Such mirrors can be seen in a train compartment. By recursion, it is very easy to calculate the dimension of a geometric object mathematically. Mathematicians have noticed that some recursive objects have a fractional (fractal) dimension. Such objects are called fractals. Let's recall that a point has dimension 0; lines (curves and straight), 1 ("length"); plane and surface, 2 ("area"); three-dimensional space and bodies, 3 ("volume"). To imagine a geometric object with fractional dimension, the reader could search on the Internet for any website with images of fractals or look at Figure 10.5.

If a computational algorithm is based on the principle of dividing the main problem into subtasks, and each of them repeats the main one, then we get a recursive algorithm. Such an algorithm solves a problem by reducing it to solving the same task one or more times, though in a simpler version. An example of such algorithm is shown in Figure 10.1. The factorial calculation begins with the value $n=5$, and then either a recursive ascent or a recursive descent occurs. Note that we could start with any natural number n whose factorial is known to us. For example, take $10!=3628800$ and $n=10$ or $n=1$ and $n!=1$ and use only a recursive lift (ascent).

A verbal description of the used *Factorial* function used is the following. If n is equal to 5, then $n!$ is equal to 120 (see the first two lines of the program in Figure 10.1). If n is greater than 5, the factorial of such an integer is $(n-1)! \bullet n$. Example $6!=5! \bullet 6=720$. If n is less than 5, then $n!$ is equal to $(n+1)!/(n+1)$. Examples: $4!=5!/5=24$, $3!=4!/4=6$, $2!=3!/3=2$, $1!=2!/2=1$ and, finally, $0!=1!/1=1$, which was to be proved.

On the Internet there is a very interesting website, http://oeis.org, "Open Encyclopedia of Integer Sequences". If the reader enters the sequence of numbers 1, 1, 2, 6, 24, 120, 720 in the window of this website and clicks the *Search* button, the website answers that there is a "factorial" hidden here, and some interesting information is provided about this mathematical operator.

It is very amusing to enter various sequences of integers on this website and to find out what regularities they obey. So, for example, if the reader enters the numbers 1, 2, 3, 5, 8, 13, 21, then the answer is that they are Fibonacci numbers—a number sequence, where each successive number is the sum of the two previous numbers (see Figure 10.2).

It is interesting to find the regularity of this or that numerical sequence autonomously or with the help of the website shown in Figure 10.2. However, it is even more interesting to fool this website by entering a numerical sequence with a known pattern and seeing that the website could not find the answer. The reader can register on this website and enter new information into it. Let's try to

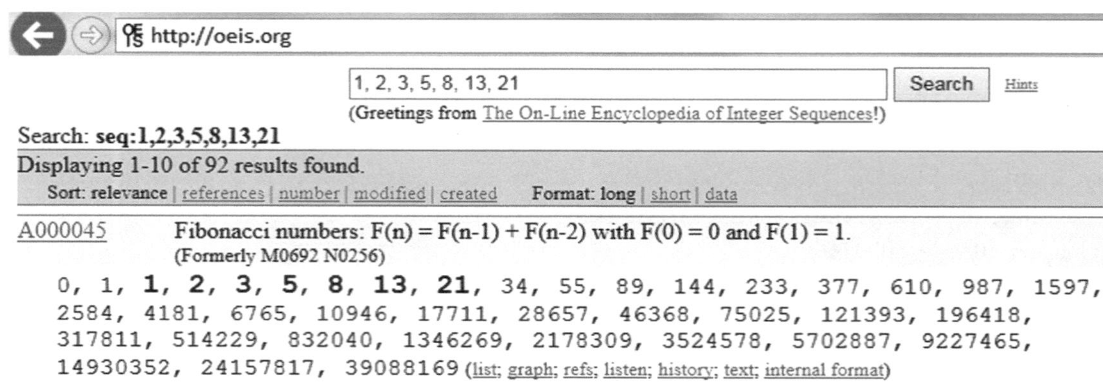

FIGURE 10.2 Website for determining regularities in integer sequences.

do this. Many people know what Fibonacci numbers are (see above). However, very few people know what *beautiful* Fibonacci numbers are. The "beautiful" numbers are 1, 3 and 7. If the reader adds them, it is possible to get another "nice" number, 11. Let's consider the epigraph to this Study: "Three, seven, ace (11)". Figure 10.3 shows a recursive function that returns the beautiful Fibonacci numbers not only for positive but also for negative arguments.* In this more complete sequence of numbers there is also a five.

So, if the reader enters numbers 1, 3, 7, 11, 21, 39 in the window of the mentioned website (Figure 10.2), he/she will "figure it out", noting that each regular number is the sum of three (and not two—see Figure 10.2) previous ones. However, if the sequence is expanded to −1, 3, 1, 3, 7, 11, 21, 39 (by introducing −1 and 3 at the beginning of the sequence), it will result in a "disservice" on this website, and an answer will not follow

We suggest readers "play" with the website *oeis.org*, introducing numeric sequences created by programming (see Figures 10.1 and 10.3) or based on their own ingenuity. The *oeis.org* website, for example, knows that 1, 2, 5, 10, 20, 50 and 100 are denominations of US banknotes, and 5, 10, 20, 50, 100, 200, 500 are euro banknotes, and it predicts future banknotes in case of hyperinflation. However, this website does not know that 1, 2, 3, 5, 10, 25, 50, 100 are denominations of banknotes of the former USSR.† It is also unknown to the website that 7, 9, 11, 13, 15, 19, 22, 28, 48 were the cost (in kopecks) of ice cream, also in the former Soviet Union. In the program shown in Figure 10.3, it is possible to change the reference numbers ("three, seven, ace" to "three, seven, queen" (another three, see the epigraph), in order to get a new sequence of integers and to pass it through the sieve of the website *oeis.org*.

However, back to our main topic, recursion.

The calculation of the factorial (Figure 10.1) and of the Fibonacci numbers (ordinary or beautiful) can be executed without recursion: knowing the previous numbers, it is possible to calculate the next one, and that next then becomes one of the previous ones, and

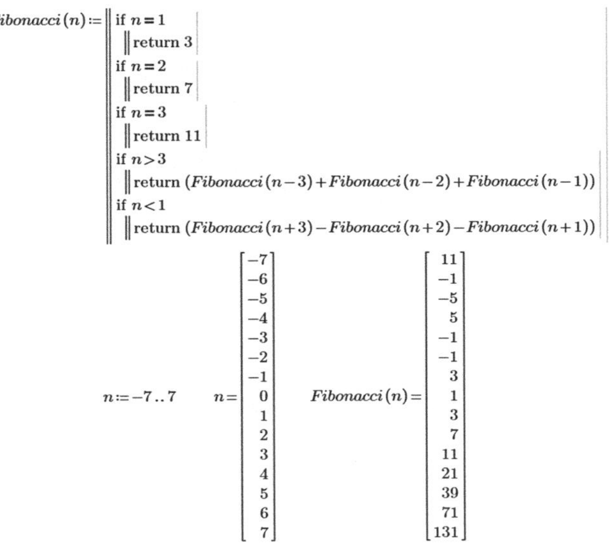

FIGURE 10.3 Beautiful Fibonacci numbers.

the operation continues until the required number is reached. This calculation (recurrence instead of recursion—the examples below in Figure 10.10 and in Figure 3.1 in Study 3) is faster and does not require significant computer memory resources. However, recursion makes it much easier to write a code. A classic example is a code that solves the puzzle of the "Tower of Hanoi".

The essence of the puzzle. There are three rods, the first of them (in our program in Figure 10.4) is marked with A, and the disks are settled in the manner of a children's pyramid: the largest disk is at the bottom while the smallest is at the top. It is proposed to move all these disks to rod C, taking them one by one. For temporary storage, it is possible to use the third rod B. It is not possible to put a large disk on a smaller one. The program in Figure 10.4 is enough to transfer all the disks in an n pyramid. After the launch, the program will return to the order of the disk's relocation:

n = 2: AB, AC and BC (three moves)

n = 3: AC, AB, CB, AC, BA, BC and, finally, AC (seven moves)

n = 4: … (15 moves), etc.

In the general case the number of permutations p is 2^n-1.‡ The problem with n disks is easily reduced to

* Once upon a time, the author of this book was fond of formalism in programming and wrote a "beautiful" program in the Pascal language for calculating the Fibonacci numbers, where the number of characters on each program line was equal to the next Fibonacci number.

† There are 7 banknotes everywhere. It is believed that a stable banking system should be based on a "beautiful" number of banknotes. In the USSR, by the way, before the monetary reform of 1961 and coins there were seven—1, 2, 3, 5, 10, 15 and 20 kopecks. The optimal number of banknotes and coins, as well as the number of weights for weighing are interesting mathematical problems.

‡ This is named nth Mersenne number, as it is possible to check in the mentioned oeis.org website.

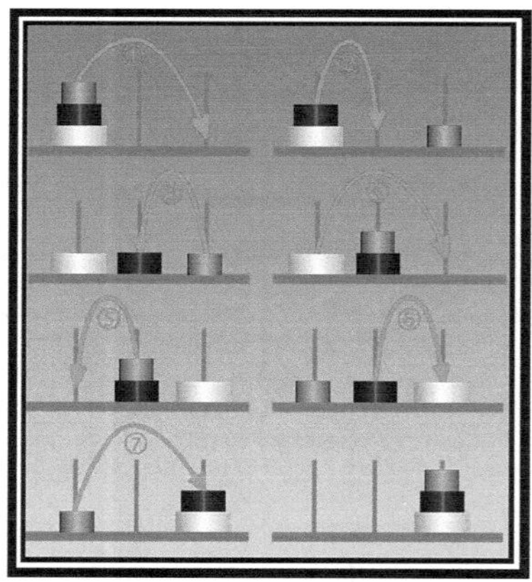

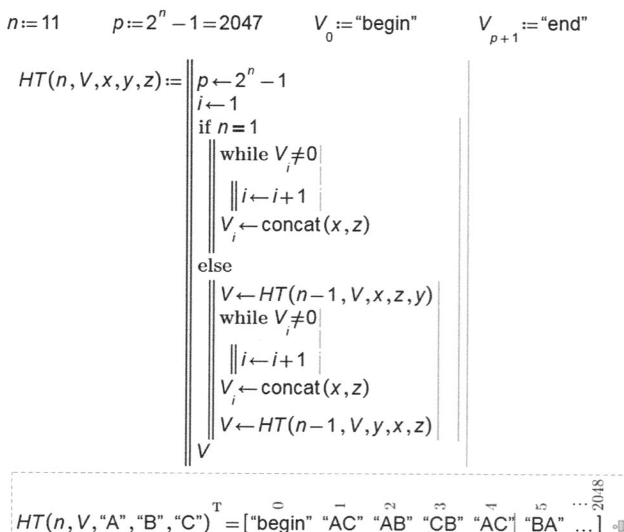

FIGURE 10.4 Hanoi Tower with recursion.

the problem with n–1 disks, and the problem with n–1 disks reduces to the problem with n–2 disks, etc., to the problem with two disks, which is solved simply: we transfer the disk from the rod A to the rod C (see in Figure 10.4 the fragment of the program If n – 1) i. Hence the program in Figure 10.4 makes use of recursion.

According to an ancient legend, for several millennia Tibetan monks shifted 64 gold disks, passing them through diamond rods. When the puzzle is solved and all disks are on rod C, the end of the world will come. It only saves us that, with 64 disks, in order to solve the puzzle it requires over a trillion years, on the basis of spending a single second for each turn. This estimate is without taking into account the false moves and the time necessary for manual or computer calculation of the order of permutation of the disks. On the Internet, by the way, there are virtual games "Hanoi Tower". The reader could open it and play with 5–10 disks.

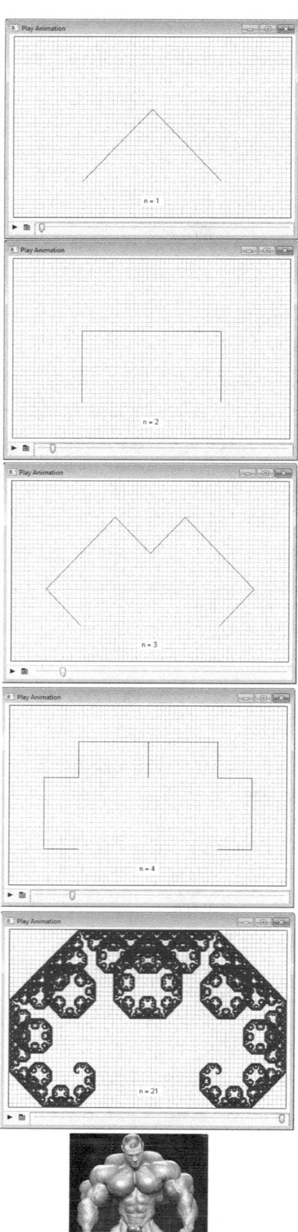

FIGURE 10.5 Fractal bodybuilding.

Now about the previously mentioned fractal. Fractal is a kind of visible picture of recursion. The development of a fractal from a simple form to a complex one is well illustrated by means of Mathcad animation. So Figure 10.5 shows five frames of a fractal development, called the Levy fractal by the name of the researcher who invented it: a straight line segment i ($n = 0$, the zero frame of the animation, not shown in Figure 10.5), which

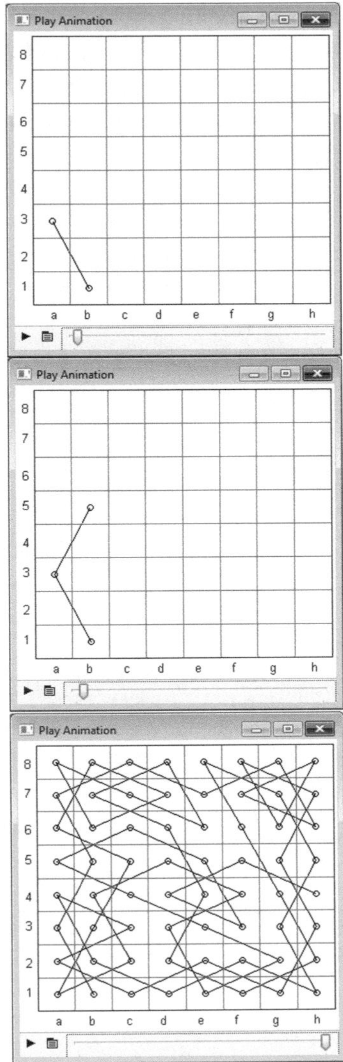

FIGURE 10.6 Knight on the chessboard.

is divided in half at a right angle ($n = 1$, the first frame of the animation). Further, this operation is repeated again and again with decreasing lengths of the segments until a certain "curly" figure appears ($n = 21$), reminiscent of a bodybuilder, as it used to be called, the modern bodybuilder (the second name of this fractal).

In Mathcad recursion helps to solve the old problem of moving a knight on a chessboard such that the knight visits every square only once. The problem is similar to the traveling salesman problem, which we will consider in Study 15, though in the chess problem it is not required that the knight return to its original position.

Chess problems played an important role in the fate of the author of this book. Back in 1983, the author was on a scientific internship in Germany, where he bought a personal computer. The seller, as a "divertissement"* (and there are a lot of such additional "divertissements" on the website of this book), showed the author a cassette with a chess program. The author immediately tested it on a chess problem from the Russian magazine *Science and Life*, "Mate in three moves." The computer solved the problem, not in three, but in two moves. This fact surprised the author, who then wrote a letter to the editors of the journal. It turned out that it was an old chess problem with a restriction: "Checkmate in three moves, but moves on a particular region of the chessboard were forbidden." However, this condition was omitted from the journal, and this was the reason for the above occurrence.

As a consequence of this chess-computer history, the editors of the journal suggested that the author supervise the journal sections "Man and Computer" and "Tips for a novice programmer", which the author led together with Yuri Pukhnachev from 1986 to 1992.

The result of this work was the publication of 50 articles in various computer magazines and the following books: *24 Studies in BASIC* (1990), *128 Tips for a Novice Programmer* (1992), *Lessons for IBM PC Users* (1992), *Mathcad for Students and Engineers* (several editions from 1996 to 2007), *Tips for Users of Mathcad* (2001), etc. Four books were translated into foreign languages (English, German and Ukrainian). The author has become a "computer writer".

The author is by education and practice a thermal power engineer who teaches students and creates computer programs for the training of heat and power engineers. Once, during a public talk at the academic council of his university, he spoke about his innovative software development and noted that the university did not teach any course on informatics for the students. The university rector responded that if the author is so clever then he could teach this informatics course himself, basing the course not on programming languages, but on modern mathematical programs. The author agreed and for almost 30 years has been teaching this course at the Moscow Power Engineering Institute (MPEI, National Research University).

During this time, the author has accumulated a lot of interesting problems with *interdisciplinary connections*,

* *Divertissement*, in ballet: a short dance that displays a dancer's technical skill without advancing the plot or character development.

where, in the same class, problems in mathematics, physics, chemistry, mechanics, literature, visual arts, etc., are solved with a computer. This educational method later became known as STEM, Science, Technology, Engineering and Mathematics. A Russian publishing house published the book *Physical and Mathematical Studies with Mathcad and the Internet* (2016 and 2018), of which this is an extended English translation.

Yes–No in Mathcad Or

Boole's Origami

ABSTRACT

In Study 11, you can learn about Boolean and fuzzy logic, binary and infinite-value logic, about the features of its application in programming. We consider new three-dimensional surfaces that imitate the laws of clear logic. A machine for fuzzy jury voting is described.

Mathematics: Boolean function, Boolean algebra, Boolean function extremum, fuzzy logic.

Physics: Switches of electrical circuits, voting machine.

IT: Boolean operators, tree operator, vectorization of computer calculations.

Art: Origami.

Study website: https://community.ptc.com/t5/PTC-Mathcad/Study-11/m-p/617467.

In Mathcad 15 there are operators implemented as prefix and infix operators, see the right-hand side of Figure 11.1, which returns only two values: 1 (not zero) or 0 (zero): Yes–No, True–False. These operators are collected on the *Boolean* toolbar in Mathcad 15 (see the left-hand side of Figure 11.1) and Comparison in Mathcad Prime (see Figure 11.2).

Sometimes Boolean* operators (or, otherwise, Boolean functions) are called *binary* or *logical.* The operands of these operators (the arguments of these functions) in Mathcad can be real numbers, however a number different from zero is perceived as true while zero as false.

In some programming languages (Pascal, for example), you can start working with Boolean and other types of variables only after they are listed in the variable type declaration. This is done to save computer's memory: much less space is allocated for storing the Boolean variable than for storing an integer, and even more so material. In the present day the memory of a computer, as a rule, is not a limiting factor in solving problems, and so such a pre-declaration of variable types can be abandoned. So, at least, it is in Mathcad.

We will talk about Boolean functions, starting from the problems that arise when working with them in Mathcad or in other mathematical programs and programming languages, and, in general, when using digital computer technology, which is based on the binary "atom": the basic element of memory, located in one of two states (charged-discharged, magnetized-demagnetized, etc.). From "atoms" (bits) are made "molecules" (bytes), which in turn form new "connections"—variables, arrays of variables (see Study 2), etc.—everything that programmers use. Arithmetic and other actions on numbers, variables and arrays of variables are essentially nothing but a variety of Boolean (bit-wise) operations.

BOOLEAN FUNCTIONS OF ONE ARGUMENT

There are *four* basic Boolean functions of one argument: see Table 11.1, but in practice only one, f_1, which is called

* The word "Boolean" comes from the name of George Boole—English mathematician and logic professor of mathematics at the Royal College of Cork (now Cork University College). G. Boole is the creator of symbolic logic.

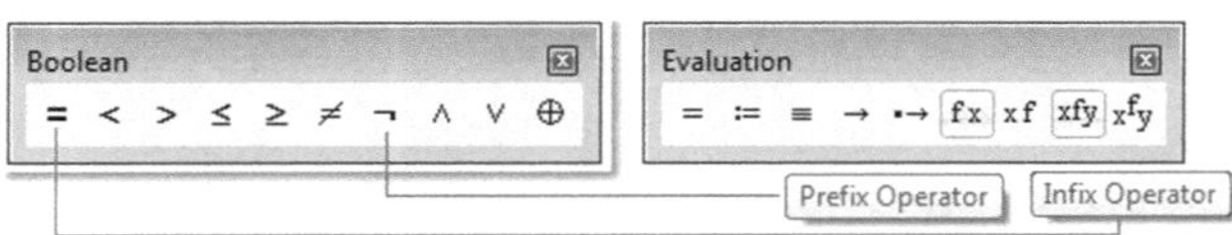

FIGURE 11.1 Panel of Boolean operators in Mathcad 15 (left).

FIGURE 11.2 Panel of Boolean operators in Mathcad Prime.

negation (*inversion*), works. The other three functions return either their argument (function f_2), or the constants 1 (f_3) and 0 (f_4).

The first row of the notation graphics in Table 11.1 (and Table 11.2), shows the Mathcad operators that are used to implement the Boolean function. The symbol (prefix* operator of negation or inversion) is present on the panels shown in Figures 11.1 and 11.2.

From these functions it is possible to compose more complex functions of one argument, but they, as a rule, are not required. Let's see how, for example, the function $f_2\left(f_4\left(f_1(a)\right)\right)$ operates. If $a=1$, then $f_1(a)=0$, $f_4(f_1(a))=0$, $f_2\left(f_4\left(f_1(a)\right)\right)=0$. If $a=0$, then $f_1(a)=1$, $f_4(f_1(a))=0$, $f_2\left(f_4\left(f_1(a)\right)\right)=0$. The table of values tells us that $f_2\left(f_4\left(f_1(a)\right)\right)=f_4(a)$ QUOTE f2f4f1a=f4a , so it is meaningful to replace a simple function with a basic Boolean function.

BOOLEAN FUNCTIONS OF TWO ARGUMENTS

The basic Boolean functions of two arguments are sixteen: see Table 11.2.

Table 11.2 is provisionally divided into two halves—the "nominal" (f_1—f_8) and the nameless (f_9—f_{16}). Here are the names of the first eight functions. Rather, seven (another "magnificent seven", see Study 4): two functions

* A *prefix* by definition means that the operator symbol is placed *before* the operand. In Mathcad, there are *postfix* built-in operators also, the factorial operator, for example, (5!), where the operator symbol (!) is placed after the operand (5). The *infix* operator has two arguments, and its symbol is between them (an example is an addition operator with two operand terms). In Mathcad 15 there is also a *tree-like* user operator, see Figures 11.7 and 11.12.

TABLE 11.1 Boolean Functions of One Argument

a	f_1	f_2	f_3	f_4
0	1	0	1	0
1	0	1	1	0
Notation	¬a	a	1	0
	Not(a)			
	!a			
	¬a			
	ā			

(f_5 and f_6) are called the same—implication or logical consequence.

- f_1—conjunction (logical multiplication);
- f_2—disjunction (logical addition);
- f_3—equivalence (equivalence, identity);
- f_4—(non-equivalence, separating disjunction, sum modulo 2);
- f_5 and f_6—implication (logical consequence): f_5—implication from a to b; f_6—implication from b to a;
- f_7—Pierce function (Webb function, the Dagger function, the anti-disjunction—we'll talk about it in detail: see Figures 11.7 and 11.8);
- f_8—function (prime) of Scheffer (anti-conjunction);

The other eight functions of Table 11.2 (f_9—f_{16}, as well as the last three functions of Table 11.1), we repeat, have neither names nor practical applications.† These are either constants (f_{15} and f_{16}), or functions of only one argument (f_{11}—f_{14}). The name, and even then conditionally, can only be given to the functions f_9 and f_{10}: the inversion (negation) of the implication. But these two functions, as a rule, apply only to real arguments: the example $\pi > e = 1$ (yes, the numerical value of the ratio of the length of the circle to its diameter is greater than the numerical value of the base of the natural logarithm).

Figure 11.3 shows the calling operators in Mathcad 15 of some basic Boolean functions of two arguments from Table 11.2. These operators are implemented in the form

† In Mathcad, there is one more operator that returns zero or one: It is , the set *membership operator*. This infix operator determines whether its first operand belongs to one of the possible numeric sets (the second operand): C is a complex number, Q is a rational number (this operator belongs to the symbolic math), R is a real number, and Z is an integer (see Figure 12.1 in Study 12).

TABLE 11.2 Boolean Functions of Two Arguments

<table>
<tr><th>ab</th><th>f1</th><th>f2</th><th>f3</th><th>f4</th><th>f5</th><th>f6</th><th>f7</th><th>f8</th><th>f9</th><th>f10</th><th>f11</th><th>f12</th><th>f13</th><th>f14</th><th>f15</th><th>f16</th></tr>
<tr><td>00</td><td>0</td><td>0</td><td>1</td><td>0</td><td>1</td><td>1</td><td>1</td><td>1</td><td>0</td><td>0</td><td>1</td><td>1</td><td>0</td><td>0</td><td>1</td><td>0</td></tr>
<tr><td>01</td><td>0</td><td>1</td><td>0</td><td>1</td><td>0</td><td>1</td><td>0</td><td>1</td><td>0</td><td>1</td><td>1</td><td>0</td><td>0</td><td>1</td><td>1</td><td>0</td></tr>
<tr><td>10</td><td>0</td><td>1</td><td>0</td><td>1</td><td>1</td><td>0</td><td>0</td><td>1</td><td>1</td><td>0</td><td>0</td><td>1</td><td>1</td><td>0</td><td>1</td><td>0</td></tr>
<tr><td>11</td><td>1</td><td>1</td><td>1</td><td>0</td><td>1</td><td>1</td><td>0</td><td>0</td><td>0</td><td>0</td><td>0</td><td>0</td><td>1</td><td>1</td><td>1</td><td>0</td></tr>
<tr><td>Notation</td><td>^</td><td>˅</td><td>=</td><td>⊕ ≠</td><td>≥</td><td>≤</td><td></td><td></td><td>></td><td><</td><td>¬a</td><td>¬b</td><td>a</td><td>b</td><td>1</td><td>0</td></tr>
<tr><td></td><td>*</td><td>+</td><td>↔</td><td><></td><td>→</td><td>→</td><td>↓</td><td>|</td><td></td><td></td><td></td><td></td><td></td><td></td><td></td><td></td></tr>
<tr><td></td><td>×</td><td>o</td><td>≡</td><td>><</td><td></td><td></td><td>¬And</td><td>¬Or</td><td></td><td></td><td></td><td></td><td></td><td></td><td></td><td></td></tr>
<tr><td></td><td>·</td><td>Or</td><td></td><td>Xor</td><td>⇨</td><td>⇨</td><td></td><td></td><td></td><td></td><td></td><td></td><td></td><td></td><td></td><td></td></tr>
<tr><td></td><td>e</td><td>||</td><td>Eqv</td><td>!=</td><td>Imp</td><td>Imp</td><td></td><td></td><td></td><td></td><td></td><td></td><td></td><td></td><td></td><td></td></tr>
<tr><td></td><td>And</td><td>max</td><td>==</td><td></td><td></td><td></td><td></td><td></td><td></td><td></td><td></td><td></td><td></td><td></td><td></td><td></td></tr>
<tr><td></td><td>&</td><td></td><td></td><td></td><td></td><td></td><td></td><td></td><td></td><td></td><td></td><td></td><td></td><td></td><td></td><td></td></tr>
<tr><td></td><td>&&</td><td></td><td></td><td></td><td></td><td></td><td></td><td></td><td></td><td></td><td></td><td></td><td></td><td></td><td></td><td></td></tr>
<tr><td></td><td>min</td><td></td><td></td><td></td><td></td><td></td><td></td><td></td><td></td><td></td><td></td><td></td><td></td><td></td><td></td><td></td></tr>
</table>

of infix operators: operators, the symbols of which, as already noted, are between two operand arguments. The arrow over the vector operands in Figure 11.3 is a special vectorization operator, i.e. element-wise execution of the action on vectors.

For some operators shown in Figure 11.3, there can only be Boolean operands (0 or 1), and for others, also real ones: 0.1, 0.9, 2.5, etc. Figure 11.4 shows the operation of the operators Exclusive or* (*Xor*) and Not equal to real operands: not with 0 and 1, but with 0.1 and 0.9. These operators for Boolean operands (Figure 11.3) gave identical answers, while for real operands (Figure 11.4) the answers were different.

The answer of the first operator in Figure 11.4 (only zeros) is due to the fact that numbers other than zero are perceived as units (Truth) by Boolean functions of Mathcad. This is due to the fact that in earlier versions of Mathcad there were no *And* and *Or* operators, and Boolean operations were implemented through arithmetic operators of multiplication and addition, see Figure 11.5. If you add two units (the *Or* operator), you get a two (true), a number that is different from zero.

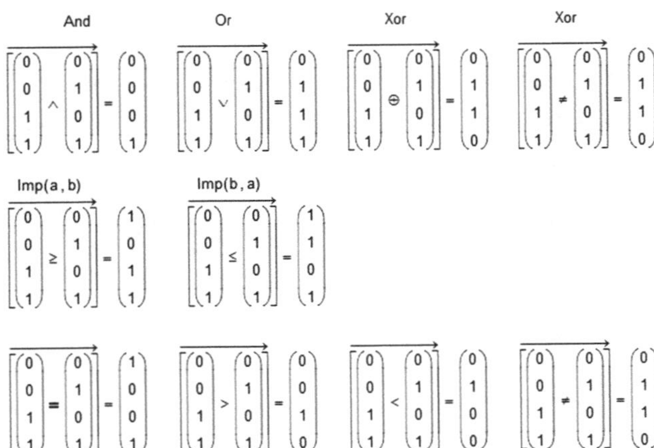

FIGURE 11.3 Operators for calling Boolean functions in Mathcad 15.

FIGURE 11.4 Operator "Exclusive" or "Not equal" with real arguments.

FIGURE 11.5 The implementation of logical multiplication and logical addition in earlier versions of Mathcad.

* By the way, our conversational "or" is often not simple "or", namely "exclusive or". When we ask, for example: "Did you go to the cinema or stay at home?", we put this question in the "exclusive or" because a person cannot be at the cinema and at home at the same time. And further, in written texts (including this book) a hybrid "*and/or*" ("*and*" and "*or*"), which editors have only recently ceased removing from manuscripts, is now used.

Boolean functions *And* and *Or* are often illustrated by an electrical circuit: the serial connection of switches (devices in two possible positions 0 and 1) is a conjunction (logical multiplication) and the parallel one is a disjunction (logical addition). Figure 11.6 shows a less trivial example, the electrical analog of the equivalence (*Eqv*): a circuit for connecting two switches, so that the light is independently switched on and extinguished from two different places. Such switches will be useful in a long corridor or on the stairs of a multi-story building, in a suburban area and in other places.

Tables 11.1 and 11.2 attempt to collect all the names of the base functions and the symbols of the basic operators used to implement the Boolean algebra. The list is, of course, incomplete. You can extend it with examples from other programming languages (Pascal, Fortran, etc.) and mathematical programs (Maple, MATLAB®, Mathematica, etc.).

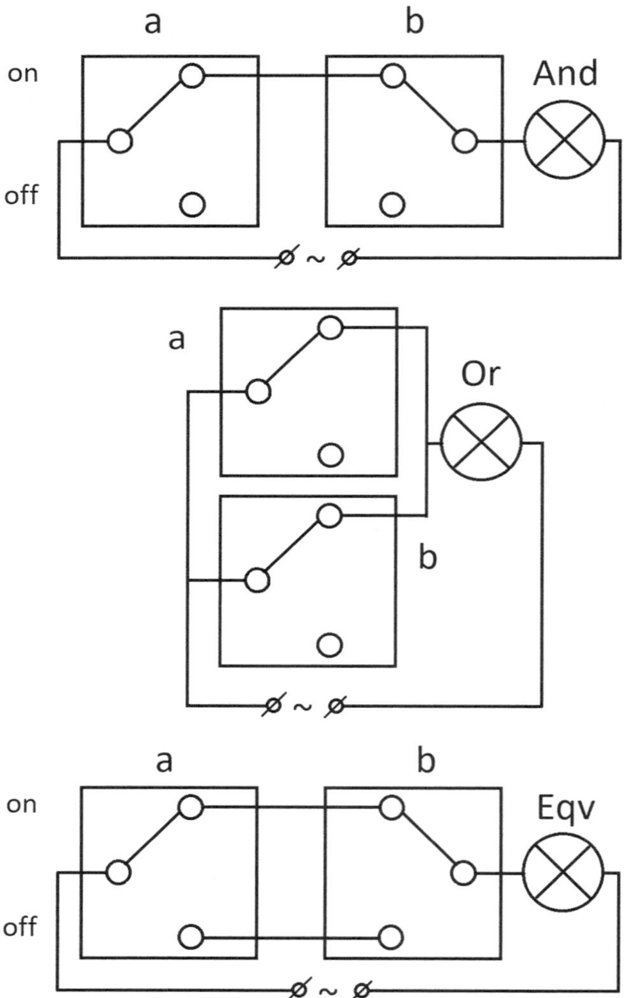

FIGURE 11.6 The electrical analogs of the three Boolean functions And, Or and Eqv.

It is possible to note some *redundancy* of the functions collected in Tables 11.1 and 11.2. In specific programming languages, there is a certain limited set of built-in Boolean functions and operators, with which you can make any other Boolean functions. To prove the sufficiency of such a set for produce any Boolean operators is a difficult mathematical task. If a finite number of functions are chosen, the actions that are subordinate to special rules are described with them and the possibility of realizing all other functions by means of these functions and actions is proved, then it is said to be an algebra.*

Here is a list of such functions and operators embedded in popular software:

- BASIC programming language: *Not*, *And*, Or, *Xor* and *Imp*
- C programming language: !, & (logical multiplication), && (bit-wise conjunction), !=, || and ==
- Mathematical program Mathcad: ¬, ∧, ∨ and ⊕ (see Figures 11.1 and 11.2).

The programmer can enter missing binary functions (operators) into the program itself through the mechanism of user functions, and we'll show this below.

The distinction of binary functions and operators into main (basic) and auxiliary ones appeared long before computers and was legitimized in the form of binary algebras (their base notation is shown in parentheses):

- Algebra of logic (¬, &, ∨, → and ↔)
- Boolean algebra (¬, & and ∨)
- Zhegalkin algebra (&, ∨ and ⊕)
- Peirce algebra (↓)
- Sheffer algebra (|).

* An algebra is a non-empty set of elements of any nature with associated algebraic operations satisfying given axioms. The subject of algebra studies are not the elements, but the algebraic operations. There are several types of algebras. One of the most important and most studied types are *groups*, i.e. algebras with one associative binary operation, including a unit and each element having an inverse element. The most important types of algebras with two binary operations are *rings and fields*. Operations in them are usually called addition and multiplication.

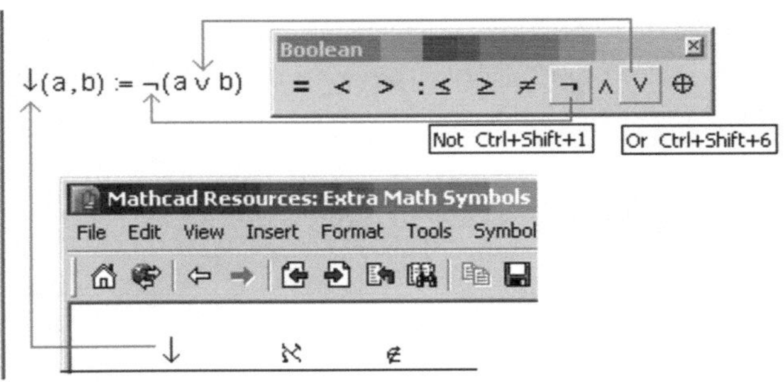

FIGURE 11.7 Creating the Boolean Peirce function in Mathcad using negation and logical addition.

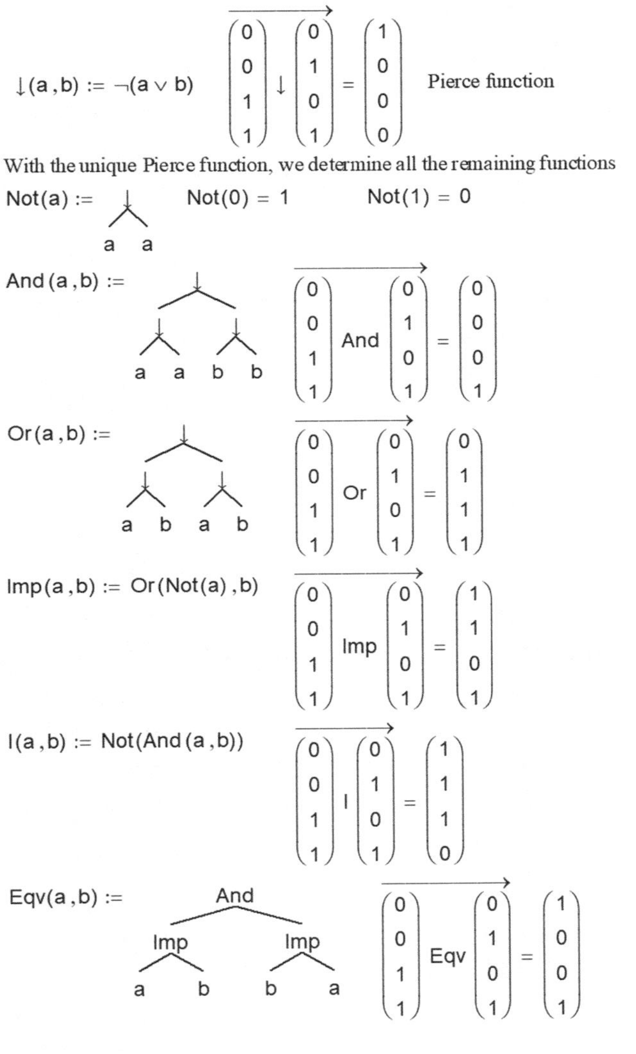

$$\downarrow(a,b) := \neg(a \vee b) \qquad \overrightarrow{\begin{pmatrix}0\\0\\1\\1\end{pmatrix} \downarrow \begin{pmatrix}0\\1\\0\\1\end{pmatrix}} = \begin{pmatrix}1\\0\\0\\0\end{pmatrix} \quad \text{Pierce function}$$

With the unique Pierce function, we determine all the remaining functions

Not(a) := ↓(a, a) Not(0) = 1 Not(1) = 0

And(a,b) := ↓(↓(a, a), ↓(b, b))

$$\overrightarrow{\begin{pmatrix}0\\0\\1\\1\end{pmatrix} \text{And} \begin{pmatrix}0\\1\\0\\1\end{pmatrix}} = \begin{pmatrix}0\\0\\0\\1\end{pmatrix}$$

Or(a,b) := ↓(↓(a, b), ↓(a, b))

$$\overrightarrow{\begin{pmatrix}0\\0\\1\\1\end{pmatrix} \text{Or} \begin{pmatrix}0\\1\\0\\1\end{pmatrix}} = \begin{pmatrix}0\\1\\1\\1\end{pmatrix}$$

Imp(a,b) := Or(Not(a),b)

$$\overrightarrow{\begin{pmatrix}0\\0\\1\\1\end{pmatrix} \text{Imp} \begin{pmatrix}0\\1\\0\\1\end{pmatrix}} = \begin{pmatrix}1\\1\\0\\1\end{pmatrix}$$

I(a,b) := Not(And(a,b))

$$\overrightarrow{\begin{pmatrix}0\\0\\1\\1\end{pmatrix} \text{I} \begin{pmatrix}0\\1\\0\\1\end{pmatrix}} = \begin{pmatrix}1\\1\\1\\0\end{pmatrix}$$

Eqv(a,b) := And(Imp(a, b), Imp(b, a))

$$\overrightarrow{\begin{pmatrix}0\\0\\1\\1\end{pmatrix} \text{Eqv} \begin{pmatrix}0\\1\\0\\1\end{pmatrix}} = \begin{pmatrix}1\\0\\0\\1\end{pmatrix}$$

FIGURE 11.8 Building the Boolean functions using the Pierce function.

The last two binary algebras are noteworthy because in their basic form there is only one binary function ↓ or |, from which it is possible to build all the remaining ones.

Figure 11.8 shows the Mathcad 15 sheet, where other Boolean functions are built using the Pierce function (↓, see Figure 11.7): one function of one argument (negation, inversion, *Not*) and five functions of two arguments: *And*, *Or*, *Imp*, Sheffer's stroke (|) and *Eqv*. The last three functions (*Imp*, Sheffer's stroke and *Eqv*) are defined using previously defined functions. This is done to make the drawing more compact, but you can refuse the mechanism of embedding user functions, for example *(Imp (a, b):=Or (Not (a), b))* and operate "for the purity of the experiment" only by Pierce's (stroke) function.

Once, high hopes were placed on the Peirce and Schaeffer algebras, in the sense of building a digital computer of the same type. Then this idea was abandoned for a number of reasons, the main one being that any computer already only consists of certain types of elements, from transistors integrated into integrated circuits (chips).

One can also note a certain *insufficiency* of the set of mathematical tools displayed in Tables 11.1 and 11.2. Take, for example, what is probably the most popular function of Boolean algebra—a conjunction (a consequence of its popularity and that it has the highest number of names and symbols for its designation, see column f_1 in Table 11.2). The truth-of-truth (the truth of conjunction) column in the truth table (as Tables 11.1 and 11.2 are called) should ideally be as shown in Table 11.3.

A dash beside zero in the first column of Table 11.3 means that if the first (*a*) argument is zero, then there is no need to check what the second argument (*b*) is equal to, and vice versa. This is provided by creating some ("not lazy") programming languages—the C language, for example. When programming the same in BASIC, the conditional jump on a conjunction can be written as follows:

If *a* And *b* Then (first method)

but it is better this way:

If *a* Then If *b* Then or If *b* Then If *a* Then (second method)

The second way of coding allows not only the speeding up of the calculations, but also the avoidance of some errors—the logical expression *b* can make sense only if

TABLE 11.3 Closed Conjunction

ab	*a And b* (f_1)
0 –	0
– 0	0
1 0	0
1 1	0

the alternative question *a* is answered positively. Here is a typical example of such a "programmer" situation:

If I > 0 Then If V(i) >V(i–1) Then ...

We can say that in the BASIC language there are two conjunctions: *And* and *Then If.*

In Tables 11.1 and 11.2, we repeat, we collected the Boolean functions of one (Table 11.1) and two (Table 11.2) arguments. But, returning to the conjunction, we can say that this function has not two but one-and-a-half arguments (see Table 11.3).

The same non-integer number of arguments can be observed for other Boolean functions.

It can be noted that the operators in Table 11.2 were originally designed to work not with binary but with real operands: ">", "<", "≥", "≤", "=" and "≠". But if we take into account the fact that zero and one are in the set of real numbers, then the inclusion of these operators in Table 11.2 is completely natural. In this series (">", "<", "≥", "≤", "=" and "≠") redundancy and insufficiency can also be noted. With redundancy, everything is more or less** clear (for example, "more", is the inversion from the operator "less than or equal to", etc.).

Insufficiency can be observed, for example, when working with real variables, since in this case instead of the "equal" operator, it is more appropriate to use the "approximately equal" operator, which is not among the operators in the built-in lists. One can also recall the existence of the concepts "much more" or "much less". These relational operators also return binary values, but in fact they have not two but three arguments (operands): a comparable pair of real numbers and some kind of contextual notion of what is "about" or "much".

Figure 11.9 shows the creation of the Boolean operator *Approximately equal* in Mathcad 15.

In Mathcad Prime, there is an option that makes it possible to turn the "equals" operator into an "approximately equal" operator, see Figure 11.10.

Boolean functions in Mathcad can be used not only for logical computations. They can also be used to solve equations.

Example 1. Solve the equation:

$$\sqrt[4]{\sin \pi x}-\sqrt{\cos 1.7x}=\sqrt[4]{16-x^2}-\sqrt{x^6+x^5-4x+9}.$$

For the solution, one of the solver operators described in Study 4 could be used. And it would be nice to first plot the graphs of both sides of the equation (note that it is sufficient to plot only the interval [–4, 4], the inequality $16 \geq x^2$) in order to locate the available roots. But you can do otherwise.

We introduce the following logical variables: $a := \sin \pi x \geq 0$; $b := \cos 1.7x \geq 0$; c: = $x^6+x^5-4x+9 \geq 0$; $p := \sqrt[4]{\sin \pi x}-\sqrt{\cos 1.7x} \geq 0$; q:= $\sqrt[4]{16-x^2}-\sqrt{x^6+x^5-4x+9}$. Clearly, the roots of the equation will be found approximately (with the exception of the obvious integer root $x=0$), so we must choose the desired step, and for each point of the above segment obtained with this step, we need to verify the truth of the statement.

$$f_{13}(a)\wedge f_{13}(b)\wedge f_{13}(c)$$
$$\wedge\left(f_{13}(p)\wedge f_{13}(q)\vee f_{11}(p)\wedge f_{11}(q)\right)\wedge(p\approx q).$$

The test statement can be simplified if we transform the original equation to the form $\sqrt[4]{\sin \pi x}+\sqrt[4]{16-x^2}=\sqrt{x^6+x^5-4x+9}+\sqrt{\cos 1.7x}$. Such a form will make it impossible to verify the truth of the statement that both sides of the equation at each admissible point have the same signs. And for r:= $\sqrt[4]{\sin \pi x}+\sqrt[4]{16-x^2}$ and t:= $\sqrt{x^6+x^5-4x+9}+\sqrt{\cos 1.7x}$, it will be required at each point to check the truth of the statement $f_{13}(a)\wedge f_{13}(b)\wedge f_{13}(c)\wedge(r\approx t)$.

Example 2. Solve the equation $\sqrt[4]{\sin \pi x}+\sqrt{\cos 1.7y}=\sqrt[4]{16-x^4}+\sqrt{9-y^2}$.

The solution of one equation with two variables can be any (including infinite) number of points in the plane. For example, the equation $\sqrt{-x^2-y^2}=0$ has a unique solution (0, 0), while

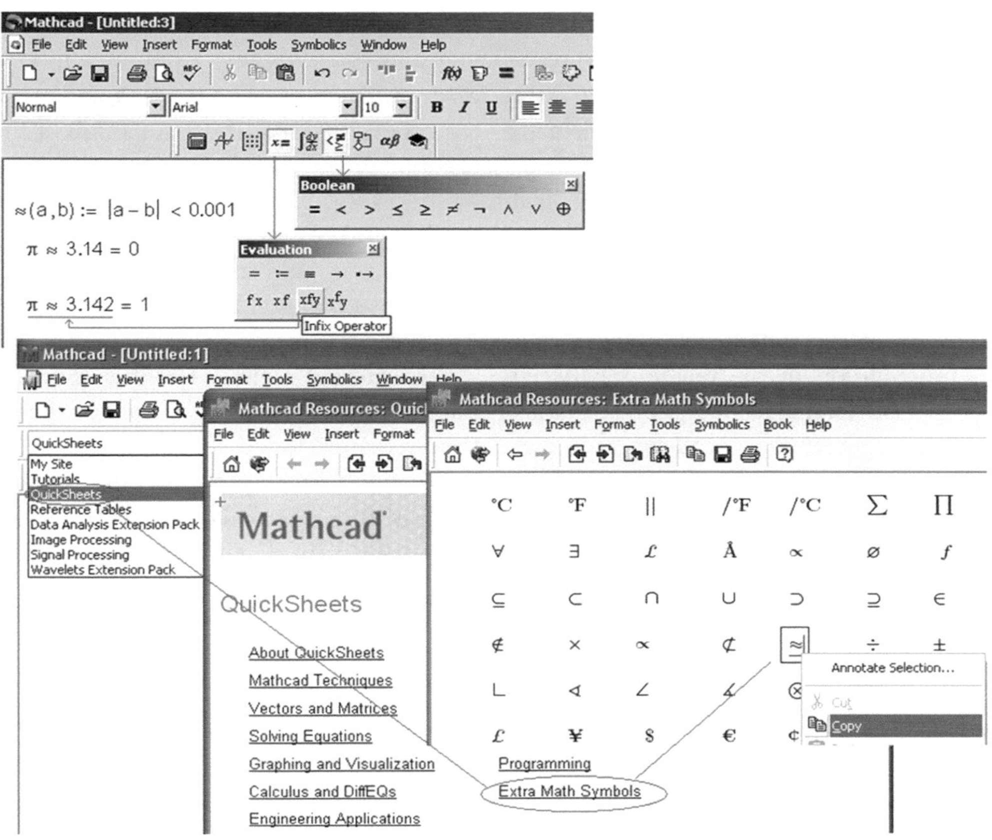

FIGURE 11.9 Creating a Boolean operator **Approximately equal to** in Mathcad 15.

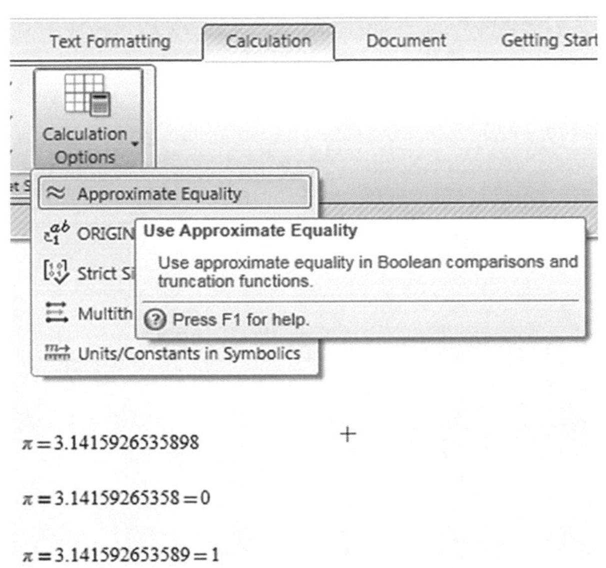

FIGURE 11.10 Approximately equal in Mathcad Prime.

the equation $x^2+y^2=1$ has infinitely many solutions—points forming a circle of unit radius and center at zero. The solutions of equations with two unknowns are discussed in Study 4. Here we again apply the logical variables: $a{:=}\ \sin\ \pi x \geq 0$; $b{:=}\ \cos\ 1.7y \geq 0$ QUOTE 4sinπx+cos1.7y ; $p{:=}\ \sqrt[4]{\sin \pi x}+\sqrt{\cos 1.7y}$; $q{:=}\ \sqrt[4]{16-x^4}+\sqrt{9-y^2}$, and we will verify the truth of the statement $f_{13}(a) \wedge f_{13}(b) \wedge (p \approx q)$ at the points of the rectangle $-2 \leq x \leq 2$, $-3 \leq y \leq 3$, defined with a certain step in each of the variables. All the points found in this way will give us an approximate solution of the original equation.

The classical Boolean algebra operates with argument-statements that can be true or false. For example, saying *a*: "The student passed the math exam", saying *b*: "The student passed the computer science exam", saying $a \wedge b$: "The student passed the exams in mathematics and computer science". Statements can be naturally related to sets.* For example, A is the set of students who passed the math exam; B is the set of students who passed the exam in computer science. The question can be posed as follows: "Is it true or false that this student belongs to the set $A \bullet B$?"

Such a natural correspondence of Boolean algebra and algebra of sets allows one to take advantage of this or that interpretation of the statements. Thus, a geometric visual representation of the relations between statements (sets) in a Boolean algebra is represented by circles† or other figures. Such figure is called a Venn diagram.‡ Element 1 of Boolean algebra is the universal set, or a true statement, it can be represented as a square and denoted by the letter Ω. To the element 0 corresponds an empty set, or a false statement, it can be denoted by $\emptyset$. Some non-empty set A, or some statement a, is represented as a circle or another figure located in a square. What lies outside the set A is the complement of A or the negation of the statement a, that is, $\neg A$. The sum (disjunction) of two sets A and B is represented as a union of the figures representing them and can be indicated by the symbol "+". The product (conjunction) of two sets A and B is represented as an intersection of the two figures representing them and can be designated by the symbol "•". The Venn diagram is used to illustrate the validity of axioms and Boolean algebra theorems, as well as to represent the relations between logical statements. As a special case of Euler circles, the Venn diagram depicts a finite Boolean algebra, clearly describing all its properties and the combinations of these properties. For $n=3$, this diagram is usually represented as three circles with centers at the vertices of an equilateral triangle and with the same radius, slightly smaller than the length of the side of the triangle. All this design is placed in a square.

With the help of Venn diagrams, you can simplify or compose several Boolean functions from table functions. It allows us to understand the validity of logical identities. Clearly, the same figures on the Venn diagram correspond to the (left and right) parts of the logical expressions. Thus, the union of the circle A and the intersection $A \bullet B$ is the circle A, therefore, we get $A+A \cdot B=A$. Note that for any statement (set) A the Boolean expression $a+(\neg a)$ is true, i.e. $A+(\neg A)=\Omega$. The same is equivalent to the property $\neg(\neg a)=a$, which is also called the "Excluded Third Law". Indeed, if A is a circle inside a square, then $(\neg A)$ is a part of a square without a circle, but then $\neg(\neg A)$ is the part of the square from which the part of the square without the circle is excluded; just the circle.

We list the main properties of Boolean algebra. They are easy to understand by giving the statements a, b and c the values "true" and "false", or by shading certain areas on the Venn diagram for the sets A, B and C.

1. $A+A=A$
2. $A \bullet A=A$
3. $A+B=B+A$
4. $A \bullet B=B \bullet A$
5. $A+A \bullet B=A$
6. $A+(B+C)=(A+B)+C$
7. $A \bullet (B \bullet C)=(A \bullet B) \bullet C$
8. $A \bullet (B+C)=A \bullet B+A \bullet C$
9. $(A+B) \bullet (A+C)=A+B \bullet C$
10. $A+\Omega=\Omega$
11. $\neg\Omega=\emptyset$
12. $\neg\emptyset=\Omega$
13. $A+\emptyset=A$
14. $A \bullet \Omega=A$
15. $A \bullet \emptyset=\emptyset$
16. $A+(\neg A)=\Omega$
17. $A \bullet (\neg A)=\emptyset$
18. $\neg(A+B)=(\neg A) \bullet (\neg B)$
19. $\neg(A \bullet B)=(\neg A)+(\neg B)$
20. $\neg(\neg A)=A$.

* Traditionally, statements, or logical variables, are denoted by lowercase letters, and sets by capital letters. Also traditionally, the notation for logical operations in Boolean algebra and in the algebra of sets is different.

† In solving a whole series of problems, Leonard Euler used the idea of representing sets by means of circles. However, this method was used by the eminent German philosopher and mathematician Gottfried Wilhelm Leibniz for geometric interpretation of logical connections between concepts even before Euler. The German mathematician Ernst Schröder used the method of Euler circles in the book *The Algebra of Logic*.

‡ In mathematical logic, such figures were introduced by the English mathematician John Venn, who described them in detail in his book *Symbolic Logic*, published in London in 1881.

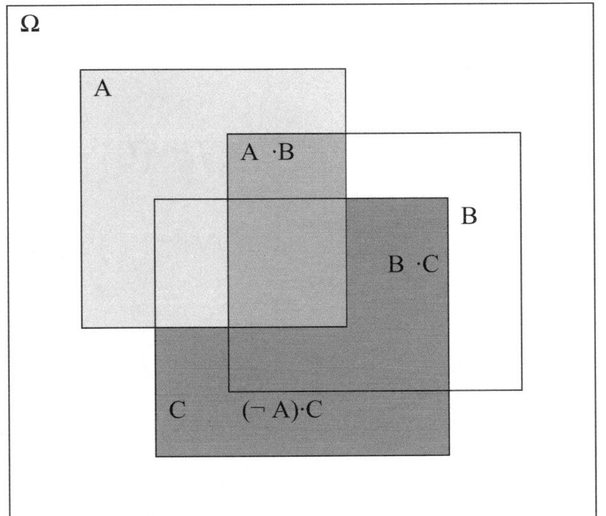

FIGURE 11.11 The Venn diagram for A + A • B + B • C + (¬ A) • C.

These properties are most often used to prove equalities that define complex statements or sets.

For example, you want to simplify a logical statement $f(a,b,c) = a \vee a \wedge b \vee b \wedge c \vee (\neg a) \wedge c$. Of course, you can, using basic logic functions, create a table of values for this function, and then look for another, simpler function with exactly the same values (if it exists). But we proceed as follows. We draw the Venn diagram in the case of the intersection of all three sets (for convenience we replaced the circles with rectangles in Figure 11.11). Given the location of the rectangles, we see that

$$A + A \cdot B + B \cdot C + (\neg A) \cdot C = A + C.$$

Now we strictly prove this equality using the properties of a Boolean algebra:

$$A + A \cdot B + B \cdot C + (\neg A) \cdot C = (A + (\neg A) \cdot C) + (A \cdot B + B \cdot C)$$

$$= (A + (\neg A) \cdot C) + B \cdot (A + C)$$

$$= (A + (\neg A)) \cdot (A + C) + B \cdot (A + C)$$

$$= \Omega \cdot (A + C) + B \cdot (A + C)$$

$$= (\Omega + B) \cdot (A + C) = \Omega \cdot (A + C) = A + C.$$

It is not always possible to obtain a rigorous proof of the assumed equality. And if the equality is not right, then it is even more difficult to find a solution, since, when carrying out the transformations, it is necessary to bring equality to a contradiction. A graphical method using Venn diagrams is not a rigorous or complete solution if the equation is verified, since in the case of a graphical solution it will be necessary to consider all possible locations of all sets. But the diagrams give an unambiguous answer in the case when the equation is not verified. Indeed, if it was possible to find at least one example when the equation is not true, then it is not exactly satisfied.

Clearly, the use of the diagram is especially convenient if there are a lot of statements.

If we talk not about classical Boolean algebra, but about the actual programming practice, it should be recognized that the variables appearing in Tables 11.1 and 11.2 can take not two (0 or 1) but three values: 0, 1 and indefinite. We have already fixed this feature in Tables 11.3 and 11.4, where instead of specific values of the arguments (0 or 1) there is a dash. In programming languages, there are tools that handle such "dashes" in truth tables. If the argument of the Boolean function is not defined, then the calculation can either be interrupted by an error message or go through the third scenario.

Arguments of Boolean functions can take not two but three, but an infinite set of real values. This set is divided into two extremely unequal parts: zero and nonzero (¬0, if we speak in the language of Table 11.1 for non-zero numbers, and we repeat, Boolean functions of Mathcad are perceived as units). It also happens that the Boolean function returns not only zeros and ones. Here, for example, how the *Or* function works in one of the implementations of the BASIC language (or rather, the version of the BASIC language that was completed with the popular PC "Iskra-226", the tracing machine "Wang-2000"): see Table 11.5.

You can allow such work for the extended operator: see Table 11.6.

TABLE 11.4 Boolean Functions of One-and-a-Half, One and Zero of the Arguments

a b	*a Or b* (f_2)	*a b*	*a* (f_{11})	¬*a* (f_{13})	*a b*	*a* (f_{12})	¬*b* (f_{14})	*a b*	*1* (f_{15})	*0* (f_{16})
0 0	0	0 –	0	1	– 0	1	0	– –	1	0
0 1	1	0 –	0	1	– 1	1	1	– –	1	0
1 –	1	1 –	1	0	– 0	0	0	– –	1	0
1 –	1	1 –	1	0	– 1	0	1	– –	1	0

TABLE 11.5 "Disjunctive Conjunction"

a	*b*	*a Or b*
0	0	0
0	¬0	1
¬0	0	1
¬0	¬0	2

TABLE 11.6 Extended Conjunction (Disjunction)

a	*b*	*a Or b*
0	0	0
0	¬0	1
¬0	0	2
¬0	¬0	3

This is one case when the first argument (operand) is not equal to zero, and the second case is when the second is not equal to zero, and the third case is when both arguments are equal to zero at the same time.

Summarizing our analysis of Tables 11.1 and 11.2, we can say that the described Boolean functions in real computer implementations can have non-binary arguments and return again non-binary results. But there is no special non-binary meaning in this. Simply, we repeat, for technological reasons, the real variables in the described implementations of programming languages (BASIC, Mathcad) "concurrently" play the role of binary ones. Boolean functions take their real arguments "binary": zero is zero ("No", "False"), and the rest is a unit ("Yes", "True").

This, one might say, "categorical" behavior of the described built-in functions contradicts the provisions of *fuzzy sets theory* (see Study 12) and fuzzy logic theory. It is necessary, for example, to process on the computer not only the "black and white" (binary) answers of the questionnaire type "Yes (1)", "No (0)", but also the "color" (real) answers: "Yes (1)", "Yes, rather than no (0.75, for example)", "No, yes, no (0.5)", "Rather no than yes (0.25, for example)" and "No (0)". If we talk not about statistics, but about electrical engineering, and return to electrical circuits, which are often used to illustrate the work of Boolean functions (see Figure 11.6), it is possible to mention the fact that now in everyday life switches smoothly change the glow of lamps from 100% to zero. Such devices were earlier used in theaters and cinemas. Doctors say that a smooth transition from light to darkness through the twilight does not spoil the eyesight. (In cinemas, light is extinguished smoothly, of course, not for medical reasons, but for other reasons—if the light is abruptly extinguished, panic may occur.)

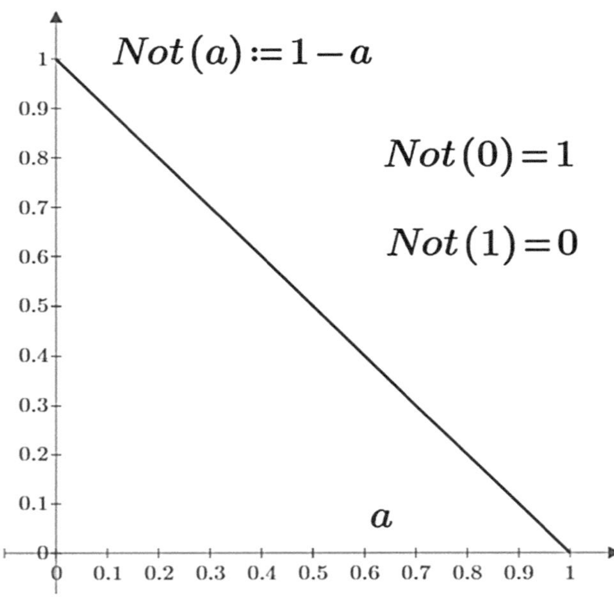

FIGURE 11.12 Smooth negation.

There are many other examples that lead to the fact that the arguments of the functions listed in Tables 11.1 and 11.2 can and should not only be binary, but also real numbers from zero to one. And the functions listed in Tables 11.1 and 11.2 must return real values that smoothly change from zero to one. Here's how, for example, you can specify a "smooth" negation function, see Figure 11.12.

"Smooth" conjunctions (*And*) and "smooth" disjunctions (*Or*)* turn out on their own if one remembers that

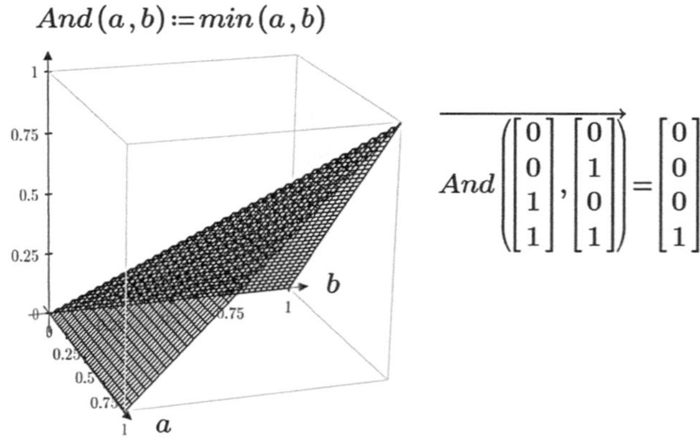

FIGURE 11.13 Fuzzy Graph **And.**

* Some Mathcad users find it difficult to remember that ^ is a logical multiplication, and ∨ is a logical addition. Therefore, it will not be superfluous to introduce into account the user-defined functions And and Or, even if we work only with Boolean logic.

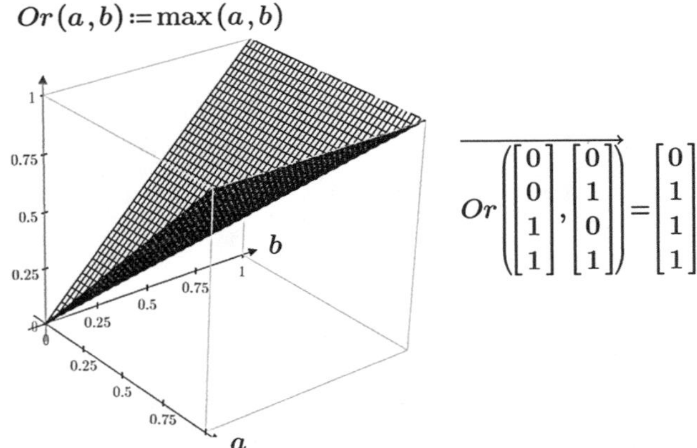

FIGURE 11.14 Fuzzy Graph **Or.**

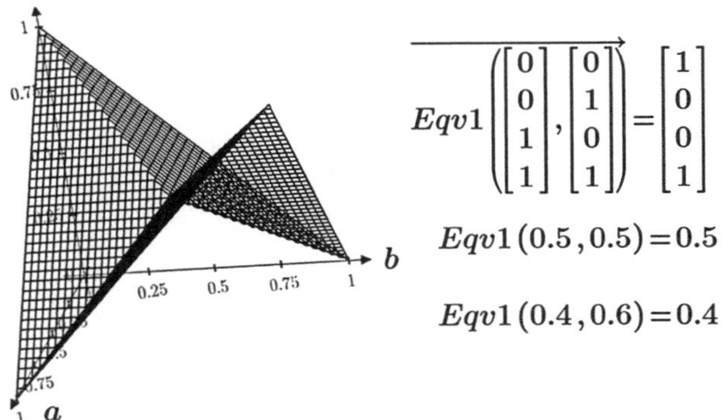

FIGURE 11.16 Fuzzy Graph **Eqv** (second option).

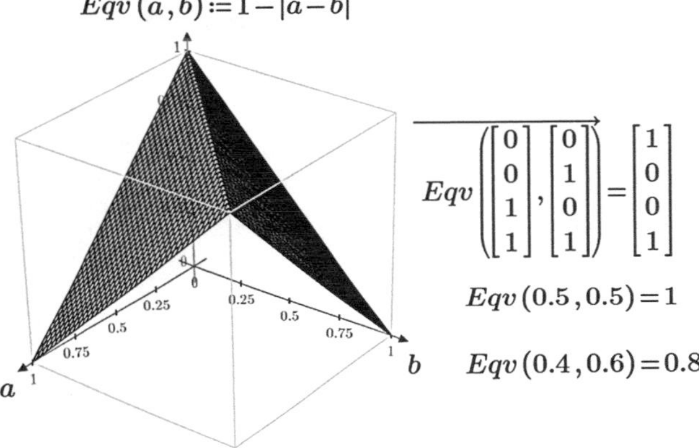

FIGURE 11.15 Fuzzy Graph **Eqv.**

one of the notations of the conjunction is *min* (minimum, see column f_1 in Table 11.2), and one of the disjunction symbols is *max* (maximum, see column f_2):

And(a, b) := min(a, b) Or(a, b) := min(a, b)

Figures 11.13 through 11.16 show a three-dimensional graphical representation of fuzzy Boolean functions *And*, *Or* and Eqv. This last function is represented in two variants of *Eqv* (Figure 11.15) and *Eqv1* (Figure 11.16).

The broken surfaces shown in Figures 11.13 through 11.16 and 11.19 below are similar to the Japanese origami figures, folded from a sheet of paper.

Figure 11.17 shows the creation of fuzzy Boolean functions and their display on the graphs of lines of the same level (on contour charts).

FUNCTIONS OF MANY ARGUMENTS

Boolean functions of three arguments are 256, and four — Reader, define it yourself!

The functions *And* and *Or*, if they are identified with the min and max functions (see above), can have a variable number of arguments, with Boolean (0 or 1) or fuzzy (from 0 to 1) values. This property is also possessed by some other Boolean functions.

What kind? Reader, again find this by yourself!

Figures 11.18 and 11.19 show the formation in Mathcad of a fuzzy function of three arguments, returning the decision of the jury, which can issue no longer "black and white" answers (guilty, innocent), but "color": guilty of 30%, innocent by 70%, etc. In the electric analog of the voting machine, the switches (0 or 1) are replaced by rheostats (from 0 to 1).

A SAD NOTE

It is said that in the United States, an electric chair is energized by several people, even though the real switch is activated by only one person. The remaining members of this execution activate fake switches. And no one knows which are the fakes, and which is the real switch. Such a somewhat sanctimonious procedure makes it possible for each such executioner to think that it was not him, but someone else was the cause of the person's death. If all the actual switches are changed with rheostats (see Figure 11.18 below),

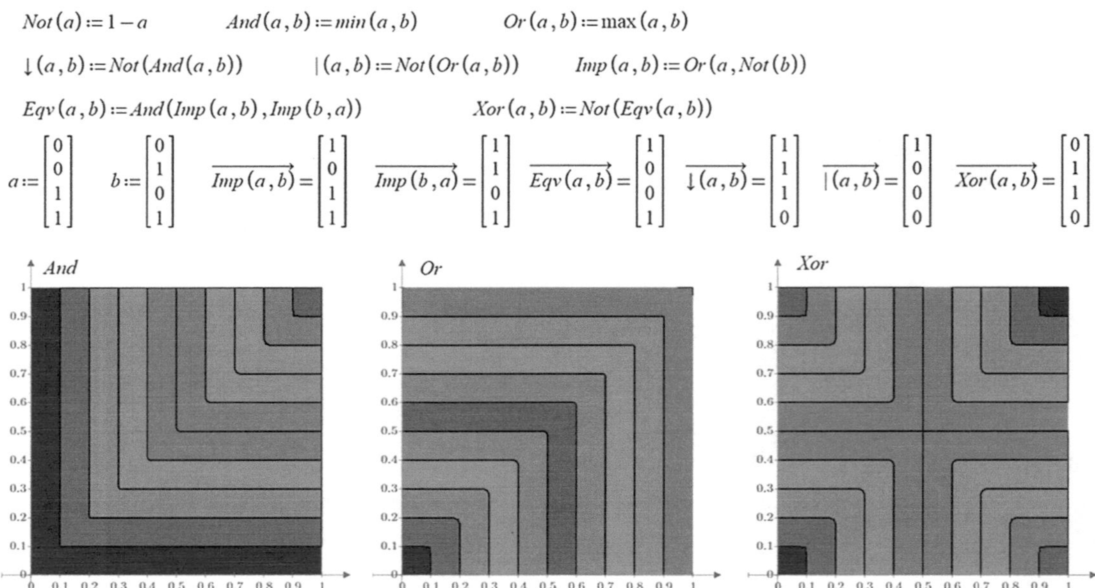

FIGURE 11.17 Fuzzy Boolean functions.

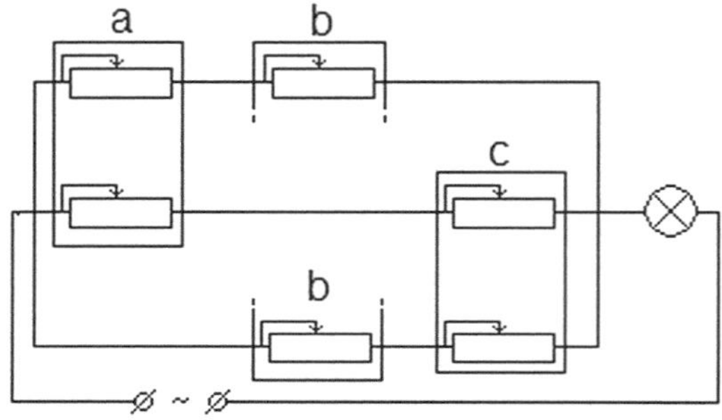

FIGURE 11.18 Voting machine: parallel (Or) connection in series (And) connected switches (rheostats).

$And(a) := min(a) \quad Or(b) := \max(b) \quad Verdict(a,b,c) := Or\left(\begin{bmatrix} And([a\ b]) \\ And([a\ c]) \\ And([b\ c]) \end{bmatrix}\right)$

$f(b,c) := Verdict(0.7, b, c)$

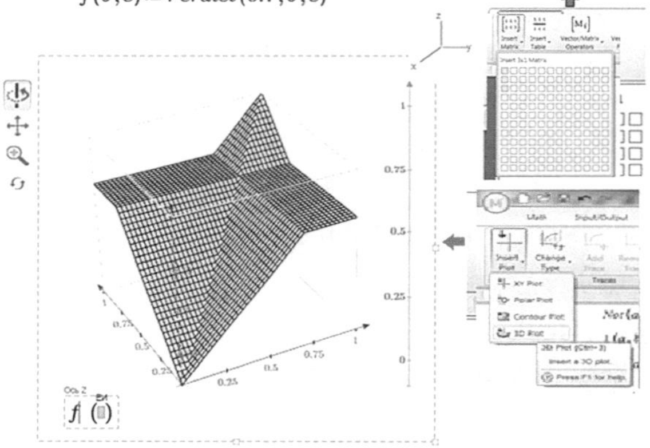

FIGURE 11.19 Function of fuzzy voting of three members of the jury.

smoothly changing the voltage, then death can be replaced by non-lethal punishment: the convicted criminal will receive an electric shock (strong or weak) but will remain alive.

The *Verdict* function, shown in Figure 11.19, with binary arguments returns a binary response, and for real arguments it returns a real one, naturally. The corresponding "Boolean die" (origami) is shown at $a = 0.7$: we see a hybrid of the conjunction with disjunction where the fuzzy (from 0 to 1) opinion of one member of the jury smoothly translates the verdict from the *Or* area (see Figure 11.14) to the And area (see Figure 11.13).

ASSIGNMENTS FOR READERS

1. Create a function similar to the one shown in Figure 11.19, but not for three, but for any number of jury members who make their verdict vague—a real number from 0 to 1.
2. Create a Boolean function to turn on the light from anywhere.
3. Find online logic puzzles and try to solve them using the tools described in this Study.

STUDY 12

Fuzzy Sets Or

Optimal Fire Bucket

ABSTRACT

In this Study, we discuss how to solve optimization problems, the theory of fuzzy sets and how it can be applied to "Boolean" digital computers. The theory of fuzzy sets and its application on a computer for the construction of membership functions are considered. It is shown how you can use computer graphics to enumerate membership functions.

Mathematics: Set, subset, cylinder volume, cone volume, ball volume, sequence limit, surface area, function minimum, derivative, normal distribution function, Gaussian curve, level line of a function of two variables, function of an element belonging to a fuzzy set, addition of fuzzy sets, multiplication of fuzzy sets, objective function.

Physics: Body volume, filtration.

IT: The operator of membership to a set, computer analytical transformations, fully dimensional function, functions for solving optimization problems with constraints.

Art: An American film, *12 Angry Men*, directed by Sidney Lumet

Study website: https://community.ptc.com/t5/PTC-Mathcad/Study-14/m-p/642397.

In mathematics there are the very fundamental concepts of *set* and *subset.*

There is, for example, the set of geometric bodies. In this set there is a subset of cones, and in this subset there is a (sub)subset of straight circular cones.

In Mathcad Prime, there is an operator that determines whether a value belongs to a certain numeric set, and we have already dealt with it in the previous Study. So, from Figure 12.1 it can be seen that the number π belongs (1, yes) to the numeric sets "Complex numbers *C*" and "Real numbers *R*", but does not belong (0, no) to the sets "Rational numbers *Q*" and "Integers *Z*". In the program shown in Figure 17.2 in Study 17, when solving the problem of fishermen and fish, we will use the membership operator for the set of *Z* integer numbers.

In Mathcad there are other functions for working with a set, returning 0 or 1. Here are some of them:

- IsString—whether the argument of the function is a string:
- *IsString*("123") = 1, but *IsString*(123) = 0
- *IsArray*—whether the argument is an array (an array is a vector or a matrix, see Study 2)
- *IsFunction*—whether the argument is a function
- *IsScalar*—whether the argument is a scalar
- *UnitsOf*—determines which set of physical quantities (length, volume, area, etc.) belongs to the argument.

We often come across with the notion of "set" in everyday life.

Example. There are a lot of buckets. In this set there is a subset of fire buckets.

Let's talk about them.

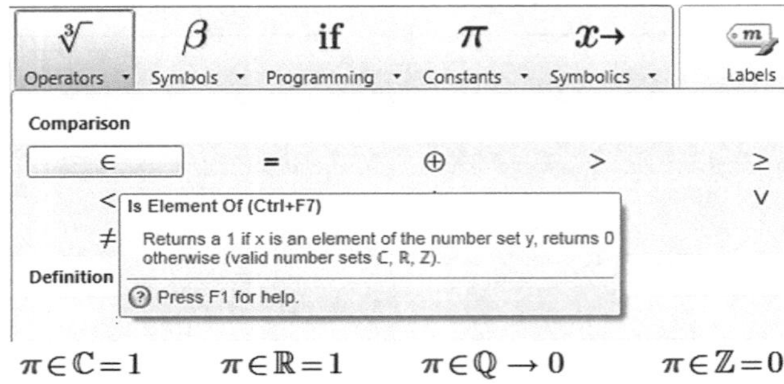

FIGURE 12.1 The operator of membership applied to sets of numbers.

The fire bucket (Figure 12.2) differs from the conventional bucket because it is made in the form of a *straight circular cone*. The conventional bucket is usually a truncated straight circular cone or a straight circular cylinder. The fire bucket is however made in the form of a cone so that it does not have a bottom on which it could stand. Water can be collected in this bucket, though it will be necessary to pour it on a fire immediately, without letting go of the bucket. To use such a bucket at home for other purposes (for cleaning the premises, for example) is almost impossible.

The author of this book during his serving time in the army used a fire bucket for cleaning the floors in the barracks (as there was no other). It was placed on an inverted stool. Hence, probably, his "love" for fire buckets, poured on the pages of this as well as his other books on Mathcad. When visiting a site at a summer cottage cooperative, the author always looks first of all for a fire shield with a fire bucket. Another attribute of the "bottomless" fire bucket is its red color. However, though important, this detail is superfluous for our further narrative.

FIGURE 12.2 Fire bucket.

$$V(r,h,n) := \left\| \begin{array}{l} \Delta h \leftarrow \dfrac{h}{n} \\ V \leftarrow 0 \\ \text{for } i \in 0..n-1 \\ \quad \left\| \begin{array}{l} R \leftarrow r \cdot \left(1 - \dfrac{i \cdot \Delta h}{h}\right) \\ v \leftarrow \pi \cdot R^2 \cdot \Delta h \\ V \leftarrow V + v \end{array} \right. \\ V \end{array} \right.$$

$r := 10\ cm \qquad h := 20\ cm$

$n := 7 \qquad V(r,h,n) = 2.565\ L$

$n := 1000000 \qquad V(r,h,n) = 2.094\ L$

$$\frac{1}{3} \cdot \pi \cdot r^2 \cdot h = 2.094\ L$$

FIGURE 12.3 Calculation of the volume of the children's pyramid/cone, consisting of *n* disks or cylinders.

The cone, along with the spherical ball and the cubes, is one of the first geometric bodies a child encounters. Let's recall the pyramid toy consisting of separate disks around a central rod, or the puzzle "Hanoi towers", which is described in Study 10 (see Figure 10.4). These toys suggest how it is possible to prove that the volume of the cone is exactly one-third of the volume of the cylinder into which the cone is inscribed. Figure 12.3 shows a Mathcad-sheet for calculating the volume of the children's pyramid with radius of the base `r` and height `h` consisting of `n` disks—straight circular cylinders with no hole in the middle.

A more complicated part of this calculation is the determination of how the radius *R* of the next *i*th disk will change. In Figure 12.4, the loop operator with

$$V(r,h,n) := \sum_{i=0}^{n-1} \pi \cdot \left(r \cdot \left(1 - \frac{i \cdot \dfrac{h}{n}}{h}\right)\right)^2 \cdot \frac{h}{n}$$

$$\lim_{n \to \infty} V(r,h,n) \to \frac{\pi \cdot h \cdot r^2}{3}$$

FIGURE 12.4 Summation and limit for calculating the volume of a child's pyramid.

parameter i for the sum calculation is replaced by the sum operator itself. Then, using the limit operator (if the number of disks tends to infinity, then the stepped pyramid turns into a smooth cone), we did what we had to prove—the volume of the cone is one-third of the volume of the cylinder into which the cone is inscribed. When a child assembles a pyramid, then we can assume that he/she is preparing to study the basics of integral calculus. And what is the volume of a ball equal to? It is possible to refer to the handbooks on a bookshelf or on the Internet and to find the answer, though it is also possible to derive the specific formula, imagining the same children's pyramid, built not as a cone, but in the form of a hemisphere.

A straight circular cone (hereinafter simply a cone) has two parameters: the radius of the base r and the height h, and with them it is possible to calculate not only its volume V (see Figure 12.4), but also the total surface area S, consisting of the lateral surface and of the surface of the cone base.

Many interesting optimization problems are connected with the cone. Let's solve some of them.

PROBLEM 1

Determine the ratio of the radius of the cone base r to its height h, so that minimizes its total surface area S. Children like to blow bubbles, or to watch adults do it. Imagine such a trick. Soap bubbles start, but not in the form of spheres (closed surfaces with a minimum value of the surface area for a given value of the volume), but in the form of cones with the same property.

Figure 12.5 shows, let's say, *an engineering approach* for solving this problem for a cone with a volume of 100 m^3. To an engineer was given the task of designing a storage tank for some loose material (sand or cement, for example) in the form of an inverted cone with a lid and of using the minimum of metal material for it.

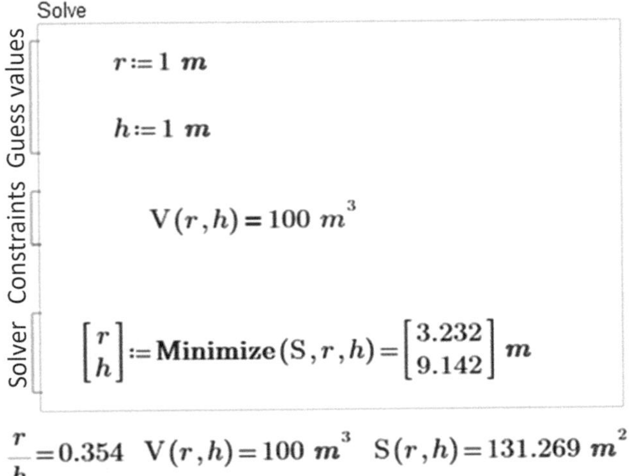

FIGURE 12.5 Cone with minimal surface area: solution 1.

The conic capacity of a given volume can be constructed with different r and h ratios. In the Mathcad calculation, shown in Figure 12.5, we determine the values of r and h so that the volume of the cone is kept equal to 100 m^3 (constraint) and its surface area is minimal. In the calculation with the Solve block three areas are used: the area of *Guess Values* to the solution, the area of *Constraints* and the *Solver* area with the built-in *Minimize* function. This function, by use of a specific numerical algorithm (see Study 13), changes the values of the variables r and h, starting from the initial approximations, so that the function S takes a minimum value, and the constraints are not violated. In our problem, the restriction is that the volume of the cone must be equal to the given 100 m^3. However, in this area it is possible to add additional constraints, for example, the following: in order to store the bulk material (remember the hourglass), the angle of inclination of the lateral surface of our inverted cone should be no less than a certain value. Otherwise, the loose material will get stuck in the cone.

One can think a little and figure out* that our problem about the minimum surface area of a cone with a given volume has not two (r and h), but one (r or h) parameter. If the volume of the cone V is given and, for example, its radius is r, then it is not difficult to determine its height h. This approach to solving our problem of optimizing cone sizes is shown in Figure 12.6.

In the solution shown in Figure 12.6, after some simple manual analytic transformations, we could see that the user function with the name S had not two (r and h, as in the solution in Figure 12.5), but one (r) argument. This allowed us to construct a Cartesian graph of the change in the value of the function S depending on the argument r for a given volume of the cone V, which clearly shows a minimum point whose value is refined with the same *Minimize* function.

Strictly speaking, in this case the function S is a function of two variables r and V (see Figure 12.5). Fixing the value $V :=$ 100 m^3 it allows us to plot the section of the surface S = S(r, V) through the plane $V =$ 100 m^3 (see

* "One can think a little and figure out" that our paradoxical transport problem, which we will solve in Study 15 (see Figures 15.15 and 15.16), also has a smaller number of unknowns—not four, but only one.

$V:=100\ m^3 \qquad V=\frac{1}{3}\cdot\pi\cdot r^2\cdot h \qquad h=\frac{3\cdot V}{\pi\cdot r^2}$

$S=\pi\cdot r\cdot\sqrt{r^2+h^2}+\pi\cdot r^2 \qquad S(r):=\pi\cdot r\cdot\sqrt{r^2+\left(\frac{3\cdot V}{\pi\cdot r^2}\right)^2}+\pi\cdot r^2$

$r:=2\ m, 2.1\ m..5\ m$

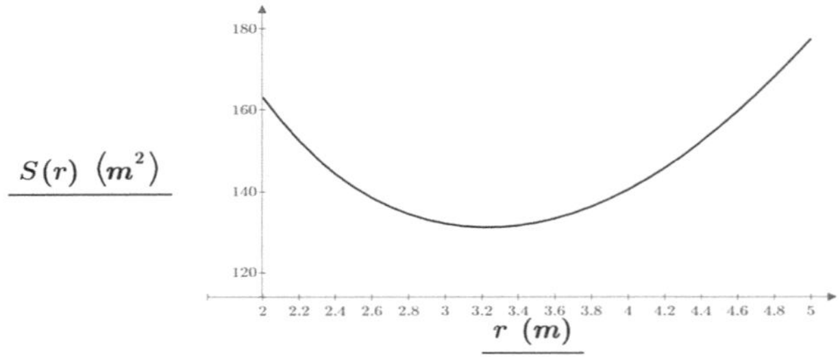

$r:=3.2\ m \qquad r:=\text{Minimize}(S,r)=3.232\ m$

$h:=\frac{3\cdot V}{\pi\cdot r^2}=9.142\ m \qquad \frac{1}{3}\cdot\pi\cdot r^2\cdot h=100\ m^3 \qquad S(r)=131.269\ m^2$

FIGURE 12.6 Cone with minimum surface area: solution 2.

Figure 12.6). To understand where the minimum points of the indicated function are located, it is sufficient to construct its three-dimensional graph.

However, the solution implemented in Figure 12.5 also has a right to life. It implements the KISS principle. KISS is an acronym of the English phrase: "Keep it simple, stupid!" It calls for solving tasks simply and resorting to sophisticated algorithms only when simple methods fail or are difficult to implement. We will need "refinements" in another method to solve this optimization problem, see Figure 12.7.

Figure 12.7 shows a *mathematical* approach to solving the problem, rather than an engineering approach. Let's return to Figure 12.5. It calculates not only the specific values of r and h, but also their ratio, which is equal to 0.354. It is possible to change the value of the given volume of the cone V and see that the value of this ratio does not change. However, this can be shown and, most importantly, proven in a different way. The mathematical approach to the solution of the problem is based on the fact that at the minimum point the derivative of a smooth continuous function is zero. This is in particular expressed by the fact that in the graph of Figure 12.6 the tangent to the curve $S(r)$ at the minimum point (r = 3.232 m) is parallel to the abscissa axis.

In the solution in Figure 12.7, four operators of Mathcad symbolic mathematics are involved:

1. The *solve* operator searches for the root of the equation or the zero of the function. This operator is used twice in our solution: to solve the equation (to derive the formula for calculating the height

$V(r,h)=\frac{1}{3}\cdot\pi\cdot r^2\cdot h \qquad S(r,h)=\pi\cdot r\cdot\sqrt{r^2+h^2}+\pi\cdot r^2 \qquad V=\frac{1}{3}\cdot\pi\cdot r^2\cdot h \xrightarrow{solve,h} \frac{3\cdot V}{\pi\cdot r^2}$

$\pi\cdot r\cdot\sqrt{r^2+h^2}+\pi\cdot r^2 \xrightarrow{substitute,h=\frac{3\cdot V}{\pi\cdot r^2}} r\cdot\left(\pi\cdot r+\sqrt{\frac{\pi^2\cdot r^6+9\cdot V^2}{r^4}}\right)$

$\frac{d}{dr}r\cdot\left(\pi\cdot r+\sqrt{\frac{\pi^2\cdot r^6+9\cdot V^2}{r^4}}\right) \xrightarrow{simplify} \frac{\sqrt{\pi^2\cdot r^2+\frac{9\cdot V^2}{r^4}}\cdot\left(2\cdot\pi^2\cdot r^6-9\cdot V^2+2\cdot\pi\cdot r^5\cdot\sqrt{\pi^2\cdot r^2+\frac{9\cdot V^2}{r^4}}\right)}{\pi^2\cdot r^6+9\cdot V^2}$

$\frac{\sqrt{\pi^2\cdot r^2+\frac{9\cdot V^2}{r^4}}\cdot\left(2\cdot\pi^2\cdot r^6-9\cdot V^2+2\cdot\pi\cdot r^5\cdot\sqrt{\pi^2\cdot r^2+\frac{9\cdot V^2}{r^4}}\right)}{\pi^2\cdot r^6+9\cdot V^2} \xrightarrow{solve,r} \begin{bmatrix}\frac{8^{\frac{5}{6}}\cdot 9^{\frac{1}{6}}\cdot V^{\frac{1}{3}}}{8\cdot\pi^{\frac{1}{3}}}\\ -\frac{8^{\frac{5}{6}}\cdot 9^{\frac{1}{6}}\cdot V^{\frac{1}{3}}}{8\cdot\pi^{\frac{1}{3}}}\end{bmatrix}$

And other 8 zeros

$\frac{r_{opt}}{h_{opt}}=\frac{\frac{8^{\frac{5}{6}}\cdot 9^{\frac{1}{6}}\cdot V^{\frac{1}{3}}}{8\cdot\pi^{\frac{1}{3}}}}{\frac{3\cdot V}{\pi\cdot\left(\frac{8^{\frac{5}{6}}\cdot 9^{\frac{1}{6}}\cdot V^{\frac{1}{3}}}{8\cdot\pi^{\frac{1}{3}}}\right)^2}} \xrightarrow{simplify} \frac{r_{opt}}{h_{opt}}=\frac{\sqrt{2}}{4}$

FIGURE 12.7 Cone with minimal surface area: solution 3.

of the cone by its volume and the base radius, see the top of Figure 12.7) and to find the zeros of the derivative of the function $S(V, r)$ with respect to r for V = constant. In the first case, the root of the equation can also be found mentally (incidentally, we did this in the second version of the solution of the problem in Figure 12.6). However, the third solution variant (Figure 12.7, search for the roots of the derivative) is unlikely to be done without a computer. Or, rather, it would be possible for a good mathematician, but not for an engineer.

2. The *substitute* operator replaces a variable for another variable or expression. In the solution in Figure 12.6, we manually changed the variable h to the expression $(3 \cdot V/\pi)/r^2$. Figure 12.7 shows how this is automatically done in Mathcad.
3. The operator d■/d■ takes the derivative.
4. The *simplify* operator rearranges an expression in a more compact form. We reduced the derivative so that it fits into Figure 12.7.

The derivative of the function of the cone surface area with respect to the base radius has 10 zeros—two real opposites in sign (they are shown in Figure 12.7) and eight complex ones with imaginary parts. The first zero is taken (copied and pasted) in order to prove that for any volume of the cone the ratio of the radius of the base of the cone of the minimum surface area to its height is equal to the root of two divided by four, that is, approximately 0.354 (see Figure 12.5).

As an exercise, the reader can try to find similar optimal ratios of radius r to height h for other geometric bodies: for a cylinder with a flat and round bottom, but without a round flat lid; for a cylinder with a round flat bottom and with a round flat lid; for a cone with a lid in the form of a hemisphere (but not with a round flat lid) (Figures 12.5–12.7), etc. In similar, but already hollow, bodies (shells), it is possible to put horizontal or vertical partitions that increase the total surface area of such a body and to change the optimal ratio `r` to `h`. All these problems are easy to solve in an "engineering" sense (numerically) for particular cases with a given volume, but only some of them can be solved "mathematically" (symbolically) for the general case without fixing the volume. Good luck!

If the reader does not get special luck, he/she can go to the PTC Community forum, where these problems are solved not only with the help of Mathcad, but also with the help of the Maple and the Mathematica packages

PROBLEM 2

There is a conical funnel into which a filter is rolled, folded from round filter paper. Many of us have worked with such filters in chemistry lessons at school and university. The question is what should be the angle of the "socket" of the funnel cone (β, see Figure 12.8), at which the filtering rate would be maximal?

The solution to this optimization problem is shown in Figure 12.8.

It is assumed in the solution that the filtration rate (flow rate q) is proportional to the product of the height of the liquid column h and the filtering area S. The "trade-off" between the height of the liquid column and the filtering area is 90° (right angle). By the way, almost all conical funnels are made with this angle of socket, for the overflow of the liquids. The reader can, if desired, verify this.

In our problem of a filter in a conical funnel, the parameter R is the radius of the round paper blank, whose a circular sector is used to make the conical filter cap of base radius r and height h.

Using the last two operators (shown in Figure 12.8), it is easy to evaluate the sector angle α that must be cut out from the round billet, so that a cone with a 90° socket can be twisted from the remaining sector. This angle (α_{opt}) is approximately 105.5 angular degrees (*deg*).

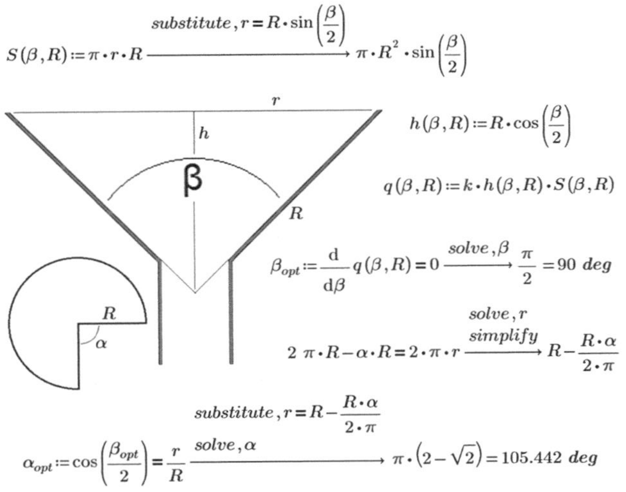

$$S(\beta,R) := \pi \cdot r \cdot R \xrightarrow{substitute,\, r = R\cdot\sin\left(\frac{\beta}{2}\right)} \pi \cdot R^2 \cdot \sin\left(\frac{\beta}{2}\right)$$

$$h(\beta,R) := R\cdot\cos\left(\frac{\beta}{2}\right)$$

$$q(\beta,R) := k\cdot h(\beta,R)\cdot S(\beta,R)$$

$$\beta_{opt} := \frac{d}{d\beta} q(\beta,R) = 0 \xrightarrow{solve,\beta} \frac{\pi}{2} = 90\ deg$$

$$2\ \pi\cdot R - \alpha\cdot R = 2\cdot\pi\cdot r \xrightarrow[simplify]{solve,r} R - \frac{R\cdot\alpha}{2\cdot\pi}$$

$$\alpha_{opt} := \cos\left(\frac{\beta_{opt}}{2}\right) = \frac{r}{R} \xrightarrow[solve,\alpha]{substitute,\, r = R - \frac{R\cdot\alpha}{2\cdot\pi}} \pi\cdot\left(2-\sqrt{2}\right) = 105.442\ deg$$

FIGURE 12.8 Solution of the problem of an optimal paper conical filter.

And what sector needs to be cut from a round billet of radius *R* in order to obtain a conical capacity (our fire bucket, for example) of the maximum volume?

The solution of this **third** optimization problem for the cone is shown in Figure 12.9. We immediately abandoned the "engineering" approach and took advantage of the mathematical.

Three user functions *r(α, R), h(α, R) and V(α, R)* (optimization objective function) are created when solving the problem of the maximum volume of a cone folded from a round billet. The last function takes the derivative with respect to *α*. One of the zeros of the derivative will be the solution of our problem: from the round billet it is necessary to cut a sector with approximately 66° in the top, the remainder (the larger sector) is rolled into a cone with 109.5 angular degrees in the "socket". It is also possible to roll one more cones from the yet-to-be-carved sector. A question is asked (a new problem for the reader), what will be the optimal cutting of the round billet into two sectors, from which it is possible to collapse two cones with the maximum total volume. And if the round billet is cut not into two, but into three and more sectors, billets for cones? Will the total volume of these cones exceed the total volume of the two cones? We suggest readers to conduct these calculations autonomously by numerical and symbolic methods.

Here we are, in the manufacture of cones, come to the solution to the problem of our fire bucket.

The fire bucket, we repeat, is made in the form of a cone so that it cannot be put on the floor and used for a not-intended purpose (for washing, for example)—such a bucket will fall on its side.

$$2\pi \cdot R - \alpha \cdot R = 2\cdot\pi\cdot r \xrightarrow[\text{simplify}]{\text{solve}, r} R - \frac{R\cdot\alpha}{2\cdot\pi}$$

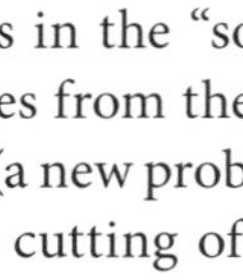

$$r(\alpha, R) := R - \frac{R\cdot\alpha}{2\cdot\pi} \qquad h(\alpha, R) := \sqrt{R^2 - r(\alpha, R)^2}$$

R

α

$$V(\alpha, R) := \frac{1}{3}\cdot\pi\cdot r(\alpha, R)^2 \cdot h(\alpha, R)$$

$$\frac{d}{d\alpha}V(\alpha, R) \xrightarrow{\text{simplify}} -\frac{R^2\cdot\sqrt{-(R^2\cdot\alpha\cdot(\alpha-4\cdot\pi))}\cdot(18\cdot\pi\cdot\alpha^2 - 3\cdot\alpha^3 - 28\cdot\pi^2\cdot\alpha + 8\cdot\pi^3)}{24\cdot\pi^2\cdot\alpha\cdot(\alpha-4\cdot\pi)}$$

$$\frac{R^2\cdot\sqrt{-(R^2\cdot\alpha\cdot(\alpha-4\cdot\pi))}\cdot(18\cdot\pi\cdot\alpha^2 - 3\cdot\alpha^3 - 28\cdot\pi^2\cdot\alpha + 8\cdot\pi^3)}{24\cdot\pi^2\cdot\alpha\cdot(\alpha-4\cdot\pi)} \xrightarrow{\text{solve}, \alpha} \begin{bmatrix} 2\cdot\pi \\ \frac{2\cdot\pi\cdot(\sqrt{6}+3)}{3} \\ -\frac{2\cdot\pi\cdot(\sqrt{6}-3)}{3} \end{bmatrix}$$

$$\alpha_{opt} := \frac{2\cdot\pi\cdot(3-\sqrt{6})}{3} = 66.061\ deg$$

$$\beta_{opt} := 2\cdot\text{asin}\left(\frac{r(\alpha_{opt}, R)}{R}\right) \xrightarrow{\text{simplify}} 2\cdot\text{asin}\left(\frac{\sqrt{6}}{3}\right) = 109.471\ deg$$

FIGURE 12.9 The maximum volume of a cone folded from a round billet.

When solving the problem of a conical fire bucket (see Figure 12.9), we understood by optimizing the *maximization* of the bucket volume with the chosen technology of its manufacture—cutting the sector from the round billet and then folding this sector into a cone. However, at the same time the bucket lost not only its *optimality* but also its name "bucket": it turned into something like a basin (a fire-basin?). Or an inverted Vietnamese hat. We chased the *number* (bucket volume) and lost the *essence* (the function, the hypostasis of the bucket)—a container for carrying a liquid easily.

Nature does not like angles but also *calculations* do not like, which in some cases simply kill her. This can be seen not only in biology and physics, where the research tools often unrecognizably spoil or even kill the object of investigation, but also in *computer science*. And not only in the application of computers (*counting*, roughly speaking) to solving natural science problems, but also in the field of applying computers to the computers themselves.

The doctor does not need to know the exact *numerical* value of the temperature of the patient's body, e.g. 36.6, 38.9, to give a diagnosis. It is enough to express the thermometer readings by ranges (separate *sets* of temperatures) that physicians agreed in advance, e.g. "low temperature", "normal temperature", "elevated temperature", "high temperature", "very high temperature (fever)". The boundaries of these estimates (sets of temperatures), although intuitive, are, nevertheless, blurred—"bushy". This is determined not only by modern ideas (the theory of fuzzy ("bushy") sets, TFS), but also by the error of the thermometer itself, the temperature measurement technique, etc. A fresh graduate of a medical college, without hesitation, will tell you where the boundary between high and very high temperature lies. An experienced doctor may, however, not know this, even though he makes diagnoses that are no worse than a beginning doctor. An experienced doctor knows that the boundaries are not only blurred, but also context dependent. Even the patient's parameters, expressed not in material but in a Boolean form (Wasserman's reaction, Koch's wand, analysis for AIDS, etc.) also have "bushy" borders. This is well known to laboratory technicians conducting tests. If the reader looks into any handbook of a therapist, which describes the symptoms of illness, then, as a rule, the reader will not see specific numbers (body temperature, blood pressure, hemoglobin content in the blood, etc.), but rather swords—"raised", "lowered," etc. The programs for making a diagnosis by entering certain patient parameters into a computer did not receive wide practical application. One of the difficulties in this matter is the translation of the number into the linguistic category.

It is customary to distinguish three revolutions that brought programming to new levels: *structural*, *object-oriented* and *visual*. However, these revolutionary aspects were more focused on *programs* (art for the sake of art) and almost did not touch the *object* of programming—those models of the real world, whose properties and events are programmed to simulate. Moreover, a retrospective view can also catch some counterrevolution in the refusal, for example, of the transition from analog computers to digital technology. Although recently there is some kind of renaissance here—the revival of the principles of analog modeling on modern digital computers. This can be mentioned in the technology of visual programming, where the previous controls are recreated in a similar way to some analog adders, integrators, etc. However, the virtuality of these neo-analog machines implies their strict determinateness (Boolean), which entails not only positive but also negative consequences. The very principles of TFS have long been at the fingertips of programmers. In a literal sense. The mouse of the computer reacts to two events—click and double click. And then, actually, why one double click differs from two single ones? Because of the length of the pause between the two clicks. If it is expressed in the language of human communication (very short, short, long, etc.), then this will be a typical example of sets with "bushy" borders. How long does it take to increase the short pause so that it turns into a long one and that the double click is perceived by the computer as two single ones? And how many seeds need to be added to a handful to make it a heap? And how many centimeters is the lowest mountain above the highest hill?

At the peak of the structural revolution, when the keyword *GOTO* was anathematized in all programming "temples", one could often hear the following: "It is almost impossible to teach well programming students oriented initially to BASIC: as potential programmers, they are mentally stolen without hope of healing." There were also more categorical warnings of the type: "Caution! The programming exercise can deprive you of a future. Do not think that by learning how to program, you will achieve something in life!" Traditional programming makes the programmer look at the multicolored world through black-and-white glasses: the Boolean variable can take only two values (yes/no), and the real variable is strictly defined in the specified range with a fixed length of the mantissa, etc. The truth, as always, lies in the middle. However, extreme points of view are not useless—they seem to prop up the truth from the two sides, not letting it slip to extremes. And the truths themselves are "bushy" in some things. One of them might sound like this: "If you want to know the world (fuzzy, "bushy", non-numerical) and manage it, then be afraid of traditional programming languages and mathematical programs with their strict determinism."

However, let's return to our task about the fire bucket and try to solve it with the help of the TFS apparatus. We will conduct a public opinion survey and learn as much as possible about the parameters of the optimal fire bucket: its convenient geometry (the radius of the base of the cone and the height) and its optimum volume (about the weight of a bucket of water). This is where the TFS position will manifest itself in all its glory. How much water should be added to a bucket to turn it from light to heavy? How much is it necessary to increase or to decrease the radius or the height of the bucket so that it is no longer convenient to carry water in it? These "*how many*" questions are typical aspects of fuzzy sets. In Mathcad, as in other popular packages, there are no variable types for storing such quantities, but we will nevertheless try to solve the posed problem.

People's views about the optimal (convenient) radius of the cone base of a fire bucket can be obtained as follows: after making many buckets of various geometries, allow people to add water to them and to evaluate them on such a scale:

- convenient (1)
- more convenient than inconvenient (0.67)
- difficult to answer (0.5)
- more inconvenient than convenient (0.34)
- inconvenient (0).

Other estimates can be taken into account in the range from zero to one. Figure 12.10 shows the beginning of a Mathcad-sheet with the solution of the optimal fire bucket problem with elements of the fuzzy set theory. If in the theory of Boolean sets an element can (value

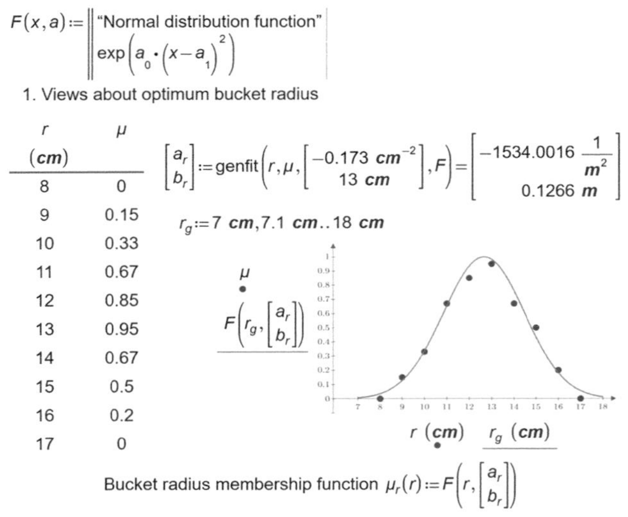

$F(x,a):=\left\| \begin{array}{l}\text{"Normal distribution function"}\\ \exp\left(a_0\cdot(x-a_1)^2\right)\end{array}\right.$

1. Views about optimum bucket radius

r (cm)	μ
8	0
9	0.15
10	0.33
11	0.67
12	0.85
13	0.95
14	0.67
15	0.5
16	0.2
17	0

$\begin{bmatrix}a_r\\ b_r\end{bmatrix}:=\text{genfit}\left(r,\mu,\begin{bmatrix}-0.173\ cm^{-2}\\ 13\ cm\end{bmatrix},F\right)=\begin{bmatrix}-1534.0016\ \frac{1}{m^2}\\ 0.1266\ m\end{bmatrix}$

$r_g:=7\ cm, 7.1\ cm..18\ cm$

Bucket radius membership function $\mu_r(r):=F\left(r,\begin{bmatrix}a_r\\ b_r\end{bmatrix}\right)$

FIGURE 12.10 The problem of the optimal fire bucket: beginning of the calculation.

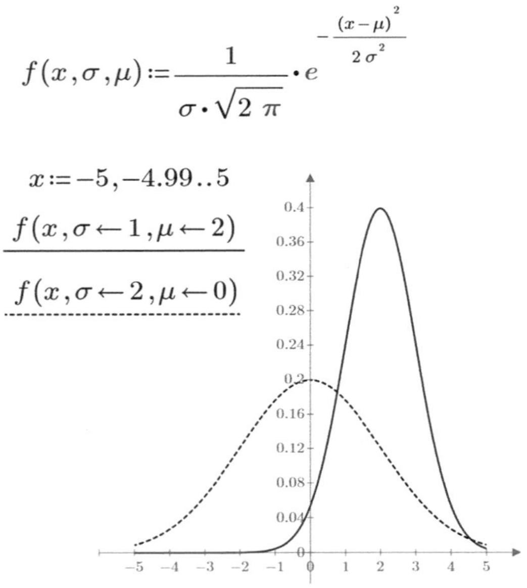

$$f(x,\sigma,\mu):=\frac{1}{\sigma\cdot\sqrt{2\pi}}\cdot e^{-\frac{(x-\mu)^2}{2\sigma^2}}$$

$x:=-5,-4.99..5$

FIGURE 12.11 The normal distribution function.

1) or cannot (value 0) belong to some distinct set (see Study 11), then in the theory of fuzzy sets this element can belong to some fuzzy set with probability from zero to one. This parameter is given by the *membership function*—the central concept of the theory of fuzzy sets.

The first operator in the calculation in Figure 12.10 is the function F with the argument x and the parameter a. The argument x is a scalar, and the parameter a is a vector with two elements a_0 and a_1. This unusual definition of the function is due to the fact that later it will be invoked as an argument to the built-in function *genfit*, whose features are explained below.

The function F is a simplified version of the normal distribution function, whose standard form is shown in Figure 12.11. We have already used a similar function in Study 9 for laying rails, according to them a train or a tram will rotate without sharp side shocks, see Figure 9.35.

In the formula in Figure 12.11, the variable μ is the mathematical expectation (mean value), and the variable σ is the standard deviation.

In the solution in Figure 12.10, we repeat, the formula for the normal distribution is somewhat simplified—it was verified that the value of μ always remained in the range from zero to one.

In the calculation in Figure 12.10, people's estimates for the optimum radius of the fire bucket are entered into a table with two columns, using two vectors with the names r and μ. We limited ourselves to ten points (people's estimates of the bucket radius), but they can be much more: how many people—so many opinions (*tot capita, tot sententiae*). The reader, if desired, can interview his friends and supplement the vectors r and μ with new pairs of numbers. The survey data are processed by the method of least squares, where the curve of the normal distribution is taken as the approximating curve. This is performed using the Mathcad built-in function *genfit*, intended for "**fit**ting" points by a user-defined function of the **gen**eral type. In our case, this function F is a simplified normal distribution function with two arguments. However, the fitting function can have any number of arguments. Therefore, these arguments are stored as a vector with a variable number of elements. The function *genfit* has four arguments: two vectors with the coordinates of the initial processing points (r and μ), the vector of the first approximations and the name of the fitting function (F). It is very important to set the correct values of the first approximation with the correct units of measurement. Otherwise, the function *genfit* will return no response (vector values of the coefficients of the fitting function), but an error message. The correct unit of measurement is easy to determine—it is enough to analyze the behavior of the fitting function. To set the value of the first approximation for the coefficient a_1 is also not very difficult—it is the value of the element of the vector r at the maximum value of the vector μ. It is more difficult to set the value of the coefficient a_0. Here it is possible either to enter random numbers and to see how the graph looks, or, without relying on the case, to proceed as shown in

Figure 12.12: by determining the value of the coefficient a_0 for the case when the fitting curve passes through one of the source points—through the point r = 15 cm and μ = 0.5, for example. Figure 12.12 creates an auxiliary function named *f*, with argument *x* and parameter *a*. This function uses the built-in function *root* to search near the point of the initial assumption (1/cm²) the solution, which will be the initial assumption for the function *genfit*. Without this tricky method it will be quite difficult to set the correct value of the initial assumption for the coefficient a_0.

Based on the function *F* and on the optimal coefficients a_r and b_r, the function $\mu_r(r)$ is created, this is the function of membership of a particular value of the base radius of the fire bucket to the fuzzy set "optimal radius of the fire bucket base", see Figure 12.10.

The same "public opinion polls" with their subsequent statistical processing were carried out in relation to the height of the fire bucket (Figure 12.13) and its volume (Figure 12.14).

Our fire bucket should be optimal **And** in reference to the radius of the base, **And** in height, **And** in volume. In the mathematics of fuzzy sets, the operation of logical multiplication (And) is performed by the command

$$f(x,a) := e^{a\cdot(x-13\,cm)^2}$$

$$a := \frac{1}{cm^2} \qquad \text{root}(f(15\,cm,a)-0.5,a) = -0.173\,\frac{1}{cm^2}$$

FIGURE 12.12 The determination of the initial approximation for a coefficient.

2. Views about optimum height of the bucket

h (cm)	μ
26	0
28	0.1
30	0.34
32	0.67
34	1
36	0.9
38	0.63
40	0.4
42	0.2
44	0

$$\begin{bmatrix} a_h \\ b_h \end{bmatrix} := \text{genfit}\left(h,\mu,\begin{bmatrix} -0.1\,cm^{-2} \\ 35\,cm \end{bmatrix},F\right) = \begin{bmatrix} -419.8702\,\frac{1}{m^2} \\ 0.3501\,m \end{bmatrix}$$

$h_g := 23\,cm, 23.1\,cm..43\,cm$

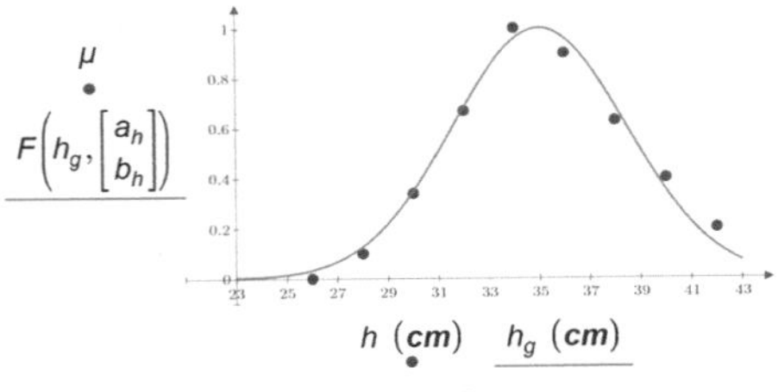

Bucket height membership function $\mu_h(h) := F\left(h,\begin{bmatrix} a_h \\ b_h \end{bmatrix}\right)$

FIGURE 12.13 The problem of the optimal fire bucket: continuation of the calculation.

3. Views about optimal volume of water in the bucket

v (L)	μ
8	0.2
9	0.4
10	0.8
11	0.95
12	0.9
13	0.7
14	0.2
15	0.1

$$\begin{bmatrix} a_v \\ b_v \end{bmatrix} := \text{genfit}\left(v,\mu,\begin{bmatrix} -0.01\,L^{-2} \\ 10\,L \end{bmatrix},F\right) = \begin{bmatrix} -165.7121\cdot 10^3\,\frac{1}{m^6} \\ (11.2689\cdot 10^{-3})\,m^3 \end{bmatrix}$$

$v_g := 7\,L, 7.1\,L..16\,L$

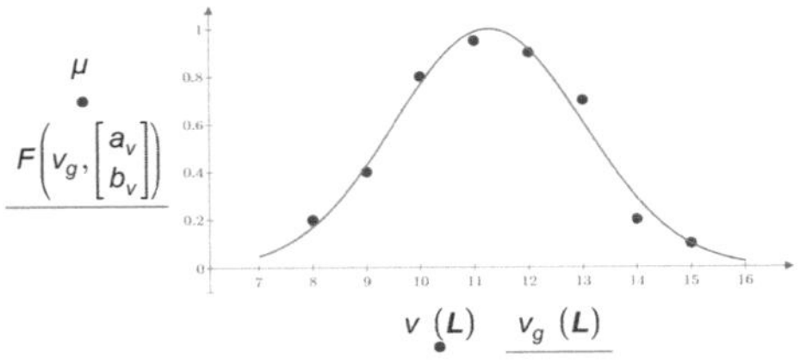

Water volume membership function $\mu_v(v) := F\left(v,\begin{bmatrix} a_v \\ b_v \end{bmatrix}\right)$

FIGURE 12.14 The problem of the optimal fire bucket: continuation of the calculation.

for determining the *minimum*, and we have already mentioned it in Study 11 (see Figure 11.12). In Figure 12.15, the membership function $\mu_{rh}(r, h)$ for a particular bucket with parameters *r* and *h* is formed for the fuzzy set "optimal fire bucket". To do this, the built-in Mathcad function *min*, is used in the calculation, having as its argument a vector of three elements: the previously given functions μ_r, μ_h and μ_v (see Figures 12.10, 12.13 and 12.14). The function $\mu_{rh}(r, h)$ itself will have only two arguments, since the bucket volume depends on their size.

We need somehow to "add up", or rather, sum up the opinions of people about the optimal parameters of the fire bucket, shown in Figures 12.10, 12.13 and 12.14. In TFS there are no concepts of addition, subtraction, multiplication, etc., underlying the traditional "Boolean" mathematics and implemented in Mathcad by infix operators +, –, ·, etc. In TFS, *multiplication* (*intersection* of sets, *And*), we repeat, is replaced with the operation of finding the *minimum* (**min**), and *addition* (*merging* sets, **Or**), by searching for a *maximum* (max). The mathematics of Boolean sets is a special case of the mathematics of fuzzy sets. In programs, instead of the function (operator) *And*, one can use the function of finding the minimum, and instead of the function *Or*, the function of finding the maximum, as we already mentioned in Study 11. In our problem, the membership function μ_{rh} is obtained by *fuzzy multiplication* (*min*) of the functions μ_r, μ_h and μ_v—the fuzzy set "optimal bucket" lies at the *intersection* of three other fuzzy sets: "optimal bucket radius" (Figure 12.10), "optimal bucket height" (Figure 12.13) and "optimal bucket volume" (Figure 12.14). It was possible to write the μ_{rh} function as a two-argument function with r and h, rather than a three-argument one (arguments r, h and v) by considering the formula for the volume of a cone that links all these three arguments.

In step 5 in Figure 12.15, using the Mathcad built-in function *Minerr,* the parameters of the optimal fire bucket are searched.

Figures 12.16 and 12.17 show the contour lines of the two-argument (of arguments r and h) functions of membership of a specific fire bucket to the fuzzy set "optimal fire bucket".

In Figure 12.17 it is possible to see a fire bucket, crushed due to its use not for its intended purpose (to extinguish the fire), but for washing the floors by inserting a bucket into an inverted stool.

At the beginning of this chapter, a fire shield at the summer residence of the author was mentioned. So, after the calculations outlined above, this author decided to measure the parameters of a real fire bucket. However, when he came up to this shield with a ruler, it turned out that there was no fire equipment on the shield, and all this (a crowbar, a shovel, an ax and a fire bucket) was simply painted with red paint on a green fire shield. However, a query in the Internet search engine led the author to a fireman's store, where conical fire buckets with dimensions of 280 by 380 mm, a volume of 8 liters and a weight of 1.1 kg are sold at a price of 110 rubles apiece. This (except for the price) was pretty close to what we got with our "fuzzy" calculations.

The website of this store contains the most common answers to the question why the fire bucket has the shape of a cone.

$$\mu_{rh}(r,h) := min\left(\begin{bmatrix}\mu_r(r)\\ \mu_h(h)\\ \mu_v\left(\frac{1}{3}\cdot\pi\cdot r^2\ h\right)\end{bmatrix}\right)$$

Solve

Solver Constraints Guess values

$r := 15\ cm \quad h := 40\ cm$

$\mu_{rh}(r,h) - 1 = 0$

$$\begin{bmatrix}r\\h\end{bmatrix} := \mathbf{Minerr}(r,\mathbf{h}) = \begin{bmatrix}0.1501496\\0.3814816\end{bmatrix} m$$

$$v := \frac{1}{3}\cdot\pi\cdot r^2\cdot h = 9.01\ L$$

$\mu_{rh}(r,h) = 0.4281552$

$\mu_r(r) = 0.42815524$

$\mu_h(h) = 0.66152119$

$\mu_v(v) = 0.4281553$

FIGURE 12.15 Determination of the parameters of the optimum fire bucket.

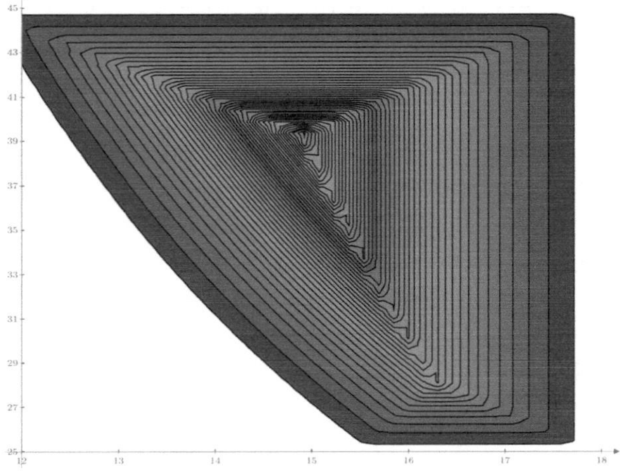

FIGURE 12.16 Contour lines of the membership function "optimum fire bucket".

7. Crushed fire bucket (surface by membership function)

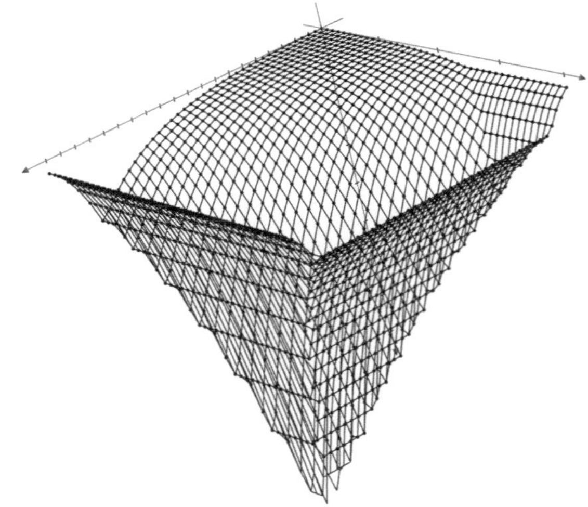

FIGURE 12.17 The surface of the membership function is the "optimum fire bucket".

1. From the production point of view: This form allows to not solder the bottom, therefore, the production process is noticeably simplified, and the bucket becomes cheaper.

2. From the domestic point of view (the most common opinion): This bucket will not be stolen from the fire shield, since it is not suitable* for domestic use.

3. The most basic opinion. From the point of view of fire science, this form of bucket can quickly cope with the fire: first, the cone-shaped bucket allows in winter to punch holes in fire reservoirs; secondly, a special form of a fire bucket allows to avoid splashing of water during firefighting. The water does not pour out evenly as from the usual bucket, but the water from the fire bucket is poured out only by a deliberate stream.

However, back to the main topic.

Even a not very attentive reader will notice inaccuracies made in solving the problem of the optimal fire bucket. Here are three of them:

- the bucket never fills up to the brim;
- the author is very free to treat such concepts as volume, weight and mass of the bucket, confusing them;
- the weight of the empty bucket, as well as the material from which it was made, is not taken into account.

However, let's look once again at the graphs illustrating the fuzzy sets in Figures 12.10, 12.13 and 12.14 in order to understand the most important feature of solving problems involving the TFS apparatus. Our solution isolates, if one may put it this way, the *essence* of the problem, ignoring various *trifles*: the density of water, the weight of the empty bucket, the degree of its filling.

This feature is currently implemented, for example, in automatic control systems, where the controllers tuned to the positions of the TFS are more "attentive" to the main signal and less susceptible to noise. After all, to control the work at a power plant, for example, we do not need to know the specific value of the vapor pressure in the boiler. We need to know whether it is elevated or not, and proceed from this decision. The concept of "low blood pressure" or "high blood pressure" (as, indeed, the concept of "low temperature" or "high temperature", see above) are typical representatives of fuzzy sets. It turned out, although it seems paradoxical, that traditional "clear" control algorithms are qualitatively inferior to "fuzzy" or they are just particular cases. In the theory of automatic control, there was a certain stagnation, because no new algorithms could compare with the good old proportional integral differential (PID) control algorithm. The principles of PID regulation can be seen, for example, in the procedure for making a decision to grant a loan to a bank customer, when the decision-maker takes into account, first, the amount of money in the current account of the applicant (the *proportional* component—the richer the client, the more money can be given as a loan), second, the dynamics of the change in the client's current account (the *differential* component, e.g. is the client's business on the rise or in decay) and, third, the average amount of money the client has handled for the last five years (for example, the *integral* component whether the client has just deposited money to create an appearance of well-being). It is possible to consider other components, but three is a beautiful number. In addition, in the end, the decision to issue the same loan is usually made "by inspiration", taking into account the lack of even blurry, "bushy" information. In Study 15 it is also possible to read about bankers, good and bad.

The PID algorithm of regulation was somehow imperceptibly fetishized. The ideas of fuzzy control are a fresh stream in the theory of automatic regulation, whose main provisions are currently being assessed. True, there is another opinion. Some scientists believe that the use of the TFS apparatus in the theory of automatic regulation and in cybernetics in general is an attempt to replace one *uncertainty* with another. Some researchers believe that TFS (it is only 40 years since it was disclosed to the world by L.A. Zadeh—an American of Iranian origin in 1965) is a "well-forgotten old". By tradition, it is customary to illustrate Boolean sets with circles with sharply contoured borders. Fuzzy sets are circles formed by separate points: there are many points in the center of the circle, but closer to the periphery

* You might think that this is related to the peculiarities of the Russian mentality, but on the Internet you can find pictures of foreign fire buckets with the inscription "fire", in which the bottom is also not flat, but, for example, spherical with an additional handle soldered to it for easy pouring out of water.

their density decreases to zero; the circle seems to be blurred (becomes "bushy") on the edges. These "fuzzy sets" can be seen in a dash—on the wall where the targets are hung. Traces from the bullets form *random* sets with known mathematics. It appears that for operations with fuzzy sets the already well-developed apparatus of random sets is suitable.

We speak about a *fuzzy set*. In order to be consistent, we have to state that the element of a fuzzy set is a new fuzzy set of new fuzzy sets, etc. Let's return to the classical example, to the *heap of grain*. The element of this fuzzy set will be a *million grains*, for example. However, a million grains are not a Boolean *element*, but a *new fuzzy set*. After counting the grains (manually or automatically), it would be no wonder if 999,997 grains, for example, were mistakenly taken for a million. Here we can say that the element 999 997 has a value equal to 0.999997 for its membership function to the set "million". In addition, the grain itself is not an element, but a new fuzzy set: there is a full grain, and there are also two fused grains, an underdeveloped grain or just a husk. When counting grains, a person might somehow reject one or take two grains for one, or perhaps, in another case, take one grain to be two. A fuzzy set is not so easy to implement in a digital computer with classical languages: the elements of the array (vector) must be new arrays of arrays (nested vectors and matrices, if we talk about Mathcad). The classical mathematics of Boolean sets (number theory, arithmetic, etc.) is a hook by which a *reasonable person* fixes (determines) himself in a slippery and fuzzy surrounding world. A hook, as known, is a rather rough tool, quite often spoiling what is clinging to it. It is also difficult to force terms that reflect fuzzy sets (and there are enough of them in this book—"a lot", "slightly", "a little bit", etc.) into the computer because they are *contextually dependent*. It is one thing to say "Give me some seeds (grain)" to a person who has sunflower seeds in his pocket, and another thing to say this a person sitting at the wheel of a truck laden with seeds.

Is it possible to see a certain *crisis* in the theory and practice of programming, relating to the contradiction between the Boolean structure of programs (data) and the fuzzy world? Is it necessary to develop "fuzzy" programming languages for implementing "fuzzy" algorithms and for placing "fuzzy" data? Opinions here are different. Programmers have at least learned to "push" the fuzzy world into a strictly deterministic computer.

Our approach to the selection of the optimal fire bucket can be applied to the choice of the optimal sagging chain, which was described in Studies 4 and 7. At the end of Study 4, the parameters of the chain with the minimum support forces at anchor points located on a given level were determined. It was shown that for such a chain, the ratio of its length to the distance between the attachment points should be 1.258. However, will such a chain also be beautiful? After all, chains are often used purely for decorative purposes—they, for example, may enclose various monuments.

It is possible to hang chains of different shapes on the street (with different values of the ratio of the length of the chain to the distance between the attachment points) and ask passers-by (experts) to give their estimates of "beauty" on a similar scale to that we used for the fire bucket: the chain is ugly; the chain is rather more ugly than beautiful; I cannot determine; the chain is rather beautiful and the chain is beautiful. We can expect that the chain with the ratio of its length to the distance between the attachment points, equal to 1.258, will turn out to be beautiful. After all, it often turns out that a well-built bridge turns out to be beautiful. And vice versa!

In Study 14, we will return to the theory of fuzzy sets, relating it to the growth of a man.

And at Last

Mathcad's built-in tools for solving optimization problems sometimes prove powerless in the presence of a large number of optimization variables and of a complex objective function. Here it is necessary to develop an individual solution scheme or to look for a solution by a provider. In this case the company "Sigma Technology", for example, offers users of Mathcad and of other popular calculation programs user-defined functions for solving complex optimization problems.

For more information, see http://www.iosotech.com.

ASSIGNMENTS FOR READERS

Try to solve optimization and other problems in this book with the help of the tools of fuzzy sets.

The Journey of a Circle in a Triangle, and of a Triangle in a Hollow Or

A Self-Directed Computer Director

ABSTRACT

In this Study, readers learn how to animate numerical methods for solving problems on a computer. The methods of step-by-step tracing of the process of numerical solution of problems on a computer and the visualization of the results are shown. Living names of methods for solving problems on a computer are discussed.

Mathematics: Cartesian coordinates, triangle, inscribed circle, properties of the bisectors of the triangle, Pythagoras theorem, Heron's theorem, triangle area, system of equations, equation of a straight line on the plane, level curves (contour lines) of a function of two variables, Bernoulli lemniscate.

Physics: The Coriolis force, water swirling.

IT: Tracing of the intermediate data of numerical methods for solving equations, symbolic mathematics, animation, simplex method, "Two steps" method, pretest cycle, loop with parameter, conditional operator.

Art: Cinema.

Study website: https://community.ptc.com/t5/PTC-Mathcad/Study-13/m-p/483810.

Problem. A triangle is given by the Cartesian coordinates of its vertices (x_1, y_1), (x_2, y_2) and (x_3, y_3). Find the following parameters of a circle inscribed in this triangle: the radius r and the Cartesian coordinates of the center (x_r, y_r) (see Figure 13.1).

It is possible to try to search for ready-made formulas in reference books or on the Internet, and by using them the values of the unknowns r, x_r and y_r can be calculated from the values of the given variables x_1, y_1, x_2, y_2, x_3 and y_3. However, ready-made formulas for all three quantities r, x_r and y_r cannot be found in reference books or on the Internet (they cannot be, or rather, cannot yet be found: read the Study further). It is possible to find only a very elegant formula, according to which the radius of a circle inscribed in a triangle is calculated depending on its area (S) and on its half-perimeter (p)* or depending on the lengths of the sides of the triangle (a, b and c) and on its half-perimeter (the area of the triangle is expressed by the Heron formula $S_\Delta = \sqrt{p(p-a)(p-b)(p-c)}$, that is studied in school):

$$r = \frac{S}{p} = \sqrt{\frac{(p-a)(p-b)(p-c)}{p}}$$

The quantities S, p, a, b and c will need to be calculated by other formulas, which again need to be searched

* Most people will understand the variable p to be the perimeter of the triangle: the sum of the lengths of its sides. However, in this formula for the radius of the inscribed circle the variable p is reserved to be the semi-perimeter—for the sake of mathematical simplicity, while the perimeter itself is denoted by the capital letter P, just as the area is denoted by the capital S. And it is very confusing for many. Truly it is said: "Simplicity is worse than theft."

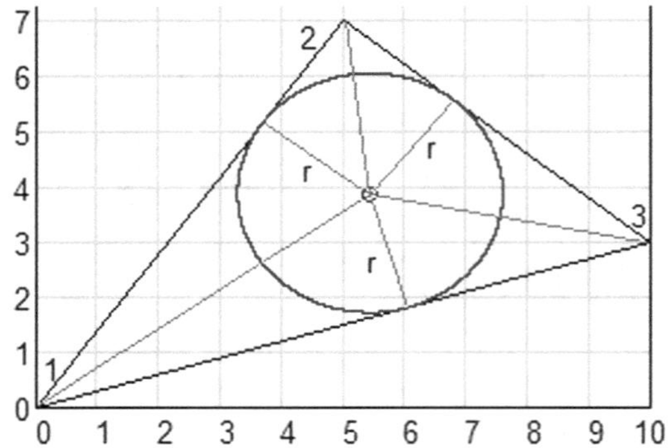

FIGURE 13.1 The problem of a circle inscribed in a triangle.

somewhere or derived by the reader. The lengths of the sides of a triangle can, for example, be calculated from the values of the Cartesian coordinates of the vertices using the theorem of another ancient Greek, Pythagoras, see Figure 13.2.

Figure 13.2 is a procedure for calculating the radius of a circle inscribed in a triangle, made in Mathcad. The coordinates of the vertices of the triangle in Figure 13.2 are given in meters. Many "pure" mathematicians will mock this approach. They will immediately say, and where are the units of length?* Why are there just meters, instead of centimeters, kilometers or feet with inches?! And here's why! Inputting units of measurement in the calculation allows control of the dimensions and the identification of possible errors when entering formulas. Someone, for example, may forget to square a variable or mistakenly raise another to a cube. Calculations without meters "swallow" this oversight, while by entering dimensional values, then this computational error will be immediately noticed and corrected.

When solving the problem of a triangle inscribed in a circle, one cannot look for ready-made solutions, but should proceed after having analyzed the problem. Recall, for example, that the radius drawn at the tangent point is perpendicular to the tangent (see below). As result of this analysis, it is not difficult to write down a system of three algebraic equations describing the position of a circle in a triangle, and to solve this system analytically on a computer (symbolically, see Figure 13.3) or numerically (approximately, see Figure 13.4). Previously, for obvious reasons, it was impossible to think about this "frontal attack" on the problem—there were no available computer solutions for systems of non-linear algebraic equations.† Now it's different! Almost every student or schoolchild has a computer at his fingertips, in which it is easy to download the appropriate software for solving such problems!

Three algebraic equations are obtained if the original triangle is divided into six right-angled triangles (we return to Figure 13.1). In each of them one of the vertices rests against the center of the circle, and one of the legs is the radius of the circle. Hypotenuses of these right-angled triangles are segments of straight lines connecting the center of the circle with the vertices of the triangle. Our equations, shown in Figures 13.2 and 13.3, fix the fact that the lengths of the sides of the original triangle are the sums of the lengths of pairs of legs of adjacent rectangular triangles.

$$x_1 := 0\text{m} \quad y_1 := 0\text{m} \qquad x_2 := 5\text{m} \quad y_2 := 7\text{m}$$

$$a := \sqrt{(x_1 - x_2)^2 + (y_1 - y_2)^2} = 8.602\,\text{m}$$

$$x_3 := 10\text{m} \quad y_3 := 3\text{m}$$

$$b := \sqrt{(x_2 - x_3)^2 + (y_2 - y_3)^2} = 6.403\,\text{m}$$

$$c := \sqrt{(x_1 - x_3)^2 + (y_1 - y_3)^2} = 10.44\,\text{m}$$

$$p := \frac{a + b + c}{2} = 12.723\,\text{m}$$

$$S := \sqrt{p\cdot(p - a)\cdot(p - b)\cdot(p - c)} = 27.5\,\text{m}^2$$

$$r := \frac{S}{p} = 2.161\,\text{m}$$

FIGURE 13.2 Calculation of the radius of a circle inscribed in a triangle in Mathcad.

* In older versions of Mathcad, it was possible to operate with non-specific units (meters, seconds, kilograms, etc.), but with physical quantities (length, time, mass, etc.), which is more logical for our problem. However, this will probably disappoint the engineers who love specific values in the calculations. It is not possible to please everyone. Someone, for example, can even be disappointed by such a "purely mathematical" expression as $x + x^2$: it is not possible to add a variable and its square!

† These equations are not algebraic (consisting of polynomials), but irrational. The term "algebraic" here means that this is not a differential equation.

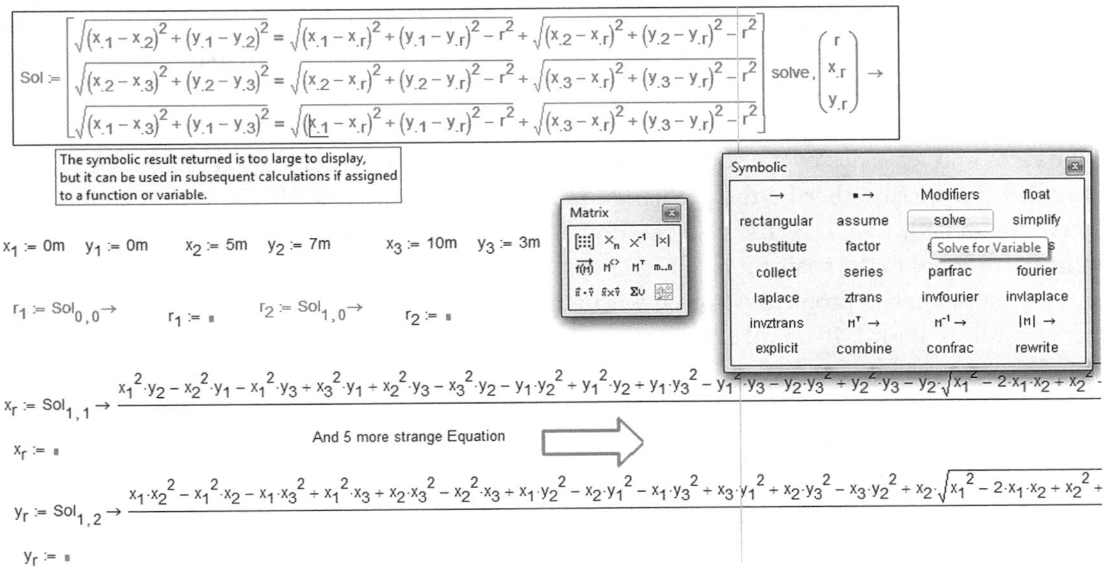

FIGURE 13.3 An analytical solution of the problem of a circle inscribed in a triangle.

Figure 13.3 shows the use in Mathcad of an analytic (symbolic) operator *solve* when searching for *formulas* for the variables r, x_r and y_r. Mathcad solved the problem, though it was not able to "display" the very large response*—a matrix with two columns (the two solutions of the system) and three rows (three formulas for three unknown systems). This error message about "fell out" from the "reddened by the effort" of the operator *solve*. Guided by this message, we have printed out the individual elements of the response matrix *Sol*, and even then only small parts of the formulas for x_r and y_r. The Mathcad package could not "display" the formula for r. These formulas can be simplified, but they will still remain too voluminous. Therefore, they are not available in reference books or on the Internet. However, in principle, despite their very large size, they can now be placed in online references in the form of, for example, very long texts, not for educational purposes, but for copying and pasting them into computer programs when it is necessary to calculate the position of a circle inscribed in a triangle from the coordinates of the vertices of a triangle.

On the three formulas (two visible and one invisible, see Figure 13.3), the values of the unknown quantities r, x_r and y_r were calculated from the values of the coordinates of the vertices of the triangle given earlier.

If we need to solve our problem no longer on a plane but in space (to find the parameters of a sphere inscribed in a tetrahedron, for example), then the analytical approaches will not work here. In this case, we must immediately resort to numerical methods for solving the problems. Let us illustrate them on the already solved "planar" problem of a circle and a triangle (see Figure 13.4).

Figure 13.4 shows the *numerical* solution of our system of three non-linear equations. In order to do this, it is necessary to input the original data (the coordinates of the vertices of the triangle), to enter the *Given* keyword, to specify the values of the first assumption of the solution, and to write the system of equations. The Mathcad built-in function *Find* starts to change the values of its arguments (the unknowns r, x_r and y_r) with

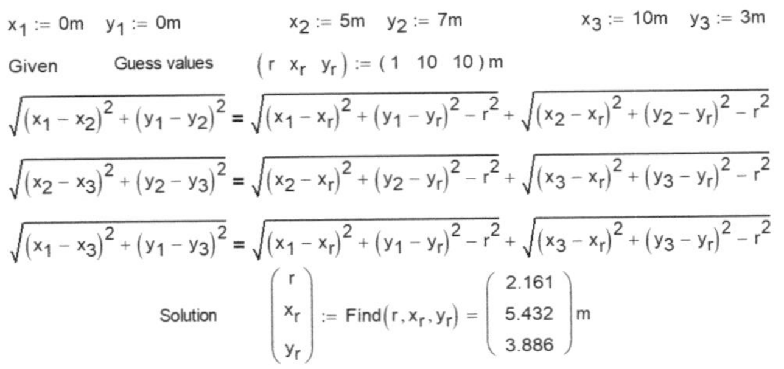

FIGURE 13.4 Numerical solution of the problem of a circle inscribed in a triangle.

* We already wrote about this issue in Study 4, see Figure 4.3.

a special numerical algorithm so as to finally come to a solution—to find the values of these variables, whose substitution into *the* initial equations turns them into identities. Rather, *almost* (just about, *approximately*) into identities: the right and left sides of the equations must differ in modulus from each other by the pre-assigned number *CTOL*. By default $CTOL = 10^{-3}$, and the units of measurement are equal to the basic units of SI meters in our case, i.e. our geometric problem is solved within a millimeter accuracy (10^{-3} m). If we measured our triangle and circle not by meters, but by centimeters, then the answer would be too rough. Here it would be necessary to change the value of the system variable *CTOL* to the value of 10^{-7}, for example.

And by what algorithm does the Find function "push" the values of the unknowns from the initial assumption given by the user to the sought solution?! It is possible to find out the name of this algorithm (method) by right-clicking on the word *Find* and by searching for information on the particular method on the Internet (see Figure 13.5). However, the reader can do something different. not so boring.

The Mathcad 15 package, in which the problem was solved numerically (Figure 13.4), has a very useful capability of recording the intermediate values of the unknown quantities when implementing some of the built-in tools for the numerical solution of problems. In our case, we can recall the catchword and say: "The goal is nothing—motion is everything!" We are not particularly interested in the goal—the place of the "inscribing" of the circle in the triangle (we have already found it analytically, see Figure 13.3, and numerically, see Figure 13.4), but how the circle is "rolled up numerically," that is, its motion.

Figure 13.6 shows the path of the center of the circle from the point of the initial assumption ($x = 10$ m, $y = 10$ m)

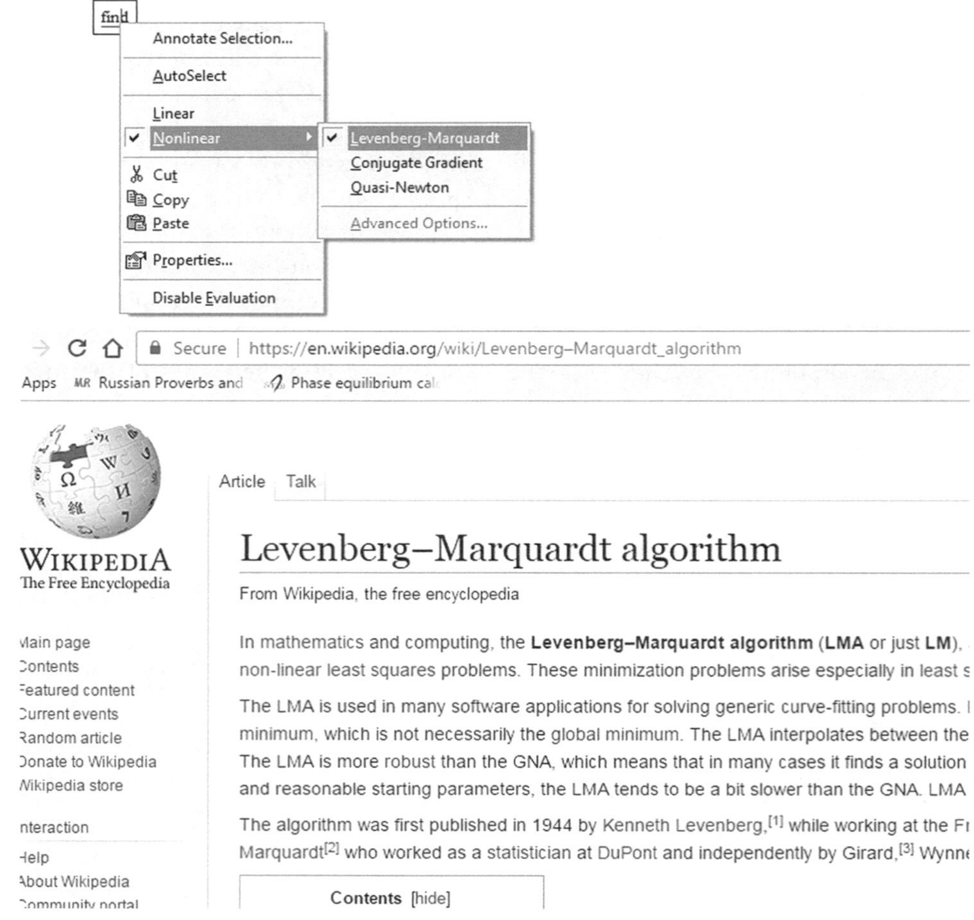

FIGURE 13.5 The selection of the method for solving the system of equations in Mathcad 15 and online help.

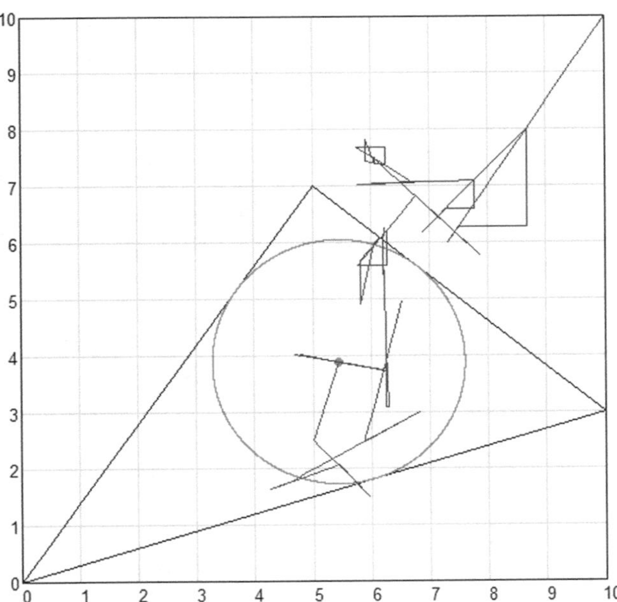

FIGURE 13.6 Trajectory of the motion of the center of a circle in a triangle.

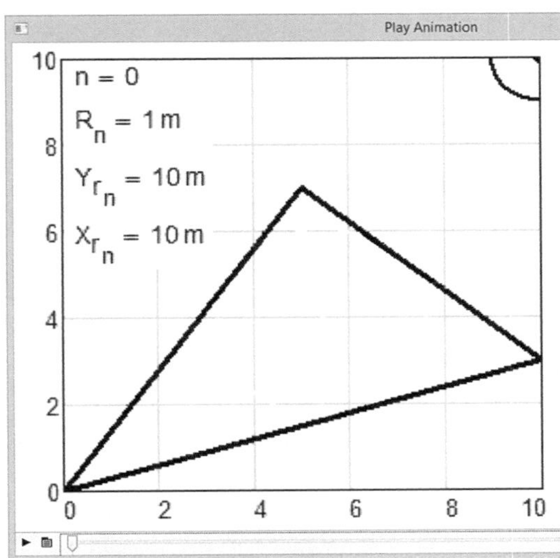

FIGURE 13.7 The initial frame of the animation of a circle in a triangle.

to the "target", to the solution ($x = 5.432$ m, $y = 3.886$ m). We specifically asked the Mathcad package to perform a rather difficult task, to put the point of the initial assumption outside the triangle to see how the package "will reach the goal": "Computer, You imagine a lot about yourself—let's see if you swallow this!" And it would be necessary, of course, to set the values of the first assumption near the solution—somewhere in the center of the original triangle. However, we repeat "The goal is nothing—motion is everything!"

However, it is interesting to see this circle motion with the **animation**, the creation tools available in Mathcad 15, and that we described in detail in Study 8. Figure 13.7 shows the initial frame of this animation ($n = 0$, and the whole frames will be 373): the circle (only its quarter-arc is visible) with a radius of 1 m is at the point $x = 10$ m, $y = 10$ m. Then (the first 94 frames of the animation), the circle breaks off and hits the nearest side of the triangle (Figure 13.8), jumps and strikes the triangle again and again, as if trying to pierce a hole in it.

However, this "frontal ram"-type of attack is not possible and the circle begins to roll along the side of the triangle, trying to somehow penetrate into it (Figure 13.8). These attempts are also unsuccessful. Then our "living and annoying" circle begins to decrease in size almost to a point (Figure 13.9), finds a hypothetical "hole" in the side of the triangle and penetrates into it (Figure 13.10).

Then, once in the triangle, the point increases to a circle and begins to "beat" inside the triangle (Figure 13.11), changing its position and radius, until it "calms down"—it does find its place in it: the solution of the problem (Figures 13.12 and 13.6).

This "living" behavior of the circle when it is "numerically" inscribed in a triangle, shown in Figures 13.7–13.12, can be seen only with a good spatial imagination. However, the reader can cut out pictures with animation frames from a magazine, sew them to a notepad, and

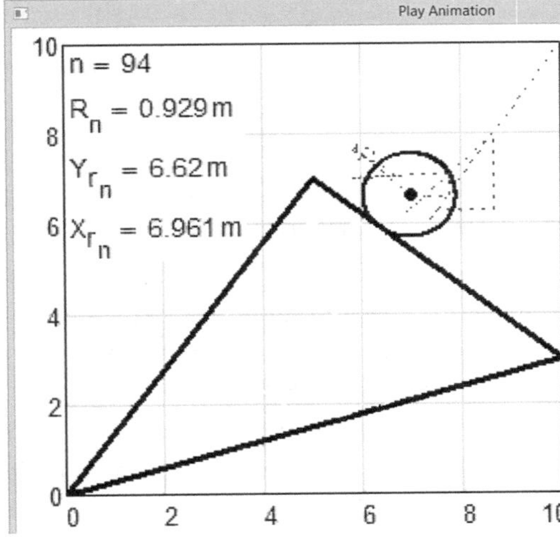

FIGURE 13.8 The 94th frame of the animation of a circle in a triangle.

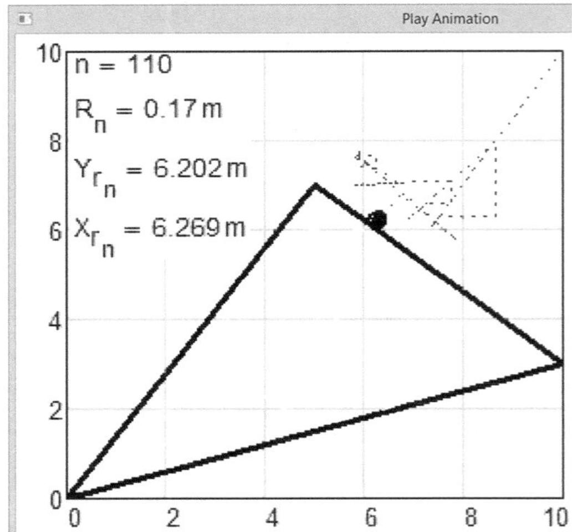

FIGURE 13.9 110th frame of animation of a circle in a triangle.

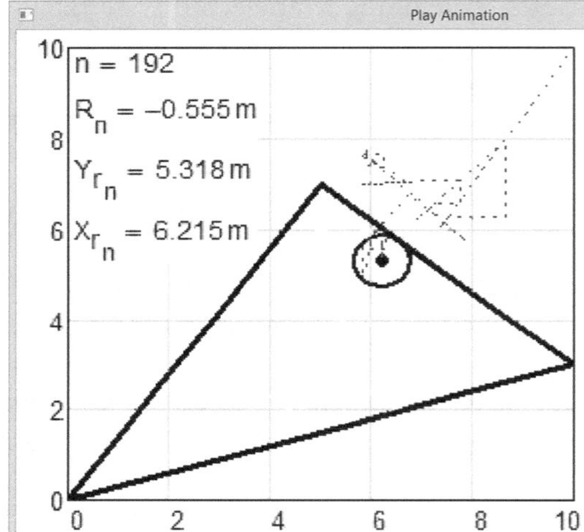

FIGURE 13.10 The 192th frame of the animation of the circumference in a triangle.

quickly scroll through it, by bending and squeezing the pages and releasing them.* Such book-toys (flip-book) with planar subjects were very popular before the invention of cinema and television. So, similarly, the Mathcad 15 does this while creating the animation, "stapling" the individual frames, and then showing them at a speed sufficient for animation.

How can I output intermediate data when running the Mathcad 15 built-in numerical solution tools?

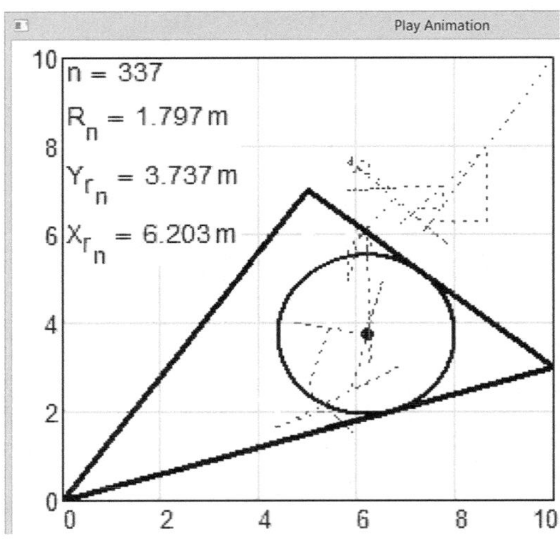

FIGURE 13.11 337th frame of animation of a circle in a triangle.

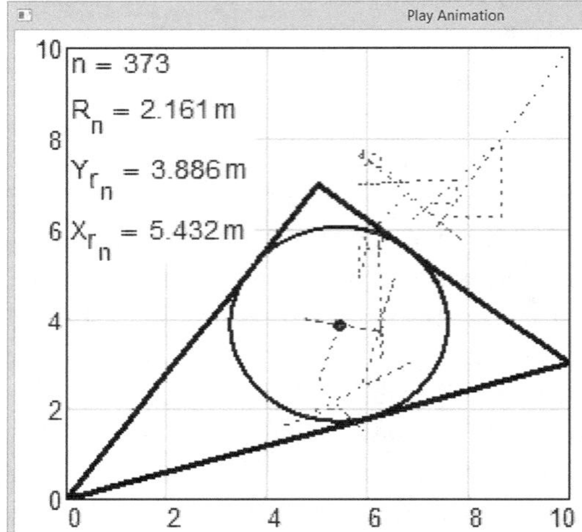

FIGURE 13.12 The final frame of the animation of the motion of a circle in a triangle.

In order to do this, Mathcad 15 has three tools: the built-in trace function, the Toggle Debugging (mode) and the Trace Window, where intermediate data are entered. All of them are shown in Figure 13.13.

The *trace* function must be inserted into one of the algebraic equations (see the upper left corner in Figure 13.13). The arguments of the *trace* function are alternately: the radius of the circle (as in Figure 13.13) and the coordinates of the center, whose initial, intermediate and final values will be automatically recorded in the Trace Window (bottom left corner of Figure 13.13), if the debugging mode is enabled (see the right side of Figure 13.13). These arrays of data must be copied, transferred

* Eureka! The reader can ask a publisher to edit this book so that the pictures 13.7–13.12 are located in the same place on different pages and they could be "revived" by quickly flipping through the pages of the book.

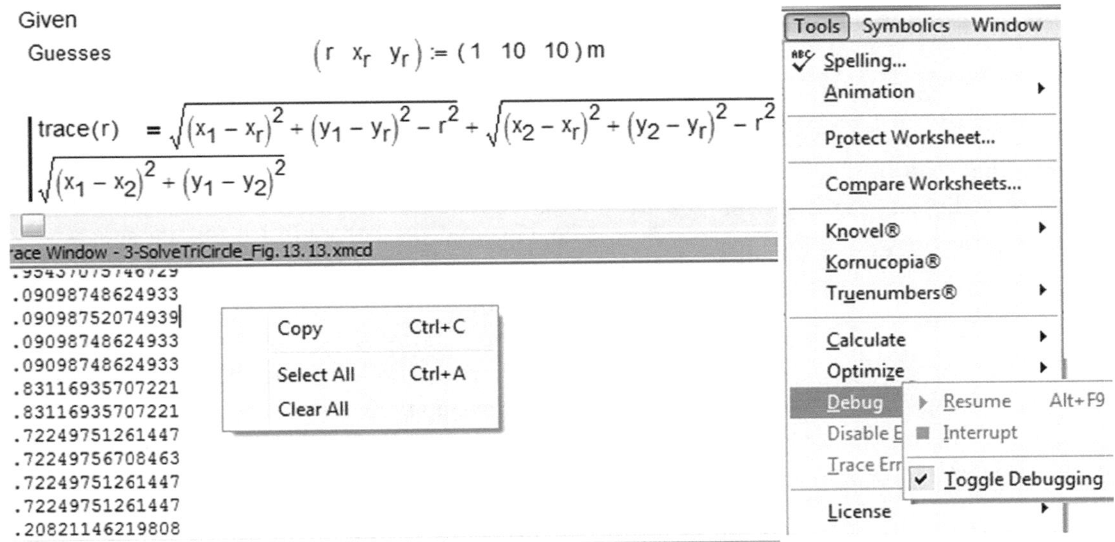

FIGURE 13.13 Tools for outputting intermediate results of a numerical solution search.

into calculation in the form of vectors, displayed on a* graph and, by using them, it is possible to create the animation of the circle from the assumed first point to the end point. In this case, the reader can change the initial size and position of the circle, as well as the method of numerical solution of the problem and the accuracy of finding the root of the algebraic equation. So, the reader can "shoot a rather fascinating movie", i.e. to engage in a kind of *computer-based visual arts*, based on applied mathematics.

In Figures 4.24 and 4.27 of Study 4, the use of the *trace* function for finding the zero of a function is shown, using the secant method and the half-division method.

On the PTC Community website, it is possible to see the animation of the numerical solution of the "reverse" problem—the problem of a circle circumscribing a triangle. The circle breaks from the place in the center of the triangle almost to the solution, and then "in doubt" (again a living organism!) begins to scour it, seeking the final solution.

Well, and if the reader does not get carried away with "filming a movie with a circle and a triangle in leading roles", and prefers to solve the problem of inscribing a circle in a triangle using traditional methods, then we need three theorems: "There is nothing more practical than a good theory!"

Theorem 1. All three bisectors of the triangle intersect at one point.

Theorem 2. The center of a circle inscribed in a triangle lies at the intersection point of the bisectors.

Theorem 3. The bisector of the angle of the triangle divides the opposite side into segments proportional to the adjacent sides.

By the way, in previous calculations we relied on a more important theorem: the radius of the circle drawn to the point of tangency is perpendicular to the tangent. This is what allowed us to draw six rectangular triangles and to write down for them a system of non-linear algebraic equations, which we solved analytically and numerically.

Thus, relying on the three theorems listed above, we outline the plan for solving the posed problem. It is sufficient for us to find the intersection point of only two of any bisectors (Theorem 1). For this we need the equations of these bisectors. We obtain the equation of a straight line (bisectors), knowing two points lying on this line.

We proceed from the fact that the lengths of the sides of the original triangle have been previously found. We denote with a_{12} side length A_1A_2, a_{13} side length A_1A_3 and a_{23} side length A_2A_3 (see Figure 13.1). Let A_3B_3 be the bisector of the angle A_3, with the point B_3 lying on the side A_1A_2. Similarly, A_2B_2 is the bisector of the angle A_2,

* ESC and F9 to activate the Trace Window.

with the point B_2 lying on the side A_1A_3. Then (Theorem 3) we have the relations $\frac{A_2B_3}{A_1B_3}=\frac{a_{23}}{a_{13}}$ and $\frac{A_3B_2}{A_1B_2}=\frac{a_{23}}{a_{12}}$.

If two points $A(x_A;y_A), B(x_B;y_B)$ of the plane are known, then the coordinates of the point $M(x_M;y_M)$, which divides the segment AB in the ratio $\frac{AM}{BM}$, are expressed by the formulas: $x_M=\frac{x_A+\lambda\cdot x_B}{1+\lambda}$, $y_M=\frac{y_A+\lambda\cdot y_B}{1+\lambda}$. QUOTE xM=xA +λ•xB1+λ QUOTE yM=yA+λ•yB1+λ

Using these formulas, we can straightforwardly calculate the coordinates of the points $B_3(x_{B3},y_{B3})$ and $B_2(x_{B2},y_{B2})$.

In the general case the equation of a straight line has the following form $Kx+Ly+P=0$. However, since at the origin A_1 there is a vertex, the sought bisectors A_2B_2 and A_3B_3 do not pass through the origin, and therefore the equation of each bisector will have the simplified form $kx+ly+1=0$. The sought coefficients are obtained from the system obtained by substituting the coordinates of two points lying on this line into the equation of the straight line. Thus, the equation for the bisector will have the form $k_2x+l_2y+1=0$ where $\begin{cases}k_2x_2+l_2y_2+1=0\\ k_2x_{B2}+l_2y_{B2}+1=0\end{cases}$, and for the bisector A_3B_3 it will have the form $k_3x+l_3y+1=0$, where $\begin{cases}k_3x_3+l_3y_3+1=0\\ k_3x_{B3}+l_3y_{B3}+1=0\end{cases}$. It remains to find the point (x, y) of intersection of these bisectors, since (Theorem 2) it is the center of the circle. The coordinates (x, y), thus, are found from a system of six linear equations

$$\begin{cases}k_2x_2+l_2y_2+1=0\\ k_2x_{B2}+l_2y_{B2}+1=0\\ k_3x_3+l_3y_3+1=0\\ k_3x_{B3}+l_3y_{B3}+1=0\\ k_2x+l_2y+1=0\\ k_3x+l_3y+1=0\end{cases}$$

, in which the numbers x_2, x_3, y_2, y_3 are determined by the condition, while the numbers $x_{B2}, x_{B3}, y_{B2}, y_{B3}$ were previously found.

Ugh! Let us return to the "film shootings" with nonlinear equations.

The system of algebraic equations shown in Figures 13.3 and 13.4, as we have already noted, has two solutions: one with a positive value of *r* and the other with the same absolute value but a negative value of *r*. Figure 13.14 shows a frame of another "motion picture"—the solution of a system of two algebraic equations: the ellipse equations and the Bernoulli lemniscate equation, which has not two but six real roots. Around these two closed curves and their intersection points (the roots of the system of two equations), the point of the first assumption rotates in a circle. Another point (the current solution of the system of equations is the point of intersection of the ellipse with the lemniscate) skips from one place to another. This authorial animation on the Planet PTC website is numbered 2418. It shows how the point of the first assumption affects the decision of the system—the choice of one or another root. This animation focuses on the importance of choosing the starting point of the solution.

Sometimes, instead of the term "Initial guess", which is very important when using numerical methods for solving problems, the term "initial **approximation**" is used, which is not entirely true (we noted this in Study 4), and the animation shown in Figure 13.14, proves this: very often the solution is not the point closest to the original, but quite different (this is fixed in Figure 13.14), and in some cases even one of the most distant ones. Everything depends on the system of equations and the method applied to its numerical solution.

In the animation shown in Figure 13.14, the point of the first assumption moves along a circle around the ellipse and the Bernoulli lemniscate. Another point that marks the root of the system of equations will jump

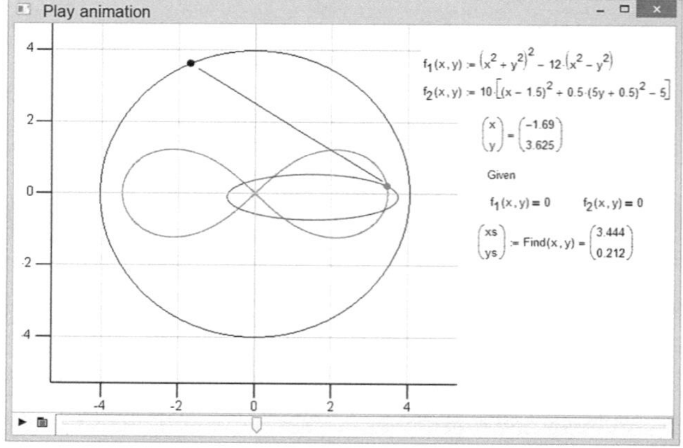

FIGURE 13.14 Animation frame on the relation between the first assumption and the root of a system of equations.

from an intersection to another of the two closed curves. In the specific case, shown in Figure 13.14, the next root is not always close to the point on the circle.

Observing the motion of the circle in the animation (Figures 13.1–13.13), we see that it behaves like a living organism, clogging its "triangular burrow". Here it is very appropriate to recall that some numerical methods for solving problems have "animal" names. For instance, one of the methods for finding the minimum of a function is called the amoeba method. The reader can understand the origin of this name by looking at how a triangle, rather than a circle, can behave like a living being and creep somewhere, while solving a mathematical problem.

The frames of the triangle creep animation are shown in Figure 13.15 (the author of the program and of the animation is V.R. Sabanin): a futuristic triangular amoeba swinging from side to side, crawling from the right bottom corner into a certain hollow (the minimum of a function of two arguments), decreases in size (shrinks), until it finds the sought point, shown in the contour lines of Figure 13.15 frame.

It is possible to find the description of this optimization method on the Internet (its other names are the Nelder-Mead method, the method of the deformable polyhedron or the simplex method) and understand its essence. The method is rather complicated in both description and implementation, but it is effective: the minimum point is reached fast (with a small number of moves) and quite accurately. If the function being analyzed does not have two, but three arguments, then the triangle is replaced by a tetrahedron. If there are more than three variables, then the corresponding polyhedron is selected.

Figure 13.16 shows the description and implementation of a less quick but rather simple optimization method of the author, which can be conditionally called "Two Steps".

The program, stressing once again, is quite simple—it took more time to find its name than to create it. It would be possible, of course, to apply the standard gradient descent method, in which steps are taken in the direction of the gradient (when searching for the maximum) or minus the gradient (when searching for a minimum), but the goal was to write the simplest program for the implementation, especially without thinking about its speed. The program was named "Two Steps". The program when looking for a minimum makes two steps from the next point of approach and transfers the reference point to the point where the value of the optimized function is minimal. These two steps of length X are made in all the coordinates (their number L) in the negative ($-D$) and positive (D) directions. As we approach the minimum, the steps are shortened. The "Dance" is repeated until the step is shorter than the specified error value ($D <$ CTOL, see Figure 13.17).

The program "prints" not only the final point, but also the entire path to this point (matrix M), which can be displayed on a graph (see Figures 13.18 and 13.19) and animated (see Figure 13.16).

In the program, it is not the algorithm that is interesting, but the trace left by the program on the X–Y plane ("in the steppe") when optimizing some function of two variables, see Figures 13.18 and 13.19.

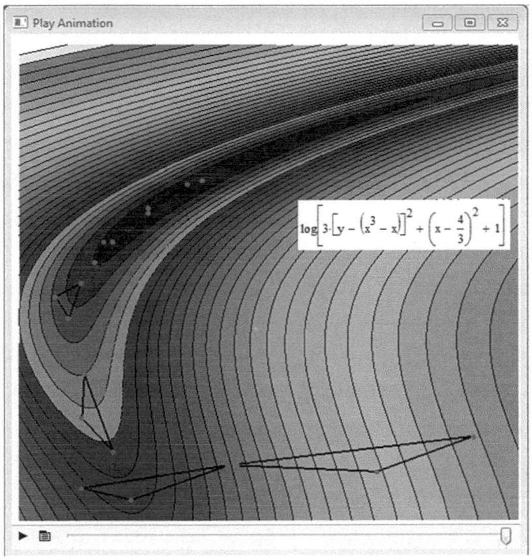

FIGURE 13.15 Climbing a triangle into a hollow.

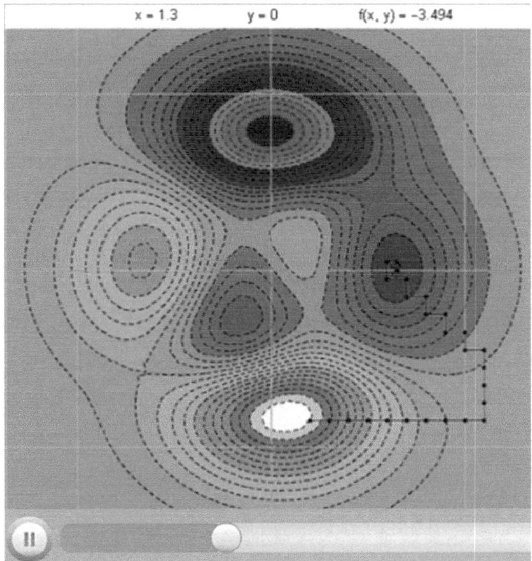

FIGURE 13.16 Animation of the "Two Steps" method.

```
MinTwoStep(f,x,D) := | (L ← last(x)  fmin ← f(x)  j ← p ← n ← 0  M<0> ← x)
                     | while D > CTOL
                     |   p ← 1
                     |   while p
                     |     p ← 0
                     |     for i ∈ ORIGIN..L
                     |       for X ∈ -D,D
                     |         (xi ← xi + X   F ← f(x))
                     |         (p ← X   j ← i   fmin ← F)  if F < fmin
                     |         xi ← xi - X
                     |     (xj ← xj + p   n ← n + 1   M<n> ← x)
                     |   D ← D/2
                     | M
```

FIGURE 13.17 The Mathcad program "Two Steps".

The form of the trace of the program "Two Steps" allowed it to be given other names. It is believed that the more names a person has, the more that person has heavenly patrons. This, perhaps, can also be attributed to programs (methods) for solving problems.

If the reader fills a tub with water, throws a feather in it, for example, and pulls out the drain stopper, then initially the feather will move calmly to the drain-hole, and then it will turn in a whirlpool. Much as shown in Figures 13.18 and 13.19, with the initial steps of approximation (Figure 13.18), the intermediate points drift to a minimum and then "twist in the whirlwind of dance" (Figure 13.19), which allows us to call this optimization method not only "Two Steps", but also "Waltz".

Modern software tools allow easy animation and the use of quick numerical methods for solving mathematical

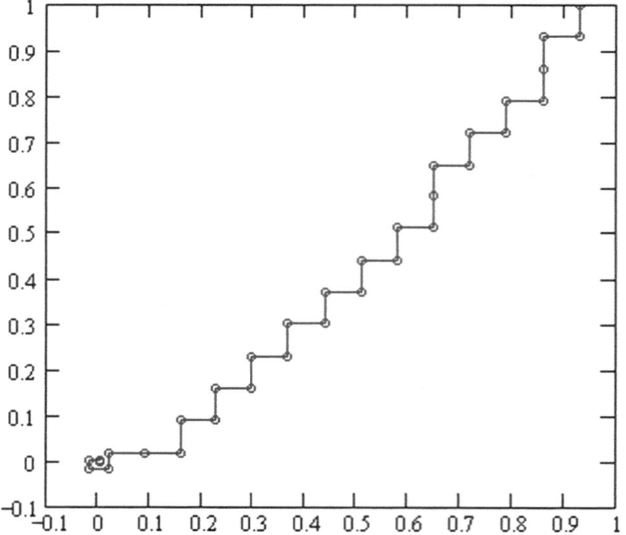

FIGURE 13.18 Initial steps to Minimize.

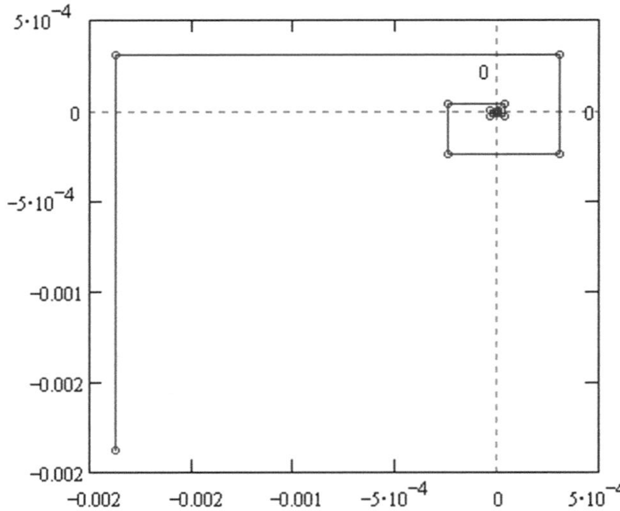

FIGURE 13.19 Final steps to Minimize.

problems. Animation allows us to better understand these methods, features and limitations when applying them to solve applications on the computer. In addition, this animation will revitalize, making it more entertaining to study these methods in seminars and lessons in informatics and applied mathematics.

The author does not tire of repeating to his students that from learning it is necessary first of all not to try to get knowledge and skills, but pleasure. Study is work, very difficult work, and without pleasure—without a love of work, even a burdensome, prestigious and highly paid work can turn into hard labor. Pleasure is unthinkable without a joke, without an easy and ironic attitude to the most serious things. And what can be "more serious" than mathematics? This book can be seen as an attempt to return students' interest in this most basic human ability—to gain knowledge. Many students are drawn to the computer, though their future engineering specialty is not always computer technology, but, for example, heating engineering. The author hopes that this book will help these students, at least through the computer and the Internet, to revive the ability to gain pleasure from studying in a technical college whose basis is mathematics. It was not for nothing that Immanuel Kant used to say that "in every natural science there is as much truth as there is mathematics in it."

ASSIGNMENTS FOR READERS

Create a 3D animation of the motion to the minimum point using the "Two Steps" method. The "planar" case is shown in Figures 13.18 and 13.19.

Interpolation, Extrapolation and Fitting Or

*Lies, Damned Lies and Statistics**

ABSTRACT

In this Study, readers learn about some statistical methods for the processing of data sets. The methods of interpolation, smoothing and the extrapolation of a set of statistical data obtained by the author in his lecture for students are discussed. A new method is proposed for the qualitative assessment of student growth using ancient length units.

Mathematics: Function, correlation, interpolation (linear and splines), extrapolation, regression analysis, statistical sampling, table of function value, calculation accuracy, scattering diagram, histogram, system of two linear equations, least squares method, derivative, spline, normal distribution function, Gaussian curve, exponential dependence, graph of a function of two variables, level (contour) lines, minimum of a function of two variables.

Physics: Statistical processing of experimental data, measurement error, heat conductivity, weight and height of a person.

IT: Database with fields and records, interactive calculations on the Internet, floating-point numbers, calculations by empirical and pseudo-empirical formulas, computer tools for analytical transformations (symbolic mathematics), cloud function, genetic optimization algorithm.

Art: Ivan Turgenev *Mumu*; *Hercule Poirot* by Agatha Christie; John Galsworthy *Forsyte Saga*.

Study website: https://community.ptc.com/t5/PTC-Mathcad/Study-14/m-p/642397

As was rightly noted "the exaggeration of the role of computer science is as, and perhaps even more dangerous, than the understatement of this role". Summing up the last decade, we can say with confidence that modern students trust Wikipedia more than their lecturer, and their belief in the possibility of computational packages should have no limits at all. Not being able to understand mathematical methods and algorithms "hidden" in the depths of mathematical computing programs, young researchers allow fundamental and unavoidable errors in their work. Overcoming this situation as early as possible (i.e. at school) could be achieved through the students' acquaintance with the basic principles of numerical methods, with the help of examples that are accessible to the understanding of schoolchildren. The solution of everyday mathematical problems, processed

* This phrase is attributed to many famous people—Google it and see who said so, and most importantly, on what occasion. This version is according to (Wikipedia) Mark Twain (*Chapters from My Autobiography*) that were attributed (unproved) to British Prime Minister Benjamin Disraeli https://en.wikipedia.org/wiki/Lies,_damned_lies,_and_statistics.

for the education of schoolchildren with the support of computer science, will be a real step toward the fundamentalization of learning computer science.

We will demonstrate a real attempt to unify informatics and mathematics within the framework of a single educational area, and we will determine the points of contact between mathematics and computer science. Let's give examples of the effective interaction of these sciences, using objects and problems that are understandable to everybody.

First, we will examine some terms that will help us understand the essence of the problem raised in this Study.

Function (from Latin *functio*, execution, implementation) is the law *f*, according to which each element *x* from one numerical set is put in correspondence to a certain definite element *y* from another numerical set, so that $y = f(x)$.

Correlation (from Latin *correlatio*, ratio) is a probabilistic or statistical dependence that does not have a strictly functional character (see above) because it is impossible to accurately take into account the influence of a set of simultaneously changing factors.

Interpolation (from Latin *interpolatio*) is the approximate or exact determination of a quantity from known individual values of this quantity, i.e. recovering (exactly or approximately) the function (see above) from its several known values.

Regression (from Latin *regressio*—reverse movement) is the dependence of the average value of any quantity on any other value or on several values, such that the same value of the independent variable *x* can correspond, depending on the case, to different values of *y*.

Examples of function and correlation were given at the end of Study 1, when the relationship between atmospheric pressure, boiling point of water and height above sea level was considered.

And now to business (the main subject).

When processing experimental data, we usually encounter two fundamentally different cases: either the data obtained are related by some functional dependence, whose formula is unknown to us, or there is no functional dependence, although a correlation is observed.

However, there is also a third option: the two quantities are actually related by a functional dependence, and theoretically to each value of the argument *x* corresponds exactly one value of *y*. However, during the experiment it is not possible to obtain the values with a sufficient degree of accuracy, and therefore the experimental results are presented in the form of tables of values (x_i, y_i), $y_i = f(x_i) \pm \varepsilon_i$ where ε_i are measurement errors. If this is the case, by simply connecting the experimental points through straight lines, we obtain a broken line that does not have anything in common with the real functional dependence. And the shape of this broken line due to measurement errors is unlikely to be reproduced during a repeated experiment.

In the first of the cases considered, interpolation is used to obtain an approximate formula for the dependence of two quantities. However, in the second and third cases, regression analysis is used. Let us give some examples to support this.

Students who solve problems in physics, chemistry and other disciplines often have to look into references, where a function is presented in the form of a table: for example, the thermal conductivity of a metal as a function of temperature or the density of an aqueous solution of a salt at a fixed concentration depending also on temperature.* In these tables, pairs of "temperature-property" numbers are shown only for certain temperature values, for example, 0, 10, 20, 30, etc. degrees on the Celsius scale. And the value of the property (the thermal conductivity of the metal or the density of the solution) should, for example, be taken at 17°C. It is understood that the values indicated in the tables are accurate (to be more precise, they were obtained with a sufficient degree of accuracy). Since physical laws allow us to consider the analyzed quantities to be functionally dependent, interpolation is used. The simplest is the linear interpolation while the interpolation by polynomials or splines is more complex. With linear interpolation, we fix two points, mentally, on paper or on the computer display, draw a straight line through them and find the desired intermediate value on that line. In non-linear

* Previously, we had to determine from the tables the value of trigonometric, logarithmic, exponential and other elementary functions. Now the tables of values of special functions, for example, the differential and integral Laplace functions, are actively used.

interpolation, a function having a certain smoothness is sought, the graph of which passes through the indicated points (interpolation knots).

However, we should always remember that in the absence of any specific functional relationship between the two quantities, interpolation can lead to a false result.

Once, before a lecture on "Regression Analysis" on a course on "Information Technology", the author selected the example of a statistical sample for such analysis. However, when he "climbed the pulpit" and looked at the audience (Figure 14.1), he realized that a sample was right in front of his eyes. A roll call of students was conducted. The young men got up and reported their *weight** and *height* (female students were exempted from this procedure, for obvious reasons). The numbers were recorded in the two vectors with names *Weight* and *Height* in Mathcad. Fifty elements were obtained in the vectors. These arrays of numbers served as a good example for the lecture.

As already noted, there are two main methods for the initial processing of the obtained data: approximation (by conducting a line near the points) and interpolation (by setting a line through the selected points). Each method will provide some insight into the dependence (functional or statistical) of the two quantities. Each method has strengths and weaknesses, and also its own features for the computer implementation.

Let's start with interpolation. It (theoretically) can be implemented with the help of any known function, but in practice it is usually done by polynomials. And first

FIGURE 14.1 Students: "experimental statistical rabbits" (author's photo).

* It was, of course, the mass of a student, not the weight.

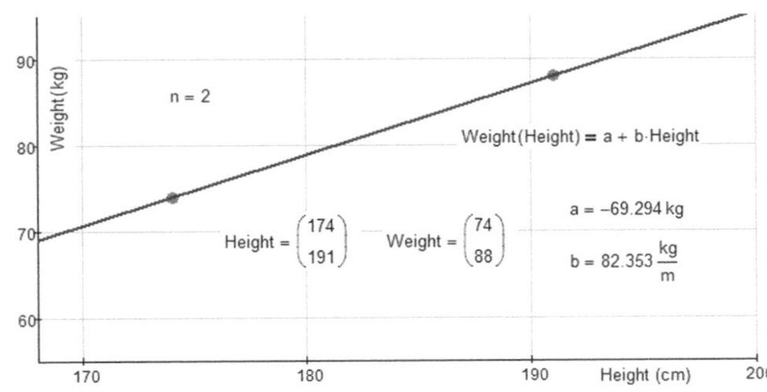

FIGURE 14.2 Linear interpolation for two students.

of all, polynomials of the first degree, whose graphs are straight lines on the plane. This interpolation is called linear.

Linear interpolation, formally applied to the pair "Height-Weight" values, is shown in Figure 14.2. Data from two students (a "small student" and a "big student") were taken and a straight line was drawn through the corresponding points (interpolation nodes). The parameters of this line were also determined—the coefficients a and b of the equation $y = a + b\,x$.

How the coefficients a and b were calculated is shown in Figure 14.3. The reader can search in reference books for the corresponding formulas, or he/she can solve a couple of linear equations in Mathcad (Figure 14.3) and get the answer.

In Figure 14.3, the weight and the height of "small and big students" are entered in a table, then the system

Height (*cm*)	*Weight* (*kg*)
174	74
191	88

$$a + Height_1 \cdot b = Weight_1$$

$$a + Height_2 \cdot b = Weight_2$$

$$M := \begin{bmatrix} 1 & Height_1 \\ 1 & Height_2 \end{bmatrix} \qquad v := \begin{bmatrix} Weight_1 \\ Weight_2 \end{bmatrix}$$

$$\begin{bmatrix} a \\ b \end{bmatrix} := \mathrm{lsolve}(M, v) = \begin{bmatrix} -69.294\ kg \\ 82.353\ \dfrac{kg}{m} \end{bmatrix}$$

$$a = -69.294\ kg \qquad b = 823.529\ \frac{gm}{cm}$$

FIGURE 14.3 Numerical solution of a system of linear equations for a straight line passing through two points on the plane (see Figure 14.2).

$$\begin{bmatrix} a + Height_1 \cdot b = Weight_1 \\ a + Height_2 \cdot b = Weight_2 \end{bmatrix} \xrightarrow{solve, \begin{bmatrix} a \\ b \end{bmatrix}} \begin{bmatrix} \dfrac{Height_1 \cdot Weight_2 - Height_2 \cdot Weight_1}{Height_1 - Height_2} & \dfrac{Weight_1 - Weight_2}{Height_1 - Height_2} \end{bmatrix}$$

$Height_1 := 174\ cm$ $\quad Weight_1 := 74\ kg$

$Height_2 := 191\ cm$ $\quad Weight_2 := 88\ kg$

Copy $\quad$ Copy

$$a := \frac{Height_1 \cdot Weight_2 - Height_2 \cdot Weight_1}{Height_1 - Height_2} = -69.294\ kg \qquad b := \frac{Weight_1 - Weight_2}{Height_1 - Height_2} = 823.529\ \frac{gm}{cm}$$

FIGURE 14.4 A symbolic solution of a system of linear equations of a straight line passing through two points on the plane (see Figure 14.2).

of two linear algebraic equations for a straight line passing through two points is written, by creating the square matrix of the M coefficients of the unknowns a and b and a vector v of free terms. The solution of the system is found by using the Mathcad built-in function *lsolve*. When specifying the equations of the straight line passing through two given points with coordinates (x_1, y_1) and (x_2, y_2) our system of two linear equations can be also solved analytically (Figure 14.4) to obtain the formula for determining the coefficients a and b.

If we choose the pair of students randomly, then our straight line can be oriented in different directions in the "Height-Weight" plane, so in this case this approach for describing the dependence of the quantities is not reliable. Linear interpolation here is not adequate: it once again "brings grist to the mill" of those who jokingly or seriously claim that these are "lies, damned lies and statistics".

Now, to process the results, we will try to use linear approximation, i.e. to draw a straight line near the points. In statistics, this approximation, constructed as a result of applying a certain optimization method, is called linear regression. To find the linear regression, we need to construct a scattering diagram, i.e. to "pour out" on the Cartesian plane all 50 points with the statistical data of the students, and then try somehow to draw a straight line showing the interdependence (correlation) of the height and weight of a person, not contradicting the basic trends in the behavior of the measured values (see Figure 14.5).

Once upon a time in the pre-calculator and pre-computer era, this work was done approximately like this: the points of the original statistical data were put on millimeter graph paper, and then with a transparent ruler a straight line was drawn, which best described the relationship of the height of a person and his weight. At the same time, when drawing this line everyone could be guided by some subjective notions about the optimality of the position of this line on the plane, that is, the values of the coefficients a and b that determine the equation of the straight line $y = a + b \cdot x$ (the constants a and b are calculated and written in Figures 14.2 and 14.5). On a computer, this work can be automated if some objective optimization criterion is selected. A possible criterion is the minimum sum of the squares of the "vertical" deviations of the points from the curve (in our case, from a straight line). This method of calculation is called the *least-squares method* (LSM). If there is a dispute about which straight line drawn through the array of points is "more correct", then it is possible to measure the vertical deviations of the points from the straight line with a compass and a ruler, to raise to the square each deviation, to sum these squares and then to compare the amounts obtained. The straight line whose sum turns out to be smaller, also correctly represents the statistical dependence being investigated. If we only have

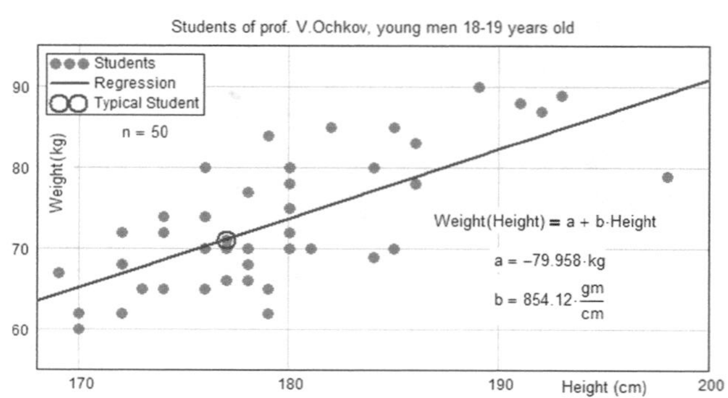

FIGURE 14.5 Linear fitting (approximation).

two points (Figure 14.2), or even more than two points, but they all lie on the same line, then obviously the sum of the squares of the deviations will be zero. However, if not all the points lie on the same straight line (Figure 14.5), then we can assume that there is a position of the line (such values of the coefficients a and b) such that the sum of the squares of the deviations will be *minimal*. Figure 14.6 shows a Mathcad-sheet that solves this optimization problem.

In the Mathcad-sheet, shown in Figure 14.6, a user function named *SSD* (i.e. the sum of the squares of the deviations is the objective optimization function) is formed, which has two arguments—a and b. After the guess values of these optimization variables are assigned, based on the empirical formula that states that the height of an adult in centimeters is equal to his/her weight in kilograms plus one hundred. Our problem is to test this hypothesis on this *statistical sample*—on 50 students. Then, as shown in Figure 14.6, the Mathcad built-in function *Minimize* is called, which starts changing the values of the variables a and b initially set by us, so that the *SSD* function reaches a minimum value. How this operation is performed is another matter. We see that with these initial values of the coefficients a and b, the *SSD* function returned a value of 3396 kg², instead of the optimal values of 1505 kg², and we also believe that a smaller value within the specified accuracy cannot be found. In Figure 14.5, it is possible to see these optimal values of the coefficients a and b and an "optimal" straight line that "pierces" the array of points. We shall return to the optimality of this solution below.

The problem formulated in Figure 14.6 can be solved analytically, if we recall that for a differentiable function of two arguments at the minimum point,

$$SSD(a,b) := \sum_{i=1}^{n} \left[Weight_i - (a + b \cdot Height_i)\right]^2$$

$$a := -100kg \qquad b := 1\frac{kg}{cm} \qquad SSD(a,b) = 3396\,kg^2$$

$$\begin{pmatrix} a \\ b \end{pmatrix} := Minimize(SSD, a, b)$$

$$a = -79.958 \cdot kg \qquad b = 854.12 \cdot \frac{gm}{cm} \qquad SSD(a,b) = 1505\,kg^2$$

FIGURE 14.6 The least-squares method as an optimization problem.

the partial derivatives with respect to both arguments are equal to zero (a necessary condition, but not sufficient). Figure 14.7 shows the symbolic solution of the system of two equations obtained by the partial derivatives of the function *SSD*(a, b) with respect to arguments a and b.

Formulas for the constants a and b, derived in Figure 14.17, are useful for working in Mathcad Express, where the *intercept* and *slope* functions are deactivated.

The *SSD* function sums up the *squares* of the "vertical"* deviation of the points from the straight line. And this is understandable: by removing the square, the summation leads to an incorrect statement of the problem: the deviations with different signs will be added together. Is it possible to replace squaring by taking an absolute value? To do this, it is sufficient to make a corresponding change to the first calculation operator shown in Figure 14.6. However, in this case the *Minimize* function will have poor convergence and will provide completely different answers for different initial values of the variables a and b. The fact is that the absolute value function is angular, if one can put it this way. And nature does not like sharp corners. In addition, it is worth looking at Figure 14.8, where an unsuccessful attempt at an analytic solution of a system of two equations was made—not the partial derivatives of the function *SSD*(a, b), which are equal to zero, but rather from the function of the *SAD*(a, b): the sum of the absolute (values) deviations. The fact is that the function $y = |x - x_0|$ does not have a derivative at the point x_0, and any other functions containing the module have the same property. In addition, the operation of numerical differentiation itself is incorrect, i.e. it can produce incorrect results.

The website http://www.math.wpi.edu/Course_Materials/SAS/lablets/7.3/7.3c/index.html shows a description and comparison of the least squares and the least modules the methods.

A point in Figure 14.5 is circled—this, it is possible to say, is a typical student whose weight deviation from the center line (standard) is minimal.

* The reader can try to sum the squares of "horizontal" deviations—not the values of the masses of students, but their heights. It is not possible to work with "diagonal" deviations, because our two vectors with student data have different dimensions—mass and length. However, the "vertical" deviations here are more appropriate—the height of a person (the argument) affects its weight (function), and not vice versa.

$$SSD(a,b) := \sum_{i=1}^{n} \left(Weight_i - (a + b \cdot Height_i)\right)^2$$

$$\begin{bmatrix} \frac{d}{da} SSD(a,b) = 0 \\ \frac{d}{db} SSD(a,b) = 0 \end{bmatrix} \xrightarrow{solve, \begin{bmatrix} a \\ b \end{bmatrix}} \left[\frac{\left(\sum_{i=1}^{n} Weight_i\right) \cdot \sum_{i=1}^{n} Height_i^2 - \left(\sum_{i=1}^{n} Height_i\right) \cdot \sum_{i=1}^{n} Height_i \cdot Weight_i}{n \cdot \sum_{i=1}^{n} Height_i^2 - \left(\sum_{i=1}^{n} Height_i\right)^2} \quad -\frac{\left(\sum_{i=1}^{n} Height_i\right) \cdot \sum_{i=1}^{n} Weight_i - n \cdot \sum_{i=1}^{n} Height_i \cdot Weight_i}{n \cdot \sum_{i=1}^{n} Height_i^2 - \left(\sum_{i=1}^{n} Height_i\right)^2} \right]$$

$$Height := M^{\langle 1 \rangle} \cdot m \qquad Weight := M^{\langle 2 \rangle} \cdot kg \qquad n := last(Height) = 50$$

$$a := \frac{\left(\sum_{i=1}^{n} Weight_i\right) \cdot \sum_{i=1}^{n} Height_i^2 - \left(\sum_{i=1}^{n} Height_i\right) \cdot \sum_{i=1}^{n} Height_i \cdot Weight_i}{n \cdot \sum_{i=1}^{n} Height_i^2 - \left(\sum_{i=1}^{n} Height_i\right)^2} = -79.958\ kg \qquad b := -\frac{\left(\sum_{i=1}^{n} Height_i\right) \cdot \sum_{i=1}^{n} Weight_i - n \cdot \sum_{i=1}^{n} Height_i \cdot Weight_i}{n \cdot \sum_{i=1}^{n} Height_i^2 - \left(\sum_{i=1}^{n} Height_i\right)^2} = 854.117\ \frac{gm}{cm}$$

FIGURE 14.7 A symbolic solution of the optimization problem—to find the minimum of a function of two arguments.

In this regard, there is an interesting proposal to summarize the results of various beauty contests. Now they have too much subjectivity, which means—resentment, tears and even litigation. In the finals of such contests "beautiful beauties" usually fall (see, for example, Figure 14.1), of which it is rather difficult to choose the most optimal, sorry, the most beautiful miss or Mrs. So, it is possible for these finalists to measure weight and height or other sizes (notorious 90-60-90, for example), to draw a line through the points and to select the winner as shown in the circle in Figure 14.5.

However, let's return to our more serious cases. The built-in Mathcad function *Minimize*, shown in Figure 14.6, allows us to implement the least-squares method in its general formulation: the user (objective) function named *SSD* can in principle have any number of arguments, and its right side—any functional dependence for fitting out points. However, for specific cases in Mathcad, there are simpler tools for solving the problem of statistical processing of data by the method of least squares. For example, Figure 14.9 shows a call of the two Mathcad built-in functions *intercept* and *slope* that return the value of the coefficients a and b of the fitting line $y = a + b \cdot x$. The formulas we previously derived in Figure 14.7 are implemented in these two functions. They are also used in the calculation in Figure 14.9. In addition, Figure 14.9 shows the operation of another function that returns linear regression coefficients, the *line* function. The value of the coefficient a is the ordinate value of our straight line at the intersection with the Y-axis, and the value of the constant b is the tangent of the slope of the line with respect to the X-axis. The tangent and the angle could be mentioned if our initial vectors were dimensionless. With dimensional vectors (mass and length, as in our problem), this is simply the ratio of the vertical increment (in grams) to the horizontal increment (in centimeters).

Through the points shown in Figure 14.5, it is possible to draw not only a straight line. In fact if a researcher sees in the original scattered points not a straight line, but, for example, a parabola or a higher degree polynomial with coefficients a, b, c, d, etc., then he/she can call the Mathcad function *regress*, which no longer has two arguments (see Figure 14.9), but three: the ordinate of the points, the abscissa of the points and the degree of the polynomial: 1 (our case, shown in Figures 14.5, 14.7, 14.8 and 14.9), 2 (parabola), 3, a polynomial of third degree, 4, fourth (see Figure 14.10 below), etc., until the degree of the polynomial becomes equal to the number of points minus one.

If the degree of the polynomial is equal to the number of points minus one, then the approximation (by determining a curve near the points) becomes an interpolation (by determining a curve through the points). However, it is not so easy. The author's animation (Figure 14.10) shows how the shape of the approximating curve changes near 15 points and how new coefficients k_i are added as the degree of the polynomial increases from 1 to 14.

When the degree of the regression polynomial approaches the value of the number of given points (Figure 14.11), an *oscillation* (Latin, *oscillatio*) can be observed. We see how the regression curve approaches the points (for $n = 14$, we repeat, the approximation becomes an interpolation), though in the intervals between the points the curve can "swing"—far away from the points, invalidating the entire interpolation and approximation.

In the intervals between the three right-hand points in Figure 14.11 because of the oscillations, we see not

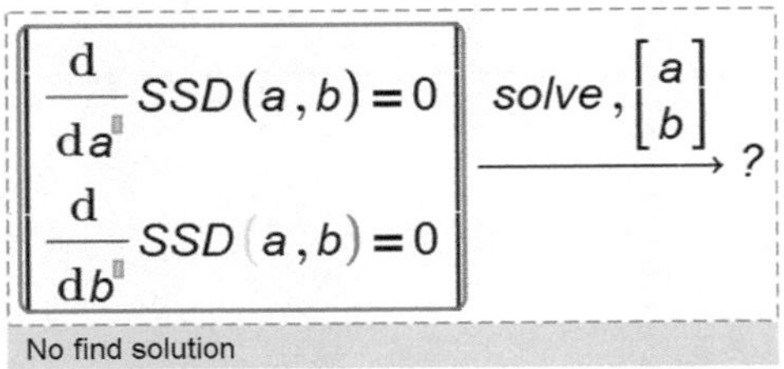

$$\left| \begin{array}{l} \dfrac{d}{da} SSD(a,b)=0 \\ \dfrac{d}{db} SSD(a,b)=0 \end{array} \right| \xrightarrow{solve, \begin{bmatrix} a \\ b \end{bmatrix}} ?$$

No find solution

FIGURE 14.8 Unsuccessful attempt to solve the system of two equations.

$$a := \text{intercept}(Height, Weight) = -79.958\ \mathbf{kg}$$

$$b := \text{slope}(Height, Weight) = 854.12\ \frac{\mathbf{gm}}{\mathbf{cm}}$$

or

$$\begin{bmatrix} a \\ b \end{bmatrix} := \text{line}(Height, Weight) = \begin{bmatrix} -79.958\ \mathbf{kg} \\ 85.412\ \dfrac{\mathbf{kg}}{\mathbf{m}} \end{bmatrix}$$

$$a = -79.958\ \mathbf{kg} \qquad b = 854.12\ \frac{\mathbf{gm}}{\mathbf{cm}}$$

FIGURE 14.9 Tools of Mathcad to solve a linear least-squares regression problem.

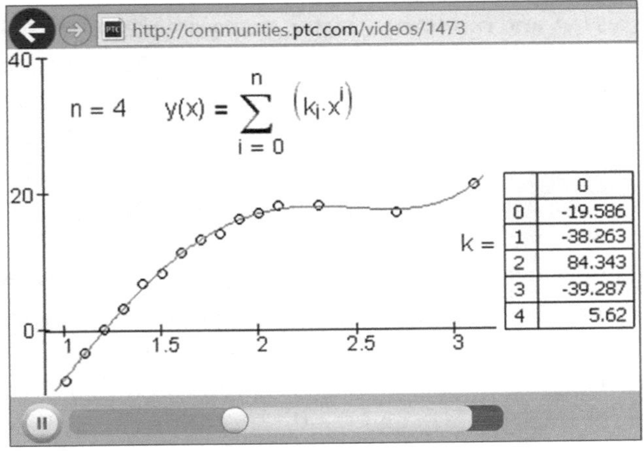

FIGURE 14.10 A frame of the animation of the **regress** function: a polynomial of the 4th degree through 15 points.

"statistics" and not even "Lies, damned lies and statistics" (see the second title of this Study).

The *regress* function can be applied not only to a line, but also to a surface, processing two vectors (or rather, a matrix with two columns) or a square matrix.

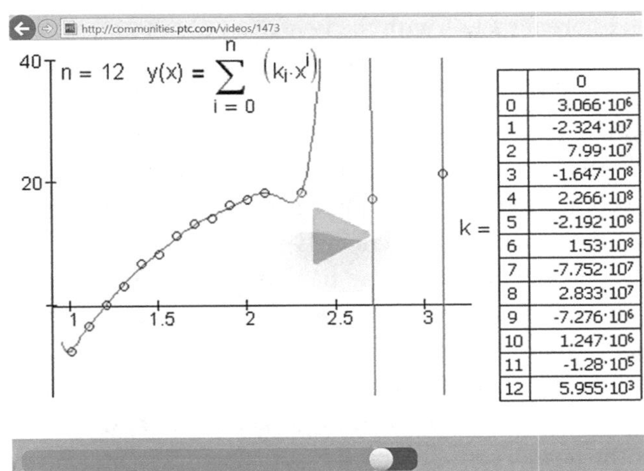

FIGURE 14.11 A frame of the animation of the **regress** function: a polynomial curve (part of it is not shown) of the 12th degree through 15 points.

However, let us return to the definitions provided at the very beginning of this Study. The relationship between the height and weight of a person is a typical example of *correlation*, while our search for the values of the coefficients *a* and *b* of a linear *function* is a classic example of *regression analysis*.

In traditional mathematics, as a rule, we are dealing not with correlation, but with clear functional dependencies. Each value of a flat angle, for example, corresponds to a clear value of the sine, and to each value of the radius of the circle, to a value of its length, etc. However, in real life, everything is much more complicated. How does the thermal conductivity of a metal, which we mentioned above, depend on temperature? On the one hand, there is a clear functional dependence, tabulated in many reference books* though, on the other hand, some unaccounted factors can influence this physical property of a metal, for example, the presence of impurities, the shape of the sample used for measuring the thermal conductivity, the age ("fatigue") of the metal, etc.

On the website http://twt.mpei.ac.ru/MCS/Worksheets/Thermal/T-T-2-Tab-3-04.xmcd it is possible to approximate tabulated data of the thermal conductivity

* In the traditional manuals, dating back to pre-computer times, there were reported not the formulas for computer calculation, but tables for manual calculations. In modern Internet directories, some of which are created, among other things, by the author of this book, a compromise variant is applied: a table is given, whose intermediate values are found by interpolation.

of different metals with the help of a polynomial of variable degree and to calculate the sought value. The density of a salt solution can also be uncertain depending on the temperature and on the concentration: the salt may have impurities, the dissolving water may also not be completely pure, etc. On the website http://twt.mpei.ac.ru/TTHB/1/Water-chem/C-NaCl.html there are online calculations of the density of an aqueous solution of this salt as a function of temperature and concentration. Here the statistical processing is conducted not on the plane, but in space: the corresponding function has not one, but two arguments. The first chapter of the book *Thermophysical Properties of Thermal Power Engineering Working Substances* (http://twt.mpei.ac.ru/rbtpp) is devoted to techniques for obtaining these dependences from tabular and other data.

Seeing a table with pairs of values in a book or on the Internet, we repeat, it is not so easy to determine if the data in the table are from a function or from a correlation, and if interpolation (holding the line strictly through the dots, Figure 14.1) or approximation (smoothing, the line near the points is shown, Figures 14.5 and 14.10) apply to these data.

We have already mentioned the linear interpolation, see Figure 14.2. In Mathcad, there is the built-in function *linterp(X, Y, x)*, that creates a user function with argument *x* for the interpolation (interp) by straight line segments (l) of points whose coordinates are stored in the vectors *X* and *Y*. For this piecewise-linear interpolation it is necessary that the vectors of the original data vectors are of the same size, real numbers are stored and the elements are arranged in ascending order in the vector *X*.

Linear interpolation is very easy to implement. However, "Simplicity Is Worse Than Theft": with this simplified interpolation, the function turns out to be angular, and nature, we repeat, does not tolerate sharp angles and tries to smooth them. Let us recall how the sea waves turn sharp fragments of stones into smooth pebbles. However, the main thing is that the "broken" function, consisting of segments of straight line, has as a first derivative a set of constants. The second derivative of this function is generally zero (stolen!) at all points except at the "angular" ones, in which the second derivative simply does not exist. This characteristic often makes it difficult to apply some numerical methods to this function, for example, to search for a zero or for a minimum—see the example in Figure 14.8, where an "angular" function was involved.

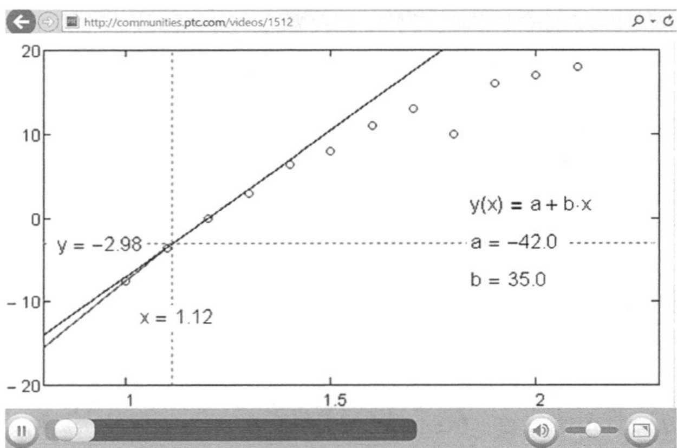

FIGURE 14.12 A frame of animation of a linear interpolation.

If a straight line can be drawn through two points (see Figures 14.2 and 14.12), then through four points, we have a cubic parabola. This is the basis for the cubic spline interpolation for tabular data dependencies. The computer implementation is more complex than in the case of linear interpolation, but it lacks the "angular" flaws noted above.

The website https://community.ptc.com/t5/PTC-Mathcad-Questions/cspline/m-p/541853 shows the author's animation of an interpolation with cubic splines. Three frames of this animation are shown in Figure 14.13.

In Mathcad, spline interpolation can be implemented using the *lspline*, *pspline* and *cspline* functions. The prefixes *l*, *p*, and *c* indicate which *extrapolation* will be used (in order to find the values of the reconstructed function outside the interval of the discrete values of the vector *X*). Namely: *l* is a line based on two extreme points, *p* is a parabola based on three extreme points and *c* is in a cubic dependence based on four extreme points.

Spline interpolation is widely used in computer graphics, when, for example, it is necessary to draw a smooth line or surface through fixed points of an object. There is such a technology for creating computer cartoons: there are fixed luminous points on a character, the character moves and the three coordinates of these points are determined by the computer. Then the computer passes through these points a curved smooth surface or pieces of surface and recreates (animates) the movement of the virtual person.

If interpolation (from Latin, *inter*, among, between; and *polio*, correct, change) means finding intermediate values for a number of statistical data, and it is mathematically expressed in the construction of a function passing through given points (interpolation nodes),

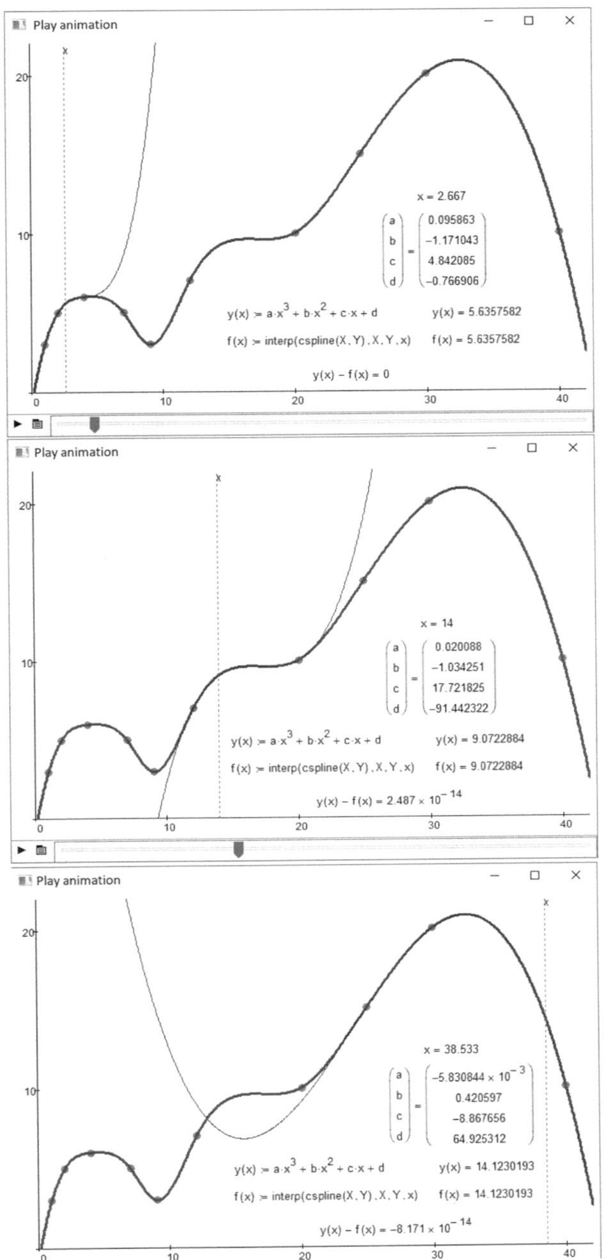

FIGURE 14.13 Animation of the spline interpolation.

extrapolation (from Latin *extra*—over, out; and *polio*, correct, change) is the procedure of extending (to the future) the conclusions drawn about an object on the basis of its present and (or) past states. In mathematics, extrapolation is expressed by the continuation of a series of data based on certain formulas or by constructing a function that describes the behavior of an object in the present and presumably describes the behavior of the object in the future.

Extrapolation in Mathcad can also be implemented with the built-in function *predict*(prediction), which contains a very complex algorithm for autocorrelation of successive values in a sliding window.* Figure 14.14 shows how this built-in function predicts the values of the user's function *y(x)*: at the first points, the values of the given function are fairly well predicted, but then a stall occurs.

The reader could try with the help of the author's website at http://twt.mpei.ru/mcs/Worksheets/rbtpp/predict.xmcd, which uses the *predict* function, to forecast the air temperature or the dollar rate for the coming week, based on the data of the previous week, month, year, etc. Then, when the future becomes the past, the reader can estimate the quality of the forecast. Some considerations about prediction, or rather about forecasters follow.

In the media it is often possible to see the advertisements of various kinds of predictors, conventional and genetic. So, for example, these people can give predictions for the dollar or the euro exchange rate, the

Predicted Function $y(x) := \exp\left(\frac{-x}{100}\right)\cdot\sin\left(\frac{x}{10}\right)$

Function values in 100 points $k := 0..99$ $data_k := y(k)$

Prediction (extrapolation) function values for the subsequent hundred points according to n function values $n := 50$

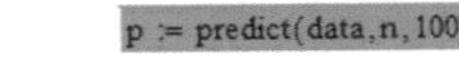

Graphical test extrapolation $i := 0, 5..99$

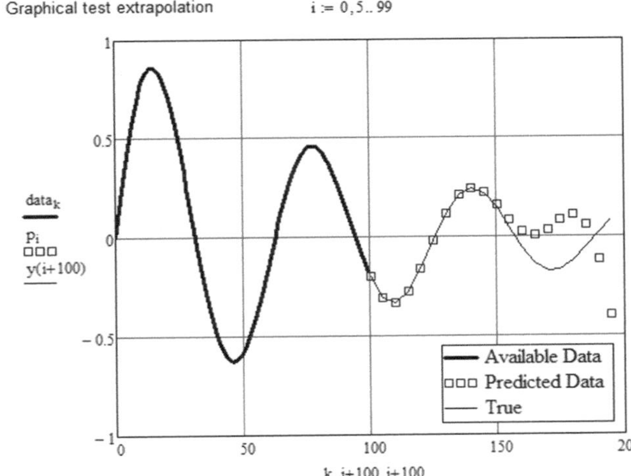

FIGURE 14.14 Sine prediction in Mathcad.

* From Mathcad Help: The **predict** function uses Burg's method to calculate autocorrelation coefficients for the last m points in **v**. These are used to predict the value of the (m+1)th point; this procedure repeats in a sliding window. Linear prediction can be used for extrapolation but should not be confused with linear or polynomial extrapolation. This algorithm is useful when data are smooth and oscillatory, though not necessarily periodic.

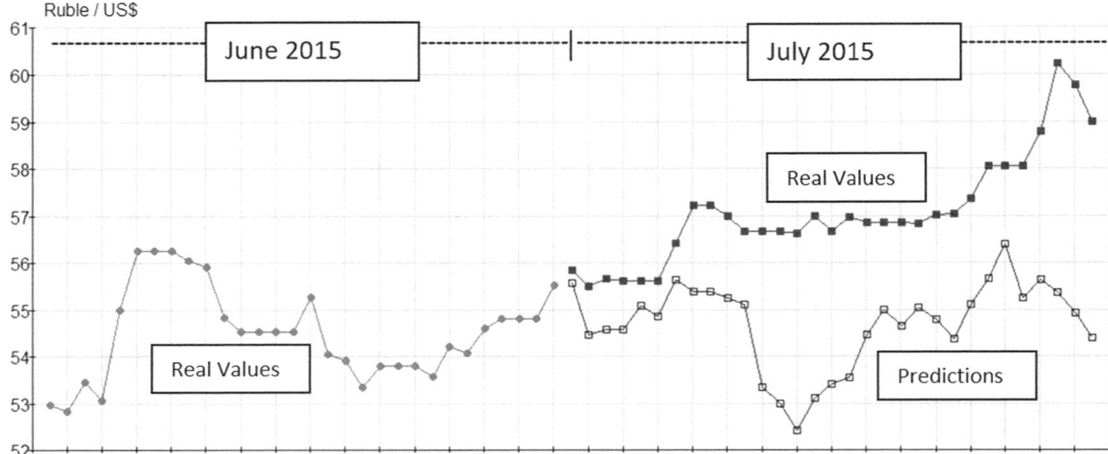

FIGURE 14.15 The prediction of the ruble/dollar in Mathcad.

shares of companies or how to make money with the exchange of securities. However, it is probable that only the fortune-teller is enriched in this way. Figure 14.15 shows how the *predict* function, using data on the dollar-ruble exchange rate for June 2015, "predicts" the dollar exchange rate for July 2015.

It would be good to publish in the media these curves comparing prediction data and real data for air temperature, for example. Based on them it would be possible to evaluate the quality of the work of various forecast offices.

In Mathcad, there are many functions that allow processing discrete random or regular data—to fit them (approximate) or to build interpolations or extrapolations from them. Figure 14.16 shows an Internet website created using Mathcad, on which table data are approximated using an arbitrary function of three arguments. The website shows the work of the *logistics** function. With its help, it is possible to describe the three main phases of the change in many objects and phenomena of animate and inanimate nature: the emergence, rapid development and dying (attenuation). A concrete example: the number of steam† locomotives in the world initially grew slowly, then there was a rapid growth in their production, and then they were gradually replaced by diesel locomotives and electric locomotives.

In Mathcad the following built-in functions with the postfix *fit* (fitting), are designed for this purpose:

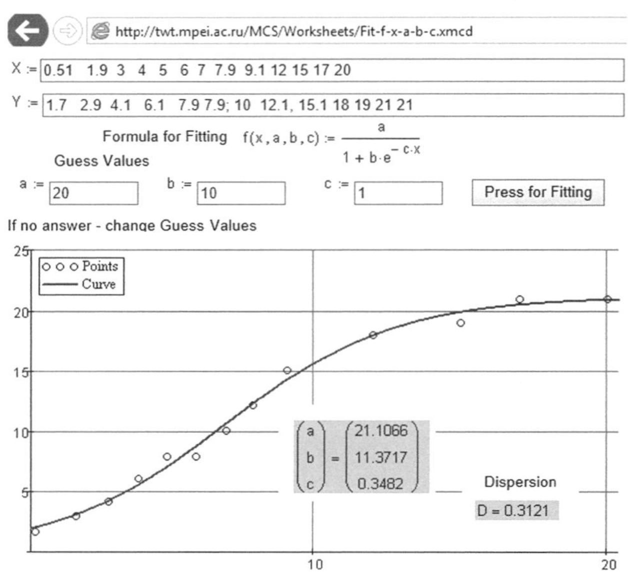

FIGURE 14.16 Fitting tabular dependence by an arbitrary formula.

expfit(*vx*, *vy*, [*vg*])‡ – approximation by formula (dependencies) $a \bullet e^{b \cdot x} + c$ (we will use this function later to solve a specific practical problem, see Figure 14.25).

lgsfit(*vx*, *vy*, *vg*) – $1 + b \bullet e^{-c \cdot x}$

lnfit(*vx*, *vy*) – $a \bullet ln(x) + b$

logfit(*vx*, *vy*, *vg*) – $a \bullet ln(x + b) + c$

* The logistic equation, also known as the Verhulst equation (in honor of the Belgian mathematician who first formulated it), originally appeared in relation to the population growth model.

† We animated the steam engine in Study 8.

‡ The argument *vg* is the vector of the first assumptions during the solution of the given regression problem. Vectors *vx* and *vy* store the coordinates of the nodes of the sample being analyzed. The square brackets mean that the parameter is optional.

medfit(vx, vy) – $a + b \bullet x$—an alternative algorithm that does not minimize the sum of the squares of deviations (see Figure 14.6), but the *median-median linear regression* for calculating the coefficients a and b (see Figure 14.17)

pwrfit(vx, vy, vg) – $a \bullet x^b + c$

sinfit(vx, vy, vg) – $a \bullet \sin(x + b) + c$

linfit(vx, vy, F)—linear combination of elementary functions

genfit(vx, vy, vg, F)—regression analysis in general form

From this list of "fitting" functions, the *medfit* function deviates a little, in its name the shortening *med* means not the *type* of approximating function, as for the other functions in the list, but the *method* of linear approximation. The *medfit* function as well as the line function (see Figure 14.9) is intended to solve linear regression problems. However, the function *medfit* is not based on the method of least squares, but on the method of median-median regression that minimizes the sum of the absolute values of the *median* errors in the nodes.* Figure 14.17 shows a comparison of the regression analysis of the parameters of our students, conducted by the method of least squares (a more sloping straight line) and by the median-median method.

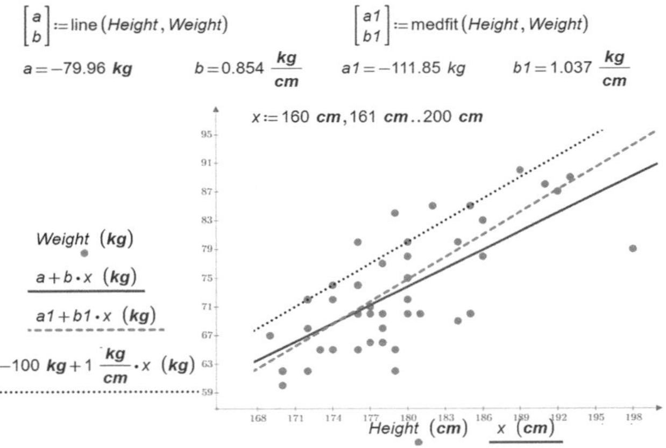

FIGURE 14.17 Comparison of the results of the data processing by the median-median regression (**medfit** function is used) and by the least-squares method (**line**).

Which straight line in Figure 14.17 is closer to the truth? Or we will ask a little differently, which straight line lies less (see the second name of the Study)? The answer to this question,† if, of course, it exists at all, depends, specifically, on which *distribution* is embedded in the analyzed random variables. In Mathcad, by the way, there is a set of built-in functions for generating random variables according to different (probability) distribution laws.

However, there are cases where there is no correlation of the data, for example, as it was revealed in a study of the statistical relationship between the USE (Unified State Examination) scores in mathematics and the results of work tests. The scattering diagram (Figure 14.18) makes it clear that the residual level of school mathematical knowledge depends little on the points obtained on the USE, and further regression analysis of the data does not make sense, since the regression line will be located almost horizontally.

On the other hand, a preliminary analysis of the data can help to identify weak correlations. Thus, for example, an insignificant but negative correlation was found between the time spent by students at the computer for entertainment purposes and the success of passing the USE in mathematics. For students from small towns it was −0.04, and for students from large cities it was −0.25.

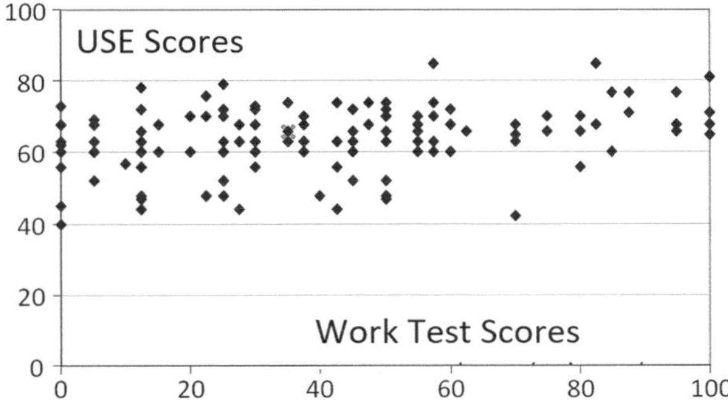

FIGURE 14.18 Scattering diagram between points of work test and points of the USE on mathematics.

* The median in statistics is the average index in a series of numbers. The median for a series of N numbers is simple. It is necessary to put the numbers in ascending order. If N is an odd number, then the median is equal to the central number. If N is an even number, then the median is equal to the arithmetic mean of the central pair. For example, for a series of 1, 2, 4, 4, 6, 9, 12, 21, the median is 5 = ((4 + 6)/2).

† Our idea of an objective evaluation of the results of a beauty contest (see above) can "stumble" on disputes about which regression analysis method shall be used here. The reader can propose to determine the winner of the contest according to one method (one straight line), and the vice-winner to another method.

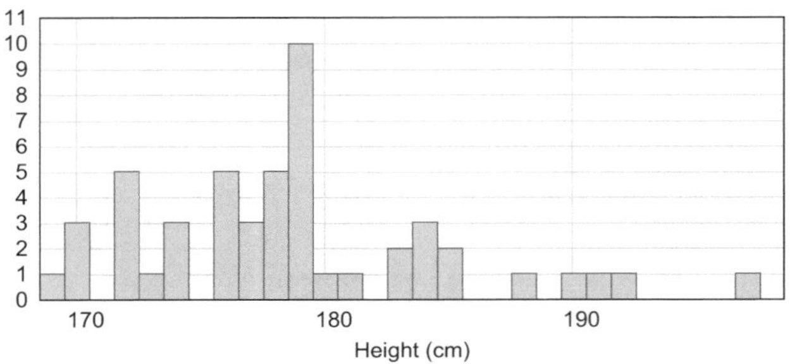

FIGURE 14.19 Example of a histogram in Mathcad.

We see that, when beginning to statistically process random variables, it is desirable to pre-build graphs and diagrams for visual analysis of data. Thus, the histogram function built into Mathcad allows us to build a *frequency characterization* (histogram) of the sample. Figure 14.19 shows graphically the frequency of occurrence of a given growth value in our fifty students with an accuracy of one centimeter.

So, from the histogram in Figure 14.19 it can be seen that the majority of students (10 people) have a height of 179 cm. However, it is difficult to see some statistical regularity in Figure 14.19. If our sample of students was much larger (students of the whole university or of the whole country), then we would see that on the histogram a certain bell-shaped curve would be drawn—the curve of normal distribution: very few people have very low or very high height, while a large number of people will have a certain average (statistical average) height.

Figure 14.19 shows that our histogram is slightly shifted with respect to the normal distribution. However, let's remember for the second time the second title of this Study: "There are lies, damned lies and statistics". People at polls often freely or involuntarily slightly distort information about themselves. It can be assumed that some students, naming their height aloud, added 1, 2 or even 3 cm to it. Let's say this a little more softly. Short students rounded their height to a centimeter higher (for example, 156.3 cm became 157 cm), while tall students rounding was less pronounced (for example, 179.3 cm became 179 cm). In such surveys (in the case of the population census, for example) women often dissimulate: they reduce their age and/or weight, attribute themselves to a non-existent husband, etc.

On the Internet, the reader can find a not very high-quality, but a very interesting photograph of a "living" histogram of people's height. The author does not place this photo here—it is of poor quality and its author is unknown. The author places here an Excel spread-sheet showing this photo, see Figure 14.20.

A group of players assembled on a football field according to their height, not in a line, but in a column with a different number of people in each row. Each row corresponds to a certain height: five feet (152.4 cm), five feet one inch (154.94 cm) and so on to 6 feet 5 inches (195.58 cm). In a "poll" like this, cheating when answering about height is more difficult: overestimating this parameter by 2–3 cm would be immediately noticeable in the corresponding row, unless the player were to stand on tiptoe. In the diagram shown in Figure 14.21, there are no empty rows in the middle, as we observed in Figure 14.19, because the steps on these histograms are of different sizes—not one centimeter but one inch (2.54 cm).

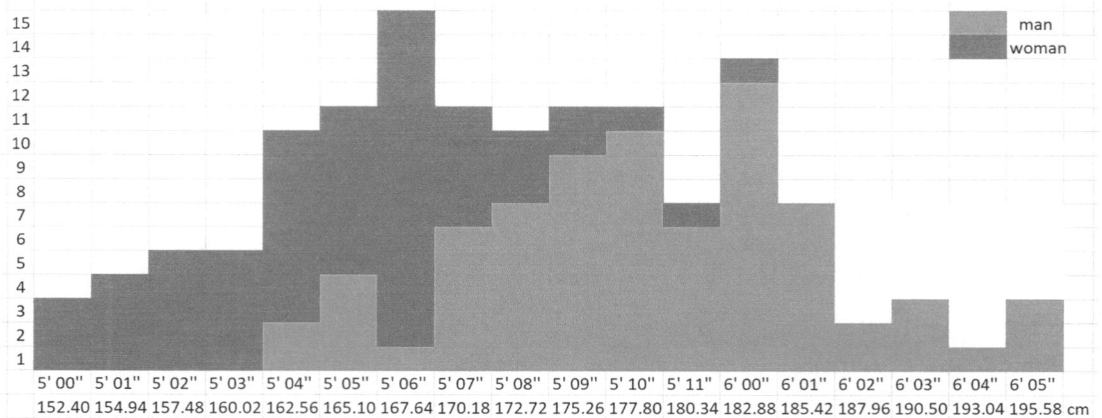

FIGURE 14.20 A living human height histogram (Excel variant).

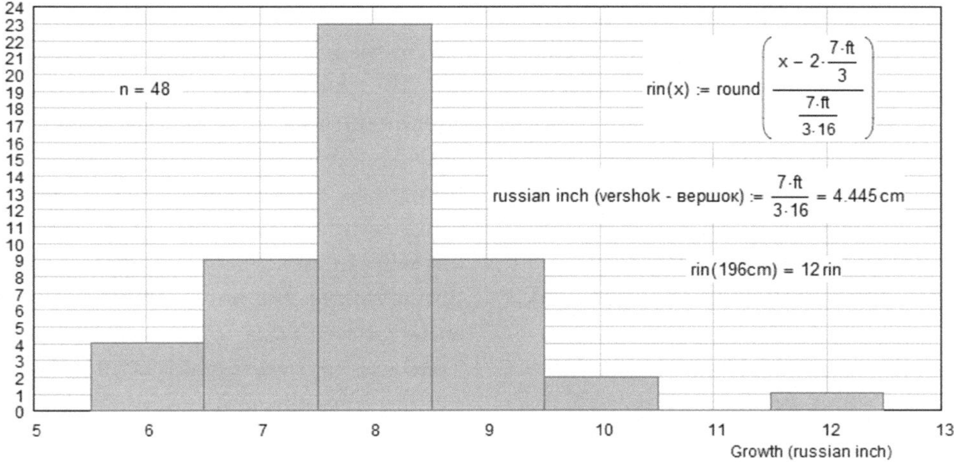

FIGURE 14.21 "Gradual" student growth (height) histogram (see one more similar gigagram in Figure 27.12 in Study 27).

In the United States and the UK, a person's height is measured in feet and inches,* and in Old Russia arshins and vershiks were used for this,† but arshins (specifically *two arshins*) by default did not specify any value. The fact is that the height of an adult person was very rarely less than two arshins (approximately 142 cm) or more than a sazhen (three arshins—about 213 cm). We read (and remember, see below) in the story "Mumu" of Turgenev: "Among all her servants, the most remarkable person was the janitor Gerasim, a man of twelve vershok (Russian inches) of height, a hero and deaf-mute from birth." It is easy to calculate the "heroic" height of Gerasim, taking into account the unspoken two arshins—approximately 196 cm. Figure 14.21 shows a histogram of the height (growth) of "author" students (see Figure 14.5) in increments not of one centimeter nor of one inch, but of one "good Old Russian" inch (vershok).

On the "vershok" histogram (Figure 14.21), a certain "statistical bell" is clearly visible, it is the curve of the normal distribution. Students are divided into five restricted groups based on height: short (4 people with a height of 6 vershoks), rather shorter than average (9 people, 7 vershoks), average (23 people, 8 vershoks), rather taller than average (9 people, 9 vershoks) and tall (2 people, 10 vershoks). A student fell out of our "statistics" and found himself with an anomalous height of 12 vershok, as Turgenev's Gerasim: the exception only confirms our statistical rule. By the way, here we involuntarily touched one of the most important questions for data processing: how to determine the optimal step when working with samples. After all, it is possible to group the data in some other way. For example, according to descending growth values in intervals of 1 meter in length, starting from zero (then the histogram would have the obviously false form of the rectangle).

Looking at Figure 14.21, the reader can go back to Study 12 and say that the height of a person can be "clearly" measured in centimeters, feet, inches, arshins and vershoks, though these data say little about the subject. However, if we refer to the categories of fuzzy sets "short", "rather shorter than average", "average", "rather taller than average" and finally, " tall", the analysis will be much more informative.

There are many factors that influence the relationship between height (growth) and weight of a person: for this, their relationship is a correlation, not a function. An important factor is the age of the person. Few people manage to keep their weight the way it was in their youth. The weight of a person is also greatly influenced by his/her genetics (heredity) and, of course, by their way of living. For a person, like a car (see below), the age is easy to determine. However, the "mileage" of a person, specifically speaking, the number of steps he did in his life, would be also a useful measurement

* In a novel of Agatha Christie, it is possible to read that the height of Hercule Poirot was five feet and five inches. The Forsyte ancestor described in the famous "Forsyte Saga" by John Galsworthy had the same height. Knowledge of ancient and foreign units of measurement helps to feel the "spirit" of classical literature. We emphasized this aspect in Study 6.

† See Study 6. The size of the Russian sazhen was a variable for a long time, but by the middle of the nineteenth century, the Russian sazhens were equated to seven English feet (seven feet of water under the keel!). In the sazhen there are 3 Russian arshins, and in the arshin are 16 Russian vershoks (not a decimal system of calculation). The "charms" of such a system can be fully felt by reading the books of Sergei Aleksandrovich Rachinsky "1001 task for mental account" (see Study 6).

to determine more accurately the interdependence of weight and height (growth). The mileage is measured by the odometer of a car. Some people use a pedometer when walking or jogging.

The author here recalls a story, something that he witnessed in West Germany in the 1980s. A German professor (the author was on an internship at a German university) wanted to sell his car quickly for 2000 marks and to put it at that low price on the local automotive flea market. Potential buyers approached him for quite some time, examined the car, asked questions, but all these contacts proved to be ineffective. However, then a pretty smart man came to him, bought it, as they say, without looking at the car, and changed the price tag. Immediately the man resold it for 4000 marks before the eyes of the amazed professor. This secondhand dealer knew how to estimate the cost of a used car right away and also understood that an understated price is also bad for bargaining, as well as overpriced. Projecting this thesis onto our students, we can say that too low height is also bad for life, as well as too high.

So, the car with its years and mileage!

Once upon a time a visitor to Mathcad forum posted a table on it (see Table 14.1) and asked the members of the forum to tell how it is possible to derive a calculation formula for estimating the cost of a used car by its mileage and years from the table.

Let's try to make this regression analysis.

Table 14.1 is transferred to a Mathcad Prime sheet with three vectors *Years*, *Mileage* and *Price*, combined in a table, see Figure 14.22.

Statistical analysis of data usually begins with the construction of various kinds of graphical dependencies, which can suggest possible ways of data processing. In Figures 14.23–14.25 one can see graphs constructed from the points of the "automotive" (Table 14.1).

If the car's mileage is divided by its years, then we will obtain the average speed of the car during its operative life. The three built-in functions *max*, *mean* and *min*, which return the maximum, average and minimum values of a vector or a matrix, can be applied to this fraction. It can be seen from Figure 14.23 that the correlation of the vehicle's mileage and years is concentrated on a straight line with a slope of 1.445 kph passing through the origin. In our sample, the minimum and the maximum speeds are as follows: 1.004 kph (7.5 years and 41,000 miles) and 2.046 kph (3.5 years and 39,000 miles). If we draw two straight lines through the origin and these two points, we get a "operating" wedge, inside which our 30 sampling points lie. Hence the conclusion: if you are offered to buy a used car, the parameters of which do not fall into this "wedge" (1–2 kph), then you need to think before you buy a car that was either too heavily exploited (speed more than 2 kph), or it is not clear why it was idle (speed is less than 1 kph), or the odometer was adjusted. The years of the machine are much harder to hide (to counterfeit).

The price of the car depends on years linearly $a + b\ x$, see Figure 14.24. No other more complex dependence can be caught on the points scattered in Figure 14.24. In our case, after every year of use, it is necessary to drop an average of $393 from the price of the car. The price of the new car is about 5750 dollars.

The dependence of the car's price on the mileage (Figure 14.25) is more complicated—it varies exponentially: $a \cdot e^{b\,x} + c$. With this dependence ($b < 0$), the price of a car with mileage will never fall to zero, however this cannot be said about the price dependence on years (Figure 14.24), once in the region of 14–15 years the car becomes "priceless" in one of two senses: or it has now a zero price (our sense) or the car becomes a "priceless" museum exhibit.

TABLE 14.1 Data on the Cost of a Used Car of a Certain Brand

Years	Mileage	Price (US$)	Years	Mileage	Price (US$)	Years	Mileage	Price (US$)
11.5	88000	1195	13.5	120000	495	7.5	41000	3495
13.5	103000	750	11.5	101000	895	13.5	124000	850
10.5	82000	1295	3.5	39000	4995	10.5	77000	1595
10.5	65000	1495	10.5	78000	1295	6.5	56000	3495
12.5	97000	800	6.5	52000	2695	12.5	83000	895
10.5	70000	1495	9.5	84000	1995	9.5	67000	2495
8.5	51000	2295	4.5	39000	3995	4.5	38000	3990
10.5	80000	1495	4.5	46000	3675	6.5	43000	3400
9.5	79000	1995	12.5	92000	795	13.5	92000	795
6.5	57000	2695	11.5	108000	975	11.5	78000	1295

US$	*Mileage* (*mi*)	*Years* (*yr*)	*Price* (*US$*)
¤	88000	11.5	1195
	103000	13.5	750
	82000	10.5	1295
	65000	10.5	1495
	97000	12.5	800
	70000	10.5	1495
	Etc		

FIGURE 14.22 Mathcad Prime table with data on used cars.

$v_{max} := \max\left(\frac{Mileage}{Years}\right) \quad v_{mean} := \text{mean}\left(\frac{Mileage}{Years}\right) \quad v_{min} := \min\left(\frac{Mileage}{Years}\right)$

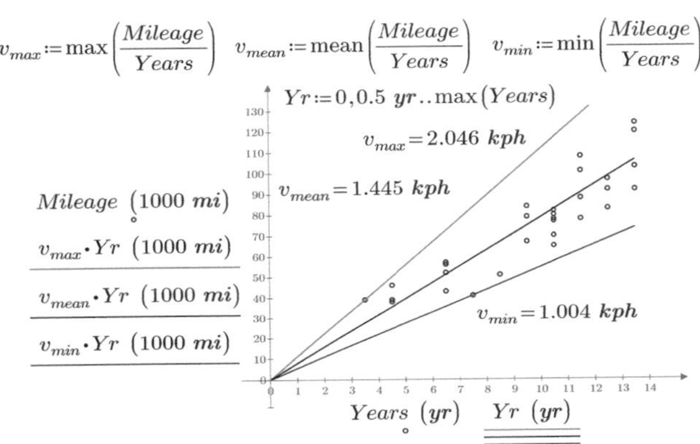

FIGURE 14.23 Correlation of cars: Mileage and Years.

$\begin{bmatrix} a \\ b \end{bmatrix} := \text{line}(Years, Price) \quad a = 5757.594\ US\$ \quad b = -392.718\ \frac{US\$}{yr}$

$y(x) := a + b \cdot x \qquad x := 0\ yr, 0.1\ yr..15\ yr$

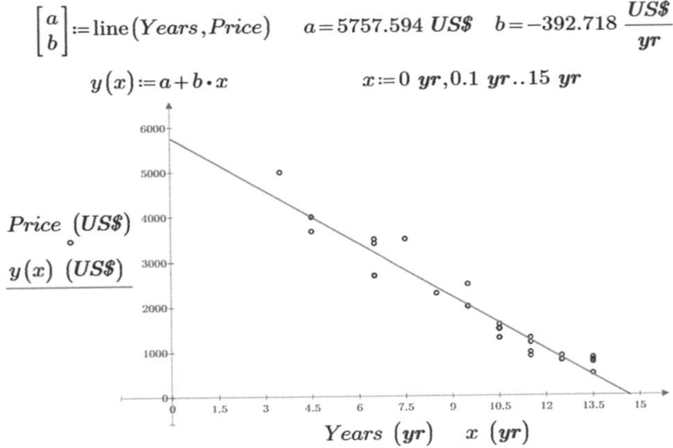

FIGURE 14.24 Correlation of the prices of the cars and their years.

Now, let us try to create our desired function of two arguments, years and mileage, by which it will be possible to estimate the value of a used car. Its form can be suggested by two separate functions of one argument

$$\begin{bmatrix} a \\ b \\ c \end{bmatrix} := \text{expfit}\left(Mileage, Price, \begin{bmatrix} 10000\ US\$ \\ -0.001\ \frac{1}{mi} \\ 100\ US\$ \end{bmatrix}\right)$$

$a = 12291.482\ US\$ \qquad b = -2.991 \cdot 10^{-5}\ \frac{1}{mi} \qquad c = 293.091\ US\$$

$y(x) := a \cdot e^{b \cdot x} + c \qquad x := 30000\ mi, 31000\ mi..130000\ mi$

Price (US$)
y(x) (US$)
Mileage (1000 mi) x (1000 mi)

FIGURE 14.25 Correlation of the prices of the cars and their mileage.

(see Figures 14.24 and 14.25). The sought function will have four parameters (*a*, *b*, *c* and *d*—see the first line in Figure 14.26) and will represent a combination of linear and exponential dependencies. It is only necessary to calculate the values of the parameters *a*, *b*, *c* and *d* so that the points shown in Table 14.1 and on the graphs in Figures 14.23–14.25 would be near a surface rather than a line. To do this, an auxiliary function is created that returns the sum of the squares of the deviations of the points from the surface—see the second operator in Figure 14.26. We have already performed this operation

$$z(Years, Mileage, a, b, c, d) := a + b \cdot Years + c \cdot e^{d \cdot Mileage}$$

$$n := \text{last}(Price) = 30$$

$$\Sigma\Delta2(a,b,c,d) := \left(\sum_{i=1}^{n}\left(Price_i - z\left(Years_i, Mileage_i, a, b, c, d\right)\right)\right)^2$$

$$\begin{bmatrix} a \\ b \\ c \\ d \end{bmatrix} := \begin{bmatrix} 4000\ US\$ \\ -300\ \frac{US\$}{yr} \\ 2000\ US\$ \\ 0\ \frac{1}{mi} \end{bmatrix} \qquad \Sigma\Delta2(a,b,c,d) = \left(1.154 \cdot 10^{9}\right)\ ¤^2$$

$$\begin{bmatrix} a \\ b \\ c \\ d \end{bmatrix} := \text{minimize}(\Sigma\Delta2, a, b, c, d) = \begin{bmatrix} \left(4 \cdot 10^{3}\right)\ ¤ \\ -9.507 \cdot 10^{-6}\ \frac{¤}{s} \\ \left(2 \cdot 10^{3}\right)\ ¤ \\ 0\ \frac{1}{m} \end{bmatrix}$$

$$\Sigma\Delta2(a,b,c,d) = \left(1.154 \cdot 10^{9}\right)\ ¤^2$$

$a = \left(4 \cdot 10^{3}\right) US\$ \qquad b = -300\ \frac{US\$}{yr} \qquad c = \left(2 \cdot 10^{3}\right) US\$ \qquad d = 0\ \frac{1}{mi}$

FIGURE 14.26 Unsuccessful attempt to work with minimize.

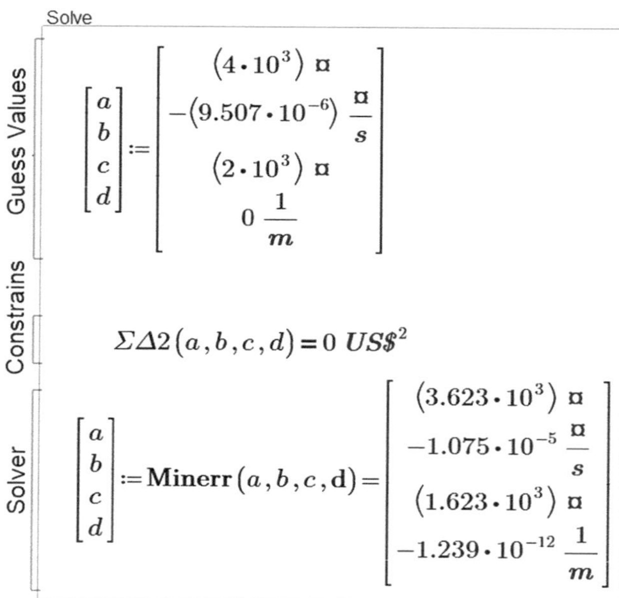

FIGURE 14.27 Successful attempt to work with Minerr function.

with the regression analysis of the height and weight of the students, see Figure 14.7.

Figure 14.26 attempts to calculate the parameters *a*, *b*, *c* and *d* by searching for the minimum sum of the squares of the deviations of the points from the surface. It turned out to be unsuccessful—the *minimize* function returned the values of the first approximation. The fact is that Mathcad Prime was built with a new algorithm for the *minimize* function. And the new is often unfinished. The author had to refer to the function *Minerr*, which has been available in Mathcad for a long time and in which possible errors and shortcomings have been identified, see Figure 14.27. We used this function in Study 4 and showed how it can be replaced by the *minimize* function, see Figures 4.17 and 4.20.

$\Delta := 1\%$ $\quad av := a \cdot (1-\Delta), a \cdot (1-\Delta) + \frac{a \cdot (1-\Delta)}{3000} .. a \cdot (1+\Delta)$

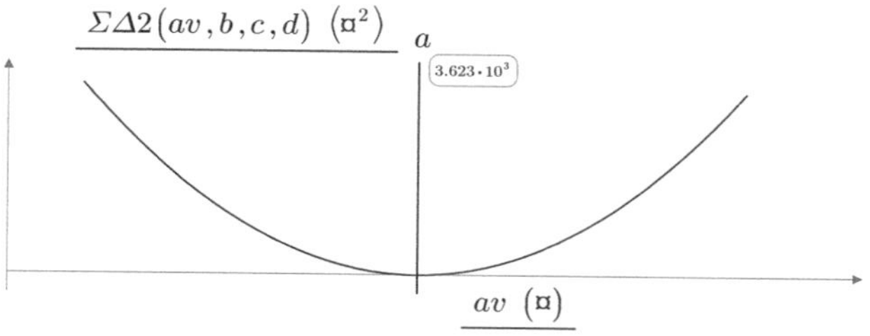

$\Sigma\Delta2(av,b,c,d)\ (¤^2)$ $\quad a$ $\quad 3.623 \cdot 10^3$

$av\ (¤)$

FIGURE 14.28 Checking the minimum point with respect to the parameter a.

However, also with the *Minerr* function it was necessary to play a bit: to provide a first assumption, to receive an answer, to see that it is not a minimum and to transfer (copy) the answer to the place of the first assumption. Control over the quality of minimization was conducted through the construction of four Cartesian graphs with

$bv := b \cdot (1-\Delta), b \cdot (1-\Delta) + \frac{b \cdot (1-\Delta)}{3000} .. b \cdot (1+\Delta)$

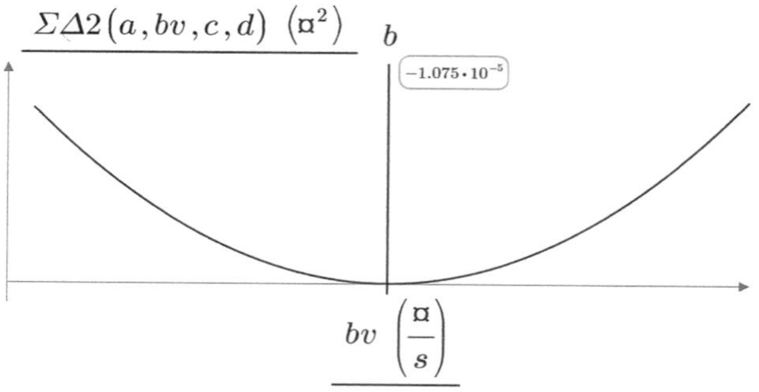

$\Sigma\Delta2(a,bv,c,d)\ (¤^2)$ $\quad b$ $\quad -1.075 \cdot 10^{-5}$

$bv\ \left(\frac{¤}{s}\right)$

FIGURE 14.29 Checking the minimum point with respect to the parameter b.

$cv := c \cdot (1-\Delta), c \cdot (1-\Delta) + \frac{c \cdot (1-\Delta)}{3000} .. c \cdot (1+\Delta)$

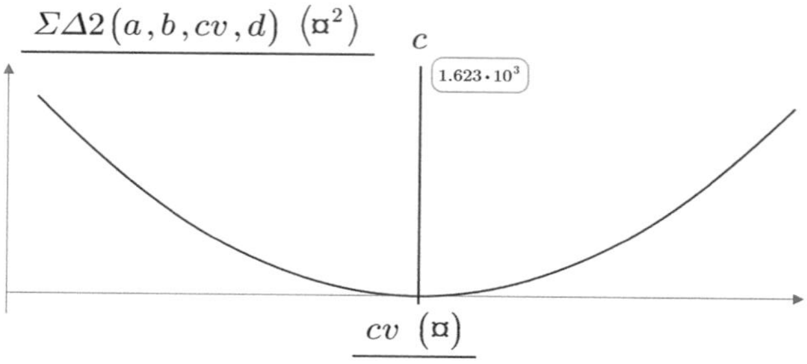

$\Sigma\Delta2(a,b,cv,d)\ (¤^2)$ $\quad c$ $\quad 1.623 \cdot 10^3$

$cv\ (¤)$

FIGURE 14.30 Checking the minimum point with respect to the parameter c.

$dv := d \cdot (1-\Delta), d \cdot (1-\Delta) + \frac{d \cdot (1-\Delta)}{3000} .. d \cdot (1+\Delta)$

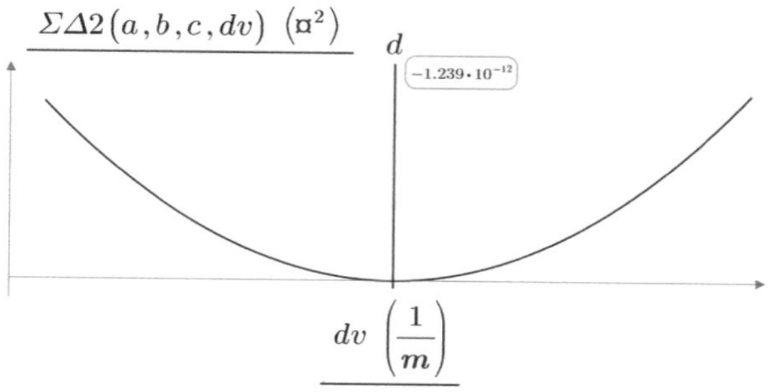

$\Sigma\Delta2(a,b,c,dv)\ (¤^2)$ $\quad d$ $\quad -1.239 \cdot 10^{-12}$

$dv\ \left(\frac{1}{m}\right)$

FIGURE 14.31 Checking the minimum point with respect to the parameter d.

$$MyYears := 9\ yr \quad MyMileage := 100000 \cdot km$$

$$z(MyYears, MyMileage, a, b, c, d) = 2191\ US\$$$

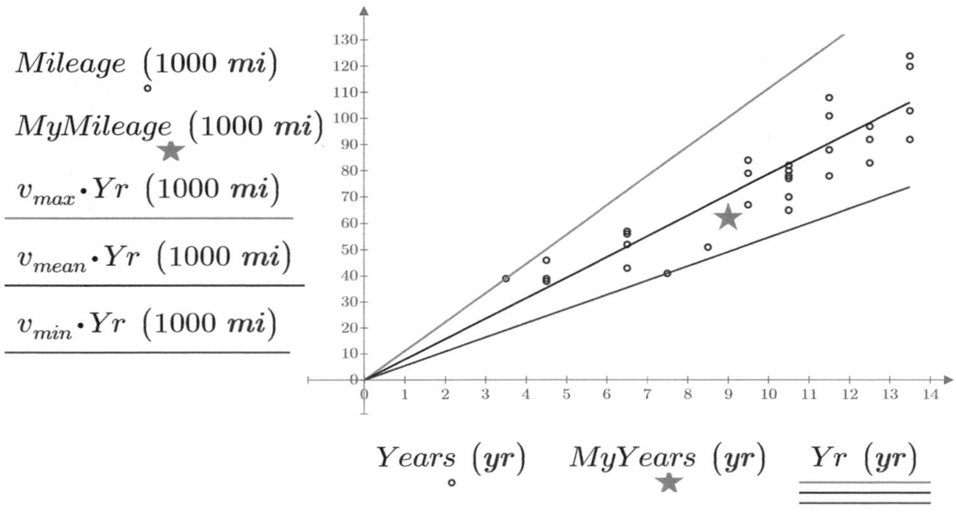

Mileage (1000 mi)
MyMileage (1000 mi)
$v_{max} \cdot Yr$ (1000 mi)
$v_{mean} \cdot Yr$ (1000 mi)
$v_{min} \cdot Yr$ (1000 mi)

Years (yr) MyYears (yr) Yr (yr)

FIGURE 14.32 Calculating the price of a used car.

horizontal markers, in which one of the four parameters is specified by the variable of the range, and the other three are fixed as constants, see Figures 14.28–14.31: at the minimum point of the function, as it is known, the values of all its partial derivatives must be zero.

Honestly, there was a desire to remove Figure 14.26 from this book and not discuss this flaw in Mathcad Prime. Moreover, we should expect that by the time the book is published, a new version of this package will appear, where this error will be corrected. However, the drawing has been left in this book and that's why.

It is not uncommon for the user of a program to see that something is wrong in it and considers that they may be the one to cause the error by default, despite the fact that everything was done correctly. And the error sits in the program itself. With such doubts, a user should contact the Mathcad forum and find out the opinions of other people. This was done, and this error in the *minimize* function was fixed and accepted by the developers. In addition, it is necessary to remember that many Mathcad tools and other mathematical programs are duplicated. This allows you to solve the problem in various ways, "dodging" from possible flaws.

Anyway, the problem of the optimal values of the parameters *a*, *b*, *c* and *d* of the regression expression for the used car price was solved and the found formula can be used in estimating the machines with mileage, see Figure 14.32, where by entering the values of the

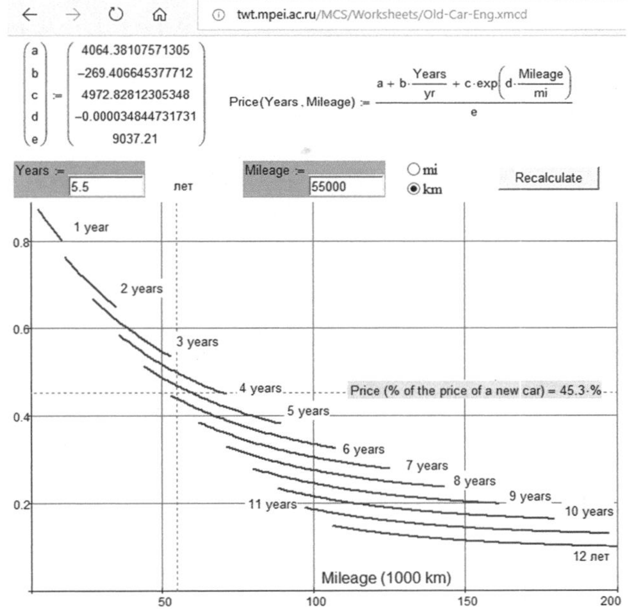

FIGURE 14.33 A website for estimating the value of a used car.

variables *MyYears* and *MyMileage*, the formula calculates the price of the car in dollars and as a percentage of the new price with zero mileage and years,* and also it shows this point among the points of the statistical sample.

* Here we must also take into account the fact that immediately after the sale, the real price of a car falls by 20–30%.

Statistical processing of data on the price of a used car allowed the creation of a Mathcad-website for estimating this correlation, see Figure 14.33.

According to the graph in Figure 14.33 it is possible not only to estimate the value of an old car, but also to cut off doubtful options, as shown in Figure 14.23 (parameters outside the speed wedge), when the mileage of the car does not match its years. It is possible to "download" the Mathcad-sheet by which calculations were made at: ftp://twt.mpei.ac.ru/ochkov/Auto.

The author of the book purchased his car twice according to a trade scheme, where the cost of the new car includes the cost of the old car, which is put on commission. So, both times the price of the old car offered by the dealer was no more than 5% of that calculated using the website shown in Figure 14.33.

ASSIGNMENTS FOR READERS

To test the solution of the regression problem for the cost of a used car, we used four Cartesian graphs, see Figures 14.28–14.31. It is possible to see this minimum on surfaces, choosing two variables and fixing two parameters. Construct these surfaces!

Go There, Knowing Where to Go Or

New Swiss Watches

ABSTRACT

In the Study an unusual solution of the travelling salesman problem and other optimization problems are described. The author fantasizes on the topic of watches in the form of a globe with the solution of the travelling salesman problem. The use of mathematical equipment to create watches of unusual shapes is discussed. The concept of the inverse travelling salesman problem is introduced.

Mathematics: Prime numbers, factorization, graph, minimum of a function, travelling salesman problem, inverse travelling salesman problem, nearest neighbor method, integer equations, integral, integro-differential equation, matrix, seven-bridge problem, Steiner problem, Steiner dimensional problem, transport problem, optimization paradoxes, financial pyramid.

Physics: Watches, bionics, pheromones, soap foam.

IT: Symbolic mathematics, animation, a cycle with a parameter, a conditional operator, matrix functions.

Art: Novel of Daniil Kharms; *Mona Lisa*, by Leonardo da Vinci.

Study website: https://community.ptc.com/t5/PTC-Mathcad/Study-15/m-p/642389

LET'S START WITH THE CLOCK

The modern *conventional* clock is a round disk with arrows. On the perimeter of this disk are located twelve Arabic or Roman numerals, forming a *dial.* However, there is no one in the world who would be left indifferent by *unusual watches*, either in form or in content. Since the invention of the watch, people have not ceased to change not only their devices (solar, sand, water, mechanical, electronic, atomic, etc.), but also their appearance, pursuing both practical and decorative purposes. If we talk about practical purposes, then we can recall the non-standard clock with a 24-hour dial. These watches (chronometers) are placed on ships (sea, air, space), in order to not confuse the day with the night. There are clocks with two dials, showing both the local and the "native" time of the traveler. In hotels and offices, it is possible to see watches with several dials, or rather, several related clocks that record time in the key capitals of the world: Tokyo, Moscow, Paris, London, New York, Rome, etc.

Mixed practical and decorative functions were brought to the *digital clock*, where time is marked not by the hands ("analog" clock), but by numbers: 14:21:37, for example. These watches were once mechanical, while now they are usually electronic. The most practical are the clocks, where time is marked by hands (it is easier to judge by the hands how much time is left before some event), while the calendar date is expressed by numbers.

However, the main "delights and refinements" in the sphere of the originality of the clock unfolds on a purely "decorative front". It is possible to see this originality, by searching on the Internet with the keywords "unusual clock". What kinds of clocks will the reader see there?

The top of designer asceticism has two hands of different lengths (thanks also for this) rotating on the wall without any numbers or other marks on the dial, which, by the way, is not present in these clocks. The other extreme is the overload of a clock with various "artistic excesses". Looking at some clocks of this type, someone could puzzle for a long time to understand what time is being shown. Often the whole design essence of such "clocks" is based on this "puzzle", at times reaching a point of absurdity.

Among all the clocks of unusual form, a clock with a specific *mathematical meaning* stands out. Of course, mathematics is present in all clocks—it is only necessary to recall the differential equation of the oscillation of the pendulum, the balance wheel or the oscillation of the quartz crystal, on whose motion modern clocks are based. In Study 9 we described a clock, showing not only the time, but also the essence of Kepler's second law. In some watches, the body is made transparent so that all their internal "mechanical-mathematical essence"* is visible. However, we will now talk about the external, so to speak, mathematics of the clock.

There are clocks that instead of the numbers 1, 2, 3, ..., 12, have mathematical expressions (radicals) $\sqrt{1}, \sqrt{4}, \sqrt{9}, \ldots, \sqrt{144}$ on their dial. In this "decoration" the reader can notice a certain "practicality"—this type of clock will be appropriate, for example, in a schoolroom for mathematics: the students, glancing at the clock, will remember the table of the squares of the natural numbers. Figure 15.1 shows the author's clock, where instead of numbers mathematical formulas and expressions are written: using these, the corresponding numbers can be calculated in the environment of the engineering calculator Mathcad. A person observing this clock can not only learn the current time, but can also evaluate his/her knowledge of mathematics and of Mathcad—are all the formulas and the notations for the numbers understandable?

It is easy to see that instead of the usual natural numbers 1, 2, ..., 12 written down with the digits from 0 to 9, it is also possible to write down the very same numbers on a dial, but written only with the help of the digits 1, 2 and 3, by making use of the addition, multiplication and power operations. Indeed: $4 = 2^2$; $5 = 2^1+3^1$; $6 = 2^1 \cdot 3^1$; $7 = 2^2+3^1$; $8 = 2^3$; $9 = 3^2$; $10 = 2^3+2^1$; $11 = 2^3+3^1$; $12 = 2^2 \cdot 3^1$.

It is worth recalling that 2 is the only even prime number, and 3 is the first odd prime number. The number 1 (unity) is not considered a prime number, although it is key in the emergence of this concept. Euclid defined prime numbers as follows: "A prime number is measured only by the unity" (for the admirers of Euclid, see *Euclid's Elements*, Book 7, Definition 11). Which kinds of measurements are we talking about here? About the most usual measurements of a segment of a numerical axis from zero to a natural number by repeatedly shifting a segment of unit length. All natural numbers (large numbers, not prime numbers), can be measured, in addition to 1, by at least one more number, i.e. it is possible to shift a segment, for example, of length 2 (even numbers), or 5 (numbers that are multiples of 5). These numbers are called composite. Speaking strictly in mathematical terms, a prime has no other divisors, except 1 and itself. It is preferred to not consider 1 (the unit) to be a prime number, not least because in this case many important theorems of number theory are easier to formulate.

In the hours designation on the dial, surprisingly, exactly 6 (half of 12) composite numbers participate: 4, 6, 8, 9, 10 and 12; 5 prime numbers and 1. Thus, prime numbers make up almost 42% of the numbers of the first dozen. Are they always located so often?†

In order to understand this frequency, mathematicians introduced the function $\pi(N)$, the number of primes not exceeding N, and a special value $\frac{\pi(N)}{N}$, for the average density of primes among the first N natural numbers. The study of the tables of prime numbers shows that with increasing N, prime numbers occur on average less and less often (although still more often than the squares of natural numbers). Euler proved that, $\lim_{N\to\infty} \frac{\pi(N)}{N} = 0$, i.e. the longer the segment $[0, N]$, the smaller the percentage of all natural numbers in this segment that are prime numbers. However, if the density limit is zero, then, is it possible that sooner or later all prime numbers "end"? It turns out that this is

* A good idea on the theme of Study 8: to create in Mathcad the kinematic scheme of a clock, transmitting the swing of a pendulum or of a balance wheel to the hands of the hour, minute and second.

† Not many years ago, a novel written by a young writer Paolo Giordano was an Italian literary best seller *The Solitude of Prime Numbers* (orig. *La solitudine dei numeri primi*, https://en.wikipedia.org/wiki/The_Solitude_of_Prime_Numbers. The novel tells of the life of a couple and plays on a similarity between the couple and the "twin prime" https://en.wikipedia.org/wiki/Twin_prime.

impossible. The theorem on the infinite set of primes can be easily proved for the reader right now.*

Suppose that there was the largest (and therefore the last) prime number. We multiply all the known prime numbers, including the largest, and add one to the product. If the number obtained in this way is divided by each of the known prime numbers, then in the remainder we get 1, i.e. the new number is not divisible by a single prime number. Therefore, it cannot be divided also by any compound number, because each composite number, in turn, is the product of some prime numbers. So, we have found a number that is larger than any known prime number and can be divided only by one and by itself, i.e. this number itself is prime, superior to all known prime numbers. We have obtained a contradiction, which means that there is no greatest prime number.

If the reader wants to know which prime numbers lie between two given natural numbers, it is possible to use the Mathcad function *IsPrime* (see "hours four" in Figure 15.1), which determines whether the argument is a prime number and/or the `factor` operator, which decomposes a composite number into prime factors, see Figure 15.2. By the way, the decomposition of a large number into prime factors is a rather complicated and lengthy mathematical operation. One of the most

$\frac{58h}{1000b}$ $\sqrt[3]{1728}$ $\pi^e < e^\pi$

log(100^5) Mathcad $\sum_{i=0}^{\infty} \frac{1}{2^i}$

x^2-30x+189 solve $\lim_{x \to 2} \frac{7x - x^2 - 10}{x - 2}$

$\left|\begin{pmatrix} 10 & 6 \\ 2 & 2 \end{pmatrix}\right|$ $\frac{\sqrt[5]{1024}}{\text{IsPrime}(23)}$

$\begin{vmatrix} a \leftarrow 3 \\ b \leftarrow 4 \\ a + b \end{vmatrix}$ $\frac{3!}{-e^{i \cdot \pi}}$ $\int_1^3 \frac{x^3}{4} dx$

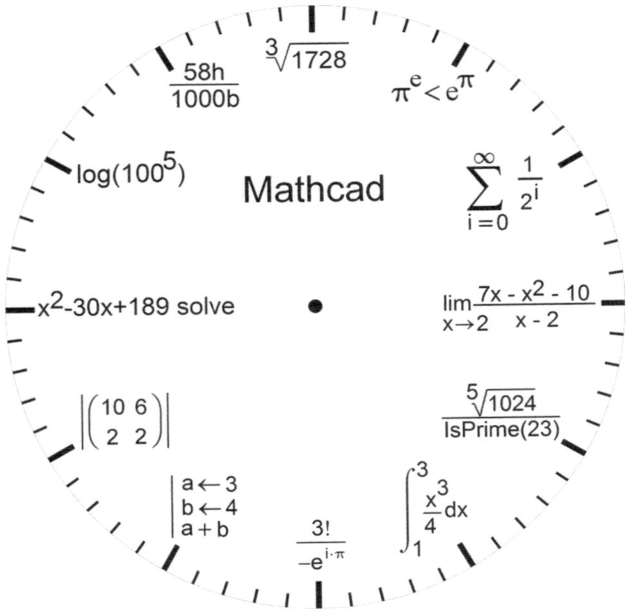

FIGURE 15.1 Author's clock with a "mathematical" dial.

* *Euclid's Elements*, Book 9, Proposition 20.

$$\text{IsPrime}(1) \rightarrow 0$$

$$11 \xrightarrow{factor} 11$$

$$11111111111111111 \xrightarrow{factor} 2071723 \cdot 5363222357$$

$$1111111111111111111 \xrightarrow{factor} 1111111111111111111$$

FIGURE 15.2 Analyzing the number "for primality" in Mathcad.

common algorithms of text encryption is based on the factorization of the number into prime factors.

From Figure 15.2 we see that 1 is neither a prime number nor a composite one. The first two prime numbers with 2 and 19 digits-units are also shown. The following prime number is made up of 23, 317, 1031, etc. digits-units. By the way, the search for the prime numbers of the year ("etc.") is a separate interesting task.

In the design of a clock there can sometimes lie complex mathematical algorithms. A Korean company produces clocks (Figure 15.3) in the form of solving the *traveling salesman's problem*. Here's what it is possible to read about it in Wikipedia (www.wikipedia.org): "The traveling salesman's problem is one of the most well-known problems of combinatorial optimization. The problem is to find the most advantageous route passing through all the cities and then returning to the original

FIGURE 15.3 A clock-globe with the traveling salesman's problem.

city. Under the constraints of the problem, the criterion of the route's advantage (shortest, cheapest, etc.) and the corresponding data on distances, costs, etc." are specified. The traveling salesman must visit each city only once. The traveling salesman's problem has many practical applications. First of all, the optimization of the routes—optimal logistics, as it is customary to say. The classic salespeople (peddlers) have now practically disappeared, although on the doors of many offices it is still possible to a sign such as "Commercial agents are not allowed to enter!". The problem of the traveling salesman is replaced these days by the courier, who is going to deliver the orders of the clients of a network store, which sells electronics, medicines, books and other things. There are also not quite ordinary applications of the traveling salesman's problem. So, for example, it is impossible to decipher the genome of a living organism without solving a rather complex traveling salesman's problem.

On the clock-globe with the traveling salesman's problem (Figure 15.3), 1440 settlements of the planet are selected every day exactly at midnight, selected at random from almost 2 million stored in the clock's memory (see Figure 15.8). Inside the globe a miniature planetarium is installed, which flashes these points on the surface of the globe. The number 1440 is the number of minutes in a day. The larger light points on the globe are the 24 nodal settlements ("cities"). They will be the hours in a day. Next, the first random city is chosen, into which the "traveling salesman" is placed, assigned with the task of bypassing all the settlements, guided by *the algorithm of the nearest neighbor*. From each city the traveling salesman goes to the nearest city that he has not yet visited. Two numbers are displayed on the digital panel of these clocks: the number of the city that our traveling salesman left (0, 1, ..., 23 hours) and the number of the settlement that he just passed (0, 1, ..., 59 minutes). The route passed on the globe is highlighted by lines connecting individual points. A person looking at this clock can not only find out the current time, but also the current position of the traveling salesman, and evaluate where the salesman will go next. On the panel of the clock, besides the current time, the name of the visited settlement is highlighted, and typical melodies and songs of the relevant region of the Earth are heard from the speakers. The dynamics of the clock can, of course, also fulfill its main function—to broadcast radio programs. The clock-globe can also be used for its intended purpose—to study the political map of the world.

Another similar clock, though not a desktop clock but a wall clock, hangs at the station in the Swiss city of Lucerne. Switzerland is a country of watches and travelers and these clocks are very appropriate there.* On this clock (flat plasma panel) is drawn the outline of Switzerland. Every night at midnight, this map shows 1,440 random cities, villages and freestanding hotels of this alpine country. A person who has just arrived at the station of Lucerne can not only determine the time by this clock, but also, if desired, he/she can follow the motion of the "traveling salesman", guess where he will turn in the next minute. This pastime will brighten up the wait for the train, make a person thinking about the problem of the traveling salesman, and about optimizing his/her own travel in Switzerland, Europe, the world.

In Figures 15.4 and 15.5 a simplified graphical display of the "dial" of these clocks is shown. In a rectangular area, 1440 squares are randomly placed, 24 of them are covered with cups, which are numbered from 0 to 23. At midnight the "traveling salesman" appears from the zero city (see the upper left corner in Figures 15.4 and 15.5) and for 24 hours visits all cities. Figure 15.4 shows the "noon" of our traveler (12:00:00) and Figure 15.5 one minute to midnight.

Animation of these unusual clocks can be seen on the author's website https://community.ptc.com/t5/PTC-Mathcad/Travelling-Salesman-Problem/m-p/490031.

In the traveling salesman's problem, schematically shown in Figures 15.4 and 15.5, different algorithms

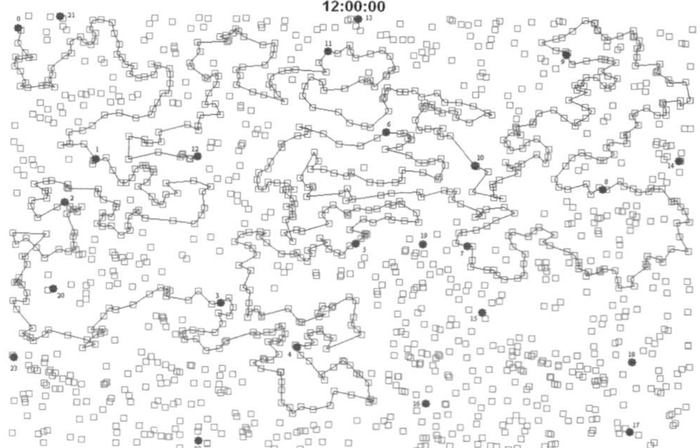

FIGURE 15.4 A clock with the traveling salesman's problem, noon.

* Here we could recall the scene about Italy versus Switzerland in the Orson Welles movie *The Third Man*: https://www.youtube.com/watch?v=nyuJQ_UO7OE.

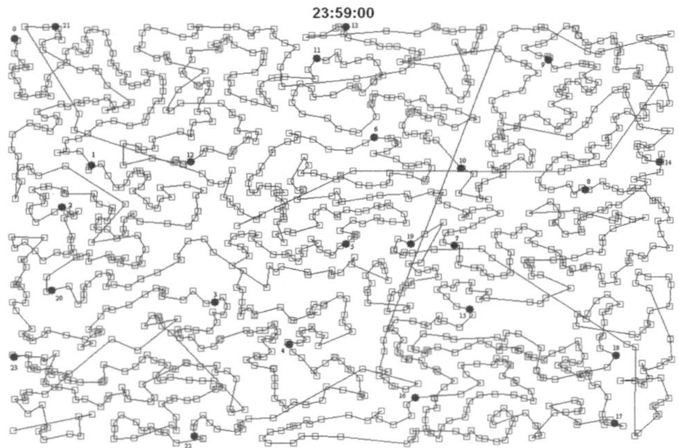

FIGURE 15.5 A clock with the traveling salesman's problem, one minute to midnight.

are implemented. One of them is the algorithm of the *nearest neighbor*: from the current city the traveling salesman is going to the nearest city that he has not yet visited. The algorithm of the nearest neighbor belongs to the category of "greedy" algorithms. Because of this "greed", our traveling salesman has to loop, and at the end of his tour jumps to distant cities (see the long stretches of the straight lines in Figure 15.5) previously recklessly missed. All this increases the total length of the route.

Visitors to the station in Lucerne, looking at these clocks, can think not only about "time and space", but also about something else.

Switzerland is not just a country of clocks and travelers, but also of bankers,* one of their qualities is their professional "greed", expressed in particular in the fact that bankers try to take money from their clients by keeping it in the bank at the minimum interest, while they give loans at the maximum rate. Professional greed forces bankers to give out loans even to those who cannot return money. This is one of the reasons for the financial crisis that has recently engulfed almost the entire world, that has destroyed many banks and has put some countries on the verge of a financial collapse.† A traveling salesman, skipping the necessary city and "eagerly" heading to the nearest, increases his route. The banker, chasing after a momentary profit, reduces the overall profit of the bank. Banking that is related to lending is dangerously bordered by usury. Another unseemly activity, similar to banking, is the construction of financial pyramids, a mathematical model of one of them will be considered at the end of this Study.

The algorithm of the nearest neighbor is corrupted by the simplicity of its implementation: the problem can be solved with the help of a compass, a pencil and a ruler. After all the cities are connected by lines, the reader can try to unravel some loops, thereby reducing the total length of the route. However, mathematicians have developed better algorithms for solving the traveling salesman's problem. One of them was stimulated by wildlife. People noticed that ants, having found a new source of food, first drag the food out of it into their anthill by random scattered routes. However, the spread of routes is gradually reduced into one single route, the shortest. Here is what it is written in Wikipedia (https://en.wikipedia.org/wiki/Ant_colony_optimization_algorithms):

> *In the real world, ants initially walk in a random order and find food to return to their colony, paving pheromone‡ trails. If other ants find such trails, they will most likely follow them. Instead of tracking the chain, they reinforce it when they return, if they eventually find the food source. Over time, the pheromone path begins to evaporate, thereby reducing its attractive strength. The more time is required for the passage of the path to and from the target, the more the pheromone path will evaporate. On a short path, for comparison, the passage will be faster and as a consequence, the density of pheromones remains high. Evaporation of pheromones also has the property of striving for a locally optimal solution. If pheromones did not evaporate, then the path chosen first would be the most attractive. In this case, the study of spatial solutions would be limited. Thus, when one ant finds (for example, a short one) the path from the colony to the source of food, other ants will most likely follow this path, and positive feedback ultimately leads all the ants to the shortest path.*

* As well as chocolate and knives. However, this detail is superfluous for our narration.

† To these countries, journalists once gave an offensive abbreviation PIGS (piglets): Portugal (P), Ireland (I), Greece (G) and Spain (S).

‡ Pheromones (Greek φέρω, "to carry" + ορμόνη, "to induce") is the collective name of substances—products of external secretion, allocated by some species of animals and providing chemical communication between individuals of the same species.

FIGURE 15.6 The path of a traveling salesman in Italy.

Figure 15.6 shows the path of the traveling salesman, laid with the help of an ant algorithm in 16,862 localities in Italy. Naturally, a traveler, who wished to pass through all the "cities and territories" of any country in the shortest way, cannot always go in a straight line from one settlement to the other. He/she will often have to take roundabout ways along existing roads and trails. Recently Google has developed a special service that allows a person to optimize the travel routes not only from point A to point B, indicated on the world map, but also on circular routes, using car, train, plane and other means of transportation.

The solution of the traveling salesman's problem, as in the case with the design of a clock, can be without practical applications but for purely aesthetic purposes. There is a technique in painting known pointillism,* when the artist paints an image with dotted strokes. We wrote about portraits in Study 5. Figure 15.7 shows a version of the famous painting by Leonardo da Vinci, where the *Mona Lisa* is drawn by 100,000 points connected by a "traveling salesman's path". The route of the traveling salesman through 1,904,711 settlements of the Earth (see Figure 15.8) can also be considered to be purely decorative.

However, back to the station clock in the Swiss city of Lucerne.

A traveler, unlike a traveling salesman, often intends to take not the shortest route, but a route that meets some other criteria. For the traveler, the preferred route can even be the longest. In this regard, we can talk about the *inverse problem of a traveling salesman*, the problem of finding the longest path.

FIGURE 15.7 The problem of the traveling salesman and the *Mona Lisa*.

Figure 15.9 shows the route of a traveling salesman through 979 cities and villages in Luxembourg using a greedy algorithm to find the shortest route. Animation of this route is placed on the author's website on PTC Community forum. We stayed in Luxembourg, not in Switzerland, because Switzerland does not "fit" into the computer when implementing the nearest neighbor algorithm—the matrix M (see Figure 15.10) will be too big. However, Luxembourg is also a country of bankers.

From Figure 15.9 (as, indeed, from Figures 15.4 and 15.5) it is clear that the chosen route is not optimal, because the traveling salesman, we repeat, because of his/her "greed" has to make loops. In addition, it is easy to show that the length of the "greedy" route depends heavily on the first city chosen. In Figure 15.9 the first and last city are marked. The path shown in Figure 15.9, is optimal in the sense that all cities were set as starting ones and a city was determined so that the path by the greedy algorithm was minimal. Another sign of the non-optimality of the routes of the traveling salesman, shown in Figures 15.4, 15.5 and 15.9, is that the lengths of the first and second halves of the path are very different. Figure 15.10 shows a Mathcad program for finding the route of the traveling salesman by the nearest neighbor method. The program returns a route vector, storing the numbers of the cities, the order of their passage to minimize the route.

* At the National Gallery, Trafalgar Square, London, there is for instance the painting of Georges Seurat, *Bathers at Asnières*.

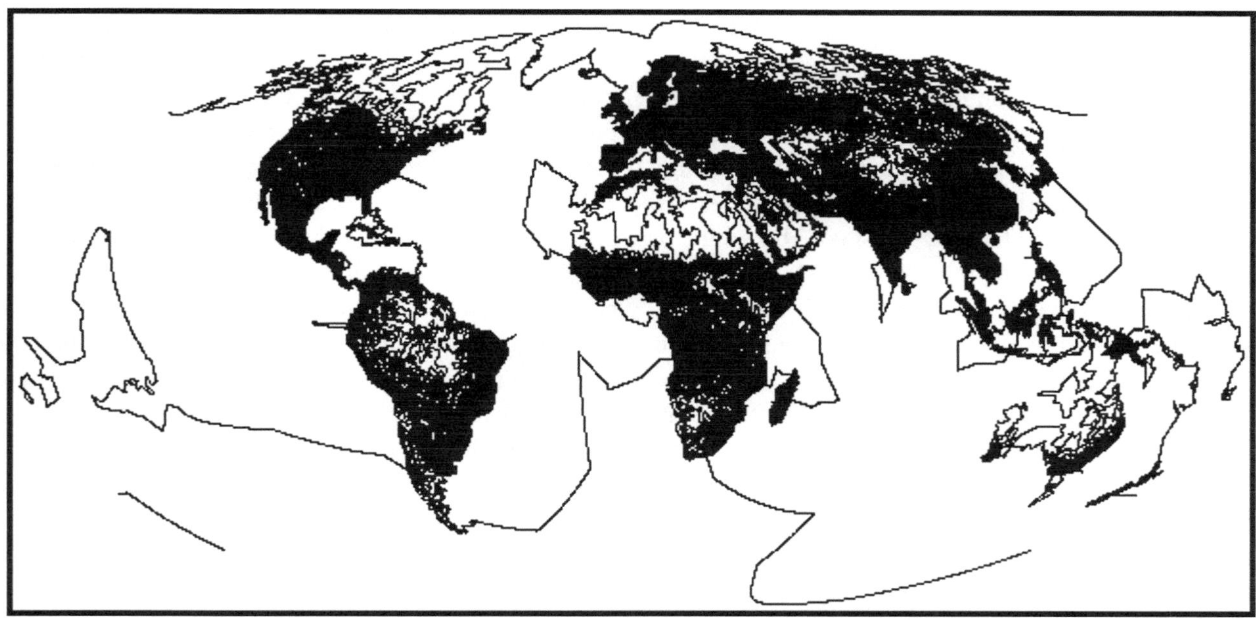

FIGURE 15.8 World tour of the traveling salesman.

The input data of the program are the vectors X and Y, with the Cartesian coordinates of the cities. The calculation begins with the creation of the square matrix M that stores numbers—distances between cities outside the main diagonal or the "infinity value" (∞, a super-large computer number) on the main diagonal. "Infinity values" will also be assigned to the elements of the matrix $M_{i,j}$ and $M_{j,i}$, if the traveling salesman has already gone from ith to jth city or vice versa. The "core" of the program are the *min* and the *match* functions. The function *min* returns the minimum element of its argument vector, while the function match returns the place of this element in the vector. The *match* function always returns a vector, even if there is only one element in it. Therefore, when somebody calls it in the program in Figure 15.10, the return operator of the first element

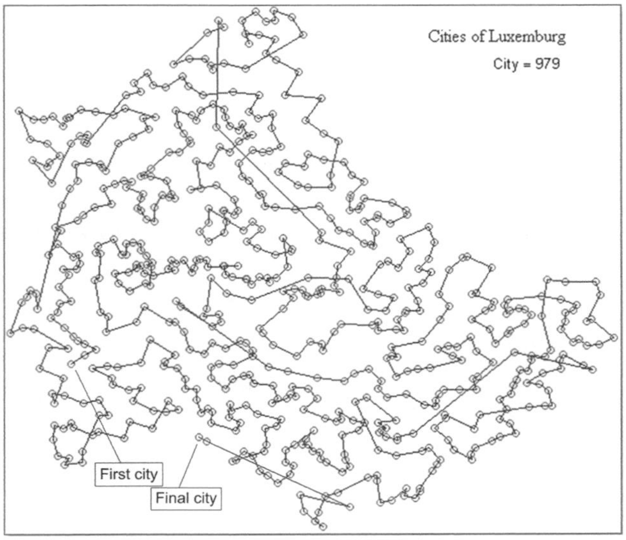

FIGURE 15.9 The route of the traveling salesman using a greedy algorithm for finding the shortest route.

```
way := | M ← | for i ∈ 1.. n
       |     |   for j ∈ 1.. n
       |     |     M[i,j] ← if[i = j, ∞, √((X[i] − X[j])² + (Y[i] − Y[j])²)]
       |     | M
       | way[1] ← np
       | for i ∈ 2.. n
       |   | for j ∈ 1.. n
       |   |   s[j] ← M(way[i−1], j)
       |   | way[i] ← match(min(s), s)[1]
       |   | for j ∈ 1.. i − 1
       |   |   | M[way[i], way[j]] ← ∞
       |   |   | M[way[j], way[i]] ← ∞
       | way[n+1] ← way[1]
       | return way
```

FIGURE 15.10 A Mathcad program of the nearest neighbor method.

of the vector, which converts such a unit vector into a scalar, is used.

However, if the function *min* is replaced by the function *max* in the program, then it will return the maximum path of the traveling salesman, where the traveling salesman is sent from the current city not to the nearest, but the furthest that he hasn't yet visited. This route, or rather its last part, is shown in Figure 15.11. The entire route cannot be shown, as it turns out all the country Luxembourg would be shaded. In Figure 15.11, only its central part is shaded.

Let's call this algorithm "generous" and note that its length also depends on the selected initial city. In Figure 15.11 this city is the same as was chosen for the implementation of the greedy algorithm (see Figure 15.9), but it is possible also to select a city, so that the path of the traveling salesman by the "generous" algorithm will be the longest. We delegate the solution of this problem to the reader. An even longer route can be obtained if, for example, the "ant" algorithm (described above) is applied to a matrix storing the distances between the cities.

The first and best-known problem of route optimization is the problem of the seven bridges of Königsberg (now the city of Kaliningrad). For a very long time among the inhabitants of this city, such a mystery was common: how to pass through all the bridges without going

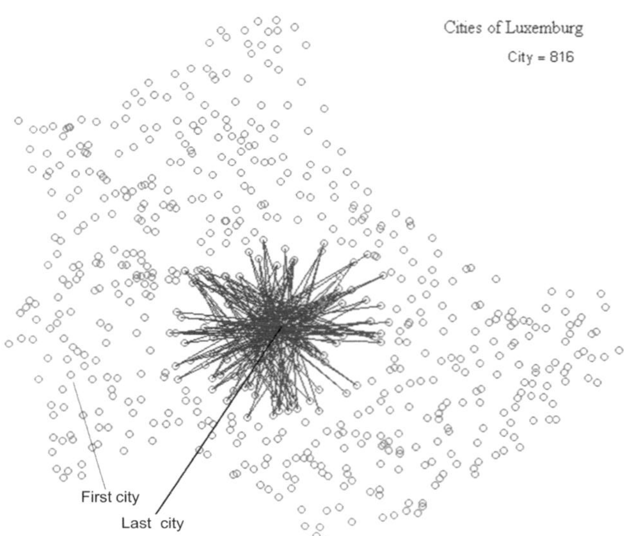

FIGURE 15.11 Completion of the traveling salesman's path through the "generous" algorithm (see the animation at https://community.ptc.com/t5/PTC-Mathcad-Questions/Reverse-Travelling-Salesman-Problem/m-p/490166).

through one of them twice? For a long time nobody succeeded in solving this problem, either theoretically or practically, but at the same time it was not possible to prove that this problem was unsolvable. In 1736, the problem of the bridges interested Leonard Euler (see Figures 27.7 and 27.8 in Study 27), an outstanding mathematician and member of the Petersburg Academy of Sciences. He was able to establish general rules, and by using them it becomes easy to determine whether it is possible to pass through all the bridges of any city, without going twice through one of them. It turned out that in the case of the seven bridges of Königsberg this is impossible.

Euler built a graph in which the bridges correspond to the lines (edges of the graph), and the parts of the city to the points (the vertices of the graph). In the course of solving this problem, the scientist identified and proved the following.

1. The number of odd vertices (vertices to which leads an odd number of edges) of a graph is always even. There cannot be a graph that has an odd number of odd vertices.
2. If all the vertices of the graph are even, then, it is possible to draw a graph starting and ending in the same pre-selected vertex, without taking the pencil from the paper.
3. A graph with more than two odd vertices cannot be drawn with one stroke.

The graph of the Königsberg bridges had four odd vertices (see Figure 15.12), therefore, it was impossible to pass through all bridges, without passing through either of them twice.

The theory of the graphs created by Euler (thanks to this problem) has found very wide application: for example, it is used in studying transport and communication

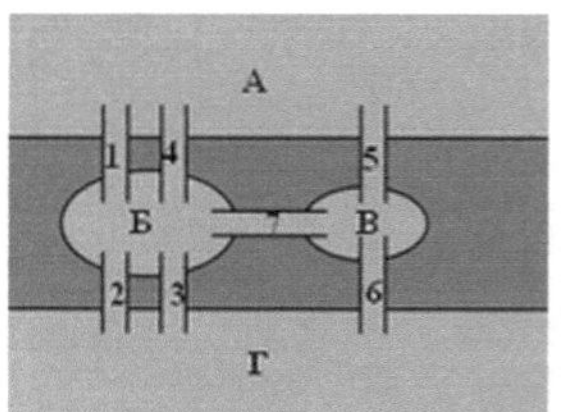

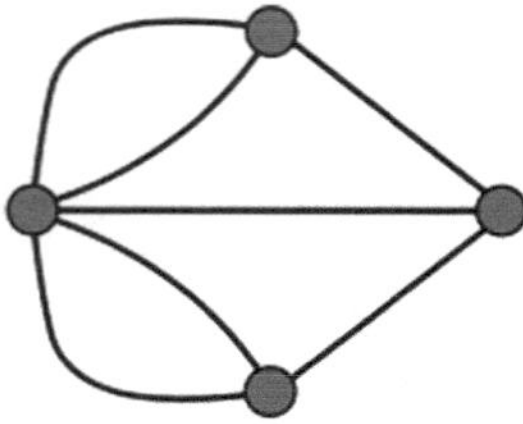

FIGURE 15.12 Simplified diagram and graph of Königsberg bridges*.

* http://eulerarchive.maa.org//docs/originals/E053.pdf.

systems and, in particular, for routing data on the Internet. Note that the traveling salesman's problem also admits a solution because its graph (see Figures 15.9 and 15.11) has all the vertices even (the entrance to the city along one road, the exit along the other, that is, exactly two edges). This graph obeys the second assertion of Euler. However, the presence of a solution does not mean at all that this solution will be perceived as natural and intuitive. For example, the traveling salesman's problem contains a paradox: often he should not go to the nearest city!

And here there are three more problems, similar to the traveling salesman's problem and also having elements of paradox.

PROBLEM 1

You are at point A (see diagram in Figure 15.13) at the border of arable land (top) and of a lawn (bottom). A thunderstorm starts, and you need to reach the house quickly, located at point B.* The speed of your running on the plowed land is 5 kph while on the lawn it is 10 kph. How will you run into the house? The first thing that comes to mind, especially in a stressful situation—is to run in a straight line from point A to point B. But the optimal route is a certain "knight's move" (see Figure 10.6 in Study 6): first you need to run on a lawn with a speed of 10 kph, and then turn to a plowed field where your speed is lower. The right-hand side of Figure 15.13 shows the schedule for changing the time of such a rescue run, depending on the coordinate of the rotation from the lawn to the arable land. The graph clearly shows the minimum point at about 40 meters.

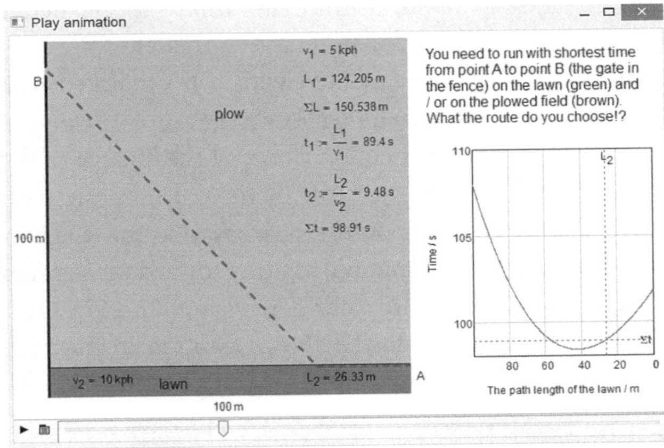

FIGURE 15.13 Optimization of running from a thunderstorm (see Study 20).

But imagine that near the house there is an asphalted path to the lawn and further down (see the left edge of Figure 15.13), along which you can run at a speed even greater than when running along the lawn—for example, 15 km/hr. Then, to reduce the travel time, you can run along the plowed field not directly to the house, but to the asphalt path, and then along it into the house. In this case, we will already have two optimization parameters—the coordinate of the point of departure from the lawn to the arable land and the coordinate of the descent from the arable land to the asphalt path. And maybe you do not even need to run on the plowed land, but only on the lawn and along the path? What is the minimum running speed on the asphalt path in this case? We ask the readers to solve this very complicated problem by themselves.

But back to the traveling salesman. Figure 15.9 shows that this wandering merchant walked between cities (points) strictly in a straight line. However, it's almost impossible to go like that on a real journey—every segment of the path should be optimized, taking into account the various restrictions. An example of this optimization is shown in Figure 15.13.

PROBLEM 2

It is not only necessary to walk around the points shown in Figure 15.9 offering the product, but also to connect the tracks so that their total length is minimal. This problem is called the Steiner problem,† from the name of a mathematician who first formulated it. It is an order of magnitude more difficult than the traveling salesman's problem. But, nevertheless, we will present it not only for the plane (Figure 15.14(a)), but also for space (Figure 15.14(b)). If somebody wants to connect the vertexes of a square with lines of minimum length, then many connect these four points by two straight lines passing through the center of the square. However, this is not the optimal solution, see Figure 15.14(a). In the Steiner problem, the solution uses three straight lines only with

* Another option for this problem. You walk leisurely along the edge of a wide beach by the seashore. 100 meters ahead to the right/left you see a free bench distant 30 meters from the sea, on which you want to sit down. You need to approach the bench as quickly as possible, otherwise it will be taken before you get there. How do you go or run to the bench, if the sea has dense sand, allowing you to move quickly, and on the beach—loose, breaking movement? In this optimization task there will be one more restriction: you need to move so that it is not noticeable that you are trying to take a bench before someone else.

† Also known as Euclidean Steiner tree problem (see Wiki).

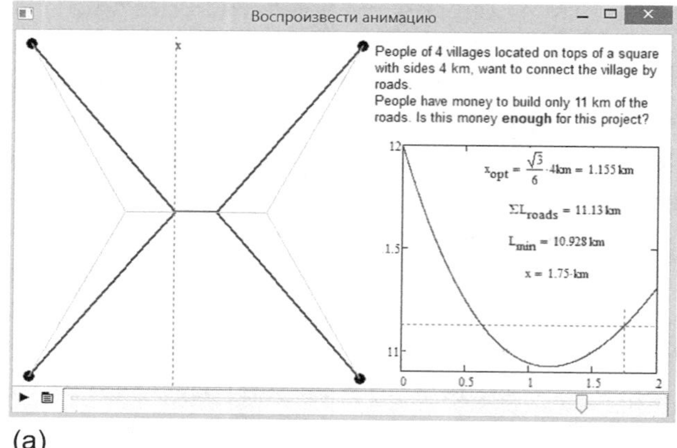

(a)

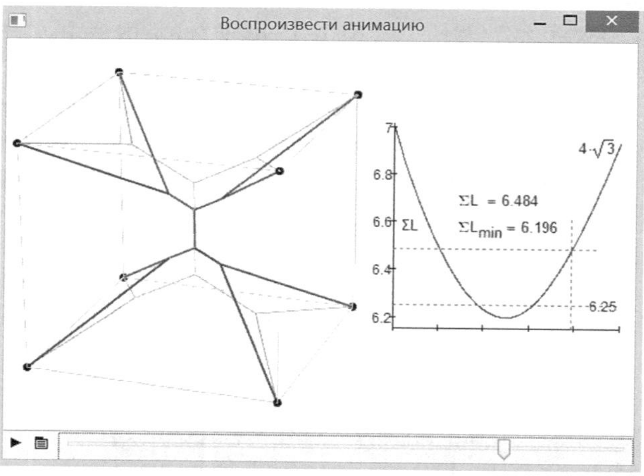

(b)

(c)

FIGURE 15.14 The Steiner problem: (a) animation on the plane; (b) animation in the volume; (c) the physical model.

an angle of 120° to each other (a Mercedes star) converging to a point.

However, there is also a volumetric version of the problem: moles live underground in burrows located on the vertices of the cube. They decided to join their homes so that the length of the moves was minimal. How should they dig their burrows? Figure 5.13(b) shows the animation frame of the solution of this problem with the display of the graph of the change in the total length of the moves (objective function) depending on the configuration of their moves. The optimal variant also implies "only three straight lines with an angle of 120° to each other converging to a point".

Steiner's problem is usually illustrated with a soapy foam (see photo in Figure 15.14(c)), which, due to surface tension, minimizes its surface.

A practical version of the Steiner problem can be seen in this example: it is necessary to bring gas (electricity, Internet, cable TV) to two houses from the main pipe with the minimum total length of pipes. The houses are located on one side of the utility line. There may be solutions such as: (a) it is possible to bring individual pipes to each of the houses perpendicular to the main; (b) it is possible to bring the pipe to the first house, and then make a withdrawal from it to the second house; and (c) it is possible to install a T-junction somewhere outside the house. For two houses this problem is solved and animated quite simply. But for three, four or more houses, the problem becomes much more complicated. A general method for solving the Steiner problem has not yet been found. The reader can try to draw a few points on a paper and to connect with straight lines, having a triangle with an angle of 120°. Then measure the lengths of these segments, sum them and draw conclusions about minimizing this amount. Better still, try to solve this problem on the computer.

In Study 9 we experimented with a five-pointed star, launching satellites from its vertices (see Figure 9.5). Figure 15.15 shows the last animation frame of the change of the configuration of line segments with a minimum total length, connecting five points that are wandering (the previous frames of the animation) tending toward the vertices of a five-pointed star. This star here was not accidental—the book was written during the celebration of the 70th anniversary of the Victory in the Great Patriotic War.*

PROBLEM 3

There are two mines where coal is excavated, and two CHP (combined heat and power), where this coal is burned without any storage. The first mine "produces" 50 tons of coal per day, and the second, 70. The first power plant burns 40 tons of coal per day, and the second, 80. Transportation of coal from the first mine to

* 9th May 2015: The Victory Day is a holiday that commemorates the victory of the Soviet Union over Nazi Germany in the Great Patriotic War. Known also as Victory in Europe Day.

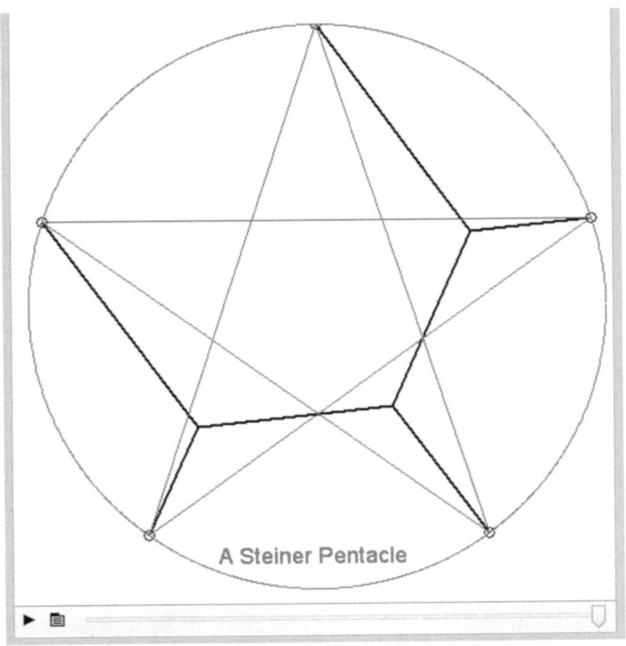

FIGURE 15.15 The Steiner problem for a five-pointed star.

the first power plant costs 1200 euros per ton, from the first mine to the second power plant, 1600, from the second mine to the first power plant, 800 and from the second mine to the second power station, 1000 euros per ton. It is necessary to organize the transportation of coal so that its cost is minimal. In Figure 15.16 the conditions of this transport problem and its "manual" solution are shown: the cheapest route (the second mine, the first thermal power station: 800 euros per ton of transported coal) is fully loaded (see the first term in the expression

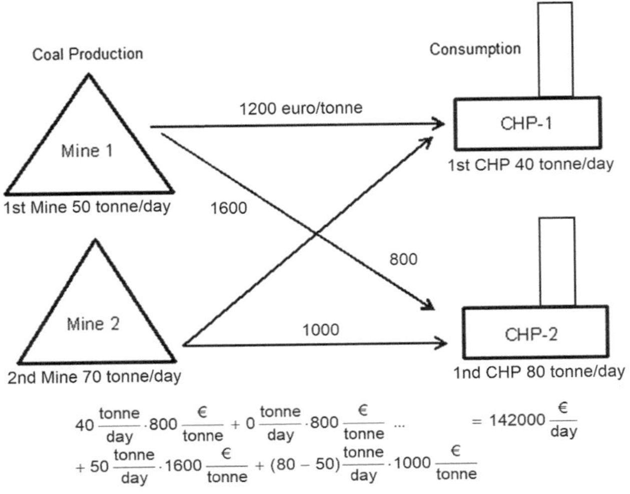

FIGURE 15.16 Scheme and incorrect manual solution of the transport problem.

$$TC(M1C1, M1C2, M2C1, M2C2) := 1200\frac{€}{tonne}\cdot M1C1 + 1600\frac{€}{tonne}\cdot M1C2 \ldots + 800\frac{€}{tonne}\cdot M2C1 + 1000\frac{€}{tonne}\cdot M2C2$$

Objective function

Guess values

$$M2C1 := 40\frac{tonne}{day} \quad M1C1 := 0\frac{tonne}{day} \quad M1C2 := 50\frac{tonne}{day} \quad M2C2 := 30\frac{tonne}{day}$$

Given

$$M1C1 + M1C2 = 50\frac{tonne}{day} \qquad M2C2 + M2C1 = 70\frac{tonne}{day}$$

$$M1C1 + M2C1 = 40\frac{tonne}{day} \qquad M1C2 + M2C2 = 80\frac{tonne}{day}$$

$$M1C1 \ge 0\frac{tonne}{day} \quad M1C2 \ge 0\frac{tonne}{day} \quad M2C1 \ge 0\frac{tonne}{day} \quad M2C2 \ge 0\frac{tonne}{day}$$

$$\begin{pmatrix} M1C1 \\ M1C2 \\ M2C1 \\ M2C2 \end{pmatrix} := Minimize(TC, M1C1, M1C2, M2C1, M2C2) = \begin{pmatrix} 40 \\ 10 \\ 0 \\ 70 \end{pmatrix}\frac{tonne}{day}$$

$$TC(M1C1, M1C2, M2C1, M2C2) = 134000\frac{€}{day}$$

FIGURE 15.17 Automatic correct solution of the transport problem.

at the bottom of Figure 15.16), while the coal remains "dispersed" on the remaining routes.

Figure 15.17 shows the automatic and non-manual solution of this problem by the standard tools of Mathcad 15—the objective function is created (*TC* is the transportation cost), a first approximation is provided (we took the values found in the manual solution of the problem shown in Figure 15.16), the keyword *Given* sets the constraints and the *Minimize* function finds unknown values that minimize the objective function verifying the constraints. The automatic solution of the problem turned out to be *paradoxical*: nothing needs to be transported along the cheapest route.

We can see in Figure 15.18 a solution to the transport problem using vectors and matrices. This technique

Mines (m) — Power Plants (p) — Price (P) of coil transportation (T)

$$m := \begin{pmatrix} 50 \\ 70 \\ 0 \\ 20 \end{pmatrix}\cdot\frac{tonne}{day} \qquad p := \begin{pmatrix} 50 \\ 10 \\ 80 \end{pmatrix}\cdot\frac{tonne}{day} \qquad PT := \begin{pmatrix} 1100 & 1600 & 700 \\ 500 & 1000 & 1200 \\ 1000 & 1600 & 1550 \\ 750 & 900 & 1150 \end{pmatrix}\cdot\frac{€}{tonne}$$

$$i := ORIGIN..\, last(m) \qquad j := ORIGIN..\, last(p) \qquad \sum_i m_i = 140\cdot\frac{tonne}{day} \qquad \sum_j p_j = 140\cdot\frac{tonne}{day}$$

$$Price(mp) := \sum_i \left[\sum_j (mp_{i,j}\cdot PT_{i,j})\right] \qquad m1_i := 1 \qquad p1_j := 1$$

$$mp_{i,j} := 0\cdot\frac{tonne}{day} \qquad Price(mp) = 0\cdot\frac{€}{day}$$

Guess values mp - from mine to plant

$$\text{Given} \qquad mp \ge 0 \qquad m = mp\cdot p1 \qquad mp^T\cdot m1 = p$$

$$mp := Minimize(Price, mp) = \begin{pmatrix} 0 & 0 & 50 \\ 50 & 0 & 20 \\ 0 & 0 & 0 \\ 0 & 10 & 10 \end{pmatrix}\frac{tonne}{day} \qquad Price(mp) = 104500\cdot\frac{€}{day}$$

FIGURE 15.18 Solution of the transport problem with using vectors and matrix.

allows the user to easily change the number of mines and power plants when solving the transport problem. Features of this solution were discussed on the website https://community.ptc.com/t5/PTC-Mathcad-Questions/One-transport-problem-linear-programming-in-matrix-form-Why/m-p/485675.

Now let's go back to the bankers whose professional greed was previously highlighted by us.

Banking turns into usury when the payment for a loan is too high. On the other hand, a too high interest on a deposit is also not welcomed—because behind such "banking products" can hide an elementary financial pyramid. Figure 15.19 shows the simplest mathematical model.

The correct banking system rests on *three* numbers, as on *three* whales. The first number N_1 is the payment for a loan. I took a hundred rubles from the bank—be kind enough at the end of the year to return 100 + N_1 rubles. The second number N_2 is the percentage of a deposit. I put a hundred rubles into the bank—receive 100 + N_2 rubles at the end of the year. The difference between the first and second number ($N_1 > N_2$) allows

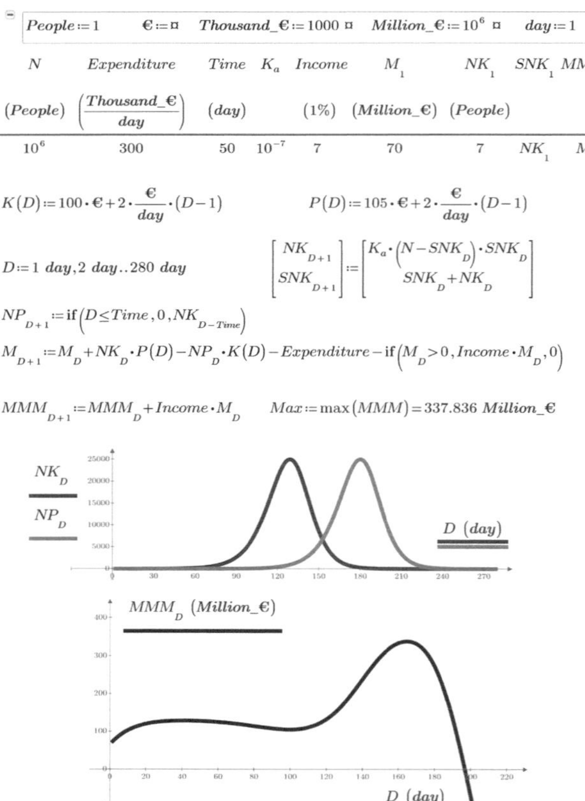

FIGURE 15.19 Mathematical model of the financial pyramid.

banks to work profitably. The third number N³, backing up the two previous ones and forcing people to bring money to the bank, is the magnitude of inflation. In a normal economic situation, a low level of inflation and a not very high fee for a loan keep the interest on the deposit within a narrow range:

$$N_1 > N_2 > N_3.$$

If inflation is high, many people, forgetting about the abnormality of this situation, easily expect 20, 30, 50 or more percent per annum on the deposit (after all, the value of N_2 should be greater than the value of N_3) and fall into the foundation of the next financial pyramid. There are also less naive people who understand that a pyramid is a special kind of game where a player must be able to *"time out"*.

So, we are building a financial pyramid (Figure 15.19).

In a city of one million residents (variable *N*), seven of them bought one share at a price of 100 rubles on the first day (D = 1). Then the number of these people will increase (vector NK). If people sell this share in 50 days (Variable Time), they will receive 200 rubles for each share transferred. These dynamics (selling rate and buying price of shares) are determined by two functions *K(D)* and *P(D)*. In the city there is a certain excitement, which is determined by the variable K_a. From the cash desk every day a certain amount of money is taken (variable *Expenditure*) for renting premises, advertising, bribes to officials, etc., as well as 7% for the income of the organizers of the pyramid (variable *Income*). A more complicated operator of the problem shown in Figure 15.19, is the definition of the number of people who bought shares in a certain day. This value is proportional (with the coefficient K_a) to the number of people who have not yet bought shares multiplied by the number of people who have already bought shares. This relationship determines the shape of the two waves of buyers and sellers of shares shown in Figure 15.19.

The last graph in Figure 15.19 shows the change in the amount of money in the cash register of the organizers of the financial pyramid. On the 164th day, the pyramid should be closed—stop buying shares and leave for somewhere outside the city.

We do not leave anywhere, we stay at our computer and, intending to invest money in some dubious enterprise, first we will calculate what can come of it. So, we

Number of inhabitants in the city $N := 1000000$

Coefficient of excitement $KA := 10^{-7}$

Number of people who bought shares on the first day $NK_1 := 7$

y(x) - is the number of shares purchased at time t

Given

$$y(1) = NK_1 \qquad y'(t) = \left[KA\cdot\left(N - \int_1^t y(t)\,dt\right) - KA\cdot\int_1^t y(t)\,dt\right]\cdot y(t)$$

$y := \text{Odesolve}(t, 365)$

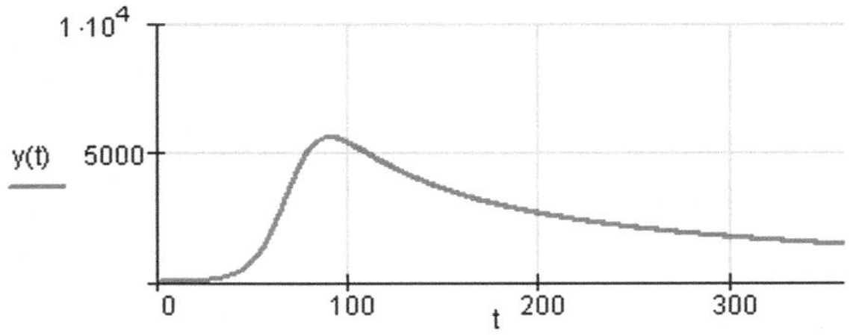

FIGURE 15.20 Simplified mathematical model of the financial pyramid (only Mathcad 11).

ORIGIN := 1 $\quad n := 70 \quad X := \text{runif}(n, 0, 100) \quad Y := \text{runif}(n, 0, 100)$

$Way := \text{anneal}(X, Y) \quad Way_{n+1} := Way_1 \quad i := 1..n+1$

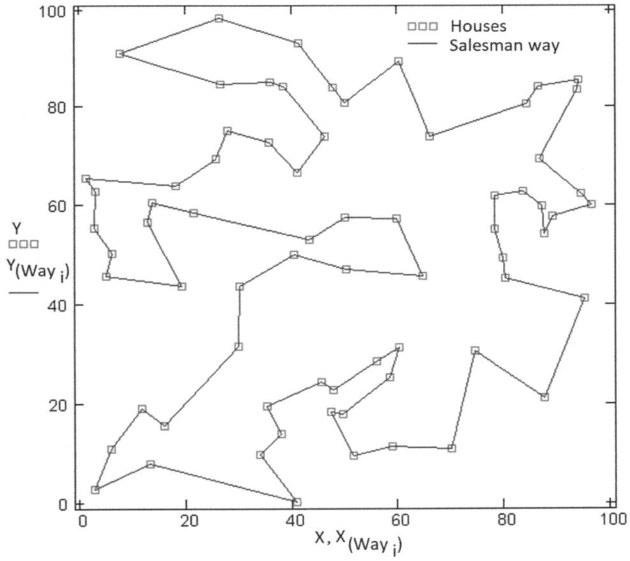

FIGURE 15.21 The solution of the traveling salesman's problem with the anneal functions.

can easily return and even increase the money spent on purchasing the computer and its legal software.

If the model of the financial pyramid is extremely simplified, its solution can be reduced to solving a differential integral equation, see Figure 15.20, where the initial growth of the incomes of the organizers is also visible, and then their decline.

Trading bitcoins and other crypto-currencies, by the way, has signs of a financial pyramid. On the other hand, all this is connected with attempts to find a substitute for the US dollar. The world's financial equivalent could become energy resources.

ASSIGNMENTS FOR READERS

1. Think of the operators for the numbers 13, 14, 15, ..., 24, which can be placed on the dial of a clock with a daily cycle, rather than a semi-day one (see Figure 15.1).

2. Find in handbooks or on the Internet a description of the solution of the traveling salesman's problem by the ant algorithm or by the annealing method (anneal is an application of the old versions of Mathcad that solves this problem with the same function, see Figure 15.21).

3. Create a function in Mathcad that returns the amount of money in a bank deposit with compound interest. This function, however, already exists in the list of the Mathcad built-in financial group of formulas.

4. Create an anti-animation for the inverse problem of the traveling salesman (Figure 15.11). The principle of anti-animation will be described in Study 26. It will allow us to more clearly trace the path of the traveling salesman maximizing the distance traveled.

5. Create a model for the exchange rate of the bitcoin. At the same time, on one hand, take into account a certain "pyramidality" of this crypto-currency, and, on the other, the world economy's need for a new "gold equivalent", which, by the way, could be energy, energy resources (see Study 29).

6. Create a Mathcad calculation of the transport problem (see Figures 15.16 and 15.17 in this Study), based on the matrix (the cost of transportation from point *i* to point *j*) and vectors (coal production at the *i*th mine and coal consumption at the *j*th power station). Research and discussion of this solution were conducted on the website https://community.ptc.com/t5/PTC-Mathcad-Questions/One-transport-problem-linear-programming-in-matrix-form-Why/td-p/485675.

Heads or Tails Or

A Three-Way Duel in Monte Carlo

ABSTRACT

In this Study we talk about the computer solution of probability theory problems and about statistical tests methods. The mathematical foundations of coin tosses and dice are discussed. A new solution to the old problem of a Three-Way Duel (a tripartite duel) is given.

Mathematics: Monte Carlo method, π number, calculation error, probability, fuzzy sets.

Physics: Throwing a coin and throwing a die.

IT: Random and pseudo-random numbers, cycle with parameter, loop with pretest.

Art: Jokes.

Study website: https://community.ptc.com/t5/PTC-Mathcad/Study-16/m-p/642393

There is an old anecdote, a joke. A student wakes up in the morning, tosses a coin and makes a guess: "If heads—I'll sleep on my right side, if tails—I'll lie on my left side, if it stands on its edge—I'll go to university, and if it hangs in the air—I will study the term paper".*

Jokes are jokes, but we often have to flip a coin—real or virtual, in order to randomly choose one of two equally probable alternatives. The football referee, for example, tosses a coin before the match to determine which team gets the ball and which gets to choose ends. A chess player starts by hiding a white pawn in one hand and a black pawn in the other and gives the opponent the choice of "one opportunity from two equi-probable ones"—to determine who will play with the white set and who will play with the black. Shuffling a deck of cards, mixing dominoes, spinning the roulette wheel, shuffling the lotto kegs in the bag (reference to Russian Lotto), we give ourselves a chance. And on this a great deal can depend even the life of a person, if you recall the name of this Study. And, instead of a simple (two-sided) duel, why should we not consider an unusual (three-way) duel (a *truel*, neologism)? The point is that the problem of the three-way duel is described in many books, where a particular solution of this problem, obtained by the method of logical reasoning, is given. We will consider a more complete solution of this problem on a computer by the Monte Carlo method (by statistical tests): we simulate a single duel, we will conduct it a sufficiently large number of times, and for each participant we will calculate the number of victories in these duels. If these numbers are divided by the total number of duels, the result will be the desired probability of winning.

However, we will start with simple problems, whose solution is known without the help of a computer. Let's remember our negligent student from the anecdote. With the help of a computer we'll toss a coin many times and count up how many times heads shows, and how many times tails respectively. Figure 16.1 shows the corresponding program for the computer, the function

* "I'll post a coursework assignment at the PTC Community forum and I will wait until someone does some work for me!"

$$HeadsorTails(n) := \left\| \begin{array}{l} Head \leftarrow 0 \\ Tail \leftarrow 0 \\ \text{for } i \in 1..n \\ \quad \left\| \begin{array}{l} \text{if } \mathrm{rnd}(1) < 0.5 \\ \quad \| Head \leftarrow Head + 1 \\ \text{else} \\ \quad \| Tail \leftarrow Tail + 1 \end{array} \right. \\ \dfrac{\begin{bmatrix} Head \\ Tail \end{bmatrix}}{n} \end{array} \right.$$

$$HeadsorTails(10000000) = \begin{bmatrix} 49.99333\% \\ 50.00667\% \end{bmatrix}$$

FIGURE 16.1 The program "HeadsorTails".

*HeadsorTails** with the argument n (the number of coin tosses), returns a vector with two elements that hold the probability of showing heads or tails.

The core of the program shown in Figure 16.1 is the Mathcad built-in function *rnd*, which returns a random number† in the interval from 0 to the value of the argument of the function *rnd*, in our case to 1. If the function *rnd*(1) returns a number less than 0.5, then we assume that heads shows, else tails shows. It remains only to count what shows in the cycle for throwing the coin: *Head* ← *Head* + 1 or *Tail* ← *Tail* + 1.

It is possible, of course, to ask why in the program in Figure 16.1 states $rnd(1) < 0.5$, and not $rnd(1) \leq 0.5$. Replacing the operator "less than" by the operator "less than or equal to" will not change anything in the calculation, because the fulfillment of the condition $rnd(1) = 0.5$ is equivalent to the fall of the coin on the edge: the probability that the function *rnd*(1) generates the value 0.5 is practically zero. This can be verified by replacing the "less than" operator by the "equals" operator in the program in Figure 16.1 and counting how many times this "head" shows with a sufficiently large number of coin tosses. The equality of the function *rnd*(1) to zero or one in general (theoretically) is impossible ("hanging of the coin in the air"), because function *rnd*(1) generates random numbers in the interval (0, 1), and not in the interval [0, 1].

If necessary the integer random numbers may represent not just two variants `0/1` (head/tail, even/odd, yes/no, etc.), but, for example, in six variants, such as required for a die (a cube, with six numbered faces). A roulette wheel‡ generates integer random numbers from 1 to 36, while numbers from 1 to 90 are present in a bag with "kegs" for playing (Russian) lotto, when the first keg is taken from it.

Figure 16.2 shows a program using the Monte Carlo method to calculate the probability of the numbers 2, 3, 4, ..., 12 showing when throwing two dice. In the column on the right, the theoretical (classical) probability of this result is calculated, which, for convenience of comparison, is also represented in percentages (which, generally speaking, are not accepted in probability theory, whose

$$\mathbf{ORIGIN} := 2$$

$$\mathrm{TwoDice}(n) := \left\| \begin{array}{l} \text{for } i \in 2..12 \\ \quad \| S_i \leftarrow 0 \\ \text{for } i \in 1..n \\ \quad \left\| \begin{array}{l} Die1 \leftarrow \mathrm{ceil}(\mathrm{rnd}(6)) \\ Die2 \leftarrow \mathrm{ceil}(\mathrm{rnd}(6)) \\ S_{Die1+Die2} \leftarrow S_{Die1+Die2} + 1 \end{array} \right. \\ \dfrac{S}{n} \end{array} \right.$$

$$\mathrm{TwoDice}(10^7) = \begin{bmatrix} 2.78258\% \\ 5.54375\% \\ 8.33525\% \\ 11.09799\% \\ 13.88425\% \\ 16.68185\% \\ 13.90013\% \\ 11.10624\% \\ 8.33167\% \\ 5.55403\% \\ 2.78226\% \end{bmatrix} \qquad \begin{array}{l} \frac{1}{36} = 2.77778\% \\ \frac{2}{36} = 5.55556\% \\ \frac{3}{36} = 8.33333\% \\ \frac{4}{36} = 11.11111\% \\ \frac{5}{36} = 13.88889\% \\ \frac{6}{36} = 16.66667\% \end{array}$$

FIGURE 16.2 Game of dice.

* A linguistic digression: In Italian, "Heads or Tails" is said "Testa o Croce" (Head or Cross), Romans called it *navia aut caput* ("ship or head"), as some coins had a ship on one side and the head of the emperor on the other. In England, this was referred to as *cross and pile*. The expression *heads or tails* results from heads and tails being considered complementary body parts: in Germany *Kopf oder Zahl* (Head or Number), in Ireland heads or harps, in Brasil *cara ou coroa* (head or cross), in Mexico *águila o sol* (eagle or sun), in Hong Kong 公定字 (head or word) and in Russian орёл или рéшка (eagle or tail).

† Rather, a pseudo-random number. The fact is that the *rnd* function returns the same series of random numbers every time it is called. When a person throws a real coin, it never happens, but when somebody throws a virtual coin on a computer, it's so specially programmed so that it is possible to debug programs. In Study 5 we considered a non-standard way of obtaining truly random numbers, see Figure 5.15.

‡ European roulette. Other variants of this "wheel of fortune" may have a different number of divisions. European roulette is sometimes called the "Ferris wheel" not because the sum of all the numbers on it is 666, but rather because this gambling has ruined many people who do not know the measures in the game. No wonder they say: "If you want to win in a casino, buy it!"

basis is given by the systems of axioms given by A.N. Kolmogorov).

The central element of the program shown in Figure 16.2, is also a function *rnd*, which returns random real (non-integer) numbers in the range from 0 to 6 (the argument of the function *rnd*). Another Mathcad built-in function is the *ceil* function* that rounds up these numbers to the next largest integer: 0.3 becomes 1, 5.7 becomes 6, etc. The inclusion of the function *rnd* in the function *ceil*(*rnd*(6)) allows us to simulate the throwing of one die—the generation of integer random numbers from the set 1, 2, 3, 4, 5 and 6.

In the program in Figure 16.2, 2 dice are thrown 10 million times (as in the coin problem) and the number of times 2, 3, 4, 5, …, or 12 are shown are recorded. These results are summed in a vector with the name s, whose first element is numbered 2 (this is defined by the operator *ORIGIN* := 2), and the last is numbered 12. This is the only program in the book where the *ORIGIN* system variable is neither 0 nor 1 (see Study 2).

Let us analyze the results obtained. Figure 16.2 shows the probabilities of showing 2 (2.77%), triples (5.55%), etc. after tossing 2 dice 10 million times. These probabilities can be counted without a computer: a die has 6 faces, while 2 dice have 36 (6^2) "faces". For 2 be achieved, there is only one combination: 1 + 1; for 3, there are two combinations: 1 + 2 and 2 + 1; for 4, there are three: 1 + 3, 2 + 2 and 3 + 1; for 5, there are four: 1 + 4, 2 + 3, 3 + 2 and 4 + 1; for 6 there are five: 1 + 5, 2 + 4, 3 + 3, 4 + 2 and 5 + 1; for 7, there are six: 1 + 6, 2 + 5, 3 + 4 , 4 + 3, 5 + 2 and 6 + 1; for 8, there are five: 2 + 6, 3 + 5, 4 + 4, 5 + 3 and 6 + 2; for 9, there are four: 3 + 6, 4 + 5, 5 + 4 and 6 + 3; for 10, there are three: 4 + 6, 5 + 5 and 6 + 4; for 11, there are two: 5 + 6 and 6 + 5; and, finally, for 12 (like 2), there is one combination only: 6 + 6.

From this it is clear that when two dice are thrown, 7 has the greatest probability of being displayed (one-sixth), while 2 and 12 have the lowest probability (one thirty-sixth). This has been known or suspected by experienced dice players for a long time and they choose the right strategy for the game accordingly. Furthermore, although the values of the classical probability of displaying six or eight points, for example, coincide, the computer experiment yields inconsistent results. And it is quite understandable why.

Running the computer programs shown in Figures 16.1 and 16.2, we did not actually simulate the throwing of a coin or of two dice, but checked the quality of the random number generator of the Mathcad program, the quality factor of the *rnd* function.† This "consumer quality" of the random number generator can be assessed visually. Figure 16.3 shows the definition in Mathcad Prime of the value of π (the ratio of the circumference to its diameter) with the Monte Carlo method. The problem is solved in this way: a square is considered, in which points are randomly "darted" and counted, and the number of points that hit the circle inscribed in this square are recorded. The ratio of the number of points in the circle to the total number of points for a sufficiently large number of throws should

ORIGIN := 1 $n := 3000$ $X := \text{runif}(n, -1, 1)$ $Y := \text{runif}(n, -1, 1)$

$i := 1..n$ $Xo_i := \textbf{if}(X_i^2 + Y_i^2 < 1, X_i, 0)$ $Yo_i := \textbf{if}(X_i^2 + Y_i^2 < 1, Y_i, 0)$

```
No := | No ← 0
      | for i ∈ 1..n
      |   if X_i^2 + Y_i^2 < 1
      |     No ← No + 1
      | No
```
$= 2358$

$\pi := 4\dfrac{No}{n} = 3.144$

Y
Yo
X
Xo

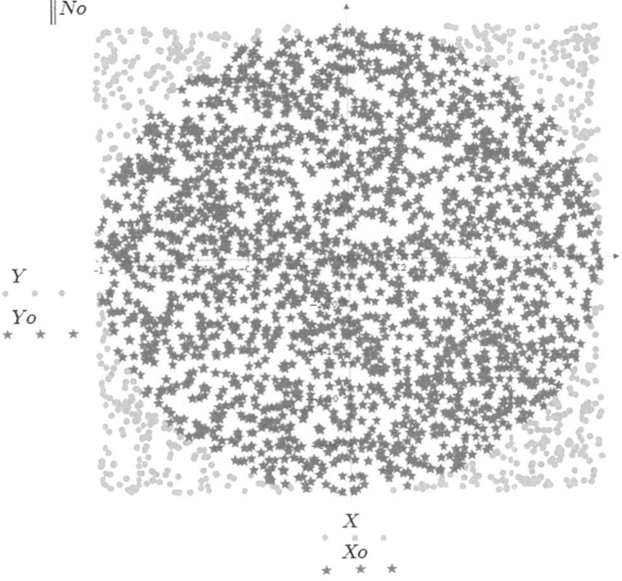

FIGURE 16.3 Number π and Monte Carlo.

* Ceil is for ceiling. In non-integer numbers there is also a "floor". The function floor (floor, the "partner" of the function *ceil*) rounds the numbers to the smallest integer—if it was 1.2, it becomes 1, if it was −4.5, it becomes −5, etc.

† In Mathcad, there are other functions that generate random numbers on a given interval, not only uniformly, but also according to other distribution laws, according to the law of normal distribution, for example, whose graph is usually drawn as a bell or a hill: at the center of the interval the probability of occurrence of an event is the highest , and at the edges of the interval, the lowest.

tend to the ratio of the area of the circle ($\pi d^2/4$) to the area of the square (d^2),* that is, to the quantity $\pi/4$.

The example with circle and square illustrates a geometric probability: if one geometric object is located within another, then the probability of a randomly discarded point falling into the smaller object is equal to the ratio of the measure of the smaller object to that of the larger object. By measure, we understand, in the case of a line, the length, in the case of a plane figure, the area, in the case of a body, the volume.

Once, students were assigned the problem of calculating the area of a figure with a complex boundary configuration in two ways: by direct integration (exact calculation) and by Monte Carlo (approximate calculation), using a table of random numbers to determine the coordinates of points "falling" on the figure and outside the figure. A student did not have the table at hand. So he took a felt pen and, with his eyes closed, began to draw points on the drawing. For the purity of the experiment, he asked his neighbor to do the same. As a result, the problem was carried out correctly, the relative error in calculating the area by the method of statistical tests turned out to be approximately 2%.

Figure 16.3 shows the Mathcad calculation of the number π by the Monte Carlo method. Here, the function *runif* is used to modify the *rnd* function. The *runif* function returns a vector with n elements that contains random numbers in the interval specified by the second and third argument of the *runif* function. When solving the problem of the approximate value of the number π, we generated two vectors *X* and *Y*, the coordinates of *n* points randomly thrown into a square with side 2. It remains only to count the number of points (*No*) in the circle, to extract the coordinates of these points in the vectors *Xo* and *Yo* and to draw them on the chart. It can be seen from the graph that the points are uniformly "smeared" over the square, which indicates the good quality of the random number generator built into Mathcad. On the website https://community.ptc.com/t5/PTC-Mathcad/ct-p/PTC Mathcad the reader can find the animation of the process of determining the number π by the Monte Carlo method. It was published by the author on March 14 (no matter what year). The fact is that this day in many countries is celebrated as the day of mathematics (3.14, March, the fourteenth day). On this day special mathematical lessons are conducted in schools, and the importance of mathematics in the educational process and in life is highlighted. By the way, February 6 in some places is celebrated as chemistry day: 2.06 is in fact the beginning of Avogadro's number (an alternative date is October 23, if we take into account not the mantissa, but the exponent of this constant 10^{23}).

In Russia, the day of the chemist (professional holiday) is the last Sunday in May. And which day can be appointed as the day of physics? Possibly August 9 (9.8, acceleration of free fall), but, on this day as a rule all physicists (school and university teachers, for example) are on vacation, so perhaps it should be September 8 (by reading this number in reverse order, as for the approximated number for π). However, June 6 (6.6) is better—the beginning of the gravitational constant, Planck's constant and the summer exam session! Heat engineering day can be celebrated on April 19, remembering that a calorie is approximately equal to 4.19 joules.

All these are built-in constants in Mathcad, see Figure 16.4.

However, let us leave the constants and go back to random numbers. The *rnd* function we used to toss the coin and the dice will also help us solve the problem of a three-way duel on the computer.

Sam, Bill and John agreed to fight in a three-way duel by the following rules:

- a draw determines who fires the first, second and third;
- duelists are located at equal distances from each other (at the corners of an equilateral triangle);

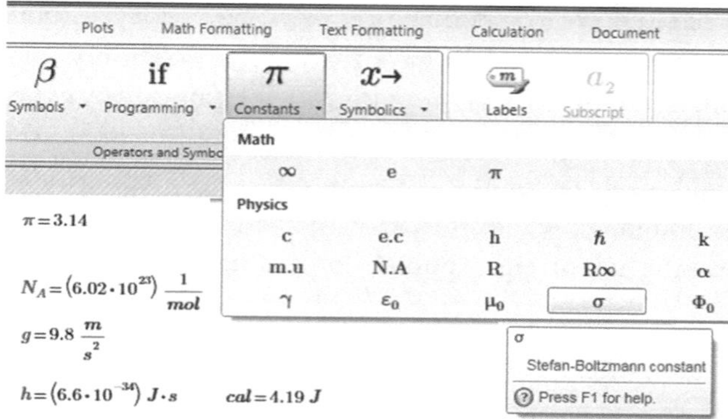

FIGURE 16.4 Built-in holiday constants, Mathcad.

* So, for example, in order to estimate the area of clouds in a photo of a piece of Earth made from space: the reader can poke a random needle into the picture and count the number of hits in the cloud. Also, on a topographic map, somebody can calculate the area of a forest or of a glade by pouring on the map, for example, millet or buckwheat and counting the grains.

- they exchange shots in turn, determined by fate, until two are killed;
- whoever is next to fire can shoot at any of the living duelists.

It is known that Sam is a sniper and never misses from this distance, Bill hits the target in 80% of cases, and John in 50%. What is the best strategy for each participant and what are the chances of their survival if they follow the best strategies?

In this duel, Sam (the sniper) and Bill (a good shooter) can adopt one of two tactics of behavior, namely, random, when the shooter knows nothing about the accuracy of the opponents and fires at the first determined by the throw of a coin (see Figure 16.1). The second tactic (most accurate), when the duelist is aware of which of his opponents shoots better, and he aims at the opponent with the higher accuracy in the hope of remaining head-to-head with the worse shooter.

John can follow another, third, "tricky" tactic. To get the highest chances of winning the duel, he must deliberately shoot avoiding the target while his two rivals are alive. After all, the next shooter, if he adheres to the tactics of "beat the most accurate", will fire not at John, but another opponent. After Sam or Bill is killed, John must show all he is capable of. In this situation, his chances of survival are 50% if he was alone with Sam, and more than 50%—if with Bill. And as a result, the chances of surviving from John will be higher than that of Sam and Bill.

Let's simulate this three-way duel on the computer. The participant of the duel, before shooting, must, first, identify the more accurate opponent, adhering to the tactics of "beat the most accurate". For this purpose, the *MostAccurate* function (Figure 16.5) is used, which returns the number of the enemy with the highest rifle qualities. In it, by looking through all the participants in a duel (the for loop), only the living opponents are taken into account, of course ($Status_{Target}$ = "alive") and not the shooter himself (*Target* ≠ *Shooter*).

```
MostAccurate(Accuracy, Status, Shooter) :=
:= | HighAccuracy ← 0
   | for Target ∈ 0..2
   |   | if Status_Target = "alive" ∧ Target ≠ Shooter ∧ Accuracy_Target > HighAccuracy
   |   |   | [MostAccurate ← Target  HighAccuracy ← Accuracy_Target]
   | MostAccurate
```

FIGURE 16.5 The "MostAccurate" function.

```
Accuracy(Accuracy, Status, Shooter) :=
:= | "Determining the number of enemies, shooting better than shooting"
   | #Accuracy ← 0
   | for Target ∈ 0..2
   |   | if Status_Target = "alive" ∧ Target ≠ Shooter ∧ Accuracy_Target > Accuracy_Shooter
   |   |   | #Accuracy ← #Accuracy + 1
   | #Accuracy
```

FIGURE 16.6 Function "Accuracy".

The duelist adhering to tricky tactics must determine before the shot how many opponents shoot better than him. This function is performed by the *Accuracy* function (Figure 16.6). It is based on the same algorithm for search opponents, as in the *MostAccurate* function itself.

Functions *MostAccurate* and *Accuracy* have as arguments the vector *Accuracy*, the vector *Status* and the scalar *Shooter*.

The *Winner* function (Figure 16.7) returns the number of the winner in a single duel. This takes into account the accuracy and the tactics of each participant in the duel (the two vectors (arguments) of the *Winner* function).

What happens in the *Winner* function?

At the beginning of the duel, all participants are alive: all three elements of the `Status` vector assume the value "alive", and the variable `Murder` (counter of the number of murders) is reset. Next, a draw is made: the first shooting (the variable `Shooter`) and the direction of the shots are determined. If the `Next` variable is equal to one, then the sequence goes

```
Winner(Accuracy, Tactic) :=
:= | "Simulation of a single three-corner duel"
   | [Status ← ["alive" "alive" "alive"]  Murder ← 0  Shooter ← floor(rnd(3))  Next ← if(rnd(1) > 0.5, 1, -1)]
   | while 1
   |   | if Tactic_Shooter = "Random"
   |   |   | while 1
   |   |   |   | Target ← floor(rnd(3))
   |   |   |   | if Target ≠ Shooter ∧ Status_Target = "alive"
   |   |   |   |   | break
   |   |   | if Accuracy_Shooter > rnd(1)
   |   |   |   | [Status_Target ← "killed"  Murder ← Murder + 1  if Murder = 2
   |   |   |   |                                                   | return Shooter]
   |   | if Tactic_Shooter = "Beat most accurate"
   |   |   | Target ← MostAccurate(Accuracy, Status, Shooter)
   |   |   | if Accuracy_Shooter > rnd(1)
   |   |   |   | [Status_Target ← "killed"  Murder ← Murder + 1  if Murder = 2
   |   |   |   |                                                   | return Shooter]
   |   | if Tactic_Shooter = "Tricky"
   |   |   | Target ← MostAccurate(Accuracy, Status, Shooter)
   |   |   | if Accuracy(Accuracy, Status, Shooter) < 2 ∧ Accuracy_Shooter > rnd(1)
   |   |   |   | [Status_Target ← "killed"  Murder ← Murder + 1  if Murder = 2
   |   |   |   |                                                   | return Shooter]
```

FIGURE 16.7 The Function "Winner".

in this direction: ... 0⟶1⟶2⟶0⟶1⟶2 ... (... Sam⟶Bill⟶John⟶Sam⟶Bill⟶John ...), if minus one, in this ... 0⟶2⟶1⟶0⟶2⟶1 ... (... Sam⟶John⟶Bill⟶Sam⟶John⟶Bill ...).

The mathematical model of the duel is based on a cycle with an exit from the middle (*while* 1 ... *return* ...): the duel continues until two effective shots are made—until the variable Murder becomes equal to two. In the body of the *while* cycle, three branches are embedded, defined by the tactics of the next shooter: "Random", "Beat most accurate" and "Tricky". In each branch of the calculation, the Target is determined—either randomly or the most accurate enemy, that is killed ($Status_{Target}$ is "killed") if, first, it is not missed $Accuracy_{Shooter} > rnd(1)$ and, second, the shooter does not fire deliberately into the air. That second option takes place only with the tricky tactic of the shooter ($Tactic_{Shooter}$ = "Tricky") and if there is more than one accurate opponent (*Accuracy*(*Accuracy*, *Status*, *Shooter*) < 2).

The definition of the next shooter is also in the cycle with the exit from the middle (*while* 1 *break*): the cycle is interrupted when, going through the next marked above (... 0⟶1⟶2⟶0⟶1⟶2 ... or ... 0⟶1⟶2⟶0⟶1⟶2 ...), "stumble" on a live participant.

The *Winner* function returns an unpredictable integer value of 0, 1 or 2, since it calls Mathcad's built-in function *rnd*, which returns a pseudo-random number in the range from zero to the value of the argument of the function rnd. The argument is equal to either one (random selection of the order of shots and imitation of a shot with a given probability of hit, proportional to the accuracy of the shooter), or three (random selection of the first shooter).

The *ChanceWinning* function (Figure 16.8) returns a vector whose elements are the ratio of the number of victories of each duellist to the total number of duels (*Duels* the third argument of the function *ChanceWinning*, the first two vector arguments are the duelists' parameters: their *Accuracy* and *Tactic*).

Now that all the necessary functions have been created, it is possible to conduct *statistical tests* (see Figure 16.8) and to record the probabilities of the victories of the participants in the duel, based on their accuracy and tactics. If the user increases the number of duels, then, with some patience, it is possible to get a result close to the theoretical.

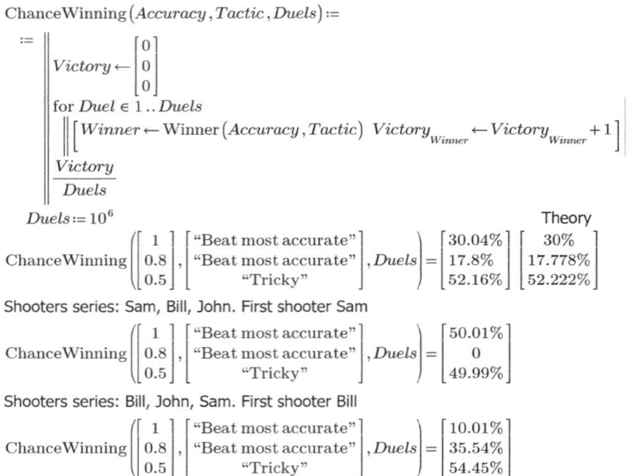

FIGURE 16.8 Conducting a series of duels.

The problem of a three-way duel, as we have already noted, is presented in many books. And what's interesting is that it's not quite right there. A priori it is believed that in this duel the weakest shooter John has the worst chance of surviving. However, if he thinks about it a little (tricky tactics), then the probability of him winning becomes the highest (52.222%).

Our solution (see Figure 16.8) suggests that John has the highest chances to survive (44-46%). By being cunning, he wins a lot, and brings Bill down.

Why such a mistake in problem setting? The fact is that the duelists have one further tactic. If the participants in the duel do not know anything about the shooting qualities of the rivals, then they try to beat the first one they encounter. Here the probability of winning can be counted immediately without a computer: Sam, 43.48% (1/(1 + 0.8 + 0.5)), Bill, 34.78% (0.8/(1 + 0.8 + 0.5)) and John, 21.74% (0.5/(1 + 0.8 + 0.5)) or 100, 43.48, 34.78). Our program can also calculate this case.

The problem was corrected, now it can be developed. The probability of winning counted by us refers to a situation when the draw for the order of shots has not yet been conducted: the variable *Next* is randomly equal to either one or minus one.

But after the draw, the chances of Sam and Bill change dramatically. Bill's affairs become very bad (10–12%), if John after his intentional miss passes the right shot not to him (*Next* = –1), but to Sam (*Next* = 1). And vice versa: Sam can lose his 30% if, after a deliberate miss of John, Bill shoots at Sam. John's probability of winning (52.2(2)%) does not depend on the order of the shots.

The reader can come up with and analyze a fourth tactic of conducting a duel: Bill and John *conspire* to target Sam, kill him, since he's so accurate, and only then will they find out the relationship between themselves. The initiator of such a conspiracy, as the reader can see, is likely to be Bill. John will go to him if he does not simulate a duel on the computer and does not know what can come of it.

Another problem for the reader is to refine the simulation program of the duel so that it is suitable for a duel with any number of participants.

Our model is not so far from real life. Dueling in its purest form is now, fortunately, not held. But you can hold a duel in the format of a paintball game, when opponents shoot each other with paintballs. Some kind of duel with the collusion of participants is observed in the markets, including financial ones. Blood does not flow there, but heart attacks occur, banks burst, companies, people and even whole countries are ruined (Greece, Spain, Italy, Cyprus, Iceland, etc., if we talk about the current situation).

Game theory and the tactics of participants' behavior are not only interesting, but also very useful. Not without reason, in 1998, the Nobel Prize winners in economics were scientists who applied the theory of games to the analysis of the work of financial exchanges.

And yet the model itself is too artificial. What is the duellist's accuracy and how to define it? Do you want to conduct real statistical tests? By the way, they are conducted with experienced firing from various types of tested weapons. There is a very well-known problem of the probability of an aircraft crash when bullets hit various parts of its hull. The highest probability of a crash, indicated there, is when bullets hit the wing. A person who is far from aviation cannot understand this situation: well, what can a small hole in the wing do? A specialist understands immediately: because the fuel tanks are placed in the wings! So, with a shot at a person in a duel—a duelist does not just have to be hit and wounded, it is necessary to be shot in a vital organ (and this is not necessarily the heart).

In addition, it is one thing to shoot at targets, and another to aim at a living person. In duels, as a rule, a duelist is not killed on the spot, but they are more usually wounded, with varying degrees of severity. The shot duelist, if he had enough strength and anger, can fire back at the rival (as in the duel of Alexander Pushkin and George d'Anthès, for example). Attempts to "land" (make close to real life) the problem of duels inevitably requires the use of the apparatus of the *theory of fuzzy sets* (TFS, see Study 12).

Nobody and nowhere measures the accuracy of a duellist by numbers, but only estimates by categories (linguistic constants): "bad", "good shooter", "sniper", etc. The status of a duellist is not a Boolean variable. Let us recall the "consultation of doctors" with Buratino, who lives without feeling: "The patient is more likely to be dead than alive"—"No, the patient is more alive than dead".*

ASSIGNMENTS FOR READERS

Simulate and solve on the computer the problem of a duel with more than three duelists.

* In Collodi's Pinocchio, Chapter 16, original in Italian: "La bella Bambina dai capelli turchini fa raccogliere il burattino: lo mette a letto, e chiama tre medici per sapere se sia vivo o morto … A vederlo, pareva morto, ma non dev'essere ancora morto perbene …"

Geese Are Flying Or

The Problem of Fishermen and Fish

ABSTRACT

This Study explains how the computer can help to solve the old English folk riddle in a new way. The solution of the extended problem of fish and fish is shown. The questions of searching for an integer solution of an algebraic equation in the manual mode and on the computer are condemned.

Mathematics: Equations in integers, a system of three linear equations.

Physics: Fishing.

IT: Loop with a pretest, loop with a parameter, flag (sign) in a program.

Art: Folklore.

Study website: https://community.ptc.com/t5/PTC-Mathcad/Study-17/m-p/642394

There is a "Russian folk" riddle (task) that goes:

> "A goose flies. A flock of geese meets him." "Hello, 100 geese," he says. They answer: "We are not 100 geese, but if we were as many as now, and so many again, and half so many and a quarter so many, and you, then we would be 100 geese. How many geese are flying in the flock?"

This task was meant to be solved mentally, mobilizing personal abilities for oral calculation. We considered similar problems in Study 6. However, these days, a calculator or even a computer is increasingly used for solving such problems.

Figure 17.1 shows the solution to this problem using the command *solve* (that solves the equation analytically) in the mathematical program Mathcad. It is sufficient for a user to create the algebraic equation ($x+x+x/2+x/4+1=100$), enter it into the computer and order the program to execute the necessary command—to solve this equation* and to obtain the answer—36 geese.

Similar riddles are common all over the world. For instance, an English analog of the geese problem is the problem of fishermen and fish:

> "Three fishermen went to bed without dividing the catch. The first fisherman, waking up at night, decided to leave, taking his share. However, the number of fish could not be divided into three. So he discarded one fish and took a third from the rest. The second and third fishermen acted in a similar way. What is the smallest number of fish that satisfies the condition of the problem?"

According to the condition of the problem, we formulate the equations, assuming all the unknowns to be natural numbers, x – 1 = 3*y*, 2y – 1 = 3z, 2z – 1 QUOTE x-1=3y, 2y-1=3z, 2z-1=3t = 3*t* and *x* is not a multiple of three (the original number of fish could not be divided

* In our "goose" equation, there is only one variable *x*. But if there are two or more variables in the equation, then after the keyword *solve*, it is necessary to specify, through a comma, for which variable (unknown) the given equation has to be solved.

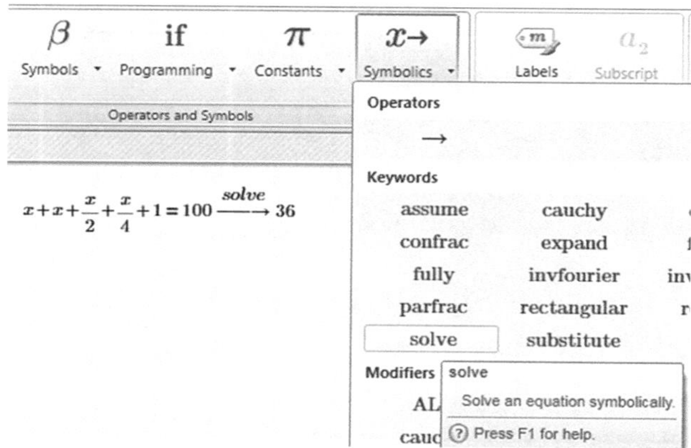

FIGURE 17.1 Solution of the geese problem in Mathcad Prime.

into three). We have obtained a system of three linear equations with four natural unknowns. If we try to find the answer for this problem through the compilation and solution of the algebraic equation in the same way as shown in Figure 17.1, it just will not work. First, you need to find not just a solution, but an *integer* solution. In the problem of the geese, the condition was chosen so that an integer solution was obtained—integer 36 geese, and not 36 geese and half a goose, for example. "Halves and other parts" are often obtained by an incorrect solution of integer problems. Let's remember how Viktor Perestukin from the cartoon *The Country of Unlearned Lessons* solved the problem of arithmetic and got one-and-a-half diggers.

Second the system obtained has infinitely many solutions.

And, third, we are interested in a solution that corresponds to the smallest number.

It is possible to proceed in the following way. If the equations are written in the form $x=3y+1$, $2y=3z+1$, $2z=3t+1$, then we see that t must be an odd number, i.e. $t=2k-1$ for $k\geq 1$. Consequently $z=3k-1$ and $2y=9k-2$, which is only true for even k. Now if we put $k=2n$ for $n\geq 1$, it follows that we can express all the unknowns as: $x=27n-2$, $y=9n-1$, $z=6n-1$, $t=3n-1$. Obviously, the smallest x value is 25, and it is obtained for $n=1$.

However, this solution can be obtained only by those who will guess the parity properties. While if we forget the parity, then we will try to solve the system by the method of search through the variable $t=3n-1$, solving for each fixed value $t\in\{1,2,3\}$ a system of three linear equations with three unknowns:

$$\begin{cases} x-3y=1, \\ 2y-3z=1, \\ 2z=3t+1. \end{cases}$$

Note that with t increasing, all the unknowns increase, while we search for the smallest number x.

For each t value, this system has a unique solution that is easy to obtain. This is left to the reader to prove.

It is much more difficult to solve a similar problem if there are more than three fishermen. After all, then the number of unknowns will increase too!

But there is another way to solve the problem, and it does not require the formulation of any equations. Moreover, it allows us to solve the problem for any number of fishermen.

Once upon a time, the author of this book taught classes to students on programming and instructed them to construct a computer program for solving the problem of the fishermen and fish according to a simple algorithm: a first guess to the solution (30 fishes, for example) is given and then it is checked whether this number satisfies the problem conditions. If not, then the assumption is decreased by one, and the test itself is repeated until the problem condition is satisfied.

In programming languages, this simple algorithm is implemented by a cycle with the verification of the condition (loop with a pretest) of the problem and a nested cycle enumerating (loop with a parameter) the three fishermen taking their share of the catch. Figure 17.2 shows how this algorithm is written in the programming language built into Mathcad Prime. The program is implemented as a user function named *Fishes* with two arguments: *Answer*, the first guess for the answer, and *fishermen*, the number of fishermen.

The algorithm of the program is simple: the words "yes" or "no" are assigned to the variable *Shared*, depending on whether the *Catch* variable (the number of fishes remaining after the departure of the next fisherman) belongs to the set of integers Z (see the operator $\in$ in Figure 17.2).* All this is arranged in a *while* loop, in which a *for* loop is nested.

Figure 17.2 also shows the results of the *Fishes* function call for different initial approximations and

* In Mathcad Prime, in addition to the set Z (integers), three more sets of numbers are defined: C, complex numbers; Q, rational numbers; and R, real numbers. This was shown in Study 12, see Figure 12.1.

$$Fishes(Answer, fishermen) := \left| \begin{array}{l} \text{“The problem of fishermen and fish”} \\ Answer \leftarrow Answer + 1 \\ Shared \leftarrow \text{“No”} \\ \text{while } Shared = \text{“No”} \\ \quad \left| \begin{array}{l} Answer \leftarrow Answer - 1 \\ Catch \leftarrow Answer \\ \text{for } fisherman \in 1..fishermen \\ \quad \left| \begin{array}{l} Catch \leftarrow Catch - 1 - \dfrac{Catch - 1}{fishermen} \\ Shared \leftarrow \text{if}(Catch \in \mathbb{Z}, \text{“Yes”}, \text{“No”}) \end{array} \right. \end{array} \right. \\ Answer \end{array} \right.$$

$\mathbb{Z} = \text{“the integers”}$

$Fishes(30, 3) = 25$	$Fishes(24, 3) = -2$	$Fishes(-3, 3) = -29$
$Fishes(300, 4) = 253$	$Fishes(252, 4) = -3$	$Fishes(-4, 4) = -259$
$Fishes(4000, 5) = 3121$	$Fishes(3120, 5) = -4$	$Fishes(-5, 5) = -3129$
$Fishes(50000, 6) = 46651$	$Fishes(46650, 6) = -5$	$Fishes(-6, 6) = -46661$
$Fishes(900000, 7) = 823537$	$Fishes(823536, 7) = -6$	$Fishes(-7, 7) = -823549$

FIGURE 17.2 Solution of the problem of fishermen and fish in Mathcad Prime.

different numbers of fishermen.* The prehistory of these answers is as follows. Schoolchildren in the above-mentioned programming group took as a first approximation 30 fish and received a traditional English response of 25 fish for the problem of three fishermen: the first fisherman took 8 fish and left 16 fish, the second fisherman took 5 fish and left 10 fish and, finally, the third fisherman took 3 fish and left 6 fish. All honestly! This solution was given by many generations of English children and adults, until Paul Dirac appeared. This English physicist became famous not only for the prediction of the antiparticle, but also because he thought of anti-fishes. Paul Dirac said that the correct solution of the problem of fishermen and fish is not 25 fish, but minus 2 fish: we throw one out of the catch—we get minus three fish, take a third, leave minus 2 fish and so on ad infinitum. To test Dirac's solution, the schoolchildren took as a first approximation 24 fish and obtained the "Dirac" response of minus two fish. And one schoolboy was not too lazy† and introduced another first approximation—minus 3 fish and received one more answer—minus 29 fish: throw one out—get minus 30 fish, take a third, leave minus 20 fish, throw one out—get minus 21 fish, to leave minus 14 and, finally, throw out one—to leave minus 15 fish. So the schoolboy corrected Paul Dirac himself! Dirac turned out to be wrong: this problem has other solutions and they constitute an infinite set. In the traditional formulation of the problem, by default, the number of fish was meant to be a *positive*‡ integer number. Dirac also crossed the zero and opened Pandora's box, depriving this problem of a single answer. Paul Dirac, in giving his answer, assumed the statement of the problem implied finding the *modulus* of the minimum number of fish. And the assumption of this "modulus" (absolute value) results in these anti-fish.

It is interesting that there was a time when negative numbers were rejected by mathematicians as unnecessary, extraneous and useless.§ Even François Viète himself, proved his famous theorem establishing the connection between the coefficients of a polynomial and its roots and published by him in 1591, only if all the roots of the polynomial are positive. He did not

* For three fishermen, the answer can be found without a computer, while for four and more fishermen it is very difficult to do so.

† Not too lazy, guided by the rule: "You, computer, you imagine a lot about yourself! Let's try and swallow this!"

‡ Ancient people knew only positive numbers—they worked only with the right-hand side of the numerical series, without any notion of negative numbers. Echoes of this "antiquity" can be seen in folklore and in our problems about geese and fish.

§ Here it is said that they were introduced in China, long before sixteenth-century Italy, but not in the context of symbolic algebra.In *Nine Chapters on the Mathematical Art* (Jiuzhang Suanshu, China, c. 100 BCE–50 CE), negative numbers were used in the chapter on solving systems of simultaneous equations. Red rods were used to denote positive coefficients, black to denote negative ones. Rules for signed numbers were given. (For more information on the history of mathematics in China, see *Mathematics in China*, http://aleph0.clarku.edu/~djoyce/mathhist/china.html) or https://web.stanford.edu/class/me161/documents/HistoryOfNegativeNumbers.pdf.

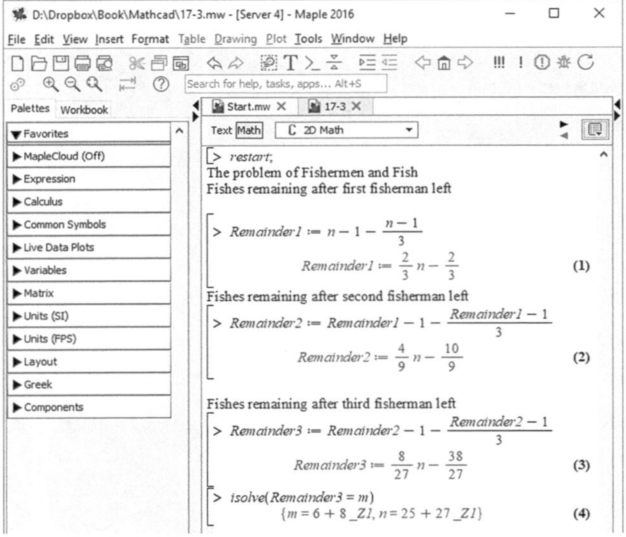

FIGURE 17.3 Solution of the problem of fishermen and fish in Maple.

recognize the negative numbers, although he was the ancestor of algebraic symbolic calculations and transformations. Italian mathematicians of the sixteenth century—Tartaglia, Cardano,* Bombelli†—were some of the first who took into account the negative as well as the positive roots of polynomials.

In Mathcad there are no tools for an integer solution of algebraic equations. But there are in another mathematical program—Maple.

Figure 17.3 shows the solution of the problem of fishermen and fish in another popular mathematical program Maple, where the conditions of the problem (actions 1–3) are formulated and the command is given to find the value of m and n, for which the remainder of the catch after its sharing would be an integer (action 4). To do this in Maple the command was given not just to *solve*, but to solve for an integer—*isolve* (*i*, integer).

The program Maple gave the answer not in the form of a specific number, but in the form of an infinite series of integers, where the variable *Z1*, already considered by us in the program in Figure 17.2, means the set of all integers *Z*, …, −3, −2, −1, 0, 1, 2, 3. If they are substituted in response (see Figure 17.3), then we get the required set of answers for m (the remainder of the fish after the third fisherman): $m = \ldots, -9, -5$ (schoolboy's answer), −1 (Paul Dirac's answer), 3 (traditional English answer), 7, 11, 15 …; and n (initial number of fish in the catch) = …, −56, −29 (schoolboy's answer), −2 (Paul Dirac's answer), 25 (traditional English answer), 52, 79, 106, ….

In the solution shown in Figure 17.2, an initial approximation, obviously *greater* than the answer, was given. Then a unit *was taken* from this number in each cycle of the program cycle. You can change the direction in which the answer is approached by setting the initial approximation to be *less* than the answer and *adding* one in the cycle. But in this case, we don't obtain the anti-fish.

ASSIGNMENTS FOR READERS

1. Fetch an old task from a book or the Internet and solve it with the help of Mathcad.

2. Create a program for the task with any number of fishermen throwing out any number of caught fishes.

---

* Cardan (Cardano), in his *Ars Magna* included negative solutions of equations and stated the basic laws of operating with negative numbers. He called positive numbers *numeri ueri* (real) and negative numbers *numeri ficti* (fictitious). He used m: as a negative sign (for example, m: 2 for −2). However, he did not allow negative coefficients in quadratic equations, since he interpreted these as partitioning squares into rectangles of smaller size, and negative coefficients would mean these rectangles would need to have sides of negative length, an absurdity. From https://web.stanford.edu/class/me161/documents/HistoryOfNegativeNumbers.pdf.

† Bombelli similarly used m. to denote a negative number, and also used p. to denote a positive number (for example, p.3 for +3). From https://web.stanford.edu/class/me161/documents/HistoryOfNegativeNumbers.pdf.

Alphabetical Puzzle Or

Help Yourself So That the Computer Helps You!

ABSTRACT

This Study explains how it is possible to combine our own intellect with the capabilities of the computer. Optimization of enumeration methods for solving a computer problem is discussed. The important problem of how to combine the mental abilities of a person and the potential of a computer is considered.

Mathematics: Decimal notation of a number, system of equations in integers.

Physics: Geographical map.

IT: Nested loop with a parameter; conditional operator; matrix; output of the result.

Art: Puzzle.

Study website: https://community.ptc.com/t5/PTC-Mathcad/Study-18/m-p/642395

Once upon a time in the Soviet magazine *Quant* the following alphabetic rebus was published.

The idea is to determine which figures are behind these Latin letters. In this case, of course, the same letters must correspond to the same numbers, and different letters to different numbers. It is clear that the letter P (leftmost in the sum) can be only one, and the letter U, nine. If so, then the letter E stores zero ($1+9=10$).* It is possible to continue such logical-arithmetical reasoning ("brainstorming" the problem) and completely solve this letter puzzle, i.e. to find out what figures correspond to the remaining unknown letters S, A, C and R. However, there is a reason to involve the computer in this work. Why!? First, the reader can rephrase† the second title of this Study and say: "Believe in yourself, and turn to the computer!" That is, again, "not bad!". It is possible in this way to "Make a mistake", i.e. to find only one solution from the set of possible solutions. We give an example of this below. Second, entertaining informatics implies the use of a computer‡ for solving puzzles with the help of a computer. It is possible with a program to substitute numbers from 2 to 8 (the numbers 0, 1 and 9 are already occupied, see above) for our undeciphered letters; to sort through all the

* This puzzle is analogous to the one in Study 3 where we examined a cipher based on the replacement of symbols.

† The name of the Study is a paraphrased Russian semi-atheistic proverb "Hope in God, but Do not Be Bad". So, we are rephrasing, so to speak, to the square. In the English version we are rephrasing to the power of 3.

‡ Honestly speaking, computer science itself exists without any computers. For simplicity, for us and for the whole world of computer science, information technology is often associated with computer technology.

```
   USA
+
  USSR
 ------
 PEACE
```

FIGURE 18.1 Alphabetical puzzle.

options and to print out the "answer". It is worth noting that three, not four, letters remained unsolved, because $A+R=10$—see the right-hand column of letters in Figure 18.1, where $E=0$. It is possible to continue "brainstorming" the problem, but from here it is worthwhile to stop and to entrust this work to the computer: designing a program for it is also a good puzzle for the mind.*

Figure 18.2 shows a program written in Mathcad Prime, which solves our alphabetical puzzle. The program has three *for* loops that sort through the letters A, C and S (cycle parameters) with a number from 2 to 8. If this is done, $USA+USSR=PEACE$, then the answer (the numbers stored in our seven variables) is written in the nth row of the matrix with the name M. In the zero line of the matrix M, the operators in the first line of the program shown in Figure 18.2, the "cap" (header) of this matrix is recorded. This line of the program serves to name the columns.

Figure 18.3 shows the result of the program in Figure 18.2. The matrix M returns four answers, but only the first one ($932+9338=10270$) is correct. In the remaining answers, $S=A$, which does not correspond to the conditions of the problem, and this is clear from a simple look at the elements of the matrix M. It is possible, of

```
M:= | M<0> ← ["#" "U" "S" "A" "+" "U" "S" "S" "R" "=" "P" "E" "A" "C" "E"]
    | [n←1 U←9 P←1 E←0]
    | for A ∈ 2..8
    |   | R←10−A
    |   | for C ∈ 2..8
    |   |   | for S ∈ 2..8
    |   |   |   | if 100 U+10 S+A+1000 U+100 S+10 S+R=10000 P+1000 E+100 A+10 C+E
    |   |   |   |   | M<n> ← [n U S A "+" U S S R "=" P E A C E]
    |   |   |   |   | n←n+1
    | M
```

FIGURE 18.2 Program for solving an alphabetical puzzle.

$$M=\begin{bmatrix} \text{“\#”} & \text{“U”} & \text{“S”} & \text{“A”} & \text{“+”} & \text{“U”} & \text{“S”} & \text{“S”} & \text{“R”} & \text{“=”} & \text{“P”} & \text{“E”} & \text{“A”} & \text{“C”} & \text{“E”} \\ 1 & 9 & 3 & 2 & \text{“+”} & 9 & 3 & 3 & 8 & \text{“=”} & 1 & 0 & 2 & 7 & 0 \\ 2 & 9 & 6 & 6 & \text{“+”} & 9 & 6 & 6 & 4 & \text{“=”} & 1 & 0 & 6 & 3 & 0 \\ 3 & 9 & 7 & 7 & \text{“+”} & 9 & 7 & 7 & 3 & \text{“=”} & 1 & 0 & 7 & 5 & 0 \\ 4 & 9 & 8 & 8 & \text{“+”} & 9 & 8 & 8 & 2 & \text{“=”} & 1 & 0 & 8 & 7 & 0 \end{bmatrix}$$

$932+9338=10270$

FIGURE 18.3 Solution the alphabetical puzzle.

course, to insert another conditional statement (if $S \neq A$) into the program shown in Figure 18.2, but we will not do that in this case since the answer is already clear in Figure 18.3.

Rebus USA + USSR = PEACE is obsolete. First, there is no such country as the USSR anymore, and second, the solution of this puzzle is described in many entertaining books and even appears on the Internet—it is only necessary to enter (USA + USSR = PEACE) in any search engine.

More recently, and still unsolved, we can consider the problem† USA + IRAN = PEACE. This is unresolved in both the direct and figurative senses. In the figurative sense: taking into account the current state of US-Iranian relations.

Figure 18.4 shows the Mathcad-program that solves this new problem. It differs from the program in Figure 18.2 quantitatively, but not qualitatively: in the new program, not three, but five nested *for* loops are written. Furthermore, an additional *if* statement is inserted into the program, which eliminates the wrong solutions when different letters correspond to the same number. Without this conditional statement, the program no longer generates 4 (see Figure 18.3), but 75 (!) solutions, which would be difficult to analyze visually and identify the wrong results. It is possible to do that here. We print out the matrix M with all 75 solutions. If it turns out that in one line a single number corresponds to different letters, then in the program shown in Figure 18.4, it would be

```
M:= | M<0> ← ["#" "U" "S" "A" "+" "I" "R" "A" "N" "=" "P" "E" "A" "C" "E"]
    | [n←1 I←9 P←1 E←0]
    | for A ∈ 2..8
    |   | N←10−A
    |   | for C ∈ 2..8
    |   |   | for S ∈ 2..8
    |   |   |   | for U ∈ 2..8
    |   |   |   |   | for R ∈ 2..8
    |   |   |   |   |   | if 100 U+10 S+A+1000 I+100 R+10 A+N=10000 P+1000 E+100 A+10 C+E
    |   |   |   |   |   |   | if U≠S∧A≠S∧N≠R∧U≠R∧S≠R∧C≠R∧N≠A∧N≠U∧C≠U
    |   |   |   |   |   |   |   | M<n> ← [n U S A "+" I R A N "=" P E A C E]
    |   |   |   |   |   |   |   | n←n+1
    | M
```

FIGURE 18.4 Program for solving a new alphabetical puzzle.

* The problem of four colors on a political map was solved approximately with this method. In mathematics, the four color theorem asserts that by painting any map located on the sphere with four colors, it will result that any two regions having a common section of the border are painted in different colors. People "brainstorming" proved that this is so for almost all cases of configuration of the map. Other cases were "explained" with the use of a computer. But even now, in the April issue of popular science magazines, cards are often placed that cannot be painted with four colors.

† This can also be checked on the Internet.

M =

"#"	"U"	"S"	"A"	"+"	"I"	"R"	"A"	"N"	"="	"P"	"E"	"A"	"C"	"E"	
1	5	3	2	"+"	9	7	2	8	"="	1	0	2	6	0	532 + 9728 = 10260
2	7	3	2	"+"	9	5	2	8	"="	1	0	2	6	0	732 + 9528 = 10260
3	5	2	3	"+"	9	8	3	7	"="	1	0	3	6	0	523 + 9837 = 10360
4	8	2	3	"+"	9	5	3	7	"="	1	0	3	6	0	823 + 9537 = 10360
5	5	7	4	"+"	9	8	4	6	"="	1	0	4	2	0	574 + 9846 = 10420
6	8	7	4	"+"	9	5	4	6	"="	1	0	4	2	0	874 + 9546 = 10420
7	7	5	6	"+"	9	8	6	4	"="	1	0	6	2	0	756 + 9864 = 10620
8	8	5	6	"+"	9	7	6	4	"="	1	0	6	2	0	856 + 9764 = 10620

FIGURE 18.5 Solutions for the new alphabetical puzzle.

necessary to insert an additional conditional statement: if $U \neq S$, for example. After that, we run the program again and output the shortened (filtered) matrix, `M`.

If it turns out that another pair of letters does not match the condition of the problem, then it would be necessary to extend the *if* statement: if $U \neq S$ becomes if $U \neq S \wedge A \neq S$ (the operator with the symbol $\wedge$ is the Boolean *And* logical multiplication, see Study 11). This operation (filtering the matrix by deleting unnecessary rows from it) would need to be continued until only the correct solutions remain in it, see Figure 18.5.

The new Rebus has not just one, but eight solutions! And what was the point of all this?! The fact that the rebus *USA + USSR = PEACE* can be solved without a computer, but the new rebus *USA + IRAN = PEACE* is very difficult to solve without a computer. We leave it to the reader to decide if this is the case, but it will be difficult to be certain that the solution found by "brainstorming" is the only solution. This problem is difficult to solve: there are nine different letters in the rebus. This means that a frontal computer attack on the problem requires nine, rather than five nested cycles, each of which need to be "chased" ten times. Such a program will run for quite a long time, the matrix M produced by it will be large and will have to undergo a lot of filtering: a long and tedious process, before producing the form shown in Figure 18.5.

But there is an opportunity to help the computer. Remembering the rules of decimal writing numbers and using the obvious values of $I = 9$ and $P = 1$, we can write equation

$$100U + 10S + A + 9000 + 100R + 10A + N$$

$$= 10000 + 1000E + 100A + 10C + E.$$

Simplifying it, we get: $1000(E+1) + 100(A-U-R) + 10(C-S-A) + E-A-N = 0$. Since $E-A-N$ is divisible by 10, then $E-A-N = 10x$, with $|x| \leq 1$ (since all variables are non-negative and not more than nine), but then, dividing by 10, we have $100(E+1) + 10(A-U-R) + C-S-A+x = 0$. Since $C-S-A+x$ is divisible by 10, then $C-S-A+x = 10y$ and $|y| \leq 2$. By dividing by 10, we get $10(E+1)+A-U-R+y=0$. Repeating the procedure again, we arrive at the equation $E+1+z=0$, where $A-U-R+y = 10z$ and $|z| \leq 2$. So, we obtain a system of equations in integers:

$$E = -1 - z \geq 0,$$

$$A + N = z - 10x + 1$$

$$S + A - C = x - 10y$$

$$U + R - A = y - 10z.$$

In addition, $0 \leq E \leq 9$, $0 \leq A \leq 9$, $0 \leq N \leq 9$, $0 \leq C \leq 9$, $0 \leq S \leq 9$, $0 \leq R \leq 9$ and $0 < U \leq 9$.

This system can be solved by brute force. And as a problem, let the reader try to prove that the rebus: "PU PIL" + "TEXTBOOK" = "INTERNET" has no solution with the help of a computer! A conclusion, which, we hope, the reader reached after solving the problems described in this Study and many other problems.

(By the way, there is an app for solving alphabetic puzzles available for Apple devices.)

For solving a problem (school, university, engineering, scientific, everyday), it is necessary to be able to combine our own intelligence with the capabilities of our computer. A computer is now used wherever possible. The time is not far off when the computer will be built right into the brain! First animals, then man. This will quickly and effectively solve any problems encountered in life.

With this very important conclusion, we finish this Study, adding here a relevant example in the form of a quotation from Wikipedia. In the Study we obtained a result for a puzzle with the names of countries (see Figure 18.4), and now let's talk about the problem of coloring countries on a geographical map, which was solved with the combination of human intelligence, accuracy and computer speed.

In mathematics, the *four-color theorem* states that any map can be painted with four colors so that any two areas that have a common border section are painted in different colors. This theorem was formulated by Francis Guthrie in 1852, but it was not possible to prove it for a long time. During this time, many attempts have been made both for evidence and for refutation, and this problem was called the *problem of four colors.*

The four-color theorem was proved in 1976 by Kenneth Appel and Wolfgang Haken from the

University of Illinois. It was the first major theorem to be proved using a computer. Appel and Haken's approach started by showing that there is a particular set of 1936 maps, each of which cannot be part of a smallest-sized counterexample to the four-color theorem. (If they did appear, there would be a smaller counterexample.) Appel and Haken used a special-purpose computer program to confirm that each of these maps had this property. Additionally, any map that could potentially be a counterexample must have a portion that looks like one of these 1936 maps. Showing this required hundreds of pages of hand analysis. Appel and Haken concluded that no smallest counterexamples exist because any must contain, yet do not contain, one of these 1936 maps. This contradiction means there are no counterexamples at all and that the theorem is therefore true. Initially, their proof was not accepted by all mathematicians because the computer-assisted proof was infeasible for a human to check by hand (Swart 1980). Since then the proof has gained wider acceptance, although doubts remain (Wilson 2014, 216–222).

And still now some popular science magazines in the April Fools' Rooms (see the end of Study 13) publish an intricate contour map that cannot be properly painted over with four colors. And many are fooled by this joke. Take four different colored pencils (real or virtual) and begin to refute the journal's statement.

The author of this book also puts every April Mathcad joke on the PTC Community website. An example of such joke is on the PTC Community forum.

ASSIGNMENTS FOR READERS

1. Solve the rebuses, available at the website http://rebus1.com, by making a program. There, incidentally, under the first number is the rebus USA + USSR = PEACE. Here is another puzzle from this website: SEND + MORE = MONEY
2. Solve a numerical rebus without or with a computer—this is a task for the amateur. Professionals themselves make up new puzzles without or with a computer. Try yourself in this field!
3. Try to apply the computer to other types of numerical puzzles—to Sudoku, for example.

   Incidentally, on the website http://twt.mpei.ac.ru/MCS/Worksheets/Sudoku.xmcdthere is a Mathcad-program for solving this puzzle.
4. Think of a good Mathcad joke for April 1.

Seven Computational Curves Or

Apollonius Bicycle and Apollo Guitar

ABSTRACT

In this Study, we talk about methods for constructing plane curves, including, among other things, the computational operators of addition (ellipse), subtraction (hyperbola), multiplication (Cassini's oval) and division (Apollonius' circle). The properties of curves on three other (remaining) computing operators with two operands are also investigated: remainder from division, exponentiation and logarithm from a given base. The theory of dimensional quantities is revised—the possibility and necessity of working with dimensional quantities in the exponent is justified.

Mathematics: Planar curves of the second order, focal and directional properties of the hyperbola and ellipse, solution of the system of two non-linear algebraic equations, vector, Cartesian and polar graph.

Physics: Trajectories of motion of celestial bodies.

IT: Symbolic mathematics, computer-aided graphics, nested loops with a parameter, conditional statement.

Art: Kozma Prutkov, *Fruits of Reflection*; gardening.

Study website: https://community.ptc.com/t5/PTC-Mathcad/Study-19/m-p/617549.

> Throwing pebbles into the water, look at the ripples they form on the surface, otherwise, such occupation becomes an idle pastime.
>
> KOZMA PRUTKOV, *FRUITS OF REFLECTION*

When the first author of the book was at school, many boys dreamed of becoming cosmonauts. In those days, the Soviet Union "lived" in space: Earth satellites were launched, probes flew over moon, a man traveled into space, etc. (see Study 24). The country did not have many other special scientific and technical achievements of global importance. All this created great interest in the theory of space flight, in particular, in celestial mechanics. Study 9 described this in some detail. In school lessons of mathematics, physics and astronomy, the teachers taught how satellites (natural and artificial) revolve around the planets in circular and elliptical orbits and how it is possible to draw an *ellipse*. It is not much more complicated than a circle (a special case of an ellipse). After one lesson, the author of this book came home, hammered two nails into the wall of the room, tied a rope between them, pulled it with a pencil and drew half of an ellipse on the wallpaper, see Figure 9.40 in Study 9. The second half could not be drawn—mother saw these physical and mathematical "arts" and gave her son a scolding.

Let's try again to draw this ellipse, not on the wall of the room, but on the computer screen in Mathcad 15, see Figure 19.1.

One could take the algebraic expression of an ellipse and use it to construct this closed curve. However, it is possible to proceed with a different approach given that an ellipse is the geometrical locus of the points of a plane such that the *sum* of distances from each point to two

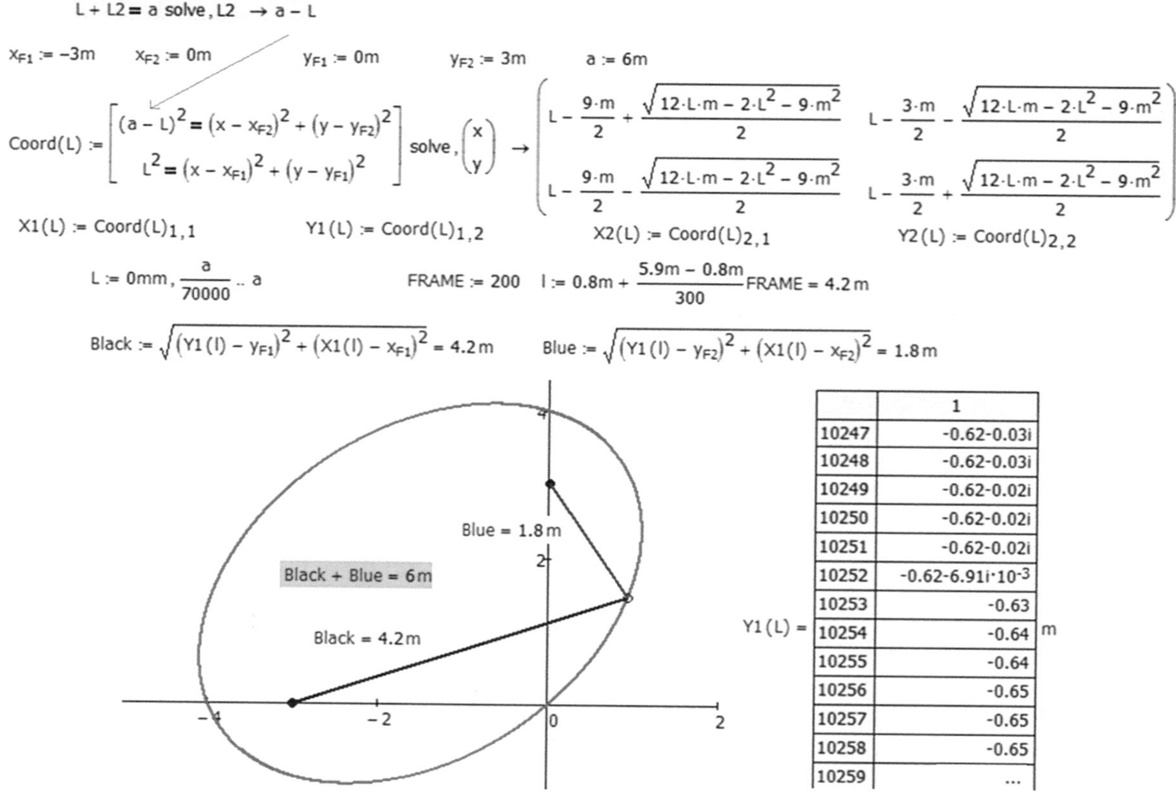

	1
10247	-0.62-0.03i
10248	-0.62-0.03i
10249	-0.62-0.02i
10250	-0.62-0.02i
10251	-0.62-0.02i
10252	-0.62-6.91i·10-3
10253	-0.63
10254	-0.64
10255	-0.64
10256	-0.65
10257	-0.65
10258	-0.65
10259	...

FIGURE 19.1 Ellipse addition.

other fixed points (foci of an ellipse) is constant. The ellipse and some other "computational" curves, which will be described below, were constructed in the Mathcad environment using this approach. A system of two nonlinear algebraic equations was solved—there was an analytic expression for the coordinates of points separated from two foci (their Cartesian coordinates x_{F1}–y_{F1} and x_{F2}–y_{F2}) at distances L (in the chart *Black*—a segment of black color) and $L2$ (*Blue*—a segment of blue color; the ellipse itself is red). The "elliptic" equation is embedded in the system of equations: $L + L2 = a$ (in the construction of other "computational" curves we will change the form of this equation). Our system has two solutions for two unknowns x and y, which form a matrix with two rows (two solutions) and two columns (two unknown systems). From this matrix four user functions with argument L and with names *X1*, *X2*, *Y1* and *Y2* are formed, through which the ellipse itself is constructed parametrically and point-wise. The variable (parameter) l is a fixed value of the length from the variable range L, for which the two segments are inscribed into the ellipse, the sum of which is constant and equal to the given value of the variable a (6 meters, see Figure 19.1: note that the variable a is not just 6, but *6 meters*). The author's calculation technique, shown in Figure 19.1, allows you to build fairly complex curves without any preliminary derivation of the analytic dependencies embedded in the curves. In this technique, analytical (symbolic) and numerical methods for solving the problem are combined.

The variable l in our calculation in Figure 19.1 is bound to the system variable *FRAME*, which controls *animation* in Mathcad. Changing the values of the variable *FRAME* from 0 to, for example, 300 (animation frames), you can show how a pencil held by a rope draws an ellipse. The animation of drawing an ellipse and other closed curves described in the Study is also located on the PTC Community and other websites. There you can download the documents allowing you to perform the corresponding calculations in Mathcad.

Figure 19.1 also shows part of the printout of the values that the function *Y1* gives for given discrete values of the variable L. There are real and complex numbers. The latter are obtained if the values of the variable L (the argument of the function *Y1*) do not allow an ellipse to be built ($L > a$, for example). Mathcad ignores these values and constructs an ellipse without problems. This feature is not present in the 3D graphics of Mathcad, thereby creating a number of inconveniences for users. The methods for constructing graphs proposed in this Study are also suitable for constructing closed surfaces that are the development of the curves under consideration: ellipsoids, hyperboloids, etc.

The ellipse in Figure 19.1 can be regarded as the orbit of rotation of one celestial body around another. This follows from the analytical solution of the corresponding differential equation, taking into account the forces acting on celestial bodies (see Study 9). During the early development of celestial mechanics attempts were made to analytically solve the problem of three or more celestial bodies obeying the law of universal gravitation. But a general analytical solution was never found—solutions were found only for a few particular cases. One of them is described in Study 9—see the round dance of the three planets in Figure 9.4. In these searches, the eyes of the researchers turned to *ellipses with more than two foci* (see Study 29), to the centers of the planets and satellites. Such closed curves are called n-ellipses or by the name of the person who first investigated them, curves of Tschirnhaus (see https://en.wikipedia.org/wiki/N-ellipse and Study 29). This philosopher, mathematician and experimenter is considered one of the inventors of European white porcelain, which, in the early eighteenth century, began to be produced in the Saxon town of Meissen near Dresden. You can visit this porcelain factory, which has been working successfully to this day, to honor Ehrenfried Walther von Tschirnhaus (1651–1708) and to make a souvenir porcelain plate with the shape and pattern shown in Figure 19.2.

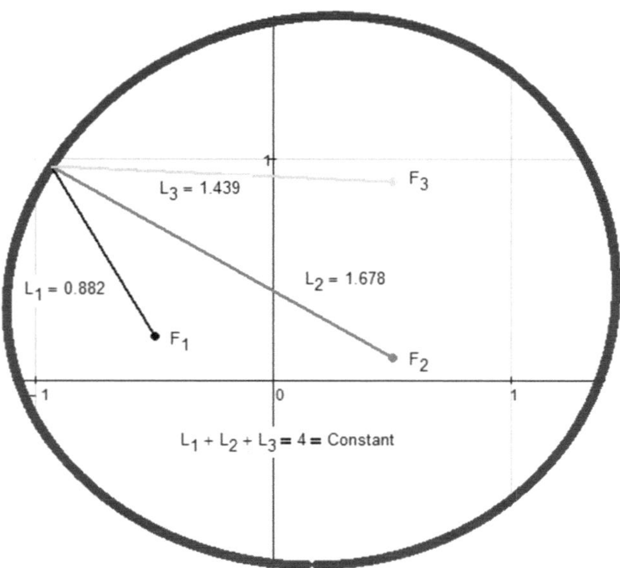

FIGURE 19.2 Ellipse with three focuses (see Figures 29.1 and 29.2 in Study 29).

On this egg-shaped plate (and these curves are also called egg-shaped*) in addition to three foci and three segments, the sum of which remains constant when drawing a blue border of the plate, you can put other information.† We offer another three-foci ellipse below, which is not only easy to draw on the computer screen, using the method in Figure 19.1, but also on the same wallpaper of the room, driving into the wall no longer two but three nail-foci (F1, F2 and F3), tying the rope between two nails (F1 and F2) and pulling it taut with a pencil and thread, using the third nail (F3), see Figure 19.3.

One may search the Internet to see if similar curves to that in Figure 19.3 have already been investigated, and whether the corresponding analytical expressions exist for Cartesian and polar coordinates. We leave this work

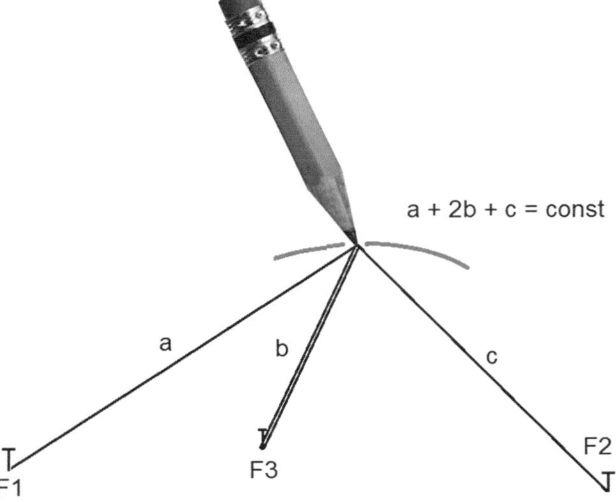

FIGURE 19.3 Drawing a three-focus ellipse (see also the Figures 29.1 and 29.2 in Study 29).

* You can "drive a few nails into three-dimensional space", "tie a string to them", and outline a certain body in the form of a real egg. To work, reader!

† The grid graphics on the plate reminds us of a prison cell. This will be a hint that the real inventor of European porcelain was not the aristocrat Tschirnhaus, but the alchemist Johann Böttger, whom Tschirnhaus kept under arrest in a "sharashka" created specifically for this "porcelain project". Tschirnhaus was the director of this closed institution—the commandant of the fortress, where this "mailbox" was located. This practice of conducting innovative scientific and technical developments two or four centuries later was widely used by the Stalinist regime. Because of this, we often do not know the names of true inventors and authors of developments, but we remember and honor only the "directors of *sharashka* (the popular name for research institutes with prison regime.)".

to the interested reader, who might also investigate the effect of increasing the number of foci and changing the order of threading the ropes through them.

If the *difference* is used instead of the *sum* of the lengths of segments, then the result will not be an ellipse, but a *hyperbola,* another plane curve of second-order: see Figure 19.4. It has two branches. One of them corresponds to the difference *Black–Blue*, and the second, the difference *Blue–Black*. Combining these two differences creates an operator of absolute value (modulus). But if you do not rely on the modulus, but work with two differences, then it is permissible to assign the parameter *a* negative values as well. In the case of a hyperbola, working with negative values of the parameter a will not change anything. But when considering the "logarithmic" curve, for example (see below), the negative values of the parameter *a* can expand the shape of the curve under study.

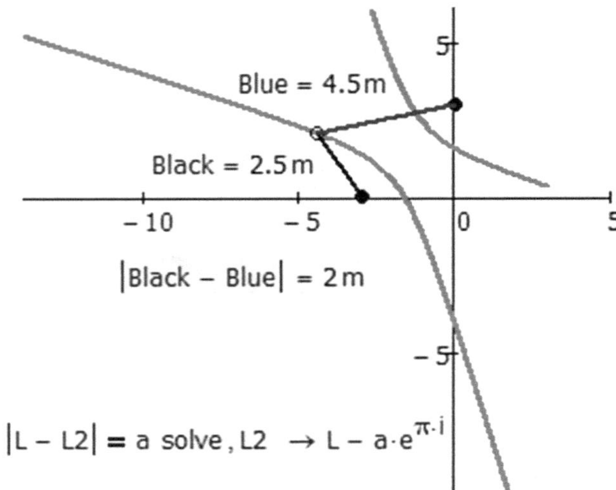

FIGURE 19.4 Hyperbola—subtraction.

We got the hyperbola, or rather, one of its branches, when we calculated the trajectory of the motion of a comet (see Figure 9.18 in Study 9 and Figure 20.4 in Study 20, the profile of the lens). Hyperbolas can also be constructed by relying not on two foci, but on one focus and the directrix, see Figure 20.12 in Study 20.

If you work with two foci, but not with the sum (see Figure 19.1) nor the difference (Figure 19.4), and with a third computational operation, namely the *product* of the values of the *Black* and *Blue* segments, it is possible to get the *Cassini oval** (see Figure 19.5.)

The closed curves in Figure 19.5 are called oval (deformed ellipse†) rather loosely. The decrease in the value of the parameter *a* (the product of the length of the *Black* by the length of the *Blue* segment) at fixed coordinates of the foci leads to the fact that this "oval" has a "waist" (see the central curve in Figure 19.5), which, with a further decrease in the value of the parameter *a*, "tears" this flat figure in two halves. Animation of this process resembles the division of a living cell. If we continue to decrease the value of the parameter *a*, then the ovals shrink to points ($a = 0$). There was a time when it was believed that the satellites rotated around the planets in orbits similar to the one shown in Figure 19.5 rather than in elliptical orbits (Figure 19.1). However, there were scientific discussions on this subject, in which Cassini himself and his oval were actively involved. But, in the end, it was proved that the ellipse "works" here. Another possible trajectory of the motion of celestial bodies is one of the branches of the hyperbola shown in Figure 19.4.

The Cassini oval, like an ellipse, can have more than two poles (foci). Such closed curves are called conditionally (see below) lemniscates ("twisted ribbons"). The most famous lemniscate is that of Bernoulli, which is a special case of Cassini's oval. This oval has a "zero waist", which occurs if the product of the lengths of the *Black* and *Blue* segments are equal to half the distance between the foci.

In Figure 19.6 one can see the evolution of a lemniscate with three fixed foci when the value of the parameter *a* changes.

Animation, four frames of which are shown in Figure 19.6, is similar to the way water is poured on an uneven surface, forming individual puddles, which gradually merge into one. If you start this animation in reverse order, you get a picture of the drying out of the puddle. Similar contours can be observed in the drying Aral Sea, for example.

Having a different number of foci at different points in the plane and setting a different value of the parameter-product *a*, you can get some pretty funny pictures.

* From the Italian, naturalized French astronomer Giovanni Domenico Cassini (1625–1712) https://en.wikipedia.org/wiki/Giovanni_Domenico_Cassini.The Cassini oval was first investigated by Cassini in 1680 when he was studying the relative motions of the Earth and the Sun. Cassini believed that the Sun travelled around the Earth on one of these oval paths, with the Earth at one focus of the oval http://mathworld.wolfram.com/CassiniOvals.html.

† The term oval may be used for any closed flat curve (in the English-language mathematical literature). In the Russian language, an oval is something completely smooth and convex.

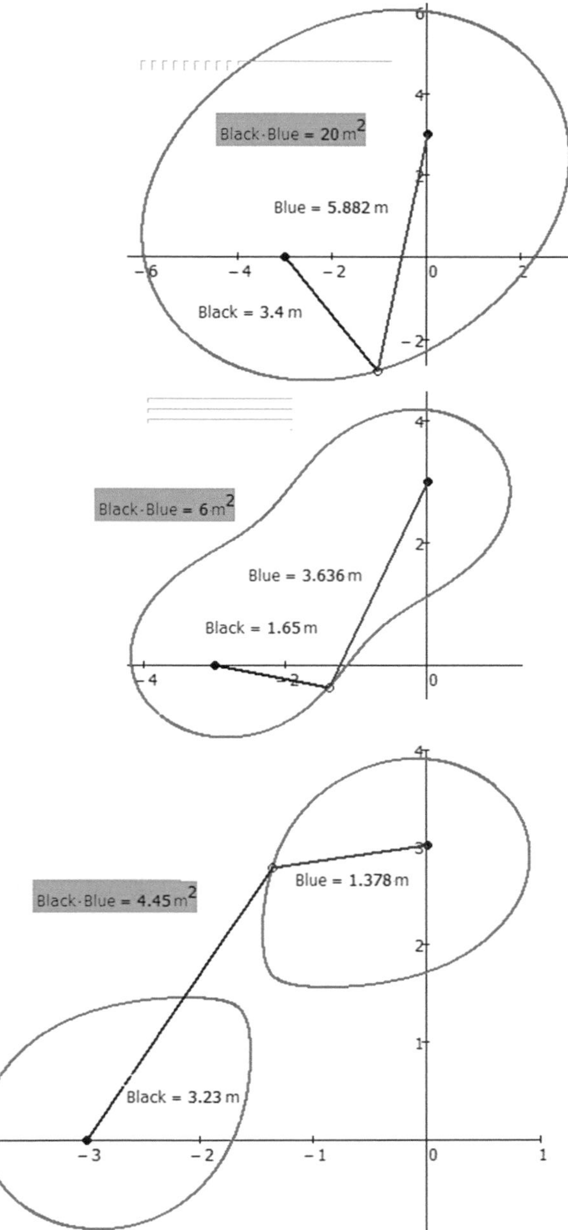

FIGURE 19.5 Oval Cassini—multiplication.

Figure 19.7 shows three stages of the transformation of Cassini's ovals, depicting a Pokémon with nine foci: ears, eyes, hands, feet and navel. Foci-eyes are additionally circled. This "virtual beast" can be caught, not only running like a madman with a smartphone in cities and towns, but also studying algebra, geometry and computer science.

The description of the lemniscate as a multi-focus Cassini oval can be found in Wikipedia. True, only in its Russian-speaking sector. In other sectors (English, German, etc.) there is no such description, which raises doubts about this interpretation of the term lemniscate. Usually they speak only about lemniscates with two foci of a special kind—Bernoulli lemniscate, Booth lemniscate, etc.

If the foci of the ellipse are placed in one single point, then a circle is obtained. However, a circle can also be constructed with two separated foci, using not the addition of segments, but their *division* (see Figure 19.8).

A circle is a special case of an ellipse, in the construction of which it is necessary to rely not on the sum of the segments (see Figure 19.1), but on their division (Figure 19.8). Such circles have a name, *the circles of Apollonius*.* The circle of Apollonius, like the hyperbola, has two "branches", shown in Figure 19.8. The two branches are circles obtained from two fractions: straight *Black*/*Blue* and reciprocal *Blue*/*Black*. The image of the circles of Apollonius, on which their forming segments *Black* and *Blue* are superimposed, resembles a bicycle with a frame and two wheels: Apollo has a chariot (or a quadriga, see the pediment of the Bolshoi Theater in Moscow or on a hundred-ruble note), and Apollonius, a bicycle—a bicycle, a two-wheeler, a "two-wheeled"—shown in Figure 19.8 and carried out in the name of this Study. Figure 19.8 also allows us to assert that the circle has a center and a focus, and these two points do not coincide.

The curves shown in Figures 19.1, 19.4, 19.5 and 19.7 (the ellipse, the hyperbola, the Cassini oval and the Apollonia circle) have been fairly well studied. But let's go into an area not yet explored, surrounded by some "red flags"—the theory of dimensional quantities.

Arithmetic operators of addition, subtraction, multiplication and division, involved in the solutions in Figures 19.1, 19.4, 19.5 and 19.7, have two operands. In Mathcad 15, in addition to these four operators with two operands, the Calculator panel (see Figure 19.9) contains three others: *exponentiation*, *a logarithm with an arbitrary base* and *a root of the nth power*. The buttons for these seven operators† are highlighted with rectangles in Figure 19.9.

* This ancient Greek mathematician, by the way, came up with the term hyperbole.

† Seven is not only a beautiful number. And the author has already written about this in previous Studies. This "beauty" also has some physical justification—remember the seven colors of the visible spectrum of light, seven notes of the European scale, seven basic SI units, seven days of the week, etc. You could also mention seven wonders of the light, seven ancient sages, etc. Another beautiful number is the triple, see http://twt.mpei.ac.ru/ochkov/Izvest_Art/art.htm.

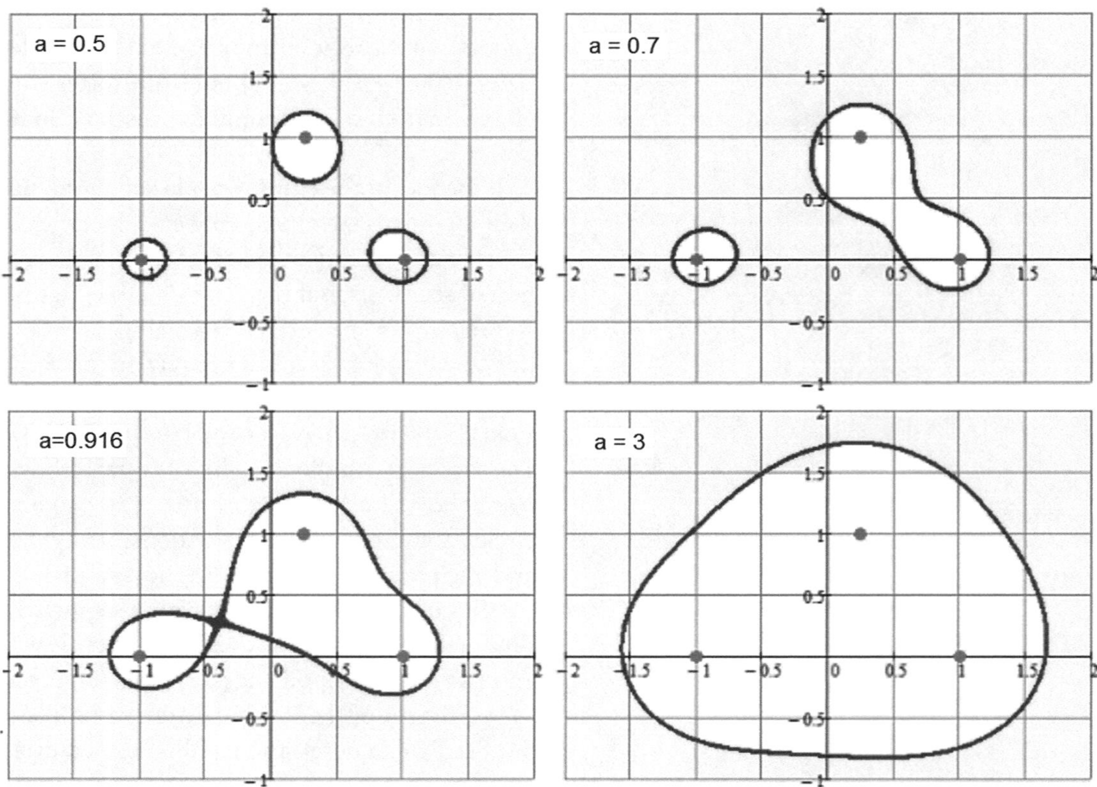

FIGURE 19.6 A three-focus lemniscate.

There was a seditious idea to go through the calculation shown in Figure 19.1, and the other three computational operators of elementary mathematics: degree, logarithm and root, *having obtained the magnificent seven computational curves*. But there was one problem that caused a heated discussion among Mathcad users, see the PTC Community website.

The point is that when constructing the curves shown in Figures 19.1, 19.4, 19.5 and 19.7, a unit of *length meter* was used. Typically, such graphics are built in dimensionless coordinates: no meters, centimeters, feet or inches. But Mathcad is equipped with a physical-value tool, which it would be a sin not to use, for example, to check the correctness of the input of formulas. As an

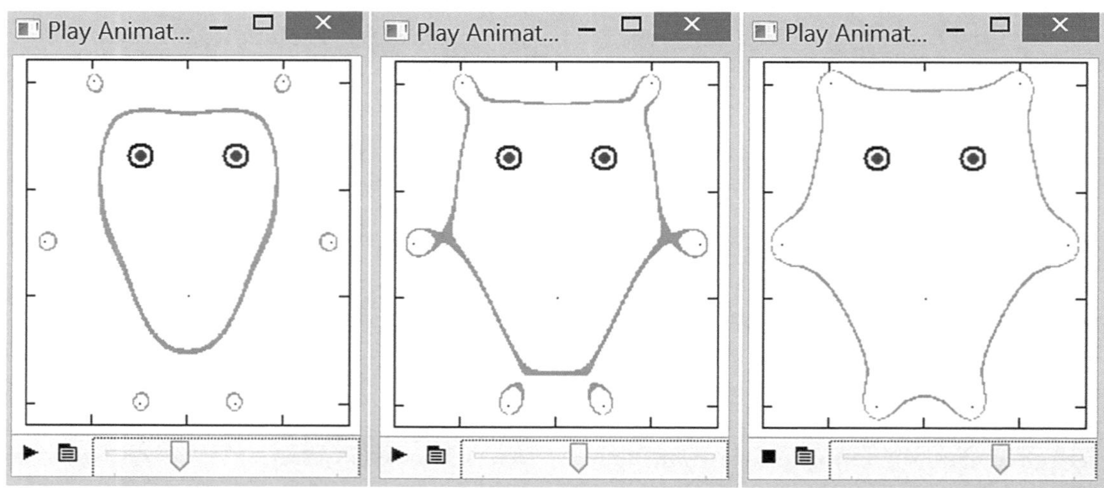

FIGURE 19.7 Pokémon Cassini.

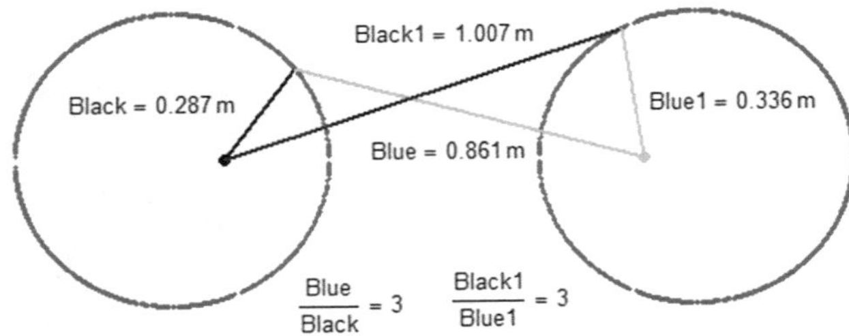

FIGURE 19.8 Apollonius circles—division.

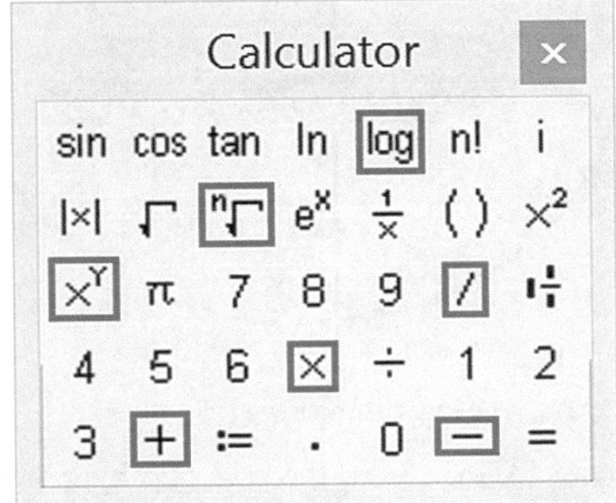

FIGURE 19.9 Mathcad 15 calculator panel.

example, if the user enters a variable raised to an incorrect power and the calculation is performed without units, it might be "swallowed"* by Mathcad, which will generate a false solution; but by using units the calculations will be interrupted resulting in an error message. In our calculations in Figures 19.1, 19.4 and 19.5 the parameter a was equal to 6 meters (Figure 19.1, an ellipse), 2 meters (Figure 19.4, hyperbola) and 20, 6 and 0.45 square meters (Figure 19.5, Cassini's ovals). Referring to Figure 19.7 the parameter a was dimensionless, or rather, had a meter dimension divided by a meter, but the straight segments along which the Apollonius circle was built still had a "meter" dimension.

If we want to construct a closed curve with a characteristic of *Black* to the *power* of *Blue*, equal to the parameter a (with units of length), then Mathcad will refuse to do this, interrupting the calculation with the error message—*the exponent cannot be a dimensional value.*

* As a participant in the discussion about computational curves, Mathcad published a curve constructed from a dimensionless calculation: in it, the difference a–L and the variable L itself (see the third line in Figure 19.1) mistakenly were not raised to the second power. A false graph was built, but another forum visitor noticed this error. It is better, of course, that the package itself should notice such mistakes and not give out false decisions. This would happen if the units of measurement were enabled.

But why should we not construct a dimensional exponent? Usually there are objections in the sense that there are no tasks that require this. But we will now show one such problem!

Let 2 meters be raised to the power of 3 meters, it will result in eight with a dimension, the essence of which needs to be discussed. For now, suppose that this unit of measure has the symbol m^m. But this is not a meter in the power of a meter, but a certain new symbol that reflects this metric pattern.† The main objection to the dimensional degree is reduced to the fact that there are no such physical quantities in nature. But our next closed curve, the curve where $Black^{Blue} = a$, suggests that this is completely wrong and that we can measure something by the unit where the meter is raised per meter.

The curve shown in Figure 19.10, is similar to the Cassini oval with the only difference being that the two halves of the new curve are not symmetrical, which is clearly seen in the lower part of Figure 19.10. This asymmetry can be viewed as a graphic illustration of the proposition that the interchange of the operands of the exponential operator changes the result. A consequence of this proposition is the fact that the closed curve in Figure 19.10 has two branches: one for the expression $Black^{Blue} = a$, and the second for the expression $Blue^{Black} = a$. Changing the shape of these two branches when changing the value of the parameter a is shown in the animation on the PTC Community website. The upper closed curve in Figure 19.10 reminds us of a guitar sound board, which justifies our giving it the corresponding name "Guitar" or "Guitar of Apollo" (see the title of this Study). In our time, Apollo would play not on the lyre, but on the guitar.‡

You can try to "stick" the meter, not only in the exponent, but also in the logarithm. Moreover, there have been precedents for taking the logarithm of a dimensional value, see Figure 19.11, which shows the calculation of the difference between the logarithms of two-dimensional quantities, pressures. And such an operation is often carried out in thermodynamics.

† That is nothing unusual. In mathematics and mathematical packages, some operators and symbols are often treated in two ways. The expression x^T, for example, depending on the situation, can mean the variable x in the power of T and the transpose of the matrix with the name x. So it will be quite reasonable to interpret m^m (or cm^{cm}, m^{mm}, ft^{in} etc.) not only as a unit of length in the power of a unit of length, but as something else, doing the correct recalculations of a physical quantity with such dimension.

‡ There is a very interesting book about the physics of music *La Fisica nella Musica* by A. Frova.

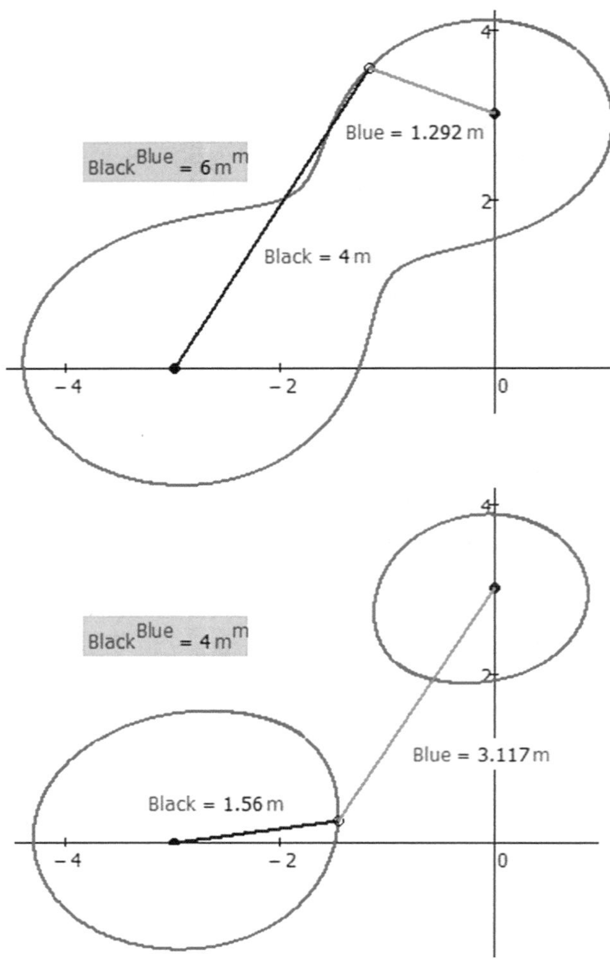

FIGURE 19.10 The exponentiation curve.

The focus of the calculation in Figure 19.11 is that the intermediate calculation of the logarithm of the fraction is not shown. But this example is not entirely appropriate here. So back to our closed curves.

Figure 19.12 shows a curve corresponding to the condition *log(Black, Blue)* = *a*. Here in the seven ($a = 7$) is a very exotic unit of measurement *log*(m, m), which also needs to be interpreted correctly (see above).

The curve shown in Figure 19.12 has two branches: one for *log(Black, Blue)* = *a*, and the second for *log(Blue, Black)* = *a*. Changing the shape of these two branches when changing the value of the parameter *a* is shown in the animation here https://www.ptcusercommunity.com/thread/137465. As in the case of a hyperbola, the parameter *a* can also have negative values.

$$log(10\ atm) - log(1\ bar) = 1.006$$

FIGURE 19.11 The logarithm of a dimensional quantity.

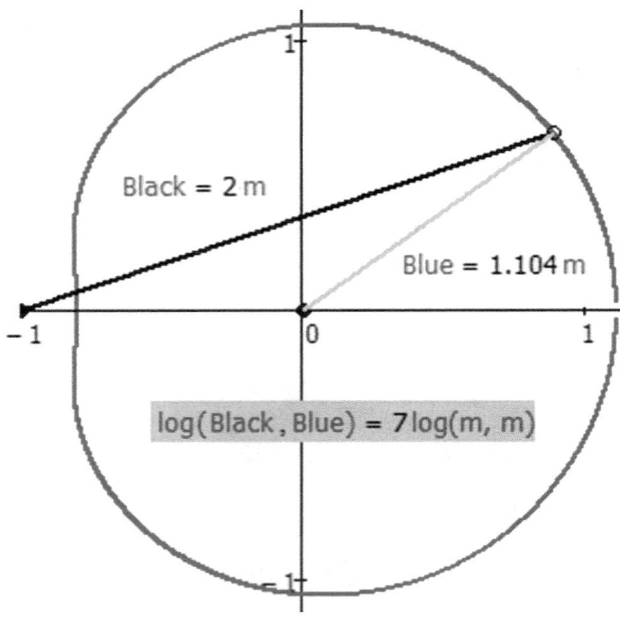

FIGURE 19.12 The logarithm curve.

And, finally, Figure 19.13 shows the seventh curve corresponding to the seventh two-operand computing operator from the panel shown in Figure 19.9—taking the *n*th root of a number. But, in Mathcad, this operator implies that *n* is an integer greater than one. Therefore, we will be a little cunning and replace the root with a degree with the inverse index: $Black^{1/Blue}$ and $Blue^{1/Black}$.

The curve shown in Figure 19.13, also has two branches: one for the expression $Black^{1/Blue} = a$ and the second for the expression $Blue^{1/Black} = a$. Changing the shape of these two branches when changing the value of the parameter *a* is shown in the animation on the PTC Community website.

Attempts by the author to find in the literature or on the Internet such "illegitimate" (from the point of view

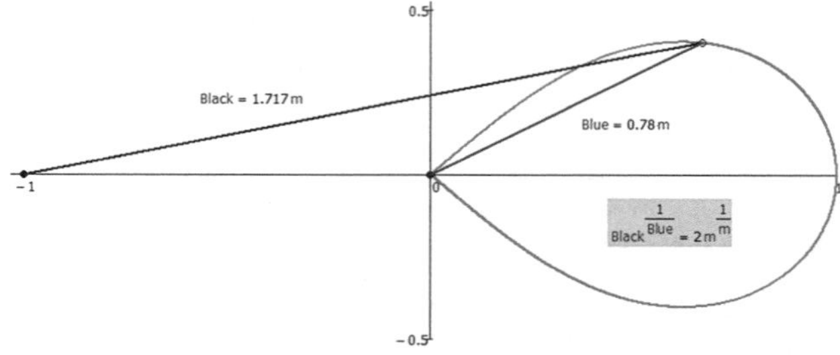

FIGURE 19.13 The *n*th root curve.

of the theory of dimensions) curves (Figures 19.9, 19.12 and 19.13) were unsuccessful. Maybe the readers will succeed. These curves require a deeper mathematical analysis—the finding of algebraic expressions describing them, constructing quadratures, etc., as was done for the ellipse, hyperbola, Cassini ovals and Apollonius circles.

The beginning of such an analysis can be as follows. If, for example, we want to determine our exponential curve in Figure 19.9, through the system of equations:

$$eq(a, b, c) = \begin{bmatrix} b^2 = x^2 + y^2 \\ a^2 = (x-c)^2 + y^2 \end{bmatrix}$$

where

$$b = \frac{\ln(\kappa)}{\ln(a)}$$

then it can be parametrized as follows:

$$x = \frac{c^2 \cdot \ln(a)^2 - a^2 \cdot \ln(a)^2 + \ln(\kappa)^2}{2 \cdot c \cdot \ln(a)^2}$$

or

$$y = \frac{\sqrt{\Delta}}{2 \cdot c \cdot \ln(a)^2}$$

where

$$\Delta = \ln\left(a^a \cdot a^c \cdot \kappa\right) \cdot \ln\left(\frac{a^a \cdot \kappa}{a^c}\right) \cdot \ln\left(\frac{\kappa}{a^a \cdot a^c}\right) \cdot \ln\left(\frac{a^a}{a^c \cdot \kappa}\right)$$

In polar coordinates this will look like this:

$$\cos(\varphi) = \frac{\rho^2 + c^2 - \kappa^{\frac{2}{\rho}}}{2 \cdot \rho \cdot c}$$

Discussion of this solution can be found at the PTC Community website.

The question of what dimension and physical meaning the length, raised to the power of length, should have remains open. Nevertheless, without waiting for the final decision of this metrological problem, the author is going to build a real, "physical" flower bed in his dacha residence in accordance with the dimensions shown in Figure 19.14.* This flower bed will have real *physical* parameters: the distance between the foci, the length of the perimeter, the area and our parameter *a* is the length to the power of the length. The physical quantity exists, but the unit of its measurement is not yet available. Unit m^m, which was mentioned above, causes controversy and doubt and requires more detailed attention. But some unit of measurement is needed here.

Imagine that someone in the US or the UK, who wants to build a similar flower bed, has a tape measure in his/her hand, marked not with *meters and centimeters*, but with *feet and inches*. What will need to replace the number 4 (see Figure 19.14)? This number must also have a dimension. We repeat the question in its form and essence, openly. Moreover, you can try to build a flower bed, where *Black* + *Blue*^{2} = *a*, but this will already be a clear violation of the law of dimensional quantities: it is not possible to add a value to its square. Our violation of this law is not so evident as in case of the "Cassini power-law oval" (Figures 19.9 and 13). And here is the other side of this problem. In mathematics, it is possible to see such problems: solve a quadratic equation of the type $2x^2 + 3x - 4 = 0$. At the same time, no one has the idea that it is impossible to add a square of the magnitude to the value itself. Here we mean that the

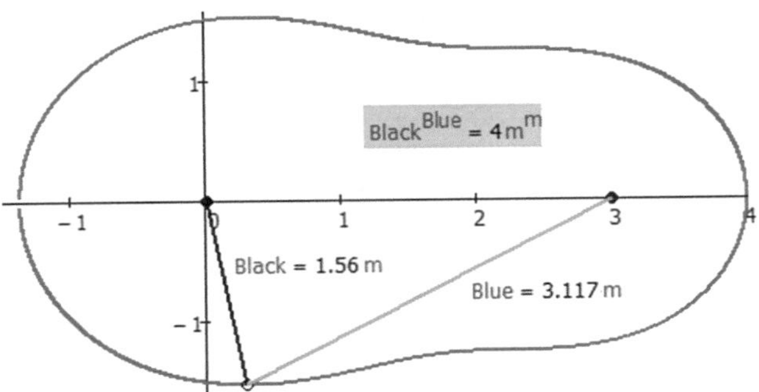

FIGURE 19.14 Exponential flower bed.

* A flower bed can be marked like this. Two pegs are hammered in, fixing the foci of this "power Cassini's oval with waist". Further with the first rolling metering tape the *Black* section is measured in meters and the value of the segment *Blue* is also calculated in meters according to the formula *ln*(4)/*ln*(*Black*). With the second rolling metering tape the distance is measured in meters and the first peg is punched, outlining the contours of the future flower bed. The operation is repeated with a new value for the length of the *Black* segment, until the contour of the flower bed marked by the pegs is closed. So, by the way, the "stupid" Mathcad dotted our ovals and other curves. A person without a computer in such cases begins to analyze the dependencies, look for asymptotes, zeros and other special points on the graph and only after that draws in the mind or on paper the curves qualitatively, not quantitatively.

coefficients for x^2 and x (two and three) contain something that is related to the recalculation of dimensions, which makes it possible to make such "illegal" addition. The expression *Black* + *Blue*^{2}, which is necessary to draw the curve, can also contain certain unitary coefficients with dimensions that allow us to add the value (length) and the square of the value (area).

The second theme of Study is this: modern computing tools allow movement away from sophisticated and incomprehensible algorithms for plotting curves and return to the origins.

Let us recall what a circle is, an ellipse, a parabola, a hyperbola, etc., as discussed above. This is the locus of points on the plane that correspond to some given characteristic. Let us, without further ado (which we did when creating the calculation in Figure 19.1), simply program this definition—we scan the bounded region of the plane where these curves can be found and note the points corresponding to the given characteristic. We have already resorted to this method by plotting a complicated function in the form of a woven napkin, see Figure 4.44 in Study 4. Figure 19.15 shows the construction of a closed curve with three foci whose distances from the curve correspond to this condition: |*Black* + *Blue* – *Green*| = *a*.

In the calculation in Figure 19.15, the region on the plane from the given values x_1 to x_2 and from y_1 to y_2, is divided into separate points, some of which are collected in two vectors X and Y and displayed on the graph. These are the points that correspond to the given property of the curve. But we use the condition "approximately equal", not "equal", because we solve the problem numerically, not analytically, and consequently approximately. As one movie character said: "It's not aesthetically pleasing, but cheap, reliable and practical." A more "aesthetic" version of the calculation, using a preliminary analysis of the dependencies, and a symbolic (analytical) solution of the system of equations is shown in the first figure of this Study.

In Figure 19.15, we got a hybrid of an ellipse and a hyperbola—some kind of *ellipsohyperbola*, in which the curve remained closed as the ellipse, and there are two branches as in the hyperbola, one of which is inside the other. This curve can also be called ellipsohyperbola because in its construction, the sum (ellipse) and the difference (hyperbola) were used. See also Figure 29.6 in Study 29 with other ellipsohyperbola. We drew in Figure 19.15 the artist's palette, a wooden plate on which paint is mixed.

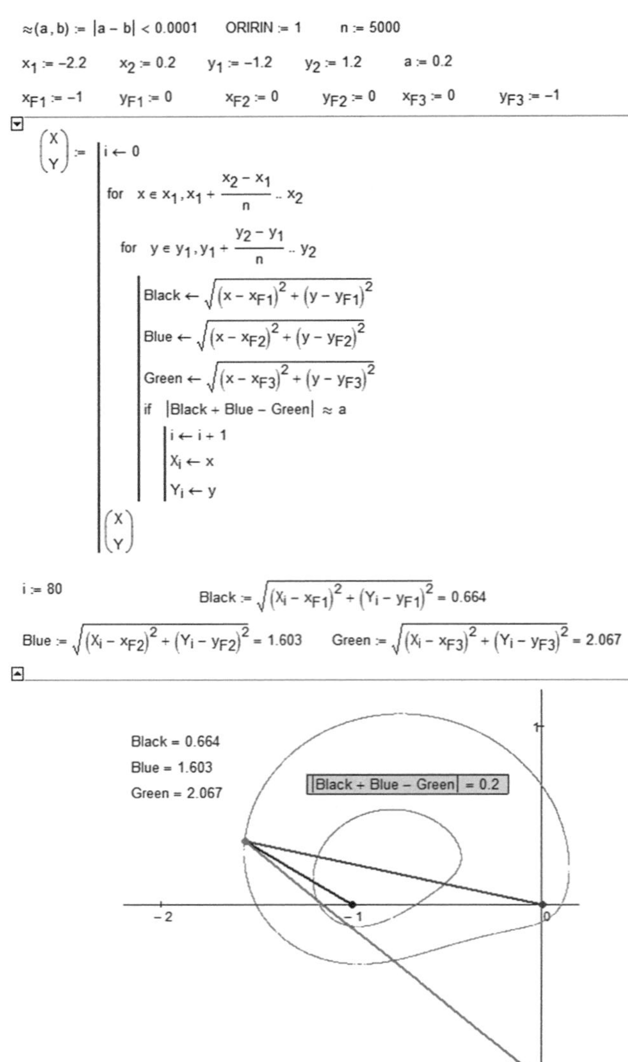

FIGURE 19.15 A universal method for drawing curves for given conditions.

On the PTC Community website you can see in the animation how this curve turns into an ellipse or hyperbola with a smooth change of its parameters in one direction or another.

The "practicality" of the calculation in Figure 19.15 is that it is not difficult to generalize it so that you can work with any number of foci and with any equality,* which sets the form of the curve. This will create very entertaining and instructive curves, not only in "static" mode, but also in "kinematics" mode with animation. Which we did: see the example in Figure 19.16.

* It is worth noting that using inequality instead of equality will allow the building of surfaces instead of lines.

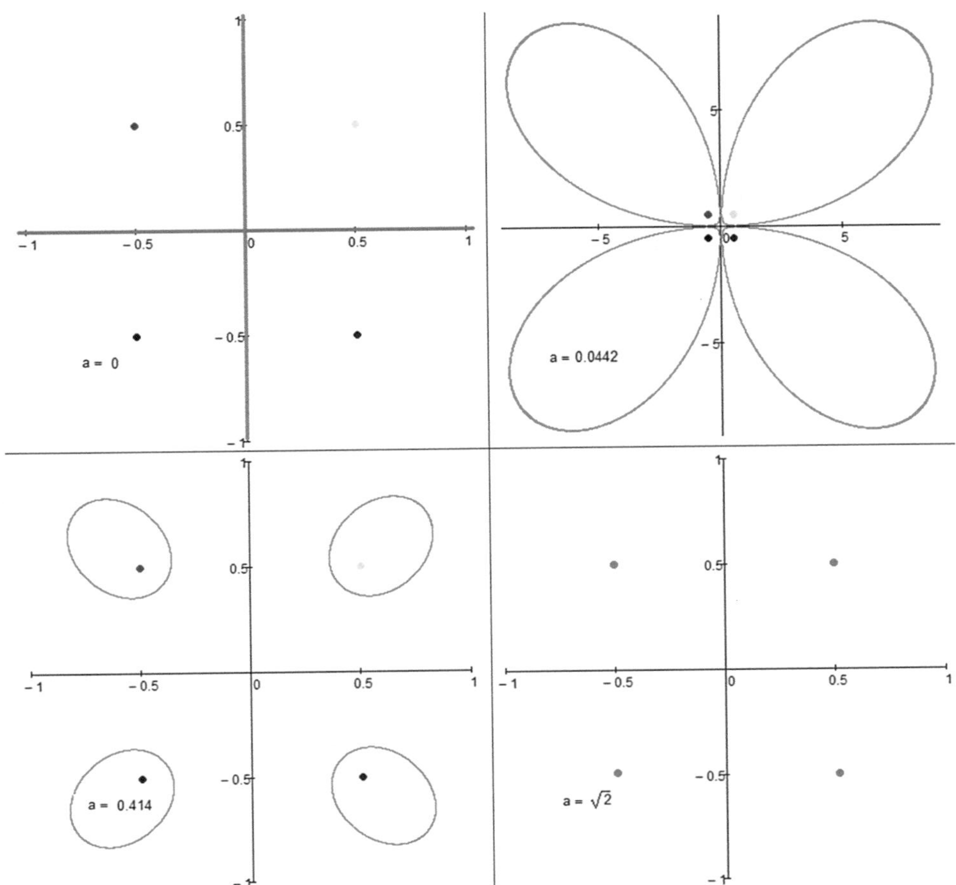

FIGURE 19.16 Four-leaf clover.

The curves in Figure 19.16 have four foci and correspond to the equality $|L1 - L2 + L3 - L4| = a$, which again makes them related to both the ellipse (sum) and the hyperbola (difference). When $a = 0$, we have four branches of the "hyperbola", which merged with the axes of the graph. If a is greater than zero, this hyperbola-cross is folded into four ovals. If a is equal to the root of the two, the leaves of this four-leaf cross are compressed into dots.

It is also interesting to display our curves not only in Cartesian, but also in polar coordinates. If we talk about the polar chart, then in Mathcad, some "red flags" are also removed: in the latest versions of this package it is possible to display negative values of the coordinate-radius, which is rather controversial from the point of view of classical mathematics. Usually, polar graphs are used where the argument changes "in a circle"—from 0 to 360 angular degrees. A typical example in Figure 19.17, which displays the simplest "circular", or rather, periodic function, a sine.

In the window of the Mathcad polar formatting chart, the show negative radii checkbox appears, the inclusion of which changes the sine curve from the circle (this is not actually a sine, but the absolute value of the sine) to an inverted heart (sine without a modulus).

If we try to construct in polar coordinates our simplest "computational" curve, the circle (see its Cartesian graph in Figure 19.7), including the option of negative radii, we can get quite interesting graphs, see Figure 19.18.

If, on the polar chart, the option to show negative radii is turned off, then the hearts and ovals (red lines) will turn into circles, and their curved radii (blue lines) will straighten, as shown on the Cartesian graph. In addition, the second "phantom" circle will disappear. And why is it a phantom!? The hyperbola is known to have two branches, see Figure 19.4. It can be assumed that the circle, ellipse and other "computational" curves also have an incomplete (real) second branch, which can be seen only on the polar chart, "jumping over

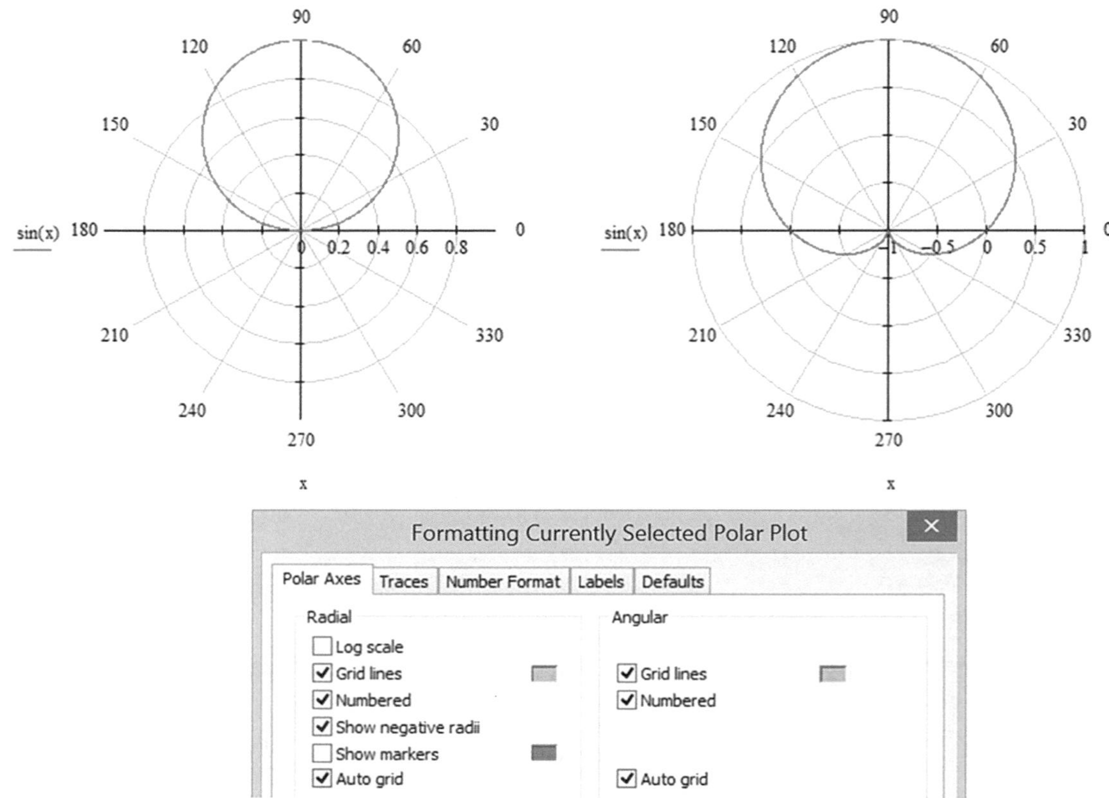

FIGURE 19.17 Sine in polar coordinates: on the left, only with positive radii-coordinates, on the right, also with negative radii-coordinates.

the red flags", allowing negative values of the coordinate-radius. We did not mention the complex values in vain—the computational operators and functions considered in the Study, in particular, the elevation to power and the logarithm have complicated graphs with several branches in the complex plane, with periodicity and so on.

The inclusion of the *n*th root operator (see Figure 19.13) in this Study, is not entirely justified to be honest, because this operator is only a modification of the exponentiation operator (Figure 19.9). Therefore, the eyes of the author turned to another important arithmetic operator with two operands—*the remainder of the division.* In Mathcad and in many other software environments, this function has the name *mod* (division by modulus). Unlike, for example, from raising to the power (Figure 19.9), the logarithm (Figure 19.11) and the root (Figure 19.12), it does not violate the generally accepted rules for working with dimensional quantities: *mod* (1.5 m, 0.4 m) = 0.3 m. Check: (1.5 m – 0.3 m)/0.4 m = 3 (integer). The equation *mod* (x, y) = a cannot be solved analytically with respect to the variable *x* or *y*. Therefore, this curve cannot be drawn by the method shown in Figure 19.1. Here you will have to "blindly" scan to draw the curves corresponding to the equality *mod(Black, Blue)* = a (first branch) and *mod(Blue, Black)* = a (second branch), see Figure 19.19 (four animation frames) and 19.20 (individual branches are highlighted in color).

For a = 0, the curves of the division function without remainder are converted into the set of circles of Apollonius—the wheels of the bicycle presented in the title of the Study.

The "magnificent seven" of the computational curves can look like this: (1) addition (ellipse), (2) subtraction (hyperbola), (3) multiplication (Cassini oval), (4) division (Apollonius circle), (5) remainder of division (?), (6) raising to the power of (?) and (7) the logarithm from a given base (?). The last two curves can be considered candidates for the "batch of computational curves" because of the unresolved "metric issue". But

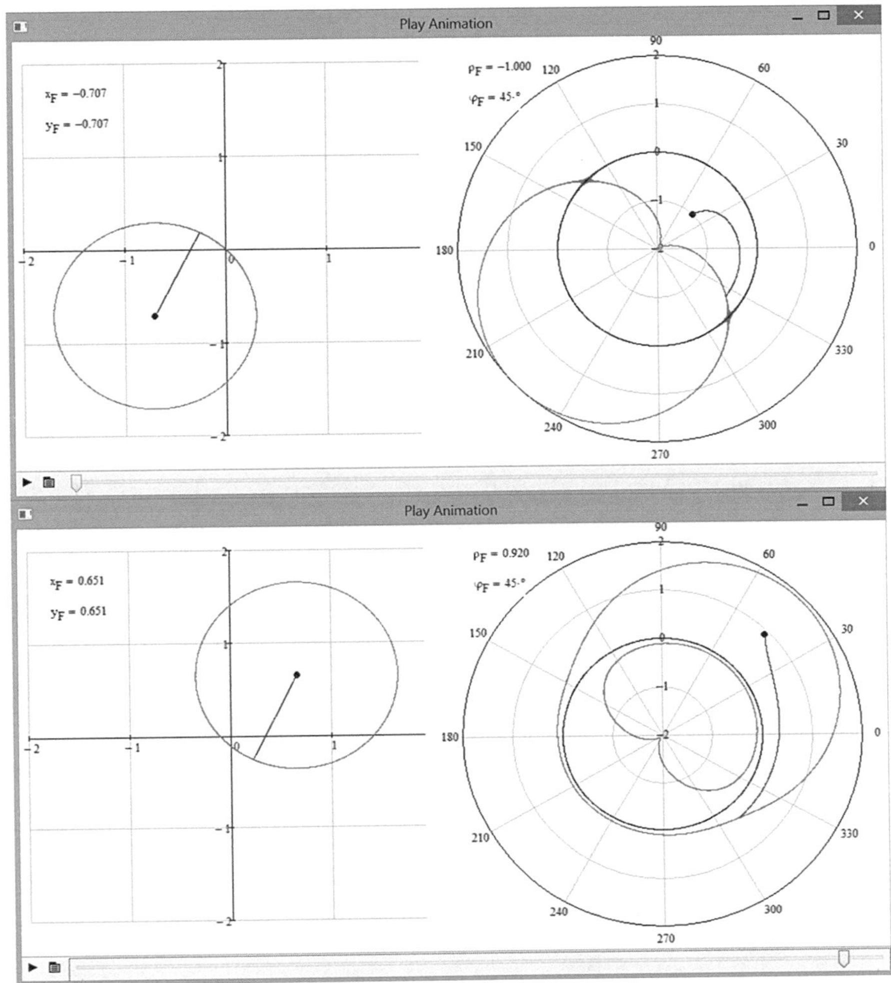

FIGURE 19.18 Circle with different center coordinates on Cartesian and polar graphs.

if the exponentiation can somehow still be left in the list of computational curves, then the logarithm here is completely irrelevant. Let's try to replace it with something else.

If subtraction can be considered the operation inverse of addition, and division the operation inverse of multiplication, consequently hyperbola can be considered the curve "inverse" of ellipse, and the circles of Apollonius the curves "inverse" of the oval of Cassini. Figure 19.21 and 19.22 show curves that are "inverse" to the curves of a division operation without a remainder.

The list of computational curves can be supplemented. The reader himself or with his students can experiment in this direction. In particular, find the seventh "legitimate" computational curve. In the meantime, we have only six of them: addition and "reverse" addition (subtraction), multiplication and "inverse" multiplication (division), division without remainder and "inverse" division without remainder,

The reader can try to build polar graphs of other "computational" curves. The practical use of this work can be doubted, but the graphics will be very entertaining. And this is an important aspect in the mathematical education and the use of computational packages.

The author is grateful to Werner Exinger and Alvaro Diaz for taking part in the discussion of this Study and creating some calculations and animations for them.

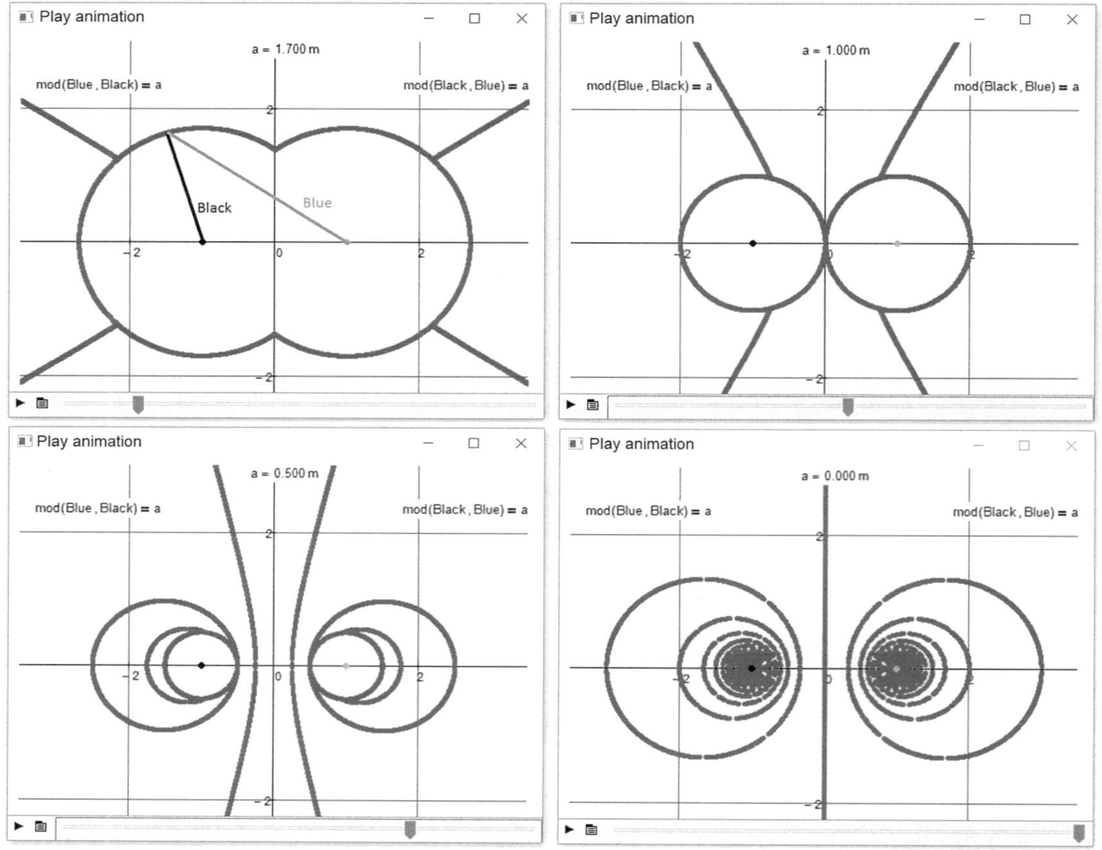

FIGURE 19.19 Curves of the remainder of the division.

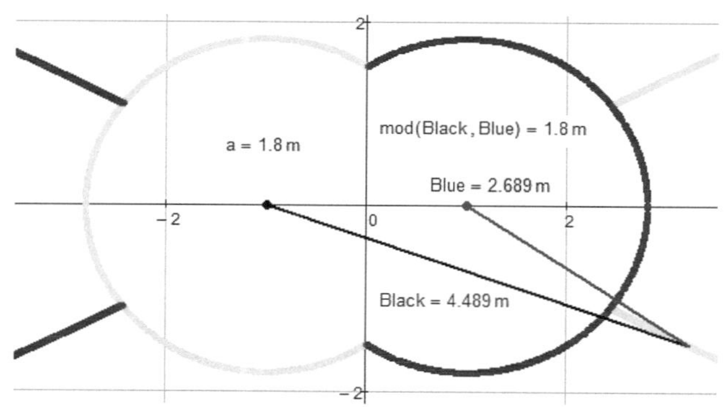

FIGURE 19.20 Two branches of the remainder of the division curve, marked with color.

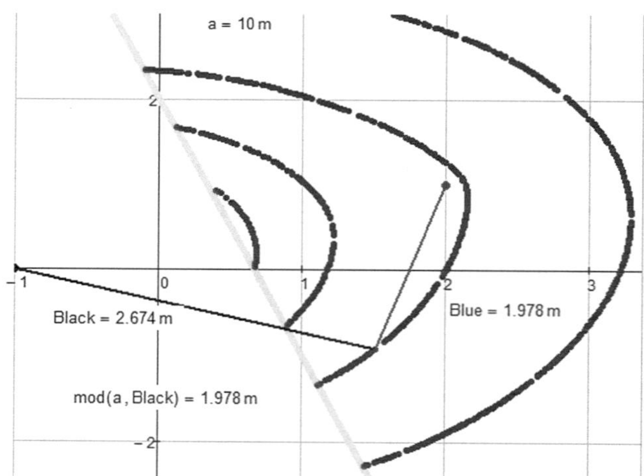

FIGURE 19.21 Curves, "inverse" of division operations without remainder.

BUG* OF THE SECOND-ORDER

If you ask what a *parabola* is, almost everyone will answer that it is a line that displays a quadratic equation on the graph. And only a few will remember that a parabola is the geometrical locus of points on the plane with a certain property. The same situation with the hyperbola. Everyone remembers his/her "school" formula $y = 1 / x$, but forgets about the geometric locus of the points. The situation is reversed with the *circle*. Everyone knows about points on the plane that are equidistant from the center of the circle, but few will immediately write an algebraic expression along which a circle is constructed on a Cartesian graph.

Let's consider the "geometric" and not the "algebraic" definition of the parabola and the hyperbola, and also of the "flattened" circle (ellipse) and construct their graphs, based not on algebraic expressions but on "point-wise" definitions.

We begin with the parabola. If a straight line is drawn on the plane and a point is placed near it, the parabola will consist of all the points equidistant from the straight line (from the parabola *directrix*) and from the given point (from *the focus* of the parabola). Such a point-like construction of a parabola is easy to make on paper, picking up a pencil, compasses and a set square: open the compass to a certain angle (Figure 19.22), place its needle at the point-focus and draw the arc of the circle. Then, with the help of a set square, draw a straight line at a right angle from the directrix to the arc so as to place on the arc of the circle a point equidistant from the focus and the directrix. Next, change the angle of the opening of the compass and repeat all over again, putting on the graph more and more points of the future parabola.

But modern schoolchildren and students, alas, are unaccustomed to working with real drawing tools, as everything can be drawn on a computer screen, and then moved onto paper by a 2D-plotter or they can even "fashion" a model on a 3D printer.

* In the context of computer science, the word "bug" could be misleading. Better the word "beetle", but there is reference to the "The Gold-Bug" of Edgar Allan Poe (see Study 3).

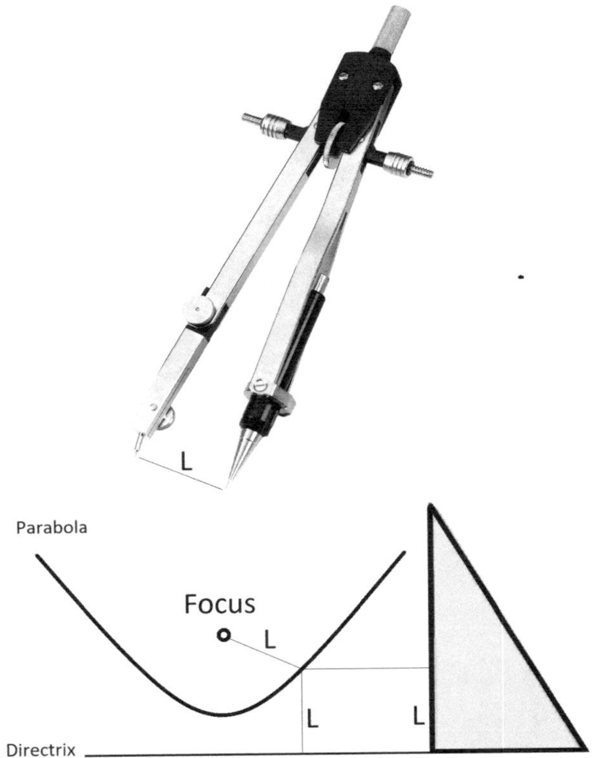

FIGURE 19.22 Constructing a parabola by a compass and a set square.

By the method shown in Figure 19.22, we can also construct a hyperbola with an ellipse. Only the segments constructed by the compass and ruler will be made, not equal, but different by a certain constant called *eccentricity* (ε). Figure 19.23 shows the construction of the hyperbola, parabola and ellipse by scanning (see Figure 19.15). The ellipse has $\varepsilon < 1$, the parabola $\varepsilon = 1$, and the hyperbola $\varepsilon > 1$.

In Study 3 we mentioned a bug, not a simple one, but a golden one, which helped the hero of the story of Edgar Allan Poe to find the treasure. We have one in Figure 19.23. The bug was also not simple, but of the second-order, made up of an ellipse, a parabola and a hyperbola—from second-order curves examined in Study 9. We will return to the eccentricity of the hyperbola in Study 20, when we consider the optical property of this curve.

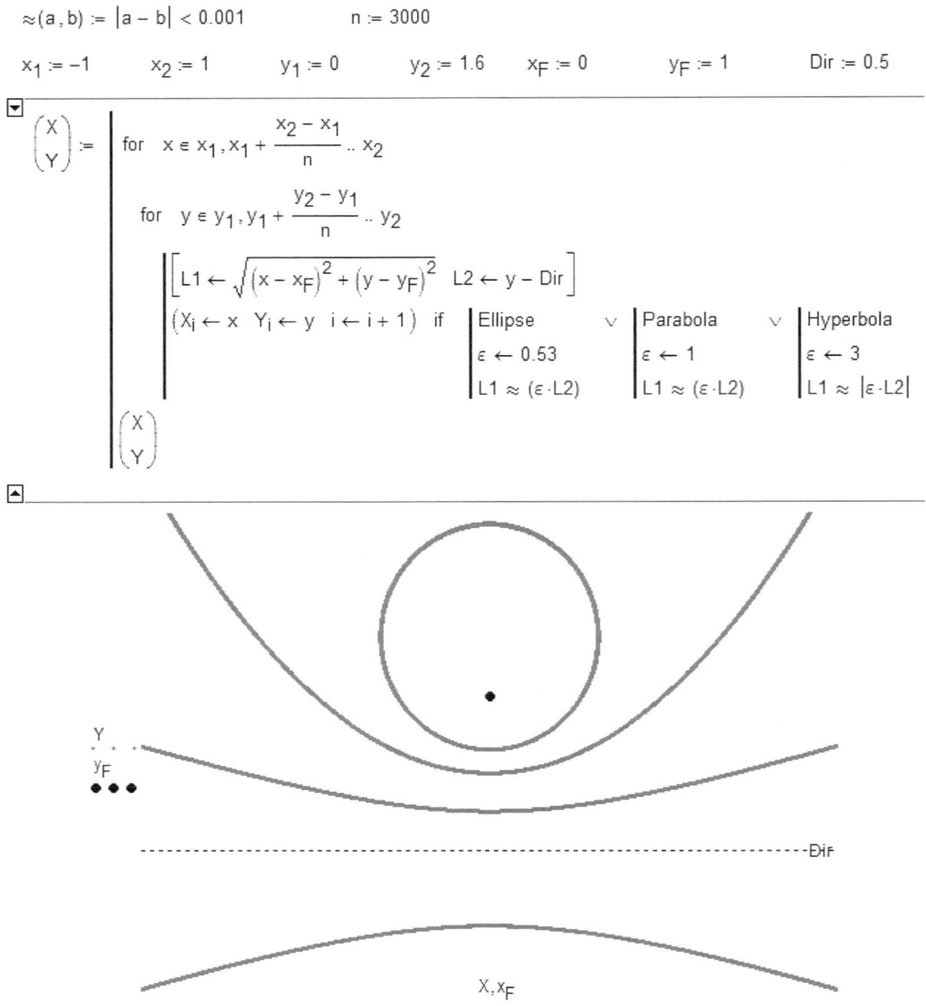

FIGURE 19.23 Ellipse, parabola and two branches of the hyperbola: bug of the second-order.

ASSIGNMENTS FOR READERS

1. Derive the analytical expressions for the new "computational" curves, considered in this Study.
2. Determine the coordinates of the centers of the two circles shown in Figure 19.8.
3. The Apollonius bicycle (see Figure 19.8) has wheel bushings placed, not in the center, but in the foci of the circles. What form should be the bicycle path, so that the bike moves strictly horizontally. Look at Figure 7.19 (in Study 7) to see such a path for a bicycle with square wheels.
4. Construct the polar graphs of the other "computational" curves described in this Study. The practical use of this work can be doubted, but the graphics will be very entertaining. And this is the main issue in mathematical education with the use of computational packages. See also the epigraph of this Study.

Journey from Home to School on the Fermat Route Or

The Second Optical Property of Hyperbola

ABSTRACT

In this Study, the solution of optimization problems is considered with reference to the problems of the reflection and of the refraction of light. A numerical and analytical solution of the algebraic equations of a lens and a focussing mirror are presented. It is shown how modern software can easily and elegantly solve optical problems. The optical property of the hyperbola, based not on the law of reflection, but on the law of refraction of light, is emphasized.

Mathematics: Solution of the equation, optimization, plane curves of the second order, Fermat's theorem.

Physics: Distance, time, speed, geometric optics, refraction and reflection of light, the Fermat principle, the principle of tautochronism, the optical path of a light beam.

IT: Numerical solution of an optimization problem, numerical solution of an algebraic equation, computer tools for analytical transformations (symbolic mathematics), computer graphics of planar curves, embedded cycle with parameter.

Art: Daniil Kharms, "Holiday"; Aleksey Nikolayevich Tolstoy, *The Hyperboloid of Engineer Garin*.

Study website: https://community.ptc.com/t5/PTC-Mathcad-Questions/Study-20/m-p/566163

The janitor nailed the flag to the house.

The passer-by asked what it meant.

The janitor answered: "This means that there is a holiday in the city."

"And what is this holiday?" asked the passer-by.

"There is a holiday because our mathematician has found one more optical property of hyperbola!" answered the janitor.

The passerby, embarrassed by his ignorance, dissolved into the air.

DANIIL KHARMS "HOLIDAY" (NEW EDITION) HTTPS://EN.WIKIPEDIA.ORG/WIKI/DANIIL_KHARMS

A schoolchild or a student is free to get from the house or the hostel to their educational institution in different ways: to walk down the street, make their way through courtyards and alleys, travel part of the way by public transport, etc. The shortest distance between two points is known to be a straight line. But it's not always possible or necessary to move in a straight line. Even if there are no obvious obstacles in the form of buildings, fences and other obstacles, then moving in a straight line, even in an open field, does not always turn out to be the optimal, if by optimization we mean not the reduction of distances, but the minimization of travel time. After all, pupils, late for school, do not think about distance, but about time.

Let's begin by solving a simple problem: a person is at point A (in an open area Figure 20.1). The clouds are gathering, thunder is heard, a thunderstorm is about to begin! It's time to get to point B where there is a certain shelter: a house, a gazebo, a canopy. How can a person run, if he is on a lawn, asphalt is laid in front of his house, and a plowed strip of arable land crosses his path? We have already solved something similar in Study 15, see Figure 15.13.

We formulated a typical *optimization* problem where the *objective function* is the time of travel from point A to point B, and the optimization *variables* are the abscissas of points C and D. The ordinates of points C and D are given and indicate straight lines for dividing the areas "lawn/arable land" and "arable land/asphalt", which are parallel to the abscissa, but these dividing lines may have another form. The speeds of running along the lawn (*l*), arable land (*al*) and asphalt (*a*) are also set and are, naturally, different: $v_a > v_l > v_{al}$. The coordinates of the initial and final points A and B are also specified.

Figure 20.1 shows the solution to this problem using the built-in *Minimize* function. It has a variable number of arguments. In our problem there are three of them: the name of the objective function *t* (time) and the names of the two optimization variables x_C and x_D. The *Minimize* function, by use of a special numerical algorithm (see Studies 4 and 13), will automatically change the values of the optimization variables x_C and x_D so that the objective function takes the minimum value. The values of the variables x_C and x_D vary in the search for the minimum from some initial values (for the first assumption, see Figure 20.1), which are set by the user himself, based on the "physics" of the problem. For x_C and x_D values taken for the first assumption, the travel time from point A to point B is 2 minutes 50 seconds, and the *Minimize* function reduces it to 2 minutes 30 seconds.

However, there are doubts, is it a real minimum, is there a better route for running along three broken straight segments? The answer is checked by constructing a contour graph and two sections of the surface of the objective function at the minimum point: all good—the minimum is found: it is surrounded by lines of one level and located at the minimum points of two curved sections with respect

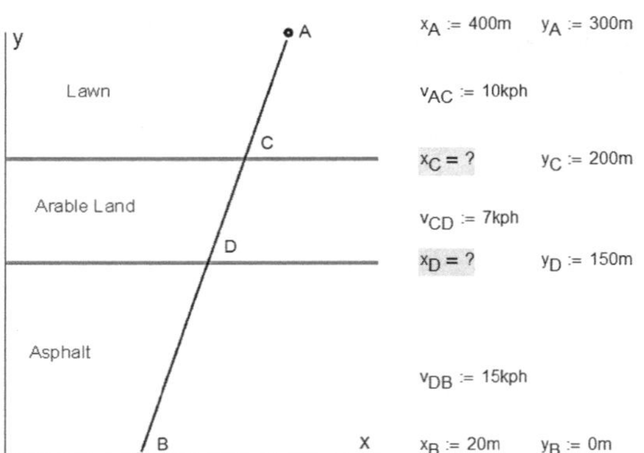

The objective function is the running time from point A to point B

$$t(x_C, x_D) := \frac{\sqrt{(x_A - x_C)^2 + (y_A - y_C)^2}}{v_{AC}} + \frac{\sqrt{(x_C - x_D)^2 + (y_C - y_D)^2}}{v_{CD}} + \frac{\sqrt{(x_D - x_B)^2 + (y_D - y_B)^2}}{v_{DB}}$$

Guess Values $x_C := 250m$ $x_D := 150m$ $t(x_C, x_D) = $ "0:2:50.037" ·hhmmss

$$\begin{pmatrix} x_C \\ x_D \end{pmatrix} := \text{Minimize}(t, x_C, x_D) = \begin{pmatrix} 326.993 \\ 304.336 \end{pmatrix} m \quad t(x_C, x_D) = \text{"0:2:29.959"} \cdot hhmmss \quad \text{Minimum}$$

(a)

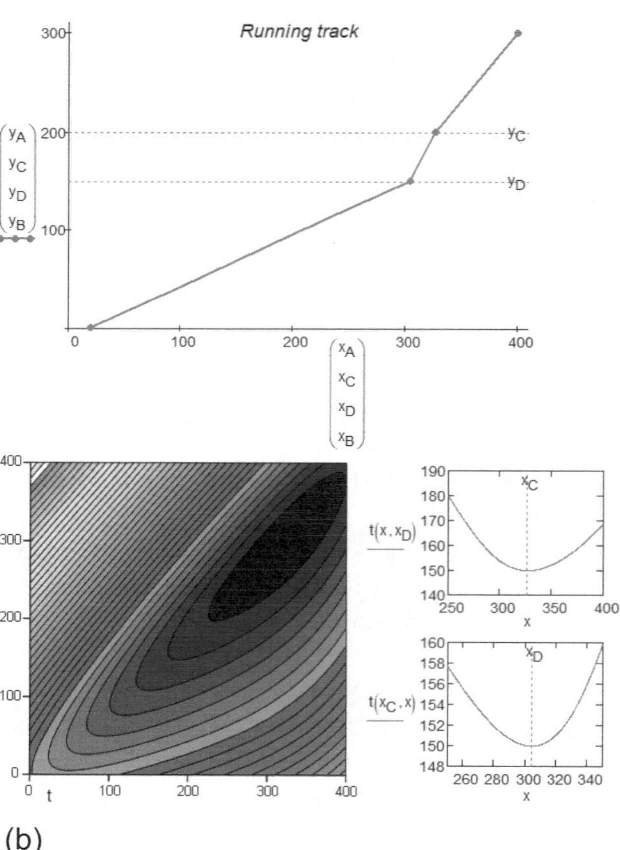

(b)

FIGURE 20.1 The task of running on lawn, arable land and asphalt: optimization: (a) solution, (b) plots.

to the two optimization variables. Section curves here seem to be superfluous, but they will be needed when there are more than two optimization variables: a person, for example, runs along a lawn, arable land, a swamp and then along asphalt (three, not two-phase boundaries, but three optimization variables). We, by the way, already drew the curves of the cross sections of the function of four arguments, when the cost of a car with mileage was determined, see Figures 14.28–14.31 in Study 14.

The graph in the center of Figure 20.1 with a broken curve resembles "a ray of light in a dark kingdom" and makes us recall the famous principle of geometrical optics formulated in 1662 by the great Fermat: the path taken from point A to point B by a ray of light is the path that can be traversed in the least time. This path is called *optical path*. Hence the kinks of straight lines, determined by the velocities of light in different media at the interface. If one speed is divided into another (adjacent), then we get the *relative refractive index n*, numerically equal to the ratio of the sine of the angle of incidence to the sine of the angle of refraction (*Snell's law*).

Note It will be interesting to relate this optic wave principle to the microscopic molecular aspect of matter. Why do lenses and mirrors behave as they do? Why does transparent material behave that way?

For some other interesting aspects of Pierre de Fermat's life, see https://en.wikipedia.org/wiki/Pierre_de_Fermat. In 1629 he gave a copy of his restoration of Apollonius' "De Locis Planis" to one of the mathematicians there, see the previous Study and https://en.wikipedia.org/wiki/Fermat%27s_principle.

Figure 20.2 shows another approach to the solution of the problem of running along the lawn, arable land and asphalt, which consists of finding not a minimum of the function of two arguments, but the root of the system of two non-linear algebraic equations: the ratio of the sines of the angles at the two interfaces is equal to the ratio of the values of the running speeds on both sides of the borders. The sine of an angle is the ratio of the length of the opposing leg to the hypotenuse, calculated by the Pythagorean theorem.

The *Find* function involved in the solution in Figure 20.2, is also based on a special numerical algorithm that, from the first assumption, will change the values of its arguments so that the system of equations written down after the keyword *Given* is transformed into a system of identities. Verification of the correctness of the solution, as in Figure 20.1, is carried out graphically: two curves

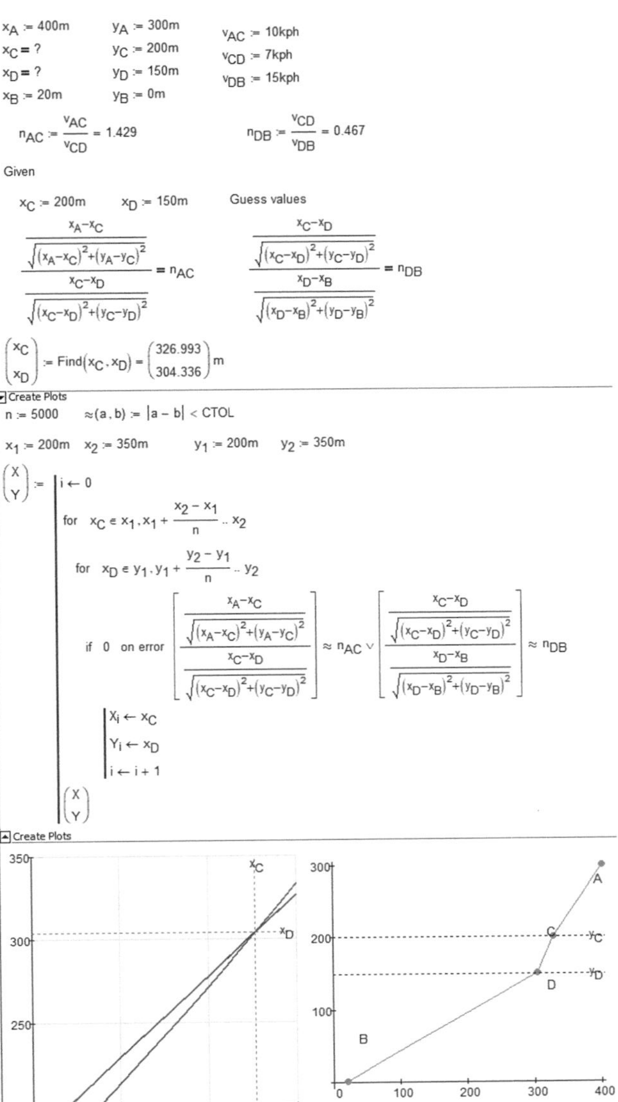

FIGURE 20.2 The problem of running on lawn, arable land and asphalt: the solution of a system of two "optical" equations.

corresponding to the two equations of the system intersect at the desired point (see the lower graph in Figure 20.2). The graph was built using the scanning method, which we described in more detail (with the program showing) in Study 19, see Figure 19.15. The scanning method is also shown in Figure 20.11.

Note The speed of light in the atmosphere and, consequently, the refractive index depends on the density of the air and its temperature. Because of this, the rays of light in the sky can "travel" through very intricate routes. Hence the mirages that can be seen in the desert

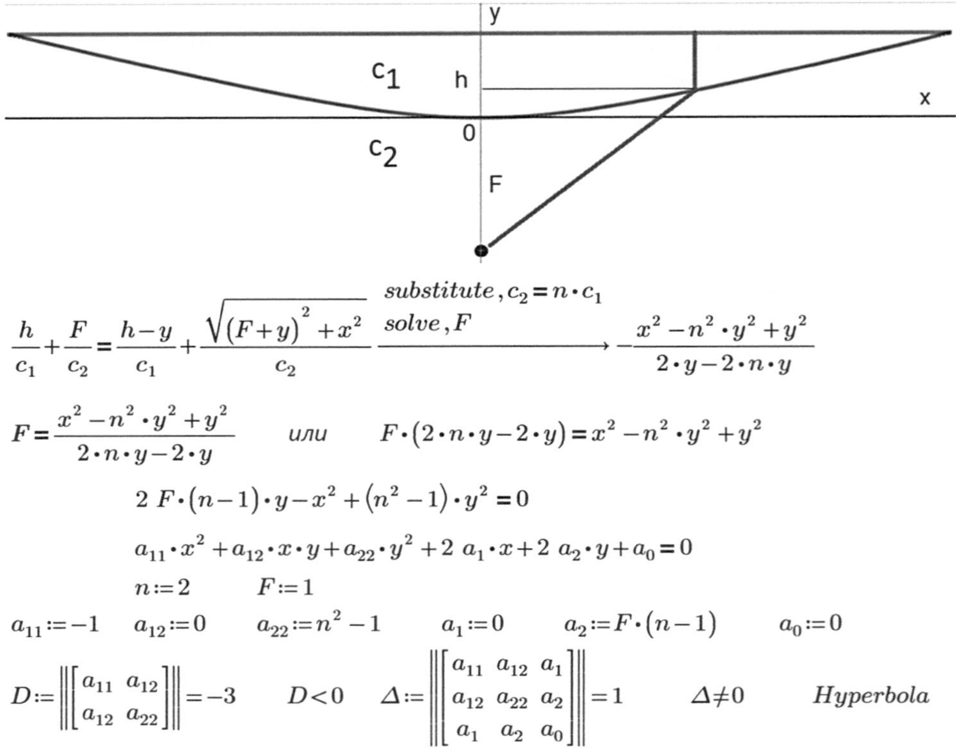

FIGURE 20.3 The derivation of the expression for a lens based on the principle of tautochronism.

or over the sea surface, where the observer on the shore sees a ship not on the surface of the water, but high in the sky, and upside down.

Consider the final moment of the traveling from home to school.

Schoolchildren enter the school gate and head to their doors, which resembles the action of a *lens* that focuses light. Let's model this optical device based on one more important principle of geometrical optics, directly following from the Fermat principle, the principle of *tautochronism*: the optical paths of light rays from a point source to its image are the same and the light spends the same time on the passage of these *optical paths*.* Mathcad can solve such, and more complicated, problems about the lens elegantly and simply, see Figure 20.3.

The plane-convex lens (see the diagram in Figure 20.3) has a thickness h in the center and a *focal length F*. The lens is made of a transparent material with a relative refractive index n. The question is: what form should the lower surface of the lens have, so that a parallel ray of light converges to the *focus*? The left-hand side of the equation in Figure 20.3 under the lens circuit, which is solved with respect to the variable *F*, is the time of passage of the light beam along the Y-axis through the glass with the speed c_1 (the first summand) and through the air with the speed c_2 (the second summand). The right part is the time of passage of the light beam no longer through the center of the lens, but in an arbitrary place, located at a distance x from the center. There are also two terms for glass and air. The solution is transformed so that it is possible to isolate the coefficients a_{11}, a_{12}, a_{22}, a_1, a_2 and a_0, forming the expression for the second-order equation. According to these coefficients, the invariant *D* and the determinant Δ are calculated, the values of which indicate that we have obtained a *hyperbola*, not a parabola or an ellipse: the value of *F* is greater than 0 (1), and the value of n is greater than 1 (2). We, by the way, have already investigated the hyperbola as a plane curve of the second-order with the help of the invariant and the determinant in Study 9, when we considered the path of a comet, see Figure 9.18 in Study 9.

In the problem in Figure 20.3, Mathcad helped us solve the initial lens equation for the variable *F* and

* The general principle of tautochronism is formulated as follows: the optical length of any beam between two wave fronts is the same. We have clarified this principle for the calculation in Figure 20.3, where a wave front is a point-focus, and the second is a flat (upper) surface of the lens.

calculate the invariant D and determinant Δ. The rest (transformation of the equation to the standard form of a plane curve of the second-order) was done manually. Such a combination of computer and human efforts is a common practice in working with mathematical programs, which allows the optimal solution of problems.

It is known that a hyperbola possesses such a "reflective" optical property: a ray emanating from a light source located in the focus of the first branch of a hyperbola, after reflection from the second branch of the hyperbola, is directed as if it originates from the focus of the second branch of the hyperbola—see Figure 20.4, showing this property. (In this figure, the hyperbola image is rotated 90 degrees and the coordinate origin is shifted as required by the canonical form of the hyperbola equation.)

Figure 20.4 shows the construction of two branches of the hyperbola according to its canonical equation, including the two coefficients a and b, along which the value of the variable c is calculated and the positions of the foci F_1 and F_2 are determined. The ray from the point F_1 (focus of the right branch of the hyperbola) after reflection from the left branch of the hyperbola is directed as if it originates from the point F_2 (focus of the left branch of the hyperbola).

The calculations in Figure 20.3 give grounds to speak about the "refracting" optical property of the hyperbola: a parallel beam of light, passing through a plane-convex lens, converges at a point if a hyperbola "sits" in the

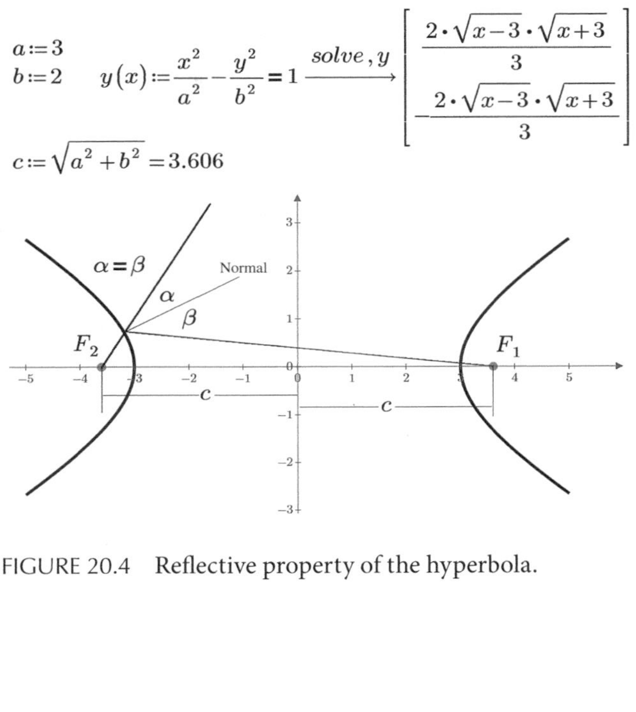

$$a := 3 \qquad b := 2 \qquad y(x) := \frac{x^2}{a^2} - \frac{y^2}{b^2} = 1 \xrightarrow{solve,\,y} \begin{bmatrix} \frac{2\cdot\sqrt{x-3}\cdot\sqrt{x+3}}{3} \\ -\frac{2\cdot\sqrt{x-3}\cdot\sqrt{x+3}}{3} \end{bmatrix}$$

$$c := \sqrt{a^2 + b^2} = 3.606$$

FIGURE 20.4 Reflective property of the hyperbola.

convex surface of the lens. Our lens is not just flat-convex, but flat-hyperbolic.

Note The "reflective" property of the hyperbola (Figure 20.4) is described in almost all the manuals (paper and electronic), where information on this second-order curve is given, although these directories are silent about the "refractive" property of a hyperbola (Figure 20.3). This can be explained by the fact that the law of light reflection is simple and understandable—the angle of incidence is equal to the reflection angle (see below). With the refraction of light, everything is not quite so simple (see Study 7). Let us repeat this once. The refractive index of light n was considered as a constant that does not depend on the wavelength of the light beam (the phenomenon of chromatic aberration), nor on the intensity of the light flux, nor on the position of the ray in the refractive substance. But it's time to remember the glass prism that decomposes white light into color components and helps, for example, to determine the composition of a substance by spectroscopic methods. This allowed, for example, helium to be found on the Sun first, and only then in the atmosphere of the Earth. It must always be remembered that real lenses reduce the beam of light not to a point, but to a kind of rainbow cluster of light energy, with which children might play on sunny days, burning out all sorts of figures on a tree. Heroes of the Jules Verne novel *The Mysterious Island*, for example, created fire with the help of two glasses taken from a clock. They filled this impromptu lens with water and clamped its edges with clay. And so an incendiary glass was made, which concentrated the sun's rays on an armful of dry moss and ignited it. Why is there little mention of hyperbola in their relationship with lenses? Because our discussion applies to huge non-spherical aberrations (chromatic aberration, coma aberration, etc.) for such lenses of large diameters ("lens apertures"). At small lens diameters, the lens surface is approximated by the first quadratic term and consequently the lens has the shape of a sphere.

In Study 7 we solved the problem of the lens through a numerical solution of the differential equation, see Figure 7.37. In Figure 20.5 this problem is supplemented by the analytical solution formula.

The solid curve in the graph of Figure 20.5 is not one curve, but three, merged into one. The first curve *y(x)* displays the numerical solution of the problem by

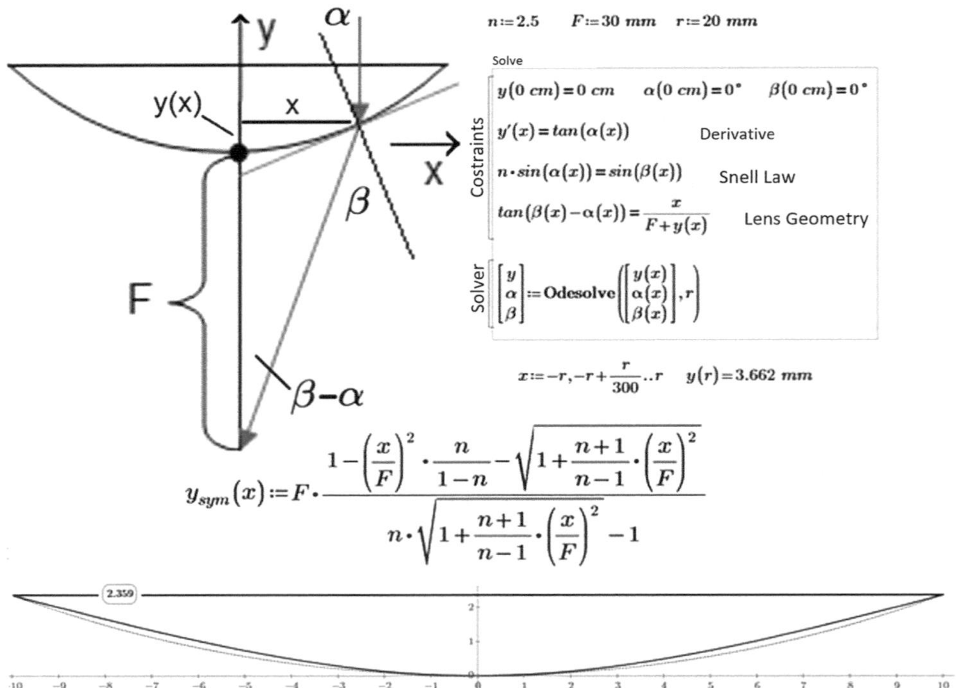

$$y_{sym}(x) := F \cdot \frac{1-\left(\frac{x}{F}\right)^2 \cdot \frac{n}{1-n} - \sqrt{1+\frac{n+1}{n-1}\cdot\left(\frac{x}{F}\right)^2}}{n\cdot\sqrt{1+\frac{n+1}{n-1}\cdot\left(\frac{x}{F}\right)^2}-1}$$

FIGURE 20.5 Numerical and analytical solution of the lens problem (the gray curve is an arc of a circle).

means of the *Odesolve* function built into Mathcad, the second is the formula (hyperbola) found by the solution in Figure 20.3, and the third $y(x)_{sym}$ is the analytic (symbolic) solution of the system of differential algebraic equations shown in Figure 20.5. This "terrible" formula for the function $y(x)_{sym}$ was derived by a Mathcad user from Holland—the homeland of the great Christian Huygens, who made a great contribution to the development of optics. This person had an good old 11th version of Mathcad with a symbolic engine from Maple (and not from MuPAD), which allowed him to solve the problem. And, of course, his intellect and skill helped. It was necessary to find the asymptotes of the continuous "numerical" curve in Figure 20.5, i.e. to find a cone in which the lens will fit in infinitely increasing its diameter with a fixed focus. Still, by the way, it was not known that this was a hyperbola. Finding the limit of the expression $y(x) / x$ turned out to be impossible, because the "numerical" function $y(x)$ is defined only in a specified range from the center of the lens to its edge. And in general, the function $y(x)$ is not a "real" function, but a kind of pseudo-function created by interpolating tabular values generated by a numerical method for solving an ordinary differential equation. The problem of the lens cone was posted on the Mathcad user forum. The question was replied by Luc Meekes, who solved the problem radically, finding an analytical solution, and the resultant expression determined the required asymptotes. The formulas in Figure 20.3 and 20.5 differ in appearance, but this is the same dependence, as we have already seen, by constructing matching graphs.

Note We flattened the lens, intentionally turned upside down (see Figure 20.3), to simplify the task. On the book's website, the reader will find a solution for the lens, turned by the convex side upward, to a parallel beam of light rays. This is usually done by burning, for example, on a sunny day on a tree with a flat-convex lens. But with this position of the lens, it will be necessary to consider the refraction of the light beam not once, but twice, at the boundaries "air-glass" and "glass-air".*

In the case shown in Figure 20.3, refraction of light occurs only on the lower surface.

And one more clarifying point. We consider not a three-dimensional, but a flat problem. The shape of the

* In the last century, in many families, there were lenses with four boundaries for the separation of media "air-glass" (or rather, plexiglas), "air-liquid", "liquid-glass" and "glass-air". They were placed in front of TV sets, the screens of which in those days were slightly larger than a postcard (see http://www.mintorgmuseum.ru/vocabulary/416). These lenses were filled with water or glycerin, which has a refractive index higher.

lens is the surface obtained by rotating the found flat curve around the y-axis. This point is often overlooked.

The lens, which should collect a parallel beam of light in focus, as it is, does not have a spherical shape, it is, as the optics say, *aspherical* (aspherical means non-spherical). Aspherical lenses are usually made as follows: the rotating spherical blank is ground to obtain the desired aspherical shape. But, nevertheless, almost all school and university optics, as part of the discipline "Physics", use the assumption that the lenses have spherical surfaces, and the learning process is not aimed at mastering the basic concepts of optics based on general physical laws, but on output or memorization of ready-made formulas for the calculation of focal lengths, diopters, magnification/reduction factors, aberrations and other purely professional optical characteristics and phenomena. School physics simply considers a priori the surfaces of lenses that ensure the convergence of a paraxial beam at a point called the focus. The problem of finding the shape of this surface for a wide beam is simply not posed. And this is understandable: when the science of geometric optics developed and the corresponding training courses were created, there were no convenient, simple and accessible means of solving the algebraic and differential equations. Now they exist and this allows us to change the methodology and content of the discipline "Optics".

The optical principle of Fermat can be applied to another important law—*the law of light reflection*. Figure 20.6 shows an analytical solution to the problem of minimizing the time for the light beam to travel from point 1 to the reflecting surface (point 2) and then to point 3. An objective function t with argument x (horizontal indentation to point 2 from point 1) is created, in which the value of the argument is sought, for which the derivative is zero. After all, if the function is smooth and continuous, then the derivative at the point of its minimum is zero. The calculations in Figure 20.6 show that the minimum of the function *t(x)* occurs when the tangent of the angle of incidence α is equal to the tangent of the reflection angle β. Consequently, the angle α is equal to the angle β. By the way, reflection and refraction "go hand-in-hand": light, falling, for example, on the glass surface is partially reflected, and partly clear with refraction deep in the glass, and the ratio of these parts depends on the angle of incidence. (Often mentioned is the critical angle, the angle for which the refracted ray does not cross the medium interfaces: $\beta_c = Arcsin(n_r/n_i)$. For any angle of incidence greater than the critical

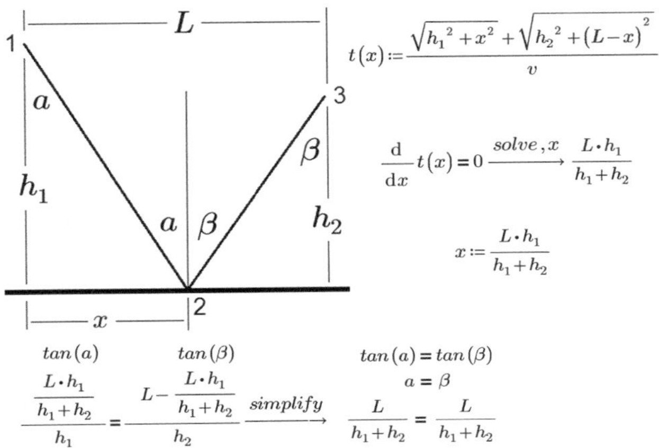

FIGURE 20.6 Proof of the law of reflection of light from the Fermat principle.

angle, light will undergo total internal reflection. Is it possible to derive it from Fermat principle?)

One of the inventors of the laser, the Nobel Laureate Charles Townes admitted that the creation of this device was inspired by the English translation of the novel by Aleksey Tolstoy *The Hyperboloid of Engineer Garin* (https://en.wikipedia.org/wiki/The_Garin_Death_Ray), translated as *The Garin Death Ray*. In Wikipedia, you can read that "the more correct name for Garin's device is the paraboloid. Tolstoy knew about this but chose the word 'hyperboloid' because of the more impressive sound." However, this device could be made not only of mirrors, but also of lenses—plane-hyperbolic, for example. Then the name of the novel and the device is completely justified. More information about this can be found at http://www.numbernautics.ru/artefact-real/838-giperboloid-inzhenera-garina-byl-i-mify.html.

We have already modeled a hyperboloid, or rather, a hyperbola, depicting the lower edge of the lens, Let us also model a paraboloid, a parabolic concave mirror (see Figure 20.7), using the already mentioned principle of tautochronism.

Figure 20.7 actually repeats the calculations shown in Figure 20.3, but not for a focusing plane-convex lens, but for a focusing concave mirror. The calculations in Figure 20.7 gave not a hyperbola, but a parabola, which has another non-optical geometric interpretation. If a straight line is drawn on the plane and a point is placed near it, then the parabola is the geometric locus of points equidistant from the given line (called the parabola *directrix*) and a given point (called the focus of a *parabola*). Hyperbolas, parabolas and ellipses, which are not considered in this

$$F+h=\sqrt{(F-y)^2+x^2}+h-y \xrightarrow{solve,y} \frac{x^2}{4\cdot F}$$

$$y-\frac{1}{4\,F}\cdot x^2=0 \qquad F:=1$$

$$a_{11}:=-\frac{1}{4\,F} \quad a_{12}:=0 \quad a_{22}:=0 \quad a_1:=0 \quad a_2:=\frac{1}{2} \quad a_0:=0$$

$$D:=\left|\begin{bmatrix} a_{11} & a_{12} \\ a_{12} & a_{22} \end{bmatrix}\right|=0 \quad D=0 \quad \Delta:=\left|\begin{bmatrix} a_{11} & a_{12} & a_1 \\ a_{12} & a_{22} & a_2 \\ a_1 & a_2 & a_0 \end{bmatrix}\right|=0.063 \quad \Delta\neq 0$$

$$f(x,y):=a_{11}\cdot x^2+a_{12}\cdot x\cdot y+a_{22}\cdot y^2+2\,a_1\cdot x+2\,a_2\cdot y+a_0$$

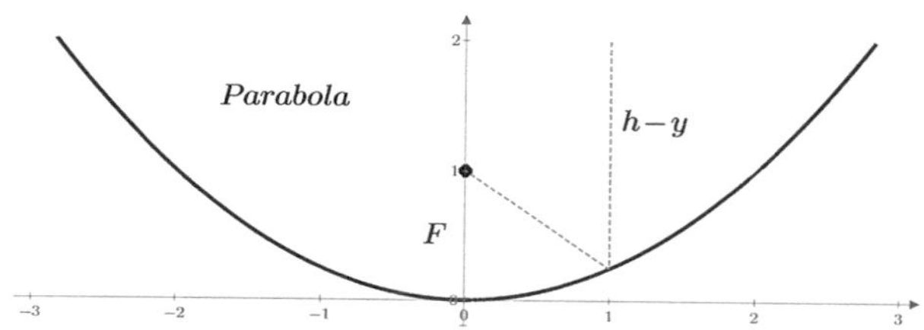

FIGURE 20.7 Derivation of the parabolic mirror equation.

Study, are called *conic curves* since they are obtained by cutting a straight circular cone with a plane. By the way, celestial bodies move along elliptical, parabolic and hyperbolic orbits (see Study 9), followed by mirror and lens telescopes. In the faculty of mathematics much attention is paid to the transformation of these curves to the canonical form, which has not six (see Figures 20.3 and 20.7 above), but only two (hyperbola and ellipse) or even one (parabola, circle, as a special case of the ellipse) coefficients.

Note A parabola, or rather, a paraboloid (the surface obtained by rotating the parabola around an axis perpendicular to the directrix and passing through the focus) can be seen not only on the walls of houses where satellite TV is watched, but also in winter on ice slides. The fact is that, once upon a time, one Soviet plant manufacturing such antennas for the needs of the military was commissioned to manufacture and convert some products. The management of the plant thought for a short time and decided to attach two leather hinges to this plate (antenna). It became very cold. And here is another possible consumer good with a "parabola inside", the idea of another "art object": a glass or a porcelain vase for fruit in the form of a paraboloid, with a round stand-head and the stem piercing in the center of the paraboloid and ending with the handle in the focus. Such a vase will look good in a mathematician's house—lest they will have nothing to talk about with the guests at the table. It can be painted not by traditional patterns, but by mathematical calculations related to the parabola.

The shape of the focusing mirror, like the shape of the lens, can also be found through solving a non-algebraic rather than a differential equation, see Figures 7.34, 7.35 and 7.36 in Study 7.

> Optics are present not only in textbooks and problem books on physics, but also in fiction. Let's remember not only Jules Verne (see above), but the Pushkin's applauses tide (https://en.wikipedia.org/wiki/Eugene_Onegin, http://www.gutenberg.org/files/23997/23997-h/23997-h.htm): Onegin enters,
> And goes over feet among the chairs,
> The double lorgnette he's aiming
> To lodges with unknown ladies.

The Krylov's fable "*Monkey and glasses*" (https://ruarrijoseph.com/publikacii/92118-basnya-krylova-martyshka-i-ochki-soderzhanie-i-moral-analiz.html). Yes, there is far to go! The eyes through which a person receives the bulk of information about the world around him/her (including, for example, reading book) are nothing more than perfect optical instruments, which we often correct and amplify with man-made optical devices: monocle, lorgnette, pince-nez, glasses, a tube, binoculars, periscope, microscope, telescope and so on. Combining mathematics, physics and literature (basic school subjects) with modern information technologies, it is possible to solve successfully and, most importantly, with interest, fairly complex optical problems, while simultaneously studying the laws of physics.

Modern computer facilities allow you to abandon many assumptions and simplifications and to make calculations for optical instruments more accurately. This can be done not only with the help of specialized programs for the calculation of optical systems (TracePro, OPTIS, LightTools, etc.), but also in Mathcad, as well as through Internet websites and professionals working in professional forums. And it is both possible and necessary to begin this study of optics with the learning of ready-made formulas, often incomprehensible because of their assumptions and simplifications, which are not spoken about, but from the generation of basic optics on the computer, then proceeding to simplified formulas. This what we have tried to do in this Study.

The author is grateful to Sergey Fedorovich, teacher of physics at the MPEI and Aleksey Sokolov, teacher of physics of the Lyceum № 1502 at MEI, for help in writing this Study.

MAGNIFIER FOR READING PHYSICAL AND MATHEMATICAL REFERENCE BOOKS

In this Study we have shown that a plane-convex lens must be plane-hyperbolic to collect a parallel beam of light into an optical point-focus (see Figure 20.3). But the hyperbola has a pair of its own geometric foci, and we discussed this in Study 19, see Figure 19.4 in Study 19. How do these two focal points of the hyperbola correlate, the optical lens and one of the geometric ones. In the parabolic mirror these foci coincide, see Figure 20.7. But can they also coincide in a hyperbolic lens?

Let's solve this problem. The hyperbola is given by the coefficients a and b entering the canonical equation $x^2 / a^2 - y^2 / b^2 = 1$. The left branch of the hyperbola (see Figure 20.8, we rotated the lens by 90°) forms a convex lens surface of thickness h. What is the refractive index of the material of such a lens, so that the optical focus coincides with the geometric focus of the hyperbola?

Figure 20.8 shows an attempt to solve this new problem based on the same principle of tautochronism. The equation of time for the passage of light from two optical paths is taken from Figure 20.3, but the lens itself, we repeat, is rotated 90° so that we can work with the canonical form of the hyperbola: $x^2 / a^2 - y^2 / b^2 = 1$. In the equation in Figure 20.8, the focal length is set equal to the sum of the two important parameters of the hyperbola $a + c$. The equation after the corresponding changes of variables is solved with respect to the variable n, the refractive index of the lens material. A formula is obtained, which includes the variable y, and which, in principle, should not affect the value of n. Numerical investigation of the function $n(a, b)$ shows that the refractive index of the material of our lens does not depend on the value of y, and in the case of the lens being formed by an isosceles hyperbola, where $a = b$, this coefficient is equal to the square root of two. In calculations based on the formula for $n(a, b)$, the value of y can be taken as a unit or per unit meter if the units of measurement in the formula are taken into account. But a further analysis of the formula in Figure 20.8 has shown that it coincides numerically with the formula for another important characteristic of the hyperbola, *eccentricity* $\varepsilon = c / a$. Figure 20.9 shows two matching surfaces constructed from the functions $n(a, b)$ and $\varepsilon(a, b)$.

Starting from the calculations in Figures 20.8 and 20.9, we can once again, and more accurately, formulate the *second optical property of the hyperbola*. A parallel beam of light, passing through a plane-convex lens, converges at one point if the convex surface of the lens is determined by one of the branches of the hyperbola. If, at the same time, the refractive index of the lens material coincides numerically with the eccentricity of the

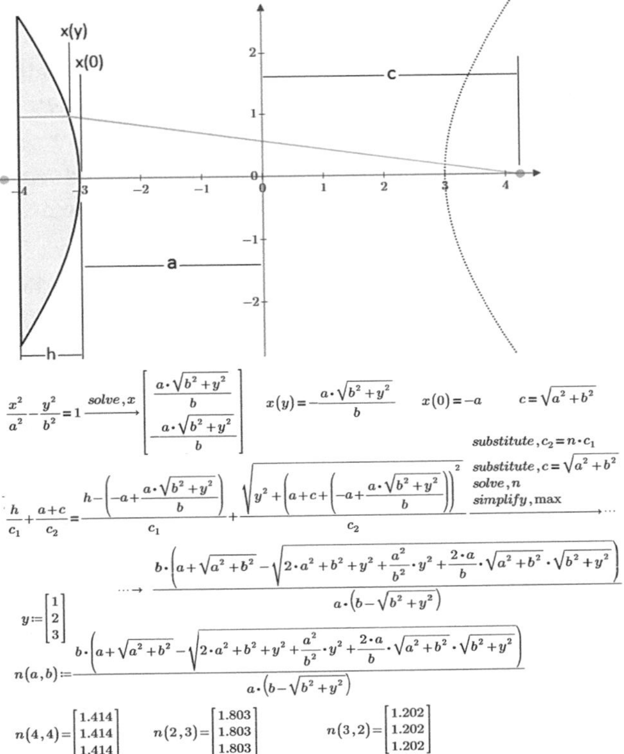

$$\frac{x^2}{a^2}-\frac{y^2}{b^2}=1 \xrightarrow{solve,x} \begin{bmatrix} \frac{a\cdot\sqrt{b^2+y^2}}{b} \\ -\frac{a\cdot\sqrt{b^2+y^2}}{b} \end{bmatrix} \quad x(y)=-\frac{a\cdot\sqrt{b^2+y^2}}{b} \quad x(0)=-a \quad c=\sqrt{a^2+b^2}$$

$$\frac{h}{c_1}+\frac{a+c}{c_2}=\frac{h-\left(-a+\frac{a\cdot\sqrt{b^2+y^2}}{b}\right)}{c_1}+\frac{\sqrt{y^2+\left(a+c+\left(-a+\frac{a\cdot\sqrt{b^2+y^2}}{b}\right)\right)^2}}{c_2} \xrightarrow[\substack{solve,n \\ simplify,\max}]{\substack{substitute,c_2=n\cdot c_1 \\ substitute,c=\sqrt{a^2+b^2}}} \cdots$$

$$\cdots\rightarrow \frac{b\cdot\left(a+\sqrt{a^2+b^2}-\sqrt{2\cdot a^2+b^2+y^2+\frac{a^2}{b^2}\cdot y^2+\frac{2\cdot a}{b}\cdot\sqrt{a^2+b^2}\cdot\sqrt{b^2+y^2}}\right)}{a\cdot\left(b-\sqrt{b^2+y^2}\right)}$$

$$y:=\begin{bmatrix}1\\2\\3\end{bmatrix} \quad n(a,b):=\frac{b\cdot\left(a+\sqrt{a^2+b^2}-\sqrt{2\cdot a^2+b^2+y^2+\frac{a^2}{b^2}\cdot y^2+\frac{2\cdot a}{b}\cdot\sqrt{a^2+b^2}\cdot\sqrt{b^2+y^2}}\right)}{a\cdot\left(b-\sqrt{b^2+y^2}\right)}$$

$$n(4,4)=\begin{bmatrix}1.414\\1.414\\1.414\end{bmatrix} \quad n(2,3)=\begin{bmatrix}1.803\\1.803\\1.803\end{bmatrix} \quad n(3,2)=\begin{bmatrix}1.202\\1.202\\1.202\end{bmatrix}$$

FIGURE 20.8 The refractive index of a lens with coincident optical and geometrical foci.

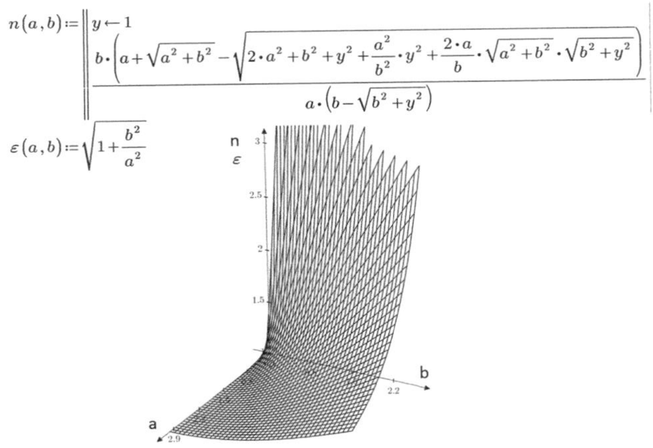

$$n(a,b):=\left\| \begin{array}{l} y\leftarrow 1 \\ \frac{b\cdot\left(a+\sqrt{a^2+b^2}-\sqrt{2\cdot a^2+b^2+y^2+\frac{a^2}{b^2}\cdot y^2+\frac{2\cdot a}{b}\cdot\sqrt{a^2+b^2}\cdot\sqrt{b^2+y^2}}\right)}{a\cdot\left(b-\sqrt{b^2+y^2}\right)} \end{array}\right.$$

$$\varepsilon(a,b):=\sqrt{1+\frac{b^2}{a^2}}$$

FIGURE 20.9 The refractive index and the eccentricity of the hyperbola.

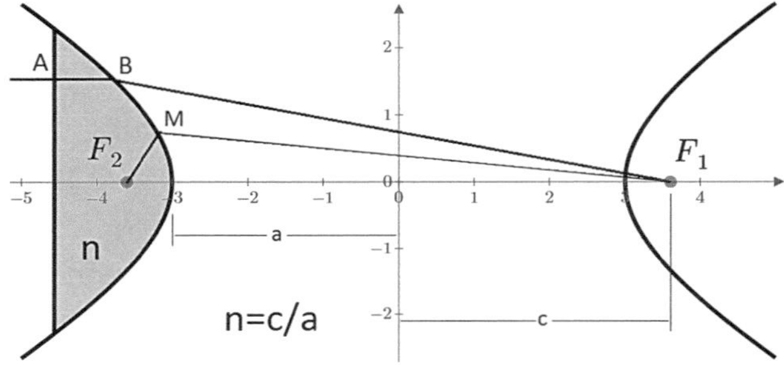

FIGURE 20.10 The second optical property of the hyperbola.

hyperbola, then the *optical* focus of such a plane-hyperbolic lens will coincide with one of the *geometric* foci of the hyperbola.

Figure 20.10 explains this new "refractive" property of the hyperbola.

On the website for this Study, the reader will find a rigorous proof of the second optical property of the hyperbola. It was obtained in this way. The hypothesis was posted on the Internet, and a nameless man from the Ukrainian city of Kharkov proved its correctness. In Study 4 we wrote that we were a little late with the discovery of a new mathematical constant. We hope that in the case of the new optical property of hyperbola, we have become pioneers.

One of the definitions of hyperbola is: the hyperbola is the geometric locus of points M of the Euclidean plane for which the absolute value of the difference of the distance from M to two selected points F_1 and F_2 (called foci of the hyperbola) is constant. In Study 19 we constructed a hyperbola according to its *focal* definition, see Figure 19.4 in Study 19.

It can be understood from Figure 20.10 that the difference between the segments $F_2 - M$ and $F_1 - M$ forms the first (left) branch, while the difference between the segments $F_1 - M$ and $F_2 - M$ is the second (right) branch of the hyperbola. If the left branch of the hyperbola forms one of the surfaces of a plane-convex lens whose material has a refractive index n equal to the eccentricity of the hyperbola c / a, then a parallel beam of light (one of its rays passes through points A – B) converges to the point F_1 coinciding with the focus of the right branch of the hyperbola.

The eccentricity (the ratio of the value of c to the value of a) is an important and determining parameter of plane curves of the second-order. If it is greater than one, then we have a hyperbola (see Figure 20.10, for example, more than one, incidentally, the refractive index of light*), and, if equal to one, a parabola, and if less than one, an ellipse (see Figure 19.22 in Study 19.) The construction of these three curves by the scanning method was described in Study 19. The construction by one focus and a directrix (a directorial, but not a focal definition of the curves) is given here.

The "extra" variable y of the function $n(a, b)$ in Figure 20.9 can be disposed of in different ways. The simplest, and we have already noted it—it is to assign a single value to the variable y, or rather the value of a unit meter, if the problem is solved using units of measurement. But it's time to remember the beginning of Study 9 with the passage on the theme "What would have happened if Newton had a computer". One of the answers to this rhetorical question is this. If Newton had a computer, then there would not be any differential calculus. However, reading about Pierre de Fermat in Wikipedia, it is said: "Regarding Fermat's work in analysis, Isaac Newton wrote that his own early ideas about calculus came directly from 'Fermat's way of drawing tangents'."

And there would be a heap of solved, and supposedly solved, problems, a lot of tables and curves on computer displays, from which any conclusions, and even more generalizing laws, would be extremely difficult—"necessity (people without a computer) is the mother of invention". If we did not have a computer, then we would be able to prove in some other way and more elegantly that the eccentricity of its hyperbola is hidden in the lens. The computer gave out "terrible" formulas, in which it is difficult to see any physical meaning.

In this book, we have repeatedly suggested ideas of objects with a hidden mathematical meaning. Here's another such idea. It is possible to produce an "isosceles" plane-hyperbolic lens from optical glass with a composition providing a refractive index equal to the root of two (the option is to use an unequal hyperbola lens with eccentricity equal to the refractive index of glass or plastic). This lens is inserted into the frame, on the case of which the foregoing calculations and explanations are engraved. A magnifying glass will be obtained for a mathematician who can use it for reading physical and mathematical handbooks with small print.

* Physicists are discussing the question of whether a material could have a refractive index less than one, or at least equal to one. How will the loupe made of such material behave?

```
N := ∞

Ans ← ["a" "b" "c" "n"]                                  = ["a" "b" "c" "n"]
for a ∈ −N..N
  for b ∈ −N..N
    for c ∈ −N..N
      for n ∈ 3..N
        if a≠0∧b≠0∧c≠0∧a^n + b^n = c^n
          Ans ← stack(Ans, [a b c n])
Ans
```

FIGURE 20.11 Fermat's last theorem.

```
N := 20

Ans ← ["a" "b" "c" "n"]                                       0  [  ⋮            ]
for a ∈ −N..N                                                 ⋮
  for b ∈ −N..N                                            1349  20 2   20  19
    for c ∈ −N..N                                        = 1350  20 2   20  20
      for n ∈ 3..N                                         1351  20 3  −20  20
        if a≠0∧b≠0∧c≠0∧a^n + b^n = c^n                     1352  20 3   20  20
          Ans ← stack(Ans, [a b c n])
Ans
```

FIGURE 20.12 The "refutation" of Fermat's last theorem.

THE THEOREM OF FERMAT AND MATHCAD

The French mathematician Pierre de Fermat (1601–1665) is known primarily not for the above-described principle of geometrical optics, but for his *Great Theorem*. In Figure 20.11, it is written not by traditional words and formulas, but by the Mathcad program.

The program in Figure 20.11 has four nested loops. We used this technique in Study 18, when the alphabetical puzzles were being solved (see Figures 18.2 and 18.4). Fermat's theorem says that for any natural number n ($n > 2$) the equation $a^n + b^n = c^n$ does not have solutions in non-zero integers a, b and c. This is what the program in Figure 20.12 says: there is a table-title ["a" "b" "c" "d"], but there is no number under the title!

But, as the reader understands, the program in Figure 20.11 could not be executed theoretically and practically. If the number N is not assigned infinity (for numerical mathematics Mathcad it is equal to 10^{307}), but any finite number, then Mathcad will disprove Fermat's theorem, see Figure 20.12.

The answers shown in Figure 20.13 are trivial—it is immediately obvious that they are incorrect, and this is due to the inaccuracy of the calculations in Mathcad. But on the Internet it is harder to find such obvious

$1782^{12} + 1841^{12} =$ 2541210258614590000000000000000000000000

$1922^{12} =$ 2541210259314800000000000000000000000000

$1782^{12} + 1841^{12} \rightarrow$ 2541210258614589176288669958142428526657

$1922^{12} \rightarrow$ 2541210259314801410819278649643651567616

FIGURE 20.13 Fermat rehabilitation (see the vertical line).

incorrect solutions or "bloopers". One of them is given in Wikipedia:

$$1782^{12} + 1841^{12} = 1922^{12}.$$

A calculator with an accuracy of not more than nine significant digits confirms this equality:

$$1782^{12} + 1841^{12} = 254121026 \cdot 10^{31}$$

$$1922^{12} = 254121026 \cdot 10^{31}.$$

But if such a test is performed in Mathcad with a higher accuracy, then Fermat will be rehabilitated, see Figure 20.13

Figure 20.12 shows that the equality $1782^{12} + 1841^{12} = 1922^{12}$ is incorrect. The equality is "true" with an accuracy of up to 13 signs of numerical mathematics Mathcad (operator =). Symbolic mathematics (operator →) gives absolute accuracy here. The "weakness" of numerical mathematics is clearly visible in Figure 20.13.

Fermat's theorem is not mentioned here by accident. This theorem of Fermat was placed without proof in the margins of one of his books. We have also put the description of the refractive property of the hyperbola without complete proof on the "margins" of this book.

ASSIGNMENTS FOR READERS

1. Find a transparent material (natural or artificial) with a refractive index closest to the root of 2. Or the root of 3. Is there any physical and mathematical meaning of the refractive index which is equal to the root of not 3, but 2.

2. Determine the parameters of the plane-convex lens, into which the equilateral hyperbola is embedded, through the solution of a non-algebraic rather than a differential equation.

3. Draw a cone in Mathcad, cut it with a plane, move this plane in the animation and get a hyperbola, a parabola and an ellipse.

4. Use Mathcad to propose a new equality of type $1782^{12} + 1841^{12} = 1922^{12}$, which can be called false only with the help of symbolic, not numerical, mathematics (see Figure 20.13).

5. "Play" on the computer with the equality $a^n + b^n + c^n = d^n$.

Optimization of Potential Energy Or

Curves of the Second-Order in Statistics

ABSTRACT

In this Study, the solution of the catenary problem is considered on the basis of the analysis of the potential energy of a mechanical system in Mathcad. Two laboratory activities at the intersection of physics, mathematics (geometry) and computer science are proposed.

Mathematics: Catenary, constrained optimization, Cartesian graph, system of linear equation.

Physics: Potential energy, thread with weights.

IT: Shortcomings of Mathcad built-in tools, the text index and the vector index.

Art: Arcs, catenary, New Year postcard.

Study website: https://community.ptc.com/t5/PTC-Mathcad/Study-21/m-p/617553.

The problem of a power cable suspended over the road (see Study 7, the chain function, the catenary) can be solved with an alternate approach, by reducing to an optimization problem.

Let's conduct a physical and mathematical computer experiment. We fix masses of equal weight on a thread so that a chaplet or a rosary is produced, see Figure 21.1. Let's see how loaded thread hangs if it is fixed at two end points. The weights will try to descend as low as possible so that their total potential energy becomes minimal. In order to prevent the weights from falling on the floor, the thread lengths from mass to mass should remain constant (our thread is inextensible and weightless). Here we see a typical *optimization* problem with *constraints*, where the *objective function* is the total potential energy of the weights to be minimized, the constraints are the constancies of the distances between the weights, and the *optimization variables* are the coordinates of the weights. Figure 21.1 shows the solution to this problem using Mathcad Prime.

So, a computational experiment! Take an inextensible and weightless (in comparison with weights) thread of length S (15 m), on which 9 beads (n) of mass 1 kg (*mass*) hang at equal distances from each other. One end of the thread is suspended at a height of 7 m (y_0), while the other at a height of 10 m (y_{10}). The horizontal distance between the suspension points is 10 m (x_{10}). How will these "beads" hang? We will solve the problem numerically, not analytically. Therefore, to all the initial values of n, S, *mass*, x_0, y_0, x_{10} and y_{10} we assign specific values (9, 15 m, 1 kg, 0 m, 7 m, 10 m and 10 m)—see the first row of the calculation in Figure 21.1. The distances here are measured in meters, not centimeters (which would be more natural for beads) because, first, it is not important, and second, the accuracy of our calculation by default is 0.001, i.e. 1 mm, if we talk about a dimensional problem where the meter is the basic unit of length. The transition to centimeters would require a change to the default tolerance—a change in the accuracy of the calculation from 0.001 to 0.00001, for example. Many people forget to do this and receive an incorrect, or rather inadequate, answer.

The potential energy is the product of the weight of the mass and its height. In the calculation, it is expressed

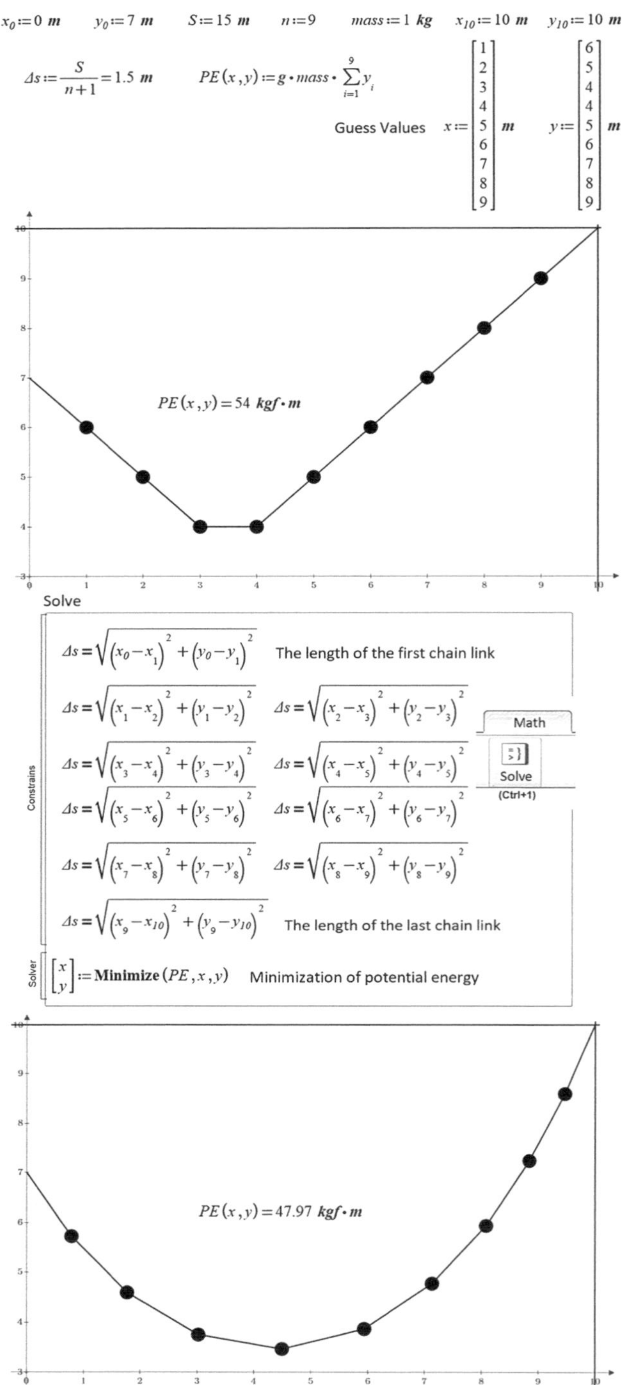

FIGURE 21.1 The solution of the problem of the weights sagging on a thread.

through a function named *PE* (potential energy). From this function, you can remove the constants *g* (acceleration of gravity) and *mass* (the mass of each weight). The solution will not change. Hence the conclusion that we have already made in this Study: a heavy metal chain and a light thread will sag equally on the Earth and on the Moon, if they have the same lengths and same coordinates of the points of suspension. However, we intentionally leave the values of *g* and *mass* in order to calculate, for example, the geometry of the chain with masses of different weights (beads, where the central bead is the largest and the others smoothly diminish toward the periphery) or a chain with a suspended massive weight (rope suspension, see Figure 7.7 in Study 7). These solutions can be found on the book's website.

The second feature of the *PE* objective function is no longer connected with the physics of the problem, but with features of its solution in Mathcad. The function *PE* has two arguments, vectors *x* and *y*, and depends only on one, on *y*. Mathcad itself requires such syntax: if optimization is carried out using two variables, then the target function must also have two arguments, no more, no less. Therefore, we had to introduce into the *PE* function an additional dummy argument *x*.* But in fact, we carry out optimization by changing the values of not two optimization parameters, but by 18—9 values in the vector *x* and 9 values in the vector *y*. The task turns out to be rather complicated. The consequence of this was that Mathcad 15 failed to solve it, but the new solver, Artelys Knitro (https://www.artelys.com/en/optimization-tools/knitro), built into Mathcad Prime, solved the problem.

We describe another computational, or rather, design nuance problem. Referring to Figure 21.1 the variables x_0, y_0, x_{10} and y_{10} are not elements of the vectors *x* and *y*, but scalar variables with text indices 0 and 10. But in the calculation there are also the vectors *x* and *y* with the same name, which store the coordinates of the 9 weights. On the website for this Study it is possible to find a solution for 25 weights. The idea of mixing of two kinds of indexes is rather doubtful—it often confuses even experienced Mathcad users.

After entering the input data and the objective function, it is necessary to give a first approximation for the optimization variables: for the two vectors *x* and *y*, the coordinates of the centers of the weights. Here it is necessary to help the computer—by providing first approximations that are close to reality—see the first graph in Figure 21.1. The center of the graph shows the calculation of the potential energy of the weights (54 kgf per meter) at the initial configuration. We want to minimize this energy while ensuring

* Mathcad has a built-in function called *time*, which has one dummy argument. This function returns the computer's time. We used it to estimate the counting time—see, for example, Figure 4.43 in Study 4.

that the constraints on the geometry of the chain are met. This minimization is performed by the *Solve Block* shown in Figure 21.1, immediately after the first graph, where 10 equalities are written in the constraint area: the lengths of the individual sections of the thread joining the weights to the chain (in the string of beads) should remain equal to Δs. The *Minimize* function uses a special numerical algorithm to change the values of the elements in the vectors x and y so that the *PE* function takes a minimum value and the constraints are satisfied. The lower graph in Figure 21.1 shows that our problem is solved: our weights have been hung on the right chain, and their potential energy has decreased from 54 to 47.97 kg-force per meter. Checking the implementation of the constraints is not difficult. It is enough to output the values of 10 expressions with square roots, written in the area of limitations of the Mathcad solver. All of them should be equal to the value of 1.5 m (Δs) to within a millimeter. It is more difficult to show that the energy of our mechanical system has reached the "bottom" of the potential well. This can be done by remembering that "there is nothing more practical than a good theory":* it is necessary to clarify what a chain function is and what its formula is, see Figure 21.2.

In Figure 7.3 in Study 7, the author's website created using the Mathcad Calculation Server technology is displayed. One can use Mathcad there without installing the package itself on a computer. It is enough that the computer has access to the Internet. On the website you need to enter the values of h_1, h_2, L and S, (7, 10, 10 and 15 meters), click the Recalculate button and get the answer, the value of the parameters of the chain function x_0, h and a, whose curve passes through two given points and has a given length. The chain function itself includes the hyperbolic cosine *cosh*. In addition, the website itself draws a catenary, in its formula h is the ordinate of the lower point of the chain, and x_0 is its abscissa. The problem is reduced to the solution of the system of three non-linear algebraic equations by the *Given-Find* block of Mathcad 15. The very same equation of the catenary (as already noted in Study 7), is derived through the solution of a differential equation obtained analyzing the forces acting on an elementary part of the chain—see, for example, the analytic calculations at https://en.wikipedia.org/wiki/Catenary. The evaluation of the correctness of the solution, based on the analysis of the potential energy of the mechanical system, is shown in Figure 21.2: weights (they are no longer 9, but 24) fill the catenary. In Figure 21.2, an additional parabola is drawn, passing through the points of suspension of the thread and through the lower point. It can be seen that the chain is sagging not along a parabola (as Galileo believed), but only close to a parabola.

Note The parameter a in the equation of the catenary, as we noted earlier, characterizes its steepness and aspect. If the value of a is reduced from a certain positive value (see Figure 21.2 to almost zero, the chain will tend to a straight line (to the string), and then (with negative values of the parameter a) turn into an arch. With this arch, by the way, it is possible to get the function *Minimize* in the calculation on Figure 21.1 replaced by the function *Maximize*, see Figure 21.3, where the

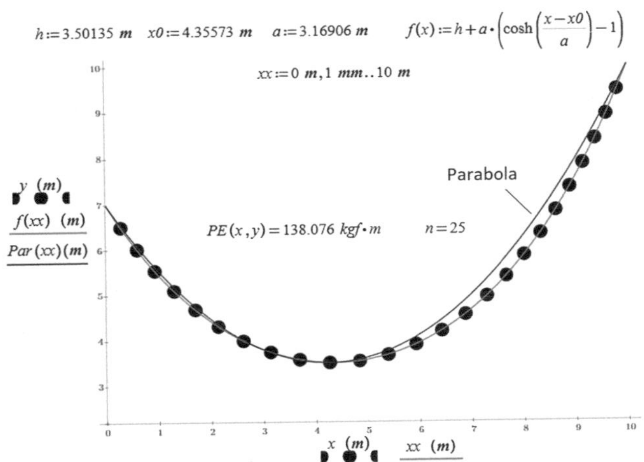

FIGURE 21.2 Checking the thread model with weights: applying a parabola and a catenary on the weights.

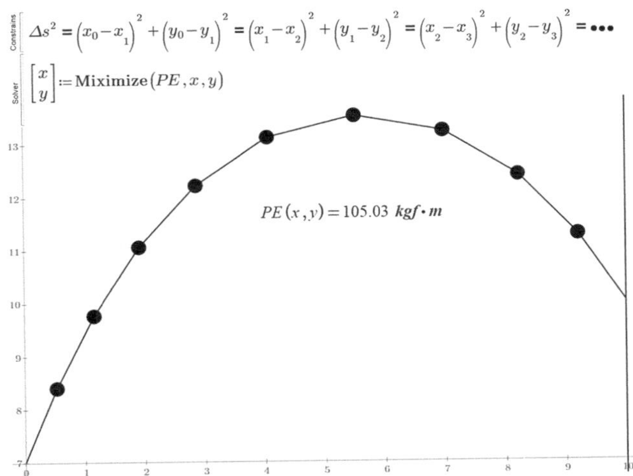

FIGURE 21.3 Arch—an inverted catenary.

* The reverse is also true: there is nothing more theoretical than a good practice.

weights will turn into balloons filled with helium, where the potential energy is not minimized, but is maximized.

In the solution in Figure 21.3 we have only one constraint not in the form of individual equations, but in the form of a chain of equality operators. The chain, so to speak, is solved by means of a chain. In addition, the square root was removed, which should speed up the calculation.

NEW YEAR'S MATHCAD-POSTCARD OR NEW MATHEMATICAL CONSTANT

At New Year, some Russian people traditionally go to the Russian baths (saunas).* The author has another, also old but more intellectual, pre-New Year's habit. He, his students and his school-age students develop and post on the Mathcad† forum (https://community.ptc.com/t5/PTC-Mathcad/ct-p/PTCMathcad) animated greeting cards with some entertaining mathematical meanings.

At the end of 2017, a New Year's postcard was published: in a square on the plane, 5 dots are thrown randomly 1000 times, and through these points a plane curve of the second-order is drawn. Anyone who sees all seven possible curves (seven is a beautiful number!), will have happiness and luck in the New Year: see https://community.ptc.com/t5/PTC-Mathcad-Questions/New-2018-Year/mp/495771.

The previous history of the Mathcad New Year's postcard is as follows. In one lesson with schoolchildren, the author showed a sagging chain (see Figure 21.4) and asked a question, namely, what function does it look like? The students answered in chorus that it was a parabola. Then the coordinates of five points (left, right, bottom and two points in the middle of the "branches" of the chain) were extracted from a photograph of the chain in the graphic editor PaintBrush, through which a plane curve of the second-order was drawn. These five points were taken from photographs of the sagging chain in different positions and with varying degrees of tension. After processing the data, it turned out that the curve of the second-order passing through the five "chain" points was, of course, not a parabola and not even one of the branches of a hyperbola, but an arc of an ellipse. The lesson ended with the description of the catenary function, which relatively accurately (taking into account measurement errors) passes through three, four, five and more points "taken" from the photograph of the sagging chain.

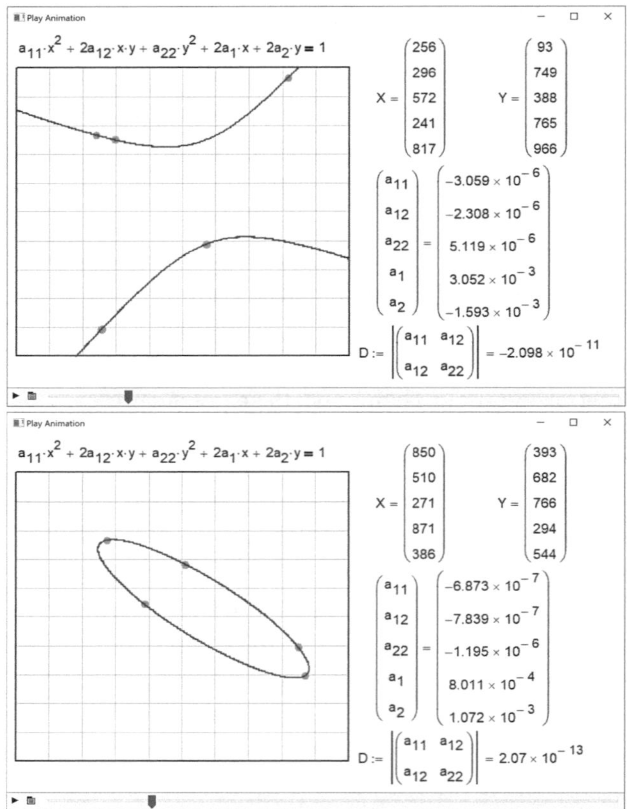

FIGURE 21.4 Hyperbola (top) and an ellipse drawn through five random points.

A question arose: which second-order curve will pass through five points randomly chosen not on the chain image, but on the whole plane. Having at hand a computer with the mathematical program Mathcad, you can try to determine this by the method of statistical tests (see Study 16).

It is known that through five points on the plane it is possible to draw the following second-order curves:

1. Two branches of hyperbola
2. Ellipse
3. Parabola (transitional case from hyperbola to ellipse)
4. Circle (special case of an ellipse)
5. Two intersecting lines (two degenerate hyperbola branches)

* See the iconic Soviet film *The Irony of Fate, or with Easy Steam!* (http://cinema.mosfilm.ru/films/film/Ironiya-sudby-ili-s-legkim-parom/ironiya-sudbi-ili-s-legkim-parom).

† It is also written MathCAD, emphasizing that the name of the package is short for mathematical CAD (Computer Aided Design) system for the automated solution of mathematical problems.

6. Two parallel straight lines
7. One straight line (a special case of 5 and 6)

Five points thrown randomly can practically (not theoretically) hold only two curves, of course: a hyperbola with two branches and an ellipse, see Figure 21.4. The author drew the remaining five curves on a New Year's card by hand, setting the desired coordinates of five points, without using a random distribution.

Figure 21.4 shows two frames of the New Year's animation with a hyperbola and an ellipse. The coordinates of the five random points (the values of the vectors *X* and *Y*) were also shown on the card, the values of the five coefficients of the equation of the curve of the second-order (a_{11}, a_{12}, a_{22}, a_1 and a_2: see the equation at the top of Figure 21.4) as well as the values of the invariant *D* corresponding to the curve obtained: $D<0$ is a hyperbola and $D>0$ is an ellipse. By the way, we used the invariant in Study 9, when we determined the shape of the trajectory of the comet's motion.

For $D=0$ (and under other additional conditions), a parabola—the transient case from an ellipse to the hyperbola—must be obtained. But this case, we repeat, took place only with artificial, and not random, assignment of the values of the vectors *X* and *Y*.

Moreover, the New Year's card recipients (for "full happiness" in the coming year) were asked to calculate how many times they saw the hyperbola and how many times an ellipse.

It turned out that a hyperbola appeared in approximately 71.84% of cases, and an ellipse in the remaining 28.16%. This was calculated, of course, not through viewing animation frames, but through a statistical computer experiment, see the corresponding Mathcad program in Figure 21.5.

Figure 21.5 shows the program for calculating the number of dropped hyperbolas (*H*) and ellipses (*E*) when throwing 5 random points 10 million times in a square of 2 by 2. At the same time (just in case!) the number of parabolas dropped (*P*) was counted.

In the Mathcad-sheet in Figure 21.5 it suffices to explain the essence of the following operators and functions:

1. The *for* statement forms a loop with the parameter *i* for throwing points into a square;
2. The function *runif* generates 5 numbers (the first argument of this function) with a random (pseudo-random) distribution in the interval from −1 to 1 (the second and third arguments of the *runif* function).
3. The *lsolve* function returns the solution of the system of linear algebraic equations (SLAE, see Study 4), the coefficient matrix for the unknowns is the first argument of the *lsolve* function, while the vector of the free terms is the second argument. The SLAE solutions are the vector of the coefficients of the required second-order equation a_{11}, a_{12}, a_{22}, a_1 and a_2).
4. The *if* function counts the "dropped" hyperbolas, ellipses and parabolas.

At the bottom of Figure 21.5, it is shown that at 10 million trials of 5 points, there were 719,484 hyperbolae (*H*), 280,516 ellipses (*E*) and no parabolas (*P*), as expected. The same approximate figures are obtained with other numbers of trials, and at different sizes of the square area where the 5 points were thrown. We will return to the form of the region below!

Many people, after receiving a greeting card, read the message, admire the picture and throw the card or postcard into a desk drawer, or even throw it away. The publication on the Mathcad user forum of the "New Year" postcard with hyperbolas and ellipses had other consequences:

```
(H E P) := | for i ∈ 1.. 1000000
           |   | X ← runif(5, −1, 1)
           |   | Y ← runif(5, −1, 1)
           |   | (a11 a12 a22 a1 a2)^T ← lsolve( [ (X1)^2  2·X1·Y1  (Y1)^2  2·X1  2·Y1 ]   (1)
           |   |                                 [ (X2)^2  2·X2·Y2  (Y2)^2  2·X2  2·Y2 ]   (1)
           |   |                                 [ (X3)^2  2·X3·Y3  (Y3)^2  2·X3  2·Y3 ] , (1) )
           |   |                                 [ (X4)^2  2·X4·Y4  (Y4)^2  2·X4  2·Y4 ]   (1)
           |   |                                 [ (X5)^2  2·X5·Y5  (Y5)^2  2·X5  2·Y5 ]   (1)
           |   | D ← | a11 a12 |
           |   |     | a12 a22 |
           |   | if(D < 0, H ← H + 1, if(D > 0, E ← E + 1, P ← P + 1))
           | (H E P)
```

H = 719484   E = 280516   P = 0   H + E + P = 1000000 $\frac{E}{H}$ = 0.389885

FIGURE 21.5 Counting hyperbolas and ellipses obtained in a square area.

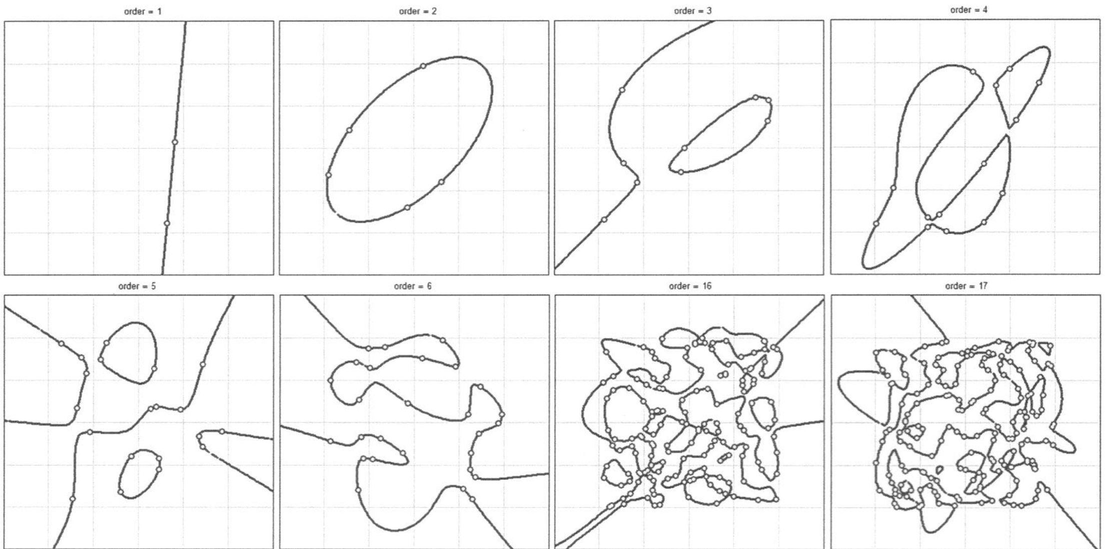

FIGURE 21.6 Curves of different orders passing through a different number of points.

1. It was suggested that a new mathematical constant of 0.2806 was discovered. It was even named jokingly (or seriously): V Points.* One colleague of the author from the department of higher mathematics asserts that he has found a way of analytically determining this constant, but he does not have time to put it all in mathematical language, since he is now completely immersed in writing a doctoral dissertation.†

2. Frank Purcell, a visitor to the forum, suggested that this constant can be determined (estimated) in another way - by solving the already well-known problem of epy IV points problem, through which two intersecting parabolas are drawn that divide the square area into certain zones . See more about this here: https://community.ptc.com/t5/PTC-Mathcad-Questions/Firecrackers-2018-or-5-th-order-curve-and-20-points/m-p/496511 and https://community.ptc.com/t5/PTC-Mathcad-Questions/Hyperbola-and-Ellipse-new-math-constant/td-p

```
N := 10^6        r := 3
(H)
(E) := | for i ∈ 1..N
       |   | while 1
       |   |   | X ← runif(5, −r, r)
       |   |   | Y ← runif(5, −r, r)
       |   |   | F ← 0
       |   |   | for i ∈ 1..5
       |   |   |    F ← F + 1  if  (X_i)^2 + (Y_i)^2 > r^2
       |   |   | break  if  F = 0
       |   | (a11)             [[(X1)^2  2·X1·Y1  (Y1)^2  2·X1  2·Y1]   (1)]
       |   | (a12)             [[(X2)^2  2·X2·Y2  (Y2)^2  2·X2  2·Y2]   (1)]
       |   | (a22) ← lsolve    [[(X3)^2  2·X3·Y3  (Y3)^2  2·X3  2·Y3] , (1)]
       |   | (a1 )             [[(X4)^2  2·X4·Y4  (Y4)^2  2·X4  2·Y4]   (1)]
       |   | (a2 )             [[(X5)^2  2·X5·Y5  (Y5)^2  2·X5  2·Y5]   (1)]
       |   | D ← |(a11 a12)|
       |   |     |(a12 a22)|
       |   | if(D < 0, H ← H + 1, E ← E + 1)
       | (H)
       | (E)

H = 701742     E = 298258     H/N = 0.70174     E/N = 0.29826
```

FIGURE 21.7 Counting hyperbolas and ellipses in a circular region.

* V. Points—this is how computer translators translate the name of this book's author: V. Ochkov.

† This man wanted to become like Pierre Fermat, who wrote down his great theorem, but did not give proof. And only a few centuries later it was found.

/495992. The fifth random point can fall into one of these zones, which determines what will be built through these five points—the hyperbola or the parabola. The calculation of the sum of the areas of these zones will give our constant. The four-point problem is described on the website http://mathworld.wolfram.com (see also E. Weisstein's article on Sylvester's Four-Point Problem).

3. A race started at the Mathcad users' forum—who will throw more points into the square and who can draw a curve of higher and higher order through them. Forum users began to throw in a square 9 points (a curve of the third order, a cube), 14 points (fourth-order), and so on to the curve of the fiftieth order (Werner Exinger)—see Figure 21.6 for some of such curves.

4. If five points are not thrown in a square, but in a circle, then a slightly different value of the constant V Points will be obtained, see Figure 21.7. What is it? Error of statistical experiment or proof of absence of the V Points constant?

By the way, schoolchildren (see the beginning of this Study) were also asked how they felt: is a plane the infinite continuation of a square or circle? Again, without thinking about the correctness of posing such a question, they answered in chorus the square. And this is understandable—in school it is "passed" the "Cartesian" coordinate system on the plane or in the volume, but polar, cylindrical and spherical coordinates are almost not touched. This lesson ended with a story about how the function describing the catenary, was discovered by Bernoulli, Huygens and Leibniz independently of each other and almost simultaneously.

VI POINTS, VII POINTS, VIII POINTS, IX POINTS, ETC. CONSTANTS. CONCLUSION

An attempt was made to discover a new mathematical constant relating to a hyperbola and an ellipse. We suggest readers try to prove or disprove the existence of the given constant, and also try to prove the presence of VI points, VII points, VIII points, IX points, etc., constants.

ASSIGNMENTS FOR READERS

1. Simulate the sagging of beads in the case where the forces of attraction are not parallel but converging to a point (as in an inhomogeneous gravitational field): for example, a long chain with weights is suspended in outer space near the Earth.
2. The beads shown in Figures 21.1 and 21.2, are strung on an extensible thread (elastic). How will they sag?
3. Determine the date with the day of the week and the exact hours when the function time, mentioned in note 1, returned a zero value.

Parallel Computing Or

Put Yourself in Order!

ABSTRACT

This Study addresses serial and parallel calculation issues, as well as how to create programs for sorting a vector and the animation of this procedure in Mathcad 15. Various laws of the distribution of random variables used in computer programs are considered. The moral problems of lining-up students according to their height in the school are considered.

Mathematics: Vector.

Physics: Chaotic motion, ordering.

IT: Sorting algorithms, parallel calculations, cycle with parameter, loop with pretest, animation, code execution speed.

Art: The art of communication in high schools and universities.

Study website: https://community.ptc.com/t5/PTC-Mathcad-Questions/My-sorting/m-p/493531.

The author of this book was chatting on the Internet with one of his former classmates, remembering good and bad school years. This classmate admitted that, for her, one of the most unpleasant moments in school were having to line-up in physical education (PE) classes (gym lessons). Girls and boys were lined up in the gym according to height and they stood in front of each other in two lines in the style of "white-top, black-bottom": in white T-shirts and black shorts. The mentioned classmate was the tallest in the class and quite full, sorry, big. In ordinary life and in ordinary clothes, it was not so noticeable but in the gym she was at the head of the line half-naked (by the standards of those times) and, as it were at the pillory. The boys looked at her and then teased her. The author also recalled an unpleasant moment in such a line-up: he had a friend of rather short stature, who always appeared at the end of the "men's" line, and the girl he liked looked at him from the female half.*

A photograph of a line-up of people according to height can be seen in Study 14 in Figure 14.20. There, the people lining up were divided into groups according to height with a step of one inch and formed a living histogram.

Yes, we must admit that the procedure for lining up according to height in school or in the army is somewhat discriminatory. In the army, too, the small soldiers are at the end of the marching column and "swallow dust". You can suggest an alternative line-up order,† for exam-

* It is said that Russian great commander Alexander Suvorov (https://en.wikipedia.org/wiki/Alexander_Suvorov) once lined up his serfs according to height in two rows: unmarried guys on the one hand, unmarried girls on the other. So there were couples whom he ordered to go to church to get married. The view of the newlyweds was of no interest to anyone. (https://lenta.ru/articles/2016/04/26/russian_slavery).

† And in general, without lining up on any basis, it is not possible to stay: schoolchildren or soldiers will argue and push, try to get closer to the right (priority) side of the ranks, or, on the contrary, "get lost in the middle of the crowd". You can suggest aligning students in the order in which they are alphabetically recorded in the class journal. But here you can also see the elements of discrimination: who is most often summoned to the board: Abramov, Yakovlev (Абрамов, Яковлев), or, conversely, some Ivanov-Petrov-Sidorov (Иванов-Петров-Сидоров) from the middle of the list. By the way, we very often resort to sorting by lining up files on the computer screen by name, type, size or time of creation. These values "in file rank" can either increase or decrease after a second click on the attribute name.

ple, by age, color of eyes, hair, or by some other attribute of schoolchildren or soldiers, which does not show any discrimination. But, one way or another, you will need to use a *sorting* procedure, which often becomes the theme of school and university lessons in informatics. The operation seems to be routine, dissected—but let's add something new to it.

Many sorting algorithms have been developed. In Wikipedia (https://en.wikipedia.org/wiki/Sorting_algorithm), you can find their names: bubble sort, sort by mixing, gnome sort, insertion sort, merge sort, binary tree sort, Timsort sort, counting sort, block sort, radix sort, sort by choice, Shell sort, comb sort, pyramidal sort, smooth sort, quick sort, introspective sort, patience sort, recursive sort, silly sort, pancake sort, permutations sort …*

These algorithms are implemented in different programming languages. We will also try to do this, but in Mathcad, which is now widely used in schools and universities.†

The basis of many sort algorithms is the exchange of numerical values in two variables. In the BASIC programming language, for example, the *swap(a, b)* operator performs this procedure. In Mathcad there is not such operator, but it is not difficult to implement it by other means.

In Figure 22.1, it is possible to see two ways to implement the *swap(a, b)* operator in Mathcad: traditional ("triangular", using an auxiliary variable) and "exotic", which many Mathcad users do not guess, not through sequential assignment, but in some parallel mode, when the assignment operators are written in a matrix form and are executed independently of each other. This technology was discussed on the Mathcad user website https://community.ptc.com/t5/PTC-Mathcad-Questions/swap-in-Mathcad/m-p/97541.

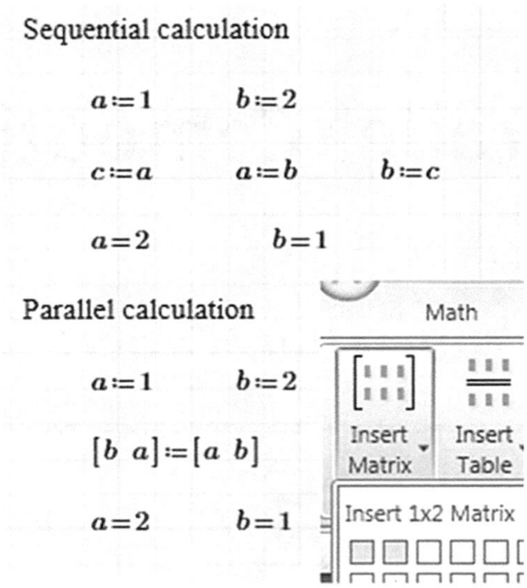

FIGURE 22.1 Two ways to rearrange the values of variables in Mathcad Prime.

If there is a permutation operator *swap(a, b)* in an explicit ("triangular") or in a matrix form, then it is easy to write a function program in Mathcad (see Figure 22.2), which probably implements the most primitive sort algorithm, whose essence is as follows.

```
MySort(V) := | Flag ← 1
             | while Flag
             |   | Flag ← 0
             |   | for i ∈ 1..last(V)
             |   |   | if V[i-1] > V[i]
             |   |   |   | [V[i] V[i-1]] ← [V[i-1] V[i]]
             |   |   |   | Flag ← 1
             | V
```

$$V := \text{runif}(10, 150, 200) = \begin{bmatrix} 163 \\ 150 \\ 190 \\ 161 \\ 178 \\ 156 \\ 188 \\ 177 \\ 172 \\ 185 \end{bmatrix} \qquad MySort(V) = \begin{bmatrix} 150 \\ 156 \\ 161 \\ 163 \\ 172 \\ 177 \\ 178 \\ 185 \\ 188 \\ 190 \end{bmatrix}$$

FIGURE 22.2 The simplest sort of a vector (Mathcad Prime).

* In his student years, the author worked part-time in a fresh food market. There the simplest work was considered ... sorting lemons. In those days (Soviet time) lemons were the only fruits that were sold not by weight, but by piece (to serve with tea) for 25, 30, 35, 40 and 45 kopeks or something like that: the larger the lemon, the more expensive it naturally was (is). On the vegetable bases they were sorted like that—they were pushed into round holes with different diameters on a special sorting table. Lemons fell into separate boxes, which were then transported to different stores, so that there was no re-sorting. The author in those years, looking at these boxes with sorted lemons, recalled the university course of mathematical statistics with its normal distribution (see for example Figure 14.35 in Study 14). One joke of those times: "Where do you work?" "At the fresh market." "Are you tired?" "Very!" "Draw boxes?" "No, I sort lemons—my head gets tired!"

† Incidentally in Mathcad, there is a sort function for this work, but for now we will forget about its existence. In addition, the Mathcad built-in functions *csort* and *rsort* allow sorting of the matrices by columns (c) or rows (r) by specifying the sort key, column or row number, of the matrix.

Schoolchildren arbitrarily line-up (in Figure 22.2, this is done by the *runif* function, which returns a vector with ten elements and random numbers in the interval from 150 to 200, the possible height of a person in centimeters), and the physical education teacher (gym teacher) passes along the line from left to right (nested *for* loop) and swaps pairs of schoolchildren if the left school student (V_{i-1}) is taller than the right (V_i). This procedure is performed until (*while* loop) the next gym teacher passing along the line reveals the disorder, and the *Flag* variable (a sign of the end of the sorting) is equal to zero.

Mathcad 15 has convenient animation tools (see Study 8), which can be used to visualize the sorting process. Figure 22.3 shows how in the program in Figure 22.2 three additional operators were inserted (they are within the framework), which allowed to show in the animation the sorting of a vector: the bars on the diagram are sequentially swapped, forming the correct line. At the same time, the next "high column" moves to the right, as if pushing and passing the baton to a column with a higher growth. This animation is posted on the website for this Study.

The system variable *FRAME*, which automatically changes its value from 0 (the initial vector is drawn in the form of chaotic bars) to 999 (the sorted vector, part of which is shown in Figure 22.3) controls the animation. Before creating the animation, it is possible to change

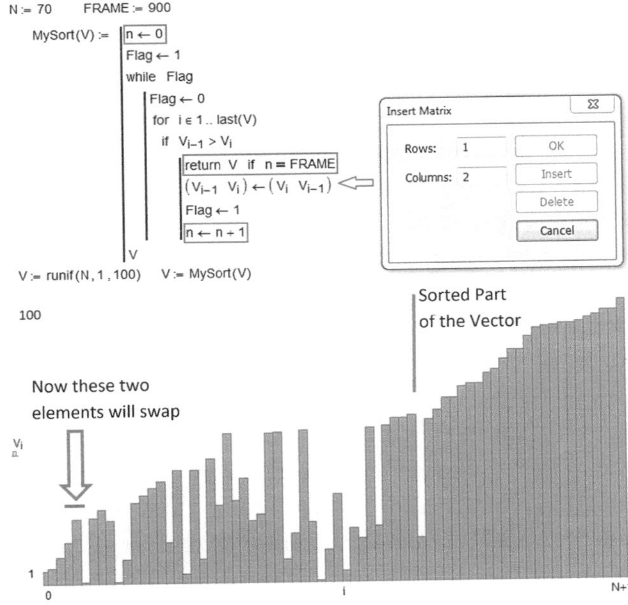

FIGURE 22.3 Creating an animation of the simplest sort algorithm (Mathcad 15).

manually the value of the system variable *FRAME* and to see the changes in the graph, in the bar chart. In this way, the program is debugged before the animation. In the program in Figure 22.3 a counter of the permutations of neighboring elements of vectors *n* is inserted. If it becomes equal to the frame number of the *FRAME* animation, the program operator *return* gives a partially sorted vector, which is plotted as a bar chart in the graph.

The nature of the animation, one frame of which is shown in Figure 22.3, could be changed drastically if at the end of the *if* operator's block (after the operator *Flag* ← 1), the *break* statement is written. It interrupts the execution of the *for* loop and transfers the control to the program: the gym teacher, rearranging the next couple of schoolchildren, does not go further with the checks along the line, but returns to its end* and proceeds to the *while* loop. With the *break* operator, the sorted vector elements will be grouped not on the right (Figure 22.3), but on the left flank of the line, but the final result will be the same: on the left will be the lowest, and on the right the highest. This animation is also available on the website for this Study.

The sort program in Figure 22.3 uses a variable number of *while-for* nested loops and this number (it is stored in the variable *n*) depends on the type of the source vector. If this program is "shoved" with the already sorted vector, then the nested *while*-for loop will be executed the minimum number of times equal to the vector size (in Figure 22.3, this is 70 times as indicated by the first operator). If this program is already given a sorted but "inverted" vector (this "inversion" is made by the built-in Mathcad function *reverse*), then the nested *for* loop will be "dull",† permuting the pairs of elements of the vector the maximum number of times.

But in Mathcad, it's easy to create a very simple program that sorts any source vector in a single cycle. In this case, the number of executions of the cycle block in

* In Mathcad, as in Pascal, there are three "interrupt" operators: return, break and continue. The first of them interrupts execution of the entire program and returns its operand (see Figure 22.3), the second operator (it is without the operand) interrupts execution of the for or while loops and transfers the control of the program to its end. The third operator also interrupts the execution of the loop but passes control to its beginning. By the way, very few people (even those completely familiar with the English language) know what the break means in the word breakfast. A breakfast is an *interruption* of the post, the breaking of which takes place on Easter or Christmas.

† If a "dull" monkey receives a peeled banana, then it will try to clean it again. So our program in Figure 22.3 will once again try to sort the already sorted vector.

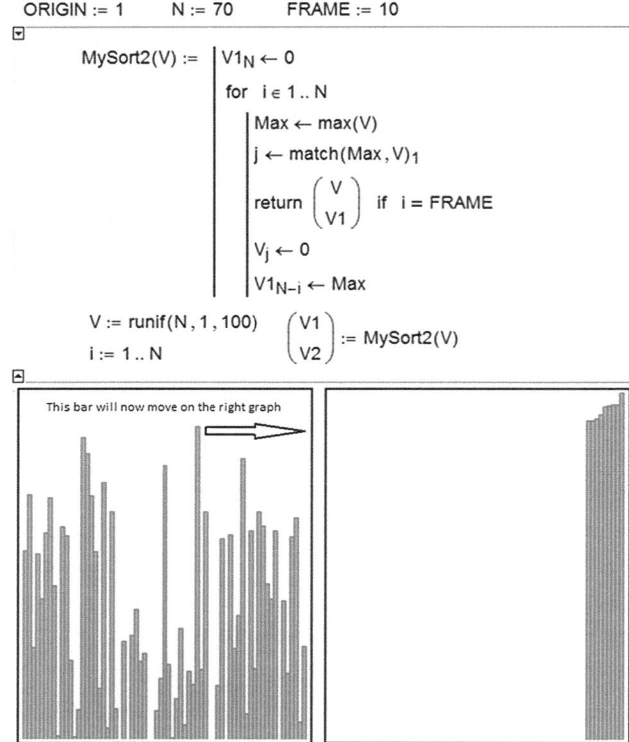

FIGURE 22.4 Creation of animation according to the second sort algorithm (Mathcad 15).

such a program will be equal to the number of elements in the source vector, see Figure 22.4.

The algorithm of the program in Figure 22.4 is that the gym teacher assesses the unsorted line of schoolchildren with a look, finds the tallest one and sends him to the left flank in a separate formed line. This operation is repeated N times, where N is the number of elements in the source vector. The bar charts in Figure 22.4 show that nine "tall" elements (see empty places in the left diagram) are transferred to the formed vector from the source vector, and another "tall" element is waiting for the next transfer. The program in Figure 22.4 executes this simple algorithm, working with two specific functions of Mathcad:

max—returns the maximum element of an array (in our case, the vector);

match—returns the number or the numbers of elements of an array of a given value (this function produces a vector with one or more elements, consequently in the program in Figure 22.4 this function is written with a single index). We have already used these two specific functions to solve the traveling salesman problem, see Figure 15.10 in Study 15.

Note The "left"-most element of a vector in Mathcad is by default the element number zero. This default value was present in the programs in Figures 22.2 and 22.3. But in the program in Figure 22.4 this default value is modified by assigning to the second important Mathcad system variable, the *ORIGIN* variable, the value one. After this assignment, the numbering of vectors will start with unity.* But in programming, zero is more commonly applied (see Study 2). Furthermore, the algorithm implemented in the program in Figure 22.4, is applicable to a source vector with positive elements. But it is not difficult to modify it for a vector with any numeric elements.

It is possible to develop the algorithm implemented in the program in Figure 22.4, by choosing from the unsorted part of the vector not only the maximum but also the minimum element and sending it to the left flank of the new vector being formed.

The animation, whose creation is shown in Figure 22.4, is available on the website of this Study. Readers can try to implement the sort algorithms described in Wikipedia (see above), or their own algorithms, in Mathcad. In this case, you can use the Mathcad built-in function time to evaluate the speed of any custom sort functions compared to the built-in function *sort*. It should be kept in mind that high-speed, sophisticated sort algorithms were created in the days of slower computers, when computation time was one of the main limiting factors. If we talk about the educational process (this book is aimed primarily at this), it will be good to recall the slow but understandable sorting algorithms. Such algorithms are good for teaching programming.

There is such a science as bionics. It allows the solution of technical problems by observing the behavior of living organisms. For example, one of the best algorithms for solving the traveling salesman problem was suggested by ants dragging their loads into the anthill by the shortest route. There is more detail about this in Study 14. When creating a sort algorithm, we can also turn to "living organisms": to line-up schoolchildren in the gym, for example, in alphabetical order, and then order them to be re-lined up according to height. All this can be recorded on a video camera, and then with a slow view on the computer monitor to analyze and try to

* A programmer was drafted into the army, he was put at the beginning of the row and received the command: "Count in order!" Our programmer first clarified in which system (binary, octal, decimal, hexadecimal or some other special) it was necessary to count, and then shouted: "Zero!"

recreate this "live" algorithm for self-sorting. If the sort performed by the gym teacher (see above) can be implemented on a single-processor machine, the self-sort of schoolchildren will require a multiprocessor system. In this case, the gym teacher can interfere with the process of such a sort and control it. Here, real technical objects or objects of living nature come to mind. In a modern aircraft, there is a central computer, and local computing devices (controllers), one of which, for example, regulates the supply of fuel to the engine and stops the fuel supply in an abnormal situation. A person has a "central" brain, but a person pulls a hand from a hot object reflexively, without using "mental abilities". All this increases the reliability of complex technical and living systems, improves their manageability.

And the second is a simpler topic of investigation. How will the nature of the sort change, if at the end of the program alternative to the one in Figure 22.2 there is not a *break* statement but a *continue* statement? And anyway, is it worth inserting them into the program? How will this affect the speed of sorting the source vectors with different height distribution?

TWO "RANDOM" MATHCAD FUNCTIONS

In Figure 22.2, it was shown how the Mathcad built-in function *runif* generates a vector containing random numbers in the range from 150 to 200. The numbers 150 and 200 are the provisionally minimal and maximum height of a person in centimeters. It is assumed that the percentage of people with a certain range of heights is the same. Figure 22.5 shows a histogram (meeting frequency) of the numbers generated by the *runif* function.

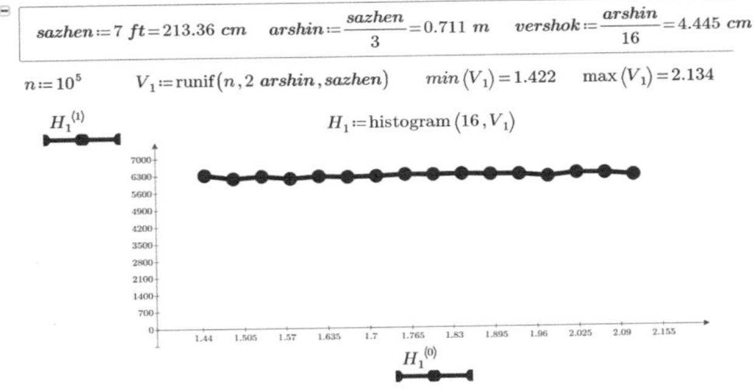

FIGURE 22.5 Histogram of a uniform distribution.

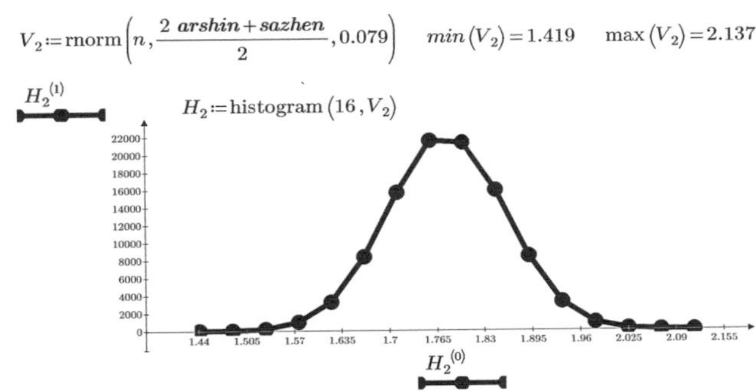

FIGURE 22.6 Generating a vector with a normal distribution in Mathcad.

At the same time, we abandoned centimeters and went over to the Old Russian arshin (7/9 yard) and vershok (tip-top), and we assumed that a person's height can vary from two arshins to a sazhen (7 feet). Why did we give up centimeters? This is illustrated in Study 14.

However, in Mathcad, there is a function named *rnorm* that allows you to generate random numbers with a *normal* distribution (see Figure 22.6), and not with a *uniform* distribution (see Figure 22.5). Selecting the values of the first (mathematical expectation) and the second (mean square deviation) arguments of this function, you can achieve the required distribution of students in terms of height. We used the normal distribution in Study 12, when the parameters of the optimal fire bucket were determined—see Figures 12.9, 12.10 (Study 12) and Figures 21.12 and 21.13 (Study 21). But the sort of the distribution of schoolchildren, students and soldiers according to height is not affected.

ASSIGNMENTS FOR READERS

1. Implement the sort algorithms described above in the Mathcad environment.
2. Take a large array of data about height of people and form a matrix from it, where the columns would have those values that are in the interval of one verst: creating a matrix similar to the histogram in Study 14 in Figure 14.20. In the columns of this matrix, place the height data with the increase. Fill in the empty places of the matrix with a constant string "empty string".

How to Calculate a Ship Or

How You Name a Boat Is How it Will Sail

ABSTRACT

This Study discusses the calculation of the sustainable navigation of ships in fresh and seawater, according to the forces acting on the ship. A laboratory work covering physics, mathematics and informatics is proposed. Engineering aspects of an old shipwreck are discussed.

Mathematics: Solution of algebraic equations.

Physics: Density, law of Archimedes, center of mass, center of gravity, center of the application of forces, stability of the ship, displacement.

IT: Cloud computing.

Art: Museums; ships' decoration.

Study website: https://community.ptc.com/t5/PTC-Mathcad-Questions/Study-23/td-p/544951.

The author of this book works at the International Association for the Properties of Water and Steam.* Its members meet every year in various cities of the world closely associated with water.† In 2015, such a meeting was in Stockholm, and the final banquet took place in the museum of the legendary Swedish sailing ship *Vasa*,‡ which was launched in 1628 and immediately sank. In 1961, the ship was raised and restored. She is currently exhibited in a museum specially built for her (Figure 23.1).

But meetings within the framework of the International Association on the Properties of Water and Water Steam (IAPWS), of course, consist not only of banquets (see Figure 24.1 in Study 24). It is necessary to work—to review and approve documents describing the procedure for calculating the properties of this important substance. These *formulas* become international standards, reference books are issued and programs for computers are created. It is noteworthy that a similar Russian reference is supplemented by Internet sites with open interactive networks and calculations of liquid water and water vapor properties, making use of the Mathcad Calculation Server technology and Elsevier/Knovel Interactive Calculation. Figure 23.2 displays one of these online calculations, through which it is possible

* www.iapws.org.

† Almost all settlements of the world arose on the banks of rivers, lakes, seas or oceans. These water areas do not divide people, but, on the contrary, give them convenient ways of communication. In 2014, for example, such a meeting took place in Moscow. (In the name of the Russian capital, according to one of the legends, there is water: *Moskva*, *akva*, water). The final banquet, without which there is not a single scientific event, took place on board a motor ship cruising under the walls of the Kremlin. In 2017, the IAPWS meeting was held in Japan in Kyoto, with the final banquet on the Lake Biwa, which is Japan's largest lake. And in 2018, the IAPWS conference took place in the Czech Republic capital Praha, where the Vltava river is running. The name of the river probably originates from the Old Germanic words "wilt ahwa", wild water (https://en.wikipedia.org/wiki/Vltava). One of the associated members of IAPWS is Egypt. So IAPWS members are looking forward to an IAPWS meeting in Egypt at some time in the future. The role of the Nile river is also often mentioned in the history of mathematics. First geometrical attempts to calculate the area of figures was associated with the Nile's flooding in Egypt. One can find a lot of information about Egyptian mathematics in the Rhind Mathematics Papyrus (1650 BC).

‡ https://en.wikipedia.org/wiki/Vasa_(ship).

FIGURE 23.1 Restored ship *Vasa* (Creative Commons ©, Attribution-ShareAlike 3.0 Unported [CC BY-SA 3.0]).

to determine the density of seawater, depending on its temperature, pressure and salinity.

If we are at the Black Sea, where one kilogram of water contains 13 grams of salt, and we dive to a depth of 150 meters, where the pressure is 1.5 MPa (about 15 atmospheres), and the temperature is, let's say, 17°C, then the density of the water will be equal to 1009.29 kilograms per cubic meter. But we can say quite readily that the density of the water will be of the order of 1010 kg/m^3, given the measurement error and density calculation technique.* In addition, it is difficult to determine the salinity of water, if we take into account the fact that the composition of water in various seas, oceans and salt lakes is different. The website in Figure 23.2 not only gives the desired value of water density but also draws three isolines showing how the density of water will change if one parameter (temperature, for example) is changed and two others (pressure and salinity of water) remain constant. For pure (fresh) water, whose thermodynamic properties are determined only by two parameters, pressure and temperature, it is possible to construct a surface or a family of curves. But seawater requires three separate Cartesian graphics. We already used this technique when, in Study 14, we decomposed a function of four arguments into four separate planar graphs.

The Internet calculation, shown in Figure 23.2, came to the mind of the author when he was sitting at a banquet in the Stockholm museum at the side of that unlucky ship (Figure 23.1). And that's why.

According to one version, the toppling of the *Vasa* ship was due to the fact that the change in the density of the water when the ship left the freshwater gulf where the shipyard was located was not taken in account. More dense water lifted the ship, a gust of wind blew, and water poured through the open cannon hatches into the holds.

So it was, or otherwise, you can argue, but the fact that there was some engineering mistake associated with incorrect calculations—this is an undeniable fact.†

Let's see how we can apply Mathcad to evaluate our hypothesis.

As a model of the ship, we take a wooden block in the form of a parallelepiped with sides of 10 mm (the width of the "deck" is variable *w*), 30 mm (ship's height, *h*) and 250 mm (length, *l*: see Figure 23.3, left). This simplification is not so far from life: real modern merchant ships have almost rectangular central cross sections (see Figure 23.3, right), and only the bow and stern are made pointed to reduce the resistance of water during movement. A "rectangular" ship is easier and cheaper to build, and then it can be filled with cargo (containers for example), which also have the form of a parallelepiped.

We can put our ship model in the water so that the sides of 10 mm (*w*) and 250 mm (*l*) form the deck and the bottom of our "ship", and the side of 30 mm (*h*), the height from the bottom to the deck. Life experience suggests that such a position of the wooden bar in the water will not be stable (see Figure 23.4, left). The force F_T (the force of gravity applied to the center of mass‡) and the force F_A (the buoyancy force of Archimedes, applied to the center of displacement, to the center of mass of the volume of fluid displaced by the floating body) will topple our "ship", and it will lie on one side.

This is explained by the fact that the center of mass was above the center of displacement, and for any small

* In the tram hangs a sign: "Places for standing 229." A question immediately arises about how accurately the number of these places was determined. You could easily round 229 to 230. What would you say if the electronic thermometer shows that your temperature is 36.589, instead of 36.6?!

† Three wise men of Gotham, They went to sea in a bowl,And if the bowl had been strongerMy song would have been longer.https://en.wikipedia.org/wiki/Wise_Men_of_Gotham

‡ Here it is more accurate to talk not about the center of mass, but about the center of gravity. But in a homogeneous gravitational field, the center of mass (the geometrical point characterizing the motion of the body or the system of particles as a whole) and the center of gravity (the geometric point through which the resultant of all gravity acts on the particles of the body at any position of the latter in space) coincide. In Study 9 we worked with a non-uniform gravitational field when we considered the motion of satellites around the planets. When we approached the Earth, we began to admit that this field is homogeneous.

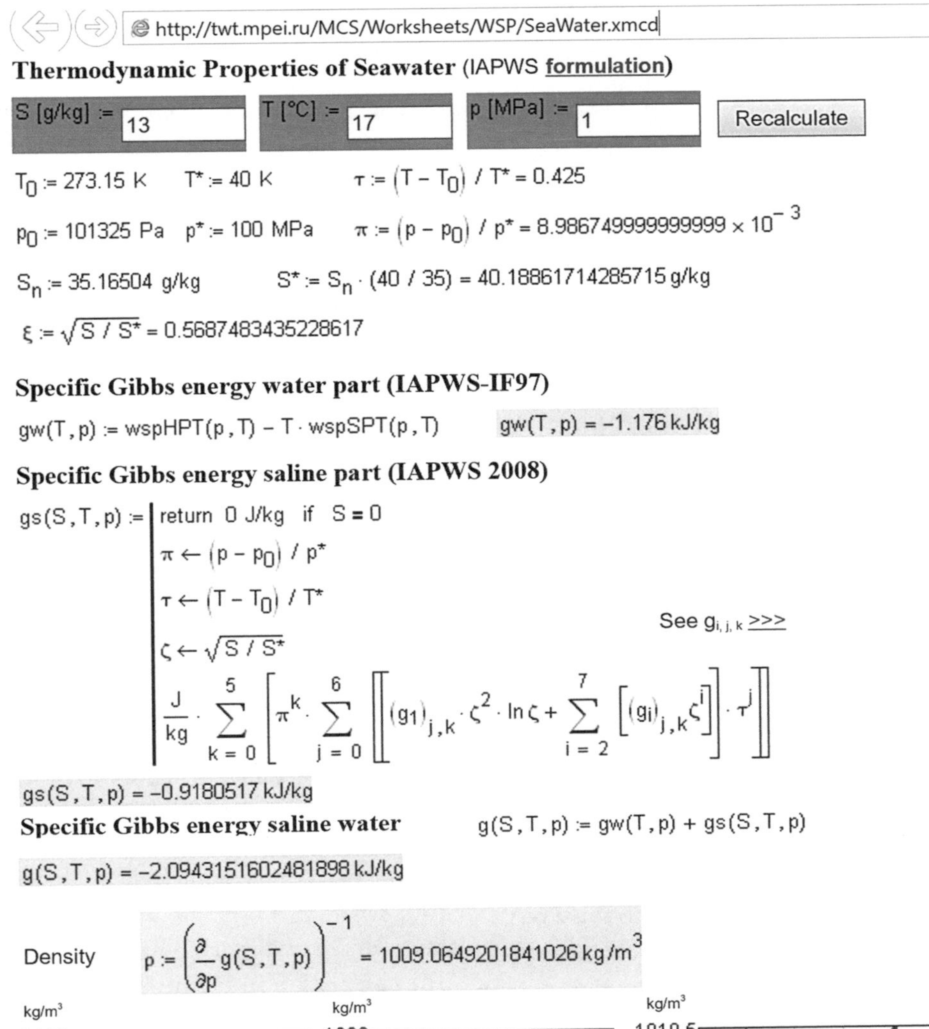

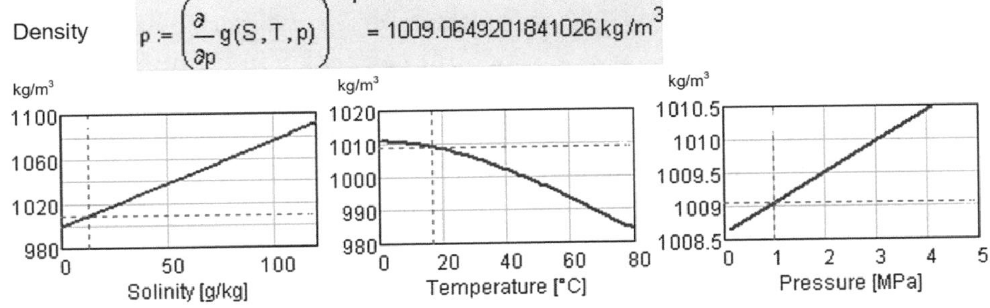

FIGURE 23.2 A website where you can estimate the density of seawater.

deviation of such a "ship" from the equilibrium position, it will turn over under the action of a pair of forces. This condition is necessary, but not sufficient, and we will note it below.

To lower the location of the center of mass, let's slightly work the bottom of our "ship" with a plane and attach a metal strip $h_м$ thick there. This will be the same *ballast* that should keep the "ship" from tipping over (see Figure 23.4, center and right). Sailing warships were built high so that the guns fired further and making it convenient to board enemy ships. Without a rather massive ballast, such a ship would not be able to stay afloat even without raised sails.*

Let's analyze stability of our "ship" using Mathcad (Figure 23.5).

The height of the displacement center (the center of the volume of fluid displaced by the floating body) can be easily determined using Archimedes law—see the upper operator in Figure 23.5, where the equation (the weight of the displaced water is equal to the weight of

* A popular Russian book about sailing: *Scarlet Sails* by Alexander Grin.

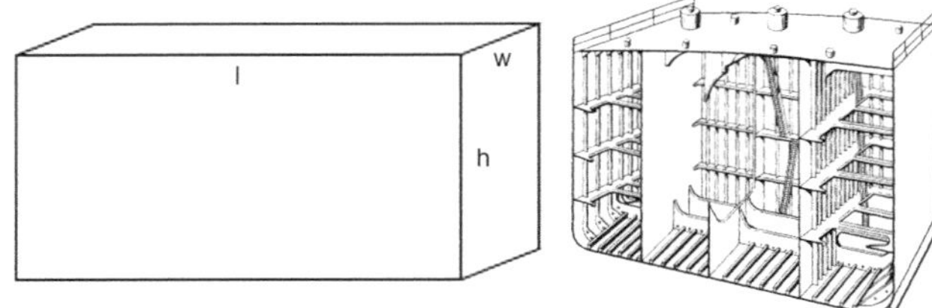

FIGURE 23.3 The ship model and the central section of a real ship.

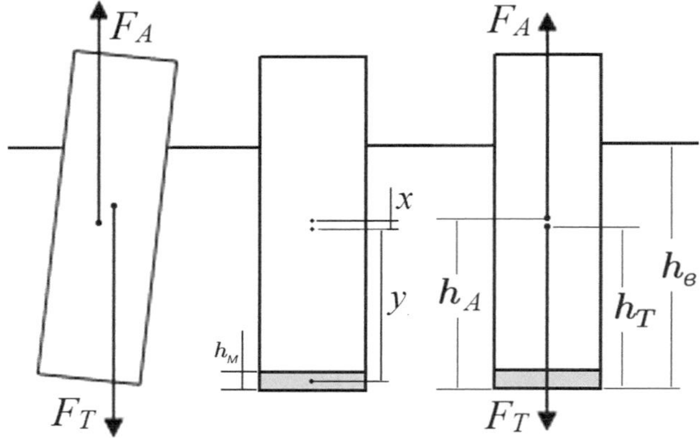

FIGURE 23.4 Forces acting on the "ship"(left, no ballast, center and right, with ballast).

the ship) is solved with respect to the variable h_d. To determine the center of mass, it is necessary to solve not one equation, but three more, see the lower operator in Figure 23.5. These equations are:

- The product of the mass of the wooden part of the ship m_w from its center to the center of mass of the entire ship (x) is equal to the product of the mass of the metal part of the ship $m_м$ by the distance from its center to the center of mass of the entire ship (y), see Figure 23.4, center.
- The height of the center of mass of the ship h_T is equal to half the ballast height $h_м$ plus the value of y.
- The height of the center of mass of the wooden part of the ship h_w is calculated in two ways:

$$h_m + (h - h_m)/2 \text{ and } h_m/2 + x + y.$$

We could derive the formula for h_T manually, but having Mathcad at hand, we can automate this work by making the balance of torque (the first equation) and the distances (the two-second equations).

Figure 23.6 shows the calculation of the stability of a ship with ballast in fresh water ($\rho = 1$ g/cm^3) using the formula for h_T, derived in Figure 23.5. The height h_A (13.55 mm) was bigger than the height h_T (13.313 mm). Our "ship" with ballast will keep upright—a pair of forces will return it to its original position with small deviations from the vertical. But this equilibrium is unstable (local potential well)—with a strong deviation of the ship from the vertical, it still lies to one side.

Let's gradually increase the density of water (take our "ship" from the freshwater harbor to the open sea) and see what happens to it. For this, let us introduce a function with the name Δ (the difference h_A and h_T) with argument $\rho_в$, construct its graph and calculate the value of the water density, where this function will change sign—where the forces h_A and h_T will start to turn our "ship" (Figure 23.7).

The calculation in Figure 23.7 shows that the critical density of water for our "ship" is 1.018 grams per milliliter. With the left graph in Figure 23.2 it is possible to estimate the salinity of water at such a density.

There is a legend about how ancient Rus (Old Russians) were saved from a raid of Vikings, the ancestors of the builders of the *Vasa*. When these pirates came on the river in their boats to the settlement of Rus, in

$$\left\| \begin{array}{l} \text{"Weight of the displaced water"} \\ \left\| \begin{array}{l} \text{"Volume of the submersed part"} \\ h_d \cdot w \cdot l \end{array} \right. \cdot \rho_d \cdot g \end{array} \right. = \left\| \begin{array}{l} \text{"Ship weight"} \\ m \cdot g \end{array} \right. \xrightarrow{\text{solve}, h_d} \frac{m}{\rho_d \cdot l \cdot w}$$

$$\begin{bmatrix} x \cdot m_d = y \cdot m_m \\ h_T = \dfrac{h_m}{2} + y \\ h_m + \dfrac{h - h_m}{2} = \dfrac{h_m}{2} + x + y \end{bmatrix} \xrightarrow{\text{solve}, \begin{bmatrix} x \\ y \\ h_T \end{bmatrix}} \begin{bmatrix} \dfrac{m_m \cdot h}{2 \cdot m_d + 2 \cdot m_m} & \dfrac{m_d \cdot h}{2 \cdot m_d + 2 \cdot m_m} & \dfrac{m_d \cdot h_m + h_m \cdot m_m + m_d \cdot h}{2 \cdot m_d + 2 \cdot m_m} \end{bmatrix}$$

FIGURE 23.5 The derivation of the formulas for calculating the height of the displacement center (h_d) and the center of mass (h_T).

$$w:=10\ mm \qquad h:=30\ mm \qquad l:=250\ mm \qquad h_m:=0.5\ mm$$

$$\rho_d:=1.0\ \frac{gm}{cm^3} \quad \text{Water density}$$

$$\rho_w:=0.8\ \frac{gm}{cm^3} \quad \text{Wood density}$$

$$\rho_m:=7\ \frac{gm}{cm^3} \quad \text{Metal density}$$

$$m_m:=w\cdot l\cdot h_m\cdot \rho_m=8.75\ gm \qquad m_w:=w\cdot l\cdot (h-h_m)\cdot \rho_w=59\ gm$$

$$m:=m_m+m_w=67.75\ gm$$

$$h_d:=\frac{m}{w\cdot l\cdot \rho_d}=27.1\ mm \qquad h_A:=\frac{h_d}{2}=13.55\ mm$$

FIGURE 23.6 Calculation of the stability of a ship with ballast in fresh water.

full armor ready to land, the Rus began to pour the most precious thing they had, salt, into the water. The density of water in the river increased, the Vikings' boats lost their stability and turned over. Some of the Vikings drowned, and some were taken prisoner.

This legend was invented by the author when working on this book. But from the point of view of physics and mathematics, it is no worse than the ancient Greek legend of how the enemy fleet was burned with the help of mirrors that focused the sun's rays on wooden ships.

However truly, we can propose such a *laboratory work*.

A "ship" is produced in the form of a ruler made of wooden and metal rails (see Figure 23.4, center and right). Such a "battleship" descends into the water—it is placed into a container of water. At the same time, the "ship" is designed (see Figure 23.6) so that it is held steady vertically in fresh water. Then table salt is poured into the water, the water is gently mixed and we watch as the "ship" starts to fall to one side.

It will then be possible to heat this aqueous solution and watch how the ship will return to its original vertical position due to the fact that the aqueous solution with a higher temperature has a lower density (see the central graph in Figure 23.2). All this can be supplemented by calculations on the ship's stability computer (solving algebraic equations—see Figures 23.6 and 23.7) and heat and mass transfer processes (solving differential equations). The shape of the ship can be made more complex, close to the real, etc. This will be a good educational theme linking math, physics, chemistry and computer science* at school. It can be written in a new Study.

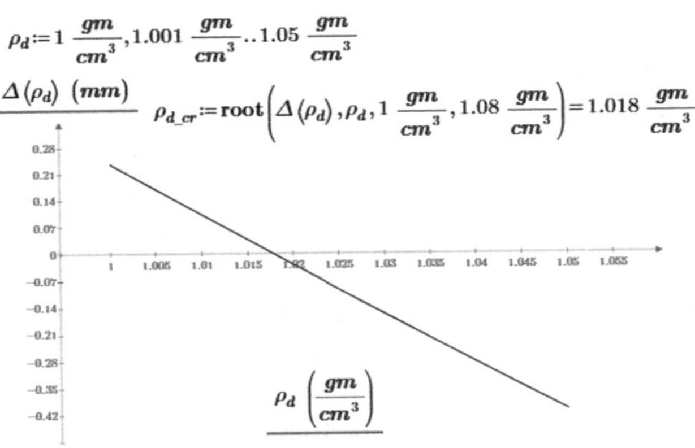

FIGURE 23.7 Stability chart of the ship depending on water density.

It is possible to try to create an animation of the turning of the ship or of its swinging together with the drawing of the vectors of the forces acting on the ship.

Of course, the calculation of the stability of real ships is much more complicated. This is a very beautiful engineering problem. Almost all great mathematicians have put their hands to try to solve it. Today we can only admire how they could have done it without computers. By the way, on modern ships, computers control special "fins" and other devices that reduce rolling even with strong sea waves and prevent a dangerous roll of the ship. Surprisingly, in a real ship, the center of displacement is still below the center of mass, but its stability is created, among other things, by the complex shape of the underwater part of the ship, which compensates for the deviation of the center of mass by offsetting the location of the displacement center. The simplest example. If our ship/bar without ballast (Figure 23.4, left) lies on one side, then it is easy to calculate that the center of mass will still be higher than the center of application of the Archimedean forces. But if the ship (or rather, the raft) is tilted by a side wind, for example, the center of application of the Archimedean forces (displacement center) will shift so as to return the ship to a horizontal position, into a global potential well.

By the way, our ship/ruler ("linear" ship) is a kind of hydrometer,† a device for measuring the density of

* Here we use the term computer science, instead of informatics because it is more commonly applied in English-speaking countries. However, informatics is the science of methods of working with information. It arose long before the advent of computers. But now work with information is inconceivable without a computer. Therefore, these two terms strongly overlap, but do not merge.

† As reported in a letter by Synesius of Cyrene, bishop of Ptolemais, the invention of the hydrometer is credited to Hypazia of Alexandria: http://www.livius.org/sources/content/synesius/synesius-letter-015.

liquid, whose principle of operation is based on the law of Archimedes. The hydrometer is a glass tube, the lower part of which, when calibrated, is filled with shot or mercury (ballast) to achieve the required mass. At the top of the tube is a scale that is graduated in terms of the density of the solution or the concentration of the dissolved substance.* A similar "scale" can be seen on the sides of ships to control its loading. Ships of the river or sea grade have several such scales. The upper mark on this scale is called a load line, which should not be submerged in water when loading a ship in a sea or river port. Otherwise, the ship can turn over. And there have been many such cases, alas.

Paraphrasing A.P. Chekhov,† we can say that everything should be fine in the ship—both the silhouette, the name and the calculations for which it was designed. Such as the small but instructive calculation we have just conducted.

CENTER OF MASS, CENTER OF GRAVITY

How to calculate the center of mass if the figure is not a rectangle (see Figure 23.4) but a more complex plane figure? If you make an appropriate request on the Internet, you will receive either formulas for simple figures (triangle, semicircle, circle sector, etc.) or a "terrible" formula with a double integral, which is practically impossible to use in the calculation for a specific case. But this is a concrete case.

The figure (say, this is the cross section of the ship) has the form of a semicircle with an outer radius of 1 m and an inner radius of 0.9 m, see Figure 23.8. It is necessary to determine the coordinates of the center of mass of this figure.

Let's do this. We can uniformly fill our figure with a set of points and determine the ordinate of the point, so that the total distance from this point to the remaining points will be minimal.‡ Whether this point will be the center of mass is a separate issue. We invite the reader to respond to it. We did something similar in Study 7, when we determined the moments of the forces on the sagging chain (see Figures 7.14 and 7.15). If this parameter

```
ORIGIN := 1   n := 10^5   X := runif(n,-1,1)m   Y := runif(n,-1,0)m

(X  Y) := | j ← 1
          | for i ∈ 1..n
          |   if (X_i)^2 + (Y_i)^2 ≤ (1m)^2 ∧ (X_i)^2 + (Y_i)^2 ≥ (90cm)^2
          |     | (XX_j ← X_i   YY_j ← Y_i)
          |     | j ← j + 1
          | (XX  YY)

n_1 := last(X) = 15101

M(y) := | M ← 0
        | for i ∈ 1..n_1
        |   M ← M + √((X_i − 0m)^2 + (Y_i − y)^2)
        | M

y := −0.5     y := Minimize(M,y) = −750·mm
```

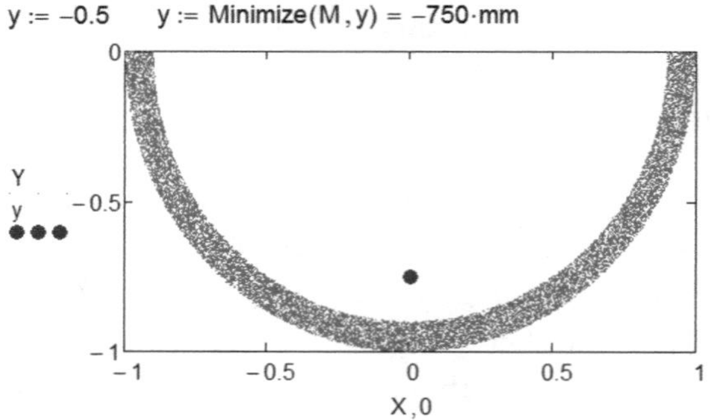

FIGURE 23.8 Approximate calculation of the center of mass of a plane figure.

is divided by the weight of the chain, then we obtain the coordinate of the center of gravity.

Figure 23.8 calculates the center of mass with a method, which can be called the Monte Carlo method. The Monte Carlo method, by the way, is often used in integrating, when determining, for example, areas of complex figures.

We take a rectangle measuring 1 m by 2 m and fill it with n points with random coordinates. We did this work in Study 16, when we calculated the number π by the Monte Carlo method, see Figure 16.3. You can also fill the rectangle with points located in the nodes of a uniform grid. Then we filter the array of points—we leave in X and Y vectors only those points that hit the "bottom of the ship"—a semicircle 10 cm thick. These points are shown in the graph of Figure 23.8, forming the given figure. In the if statement, you can add new conditions, forming a complex shape and watching the accuracy of the calculations, looking at the points of the graph. The initial points in our particular problem were 100,000 (n), and 15,101 points (n_1) hit the "ship

* See an example of the tabular and graphical dependence of density and concentration in Figure 2.3 in Study 2.

† From the play *Uncle Vanya* (1897) by Anton Pavlovich Chekhov (1860–1904), the words of Dr. Astrov (Act 2): "A human being should be entirely beautiful: the face, the clothes, the mind, the thoughts."

‡ The center of mass $x = 0$, because of the clear symmetry.

bottom". Next, a target function $M(y)$ is created that returns the total distance from the point with the coordinates (0, y) to the remaining points. Our figure in Figure 23.8 is symmetrical about the central vertical axis, and it is clear that the desired center of mass must be on this axis. But in general, a function named M must have two arguments, x and y, and in its formula, instead of zero, there must be a variable (independent optimization parameter) x. If the problem is not flat, but a volumetric one, then the function M will have three arguments: x, y and z, and then everything is done as described here. Only the memory and speed of the computer must be adequate. Further, the function $M(y)$ is found numerically near the given value (0, −0.5). *Answer*: the center of mass of our thickened semicircle is 0.75 m away from its center, see the point on the graph of Figure 23.8. As we already discussed in Study 19, remembering a famous Russian film character:* "It's not aesthetic, but cheap, reliable and practical." And did we make this determination exactly? Let us verify this in a figure where the coordinates of the center of mass are known.

Figure 23.9 shows the calculation of the center of mass of the hemisphere. It is obtained when, for example, it is necessary to determine the center of application of the Archimedean forces for our "ship" (see Figure 23.8), when our ship is as low in the water as possible. For the hemisphere, the exact formula for calculating the center of mass is known (see the expression for the variable y_t). The accuracy of our approximate calculation turned out to be satisfactory—within a fraction of a percent (5 mm, see also the point and circle in Figure 23.9).

And here is another experiment at the intersection of physics and computer science. Take a flat solid figure of complex shape, cut out of cardboard. In this, a hole is drilled, through which a nail hangs. A weight is hung on this nail along the thread (creating a plumb line). The thread draws a straight line on the figure (*option*: all this is photographed, sent to the computer and digitized). Then all the actions are repeated for another drilled hole. You can do all this for a third or fourth hole, it will be superfluous. The point of intersection of these two straight lines will be the center of gravity

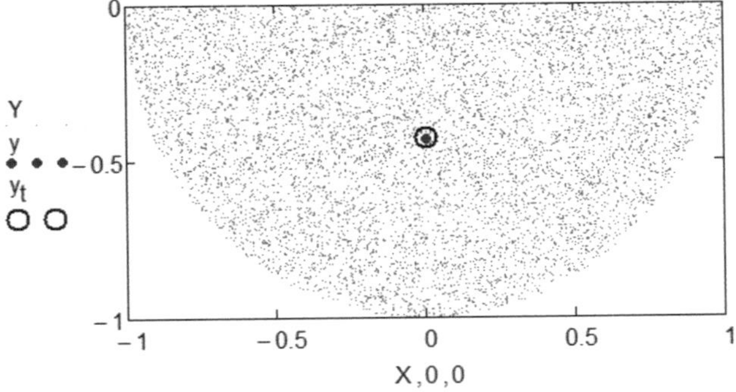

FIGURE 23.9 Verification of the approximate calculation of the center of mass of a plane figure.

(center of mass) of our flat figure. Animation of such manipulations is located, for example, in the article "center of mass" on Wikipedia. This property is due to the fact that the potential energy of a freely suspended body will be minimized—the body will try to settle as low as possible. We examined this phenomenon in Study 21 when we calculated the shape of the sagging thread with weights (see Figure 21.1). Next, a cardboard figure with two straight drawn lines is photographed, digitized to obtain an array of coordinates of its points along which lie its center of mass according to the above described computer technique. The laboratory work is ultimated by comparing the coordinates of the two points obtained from physical and computer experiments. A simplified version of the laboratory work would be as follows. A flat figure is drawn on a screen that uses the method described above to determine the center of mass. This figure is then printed on thick paper and cut from the paper with scissors. The cut is determined by the center of gravity, by double hanging it with a plumb on a stud. This can be used to verify the coincidence of the calculated and empirically determined points.

ASSIGNMENTS FOR READERS

1. Write a program for determining the center of mass of bodies of complex shapes, consisting of simple elements with a known formula.
2. Calculate how far the cannon of a ship will shoot, if you raise it from the lower deck to the top. There is a tip in Study 7.

* https://en.wikipedia.org/wiki/The_Diamond_Arm" https://en.wikipedia.org/wiki/The_Diamond_Armhttps://www.youtube.com/watch?v=9lumF8I69Io.

3. Try to prove the validity and accuracy of the method of calculating the center of mass shown in Figures 23.8 and 23.9.

4. Calculate the stability of the ship, whose cross section is shown in Figure 23.8, using different materials for the body (steel, wood, foam, etc.) and different liquids (mercury, glycerin, water with different salinity, etc.). Show how such a ship will behave if it is tilted.

5. Take a picture of your friends and acquaintances and determine their centers of "mass". It is believed that it should be located in the navel area (see the navel in the Pokémon in Figure 19.7). But is it really so?

How to Calculate a Rocket Or

"… and on Mars Will Blossom Apple Trees"

ABSTRACT

In this Study issues about the motion of a carrier rocket are considered: single- and two-stage. Tsiolkovsky's formula and the choice of a place for a cosmodrome are also discussed. Differential equations with a condition in formulas are solved numerically and analytically in this Study, and the topic of using mathematical formulas for the construction of monuments is discussed.

Mathematics: Solution of algebraic and differential equations, the manual calculations made by the characters of a feature film.

Physics: Rocket motion, Tsiolkovsky formula, Einstein's theory of relativity, acceleration of gravity, Earth rotation.

IT: Solution of an equation, the operator of choice.

Art: Soviet films and songs; monument to the conquerors of space in Moscow.

Study website: https://community.ptc.com/t5/PTC-Mathcad-Questions/Two-stage-rocket/m-p/89686.

There is a Soviet cult movie *Nine Days of One Year.** There are no chases and fights in which modern, young and not very young spectators "sat down", but there are beautiful dialogues performed by beautiful actors: Tatyana Lavrova, Innokenty Smoktunovsky, Alexei Batalov and others.

Here is one such dialogue from the film (Figure 24.1): actors Yevgeniy Yevstigneyev (theoretical physicist Nikolai Ivanovich, left) and Mikhail Kazakov (romantic physicist Valery Ivanovich, right):

—Tell me, Valery Ivanovich, how deeply are you going to penetrate the depths of our galaxy?

—At first only to the depth of 500 light years.

—At what speed?

—Close to the speed of light.

—The weight of your ship?

—One hundred thousand tonnes.

—The fuel?

—The most upgraded.

—Now we will calculate how much fuel you will need. Be kind, napkin.

—Nikolai Ivanovich, have you finished your calculations?

—Yes please. For a space ship of one hundred thousand tonnes weight at a speed close to the speed of light, in order to fly around a part of the galaxy in a reasonable time for human life, you need ten in twenty-two degrees 9710 of the most modern extra-fuel. As reference: Our planet weighs a little less, happy journey!

The dialogue is, of course, naive, but imagine that Nikolai Ivanovich does not have a napkin in his hands,

* You can see it on the website: http://kino-ussr.ru/main/217-devyat-dney-odnogo-goda-1961.html. The name of this Study is, by the way, a line from a famous song of those times. They sang it in the streets when they met the cosmonauts after another space expedition: "Cosmonauts and dreamers assert that …". Listen to the song at http://pesnifilm.ru/load/mechte_navstrechu/i_na_marse_budut_jabloni_cvesti_quot_mechte_navstrechu_quot/833-1-0-1846.

FIGURE 24.1 Shot from the movie *Nine Days of One Year.*

but a "tablet" (tablet computer) with the Mathcad program, and he makes calculations on it (Figure 24.2).

In the calculation in Figure 24.2, *Tsiolkovsky's formula* was used (by the way, Tsiolkovsky himself is mentioned in the movie*): the final velocity of the rocket v_{end} is proportional to the natural logarithm of the ratio of the starting mass of the rocket M_1 to its final mass M_2. The difference

$$v_{end} = I \cdot \ln\left(\frac{M_1}{M_2}\right) \xrightarrow{solve, M_1} M_2 \cdot e^{\frac{v_{end}}{I}}$$

$$M_2 := 100 \cdot 1000\ \textbf{tonne} = \left(1 \cdot 10^8\right)\ \textbf{kg}$$

$$v_{end} := 0.995\ c = 298.293\ 1000\ \frac{\textbf{km}}{\textbf{s}}$$

$$I := 777 \cdot 1000\ \frac{\textbf{tonnef}}{\textbf{tonne} \cdot \textbf{s}^{-1}} = \left(7.62 \cdot 10^6\right)\ \frac{\textbf{m}}{\textbf{s}}$$

$$M_{fuel} := M_2 \cdot e^{\frac{v_{end}}{I}} - M_2 = \left(1.003 \cdot 10^{22}\right)\ \textbf{tonne}$$

$$M_{Earth} := 5.97219 \cdot 10^{21}\ \textbf{tonne}$$

$$M_{fuel} > M_{Earth} = 1 \qquad \text{1 - Yes!}$$

FIGURE 24.2 The assumption of the calculation on the napkin.

M_1–M_2 is the mass of fuel and oxidant (hereinafter, simply fuel, as in the film). The proportionality factor in this equation (the value of I) is the ratio of the thrust of the engine F to the fuel mass consumption μ (the quantities F and μ will be used in further calculations).

We took I equal to 777 thousand metric tonnes (1000 kg) of force/(tonne/s). But in the film this figure is not announced, 777 thousand is our assumption (three sevens for happiness!). The value of I is also the velocity of the outflow of gases from the rocket nozzle (meters per second). The theoretical value of this velocity for nuclear rocket engines can exceed 70 km/s. The exhaust velocity for an electric motor can reach 140 km/s. Therefore, it is quite possible to dream about the "three sevens".

The Tsiolkovsky equation in Figure 24.2 is solved with respect to the variable M_1. If the final speed of the rocket is close to the speed of light (we assumed that this speed is 0.995 of the speed of light c), then the mass of fuel actually exceeds the mass of the Earth (M_{Earth}). But our calculation is, of course, very crude ("naive"). In addition, he does not take into account the important fact that, at speeds close to the speed of light, Tsiolkovsky's formula does not work. Here it will be necessary to depart from the laws of classical mechanics and resort to the theory of relativity of Einstein.

But let's come down from these galactic heights to near-Earth space, slow down the rocket a little (a hundred times) and calculate what parameters can be in real, rather than those for a semi-fantastic spacecraft. We have already done a similar calculation in Study 9—see Figure 9.14 and Figure 9.37.

Figure 24.3 makes an attempt to calculate the speed of a rocket having a starting mass of 250 tonnes, of which 10 tonnes is the payload (P), 40 tonnes is the mass of an empty carrier rocket (m) and 200 tonnes is the mass of fuel (M). The thrust of the rocket engine (F) is 2000 tonnes of force with a fuel consumption of 5 tonnes per second (tonnes/s, μ). With such initial data, the engine will operate for 40 seconds (t_{end}).

Figure 24.3 shows the numerical solution of the differential equation: the engine pushes the rocket up, and its weight pulls the rocket downward. And all this is balanced by the product of the mass of the rocket to its acceleration the first derivative of its velocity in time.

The numerical solution of the differential equation of rocket motion, the function $v(t)$ generated by points, is compared on the graph with the analytical solution (with the function $v_{sym}(t)$) found with Maple (see Figure 24.4).

* At the end of the dispute, the character of Kazakov (Valery Ivanovich) throws the following to his opponent (Yevgeny Evstigneev's character): "When Tsiolkovsky designed his missile, there were scientists-skeptics like you at the restaurant 'Yar' and you were proving insane on napkins."

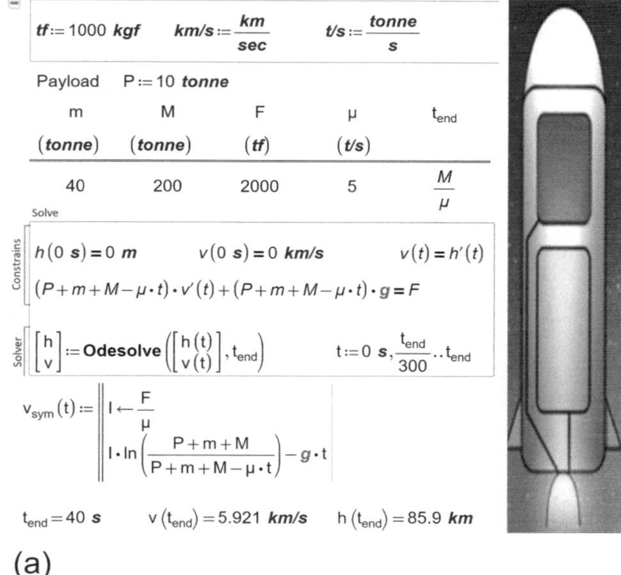

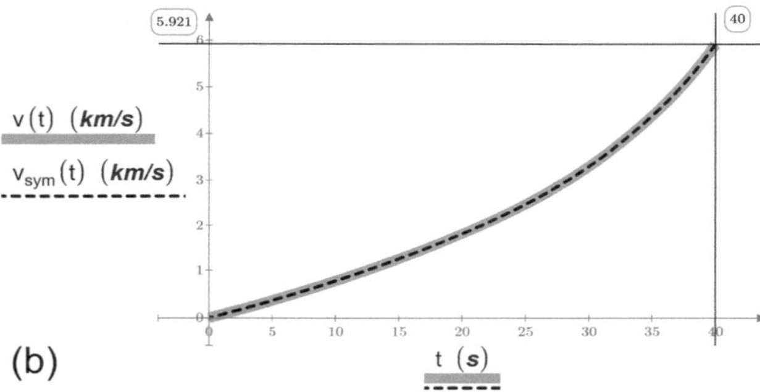

FIGURE 24.3 A single-stage rocket: (a) calculation (b) plot.

By this action, we actually derived the Tsiolkovsky formula, used in the solution in Figure 24.2, additionally inserting in it an allowance for the acceleration of gravity (the weight of the rocket). But the constant g is not a constant, but a variable, and we already wrote about this earlier (see Figure 1.27 in Study 1) and we'll talk further about this.

Note The graph in Figure 24.3 shows not one, but two curves: a thick solid pale and a thin dashed black inside a thick curve. Thus, the coincidence of the numerical and analytical (symbolic) solutions of the differential equation is verified.

$$\text{ODE} := (\mathrm{P} + \mathrm{m} + \mathrm{M} - \mu\cdot t)\cdot(v'(t) + g) = \mathrm{I}\cdot\mu$$
$$\text{Initial\_Condition} := v(0) = 0$$
$$\text{dsolve}(\{\text{ODE}, \text{Initial\_Condition}\})$$
$$v(t) = \mathrm{I}\cdot\ln\left(\frac{\mathrm{M} + \mathrm{P} + \mathrm{m}}{\mathrm{M} + \mathrm{P} + \mathrm{m} - \mu\, t}\right) - g\cdot t$$

FIGURE 24.4 The derivation of the Tsiolkovsky equation.

In the equation if the weight of the rocket $(P + m + M - \mu \cdot t) \cdot g$ is transferred to the right side with a minus sign, then Newton's second law is obtained in its pure form. Mathematically, of course, this will not change anything, but it will be more visible.

The calculation in Figure 24.3 shows that the speed of the rocket at the end of the engine operation (5.921 km/s) does not reach the first cosmic speed (≈ 7.9 km/s): after its turn by 90° (which is done by additional engines or gyroscopes), the rocket will not fly on a near-Earth orbit, but a ballistic orbit and will soon fall to the ground.

To carry our 10 tonnes of payload to the near-Earth orbit, it is still possible to increase the fuel supply. However, it is better to divide the rocket into two stages (Figure 24.5).

The differential equation of the flight of a two-stage rocket will contain the built-in operators of Mathcad *if*: if the first stage with thrust F_1 is operating, it carries the payload P, the two stages (m_1 and m_2), the fuel of the second stage M_2 and the consumable fuel of the first stage $M_1 - \mu_1 \cdot t$. The second step (the *else* operator) with thrust F_2 will carry only the payload P, the m_2 itself and the consumable fuel $M_2 - \mu_2 \cdot (t - t_1)$, where t_1 is the time of the first stage operation, and t is the total flight time of the rocket. In this scenario, the speed of the rocket at the end of the sequential operation of two stages (8.57 km/s) will exceed the first space velocity and the satellite will be put into orbit. At the same time, the rocket (or rather, its second stage) rises by 107.6 km, and not by 85.9 km (single-stage version).

The curves shown in Figures 24.3 and 24.5 are similar to the upper generatrix and edges of the monument to the conquerors of the cosmos established in Moscow (Figure 24.6, see http://www.kosmo-museum.ru/?locale=en).

I wonder if Tsiolkovsky's formula is embedded in this monument or is it an accidental coincidence? But Tsiolkovsky himself is present at the foot of the monument in the form of a statue.

In the calculations in Figures 24.3–24.5, of course, many assumptions have been made.* They do not take into account the effect of air flow on the rocket or the

* In principle, it was possible not to take into account the weight of the rocket—the product of its mass to the free-fall acceleration. This assumption, as we have already noted, is embedded in Tsiolkovsky's formula.

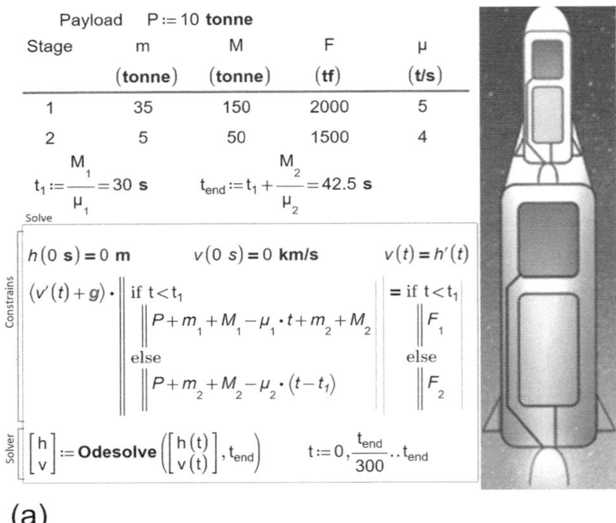

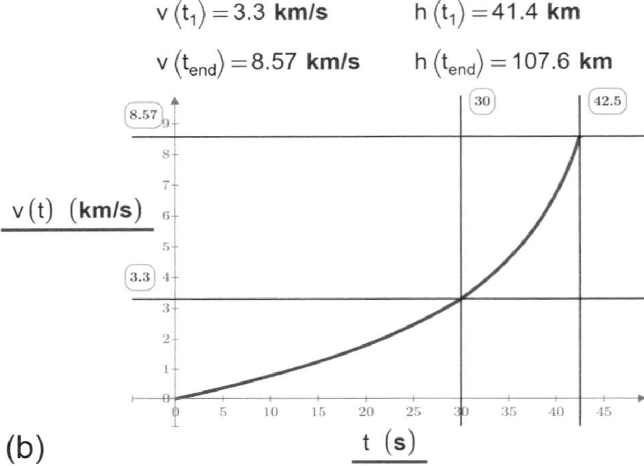

FIGURE 24.5 Calculation of the flight of a two-stage rocket: (a) calculation (b) plot.

change in the magnitude of the free-fall acceleration, depending on the height and geographical latitude. By the way, about latitude. It is known that the launching of rockets from the Earth should be carried out as near as possible to the equator and deploy the orbital station in an easterly direction. First, at the equator, the value of g is less than at higher latitudes, and second, and most importantly, closer to the equator, the rocket receives more additional acceleration due to the rotation of the Earth.

Figure 24.7 shows the value of the additional horizontal velocity the rocket will receive when it starts eastward from different points of the Earth, from five different cosmodromes as well as from the pole and the equator. If you look at the map of the location of these

FIGURE 24.6 Monument to the conquerors of space in Moscow (author's photo).

cosmodromes, it turns out that to the east of them there are places for the fall of spent rocket stages—the ocean or sparsely populated territory.

There is one more, "astronomical" factor: the orbital planes that do not coincide with the equatorial ones are unstable—the perturbing action of the Earth, the Moon and the Sun can "drop" such a satellite.

Rockets have several stages not only to save fuel or to put more useful mass into orbit for a given amount of fuel, but they also ensure that the spent rocket parts do not go into orbit but instead fall in the given area (see above) or burn in the dense layers of the atmosphere, without clogging the near-Earth space. People thought little about it at the beginning of the space age, but now

$R := 6370\ km$ $\qquad v(\phi) := \frac{2 \cdot \pi \cdot R \cdot \cos(\phi)}{day}$

$Pole := 90\ °$	$v(Pole) = 0\ \frac{km}{s}$
$Plesetsk := 63\ °$	$v(Plesetsk) = 0.21\ \frac{km}{s}$
$Baikonur := 46\ °$	$v(Baikonur) = 0.322\ \frac{km}{s}$
$Jiuquan := 41\ °$	$v(Jiuquan) = 0.35\ \frac{km}{s}$
$Cape\_Canaveral := 28.5\ °$	$v(Cape\_Canaveral) = 0.407\ \frac{km}{s}$
$Kourou := 5\ °$	$v(Kourou) = 0.461\ \frac{km}{s}$
$Equator := 0\ °$	$v(Equator) = 0.463\ \frac{km}{s}$

FIGURE 24.7 Linear velocities on the Earth's surface.

it has turned into a big problem. One often hears that the space station has changed its orbit so as not to collide with "space debris".

Discussions on the use of Mathcad for the calculation of launch vehicles can be found at the PTC Community. There are also files with the above calculations and an attempt was made to calculate the real carrier rocket "Soyuz" with three (or rather, two-and-a-half) steps.

Note The above calculations were carried out without taking into account the Coriolis forces.

ASSIGNMENT FOR READERS

Calculate the flight parameters of a real three-stage Soyuz rocket or others, taking its start and other data from Wikipedia, for example.

Calculate the flight of the rocket taking into account the Coriolis force.

STUDY 25

Hybrid Calculations on the Computer Or

One More about a Cone

ABSTRACT

The Study makes a retrospective look at the process of informatization of engineering education, in particular, on the ways of solving problems on analog, digital and hybrid computers. Binary, ternary, octal, decimal and hexadecimal formats for displaying numerical information "for printing" are touched upon, as well as paradoxes that arise. The advantages and disadvantages of analog, digital and hybrid computers are considered. We consider the simplest digital, analog and hybrid computers in conjunction with everyday items (cars, watches, etc.). The interaction with analog computers is described as a symbiosis of science and art. The analysis of architectural peculiarities and principles of work of the legendary Soviet analog computer MN-7 is given. The reasons for the decline of analog computers and the transition to digital technology are analyzed.

Examples are given of the use of hybrid computers at the present time, as well as the renaissance of analog devices and methods these days. The concept of a hybrid solution of the problem is introduced on the modern digital computer, which includes elements of the symbolic (analytical) and numerical (approximate) approaches. Using the example of solving an optimization problem (minimizing the surface of geometric bodies for a fixed volume), we examine the tools of the computer mathematics systems Mathcad (engineering calculator) and Maple: solving algebraic equations and systems, simplifying expressions, taking the derivative, etc. We conclude that modern mathematical programs give us new opportunities for solving problems in a hybrid way. We suggest ideas for registration of prize cups for winners of mathematical Olympiads.

Mathematics: Numbers with different bases, the derivative of a function, the search for zeros of a function.

Physics: Hybrid cars.

IT: Digital, analog and hybrid computers.

Art: Beautiful cup-winning mathematical Olympiad.

Study website: https://community.ptc.com/t5/PTC-Mathcad/Study-25/m-p/615013.

Nowadays when we hear the word "hybrid", we tend, as a rule, to think of hybrid cars.* These cars, when moving, use alternately or simultaneously an internal combustion engine and an electric motor fed by an electric generator and/or a battery. Due to such hybridization high economic efficiency is achieved and some environmental problems are solved: outside the city such a car can "smoke", and in the city go like a clean electric vehicle. A peculiar hybridization, which even the owners of cars themselves do not suspect, is embedded in some crossover utility vehicles (CUVs), whose front wheels are driven by a combustion engine through mechanical transmission, while, in the case of a bad road situation requiring full drive, the rear wheels are driven by

* Recently the word "hybrid" began to take a somewhat negative connotation due to hybrid wars, when military force is supplemented by other measures of influence, such as social networks on the Internet, in particular.

electric motors built into the wheel hubs. Some trolley-buses also have a hybrid feature—in the absence of overhead wires (in the historical center of a city, for example, where wires would spoil the appearance of streets and squares) they can move using accumulated energy in the batteries. Batteries can be not only electric, but may also take the form of rotating flywheels, compressed air cylinders, etc.

But there are also other hybrid "machines"—computational, the protagonists of this book. Such computers consist of *digital* and *analog* blocks. What are these?

How to add 3 to 2, not in the mind but on the computer? You can sit down at the computer, click on the keyboard and get the answer. The computer will translate 3 and 2 from the decimal to the binary format (11 and 10, see the second line in Figure 25.1, where the two binary numbers are marked with the postfix b, binary), will perform the simplest bit-wise additions on the 0s and 1s, get a binary response (101 is not one hundred-and-one but one-zero-one) and will translate it all back into the usual decimal format (5, five). The storage of 0s and 1s in digital computers is allocated in memory elements that are in one of two possible states: charged or discharged, magnetized or demagnetized, etc. From these *bits*, *bytes* and other more extensive information structures are formed. This is how electronic digital computers (PCs) work—computers with two "numbers in the head": 0 and 1. True, there were attempts to create both ternary digital computers (for example, the Soviet computer "Setun", which takes its name from the Setun River, the largest tributary of Moskva river: https://en.wikipedia.org/wiki/Setun), but they soon vanished due to the widespread release of binary electronic elements, among other things.

But you can also add 3 to 2 in another way: take 2 rheostats, connect them in parallel (see Figure 25.2), put 3 amperes on the first rheostat, with 2 amperes on the second and get the final answer: 5 amperes.

For the device shown in Figure 25.2, it is enough to connect three ammeters to get a "computer" that adds two numbers. This is the simplest analog numerical computer (ANC). The mathematical addition operation here has a direct electrical *analogy*—a parallel connection of two conductors with a current. To the scheme shown in Figure 25.2, you can connect additional conductors with rheostats and get a kind of adder, which we often use in spreadsheets. It is enough, for example, to put salaries of individual employees on such rheostats and get the final sum on the "final" ammeter. Examples of analogs of more complex operations would be the operation of differentiation contained in an electric coil and of integration in a capacitor with feedbacks. Analogs of mathematical actions can be found in pneumatic,* hydraulic, mechanical and other non-electronic devices. Analog computers are built on these principles. But there are serious drawbacks with analog technology. The maximum values of the variables used in analog computers are limited. If in the example of the calculation of the wages on an ANC we assume that 1 ruble is equal to 1 ampere, the total current (which shows how much money a certain group of workers should receive) can be very significant and exceed the capabilities of our electrical scheme. And if we assume that 1 ruble is only 0.01 ampere, for example, how to measure a penny? But the main advantage of ANCs is that they can operate easily with functions, not just variables. In order to add two functions with time as an argument on a digital computer it is necessary to perform a complex internal

3 + 2 = 5

3 = 11b 2= 10b

$$\begin{array}{r} 11 \\ +\,10 \\ \hline 101 \end{array} \quad 101b = 5$$

Result Format

Number Format | Display Options | Unit Display | Tolerance

Radix: Binary

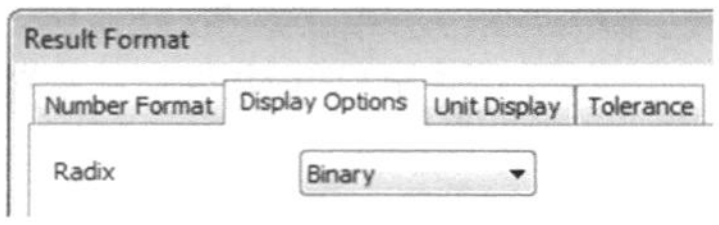

FIGURE 25.1 Imitating the addition of two numbers on a digital computer in Mathcad 15.

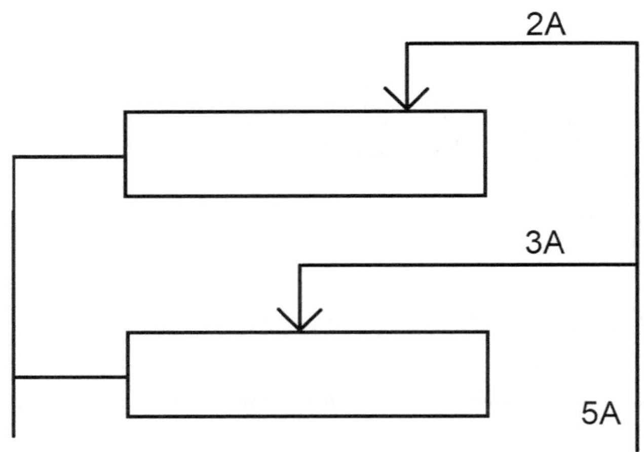

FIGURE 25.2 Addition of two numbers on an analog computer.

* Pneumatic counting and control devices were used and are used still somewhere in the explosive industries, for example.

procedure. On an analog computer, it is enough to connect two wires.

Hearing about analog computers, older generations of Russian people who have an education in electronics or control theory will immediately remember the legendary MN-7 machine, the top panel of which had complex electrical circuits assembled on it. Differential, integral-differential and other fairly complex equations and their systems were modeled on the MN-7 quite quickly and accurately, with up to the sixth order included. Digital computers of those times took days to solve similar tasks if they were solved at all. In the MN-7, the values were modeled by voltage. As already mentioned, ANCs have limitations on the maximum values of variables. For the MN-7, the maximum voltage was 100 volts. This meant that, before assembling the circuit, it was necessary to calculate the scaling coefficients, so that the result would not exceed the nominal voltage of 100 volts. There were difficulties with the representation of small quantities. What about the value that is one ten-thousandth of the maximum value? And what if the difference between the maximum and minimum values is 8–10 orders of magnitude?

The solution of systems of differential equations depends on the initial conditions. The slightest deviation of the initial conditions from the given values could lead to different solutions. Therefore, the process of specifying the initial conditions and setting the zeros of the amplifiers for modeling on analog computers was given great attention. Before running the assembled model, it was necessary to make sure that the inputs of the amplifiers were exactly the values needed (0 volts). The deviation in hundredths of a fraction during integration could give a very significant error.

MN-7 used tube operational amplifiers, characterized by low values of parasitic currents and voltages and an almost complete absence of dependence on the characteristics of the amplifier on the ambient temperature. When ANCs using transistors appeared, the procedure for setting the zeros of amplifiers turned into a whole new adventure. At times it seemed that the characteristics of the amplifiers depended not only on the environment, but also on the phases of the moon, the slenderness of the passing lab assistant and the quality of lunch in the student dining room! And if there are 128 amplifiers in the complex? On MN-7 there were 18 amplifiers.

But the speed of digital computers grew, their memory was expanded, and sophisticated, fast numerical algorithms appeared for the same differential equations, which allowed them to be solved in acceptable times on digital computers as well. Because of this, and for other reasons, analog computers became like shadows. Or, rather, they survived, but in conjunction with digital devices in the form of *hybrid* machines—as analog blocks to the already familiar digital computers. Such blocks are used, for example, in specialized computers that support simulators, where real-time calculations are needed. In airplane simulators, for example. Without analog blocks, such tasks are only possible for supercomputers whose operating time is too expensive. By the way, if the analog adder shown in Figure 25.2 is not fitted with analog keys, but with modern digital ammeters, it would represent a primitive hybrid computer.

The word "hybrid" can be applied to modern methods of solving problems in the environments of mathematical packages installed on digital computers.

Let's talk about this.

A mathematical problem, as you know, can be solved in two ways: *numerically* or *analytically* (*symbolically*). In modern computer mathematical programs (Maple, Mathcad, MATLAB®, Mathematica, etc.) special tools are provided for this purpose. Symbolic tools work with the symbols themselves,* and numerical work with numbers stored in variables. It is the skillful combination of the use of these instruments that captures the essence of the word "hybrid", which appears in the name of the Study.

Let's consider a simple concrete example of such a hybrid solution of a mathematical problem on the computer.

Let's return to cars, simple and hybrid. They, unlike pure electric vehicles, must be refueled with gasoline or diesel fuel. At large gasoline stations you can see tanks for storing fuel in the form of straight circular cylinders. And what are the proportions of these containers, if there are no special restrictions? It is easy to prove that if the diameter of the base of the cylinder d is equal to its height h, then such a geometric body will have the smallest surface S† for a given volume V. To build such a

* Or rather, if a variable stores a number, then symbolic math works with it, and if the variable does not store numbers, then with the character itself. But there are special operators that can force the number in a variable for symbolic work with it.

† As it is known, among all the geometric solids the sphere has the smallest surface for a given volume. However, a sphere-shaped tank is difficult to manufacture. They are made only for the storage of gases under pressure (gasholders, tanks for transportation of liquefied gas on specialized ships and cars).

container will require a smaller amount of metal, paint and thermal insulation, if, for example, it is used to store hot water or a refrigerant.*

And if the tank is open and does not have a top cover!? It is also easy to calculate here that in such container the ratio d / h will no longer be equal to 1, but to 2 (the height of the cylinder is equal to the radius of its base). And if the container has the shape of a straight circular cone and it is designed, for example, to store bulk materials! What is the optimal ratio of d to h? Here we have a task that possibly cannot be solved mentally—it will be necessary to attract mathematicians or a computer. We solved this problem about the cone in Study 12.

We will complicate this problem. Figure 25.3 shows the solution in Maple of the problem of the optimal ratio of the diameter and the height of a container that has the form of a kind of "glass": a straight circular cone with a base of diameter d and with height h, with a round lid (it is not visible in Figure 25.1) and a round stand. The lid and the stand have the same diameter as the base of the cone. The question is, what is the optimal ratio of d to h for such container?

Figure 25.3 shows the solution to the problem of "a glass with a lid" in Maple. This package here is chosen not by chance, as for it (along with the package Mathematica)† the main tool is symbolic, not numerical, mathematics, and we emphasized this in the introduction. Figure 25.3 analytically solves the equation of the volume of a cone with respect to the variable h. A formula is derived that relates the height of the cone to

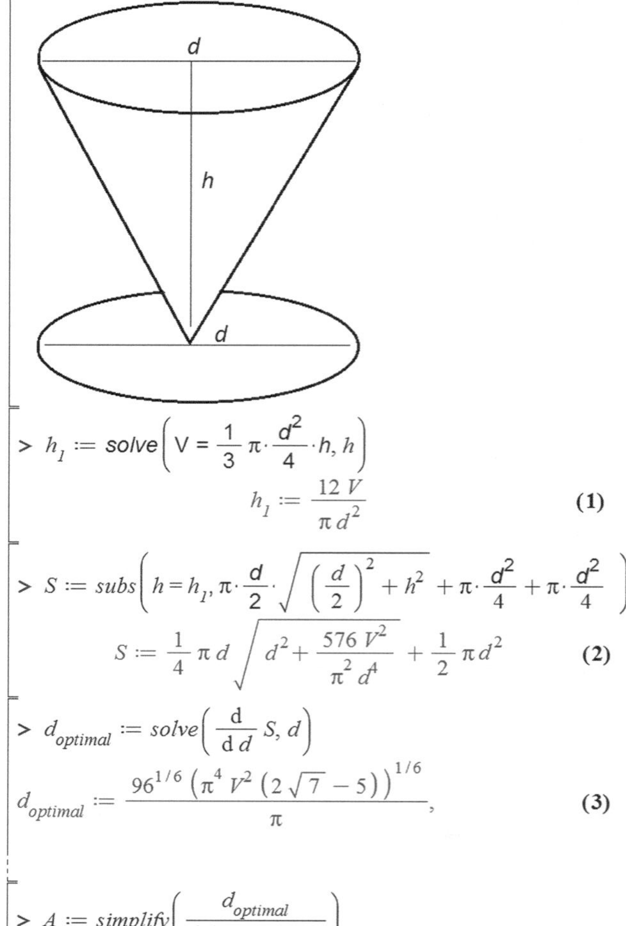

FIGURE 25.3 Symbolic calculation of the optimal "glass" with a flat lid (the Maple sheet is at https://www.maplesoft.com/applications/view.aspx?SID=154057).

its given volume V and the diameter of the base d (point 1 in Figure 25.3).‡ Further, the formula obtained for h_1 is substituted into the formula for the surface area S of a "glass with a lid" (point 2). Then the first derivative with respect to the variable d is taken of the expression for S, in order to search for the zeros§ (point 3). It is known

* These are known as heat or cold accumulators used to improve energy efficiency and energy saving. The compressor part of this air conditioner works only at night, when there is an excess of electricity and it is cheaper than in the daytime, sending the cold in the form of cool refrigerant not to the living room, but to a special tank, from which the cold can be taken later during the hot day. In the same way, it is possible to accumulate heat in winter, burning natural gas only at night, when it is in excess. In the power supply of buildings, hybrid elements are also wedged, connected, for example, with the fact that devices with different operating principles are used to produce heat, cold and electricity (trigeneration). A traditional poster of an energy profile (which hangs in many educational institutions) reads: "To give warmth and light is our profession!" It is necessary to write one more word "cold". But "letting the cold into the house" is somehow strange.

† Below we will work with Mathcad, "since its (along with the package MATLAB), main tool—it is numerical, not symbolic mathematics". The use of various packages to solve problems can also be considered another manifestation of hybridism. Formulas for the volume and surface of bodies can be sought either in the mind, or in references—paper or Internet. But it's better to do the same "hybrid" approach: remember the formula, and then check its correctness. Even better, try to derive the desired formula yourself. So you can train your brain. Some elderly people, calling their loved ones, try to get the phone number out of their memory, rather than from the phone's memory.

‡ This simple transformation can be done without a computer and even without a piece of paper—in the mind and immediately enter the answer. The combination of manual ("head") and computer character transformations is another example of hybridism.

§ There were several such solutions, but we left one of them for further transformations. Whether we took the zero of the function is a separate question for a separate analysis. Maybe our problem has several solutions?

that for a smooth continuous function (which is what we are dealing with here) the derivative is zero at a minimum point. This step also allowed us to derive a new mathematical constant,* see point 4 (the exact reproduction of the constant in radicals†) and point 5 (the calculation of the approximate value of this constant as a decimal number with floating point‡).

But if our "glass" has a flat cover replaced by a hemisphere, then it is not so easy to find the solution. Rather, it will be obtained, but in the form of a new problem—the problem of finding zeros of a polynomial of high degree. You can, of course, try to solve this problem analytically, but it's better to remember that there are numerical methods for solving similar problems.

Figure 25.4 shows the numerical solution of an even more complex problem—the problem of optimal proportions of a body composed of a hemisphere, a cylinder and a cone, which is placed on a circular support. The full volume of such body (now it is no longer a glass, but a trophy cup, the purpose of which will be described below) is 1 liter. The question is, what should be the values of the variables *d*, *L* and *h* so that the outer surface of the body (cup) will be minimal.

In Figure 25.3, we demonstrated a *mathematical* approach to the solution of the problem: we were not interested in the dimensions of a specific optimal glass in centimeters, but in the mathematical formula by which the proportions of such body of an arbitrary volume can be calculated. The solution shown in Figure 25.4, contains an *engineering* approach. A person (engineer) who designs a specific container does not have enough time to search in directories or derive the most general laws, and then consider how to apply them to his practical task§—he needs to build a concrete object. Mathcad here is chosen not by chance. It is also called an engineering supercalculator. In Mathcad, for example, you can use the units of measurement in full. This does not apply to

$$V(d,L,h) := \frac{\frac{4}{3}\cdot\pi\left(\frac{d}{2}\right)^3}{2} + \pi\left(\frac{d}{2}\right)^2 L + \frac{1}{3}\pi\left(\frac{d}{2}\right)^2 h$$

$$S(d,L,h) := \frac{4\pi\left(\frac{d}{2}\right)^2}{2} + 2\pi\frac{d}{2}\cdot L + \pi\frac{d}{2}\cdot\sqrt{h^2+\left(\frac{d}{2}\right)^2} + 2\cdot\pi\left(\frac{d}{2}\right)^2$$

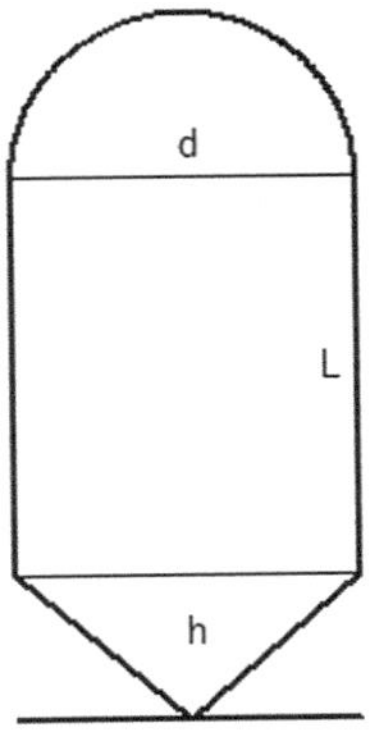

d := 1cm   L := 1cm   h := 1cm

$V(d,L,h) = 1.309\cdot cm^3$

$S(d,L,h) = 8.039\cdot cm^2$

Given

$V(d,L,h) = 1liter$

$$\begin{pmatrix} d \\ L \\ h \end{pmatrix} := Minimize(S,d,L,h) = \begin{pmatrix} 9.07099 \\ 11.09765 \\ 4.05753 \end{pmatrix} cm$$

$V(d,L,h) = 1\,liter$

$S(d,L,h) = 661.464\cdot cm^2$

$\frac{d}{L} = 0.81738$ $\frac{d}{h} = 2.23560$ $\frac{h}{L} = 0.36562$

FIGURE 25.4 Numerical calculation of the optimal cup (Mathcad 15).

Maple—and this is understandable. For a mathematician who solves abstract problems, meters, kilograms and seconds are not necessary. However, an engineer cannot do without them.

In the solution in Figure 25.4 two user functions with the names *V* (cup volume) and *S* (outer surface area of the cup) are entered, which have three (note!) arguments *d*, *L* and *h*. These are given a first approximation (assumption) equal to 1 centimeter. The Mathcad function *Minimize* will change the values stored in the variables *d*, *L* and *h* according to a specific numerical algorithm so that the function *S* (target search function, objective function) takes a minimum value, and the volume of the body *V* would remain equal to one liter (restriction when searching for the minimum of the objective function).

Note If the reader tries, as the students say, to manually "fill" the problem in Figure 25.4 on a computer with Mathcad 15, then they will get the wrong answer. Such situations lead to the bewilderment of many users of Mathcad. The fact is that the default accuracy of the numerical solution of the optimal cup problem is 0.001 m or 1 mm (0.1 cm) if the dimensional problem is solved with the support of the international SI

* There is a website on the Internet, on which are collected all sorts of mathematical constants, starting from π and e and ending with perfect exotics, little understood by many, even those knowledgeable in mathematics.

† In a simple cone without a base, the ratio *d* to *h* will be equal to the root of two, while the cone with the lid is half as large. You can remember this pattern for the design of conical containers (mathematical approach), or you can do calculations for each specific case (engineering approach).

‡ Here you can start a computational race—to announce a contest for someone who will calculate the longest decimal value of this constant. Now such a race is being held against the main mathematical constant, the number π, which appears in our formulas for calculating volume and area.

§ But one must always remember that "there is nothing more practical than a good theory".

$$V=\frac{\frac{4}{3}\cdot\pi\cdot\left(\frac{d}{2}\right)^3}{2}+\pi\cdot\left(\frac{d}{2}\right)^2\cdot L+\frac{1}{3}\cdot\pi\cdot\left(\frac{d}{2}\right)^2\cdot h \xrightarrow{solve,h} -\frac{\pi\cdot d^3+3\cdot\pi\cdot L\cdot d^2-12\cdot V}{\pi\cdot d^2}$$

$$\frac{4\cdot\pi\cdot\left(\frac{d}{2}\right)^2}{2}+2\cdot\pi\cdot\frac{d}{2}\cdot L+\pi\cdot\frac{d}{2}\cdot\sqrt{h^2+\left(\frac{d}{2}\right)^2}+2\cdot\pi\cdot\left(\frac{d}{2}\right)^2 \xrightarrow{substitute,h=-\frac{\pi\cdot d^3+3\cdot\pi\cdot L\cdot d^2-12\cdot V}{\pi\cdot d^2}} \cdots$$

$$\cdots\rightarrow\frac{\pi\cdot d\cdot\left(2\cdot L+2\cdot d+\sqrt{\frac{d^2}{4}+\frac{\left(\pi\cdot d^3+3\cdot\pi\cdot L\cdot d^2-12\cdot V\right)^2}{\pi^2\cdot d^4}}\right)}{2}$$

$$V:=1\,L \qquad S(d,L):=\frac{\pi\cdot d\cdot\left(2\cdot L+2\cdot d+\sqrt{\frac{d^2}{4}+\frac{\left(\pi\cdot d^3+3\cdot\pi\cdot L\cdot d^2-12\cdot V\right)^2}{\pi^2\cdot d^4}}\right)}{2}$$

$$d:=1\,cm \qquad L:=1\,cm \qquad \mathbf{Minimize}(S,d,L)=\begin{bmatrix}9.070739\\11.099027\end{bmatrix}cm$$

FIGURE 25.5 Hybrid calculation of the optimal cup.

measurement system, where the meter is the basic unit of length. And this is not enough for our, not "meter", but "centimeter" cup. To correctly solve the problem in Figure 25.4, you need to change the value of the built-in variable *CTOL* from 10^{-3} to 10^{-7}, for example. Because of this feature, Mathcad users sometimes disable the dimensional mechanism when solving certain tasks. Numerical calculation methods are often called *approximate* methods. And in this we have just been convinced, having solved the problem of the optimal cup approximately, with an accuracy of 100 nanometers. But the symbolic methods of solving problems can also be approximate. This happens, for example, when a complex function near the point under consideration is replaced by a series of simpler functions—see, for example, Figures 4.11 and 4.12 in Study 4. This refinement may be relevant for both analog and digital computers. It is believed that the problem should first be solved approximately, and this decision is influenced by various internal and external factors (see above). Digital machines are more deterministic and solve problems with predictable accuracy at each run of the solution.

Our problem about the trophy cup is solved (Figure 25.4), but there are doubts whether the volume of the body *V* would affect the optimal ratios *d* / *L*, *d* / *h* and *h* / *L*, shown at the end of Figure 25.4. In the problem in Figure 25.3 there were no such doubts: *d* / *h* remained constant for any volume of the "glass". That is because it is a mathematical, not an engineering solution!

In the solution in Figure 25.4, the KISS principle was used. KISS is an abbreviation for the phrase "Keep it simple, stupid".* It calls for solving problems simply, and only resorting to sophisticated algorithms when simple methods fail or are difficult to implement. And the simplicity of the problem in Figure 25.4 was that we did not analyze it deeply and did not reduce the number of arguments of function *S* from three to two. Let's do it!

Figure 25.5 shows a *hybrid* solution to the problem of our trophy cup: first using the symbolic tools of Mathcad† to get the *S* function no longer with three (Figure 25.4) but with two (*d* and *L*) arguments and one parameter *V*. To do this, we use the symbolic operators *solve* and *substitute* (in Maple it is called *subs*). One could go further along the path outlined in Figure 25.3, and try to take the partial derivatives of the function *S* with respect to the variables *d* and L and solve the reduced system of two equations. But it is possible to move to the "numerical"

* There is another popular quotation that refers to this: "Everything should be made as simple as possible, but no simpler," sometimes attributed to Einstein: "It can scarcely be denied that the supreme goal of all theory is to make the irreducible basic elements as simple and as few as possible without having to surrender the adequate representation of a single datum of experience."

† By the way, they were taken from Maple, which is included in older versions of Mathcad.

track and look for the minimum of the S function using the same built-in function *Minimize*.

On the website for this Study, you can download Maple and Mathcad files. There is also an animation of the process for searching for the optimal size of a conical glass with a spherical rather than a flat lid, but without a "leg" (a trophy cup, a cap "Drink up! Down in one!", or an ice cream cone). Figure 25.6 shows three frames of this animation with a graphical representation of the objective function.

The trophy cup we designed, shown in Figure 25.4, has the proportions found there and in Figure 25.5. However, in the cup the main thing is not the minimum of the surface,* but a certain beauty of proportions. Is there such a "beauty" in Figure 25.4? On the aesthetic aspects of the cup, views are subjective. Now we can do this: engrave on the cylindrical part of our optimal cup the mathematical calculations shown above, and hand this cup to the winner of the mathematical Olympiad. But for a hybrid, not traditional, Olympiad: its participants must use not only their knowledge and ingenuity, but also the possibilities of modern mathematical packages and even the help of the Internet to solve problems.

Hybridization can also be noted in the combination of computer characteristic transformations with similar transformations made by the user of the computer, a human being. We did this, for example, when solving the problem of the lens (see Study 20).

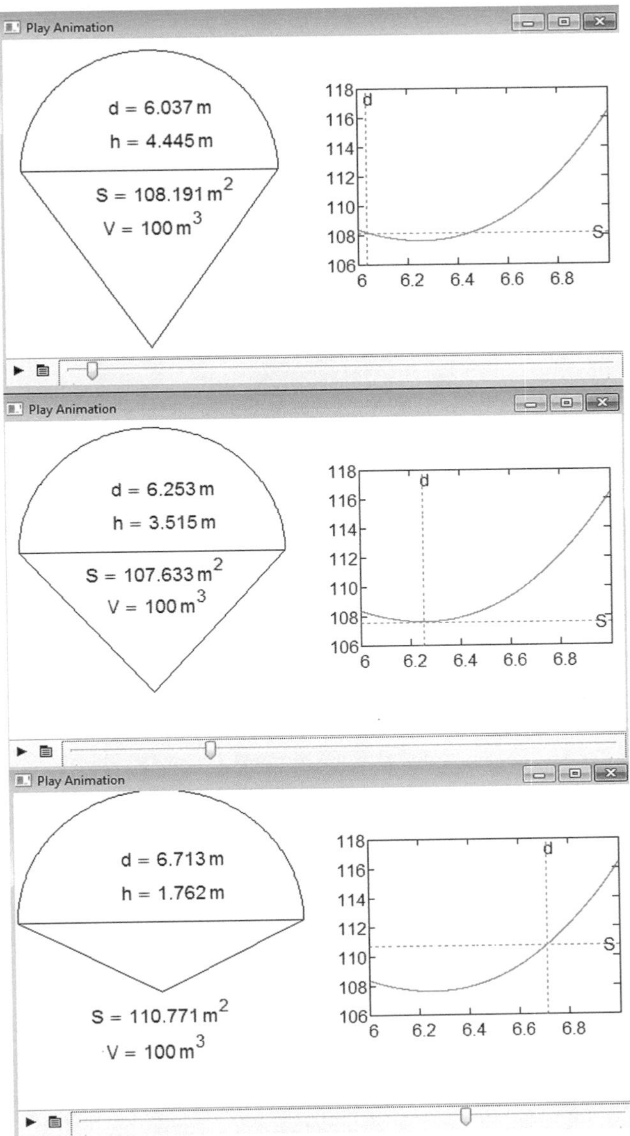

FIGURE 25.6 Three frames of the animation of the optimization problem of the size of a body, composed of a cone and a hemisphere.

CONCLUSION

The skillful use of analytical and numerical methods with several mathematical packages and the Internet increases the efficiency and accuracy of solving problems, making the process itself enjoyable by shifting routine actions to the shoulders of the computer and leaving only intellectual, creative work for a person. Such solutions can be supplemented with planar and three-dimensional graphics and animation.

ASSIGNMENTS FOR READERS

Determine the optimal sizes of different bodies: a cylinder without a top cover, a cylinder or a cone, divided in half by a vertical partition, etc.

* Why not? The cup can be made from cheap metal, and then covered with a layer of gold or silver. Our calculation minimizes the cost of this operation.

Funicular Polygons in Statics, Kinematics and Dynamics Or

Isaac Newton versus Joseph Louis Lagrange

ABSTRACT

This Study explains how it is possible to conduct real laboratory work in physics (mechanics) with the use of Mathcad. Methods for solving statics, kinematics and dynamics problems are examined by analyzing forces acting on material points (Newtonian approach), and by energy minimization (Lagrange's method). Methods for visualizing solutions using static graphics and dynamic animation with the mapping of the trajectories of the moving material points are described.

Mathematics: Solving equations and systems of equations—algebraic and differential, optimization with constraints.

Physics: Equilibrium of a mechanical system, trajectory of point motion, special types of pendulums, Newton's law, Lagrange's principle.

IT: Communication of informatics with other disciplines—with mathematics, physics, theoretical mechanics, STEM.

Art: Averchenko's story "The Focus of the Great Cinema"; painting a tulip on a computer.

Study website: https://community.ptc.com/t5/PTC-Mathcad-Questions/One-task-with-3-bobs-Forces-searching/m-p/94822.

A SHORT INTRODUCTION TO THE STUDY

There are flawed people. Some do not listen to classical music and become bored at the conservatory, while others do not understand modern art, a third group mistakes the appearance and the smell of an expensive cheese with mold, a fourth group call aged wine with a beautiful bouquet sour and prefer a glass of vodka, a fifth is blind to the beauty of mathematics. Such people should be understood—we should be sorry for them and recommend them to develop their taste and understanding of the world around them. After all, life would become much richer and more interesting. Mathematics, we repeat, is also the best gymnastics for the mind: mastering the logic of mathematics will also help master the logic of life.

It has already been said that in past, only rich and healthy people could afford to travel around the world. Now, with the development of transport, it is possible to move to any point of the Earth in a day, in comfort, and admire the natural or man-made miracles of the World. Something similar can be said about mathematics. Previously, only selected people who received a mathematical education could afford to "travel to its beauties". Now, with the advent of affordable, powerful computer-based mathematical packages, such "travels" can be made by "mere mortals" while enjoying that which earlier only the elite could have received.

This Study is an attempt to convey to readers the mechanical and mathematical pleasure that this author received by entering in contact with what at first glance appeared to be ordinary objects.

Once, the author was clearing the drawers of his desk. At the bottom of one of them were two things that almost went out of use in our electronic computer age: a bundle of millimeter-squared paper and weights for laboratory scales. In the same place were a spool of thread and a box with pins. At that time, this author was contemplating laboratory work for schoolchildren and students, where a real physical experiment would be accompanied by his computer modeling in Mathcad and by a comparison of the results of the "physics" and of the "computer mathematics". The objects in the desk (millimeter paper, weights, threads and pins) prompted the idea of activities in the three components of mechanics: *statics*, *kinematics* and *dynamics*. For these it is possible to show the solution of physical problems according to both classical (Newtonian) mechanics, as well as the Lagrange method: see the abbreviation vs. in the title of the Study (vs., versus: to counter (*Latin*)). The degree to which this work has turned out to be interesting and instructive is for the reader to judge.

PART I: STATICS

A sheet of millimeter-squared paper is attached to the wall, see Figure 26.1. With the help of pins, a thread is fixed to a paper from the two ends. To the thread are attached three weights of 20, 2 and 10 grams taken from a box with weights. The coordinates for fixing the thread

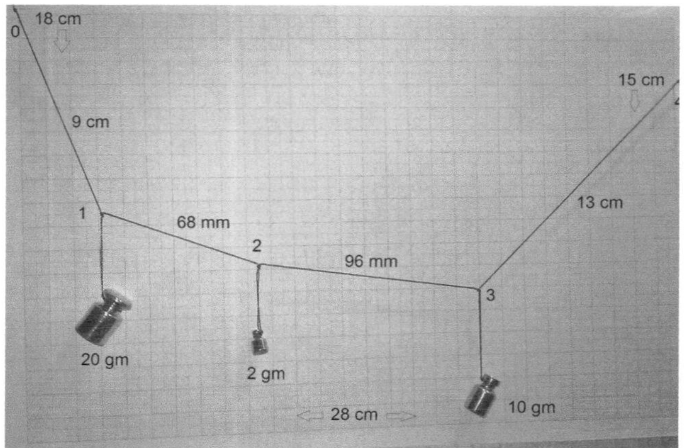

FIGURE 26.1 Physical experiment with weights on a thread.

to the paper are known: 0–180 and 280–150 mm, as well as the length of the thread segments between the nodes with weights: 90, 68, 96 and 130 mm. It is necessary to determine the coordinates of the "nodal" points 1, 2 and 3—the break points of the sagging thread. "Physically" this is easy to do—just look at Figure 26.1 and extract from the millimeter paper the coordinates of the nodes. They are approximately about: 31–95, 95–72 and 191–68 mm. But how to calculate these coordinates manually or on a computer with Mathcad?! First, of course, you need to consult about this problem on the Internet. And for this it is best to post the task on a professional forum—for example, at the Mathcad users forum (https://community.ptc.com/t5/PTC-Mathcad/ct-p/PTCMathcad). These days, in the Internet era, this is a fairly common way of solving problems and is often used by schoolchildren, students, engineers and so on.

After posting the question on the problem, answers "sprang up" from Italy, Germany, the United States, Great Britain, an Italian user with the nickname MF" suggested that our problem belongs to the class of well-known problems of classical mechanics from the statics section called *funicular polygons*. Take a weightless, inextensible, absolutely flexible thread ("rope"), which is attached at both ends and to which forces are applied at different points and in different directions. The rope takes the form of an open polygon, the parameters of which must be determined. If the rope is quite weighty (a metal chain, for example), then, under the action of gravity, uniformly distributed along the length, it will take the form of a curve that is described by a *catenary function* (see Study 7). The task of a rope polygon has to be solved by engineers, when, for example, a suspension bridge is projected. And our design in Figure 26.1 reminds us of this bridge: it is only necessary to add a second thread parallel to the first, and to replace the weights with pedestrians and transport on the roadway. See also Figure 7.6 in Study 7.

In textbooks and handbooks found on the Internet, a number of formulas and cunning algorithms are presented, with which it is possible to solve the problem of a rope polygon. But all these formulas and algorithms were created at a time when there were no computer mathematical packages with their new powerful tools for numerical and analytical problem-solving. Nowadays they are available, and this gives us the opportunity to introduce something new to the solution of this problem.

ORIGIN	gf	TOL	CTOL
1	g·gm	10^{-7}	10^{-7}

x_0 (cm)	y_0 (cm)	x_4 (cm)	y_4 (cm)	$mass$ (gm)	L_{01} (cm)	L_{12} (cm)	L_{23} (cm)	L_{34} (cm)
0	18	28	15	20	9	6.8	9.6	13
				2				
				10				

$$PE(x,y) := g \cdot \sum_{i=1}^{3} \left(mass_i \cdot y_i\right)$$

FIGURE 26.2 Calculating the balance of the weights on the thread—beginning.

Classical (Newtonian) mechanics recommends in this case that we write down the system of equations for the balance of forces and solve it—see, for example, Figure 7.9 in Study 7. We will do this—but at the end of the Study. For now, we will forget for a moment about Newton with his forces and remember Lagrange with his energies, and also remember that in Mathcad there are built-in tools for the numerical solution of *optimization problems with constraints.* Our weights in Figure 26.1 hang so that their *potential energy* is minimal. Do not let the weights fall to the floor, where their potential energy will become zero: the thread spacings between nodes—these are the constraints. Here lies the deciphering of the second title of the Study "Newton vs. Lagrange": we analyze the energy, not the force. We already did this in Study 12.

Figures 26.2 through 26.6 show a solution in Mathcad Prime of the problem of three weights on a thread using the built-in function Minimize.

Solve

Guess Values

x (cm)	y (cm)
5	10
7	8
10	7

Constraints

$$L_{01} = \sqrt{\left(x_1 - x_0\right)^2 + \left(y_1 - y_0\right)^2} \qquad L_{12} = \sqrt{\left(x_1 - x_2\right)^2 + \left(y_1 - y_2\right)^2}$$

$$L_{23} = \sqrt{\left(x_2 - x_3\right)^2 + \left(y_2 - y_3\right)^2} \qquad L_{34} = \sqrt{\left(x_4 - x_3\right)^2 + \left(y_4 - y_3\right)^2}$$

Solver

$$\begin{bmatrix} x \\ y \end{bmatrix} := \text{Minimize}\left(PE, x, y\right)$$

$$x = \begin{bmatrix} 3.09 \\ 9.44 \\ 18.93 \end{bmatrix} \text{cm} \quad \begin{bmatrix} 3.1 \\ 9.5 \\ 18.9 \end{bmatrix} \text{Data from Fig. 1} \qquad y = \begin{bmatrix} 9.55 \\ 7.11 \\ 5.69 \end{bmatrix} \text{cm} \quad \begin{bmatrix} 9.5 \\ 7.2 \\ 5.8 \end{bmatrix} \text{Data from Fig. 1}$$

FIGURE 26.3 Calculating the balance of weights on a thread (the end, while the beginning is in Figure 26.2).

Figure 26.2 shows, first, the *configuration* of Mathcad for the solution of this problem; second, the input of the constant parameters of the thread with the weights; and, third, the *creation* of a user-defined function with the name *PE*, which returns the potential energy of the three hanging weights measured from zero level—from the bottom line of millimeter paper (see Figure 26.1). The *PE* function has two arguments, *x* and *y*, although there is only one argument *y* in its description. A dummy argument *x* is introduced only for the minimization tool to work, which will be shown below in Figure 26.3 and minimizes it by two, and not by one parameter (below in Figure 26.25 the *PE* function will be created without the dummy parameter *x*, but another function, the function *PSE*, will have a dummy argument *F*). When creating a function named *PE* we used the summation operator. But you could do without it, just writing down the sum of three terms.

So, we will do this by solving a similar but simplified problem not with 3, but with 2 weights, see Figure 26.15. Above, in Study 21, we had 9 weights (see Figure 21.1), and even 24 weights (Figure 21.2). Here, without the summation operator, it was difficult to manage. In Study 21 explanations are given as to why it is not advisable to

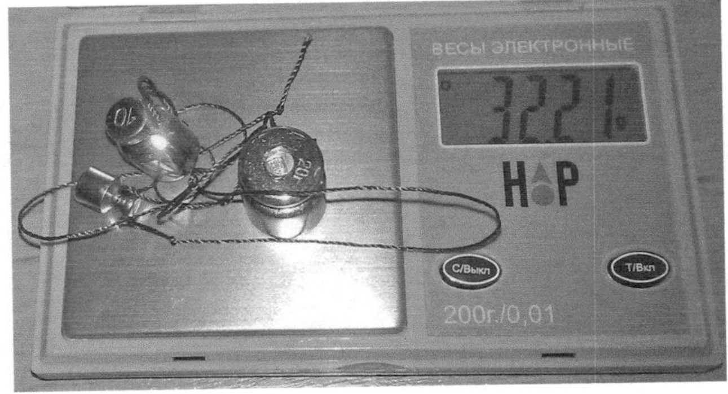

FIGURE 26.4 Weighing three weights and a thread.

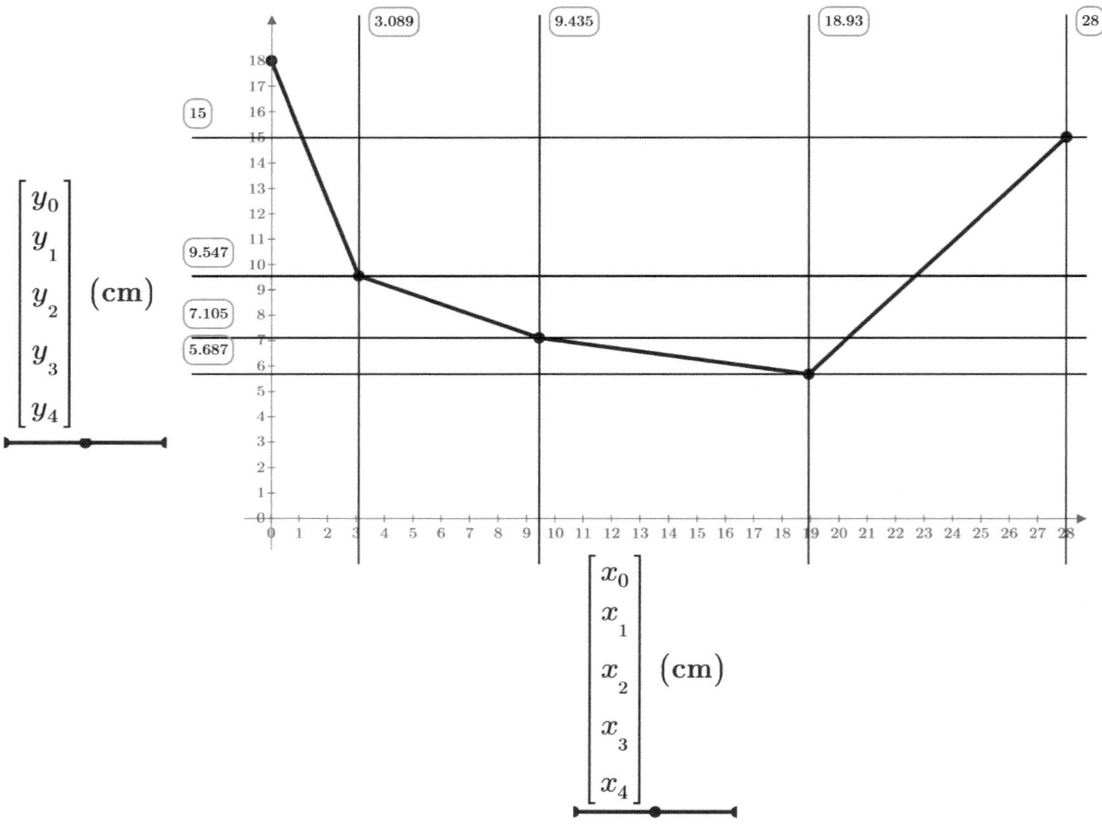

FIGURE 26.5 Graph of the sagging thread with weights.

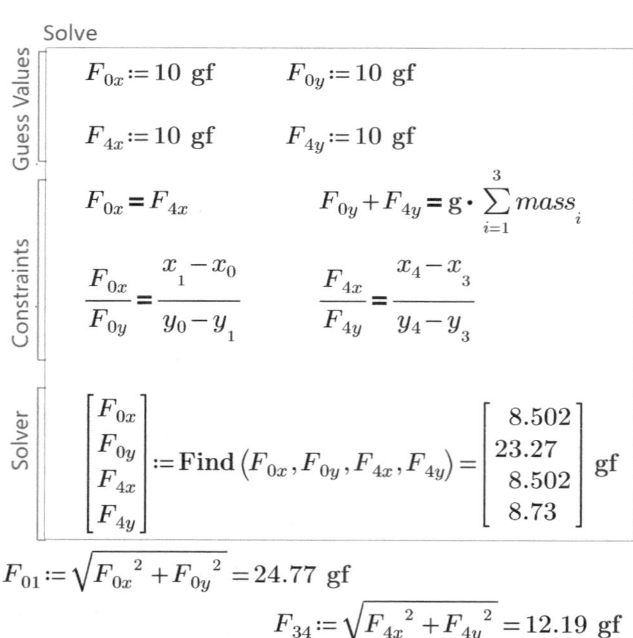

FIGURE 26.6 Calculation of forces that stretch the thread at its two ends.

represent the lengths of the segments as a vector: in the Mathcad *solve* block limits you cannot use the summation operator and other tools based on the vector index and avoid the chain of individual operators.

Mathcad configuration includes the following:

1. Conversion of the measurement system from SI (default) to the cgs system (centimeter-gram-second). The essence of the calculation does not change, but makes it convenient to work on the computer, because the answers will be displayed not in meters and kilograms, but in centimeters and grams. It goes without saying that our physical problem is solved using physical quantities—length, mass, force, time, speed, acceleration, "Evil" programming languages and spreadsheets, we repeat, played a cruel joke with us—they encouraged us to work without physical values, which is inconvenient and fraught with errors. Mathcad fixes this abnormal state of things!

2. Mathcad, tuned to the CGS measurement system (see paragraph 1 above), will give the strength in dynes (dyne, the basic unit of force in the cgs system), which is not very convenient. Therefore, the user unit gf (gram of force) is introduced into the calculation, with which we will replace the dynes shown in the answers. By the way, in Mathcad the units of force kgf (kilogram of force) and lbf (pound of force), ozf (ounce of force), *tonnef* (ton of force, see Figures 24.2, 24.3 and 24.5 in Study 24) are available, but for some reason there is no gram of force.

3. The parameters of our weights will be stored in three vectors: *mass* (mass of weight), *x* and *y* (coordinates of the attachment points of the weights to the thread). The weights are numbered 1, 2, 3, and not 0, 1, 2. Therefore, the system variable *ORIGIN* is assigned a value of 1, not 0 (default). There are also variables with indices 0 and 4 (x_0, y_0, x_4 и y_4), but these are not elements of the vectors *x* and *y*, but scalar quantities for which the index is not the element number of the corresponding vector, but the part of the variable name, shifted down. Points 0 and 4 are the fixed points for fastening the thread. In the second part of the Study, we will move point 4 to the right, solving the kinematic problem rather than the static one. The double meaning of the indexes in our solutions will confuse the inexperienced Mathcad user. It would be better, probably, to write *x0*, *y0*, *x4* and *y4*, and not x_0, y_0, x_4 and y_4. In principle, it would be possible to represent the lengths of the segments in the form of a vector named *L* and with four components with values of 90, 68, 96 and 130 mm. It could be, but it would not be worth it! Why? See above.

4. By default the accuracy of numerical calculations is 0.001. This means that the calculation will be conducted with an accuracy of one gram (one thousandth of a kilogram). And this is not enough. Therefore, the system variables *TOL* and *CTOL* are assigned values of 10^{-7}. The *TOL* variable is responsible for the accuracy of the minimum search, and the *CTOL* variable is responsible for the accuracy of the constraints.

5. The input data are entered into the calculation through a new tool that appeared in Mathcad Prime, through a table whose first line is the name of the variable, the second is the unit of its measurement, and the third is the numerical value or formula (we used this when introducing a gram of force, multiplying the acceleration of gravity by a gram of mass). In such a table, you can also enter vectors (the variable *mass*), and scalars (other variables).

The calculation of the positions of the weights at which their potential energy is minimal is shown in Figure 26.3. At the bottom of Figure 26.3, the result of calculating the values of the vectors *x* and *y* (the values found by the *Minimize* function) and the numbers taken from the millimeter paper in Figure 26.1. Deviations were less than 1%. They are related, among other things, to the fact that the weight of the thread was not taken into account in the calculation. What is this like? Weights, as we have already noted, have already disappeared from laboratories. Now electronic scales are used, which (the reader will laugh!) were also found in the drawer of the author's desk, see Figure 26.4. It was only necessary to replace the old batteries. The scales worked and showed the weight, or rather, the mass of three weights with a thread attached to them. The weight of the thread was 210 mg, which, we repeat, is one of the explanations for the differences in the data taken from the millimeter paper and the one calculated on a computer (see the lower part of Figure 26.3).

Figure 26.5 shows a graphical representation of our solution in Mathcad Prime. Markers (horizontal and vertical lines) indicate the nodal points on the graph with the display of the corresponding numerical values.

Note The graphics package in Mathcad Prime is significantly different from that in Mathcad 15, and we have already noted this aspect in the book. There are advantages and disadvantages. The main drawback of the Mathcad Prime graphics package is the absence of the possibility of drawing a grid. On the "dead" paper graphics (see Figure 26.1), this grid is needed. On it we, in particular, determined the coordinates of the nodal points of our real broken thread. On the live graph, on the graph of Mathcad Prime, the necessary points can be marked with separate "live" markers without the use of a grid. Markers move in two ways: by dragging the marker with the mouse or by inserting the desired numerical value into the label of the marker.

And what forces stretch the thread? To answer this question, one more function from the Mathcad Magnificent Seven (see Study 4), the *Find* function, designed to solve systems of algebraic equations, will help.* Minimizing energy, we repeat, is the realization of the Lagrange method in solving the problem of classical mechanics. The analysis of the acting forces is a Newtonian approach, and we will return to it again at the end of the Study.

Figure 26.6 shows the work of the *Find* function to obtain the values of the forces that pull the thread at its two ends. The *Find* function also works in the *Solve* block with the initial assumptions area, the constraint region where the system equations are written and the solution area. There are four equations in the system with four unknowns. The number of equations and unknowns can be shortened by replacing, for example, the variable F_{0x} with F_{4y}, but we will not do this, leaving the "physics" of the equations clear: the horizontal components (projections) of the forces at the reference points are equal, the sum of the vertical components (projections) of these forces is equal to the weight of suspended weights, and the ratio of the horizontal and vertical components (projections) of forces is equal to the ratio of the lengths of the corresponding legs of the right triangles into which these forces fit. The very same force, stretching the thread, is connected with the hypotenuse of this rectangular triangle. Or with the diagonal of the corresponding rectangle or parallelepiped, which often illustrates the resultant effect of two forces.

On the sites supplementing the Study and on the forum for Mathcad users, where the problem was discussed, one can see an animation showing changes in the shape of the thread with weights and the values of the forces holding the thread when moving the right reference point from left to right. At the moment when the thread is stretched to the limit and becomes almost rectilinear, it "breaks" and *Find* returns error messages. On the website of the Study, there is an animation of the shape change of the broken line (thread) when the value of the accuracy of the numerical solution changes—the values of the variables *TOL* and *CTOL*. Clearly, the wrong solution is seen at values of 10^{-3}, for example, and not 10^{-7}.

The *Find* function is intended for solving systems of equations—linear and non-linear. It is easy to understand that our system of equations is linear (see Figure 26.7) and apply the corresponding specialized tools of Mathcad to solve it: multiplication of the inverted matrix *M* that stores the coefficients for unknowns by a vector *v* that stores the values of the free terms of the system of linear equations, or function *lsolve*. The lsolve function appeared relatively recently in Mathcad. It uses a more advanced algorithm for solving systems of linear algebraic equations (SLAE) and allows us, in particular, to solve underdetermined and overdetermined problems, where the number of equations is not equal to the number of unknowns, and the matrix *M* as a consequence of this inequality is not square. Working with the operator $M^{-1} \cdot v$ or with the *lsolve* function does not require an initial assumption, but it is inconvenient that the "physics" of the problem is hidden. Before forming matrix *M* and vectors v, we had to rewrite (prepare) our system of equations in Figure 26.7 with the addition of zero coefficients and transfer of unknowns to the left side of the equations.

Our problem of the equilibrium of forces can be solved not only numerically (see Figures 26.6 and 26.7), but also analytically, *symbolically*—we automatically derive formulas for calculating the values of forces acting on the ends of the thread, depending on its configuration and mass of weights, see Figure 26.8. The "→" operator will not give a numerical answer (this is done by the "=" operator), but a symbolic one. In our case, before using it, we need to clear the numerical values of the variables by the $clear_{sym}$ operator. Otherwise, the symbolic response will not be the names of variables, but their numerical values or the result of their numerical processing (summation, subtraction, etc.). In the formulas shown in Figure 26.8, you can substitute the numerical values of the variables and get the desired numerical response. The accuracy of the calculation will be much higher than using the numerical iterative methods in the *Find* and *lsolve* functions, as well as in the inversion of the square matrix and multiplication of the matrix by the vector. But we should not forget that the symbolic solutions of more complex problems are either impossible or give very cumbersome answers.

We have already noted that in the system of equations shown in Figures 26.6, 26.7 and 26.8, we can

* By algebraic equations we refer to all types of equations (power, trigonometric and others), and not just those consisting of polynomials. We have already discussed this in Study 4. This distinguishes them from differential ones, and we talked about them in Study 9 of the book and in the third part of this Study.

$$1\cdot F_{0x}+0\cdot F_{0y}-1\cdot F_{04}+0\cdot F_{4y}=0$$

$$0\cdot F_{0x}+1\cdot F_{0y}+0\cdot F_{04}+1\cdot F_{4y}=g\cdot\sum_{i=1}^{3} mass_i$$

$$1\cdot F_{0x}+\frac{x_0-x_1}{y_0-y_1}\cdot F_{0y}+0\cdot F_{04}+0\cdot F_{4y}=0$$

$$0\cdot F_{0x}+0\cdot F_{0y}+1\cdot F_{04}+\frac{x_3-x_4}{y_4-y_3}\cdot F_{4y}=0$$

$$M:=\begin{bmatrix}1 & 0 & -1 & 0\\ 0 & 1 & 0 & 1\\ 1 & \dfrac{x_0-x_1}{y_0-y_1} & 0 & 0\\ 0 & 0 & 1 & \dfrac{x_3-x_4}{y_4-y_3}\end{bmatrix}\qquad v:=\begin{bmatrix}0\\ g\cdot\sum_{i=1}^{3} mass_i\\ 0\\ 0\end{bmatrix}$$

$$\text{lsolve}(M,v)=\begin{bmatrix}8.502\\ 23.27\\ 8.502\\ 8.73\end{bmatrix}\text{gf}\qquad M^{-1}\cdot v=\begin{bmatrix}8.502\\ 23.27\\ 8.502\\ 8.73\end{bmatrix}\text{gf}$$

FIGURE 26.7 Numerical solution of an SLAE in Mathcad according two methods.

remove the first equation, taking into account in the remaining equations that the horizontal components of the forces on the supports are equal, replacing two different variables F_{0x} and F_{4x} by one F_x. It is possible to go further in this direction and reduce the system to one equation, and then solve it symbolically or analytically. A numerical solution of such a final equation is shown in Figure 26.9. Before solving a single equation, the translation of the right-hand side to the left-hand side with a change of sign turns into a function that searches for zero—the value of the argument at which the function is zero. It is approximately equal if we take into account that we use numerical rather than analytical calculation methods.

Single equations can also be solved numerically using the *Find* function, but it is better to use specialized tools

$$\textbf{clear}_{\textbf{sym}}(x,y,mass,x_0,y_0,x_4,y_4)$$

$$\begin{bmatrix}1 & 0 & -1 & 0\\ 0 & 1 & 0 & 1\\ 1 & \dfrac{x_0-x_1}{y_0-y_1} & 0 & 0\\ 0 & 0 & 1 & \dfrac{x_3-x_4}{y_4-y_3}\end{bmatrix}^{-1}\cdot\begin{bmatrix}0\\ g\cdot\sum_{i=1}^{3} mass_i\\ 0\\ 0\end{bmatrix}\rightarrow\begin{bmatrix}\dfrac{g\cdot(mass_1+mass_2+mass_3)\cdot(x_0\cdot x_3+x_4\cdot x_1-x_1\cdot x_3-x_0\cdot x_4)}{x_0\cdot y_3-y_0\cdot x_3-x_4\cdot y_1+y_4\cdot x_1-x_1\cdot y_3+y_1\cdot x_3-x_0\cdot y_4+y_0\cdot x_4}\\ \dfrac{g\cdot(mass_1+mass_2+mass_3)\cdot(y_0\cdot x_3+x_4\cdot y_1-y_1\cdot x_3-y_0\cdot x_4)}{x_0\cdot y_3-y_0\cdot x_3-x_4\cdot y_1+y_4\cdot x_1-x_1\cdot y_3+y_1\cdot x_3-x_0\cdot y_4+y_0\cdot x_4}\\ \dfrac{g\cdot(mass_1+mass_2+mass_3)\cdot(x_0\cdot x_3+x_4\cdot x_1-x_1\cdot x_3-x_0\cdot x_4)}{x_0\cdot y_3-y_0\cdot x_3-x_4\cdot y_1+y_4\cdot x_1-x_1\cdot y_3+y_1\cdot x_3-x_0\cdot y_4+y_0\cdot x_4}\\ \dfrac{g\cdot(mass_1+mass_2+mass_3)\cdot(x_0\cdot y_3+y_4\cdot x_1-x_1\cdot y_3-x_0\cdot y_4)}{x_0\cdot y_3-y_0\cdot x_3-x_4\cdot y_1+y_4\cdot x_1-x_1\cdot y_3+y_1\cdot x_3-x_0\cdot y_4+y_0\cdot x_4}\end{bmatrix}$$

FIGURE 26.8 Symbolic solution of an SLAE.

Equation, could be solved for F_{4y}

$$\frac{F_{4y} \cdot \dfrac{x_4 - x_3}{y_4 - y_3}}{g \cdot \sum_{i=1}^{3} mass_i - F_{4y}} = \frac{x_1 - x_0}{y_0 - y_1}$$

Function, that could be nihil for a given value of F_{4y}

$$f(F_{4y}) := \frac{F_{4y} \cdot \dfrac{x_4 - x_3}{y_4 - y_3}}{g \cdot \sum_{i=1}^{3} mass_i - F_{4y}} - \frac{x_1 - x_0}{y_0 - y_1}$$

Two method to find zeros of a function - call for the function root

$$F_{4y} := \mathbf{root}\left(f(F_{4y}), F_{4y}, 0\ \mathrm{gf}, 20\ \mathrm{gf}\right) = 8.7298\ \mathrm{gf}$$

$$F_{4y} := 1\ \mathrm{gf} \quad F_{4y} := \mathbf{root}\left(f(F_{4y}), F_{4y}\right) = 8.7298\ \mathrm{gf}$$

FIGURE 26.9 Find the zero of a function.

for this. Figure 26.9 shows the operation of another function from the Mathcad Magnificent Seven, the *root* function for numerically searching for a zero of a function using the half-division method (the *root* function with four arguments) or the secant method (the *root* function with two arguments). A four-argument call to the *root* function does not need the first assumption, but it requires a zero search interval (we have 0–20 grams of force). The two-argument entry of the *root* function requires specifying the first assumption (we have 1 gram of force). But the answer in both cases is the same

The *root* function is good because it works in the truncated version of Mathcad Prime—Mathcad Express (which is free of charge), where the *Find* function and many others are muted. What is Mathcad Express? The user downloads the full version of Mathcad Prime from the website ptc.com and works with it for a month. Then it is shortened to Mathcad Express, if no payment is made (see Introduction of the book).

It is also easy to determine the values of the forces that stretch the thread between points 1–2 and 2–3. You can add six more equations to our system (see Figures 26.6, 26.7 and 26.8), and we will do this at the end of the Study "rounding" it. But one can make such a calculation outside the *Solve* block, taking into account the "geometry of right-angled triangles" of unknown forces, see Figure 26.10.

It can be seen from Figure 26.10 that our thread has the highest value in the section from the left support (point 0) to the first point with the heaviest suspended weight. We will clarify this fact in the third part of the Study.

PART II: KINEMATICS

Let's continue the lab work by switching the camera (see one of its pictures in Figure 26.1) from *shooting photograph* to *video* mode!

Let's gently pull out the right pin with the thread attached to it (see Figure 26.1) and slowly and smoothly move this knot to the right in the horizontal direction. Along which trajectories will the other vertices of the angles of our "rope polygon"—points 1, 2 and 3—move at the same time?

We will save the video again on the computer, then we will view it frame-by-frame and fix (digitize) the changing coordinates of the points, drawing them later on the millimeter paper. These trajectories can be calculated in Mathcad, and then compared with video frames.

It is clear that point 1 will move along an arc of a circle with radius L_{01}. To calculate the trajectories of points 2 and 3, the *Solve* block shown in Figure 26.3 is slightly changed, so that it returns not specific numbers combined into vectors *x* and *y*, but a user-defined function (we have it with the name *f*) whose argument has the name x_4 (see Figure 26.11).

The function with the name *f*, which returns a composite array (a vector with two elements, each of which

$$\frac{F_{12}}{F_{0x}} = \frac{L_{12}}{x_2 - x_1} \qquad F_{12} := \frac{L_{12}}{x_2 - x_1} \cdot F_{0x} = 9.109\ gf$$

$$\frac{F_{23}}{F_{4x}} = \frac{L_{23}}{x_3 - x_2} \qquad F_{23} := \frac{L_{23}}{x_3 - x_2} \cdot F_{4x} = 8.596\ gf$$

$$F_{01} = 24.775\ gf \qquad F_{12} = 9.109\ gf$$

$$F_{23} = 8.596\ gf \qquad F_{34} = 12.186\ gf$$

FIGURE 26.10 Finding the forces that stretch the thread between points 1–2 and 2–3.

Solver

$$f(x_4) := \text{Minimize}(PE, x, y)$$

$$x_2(x) := \left(f(x)_1\right)_2 \quad y_2(x) := \left(f(x)_2\right)_2 \quad x_3(x) := \left(f(x)_1\right)_3 \quad y_3(x) := \left(f(x)_2\right)_3$$

FIGURE 26.11 Forming a user function on the basis of the Solve block.

is a vector of three elements), must be decomposed into separate simple (incomplete) functions with names x_2, y_2, x_3 and y_3, Figure 26.11.

Further, it is easy to display these functions graphically by constructing a *parametric graph*, i.e. a graph where the axes are written not by a function (the abscissa axis) and its argument (ordinate axis), but by two different functions, with one argument *xx*—a *parameter* of the graph shown as a discrete variable area on the first line of Figure 26.12.

The *Solve* block is inconvenient because it cannot be inserted into Mathcad programs, and we already discussed this in Study 4. The solution here is: a user function is created based on the *Find*, *Minimize*, *Maximize* or *Minerr* functions, and it is then inserted into the program. By the way, the *Minimize* and *Maximize* functions can work without the *Solve* block if there are no optimization constraints. In this case, these two functions are inserted into the programs without any problem.

There is another way of constructing the trajectories of the displacement of points 2 and 3 when the right support is shifted. It can be conditionally called "worker-peasant", since it does not require the creation of a rather complex function that returns a composite array, but splits it into separate simple functions (see Figure 26.11). When implementing the "worker-peasant" method, we must work mainly with our hands, not with our head (see Figures 26.13 and 26.14).

Figure 26.13 shows such manual work: the Mathcad user manually changes the value of the system variable *FRAME* (it controls the animation, see Study 8) from 0 in steps of 50. Mathcad automatically calculates the values of the variables x_2, y_2, x_3, y_3, that the user enters (copies and inserts—work with hands) into the columns of the matrix *M*. Then, based on these points, functions are

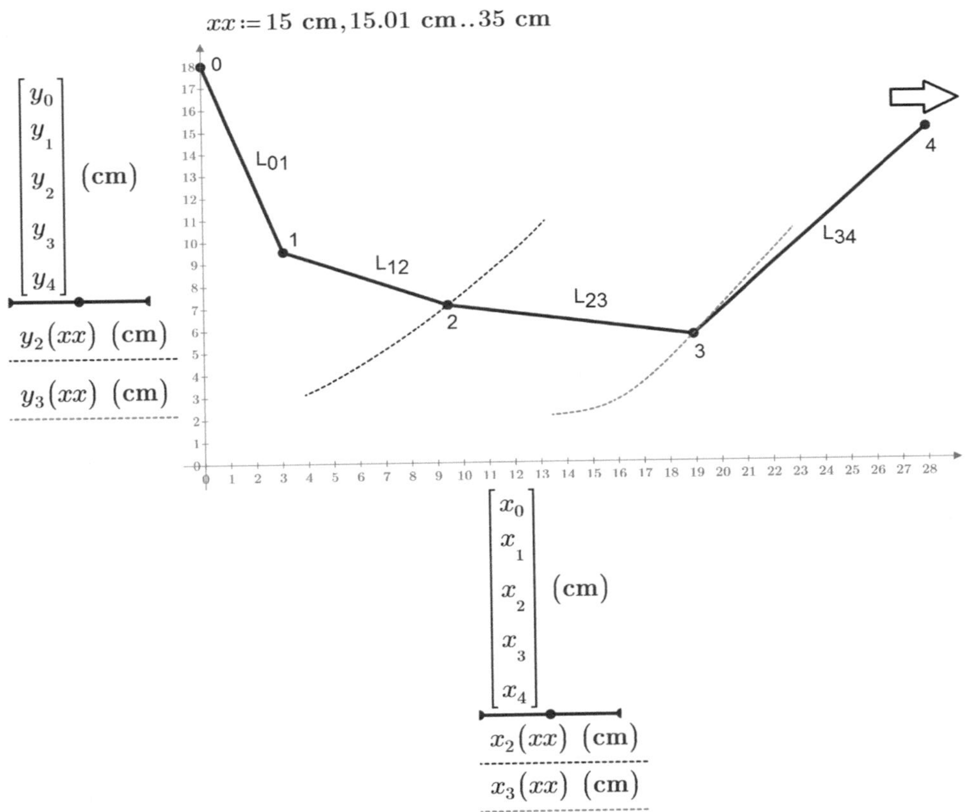

FIGURE 26.12 The construction of the trajectories of the displacement of points 2 and 3 when the right support is shifted to the right (animation https://www.facebook.com/valerij.ockov/videos/1470687612995079).

$x_2 = 1.0439\,\text{cm}$ $\quad y_2 = 2.264\,\text{cm}$ $\quad x_3 = 10.6404\,\text{cm}$ $\quad y_3 = 2.0050\,\text{cm}$

FRAME=	0	50	100	150	200		600	650	682

$$M := \begin{pmatrix} 1.0439 & 2.5004 & 3.865 & 5.0518 & 6.0183 & & 13.2276 & 14.636 & 15.7371 \\ 2.264 & 2.575 & 3.108 & 3.762 & 4.401 & \ldots & 10.892 & 13.008 & 16.594 \\ 10.6404 & 12.085 & 13.4117 & 14.53 & 15.4281 & & 22.8179 & 24.2266 & 25.3059 \\ 2.005 & 2.032 & 2.097 & 2.237 & 2.5003 & & 10.4616 & 12.5835 & 15.8201 \end{pmatrix} \text{cm}$$

$X2 := \left(M^T\right)^{\langle 1 \rangle}$ $\quad Y2 := \left(M^T\right)^{\langle 2 \rangle}$ $\quad y2(x) := \text{linterp}(X2, Y2, x)$

$n := \text{last}(X2) = 15$ $\quad x2 := X2_1, X2_1 + \dfrac{X2_n - X2_1}{300} \,..\, X2_n$

$X3 := \left(M^T\right)^{\langle 3 \rangle}$ $\quad Y3 := \left(M^T\right)^{\langle 4 \rangle}$

$y3(x) := \text{interp}(\text{cspline}(X3, Y3), X3, Y3, x)$ $\quad x3 := X3_1, X3_1 + \dfrac{X3_n - X3_1}{300} \,..\, X3_n$

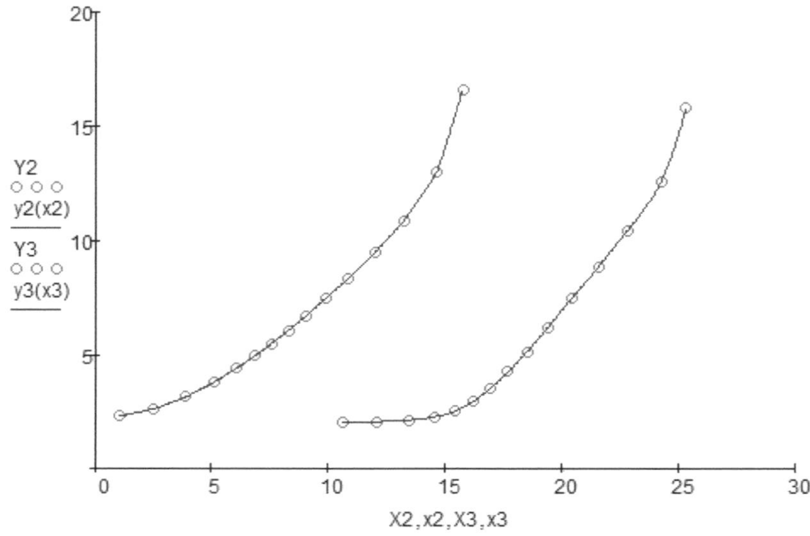

FIGURE 26.13 Piecewise linear interpolation and spline interpolation.

formed *interpolating* the discrete data from the matrix piecewise linearly (left graph) or with splines (right graph).

Interpolation (see Figure 26.13) means the passage of the curve exactly through the points. In our problem of constructing trajectories of the motion of the thread nodes, we can also apply an approximation when the curve passes near the points—see Figure 26.14, which uses the Mathcad built-in function *regress*.

The *regress* function works with a polynomial of degree k (the third argument of the *regress* function). Figure 26.14 (upper graph) shows that the trajectory of the displacement of point 3 is well described by a polynomial of the sixth degree. If you decrease the value of the variable k, noticeable deviations of the curve from the points appear. And what happens if we increase the values of the variable k? It is known that polynomials of k–1 degrees can be drawn through k points. In this case, the approximation becomes an interpolation. But it's not so simple! At high values of k, a phenomenon similar to a *hernia* is observed. See the lower graph in Figure 26.14: the curve of the polynomial passes through the points, but this is difficult to call interpolation. Scientifically, this phenomenon is called *oscillation* (not a hernia). We will talk about oscillations below.

The statistical functions of Mathcad are described in more detail in Study 14.

$$k := 6 \quad y3(x) := \text{interp}\left(\text{regress}\left(\frac{X3}{cm}, \frac{Y3}{cm}, k\right), \frac{X3}{cm}, \frac{Y3}{cm}, \frac{x}{cm}\right)$$

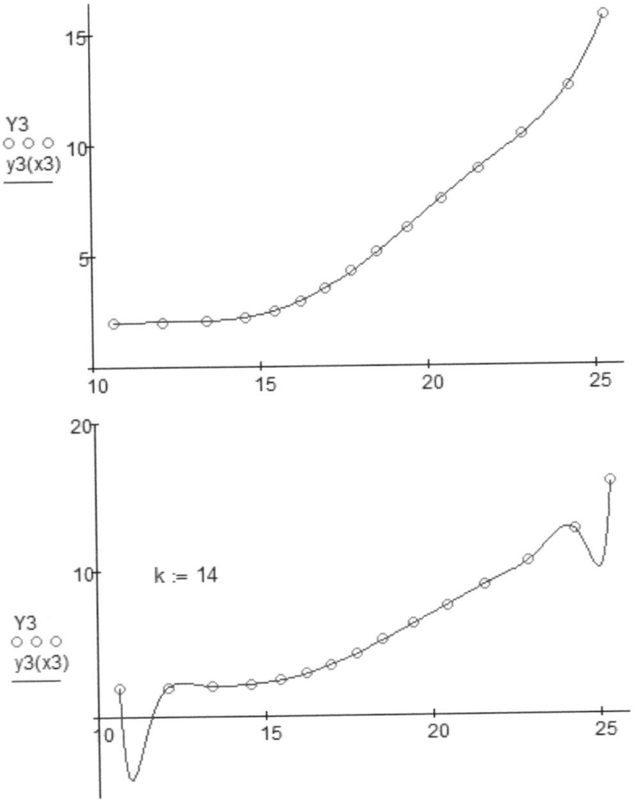

$k := 14$

FIGURE 26.14 Approximation by a polynomial.

When you move point 4 to the right at some point, our task will no longer have a solution, because the length of the thread will not suffice, and the function *Minimize* (Figure 26.11) will start returning error messages. The graph in Figure 26.12, however, will be built, but without displaying unreal sections of trajectories. The real thread with such a shift of the fourth point to the right will break in the area from the left support to the first point where the heaviest weight is suspended. What can happen here—see the second half of the third part of the Study!

Comment

This section of the Study is more correctly titled "Pseudo-Kinematics". The point is that in "real" kinematics, the displacements of the nodal points of the mechanism with rigid links are studied, which we considered in Study 8. Our thread with suspended weights (see Figure 26.1) does not have the rigid links that we use in the third part of the Study.

PART III: DYNAMICS

Our thread with weights, as we already noted at the beginning of the Study, is like a chain or a rope holding the spans of a bridge. The breakage of such a rope entails the destruction of the entire structure. But before such an anthropogenic catastrophe, the bridge, as a rule, starts to swing, to fluctuate under the influence, for example, of strong gusts of wind. On complex bridge structures, this phenomenon is continuously monitored by video recordings or special sensors. Let's simulate this process: we will move the third node of our thread (see Figure 26.1) to the right and down, and then let the weights "fly free". Calculate the trajectories of the nodal points, considering them again as material points with the mass of the corresponding weights—see the figures below. Points 1 and 3 (see Figure 26.1) will then swing along arcs of a circle, while point 2 will make an intricate path, which we will try to fix on the graph and in the animation.

Figure 26.15 shows the solution of the problem of the position of points 1 and 2, if point 3 is diverted to the right by 2 centimeters (it will fall somewhat downward). The same minimization of potential energy is being carried out, but only with two, not three, weights: the third weight is held in hand. The geometry of the thread when it deviates to the right, i.e. at the beginning of its swing (a broken dotted line) and in the equilibrium state (broken solid line) is shown in Figure 26.16. Figure 26.17 shows the area of the Mathcad Prime solve block with the algebraic and differential equations written in it, as well as the initial conditions. The tension forces of the

$$PE1(x_1, y_1, x_2, y_2) := g \cdot mass_1 \cdot y_1 + g \cdot mass_2 \cdot y_2$$

$$x_3 := x_3 + 2\ \text{cm} \qquad y_3 := y_4 - \sqrt{L_{34}^2 - (x_4 - x_3)^2} = 4.091\ \text{cm}$$

Guess Values

$$x_1 := x_1 \qquad y_1 := y_1 \qquad x_2 := x_2 \qquad y_2 := y_2$$

Constraints

$$L_{01} = \sqrt{(x_1 - x_0)^2 + (y_1 - y_0)^2} \qquad L_{12} = \sqrt{(x_1 - x_2)^2 + (y_1 - y_2)^2}$$

$$L_{23} = \sqrt{(x_2 - x_3)^2 + (y_2 - y_3)^2}$$

Solver

$$\begin{bmatrix} x_1 \\ y_1 \\ x_2 \\ y_2 \end{bmatrix} := \text{Minimize}(PE1, x_1, y_1, x_2, y_2)$$

FIGURE 26.15 Calculation of the coordinates of the node points 1 and 2 of the filament at the deviation of its third point.

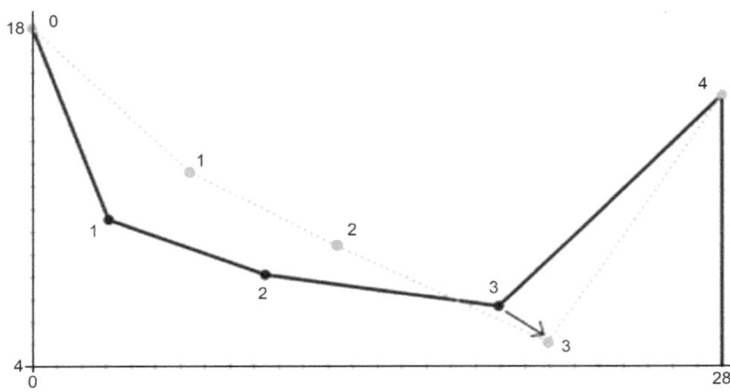

FIGURE 26.16 Geometry of the configuration of the thread with three weights at the beginning of its swaying.

two threads act at points 1, 2 and 3 (we project these forces along the abscissa and ordinate axes) while the gravity forces act only in the vertical direction. All these forces are balanced by the product of the mass of the corresponding material point by its acceleration, the second time derivative of the displacement. It is assumed that at the initial instant of time the velocities of the points are equal to zero. The forces F, stretching the threads, are equal to zero, too—we, as it were, support the canopy of these three fixed points, and then at $t=0$ we immediately release them into flight, constrained by "reins"—the stretched threads.

Solve

Constraints

Point 1

$$x1(0\ \text{s})=x_1 \quad y1(0\ \text{s})=y_1 \quad x1'(0\ \text{s})=0\ \text{kph} \quad y1'(0\ \text{s})=0\ \text{kph}$$

$$L_{01}^2=(x1(t)-x_0)^2+(y1(t)-y_0)^2 \quad F01(0\ \text{s})=0\ \text{N}$$

$$mass_1\cdot x1''(t)=F01(t)\cdot\frac{x_0-x1(t)}{L_{01}}-F12(t)\cdot\frac{x1(t)-x2(t)}{L_{12}}$$

$$mass_1\cdot y1''(t)+\text{g}\cdot mass_1=F01(t)\cdot\frac{y_0-y1(t)}{L_{01}}-F12(t)\cdot\frac{y1(t)-y2(t)}{L_{12}}$$

Point 2

$$x2(0\ \text{s})=x_2 \quad y2(0\ \text{s})=y_2 \quad x2'(0\ \text{s})=0\ \text{kph} \quad y2'(0\ \text{s})=0\ \text{kph}$$

$$L_{12}^2=(x2(t)-x1(t))^2+(y2(t)-y1(t))^2 \quad F12(0\ \text{s})=0\ \text{N}$$

$$mass_2\cdot x2''(t)=F12(t)\cdot\frac{x1(t)-x2(t)}{L_{12}}-F23(t)\cdot\frac{x2(t)-x3(t)}{L_{23}}$$

$$mass_2\cdot y2''(t)+\text{g}\cdot mass_2=F12(t)\cdot\frac{y1(t)-y2(t)}{L_{12}}-F23(t)\cdot\frac{y2(t)-y3(t)}{L_{23}}$$

Point 3

$$x3(0\ \text{s})=x_3 \quad y3(0\ \text{s})=y_3 \quad x3'(0\ \text{s})=0\ \text{kph} \quad y3'(0\ \text{s})=0\ \text{kph}$$

$$L_{23}^2=(x3(t)-x2(t))^2+(y3(t)-y2(t))^2 \quad F23(0\ \text{s})=0\ \text{N}$$

$$mass_3\cdot x3''(t)=F23(t)\cdot\frac{x2(t)-x3(t)}{L_{23}}-F34(t)\cdot\frac{x3(t)-x_4}{L_{34}}$$

$$mass_3\cdot y3''(t)+\text{g}\cdot mass_3=F23(t)\cdot\frac{y2(t)-y3(t)}{L_{23}}-F34(t)\cdot\frac{y3(t)-y_4}{L_{34}}$$

$$L_{34}^2=(x_4-x3(t))^2+(y_4-y3(t))^2 \quad F34(0\ \text{s})=0\ \text{N}$$

FIGURE 26.17 The system of equations for swinging three weights on a thread.

$$\begin{bmatrix} x1 \\ y1 \\ x2 \\ y2 \\ x3 \\ y3 \\ F01 \\ F12 \\ F23 \\ F34 \end{bmatrix} := \mathbf{Odesolve}\left(\begin{bmatrix} x1(t) \\ y1(t) \\ x2(t) \\ y2(t) \\ x3(t) \\ y3(t) \\ F01(t) \\ F12(t) \\ F23(t) \\ F34(t) \end{bmatrix}, t_e\right)$$

FIGURE 26.18 Numerical solution of the system of equations of oscillation of three weights on a thread.

Note The function *Odesolve* returns not real functions, but certain pseudo-functions (see the list in Figure 26.18), obtained by interpolating the points of the numerical solution of a system of differential and algebraic equations. On these pseudo-functions, you can build graphs (which we did in Figure 26.19) and look for special points from them, but you cannot see their analytical form.

When the point 3 is forcibly moved to the right and down (see Figure 26.16), the thread in section 0–1 can break off. Then our three weights will turn into a triple pendulum, the oscillations of which are also easy to simulate. For this, it is sufficient in the calculation shown in Figure 26.17, to expel four operators, see Figure 26.20. It is also necessary to remove the row of the vector in the Solver (see Figure 26.18), where the thread tension force *F01*, linking points 0 and 1 (the torn section) is mentioned.

Figure 26.21 shows the trajectories of the weights 1, 2 and 3 during the first second after breaking the thread at the left of the first point. The third weight moves like it and is placed along an arc of a circle and returns (almost?) to the starting point. Two other weights (apologies—material points) follow complex trajectories of motion and unfortunately return back to the starting points.

Note on Animation in Mathcad Prime

There is no happiness in life! In Mathcad Prime, there was an excellent possibility of solving differential equations using units of measurement. This is very convenient when implementing dynamic physical models.

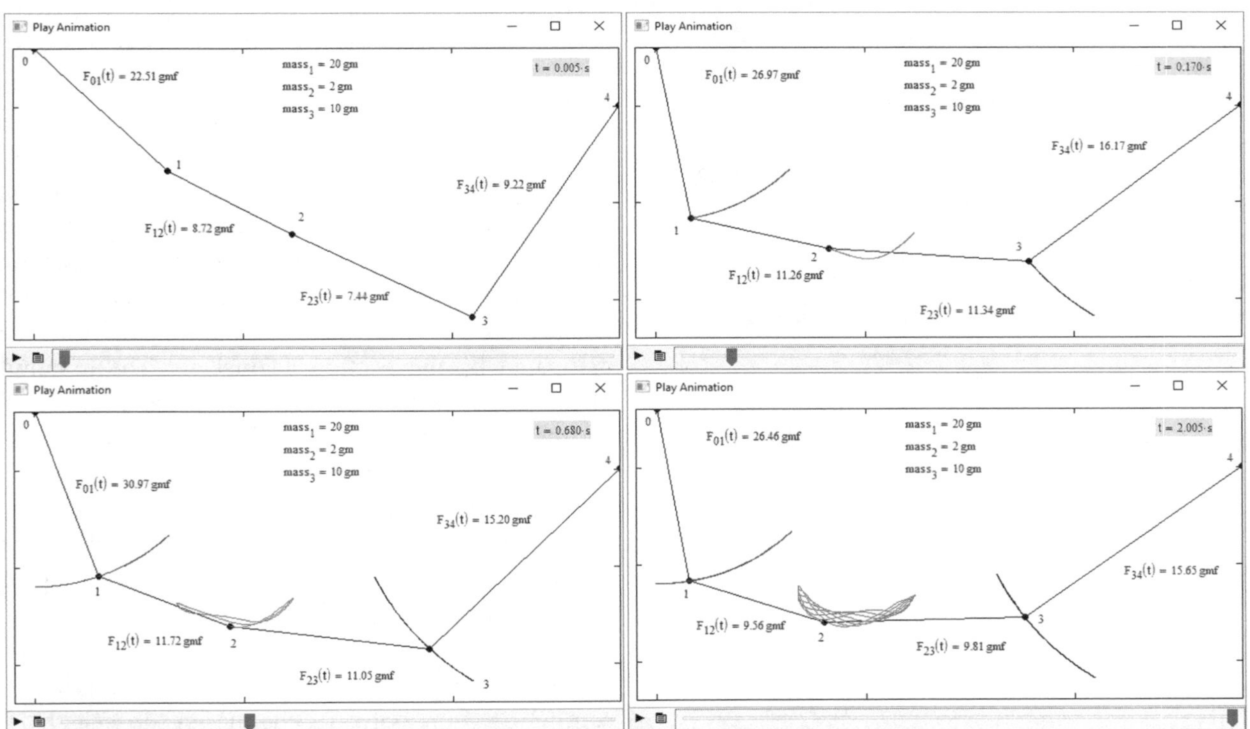

FIGURE 26.19 Four-frame animation of the swing of three weights on a thread.

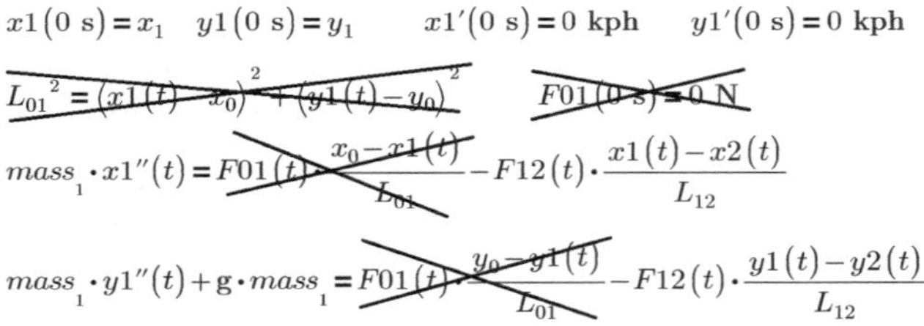

FIGURE 26.20 Equations for the first weight of the system of equations for the swinging of the triple pendulum.

We used this in both this Study and in Study 4. But this new version of the package, alas, does not yet have the means to automatically create animations, something that we wrote about in Study 8. But if you cannot, and yet really want to, then you can! Using again some kind of "worker-peasant" method—i.e. the method of work is mainly with the hands, not with the head.

Figure 26.22 shows the graphic files created in Mathcad Prime when the position of our triple pendulum is fixed after 0.02 sec (file named 002.png), 0.04 sec (file named 004.png), etc. to $t_e = 1$ s (file with name 100. png). These files were created like this: the values of the variable t_e was changed, which changed the form of the graph in Figure 26.21. This drawing was transferred to the Paint graphics editor, where it was edited and saved on a disc with the appropriate name. Further, from these files, animation was created using the website at https:// ru.toolson.net/GifAnimation/Create.

Analysis of the forces that stretch the thread (see the graphs in Figure 26.23) shows that after about one-and-one-tenth of a second (1.1 s) of the motion, the tensile force of the uppermost thread, the force *F34* (dotted on the lower graph), changes to a compression force, which does not meet our models—we have weights hanging on threads and not connected by weightless rigid rods.

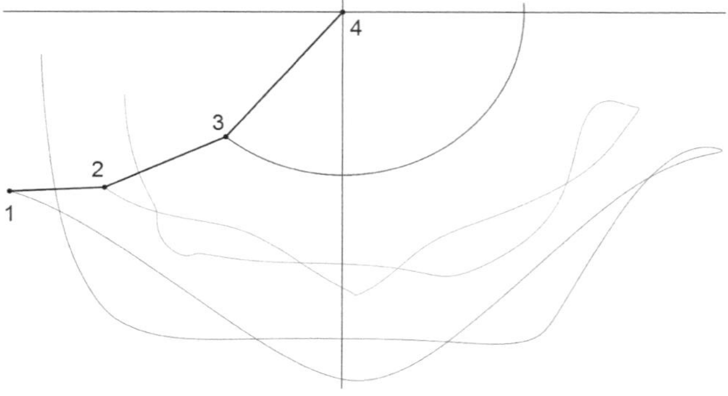

FIGURE 26.21 Trajectories of the motion of three weights after a break in the left thread.

Although such a model also has the right to exist and to be mathematically analyzed. The abscissa axis (horizontal axis) in the graph in Figure 26.23 can be called an area of weightlessness.

If in the system of equations If in the system of equations describing the dynamic model of swinging of a whole thread and shown in Figure 26.17, remove the terms with the second derivative—with acceleration, then we get a system of equations that describes our static problem, shown in Figure 26.1. Figure 26.24 shows the operation of the *Find* function, designed to solve systems of algebraic equations.

Figure 26.24 shows the solution of our problem of the weight-hanging weights (see Figure 26.1) "according to Newton" (force analysis), and not "according to Lagrange" (energy analysis, see Figures 26.2 and 26.3). The answers, of course, coincide!

Now let's replace the inextensible threads with stretchable rubber bands and take into account Hooke's law: the degree of stretching of the rubber is proportional to the force that stretches it.

Figure 26.25 shows the solution of this problem "according to Lagrange". Applying the method of minimizing the sum of two types of energy (*PSE*)—the potential energy hanging the weights (*PE*) and the potential energy of the stretched rubber bands (*SE*). In Figure 26.26 the problem is further solved "according to Newton" taking into account the balance of forces. Both calculations gave the same results. On the website for this Study it is possible to find the animation of the weights when the inextensible thread is instantly replaced by rubber bands and the resistance of the environment is taken into account. The system thus changes from one stable state to another after the damped oscillations. The potential energy of the weights is partially dissipated by the friction against air. The remainder of the energy is converted into potential energy of the weights in the new position and into potential energy of the stretched rubber bands.

The PTC Community website has solved a somewhat complicated problem of sagging weights, where inextensible threads are replaced with weightless springs (elastic bands) with different elasticity coefficients at different sites. The Lagrange and Newton methods are applied. When implementing the Lagrange method, the sum (*PSE*) of the potential energies of the weights (*PE*) and the potential energy of the springs (*SE*) were taken into account. The two calculations gave the same results.

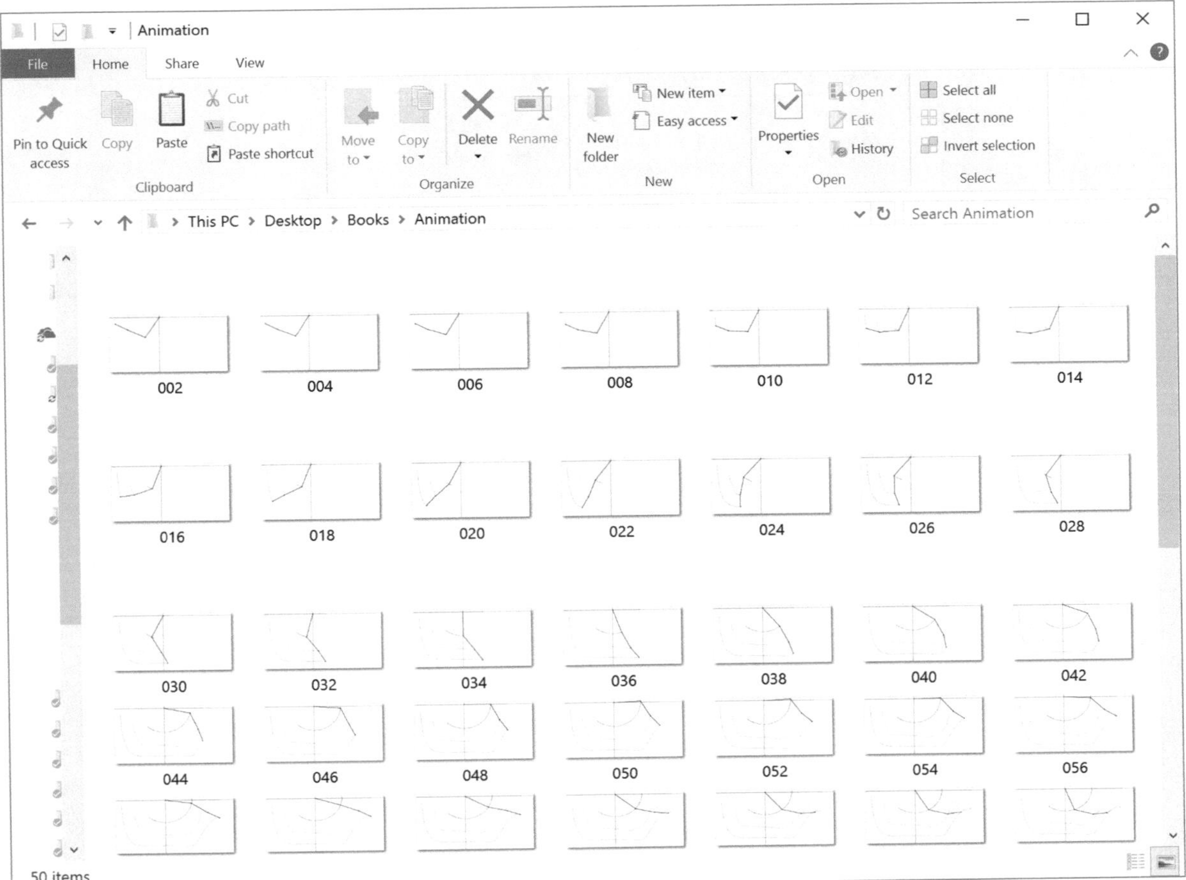

FIGURE 26.22 A set of graphic files—frames of future animation.

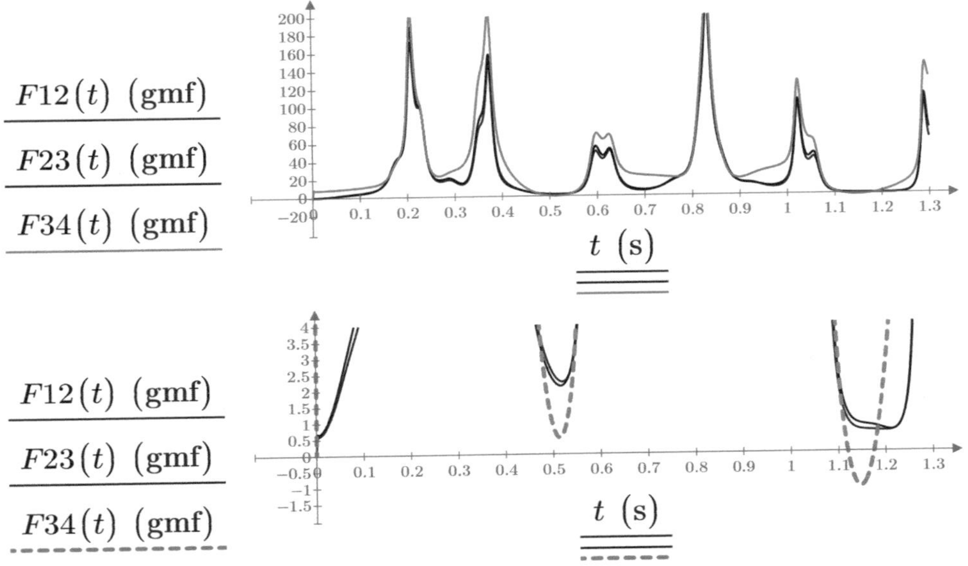

FIGURE 26.23 The graph of the change in forces acting on the pendulum thread.

L (cm)	$mass$ (gm)	x_0 (cm)	y_0 (cm)	x_4 (cm)	y_4 (cm)	gmf
9	20	0	18	28	15	$\frac{kgf}{1000}$
6.8	2					
9.6	10					
13						

Guess Values

$$x := \begin{bmatrix} 5 \\ 7 \\ 10 \end{bmatrix} cm \qquad y := \begin{bmatrix} 5 \\ 7 \\ 10 \end{bmatrix} cm \qquad F := \begin{bmatrix} 10 \\ 10 \\ 10 \\ 10 \end{bmatrix} gmf$$

Constraints

$$L_1 = \sqrt{\left(x_0 - x_1\right)^2 + \left(y_0 - y_1\right)^2}$$

$$F_1 \cdot \frac{x_0 - x_1}{L_1} = F_2 \cdot \frac{x_1 - x_2}{L_2} \qquad g \cdot mass_1 = F_1 \cdot \frac{y_0 - y_1}{L_1} - F_2 \cdot \frac{y_1 - y_2}{L_2}$$

$$L_2 = \sqrt{\left(x_1 - x_2\right)^2 + \left(y_1 - y_2\right)^2}$$

$$F_2 \cdot \frac{x_1 - x_2}{L_2} = F_3 \cdot \frac{x_2 - x_3}{L_3} \qquad g \cdot mass_2 = F_2 \cdot \frac{y_1 - y_2}{L_2} - F_3 \cdot \frac{y_2 - y_3}{L_3}$$

$$L_3 = \sqrt{\left(x_3 - x_2\right)^2 + \left(y_3 - y_2\right)^2}$$

$$F_3 \cdot \frac{x_2 - x_3}{L_3} = F_4 \cdot \frac{x_3 - x_4}{L_4} \qquad g \cdot mass_3 = F_3 \cdot \frac{y_2 - y_3}{L_3} - F_4 \cdot \frac{y_3 - y_4}{L_4}$$

$$L_4 = \sqrt{\left(x_4 - x_3\right)^2 + \left(y_4 - y_3\right)^2}$$

Solver

$$\begin{bmatrix} x \\ y \\ F \end{bmatrix} := \mathbf{Find}\left(x, y, F\right)$$

$$x = \begin{bmatrix} 3.09 \\ 9.44 \\ 18.93 \end{bmatrix} cm \qquad y = \begin{bmatrix} 9.55 \\ 7.11 \\ 5.69 \end{bmatrix} cm \qquad F = \begin{bmatrix} 24.775 \\ 9.109 \\ 8.596 \\ 12.186 \end{bmatrix} gmf$$

FIGURE 26.24 A system of equations describing a static problem without minimizing the potential energy, and taking into account the balance of forces.

L (cm)	mass (gm)	x_0 (cm)	y_0 (cm)	x_4 (cm)	y_4 (cm)	k $\left(\frac{m}{N}\right)$	gmf
9	20	0	18	28	15	0.03	$\frac{kgf}{1000}$
6.8	2					0.04	
9.6	10					0.05	
13						0.06	

$$PE(y) := g \sum_{i=1} \left(mass_i \cdot y_i\right)$$

$$SE(x,y) := \frac{1}{2}\cdot\left(\frac{\left(L_1-\sqrt{\left(x_0-x_1\right)^2+\left(y_0-y_1\right)^2}\right)^2}{k_1}+\frac{\left(L_2-\sqrt{\left(x_1-x_2\right)^2+\left(y_1-y_2\right)^2}\right)^2}{k_2}+\frac{\left(L_3-\sqrt{\left(x_3-x_2\right)^2+\left(y_3-y_2\right)^2}\right)^2}{k_3}+\frac{\left(L_4-\sqrt{\left(x_3-x_4\right)^2+\left(y_3-y_4\right)^2}\right)^2}{k_4}\right)$$

$$PSE(x,y,F) := PE(y) + SE(x,y)$$

Solve

Guess Values

$$x := \begin{bmatrix} 5 \\ 7 \\ 10 \end{bmatrix} cm \qquad y := \begin{bmatrix} 5 \\ 7 \\ 10 \end{bmatrix} cm \qquad F := \begin{bmatrix} 10 \\ 10 \\ 10 \\ 10 \end{bmatrix} gmf$$

Constraints

$$\left(L_1+F_1\cdot k_1\right)^2 = \left(x_0-x_1\right)^2+\left(y_0-y_1\right)^2 \qquad \left(L_2+F_2\cdot k_2\right)^2 = \left(x_1-x_2\right)^2+\left(y_1-y_2\right)^2$$

$$\left(L_3+F_3\cdot k_3\right)^2 = \left(x_2-x_3\right)^2+\left(y_2-y_3\right)^2 \qquad \left(L_4+F_4\cdot k_4\right)^2 = \left(x_3-x_4\right)^2+\left(y_3-y_4\right)^2$$

Solver

$$\begin{bmatrix} x \\ y \\ F \end{bmatrix} := \mathbf{Minimize}\left(PSE, x, y, F\right)$$

$$x = \begin{bmatrix} 2.926 \\ 9.466 \\ 19.313 \end{bmatrix} cm \qquad y = \begin{bmatrix} 8.738 \\ 5.945 \\ 4.436 \end{bmatrix} cm \qquad F = \begin{bmatrix} 24.245 \\ 7.941 \\ 7.388 \\ 11.498 \end{bmatrix} gmf$$

FIGURE 26.25 Calculation of the sagging of three weights on an elastic band—minimization with constraints (k is the reciprocal of the elasticity of the rubber band).

On the website for this Study you can find the animation of the weights when the inextensible thread is instantly replaced by springs. The resistance of the medium is recorded. The system thus changes from one stable state to another after the damped oscillations. The potential energy of the weights is partially dissipated by friction against air. The rest of the energy goes into the potential energy of the weights in the new position and into the potential energy of the rubber springs.

But with a decrease in the values of the coefficient k from 0.03 to, for example, 0.000001, the difference in the results of the work of the Newton method and the Lagrange method becomes noticeable, which can be explained by the peculiarities of the numerical (approximate) solution of the problems.

Which method to use to solve the problem of the sagging springs (rubber bands) with weights—according to Lagrange or according to (the easier) Newton—is for the reader to decide. Lagrange has a simpler system of equations (constraints), but a rather complex function for calculating the potential energy of the springs. In the past, we repeat, there were no simple, accurate, accessible and fast computer optimization tools, but there were only quite complex and hard-to-reach tools for solving systems of equations. Therefore, in the textbooks of mechanics, the problems of statics are solved, as a rule, only by Newton.

L (cm)	$mass$ (gm)	x_0 (cm)	y_0 (cm)	x_4 (cm)	y_4 (cm)	k $\left(\frac{m}{N}\right)$	gmf
9	20	0	18	28	15	0.03	$\frac{kgf}{1000}$
6.8	2					0.04	
9.6	10					0.05	
13						0.06	

Guess Values

$$x := \begin{bmatrix} 5 \\ 7 \\ 10 \end{bmatrix} cm \qquad y := \begin{bmatrix} 5 \\ 7 \\ 10 \end{bmatrix} cm \qquad F := \begin{bmatrix} 10 \\ 10 \\ 10 \\ 10 \end{bmatrix} gmf$$

Constraints

$$\left(L_1 + F_1 \cdot k_1\right)^2 = \left(x_0 - x_1\right)^2 + \left(y_0 - y_1\right)^2$$

$$F_1 \cdot \frac{x_0 - x_1}{L_1 + F_1 \cdot k_1} = F_2 \cdot \frac{x_1 - x_2}{L_2 + F_2 \cdot k_2} \qquad g \cdot mass_1 = F_1 \cdot \frac{y_0 - y_1}{L_1 + F_1 \cdot k_1} - F_2 \cdot \frac{y_1 - y_2}{L_2 + F_2 \cdot k_2}$$

$$\left(L_2 + F_2 \cdot k_2\right)^2 = \left(x_1 - x_2\right)^2 + \left(y_1 - y_2\right)^2$$

$$F_2 \cdot \frac{x_1 - x_2}{L_2 + F_2 \cdot k_2} = F_3 \cdot \frac{x_2 - x_3}{L_3 + F_3 \cdot k_3} \qquad g \cdot mass_2 = F_2 \cdot \frac{y_1 - y_2}{L_2 + F_2 \cdot k_2} - F_3 \cdot \frac{y_2 - y_3}{L_3 + F_3 \cdot k_3}$$

$$\left(L_3 + F_3 \cdot k_3\right)^2 = \left(x_3 - x_2\right)^2 + \left(y_3 - y_2\right)^2$$

$$F_3 \cdot \frac{x_2 - x_3}{L_3 + F_3 \cdot k_3} = F_4 \cdot \frac{x_3 - x_4}{L_4 + F_4 \cdot k_4} \qquad g \cdot mass_3 = F_3 \cdot \frac{y_2 - y_3}{L_3 + F_3 \cdot k_3} - F_4 \cdot \frac{y_3 - y_4}{L_4 + F_4 \cdot k_4}$$

$$\left(L_4 + F_4 \cdot k_4\right)^2 = \left(x_4 - x_3\right)^2 + \left(y_4 - y_3\right)^2$$

Solver

$$\begin{bmatrix} x \\ y \\ F \end{bmatrix} := \mathbf{Find}(x, y, F)$$

$$x = \begin{bmatrix} 2.926 \\ 9.466 \\ 19.313 \end{bmatrix} cm \qquad y = \begin{bmatrix} 8.738 \\ 5.945 \\ 4.436 \end{bmatrix} cm \qquad F = \begin{bmatrix} 24.245 \\ 7.941 \\ 7.388 \\ 11.498 \end{bmatrix} gmf$$

FIGURE 26.26 Calculation of the sagging of three weights on a rubber band—balance of forces.

To the previous remarks we could add that the principle of least action, the Lagrange equation and the entire “Lagrangian formalism” is derived directly from Newton’s law. The Lagrangian formalism makes it possible to reduce the number of equations in the system up to the number of degrees of freedom, taking into account the constraints. According to the principle of least action, the system on the time interval $[t_1, t_2]$ always moves so that its action takes the smallest possible value. The action over the time interval $[t_1, t_2]$ is the integral of the Lagrange function with respect to time (the difference between the kinetic and potential energy of the system).

CONCLUSIONS

The author of this book, as has already been noted many times, conducts classes on numerical methods in Mathcad with students at the advanced Moscow Lyceum

No. 1502 at the Moscow Power Engineering Institute. In the Lyceum, physical cabinets are located next to the offices of computer science. In computer science lessons students learn the Pascal language. Such lessons are actually not computer science lessons, but programming lessons, which, as the reader understands, is far from the same. At the same time, modern computer mathematical packages, which can help students not only learn computer science, but also solve both simple and complex school problems in mathematics, physics, chemistry, are completely ignored. Now, in computer science lessons, abstract problems that are not related to the educational process are solved. And such, interdisciplinary communication is very important.

So that's it. Schoolchildren or students, after completing laboratory work in the physics lab, can go to the informatics class and implement a mathematical model of the physical phenomenon on the computer, which, the reader understands, will significantly improve the efficiency of the class. An example of such a complex work was given above.

At the end of the book there is a list of similar laboratory work combining physics, mathematics and computer science.

AND THE LAST

In numerous articles and books using the Lagrange method, an example of a solution of the problem of a point bead of mass m moving without friction along a stationary vertical ring is given. It is concluded that this problem is solved much more easily if we take into account energies, not forces. We gave another example. And most importantly, they showed that modern mathematical programs with their numerical methods can radically change the idea of which solution is simpler and which is more difficult. The same bead, "rolling" from a textbook into another textbook, can be rolled with friction and with an allowance for air resistance. Here Lagrange cannot help! I'll have to call for help from the good old Newton.

Pendulum-Tulip

Let us return to the very first drawing of this Study with the hanging weights.

If the thread at the two ends of the suspension is not tied with a pin but sent through wheels fixed by the pins (through the blocks) and a couple more weights are connected to the ends of the sagging threads, we get a new more complicated balance problem. Let's first simplify it somewhat—let's leave only one weight between the blocks (Figure 26.27).

The extreme weights will be placed in air cylinders (shock absorbers), restraining the movement of the weights with a force proportional to the square of the speed of the weights. We designate the proportionality coefficient as k. We wrote about such shock absorbers at the end of Study 8, when we considered plantigrade machines. The shock absorber, unlike a rubber band or a spring, cannot store energy in a static state. The shock absorber can "hold" energy in the dynamics for a short time.

Figure 26.28 shows the trajectory of the central weights after they have been pulled down and released into the free flight without taking into account the resistance forces in the shock absorbers ($k=0$).

We get a pendulum with a period of oscillation of about 15.37 seconds. In Figure 26.28 seven "loops" of this pendulum are fixed. Our central weight (we again consider it as a material point) will return to its initial position every 15.37 seconds, at the bottom of the trajectory shown in Figure 26.28. In the center of the figure, the intersection of two straight lines with the coordinates $x=1.55$ $y=-0.689$ marks the equilibrium point, from which we pulled the central weights down and to the right. There is no friction, hence no energy dissipation ($k=0$), so our pendulum will oscillate infinitely often around the equilibrium point, converting potential energy into kinetic energy and back again.

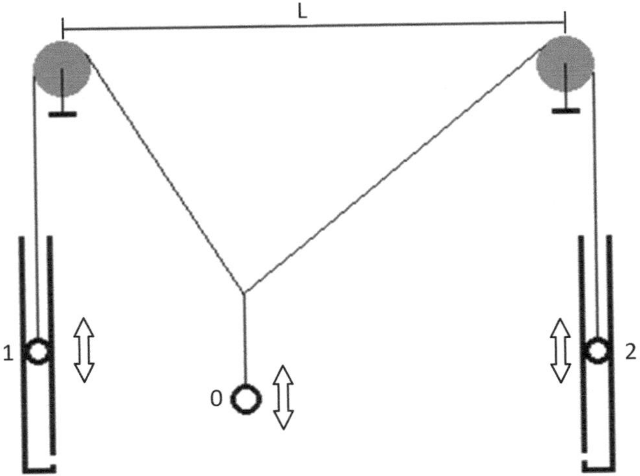

FIGURE 26.27 Scheme of a pendulum with two blocks and two shock absorbers.

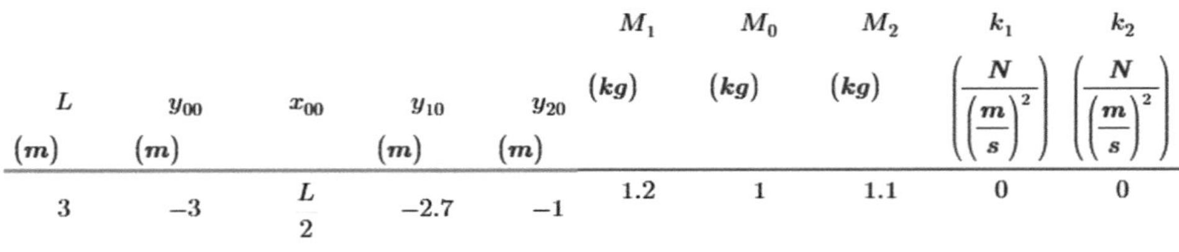

L (m)	y_{00} (m)	x_{00}	y_{10} (m)	y_{20} (m)	M_1 (kg)	M_0 (kg)	M_2 (kg)	k_1 $\left(\frac{N}{\left(\frac{m}{s}\right)^2}\right)$	k_2 $\left(\frac{N}{\left(\frac{m}{s}\right)^2}\right)$
3	−3	$\frac{L}{2}$	−2.7	−1	1.2	1	1.1	0	0

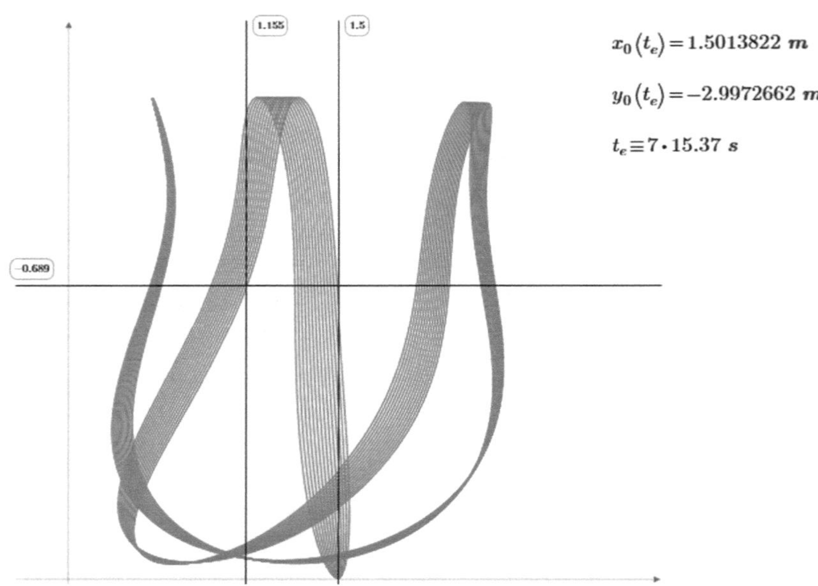

FIGURE 26.28 The trajectory of the pendulum (point 0) with disconnected air shock absorbers.

Figure 26.29 shows the trajectory of the motion of the central weights, taking into account the resistance forces in the shock absorbers ($k>0$): the intricate trajectory has a beginning (the point on the X-axis where we pulled the weight down) and the end (the point of equilibrium, the point of intersection of the two straight lines).

Near the equilibrium point, the trajectory of motion is superimposed on itself, which makes it impossible to track it. But we will return to this stage of the motion of the point at the very end of this Study, when we will work with anti-animation.

Figure 26.30 shows the equations that are solved using the built-in function *Odesolve*. The equations set the constancy of the lengths of two segments of the threads L_{10} and L_{12}, thrown over the blocks, and the balances of the forces acting on three material points. Forces acting on the central material point are projected horizontally and vertically. For the other two points, of course, only the "vertical" are considered. To the gravitational force of these weights and to the tension of the thread is added the force of resistance to motion in the air cushions, which leads to the damping of the motion of the pendulum that is recorded in Figure 26.29.

Under certain initial conditions, the trajectory of the motion of the central pendulum turns into a tulip covered with dew, the "dissolution" of which is shown in Figure 26.28, and the "flowering" in Figure 26.31. Figure 26.31 is a bit rough, but interesting from the standpoint of mathematics.

Animation of the tulip pendulum can be seen on the website of this Study. The solution embedded in the animation is not based on the coordinates of the points, but on the angles of deflection from the vertical of the threads supporting the central load. It was suggested by Alan Stevens from Great Britain. By the way, he showed that the frictional force cannot be decomposed into vertical and horizontal components by multiplying the friction coefficient, the cross-sectional area of the body, the density of the medium and the square of the corresponding velocity projection. Here you need to calculate the absolute value of the frictional force, which can then be decomposed into vertical and horizontal components.

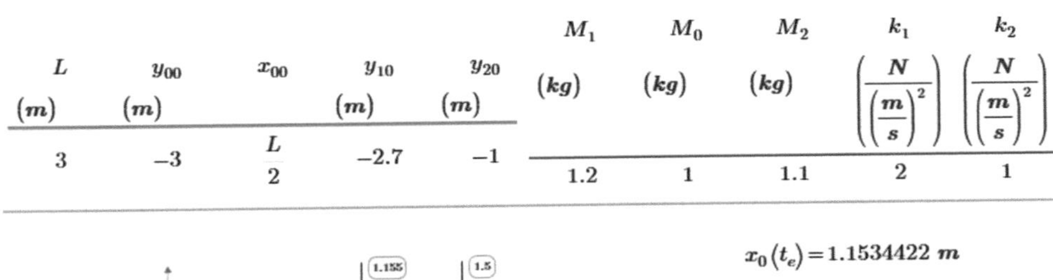

L (m)	y_{00} (m)	x_{00}	y_{10} (m)	y_{20} (m)	M_1 (kg)	M_0 (kg)	M_2 (kg)	k_1 $\left(\frac{N}{\left(\frac{m}{s}\right)^2}\right)$	k_2 $\left(\frac{N}{\left(\frac{m}{s}\right)^2}\right)$
3	−3	$\frac{L}{2}$	−2.7	−1	1.2	1	1.1	2	1

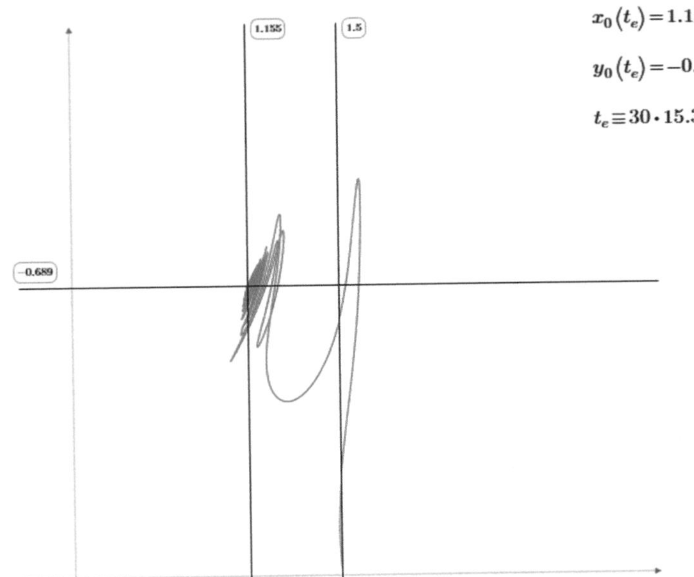

$x_0(t_e) = 1.1534422\ m$

$y_0(t_e) = -0.6989096\ m$

$t_e \equiv 30 \cdot 15.37\ s$

FIGURE 26.29 The trajectory of the pendulum (point 0) with the shock absorbers turned on.

The reader can download the Mathcad-sheet from the website, change the initial data in it and get new beautiful figures. Such a pendulum can also swing in the three dimensions. Not only interesting plane figures can appear here, but also interesting solid bodies.

AND THE VERY LAST THING

Often one hears the rhetorical question: how would the great mathematicians of modern times solve certain problems if they had at their fingertips not a quill pen, inkstand and paper, but a computer with Internet access? We have already asked this question at the beginning of Study 9.

We take the simplest "dynamic" problem: a stone with mass *m* is thrown up from zero height with velocity *v0* in the gravitational field characterized by the acceleration of gravity *g*. How will the flight altitude of the stone *h* change with the passage of time *t*? We solved similar problems in Study 9.

One can imagine how Newton (Figure 26.32) and Lagrange (Figure 26.33) would have solved this problem on a computer with an Internet connection, at what is probably the most popular mathematical website, wolframalpha.com.

In both cases the answer turned out to be the same: a well-known school formula for the stone flying along a parabola was derived (this parabola can be seen with one's own eyes, if the stone is not thrown vertically, but at an angle to the horizon). For Lagrange, an ordinary differential equation of the first order (Figure 26.33) is used, while, for Newton, one of second-order (Figure 26.32). If, on the other hand, the equations are complicated, taking into account, for example, the resistance of the medium to the flight of a stone deforming the parabola, and other factors that bring the mathematical model closer to reality, then we must abandon the symbolic (analytical) solution (Figures 26.32 and 26.33) and proceed to numerical methods. In this case, the Lagrange method, based on the balance of energies, rather than forces, often turns out to be more effective (more precise, faster, etc.).

The Englishman Newton and the Frenchman* Lagrange, can also be joined by the German Leibniz,

* Lagrange was born in Turin (at that time in the kingdom of Sardinia) and had the Italian name "Giuseppe Lodovico Lagrangia", although the family was originally from the French region of Tours, He was very successful abroad (e.g. in Berlin and in Paris).

$$L_{10} := -y_{10} + \sqrt{x_{00}^2 + y_{00}^2} = 6.054\ m \qquad L_{02} := -y_{20} + \sqrt{(L - x_{00})^2 + y_{00}^2} = 4.354\ m$$

$$L_{10} = -y_1(t) + \sqrt{x_0(t)^2 + y_0(t)^2}$$

$$y_1(0\ s) = y_{10} \quad y_1'(0\ s) = 0\frac{m}{s} \quad F_1(0\ s) = g \cdot M_1 \quad F_1(t) - g \cdot M_1 = y_1''(t) \cdot M_1 + k_1 \cdot y_1'(t) \cdot |y_1'(t)|$$

$$x_0(0\ s) = x_{00} \quad x_0'(0\ s) = 0\frac{m}{s} \quad F_2(t) \cdot \frac{L - x_0(t)}{\sqrt{(L - x_0(t))^2 + y_0(t)^2}} - F_1(t) \cdot \frac{x_0(t)}{\sqrt{x_0(t)^2 + y_0(t)^2}} = x_0''(t) \cdot M_0$$

$$y_0(0\ s) = y_{00} \quad y_0'(0\ s) = 0\frac{m}{s} \quad F_1(t) \cdot \frac{-y_0(t)}{\sqrt{x_0(t)^2 + y_0(t)^2}} + F_2(t) \cdot \frac{-y_0(t)}{\sqrt{(L - x_0(t))^2 + y_0(t)^2}} - g \cdot M_0 = y_0''(t) \cdot M_0$$

$$y_2(0\ s) = y_{20} \quad y_2'(0\ s) = 0\frac{m}{s} \quad F_2(0\ s) = g \cdot M_2 \quad F_2(t) - g \cdot M_2 = y_2''(t) \cdot M_2 + k_2 \cdot y_2'(t) \cdot |y_2'(t)|$$

$$L_{02} = -y_2(t) + \sqrt{(L - x_0(t))^2 + y_0(t)^2}$$

$$\begin{bmatrix} y_1 \\ x_0 \\ y_0 \\ y_2 \\ F_1 \\ F_2 \end{bmatrix} := \mathbf{Odesolve}\left(\begin{bmatrix} y_1(t) \\ x_0(t) \\ y_0(t) \\ y_2(t) \\ F_1(t) \\ F_2(t) \end{bmatrix}, t_e, 100000 \right)$$

$$t := 0\ s, \frac{t_e}{30000} .. t_e$$

FIGURE 26.30 A system of equations describing a pendulum with three weights, two blocks and two shock absorbers.

reminding the first two of the fundamental theorem of calculus—that the anti-derivative has another name—the anti-product. Figure 26.34 shows the solution in Maple of the slightly modified problem of the flight of a material point in the gravitational environment of the Earth: Newton sits under an apple tree and an apple falls on his head. How will the speed of the fall of the apple change over time?

Here we are solving not a differential but an *integral* equation: the potential energy of the apple decreases and smoothly converts into kinetic one. In this case, it is taken into account that the displacement is the time integral of the velocity. The equation is solved analytically. But, as is well known, more complicated similar equations have to be solved numerically. And numerical work with the integral is much simpler and more accurate than numerical work with the derivative. Moreover, "pure" mathematicians do not tire of repeating to "unclean" applied mathematicians, that the numerical finding of a derivative is generally a completely flawed operation—an "absurdity". On this "optimistic note" we finish the Study.

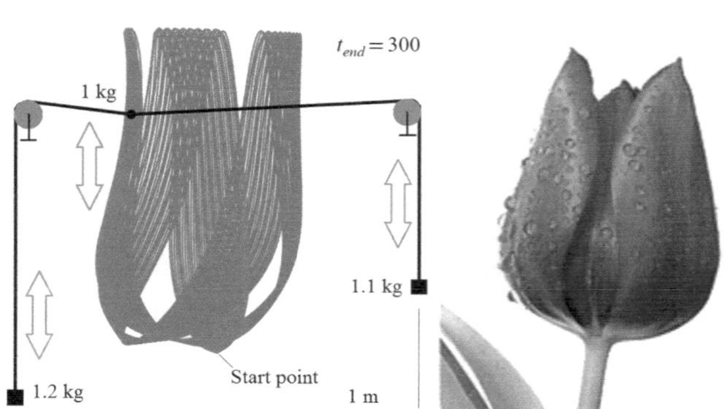

FIGURE 26.31 Tulip pendulum.

AND MOST MOST-RECENT

Now we have touched upon the anti-derivative, we will move on to talk about anti-animation. This idea arose after reading Averchenko's story "The Focus of the Great

```
h(0)=0, h'(0)=v0, m*h"(t)=-g*m
```

$$h(t) = t\,v0 - \frac{g\,t^2}{2}$$

FIGURE 26.32 Newton's solution: the sum of the forces acting on the material point is equal to the product of the mass of the point by its acceleration.

h(0)=0, (m*v0^2)/2=g*m*h(t)+(m*h'(t)^2)/2

$$h(t) = t\, v0 - \frac{g\, t^2}{2}$$

FIGURE 26.33 The Lagrange solution: the sum of the kinetic and potential energies of a moving material point remains constant.

$$IE := g * mass * h_0 = g * mass * \left(h_0 + \int_0^t v(T)\, dT\right) + (mass * v(t)\hat{}2)/2$$

$$g\, mass\, h_0 = g\, mass\left(h_0 + \int_0^t v(T)\, dT\right) + \frac{1}{2}\, mass\, v(t)^2$$

$intsolve(IE, v(t))$

$$v(t) = -g\, t$$

FIGURE 26.34 Analytical (symbolic) solution of the integral equation (IE) in Maple.

Cinema". Rather, it was so: this story was "twisted" in the head of the author as a pendulum near the equilibrium point. The story was "pulled out" from the Internet. This is its beginning.

> Once in a movie I saw an amazing picture:
>
> "Sea. Coast. A tall, steep rock, ten yards high. Suddenly the water boiled at the rock, a human head popped up, and now a man, like a giant ball, pushing away from the ground, flew ten sazhens* up, stood on the platform of the rock—completely dry and created the sign of the cross: first his fingers touched the left shoulder, then the right, then the chest and, finally, the forehead.
>
> He quickly dressed and walked away from the sea, back to front, backing off like a crayfish. He waved his hand, and the cigarette butt, lying on the road, jumped up and climbed into his fingers. The man began to smoke, drawing in itself the smoke that was born in the air. As smoking, the cigarette became more and more and finally became quite fresh, just lighted up. The man put a match to her, jumped into his hand from the ground, took out a box of matches, struck a lighted match on the box, why the match went out, put the match in the box; a cigarette sticking out in his mouth, he thrust it back into his cigarette case, bent down—and the spittle from the ground jumped into his mouth. And he went further also backward, backing like a crayfish. At home he sat down in front of an empty plate and a glass, poured a few sips of red wine from his mouth into the glass and began to drag the pieces of chicken from his mouth, putting them back on the plate where they were knit together under a knife. When the chicken came out entirely from his throat, the footman approached and took a plate, carried this chicken to the kitchen—about frying The cook put it in a frying pan, then removed the raw, stuck with feathers, led a knife over his throat, making the chicken come alive and then ran cheerfully around the yard."

* * *

Is not it true that you understand why this is the case: this is an ordinary film depicting ordinary human acts, but launched in the opposite direction.

Ah, if our life was like an obedient cinematic tape!

Turned the knob back—and went-went (https://www.youtube.com/watch?v=9p0HI9t5IB0)

Let's look again at the last animation frame of the tulip-pendulum motion in Figure 26.29 (upper right frame). There, the most interesting moment—the moment of damping of the pendulum motion near the equilibrium point—is smeared: the trajectory of motion is superimposed on itself and therefore practically invisible. Let's run the animation of the motion of this pendulum in the reverse order: not "time-forward" but "time-backward". We will draw the trajectory of the point in reverse order so that it can be clearly seen at the beginning, or rather, at the end of the path. Figure 26.35 shows frames of animation and anti-animation of the triple pendulum motion in a viscous medium.

Scrolling the animation of our unusual pendulum in the opposite direction allows us to give it another name, associated not with the flora, but with the fauna: the decaying pendulum at the end, or rather, at the beginning of its reverse movement, draws a fish—see the lower right frame of the animation in Figure 26.35. With direct animation, you cannot see it—see the upper right frame of the animation.

* Russian sazhen (сажень) is equal to seven feet.

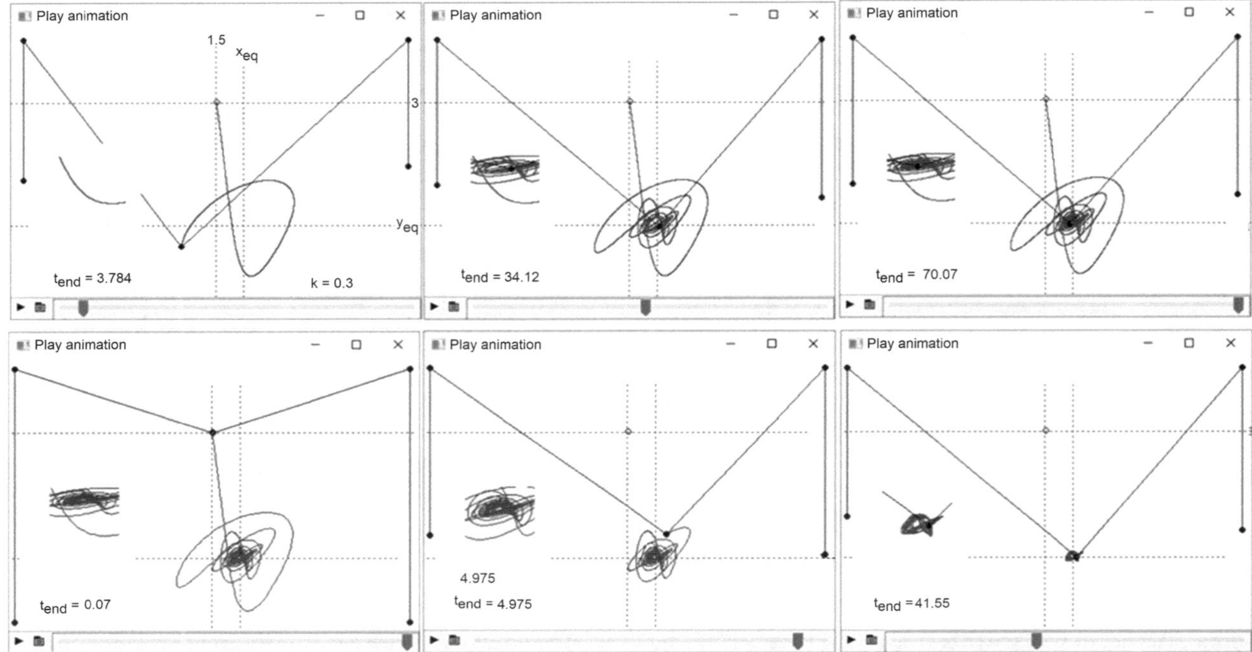

FIGURE 26.35 Animation frames (top) and anti-animation (bottom) of the pendulum-tulip movement in a viscous environment.

AFTERWORD TO THE STUDY

The author "had the impudence" to send this Study in the form of an article to the leading Russian journal on theoretical mechanics, produced at Moscow State University. It was expected that there would be no answer, or it would be in the spirit of the saying "keep your homespun nose out of the business of thoroughbreds!". A similar response was obtained when an article on celestial mechanics (Study 9) was sent to another journal. But the answer from the journal on theoretical mechanics came and was very good. It was reported that for the magazine the article does not fit both in volume and in the form of the presentation of the material, but it is very suitable for the collection "Theoretical Mechanics", which is published by the Moscow State University. The article was published in the collection "Theoretical Mechanics" and opened a new section "Theoretical Mechanics in School".

ASSIGNMENTS FOR READERS

1. Take into account the weight of a thread or, better, of a metal chain to which the weights are hung. It is possible to tie an air ball to a thread or to a chain that pulls the thread up or to create such force with the help of a block. Approaches to the solution of such a problem using the catenary function are set forth in Study 7.
2. Some or all segments of the inextensible thread shown in Figures 26.1 and 26.27, can be replaced by tensile springs or elastic bands, which obey not only Hooke's law, but also another more complex (non-linear) law of expansion-contraction. In this case, it is necessary to resort to programming, but in Mathcad, not Pascal.
3. The weights hanging on the thread can be moved not only to the right, resulting in this particular flat pendulum, but also right-forward, modeling a three-dimensional pendulum that swings not in the plane, but in volume. While it fluctuates, translating potential energy into kinetic and inverse, one can go to the computer class and simulate such a pendulum.
4. Simulate the oscillation of the pendulums, relying not on the second law of Newton, but on the principle of Lagrange. We applied this principle (the minimization principle) in the static problem when we minimized the potential energy of the system (see Figure 26.3) and the sum of the potential energy and of the energy of the stretched spring (Figure 26.25). If this principle is

applied to a dynamic model, then the difference between the potential and kinetic energies must be minimized.

5. On the website for this Study, we repeat, it is possible to find a simulation of the oscillation of pendulums taking into account the resistance of the medium. In this case, a fairly simple model is used: the drag force is proportional to the product of the cross section of the body, the density of the medium, and the square of the velocity. In this case, the proportionality coefficient is taken as a constant value. And in reality, this coefficient depends on the flow of air around the body—whether it is laminar or turbulent. Take into account the changes that include this law of gas dynamics!

6. In the problem where the threads are replaced by springs (elastic bands), we used Hooke's linear law: the degree of expansion is proportional to the value of the applied force. But in reality, this law is more complex (non-linear)—with considerable strength the elastic band either breaks or stops stretching. Model the oscillation of the pendulum with such an elastic band.

7. Solve the problems described in the Study, not in the plane, but in the volume, having taken into account the third dimension.

BHL
Or
*Application for the Ig Nobel Prize**

ABSTRACT

This Study explains how it is possible to attach a chain and a catenary to geography, history, statistics and even to scientific ethics. Statistical data on the great mathematicians of the world of all time are considered. A geographic-mathematical problem is solved. Some moral and national aspects of the names of mathematical theorems are discussed.

Mathematics: Catenary function, regression analysis, histogram.

Physics: Sagging chain, physical map.

IT: Communication with other disciplines, with mathematics, history, geography, ethics, numerical solution of systems of algebraic equations, computer graphics, introduction of an Excel table into a Mathcad-sheet.

Art: Leo Tolstoy novel, *War and Peace*; Mikhail Bulgakov play, *Flight*.

Study website: https://community.ptc.com/t5/PTC-Mathcad-Blog/Study-28/ba-p/483696.

The author has placed in this book a lot of physical and mathematical problems and riddles—either the body of each Study or at the end of each one in the form of Assignments for readers. The author of the puzzles and the visitors are making their guesses at the website PTC Community/Mathcad, where the files and animations of the book are placed. One such riddle was posted on the website https://community.ptc.com/t5/PTC-Mathcad-Questions/Do-you-know-this-curve/m-p/410542—see Figure 27.1.

The question is: which curve passes through three points on the plane—and what mathematical function does it describe?

Through three points, you can draw a parabola (see Figure 7.30 in Study 7), a circle and many other plane curves. There are people who can quickly multiply five-digit numbers in the mind. And there are people who, only by the form of the curve, by the "bends of its soul", are able to determine what function it describes. Frankly the author counted on these people when they put this puzzle on the PTC Community website. We have already solved a similar problem in Study 9—we carried out a curve of the second order through the points of motion of the comet, see Figure 9.18 in Study 9). We devoted Study 14 to this problem, talking about interpolation and approximation. Figure 27.1 shows interpolation: a smooth continuous unknown curve passes through points, rather than near them (approximation, approaching value).

The riddle was not solved. Then the author posted on the forum a visual hint, see Figure 27.2. With three points, but without a curve.

Figure 27.2 is a map of central Europe with three marked cities: The Hague (the Netherlands), Basel (Switzerland) and Leipzig (Germany). It was particularly emphasized in Figure 27.1 that three points

* Maybe this is a bad pun? Sorry.

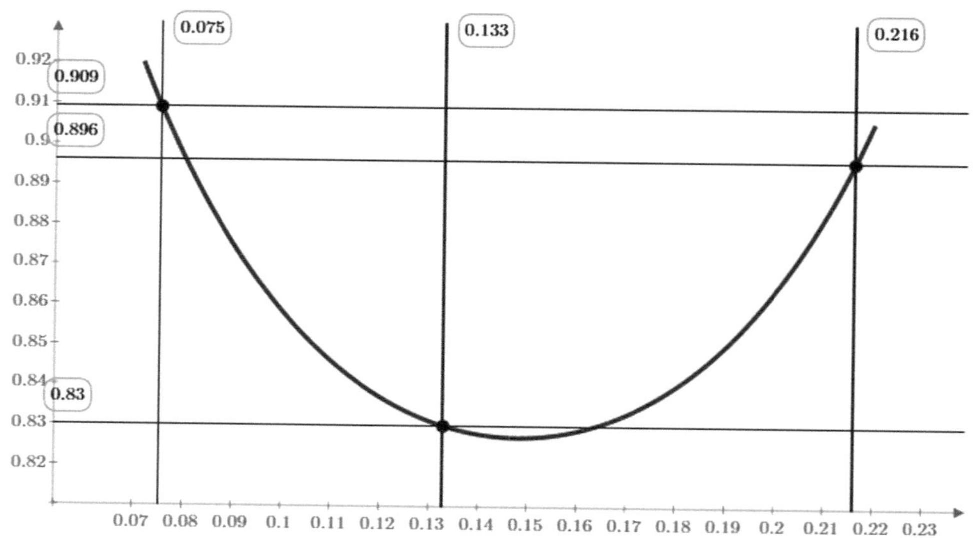

FIGURE 27.1 The first picture of the physical and mathematical riddle.

through which the curve passes have the geographical coordinates of these three cities (northern latitude and eastern longitude—see, for example, Figure 9.29 in Study 9), translated from angular degrees, minutes and seconds into radians: The Hague (upper left) 52° 05′ 03″ N and 4° 19′ 03″ E, Basel (at the bottom) 47° 33′ 30″ N and 7° 37′ 00″ E and Leipzig (upper right) 51° 20′ 00″ N and 12° 23′ 00″ E. The coordinates of the cities are taken from Wikipedia, to which we will return below.

But even with such a map as a hint, the problem of the shape of the curve was not solved. This hint only confused most people.

Then in the forum one more tip was posted—Figure 27.3, also with three points and a curve. But the coordinates of the points were of another type: 1629–1695, 1667–1748 and 1646–1716.

After this hint, Victor Korobov guessed the riddle drawings: the curves in Figures 27.1 and 27.3 give a graphical representation of the *catenary*, described in detail in Study 7.

The function describing the catenary was discovered by Bernoulli* (1667–1748, was born in Basel—see the bottom of Figures 27.1 and 27.2), Huygens (1629–1695, was born in The Hague—Den Haag—see the upper left corner) and Leibniz (1646–1716, was born in Leipzig—see the upper right corner) independently of each other and almost simultaneously. Figure 27.3 shows the years of birth and death of these three great mathematicians in the coordinates of the three points.

Figures 27.4 and 27.5 show how Mathcad Prime helped to find the parameters of the catenary passing through the native cities of its discoverers (Figure 27.1) and the years of their birth and death (Figure 27.3): a system of

FIGURE 27.2 The second figure is a hint for the physical and mathematical riddle with three cities on the map.

* Johann I. Bernoulli. And there was also Jacob, Nicholas, Johann II and Johann III Bernoulli. And they were all mathematicians and/or physicists!

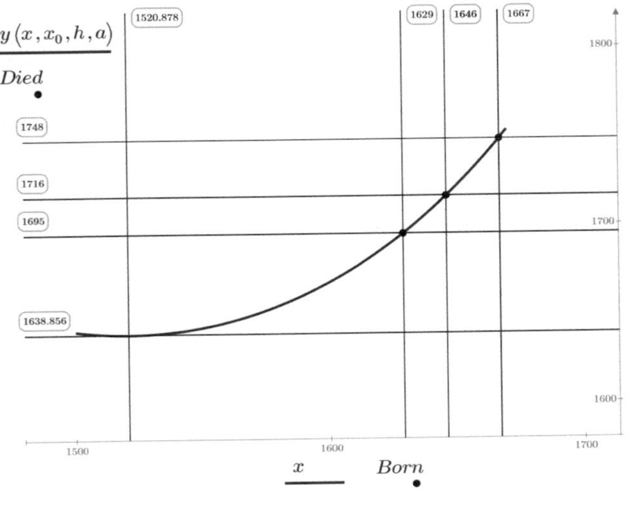

FIGURE 27.3. The second figure is a hint for the physical and mathematical riddle.

three non-linear algebraic equations are solved (see the Study 4).

Note On a geographical map hanging on the wall, the city of Basel is "below" The Hague and Leipzig. But in fact, physically Basel is at an altitude of 317 m above the sea level, while The Hague is near the sea and even just below its level. Leipzig (117 m) is one of the cities of Saxony, and, in Germany, apart from the "ordinary" Saxony with

$$y(x, x_0, h, a) := h + a \cdot \left(\cosh\left(\frac{x - x_0}{a}\right) - 1\right)$$

$$X := \begin{bmatrix} DMS\left(\begin{bmatrix} 4 \\ 19 \\ 03 \end{bmatrix}\right) \\ DMS\left(\begin{bmatrix} 7 \\ 37 \\ 00 \end{bmatrix}\right) \\ DMS\left(\begin{bmatrix} 12 \\ 23 \\ 00 \end{bmatrix}\right) \end{bmatrix} = \begin{bmatrix} 0.07535 \\ 0.13294 \\ 0.21613 \end{bmatrix} \qquad Y := \begin{bmatrix} DMS\left(\begin{bmatrix} 52 \\ 05 \\ 03 \end{bmatrix}\right) \\ DMS\left(\begin{bmatrix} 47 \\ 33 \\ 30 \end{bmatrix}\right) \\ DMS\left(\begin{bmatrix} 51 \\ 20 \\ 00 \end{bmatrix}\right) \end{bmatrix} = \begin{bmatrix} 0.90904 \\ 0.83005 \\ 0.89594 \end{bmatrix}$$

Solve

Guess Values

$$\begin{bmatrix} x_0 \\ h \\ a \end{bmatrix} := \begin{bmatrix} 1 \\ 1 \\ 0.1 \end{bmatrix}$$

Constraints

$$y(X, x_0, h, a) = Y$$

Solver

$$\begin{bmatrix} x_0 \\ h \\ a \end{bmatrix} := \mathbf{Find}(x_0, h, \mathbf{a}) = \begin{bmatrix} 0.14824 \\ 0.8272 \\ 0.04164 \end{bmatrix}$$

FIGURE 27.4 Interpolation by a catenary of the birthplaces of its discoverers (DMS function translates angular degrees D, minutes M and seconds S into radians).

$$y(x, x_0, h, a) := h + a \cdot \left(\cosh\left(\frac{x - x_0}{a}\right) - 1\right)$$

$$Born := \begin{bmatrix} 1629 \\ 1667 \\ 1646 \end{bmatrix} \qquad Died := \begin{bmatrix} 1695 \\ 1748 \\ 1716 \end{bmatrix}$$

Solve

Guess Values

$$\begin{bmatrix} x_0 \\ h \\ a \end{bmatrix} := \begin{bmatrix} 1500 \\ 2000 \\ 100 \end{bmatrix}$$

Constraints

$$y(Born, x_0, h, a) = Died$$

Solver

$$\begin{bmatrix} x_0 \\ h \\ a \end{bmatrix} := \mathbf{Find}(x_0, h, \mathbf{a}) = \begin{bmatrix} 1520.878 \\ 1638.856 \\ 112.393 \end{bmatrix}$$

FIGURE 27.5 Interpolation by a catenary of the birth dates of its discoverers.

its capital in Dresden, there is also Lower Saxony, located on the map above the "ordinary", "upper" Saxony. With the Russian two Novgorods everything is more or less in order: Nizhny Novgorod (Lower Novgorod) is located below the "upper" Veliky Novgorod (Great Novgorod) on the map and along the rivers.

This "geographical-mathematical riddle" helped us to make the first "discovery", reflected in the first title of this Study: "BHL"—the discoverers of the catenary function have names and hometowns, starting with one letter. This "scientific novelty" is worthy of the Ig Nobel Prize, noted in the second title of the Study. Why not!?

True, there was a weak hope that the curve of the catenary function, shown in Figure 27.1, would have a minimum exactly at Basel, but this did not happen: the minimum was *near* Basel. But this *near* ("about") helped us to make the second, "sculptural-linguistic" "discovery", which is also connected with mathematics, or rather with another great mathematician.

A small digression from the main topic.

There is a monument torn in two halves in two cities in Germany—in Berlin and in Frankfurt. This is a monument of an airlift between these cities dedicated to the victims who died during the Cold War (please see https://www.alamy.com/stock-photo/berlin-airlift-

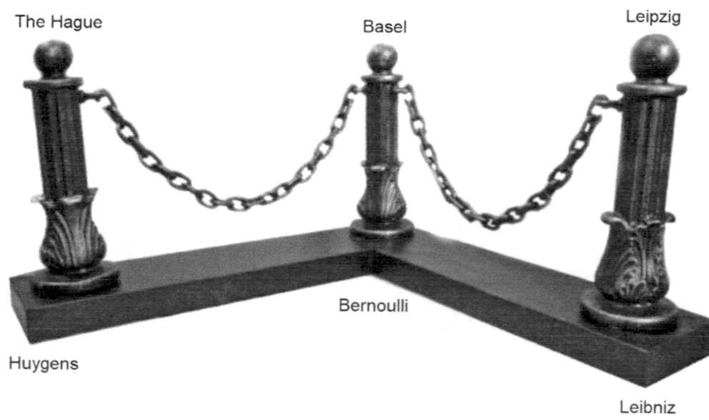

FIGURE 27.6 The project of the monument of the catenary, consisting of three parts located in three European cities (see also Figure 7.6 in Study 7).

monument.html). One "support" of this bridge is in Berlin, and the other in Frankfurt. The catenary is directly related to bridges (see Figure 7.6 in Study 7). And bridges connect people. No wonder the European currency euro shows these engineering designs!

There is an idea to build a monument to the catenary line, which will be broken not into two, but into three parts—see Figure 27.6. By the way, we have already shown the monument of the unbroken catenary in the form of an arch in the book, see Figure 7.1 in Study 7.

The monument in The Hague, Basel and Leipzig (HBL) will be a monument not so much to the catenary as to its discoverers. In the three parts of this monument in three cities, you can install stands with the map shown in Figure 27.2, and explanatory texts. It will be an excellent object, popularizing mathematics and physics.

By the way, in Basel you can see another unusual monument, a monument to a literary hero: a braying donkey.* We read in the novel "The Idiot" by F.M. Dostoevsky: "The consciousness that everything was strange weighed terribly upon me; I could understand that it was all foreign and strange. I recollect I awoke from this state for the first time at Basel, one evening; the bray of a donkey aroused me, a donkey in the town market. I saw the donkey and was extremely pleased with it, and from that moment my head seemed to clear."

The dates of birth and death of Bernoulli, Huygens and Leibniz were taken from Wikipedia. Figure 27.7 shows the beginning of the page of this most popular network encyclopedia with information about another "great mathematician"—about Euler.

Euler was a student of Johann Bernoulli, and he is the one most mentioned in this book—more than thirty times! Euler, like Bernoulli, was born in Basel but was formed as a future scientist not in this city itself, but *near* it. The author of this book, having found himself in Basel, began to scour *around* this city, seeking the "minimum of the catenary function" shown in Figure 27.1. These "yelps" led the author to the town Riehen, in which Euler spent his youthful years. There is a memorial plaque in the house where Euler lived, see Figure 27.8.

The author, examining this bas-relief, accidentally made another "physical and mathematical", sorry, "sculptural-linguistic" discovery: on the back of Euler's head the silhouette of a lion is clearly visible. And Euler was called Leonard (i.e. "Staunch lion, leon hard"†). Whether the sculptor laid this puzzle in the bas-relief of Euler, or it turned out to be an accident, the play of light and shadow, one can only guess. But the fact remains. Another "physical and mathematical" fact, a discovery worthy of the same Ig Nobel Prize!

Near Basel there is another remarkable place, "worthy" of the catenary function. This is one of the wonders, not of the whole world, but only of Europe—the Rhine Falls, which Karamzin‡ admired in "Letters of a Russian Traveller". In the vicinity, near (but the mathematical analysis is based on these "vicinities", "neighborhoods of the point") Basel and the minimum of our catenary function there is one more geographical attraction, the source of the Danube. Between these two fairly close points (the waterfall on the Rhine flows into the North Sea, and the source of the Danube flowing to the other end of Europe, the Black Sea) is a region with an interesting mathematical property. If we have this function of two arguments, a function with such topography, we seek the minimum by the numerical method of the gradient descent,§ then you can accidentally, unpredictably, fall (slip) to different ends of Europe: either the North Sea (the first local minimum) or the Black Sea (the second local minimum). Here is

* A propos. In Germany, in the city of Bremen (The "Town Musicians of Bremen") there is a monument dedicated to a donkey.

† The lion is the king of the animals. Euler with a certain degree of conventionality (see below) can be considered king, king of mathematicians. At least so it is believed in Switzerland, Germany and Russia, where he worked (Euler was buried in St. Petersburg). This "lion" was intended by the sculptor, creating a memorial plaque.

‡ Nikolay Mikhailovich Karamzin (1766–1826) was a Russian writer, poet, historian and critic. He is best remembered for his *History of the Russian State*, a 12-volume national history. See more https://en.wikipedia.org/wiki/Nikolay_Karamzin.

§ See the example of the trajectory of such a descent in Figures 13.18 and 13.19 in Study 13.

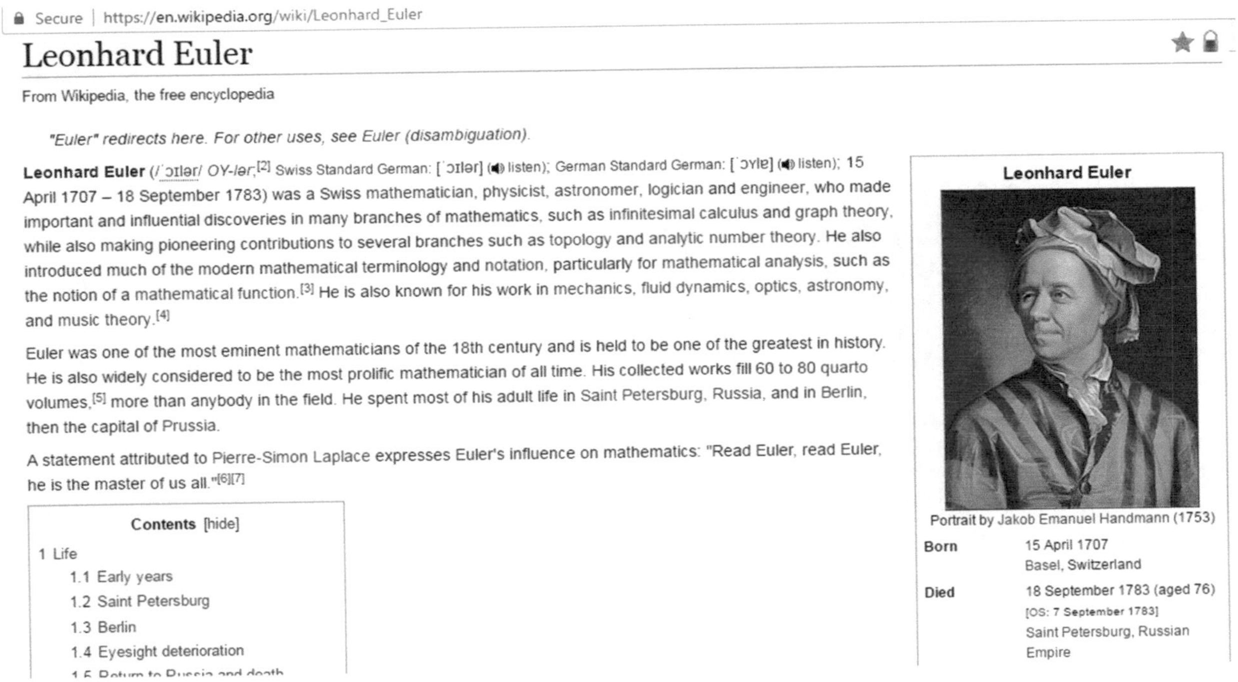

🔒 Secure | https://en.wikipedia.org/wiki/Leonhard_Euler

Leonhard Euler

From Wikipedia, the free encyclopedia

"Euler" redirects here. For other uses, see Euler (disambiguation).

Leonhard Euler (/ˈɔɪlər/ *OY-lər*;[2] Swiss Standard German: [ˈɔɪlər] (listen); German Standard German: [ˈɔʏlɐ] (listen); 15 April 1707 – 18 September 1783) was a Swiss mathematician, physicist, astronomer, logician and engineer, who made important and influential discoveries in many branches of mathematics, such as infinitesimal calculus and graph theory, while also making pioneering contributions to several branches such as topology and analytic number theory. He also introduced much of the modern mathematical terminology and notation, particularly for mathematical analysis, such as the notion of a mathematical function.[3] He is also known for his work in mechanics, fluid dynamics, optics, astronomy, and music theory.[4]

Euler was one of the most eminent mathematicians of the 18th century and is held to be one of the greatest in history. He is also widely considered to be the most prolific mathematician of all time. His collected works fill 60 to 80 quarto volumes,[5] more than anybody in the field. He spent most of his adult life in Saint Petersburg, Russia, and in Berlin, then the capital of Prussia.

A statement attributed to Pierre-Simon Laplace expresses Euler's influence on mathematics: "Read Euler, read Euler, he is the master of us all."[6][7]

Contents [hide]

Leonhard Euler

Portrait by Jakob Emanuel Handmann (1753)

Born	15 April 1707 Basel, Switzerland
Died	18 September 1783 (aged 76) [OS: 7 September 1783] Saint Petersburg, Russian Empire

FIGURE 27.7 Wikipedia page about Euler.

another "geographic and mathematical discovery" worthy of the Ig Nobel Prize!

Either way, the two branches of the catenary function line shown in Figure 27.1, can be likened to the two main rivers of Europe—the Rhine and the Danube! The chain sags under the action of gravity. This same force causes the flow of terrestrial rivers in one direction or another! However, for short path, is it possible for rivers to move to higher elevation due to inertia effects (https://worldbuilding.stackexchange.com/questions/11526/explaining-uphill-rivers-scientifically)?

There was an idea to add points on the graph in Figure 27.3: the dates of the birth and death of Euler and other mathematicians. This work was entrusted to two students who could not pass exams on the courses "information technology" and "higher mathematics". It was decided not to test their minds, but their hands, to make them "master" math and computer science: one student opened Wikipedia,* similar to the one shown in Figure 27.7, copied the name of the next mathematician into the Excel table and dictated to the second student

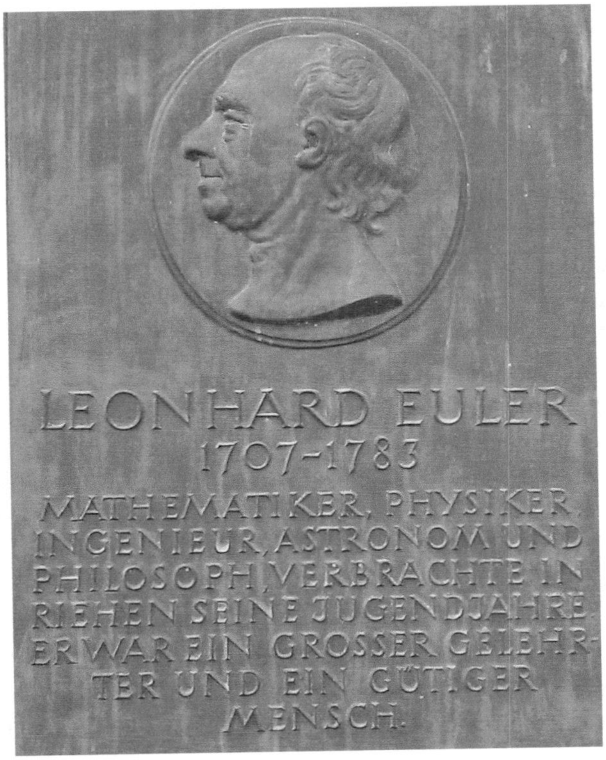

FIGURE 27.8 Memorial plaque at the Euler house near Basel (author's photo).

* And there is a Russian-language article with an alphabetical list of mathematicians and links to their personal pages.

	A	C	D	E	F	G	H	L
1	NAME	Day of Birth	Month of Birth	Year of Birth	Day of Death	Month of Death	Year of Death	Life Span
2	Abakanowicz, Bruno	6	10	1852	29	8	1905	52.9
3	Abd al-Rahman al-Khāzini							
4	Abd al-Rahman al-Sufi	7	12	903	25	5	986	82.46
5	Abel, Niels Henrik	5	8	1802	6	4	1829	26.67
6	Abramov, Alexander Alexandrovich	14	2					
7	Abramov, Alexander Mikhailovich	5	6	1946	24	5	2015	68.97
8	Abramov, Sergej Mikhailovich (scientis	25	3					
9	Abrashkin, Victor Alexandrovich	6	9					
10	Abrosimov, Alexander Viktorovich	16	11	1948	20	6	2011	62.59
11	Abū al-Wafā' al-Būzjānī				10	6	940	
12	Abu al-Hasan al-Ahwazi							
13	Abū Bakr al-Karajī			953			1029	76
14	Abū Bakr al-Rāzī			865			925	60
15	Abu Jafar Khazini (al-Khāzin)						971	
16	Abu Jafar ibn Habash							
17	Abū Kāmil			850			930	80
18	Abu Mahmud Khojandi							
19	Abū Mashar al-Balkhī	10	8	787	9	3	886	98.58
20	Abū Sahl al-Qūhī							
1960	Żurawicy, Marcin Król			1422			1460	38
1961	Zwegers, Sander P.	16	4	1975				
1962	Zwick, Uri (Israel, ?-?)							
1963	Zwicker, William S.			1949				
1964	Zworski, Maciej	8	10	1963				
1965	Życzkowski, Karol			1960				
1966	Zygalski, Henryk	15	7	1908	30	8	1978	70.17
1967	Zygmund, Antoni	25	12	1900	30	5	1992	91.49

FIGURE 27.9 Table with data about mathematicians.

the day, month and year of his birth and death. The second student entered these figures into another Excel spreadsheet. Then two of these tables were combined into one (Figure 27.9) and embedded in the Mathcad-sheet (Figure 27.10).

In the table there are 1966 mathematicians. A graphic representation of these data is shown in Figures 27.10 and 27.11. What conclusions can be drawn by looking at these points?

First, very few ancient mathematicians, whose dates of birth and death go back to the years before Christ, are clearly visible.

Second, there is a "mathematical" drop in the early Middle Ages, when they could burn the mathematician for the proof of a theorem. But from about the year 1000 onward, the number of mathematicians starts to increase with time. In these years, the number of universities opened in Europe also grew: the Renaissance happened not only in art, but also in science!

Third, the graphs show that the two World Wars took the lives not only of ordinary people, but also of mathematicians—see Figure 27.11, which has two pairs of markers: 1915–1890 = 25 years and 1950 – 1917 = 43 years. These years somehow fix the dates of birth of those who died in the two World Wars (the First World War1914–1918, the Second World War 1939–1945). Should we expect the slope of Figure 27.10 to correlate to an increase in life expectancy (length of life) from those early days until now?

Fourth, Figure 27.12 shows a histogram of the lifespan of mathematicians. On the left is the great Évariste Galois, who in his short twenty-year life (1811–1832) managed to found a modern higher algebra. On the right there is a mathematician, who lived for a long time, but who few people remember.

In Wikipedia, as you know, almost all articles are duplicated in several languages. However, these are not exact translations of articles from one language to

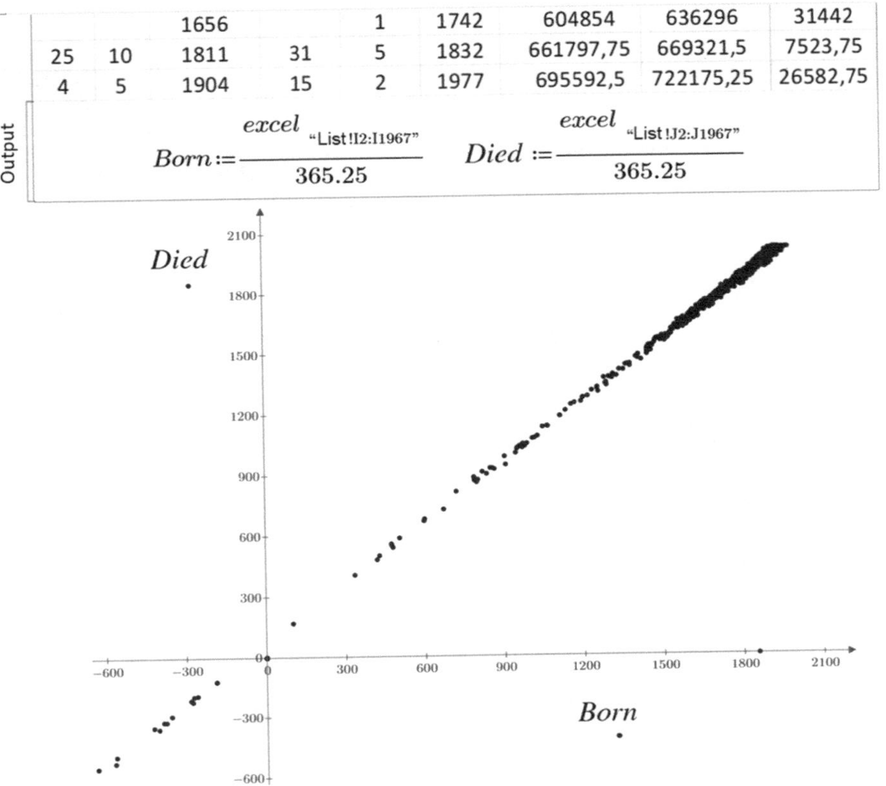

		1656		1	1742	604854	636296	31442
25	10	1811	31	5	1832	661797,75	669321,5	7523,75
4	5	1904	15	2	1977	695592,5	722175,25	26582,75

FIGURE 27.10 Mathematics in statistics.

another, but in fact different articles in terms of volume and content. This, incidentally, applies also to articles describing mathematical theorems and discoveries, and we will digress on this below. The list of mathematicians, which we analyzed statistically above, is taken from the Russian section of Wikipedia. This list includes some mathematicians who are not in other sections—English, French, German, Italian. And vice versa. Wikipedia is a very politicized resource.* It would be correct to leave for our statistics only those mathematicians that are mentioned in all significant sections of Wikipedia.

The author had wished to analyze the table in Figure 27.9 with regard to the following subject: whether there is any correlation between the degree of greatness of a mathematician and the number of years lived. But this is somewhat subjective and is not that simple. It is believed that almost all the discoveries were made in the earlier years of mathematicians, when they were not yet considered great mathematicians, but simply mathematicians. And then these people lived out their time, whether short or long, resting on their laurels. This small "Study" also fully claims the award mentioned in the second title of the Study.

And how to assess the degree of greatness† of this or that mathematician or scientist? By the number of published works? But one corresponding member of the Academy of Sciences of the USSR was awarded the Ig Nobel Prize for the world record for the number of publications. Now this person is remembered only in connection with this prize (a modern analog of the story of

* An example of explicit politicization. The mathematical institute of the Russian Academy of Sciences bears the name of a man who practically no one has heard of outside Russia. And even within Russia, many do not consider him a mathematician, worthy of having his name given to the main mathematical institute of the country. It is said that once a writer from England visited the USSR Writers' Union. He was told that in the Union of Writers of the USSR there are more than ten thousand people. The Englishman did not hear the word "thousand" and said that in England there were not ten, but only five writers. But then he recovered: "No, not five, but four. The fifth, I do not consider him a writer!"

† In Wikipedia an article on the Bernoulli family says that "9 big mathematicians and physicists (3 of them great) belong to this dynasty". Here it is time to recall the positions of the theory of fuzzy sets, which we touched on in Study 12 and which can prescribe one mathematician to refer to the set of "great mathematicians" with the membership function equal, for example, to 0.95, and the other with the value of this function equal to 0.45, e.g. consider it not a great, but simply "a major mathematician". And there are also "outstanding mathematicians", "mathematicians who have made a significant contribution to …", etc.!

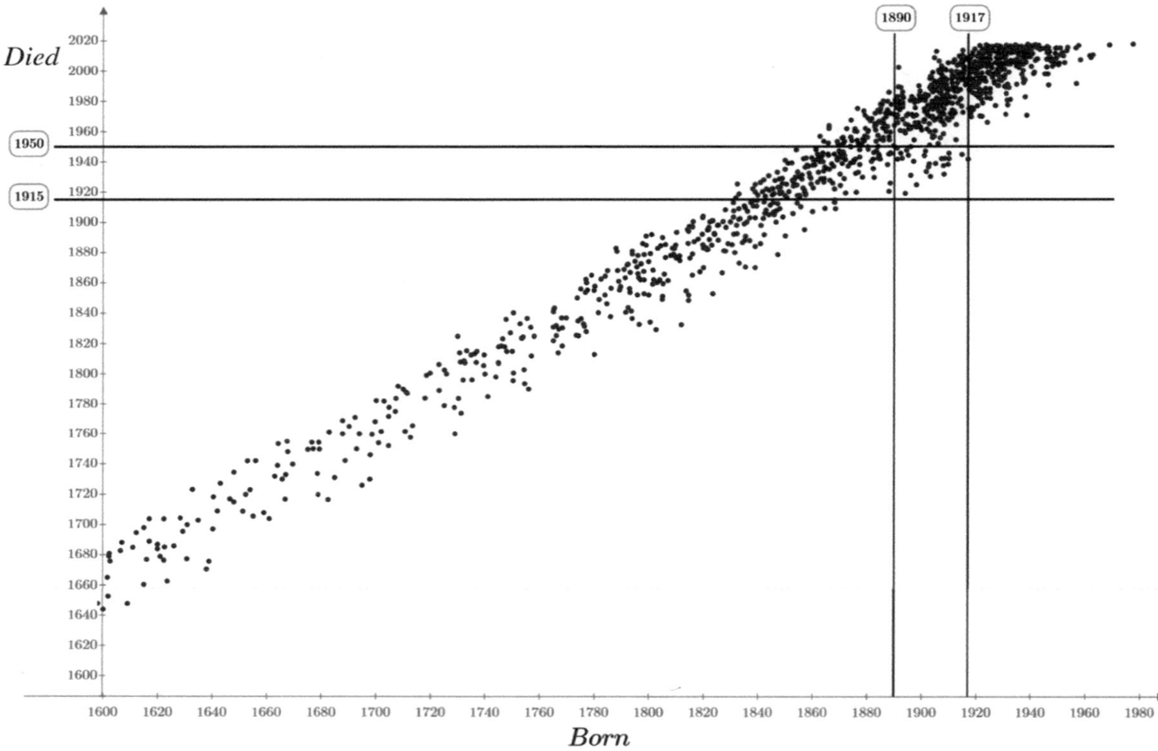

FIGURE 27.11 Mathematics and statistics in modern times.

Herostratus). The value of the scientist is currently being measured by various citation indexes.

A qualitative criterion for evaluation is the expression "Scientist with a capital letter". But the highest respect that a scientist can earn is that his name will be written with an uppercase letter. Such "names" can be found in this book. These units are physical quantities: Ampère, Ångström, Becquerel, Bell, Weber, Volta, Galileo, Gauss, Henry, Hertz, Gilbert, Gray, Dalton, Darcy, Debye, Joule, Sievert, Coulomb, Curie, Lambert, Maxwell, Newton,

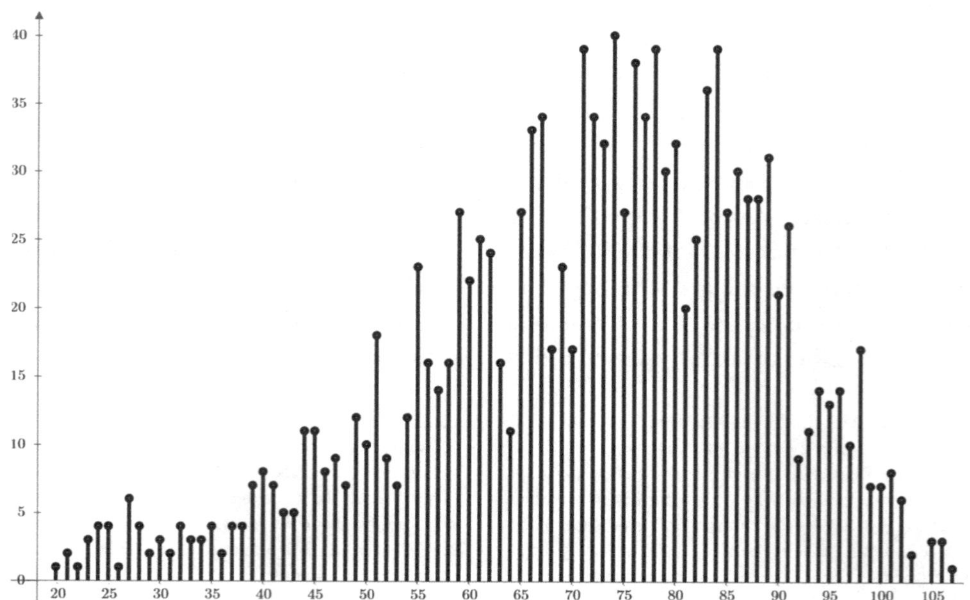

FIGURE 27.12 Histogram of the lifespan of mathematicians.

Ohm, Pascal, Planck, Poiseuille, Rutherford, Röntgen, Réaumur, Rydberg, Svedberg, Sverdrup, Siemens, Stokes, Sabin, Tesla, Thomson (Kelvin), Torricelli, Watt, Faraday, Fahrenheit, Fermi, Franklin, Hartree, Celsius, Einstein, Erlang, Ørsted, Eötvös and Jansky. But even here it's not so simple. First, these scientists can be divided into those who gave their names to the basic SI units (e.g. Ampere and Kelvin), compound units (e.g. Newton, Pascal, Watt, Joule, Coulomb, Ohm, etc.) and extra-system units (e.g. Einstein). Second, there is discrimination. No one disputes that Mendeleev, with his periodical table of chemical elements, was a great scientist, but, there was a proposal to introduce a compound (derivative) unit of Mendelian equal to J/mol/K (specific entropy, specific heat, gas constant) into calculation practice. But this proposal was not accepted because, as it is now said, of Russophobia. But there were other reasons, too, with elements of discrimination. Many scientists from the above list are overrated. So, Blaise Pascal, for example, was a writer and philosopher, fond of free time, mathematics and physics. "French" is very much among units. But this is understandable: SI is a purely French brainchild.

A NOTE ON THE UNITS

The Committee on Data for Science and Technology (www.CODATA.org) issues lists with refined values for a number of physical constants every four years. The list for 2010, for example, includes updated values for a number of constants, including those containing uncertain parameters. Thus, the uncertainty of the fine structure constant (α, the Sommerfeld's constant), which is used in atomic physics, was reduced. In addition, the Planck constant (h), which connects the magnitude of the energy of electromagnetic radiation with its frequency, was refined. The changes also affected the Avogadro number (N_A), which shows the number of particles in 1 mole of matter, and the Boltzmann constant (k), which connects energy and temperature.

Refined values of the constants, especially the Avogadro numbers and the Boltzmann constant, were perceived by a number of physicists as a signal for updating the international system of units (SI). First of all, this is connected with five units—ampere, mole, kilogram, kelvin and pascal.

Proposals to clarify the meaning of most of them have good prospects. The stumbling block was the kilogram, because there is a standard, but it is subject to changes, and this causes regular disputes in the scientific environment. A number of experts propose to determine kilograms based on the value of the Planck constant. In addition, it is proposed to give the kilogram a new name. The fact is that the kilogram is the only basic SI unit with a multiplier of 1 kilo. A contest for the best name for the updated basic unit of mass will be announced, and the unit itself, as noted above, will be tied to Planck's constant, and not to the platinum-iridium standard of kilograms stored in the Paris Chamber of Weights and Measures. The Paris standard of kilogram, which will become unnecessary in the chamber of weights and measures, can be awarded to the winner of this competition for the best name for the unit of mass. Most likely, the name *Stein*, with the abbreviation *st* and the derived units *mst*, *mcst*, etc., will win here. Stein, or rather, ein Stein, is a stone in German: an object with a mass. (Einstein, on the other hand, is also the name of the brilliant physicist Einstein, not found for various reasons in the "pantheon" of the basic and derived units of measurement. In the name of Einstein, there is only a little-valued extra-system unit for the number of photons, something that is used in photochemistry, and even then very rarely. Replacing the kilogram by the stein will correct this injustice.

A second significant change will affect not a basic, but a derivative unit, the Pascal, the unit for pressure. The point of the controversy here is that this unit is too small, and it almost always has to be used with the multipliers of kilo, mega or Giga. Because of this, in science and in technology there are alternative pressure units in use—physical atmospheres, technical atmospheres, bars, water column meters, millimeters of mercury (torr), etc. Such miscellany also often leads to errors in the calculations. It is planned to conduct some kind of "denomination" of pascal—the basic unit of pressure will be the "new pascal" equivalent to 100,000 old pascals. In this case, a transition period will be established, when it will be possible to use both old and new units of pressure, and then old pascals will be withdrawn from circulation.

Here you can note an inconvenience—working with previously released references, you will always have to correct the figures in old or new pascal. But heat engineers and thermophysicists are not accustomed to this discrepancy. After all, there are temperature scales of 1968 and 1990 with old and new Kelvin. Interactive recalculation of these scales can be conducted on the author's website http://twt.mpei.ac.ru/MCS/Worksheets/Thermal/T90-T68.xmcd. The temperature difference on these scales is very insignificant (fractions of a kelvin) but because of this there are frequent minor errors (but still errors) in the calculations. The difference in pressures in the old and new pascals will be significant, which should eliminate mistakes in recalculation—as a gross error is immediately visible.

The "weight" of a scientist is usually associated with the number of academic titles and awards. But here is a good old joke/parable about these exact titles.

One person (OP) died and went to heaven, where he was met by apostle Paul (AP) who had to decide where to send him—to paradise or to hell.

"Who are you, where did you work?" Asks AP.

"At the research institute," replies OP.

"A scientist, or what?"

"Something like that!"

AP was verifying in his "database of true scientists".

"That's right, scientist!" But, I beg your pardon, scientists are not allowed to enter paradise. They do not believe in God, they cut dogs, etc. You have a direct road to hell! There, around the corner! Proceed!" exclaims AP.

OP sadly plodded into hell. But here, behind the fence through the branches of paradise apple trees, he sees his former head of the scientific research institute, a former head of laboratory and a couple of his colleagues. He runs back and the dialogue continues.

"What do you mean, scientists are not allowed to enter paradise? Over there sit a corresponding member of the academy of science and even some academics!" exclaims OP.

"Where?"—asks AP with fright.

"There!" points the finger of OP.

"Yes, which scientists? They are not scientists." AP calms OP.

By the way, we have already raised the topic of scholarship and pseudo-scholarship (a scientist and a head (a manager) of scientists) in Study 19, mentioning the "directors of sharashka".* A vivid example. Some quite reasonably believe that if the leader of the Soviet atomic project Lavrentiy Beria was not shot when he was, he would now be considered the main Russian nuclear scientist. At present, our main nuclear scientist is an academician who has not made a single major discovery in the field of physics. These discoveries, perhaps, were, but they were so classified that nobody knows about them and will never know. In addition, in this situation it is not clear whether these discoveries were made "simultaneously and independently" from American nuclear scientists, or whether intelligence (KGB) played a role here.

The problem associated with scientific ethics concerns the history of the discovery not only of the nuclear chain reaction (see above), but also of our catenary function. The undisputed co-author here can be considered to be Newton, who, "simultaneously and independently" from Leibniz, discovered the laws of calculus, without knowledge of which neither Bernoulli nor Huygens could find the function of the catenary. In England, many believe that Leibniz was not quite "clean" and that he "borrowed" something from Newton without referring to Newton's work. In Germany, the situation is the reverse. By the way, in the same Wikipedia, in its Russian section, the fundamental theorem of calculus connecting the differential and the integral is called the Newton-Leibniz theorem. In the Italian section of Wikipedia it is also mentioned as the theorem of Torricelli-Barrow. In the English and German sectors of Wikipedia, it is simply called the Fundamental theorem of calculus and Fundamentalsatz der Analysis. By principle— neither for you nor for us! With the fundamental theorem of linear algebra (see Figures 7.25, 7.26 and 7.27 in Study 7), things are even more complicated. It is called the Kronecker-Capelli theorem, Roche-Capelli theorem, Roche-Fonton theorem and Roche-Frobenius theorem.

By the way, we can assume that these authors, "almost simultaneously and independently" from C.Y. Wang, have discovered a new mathematical constant related to the catenary function—see the note in Study 4. The author of this book first "discovered" this constant and then he learned via the Internet, that this constant was already discovered by C.Y. Wang about six months before.

ASSIGNMENTS FOR READERS

1. Download the Mathcad sheet with a built-in Excel spreadsheet and try to extract new information from it, remembering the second title of Study 14, "Lies, blatant lies and statistics".

2. Determine what in the above note on units of measurement is true, and what is an April Fools' joke.

3. Apply for the Ig Nobel Prize.

* This is what the people called the research institute with a prison regime where scientists were kept behind bars in Stalin's time.

Tomography = IT + Mathematics + Physics + Biology Or

Worker-Peasant Graphics

ABSTRACT

This Study explains how to simulate the work of medical tomography on a computer. The physical, mathematical and other foundations of different ways of studying the human body are considered. Methods of the computer graphic representation of human organs are proposed. The methods of graphical display of four-dimensional mathematical objects are discussed.

Mathematics: Function of one, two, and three arguments, a tomogram, random numbers.

Physics: X-ray, nuclear magnetic resonance.

IT: Programming cycles, two and three-dimensional graphics, animation.

Art: Computer sculpture.

Study website: https://community.ptc.com/t5/PTC-Mathcad-Blog/Tomography/ba-p/483596.

The idea of this investigation method came to a Russian doctor when he was walking through a market on Christmas Eve and saw chopped frozen pork carcasses in the meat section. On the slices of these carcasses were clearly visible outlines of bones and soft tissues. In the world's anatomical museums, it is possible to see, preserved in alcoholic preparations, such slices of human bodies. The dream of getting such cuts not only from corpses, but also from living people, has long intrigued many doctors and scientists. And only relatively recently was it realized. This was the result of important discoveries in the field of physics, the emergence of computer technology and the creation of a special mathematical apparatus.

Let's talk a little about physics, about its achievements and discoveries, which have allowed us to look inside a living person without a knife.

Firstly, there was the discovery of *X-rays*. In the city museum of Munich there is an interesting exhibit—the first X-ray photograph of a man in full growth. It was a rather dangerous experience: the evidence that X-rays are far from harmless appeared later. Nevertheless, this method of investigation has broadly entered into medical practice. In particular, X-rays helped to defeat such a terrible social disease as tuberculosis. X-rays helped to heal broken bones. X-rays are used in such a fairly routine operation as dental treatment. Today, we are obliged to undergo an X-ray lung examination every year. Soon tomography of a person will also become an obligatory procedure. It is believed that there are no healthy people, but there are not surveyed people, people who are not passed through a tomograph. Now tomography is a rather expensive procedure, prescribed by doctors only in special cases. X-rays are much cheaper and more affordable, but their images are rather fuzzy. We also note this, describing Figure 28.5 below.

There are two types of medical tomography: CT and MRI.

Computer tomography (CT) refers to methods of X-ray examination. It is based on measurement and complex computer processing of data based on the difference in the attenuation of X-ray radiation due to tissues of different density of a living organism. Magnetic resonance imaging (MRI) imposes a nuclear magnetic resonance (NMR) phenomenon. This method of investigation is based on measuring the electromagnetic response (echo) of the atomic nuclei, in particular the hydrogen atoms. The concentration of hydrogen in the human body is different in different places, which allows you to see on the computer screen "slices" of a person with the structure of internal organs.

Incidentally, initially these methods of investigation were called X-ray and nuclear magnetic tomography (NMR), but then came into use the terms CT and MRI because of people's radiophobia. Doctors often deceive their patients for good purposes. Whether it's good or bad is a difficult question. Studies on MRI instead of CT were introduced into medical practice to reduce the dose of ionizing radiation.

It is not a coincidence that the word "response" was supplemented by the word "echo". There is also a method of ultrasonic medical research (ultrasound), but it is not associated with tomography, although it also allows you to look inside the person. But not so deeply and not as clearly as with CT or MRI.

We do not go deeply into the details of these methods of investigation, since you can read about all this on the Internet. Let's just say that the scientists who made the greatest contribution to the development of the diagnostic methods described have received the Nobel Prizes. Conrad Röntgen himself was the first Nobel laureate in physics.

We have begun to associate the term "tomography" purely with medicine relatively recently, but at first it was purely a geometrical concept, which then penetrated into medicine. A brief reference description of the term "tomography" is: Tomography (Ancient Greek τομή, section, to cut)—obtaining a layered image of the internal structure of the object.

Tomography in medicine allows you to represent a three-dimensional human body in the form of a set of two-dimensional objects—planar sections, displayed on a computer screen. The doctor, moving the mouse and twisting its roller, can move along these sections and recreate the structure of the internal organs of a person, reveal anomalies in them and think about possible surgical intervention or therapeutic treatment.

And let's first create a tomogram of a three-dimensional and four-dimensional object in this geometrical example!

In Study 10, we considered a "popular" optimization problem: we take a square sheet of paper, cardboard, tin. Four equal squares of smaller size are cut out in the corners of the square. Then, from this cross-shaped workpiece, fold the box by folding the rectangular sections upward (see Figure 10.7 in Study 10). The question is, what should be the ratio of the length of the side of the small square to the length of the side of the large square, so that the volume of the box reaches its maximum?

This problem can be easily solved without a computer, remembering the basics of mathematical analysis. For the solution it is sufficient to create a function that returns the volume of the box (the product of the area of its base $(1-2x)^2$ by the height x), depending on the value of x, and then take the derivative of this cubic polynomial. We get a quadratic polynomial of the form $12x^2-8x+1$, which has two zeros: 1/2 and 1/6. It is known that for a smooth continuous function the first derivative is zero at the points of maximum, minimum or inflection. Therefore, the solution of our problem about the box of the maximum size is $x=1/6$. The other zero of the square polynomial (1/2) is a false solution: a minimum point, not a maximum. It is easy to verify this by constructing the graph of the function $(1-2x)^2 \cdot x$ or by calculating the value of the second derivative for $x=1/2$ and $x=1/6$. In the first case, it will be positive (minimum), and in the second—negative (maximum).

This solution, we repeat, can be found in many articles and Internet sources. In Study 10, we described a new and rather interesting "infinite" continuation of the problem of the boxes. It is possible to cut off 4 squares from the original workpiece, not to throw, but to start up—to make of them 4 new boxes of smaller size in the same pattern of cutting. From the 16 (4 · 4) new cut squares, it is then possible to make new boxes again. Of the 64 (16 · 4) new cut-off squares, it is possible again to make even smaller new boxes, etc., to infinity. This will be a kind of fractal, requiring optimization—determining the sizes of the sides of the square cuts, under which the total volume of all the obtained boxes will be maximum.

In Study 10, we graphically displayed a user function with the name V_5 with one argument (see Figure

$$V_1(x) := (1 - 2x)^2 \cdot x \qquad \text{One big box}$$

$$V_5(x,y) := V_1(x) + 4\cdot(x - 2\cdot x\cdot y)^2\cdot x\cdot y \qquad \text{One big box and four small boxes}$$

$$V_{21}(x,y,z) := V_5(x,y) + 16\cdot(x\cdot y - 2\cdot x\cdot y\cdot z)^2\cdot x\cdot y\cdot z \qquad \text{One big box, four medium boxes and sixteen small boxes}$$

$$\begin{pmatrix} \frac{d}{dx}V_{21}(x,y,z) = 0 \\ \frac{d}{dy}V_{21}(x,y,z) = 0 \\ \frac{d}{dz}V_{21}(x,y,z) = 0 \end{pmatrix} \text{solve}, \begin{pmatrix} x \\ y \\ z \end{pmatrix} \rightarrow \begin{pmatrix} \frac{1}{2} & \frac{1}{2} & \frac{1}{2} \\ \frac{1}{6} & \frac{1}{2} & \frac{1}{2} \\ \frac{3\cdot\sqrt{7}}{58} + \frac{9}{29} & \frac{1}{6} & \frac{1}{2} \\ \frac{9}{29} - \frac{3\cdot\sqrt{7}}{58} & \frac{1}{6} & \frac{1}{2} \\ \frac{\sqrt{270207\cdot\sqrt{7} + 5952978}}{16794} + \frac{7\cdot\sqrt{7}}{2799} + \frac{886}{2799} & \frac{3\cdot\sqrt{7}}{58} + \frac{9}{29} & \frac{1}{6} \\ \frac{7\cdot\sqrt{7}}{2799} - \frac{\sqrt{270207\cdot\sqrt{7} + 5952978}}{16794} + \frac{886}{2799} & \frac{3\cdot\sqrt{7}}{58} + \frac{9}{29} & \frac{1}{6} \\ \frac{\sqrt{5952978 - 270207\cdot\sqrt{7}}}{16794} - \frac{7\cdot\sqrt{7}}{2799} + \frac{886}{2799} & \frac{9}{29} - \frac{3\cdot\sqrt{7}}{58} & \frac{1}{6} \\ \frac{886}{2799} - \frac{7\cdot\sqrt{7}}{2799} - \frac{\sqrt{5952978 - 270207\cdot\sqrt{7}}}{16794} & \frac{9}{29} - \frac{3\cdot\sqrt{7}}{58} & \frac{1}{6} \end{pmatrix} = \begin{pmatrix} 0.5 & 0.5 & 0.5 \\ 0.1666667 & 0.5 & 0.5 \\ 0.447194 & 0.1666667 & 0.5 \\ 0.1734956 & 0.1666667 & 0.5 \\ 0.4769171 & 0.447194 & 0.1666667 \\ 0.1693997 & 0.447194 & 0.1666667 \\ 0.4462048 & 0.1734956 & 0.1666667 \\ 0.173645 & 0.1734956 & 0.1666667 \end{pmatrix}$$

FIGURE 28.1 The problem of 21 boxes.

10.8 on page 000) and a function V_5 with two arguments x and y (see Figure 10.13 on page 000). Figure 28.1 shows the solution of the problem of 21 boxes (1 large, 4 medium and 16 small), obtained after 3 steps of cutting a square billet. The function with the name V_{21} is analyzed, it already has 3 arguments.

And how to visually analyze the function of 3 arguments V_{21}, given in Figure 28.1? Here the animation of Mathcad will help us! Animation frames are three-dimensional slices of the graph of the function $f(y, z) := V_{21}(x, y, z)$, where each cut of the surface $f(y, z)$ corresponds to a certain value of the variable x in the interval from 0 (boxes are absent) to 0.5 (20 boxes—the original workpiece was cut into 4 squares, and 5 boxes were made from each square).

If the value of x is 0 (the upper left image in Figure 28.2), the function $V_5(x, y, z)$ is also 0, and we have a plane surface bounded by the y and z axes. If the value of x is 0.5 (see the left image in Figure 10.14 on page 000), we repeat, we have not 21 but 20 boxes: the original square is divided into 4 identical squares, of which 4 large boxes are made and 16 small. If we change the value of the parameter x from 0 to 0.5, then we can see the deformation of the cross sections of the four-dimensional surface of the function of 3 arguments, see Figure 28.3. So, with the help of a tomogram, you can visualize the function of 3 arguments—a four-dimensional graphic object!

So, tomography allowed us to visualize a four-dimensional object!

But let's go back to three-dimensional bodies—to the human body, for example.

Figure 28.4 shows the construction of two closed surfaces in Maple. This mathematical program has a built-in function *implicitplot3d* from the *plots* package for the graphical representation of closed (implicit) algebraic expressions: not $z = f(x, y)$, but $f(z, y, z) = 0$. The calculation applies the formula of the torus, deformed and torn into two halves:* in front of the x quadratic term there is not the unit (as for the torus), but a constant, 1.3. We will consider that these are our two human kidneys or two halves of the lungs. It is possible to introduce additional coefficients in the formulas of closed surfaces (torus, sphere, ellipsoid, etc.) and obtain other interesting closed and open surfaces similar to human organs such as the liver, stomach, spleen, etc.

Figure 28.4 also displays a dialog box for lighting a three-dimensional design: you can move the mouse with a light of a specified color (see the cone in the lighting dialogue box) and achieve maximum expressiveness of the 3D object, increasing its volumetric effect. To do

* The torus is not accidental. A patient is passed through a large torus during tomography, in the center of which either X-rays (CT) or a magnetic field (MRI) are focussed.

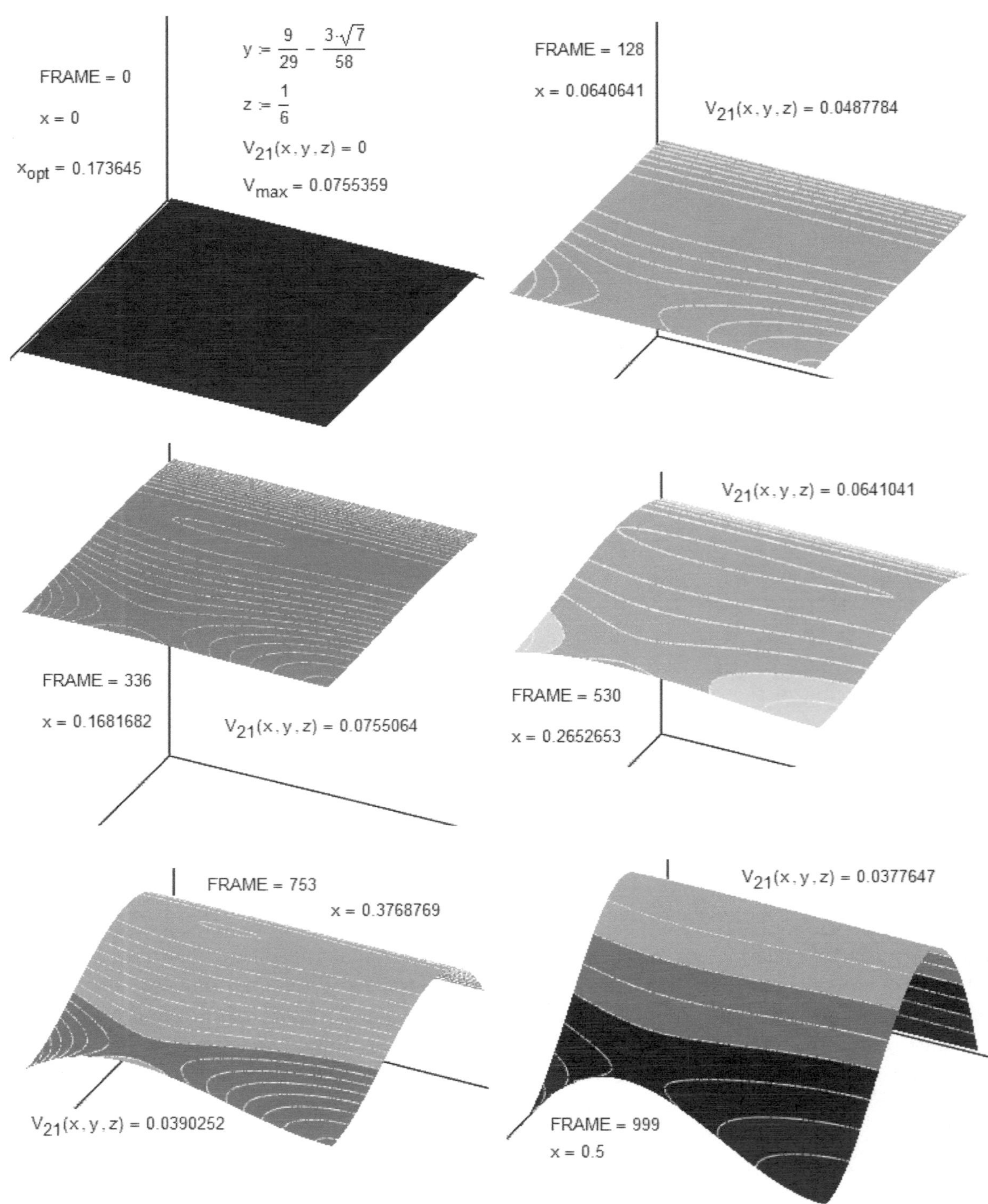

FIGURE 28.2 A four-dimensional tomogram.

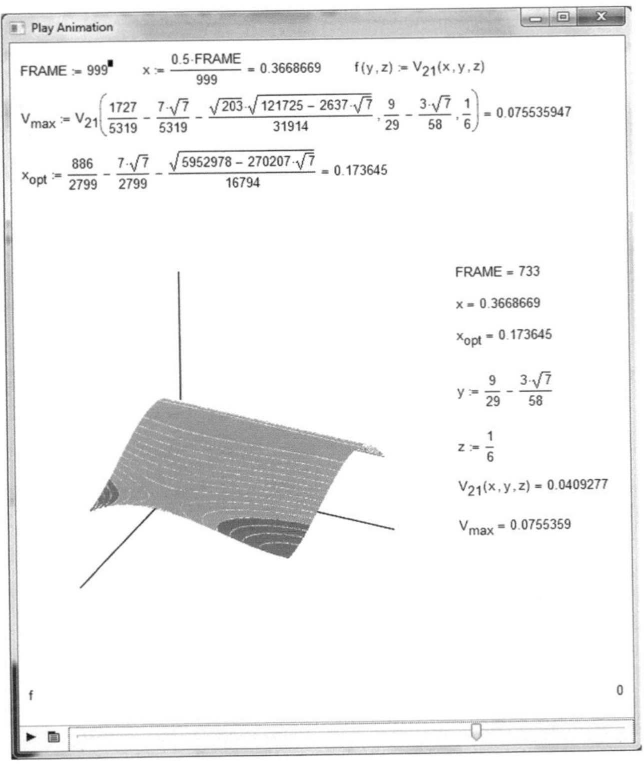

FIGURE 28.3 Video player showing a four-dimensional tomogram.

this, use fog, perspective and other tools for formatting 3D graphics.

On the website for this Study, you can see the animation of the transformation of the torus into two kidneys (in two halves of the lungs) with a smooth change in the coefficient with the argument *x* from 1 to 1.4 in the torus formula.

However, three-dimensional graphics of mathematical packages are more an advertising tool than a tool for mathematical modelling. Figures,* similar to the one shown in Figure 28.4, previously placed on boxes with a distribution of mathematical packages, and now are on the websites of software developers. If you want to design complex three-dimensional designs, then you need to work with special CAD programs. And if you want to display some mathematical object (surface or geometric body), then three coordinates often do not suffice, see above Figure 28.2. In addition, we work with a flat screen display and a flat sheet of printing paper. Therefore, Mathcad does not have a built-in function, similar to the one shown in Figure 28.4. For practical engineering calculations, volumetric graphics are not particularly needed. Therefore, its capabilities were dramatically reduced in the new version of Mathcad (in Mathcad Prime).

Nevertheless, our deformed torus, transformed into two kidneys or halves of the lungs, can also be constructed in Mathcad by the method described in Study 19—by filling the surface or volume with points whose coordinates satisfy certain conditions.

Figure 28.5 shows how our kidneys/lungs can be constructed in Mathcad by placing a small ball that will mimic a kidney stone or some darkening in the lungs diagnosed with tomography.

In a box bounded by coordinates from −0.4 to 0.4 along the X-axis, from −0.8 to 0.8 along the Y-axis and from −0.3 to 0.3 along the Z-axis, *n* random points are thrown. Those points that form the kidney/lung surface and the volume of the stone in the kidney (ball) are recorded in three vectors *X*, *Y* and *Z*. The random function is generated by the Mathcad built-in function *rnd*, which returns a random number with a value from zero to the value specified by the argument of this function. For point-wise surface shaping, the operator ≈ (approximately equal to; specified by the first operator in Figure 28.5) is used, and for the formation of a geometric body, the operator is ≤ (less than or equal to). The resulting structures (kidney/lungs with a stone or a darkening inside) are shown in the lower part in Figure 28.5.

But back to medicine. There are two methods of X-ray examination of the internal organs of a person: radiography (obtaining X-rays) and fluoroscopy (X-ray radiography), when the image of the object is obtained not on a film, but on a luminous (fluorescent) screen. In the old days, a doctor could sit near this screen and turn

* The author showed Figure 28.4 to some colleagues and asked them to describe what it looked like. They said coffee beans, cedar nuts, etc., but no one called it the kidneys or lungs of a person. Nevertheless, we will consider it to be showing a kidneys, inside of which a stone or a light dimming was formed.

the patient's torso, seeking foci of pulmonary tuberculosis, for example. The hands of the doctor would be protected with thick gloves made of special rubber with lead additives, and a heavy apron made of the same rubber was covering the body. After all, with this type of investigation, not only the patient, but also the doctor would be subjected to ionizing irradiation. However, nowadays we can safely rotate our kidneys/lungs, shown in Figure 28.5, with just one hand by using the computer mouse. Even so, at an earlier stage, when displays had television tubes there was still a risk of a small amount of ionizing irradiation. Today we use safer flat displays of a different type.

But our kidneys/lungs can be seen more clearly with the help of *tomography*! The image in Figure 28.5, we repeat, can be considered as a fuzzy X-ray. Medical tomography was then invented to improve the clarity and, therefore, the informativeness of such medical research.

Figure 28.6 shows the already described computational actions, but not in a three-dimensional space (Figure 28.5), but on a plane. In this case, the value of the variable z (vertical axis) is fixed, which is associated with the system of the variable *FRAME* that corresponds to the animation in Mathcad (animation frame counter, see Study 8). We will have 300 frames: 300 sections of the kidney/lung. By changing the value of the variable z, we can get a tomogram (section) of our three-dimensional object, see Figure 28.7.

By moving the slider of the video player shown in Figure 28.7 it is possible to identify abnormalities in our kidneys/lungs.

By the way, nothing was found on the Internet on the request for the "formula of the kidney shape". That's

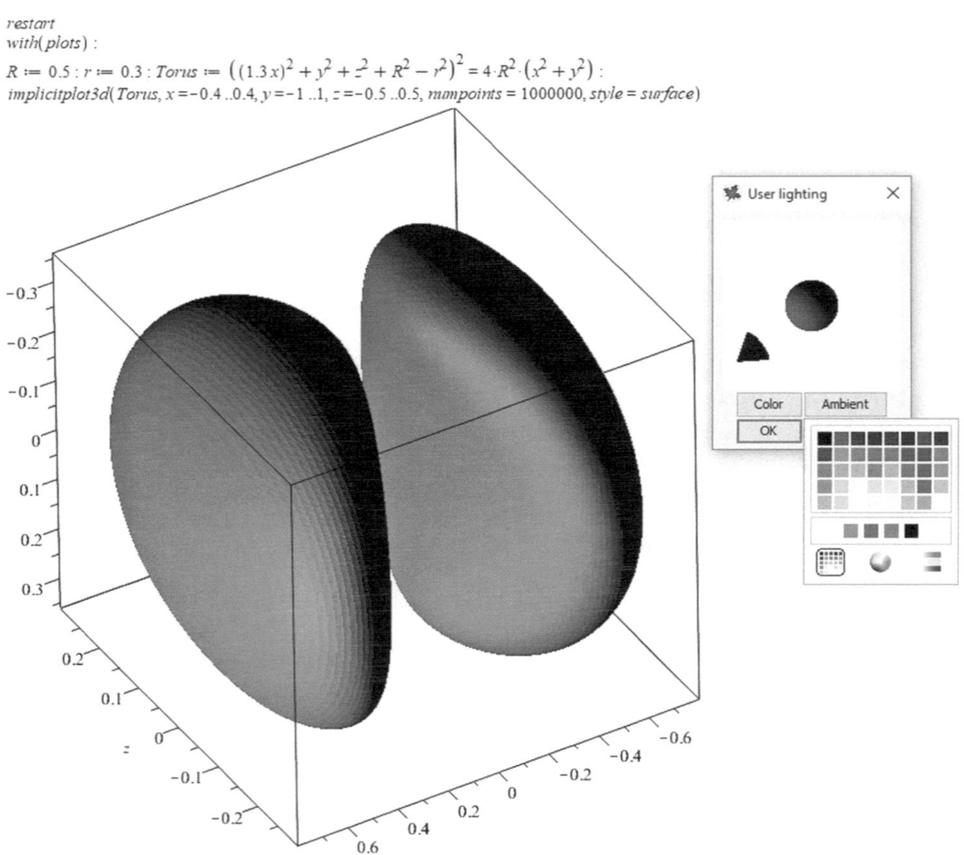

FIGURE 28.4 Building kidney/lung surfaces with Maple.

$\approx(a,b) := |a-b| < 0.0005$ ORIGIN := 1 $R := 0.5$ $r := 0.3$

$f_1(x,y,z) := \left[(1.3x)^2 + y^2 + z^2 + R^2 - r^2\right]^2 - 4R^2(x^2+y^2)$ Kidney / Lung

$r_1 := 0.05$ $f_2(x,y,z) := (x-0.2)^2 + (y-0.4)^2 + (z+0.1)^2 \le r_1^2$ Stone

$x_1 := -0.4$ $x_2 := 0.4$ $y_1 := -0.8$ $y_2 := 0.8$ $z_1 := -0.3$ $z_2 := 0.3$

```
n := 1000000     (X Y Z) := | for i ∈ 1..n
                            |   x ← x1 + rnd(x2 − x1)
                            |   y ← y1 + rnd(y2 − y1)
                            |   z ← z1 + rnd(z2 − z1)
                            |   if f1(x,y,z) ≈ 0 ∨ f2(x,y,z)
                            |     Xi ← x
                            |     Yi ← y
                            |     Zi ← z
                            | (X Y Z)
```

(X, Y, Z)

FIGURE 28.5 Pointed construction of "kidney/lungs" with a stone/dimming inside with Mathcad.

FRAME := 100 $z := \left(-0.31 + 0.62 \cdot \frac{\text{FRAME}}{300}\right) = -0.103$

```
(X Y) := | for i ∈ 1..n
         |   x ← x1 + rnd(x2 − x1)
         |   y ← y1 + rnd(y2 − y1)
         |   if f1(x,y,z) ≈ 0 ∨ f2(x,y,z)
         |     Xi ← x
         |     Yi ← y
         | (X Y)
```

x

z = −0.103

FRAME = 100

y

FIGURE 28.6 Obtaining a kidney/lung section using Mathcad.

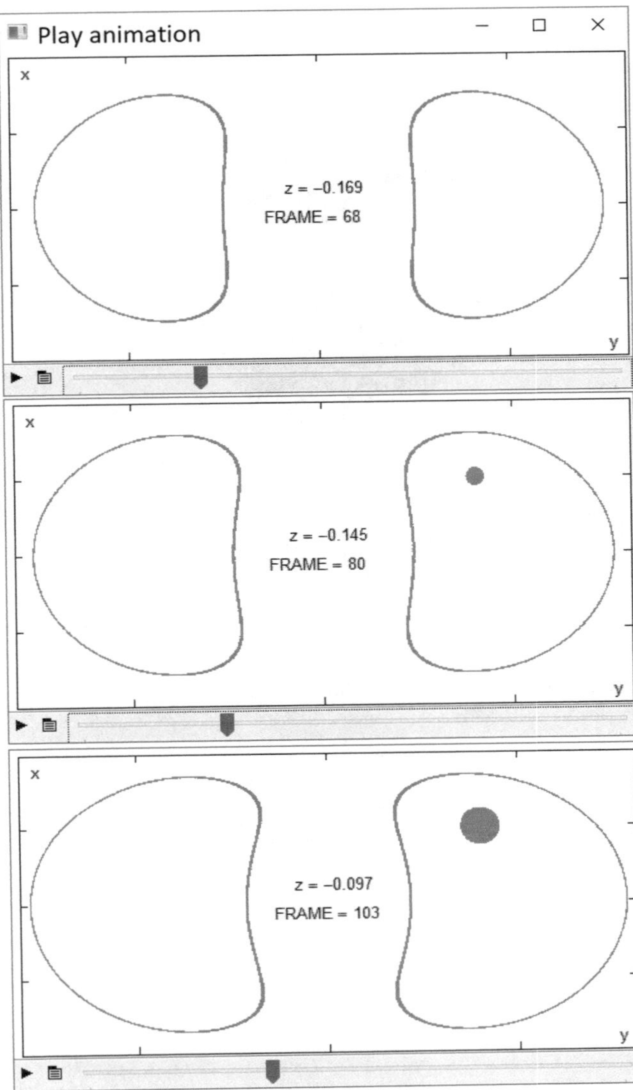

FIGURE 28.7 Frames of a tomogram of kidneys with a stone inside (dimming).

why the author had to "deform and tear" the torus. But on request, the "heart shape formula" found an algebraic expression with the variables *x*, *y* and *z*, along which you can build a heart—not in an anatomical, but in a "romantic" form (see Figure 28.8). Having said that, our Study also describes tomography in a certain "romantic" style!

It would be interesting to find on the Internet or deduce the formulas autonomously for other human organs, for example the stomach, liver or spleen. They could be placed in a certain three-dimensional space

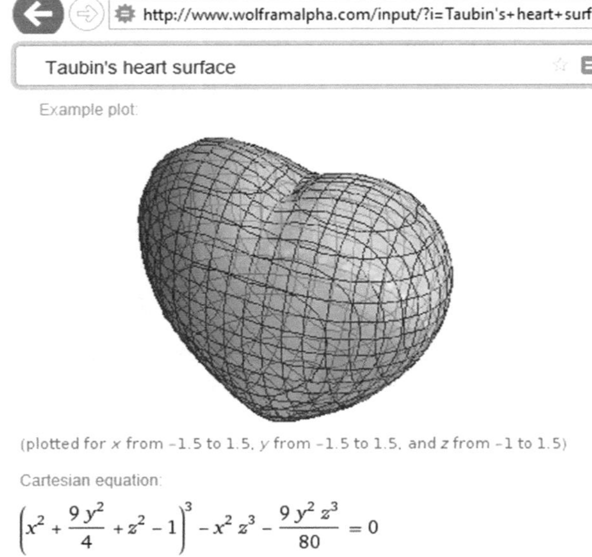

FIGURE 28.8 Building a "heart" surface using the WolframAlpha website (http://www.wolframalpha.com/input/?i=Taubin's+heart+surface).

of a person and the research conducted on a tomograph. This is a simplistic way to simulate such an investigation.

CONCLUSION

The theme of laboratory work at the intersection of informatics, physics, mathematics and biology is offered: to work with interdisciplinary connections is something that is very necessary for modern secondary and higher schools.

We live in a three-dimensional world, in which the fourth dimension is time. The described computer tools and methods make it possible, for example, to visualize the temperature of a three-dimensional surface of a heated or cooled body in time—to apply the principles of tomography to real physical objects.

ASSIGNMENT FOR READERS

Create functions that describe the different shapes of human organs.

A New Ellipse Or

Math Porcelain Service

ABSTRACT

Egglipse was first explored by Maxwell, but Descartes discovered a way to modify the pins-and-string construction for ellipses to produce more egg-shaped curves. There have been no examples of serious scientific and practical applications of three-foci ellipses until now. This situation can be changed if porcelain and ellipses are combined. In the introduced concept of the egg-ellipse, unexplored points are observed. The new three-foci ellipse with an equilateral triangle, a square and a circle as "foci" are presented for this application and can be transformed by animation. The new elliptic-hyperbolic oval is presented. The other two similar curves, hyperbola and parabola, can be also used to create new porcelain designs. Curves of the order of 3, 4, 5, etc., are interesting for porcelain decoration. An idea of combining of 3D printer and 2D color printer in the form of 2.5D printer for porcelain production and painting is introduced and listing functions in Mathcad are provided. The Study is based on the author's article in the Journal *Symmetry*.

Keywords: Ellipse; parabola; hyperbola; elliptic-hyperbolic oval; 2.5D printer; augmented reality; programming; graphics; animation.

Mathematics: Ellipse, parabola, hyperbola.

Physics: Porcelain production.

IT: Augmented reality, programming, graphics, animation.

Art: Porcelain decoration.

Study website: https://community.ptc.com/t5/PTC-Mathcad/NewEllipse/m-p/583575.

INTRODUCTION

The inventor of European white porcelain is Count Ehrenfried Walther von Tschirnhaus (1651–1708), who, at the turn of the seventeenth and eighteenth centuries conducted experiments in Saxony on the creation of porcelain, and then organized its production in Meissen near Dresden. However, some historians believe that the real inventor of European white porcelain was not the aristocrat Tschirnhaus, but the monk-alchemist Johann Friedrich Böttger (1682–1719). Tschirnhaus kept him under arrest in one fortress. Other researchers quite reasonably believe that after the death of Tschirnhaus, Böttger illegally called himself the inventor of European porcelain and was locked up because he once tried to sell the secret of making porcelain to Prussia, however this attempt was suppressed.

Böttger was an alchemist. These medieval "non-chemists" tried, in particular, to discover the philosopher's stone—a reagent necessary for the transformation of inexpensive metals into gold. The invention of porcelain somehow realized this dream—porcelain in those years, and even now is a very expensive commodity made from relatively cheap raw materials (kaolin, quartz, etc.), but bringing high profits, if it is manufactured with intelligence and talent. Prior to Tschirnhaus and Böttger in Europe there was only imported Chinese porcelain. However, following this development, porcelain was produced in Austria, France, England, Italy, Russia, the USA and in other countries. Saxon porcelain is a special porcelain because it was the first porcelain in Europe. In addition, European porcelain cannot be considered an exact copy (replica) of Chinese porcelain. Porcelain from Meissen is hard porcelain, and ancient China is soft porcelain.

Tschirnhaus left his mark in mathematics also. In Dresden there is a gymnasium with a mathematical inclination, which bears the name of Count Tschirnhaus. He, in particular, explored ellipses with three foci, which in German literature are called "the egg-shaped ellipses of Tschirnhaus" [1]. This ellipse was first explored by James Maxwell in 1846. However, Rene Descartes discovered an interesting way to modify the pins-and-string construction for ellipses in order to produce more egg-shaped curves [2, 3]. Fresh ideas for porcelain decoration can be generated due to a combination of porcelain with ellipses.

At the Meissen manufactory, all plates and saucers are made to be round only. Only large plates are made oval (elliptical). Other porcelain factories produce plates and saucers that are not only round, but also oval, square or triangular, with rounded corners and so on. Oval (elongated) plates are usually served with fish. However, Meissen porcelain is not the porcelain on which something is served at the table. Meissen porcelain is usually only admired or bragged about before the guests: "Anxious to obliterate the memory of that emotion, he could think of nothing better than china; and moving with her slowly from cabinet to cabinet, he kept taking up bits of Dresden and Lowestoft and Chelsea, turning them round and round with his thin, veined hands, whose skin, faintly freckled, had such an aged look" [4].

An ellipse is the locus of points on a plane, for each of which the sum of the distances from two other points, called foci, are constant [5–8]. There is a simple way of drawing an ellipse: two pins are fixed onto a sheet of paper, a string is attached to them, along which a pencil slides, drawing an ellipse.

If two pins stick to one point (take only one pin), then a circle will be drawn—the locus of points equidistant from the center of the circle. If one takes three pins and sticks them into three different points (foci), one can draw an ellipse with three foci—the egg-shaped ellipse investigated by Tschirnhaus (Figures 29.1 and 29.2).

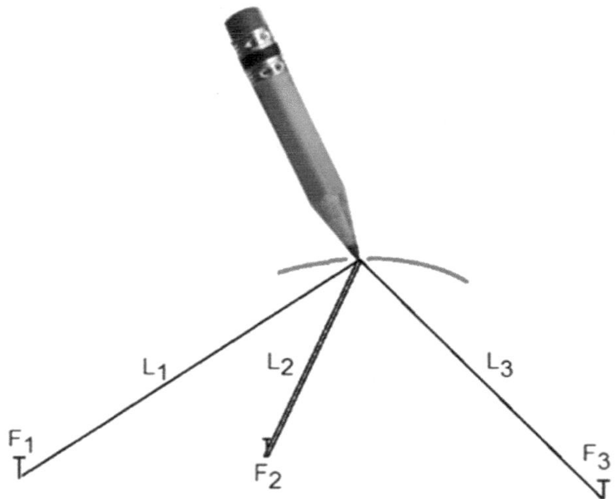

FIGURE 29.1 Drawing a three-foci ellipse, $L_1 + 2L_2 + L_3 =$ constant (see Figures 19.2 and 19.3 in Study 19).

The rope is fixed with a pin to point F_1, then the pencil lead is tossed over through the pin stuck at point F_2 and fixed to point F_3. Moving the pencil and thus changing the lengths L_1, L_2, and L_3, but keeping the sum of $L_1 + 2L_2 + L_3$ constant, it is possible to draw the closed curves shown in Figure 29.2. The author will not use the 2 before L_2 in the text that follows. This trick was found because the author used a model with a string in Figure

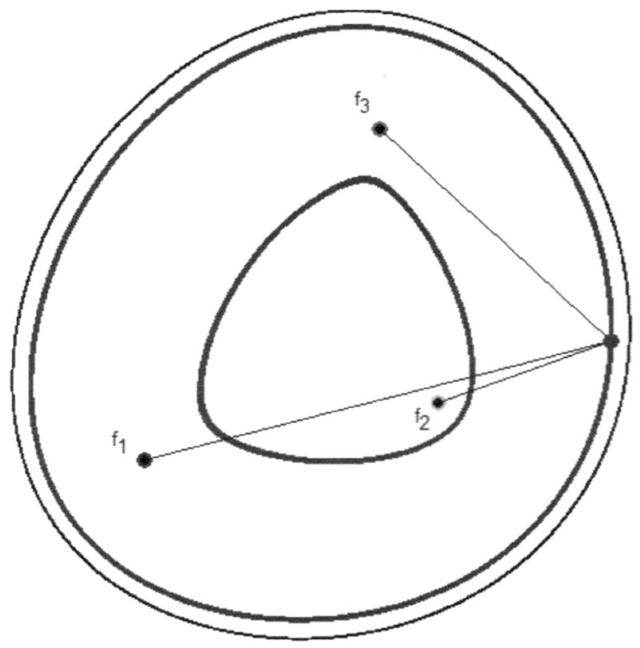

FIGURE 29.2 Sketch of a porcelain plate with two blue three-foci ellipses of Tschirnhaus (see the Figures 19.2 and 19.3 in Study 19).

29.1 with a double string between the second focus and the pencil.

Two-foci ellipses have important scientific applications—many natural and artificial celestial objects (planets and their satellites) move in elliptical orbits or in orbits close to circular ones. Three-foci ellipses, as mentioned in the introduction, do not have any serious scientific and practical application. This situation can be fixed, if you think that the mathematician Tschirnhaus attended the creation of European porcelain.

Figure 29.2 shows a sketch of a porcelain plate with two blue (in the original color picture) three-foci ellipses of Tschirnhaus. The plate itself is made in the form of such an ellipse (black contour of the plate) [9–13].

Three foci f_1, f_2, f_3, are drawn on the plate, as well as a trace of a stretched rope, the length of which remains constant when drawing the outer large ellipse of Tschirnhaus, on which one point is marked. To draw the same internal ellipse, the rope was shortened. The distances from the foci to one of the points of the outer ellipse are denoted by L_1, L_2 and L_3. These parameters can be estimated in this way:

$$\begin{aligned} L_1 &= \sqrt{(x-x_{f1})^2+(y-y_{f1})^2} = 24.59 \text{ cm}, \\ L_2 &= \sqrt{(x-x_{f2})^2+(y-y_{f2})^2} = 9.35 \text{ cm}, \\ L_3 &= \sqrt{(x-x_{f3})^2+(y-y_{f3})^2} = 16.06 \text{ cm}, \\ L_1+L_2+L_3 &= \text{const} = 50 \text{ cm}. \end{aligned} \tag{29.1}$$

The sum of these distances remains constant. For the outer ellipse, this distance is 50 cm, for the inner ellipse it is 33 cm and for the edge of the plate it is 52.5 cm.

MATERIALS AND METHODS

The first idea of this Study was to show a scanning method, which uses one flat rectangular area and allows the generation of new, unusual, interesting curves with very simple properties. The traditional mathematical methods for analysis, based on the symbolic solution of equations and equation systems, cannot always provide a solution method to generate curves with given properties. The method proposed by the author is applicable for the generation of not only curves, but also for geometric figures, i.e. if not equations, but inequalities are considered. This method is also applicable for the generation of geometrical bodies with given properties.

The second idea of this Study was to provide a scanning method for the generation of geometric objects, which is optimally correlated with the currently widely used 2D and 3D printing technology. The author introduced a concept of a 2.5D printer.

The inventors and researchers of new curves used, as a rule, analytical methods for solving problems. This significantly limits the search capabilities and is very time consuming. Numerical methods, one of which is proposed by the author, opens new possibilities in this search, but does not deny analytical methods.

In Figure 29.2, three foci are marked, not by points, but by small circles. And this is understandable—the point in its mathematical representation will not be visible on the plate. In this connection, there was an idea that broadens the notion of an ellipse. What if, in constructing ellipses, one does not rely on point-foci but on circles and draws an ellipse with such a property. Two, three, four or more circles are drawn on the plane, and then a curve is drawn, the sum of the distances from the curve-points to the circles-foci (guiding circles) are constant. This is a directrix curled into a circle. The distance from a point to a circle is easy to determine—the length of a straight line between the point itself and the point on the circle lying on the line joining the point to the center of the circle. The center of the circle, the point on its circumference and the point on our new ellipse must lie on one straight line. Instead of circle-foci, other closed curves: ellipses, triangles, squares, rhombuses, rectangles (polygons), straight lines, etc., can be used. The distance from a point to these curves is also easy to determine. For example, a point on the contour of a triangle or of a square that is closest to a point that is not on this contour, can be either on the vertex of the triangle or of the square (see the triangle in Figure 29.3) or on a segment perpendicular to the side of the triangle or of the square (Figure 29.3). It is easy to create appropriate procedures or functions and use them when building our new ellipses. A Mathcad-sheet with functions returning the coordinates of various points on segments of lines, circles, squares and equilateral triangles closest to the given point is stored as supplementary material (File_S1.xmcd). The codes are shown below.

RESULTS AND DISCUSSION

New Type of Ellipses

Using a circle instead of a point as a focus of the ellipse does not change anything in the form of the ellipse if the ellipse does not intersect the circle-focus. This occurs

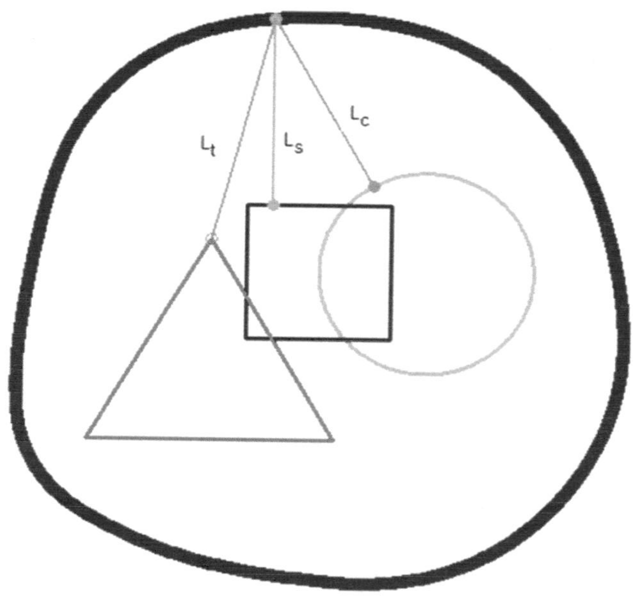

FIGURE 29.3 New three-foci ellipse $L_c + L_s + L_t = 1.80$ m.

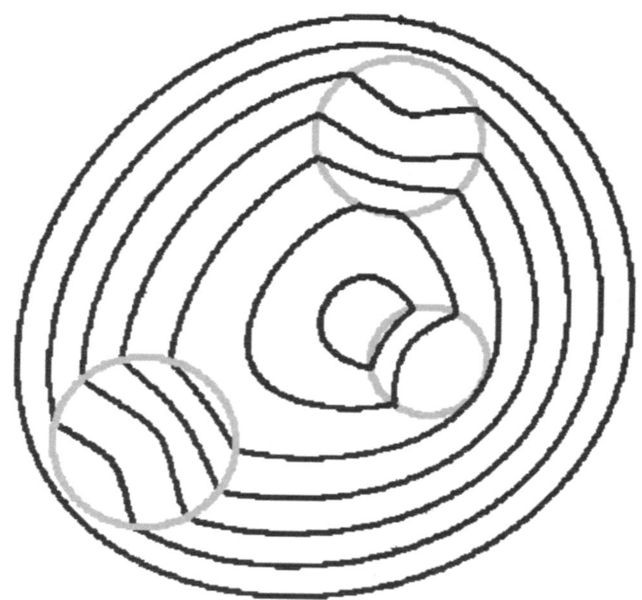

FIGURE 29.4 Sketch of a porcelain plate with the modified ellipses of Tschirnhaus (different blue ovals correspond to different values of the sum $L_{\text{circle1}} + L_{\text{circle2}} + L_{\text{circle3}}$).

when the value of the sum of $L_1 + L_2$ (bifocal ellipse) or the sum of $L_1 + L_2 + L_3$ (three-focal ellipse) is sufficiently large. Therefore, the author conducted a study not only of a circle, but of an equilateral triangle and a square as the focus of the ellipse. The animation of this study is provided within the supplementary material (PTC Community).

Figure 29.3 shows a new previously unseen three-foci ellipse, where an equilateral triangle, a square and a circle act as "foci" (directing closed curves, directrixes). The square in Figure 29.3 seems distorted, but this is an optical illusion arising from the intersection with the triangle and the circle.

By changing discretely, the sum of the distances from the points of the ellipse to fixed "foci", families of ellipses can be obtained. Figures 29.4 and 29.5 show modified Tschirnhaus ellipses in blue (in the original color picture). The blue closed curves have different sums of distances. The three "foci" of these ellipses are the three red circles.

In Figure 29.4, the outer, largest ellipse is the traditional (normal) egg-shaped Tschirnhaus ellipse with three focus points, located at the centers of the three circles. The remaining six ellipses have deviations from the traditional ellipse inside the circles-foci, which cause a specific visual effect. This is one of the goals of fine art in general and of the decoration of porcelain in particular.

Figure 29.5 shows two sketches of a Tschirnhaus mathematical plate obtained with a circle and two squares in the role of foci. On the right-hand figure, a kind of horse or zebra with longitudinal, instead of transverse, strips can be seen. The foci in Figures 29.4 and 29.5 can be likened to certain lenses, which distort the images.

The task of determining the distance from a given point to an ellipse is itself quite difficult even for a simple two-foci ellipse. However, the methods for constructing curves with given properties, described below, allow us to solve this problem in a simple way, which consists of finding the minimum element of a vector.

Combining figures (closed curves), which are used as "foci", and their mutual arrangement, it is possible to manufacture and to color plates with a non-repetitive mathematical pattern.

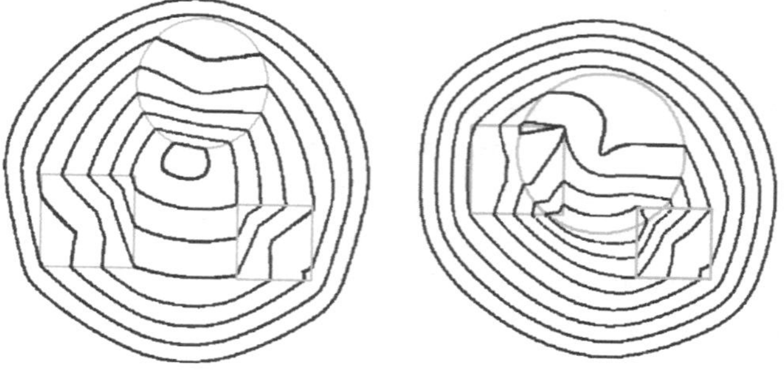

FIGURE 29.5 Sketches of the drawing of two porcelain plates obtained with a circle and two squares as "foci" (different blue ovals correspond to different values of the sum $L_{\text{circle}} + L_{\text{square1}} + L_{\text{square2}}$).

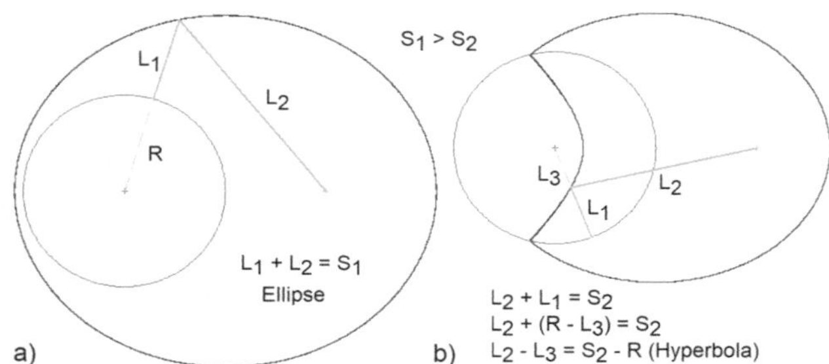

FIGURE 29.6 An ellipse: (a) the blue oval and an elliptic-hyperbolic oval; (b) the blue oval, the brown circle is the second focus, the first focus is a point.

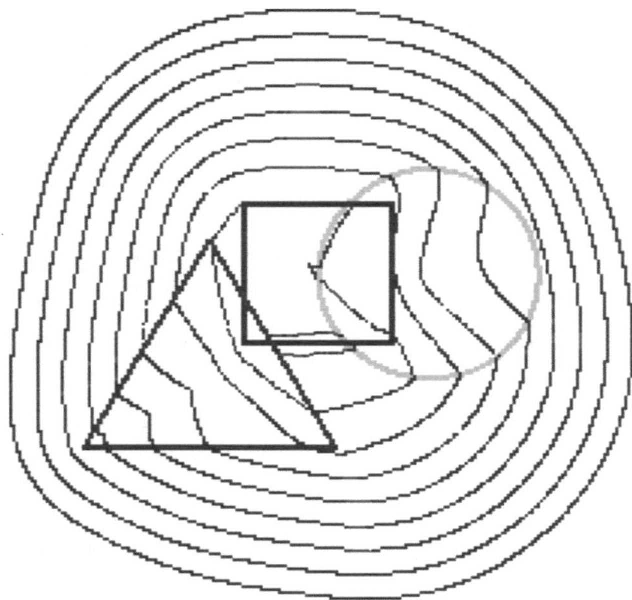

FIGURE 29.7 Another sketch of a Tschirnhaus plate; the green triangle is a first focus, the dark blue square is the second focus, the red circle is a third focus; blue ovals have different sum $L_c + L_s + L_t$.

If a circle is used as a focus instead of a point, a classic ellipse (see Figure 29.6(a)) or a new closed curve (elliptic-hyperbolic oval), which consists of a part of an ellipse and a part of one hyperbola branch (see Figure 29.6(b)), can be obtained. Such an elliptic-hyperbolic oval is observed if the sum of distances from its points to the focus point and to the focus circle is sufficiently small and the elliptic-hyperbolic oval crosses the circle.

Animation and Augmented Reality

Figure 29.7 shows a Tschirnhaus plate with the three-foci ellipses formed by a red circle (in the original color picture) , a blue square, and a green equilateral triangle (see also Figure 29.4). However, on the plate only one modified three-foci ellipse can be included in the animation.

At the moment, a new trend of information technology called augmented reality is booming: museum visitors look at an exhibit through their tablets or smartphones and receive not only additional audio and video information, but also see the object in a new perspective or in a new way.

A person who sees a mathematical Tschirnhaus plate on the wall, on a table or in a museum display, directs his tablet or smartphone to the plate and sees on his screen not only information about the mathematics and the porcelain art object, but also an animation, five frames of which are shown in Figure 29.8 as an example.

Figure 29.8 shows only one blue (in the original color picture) Tschirnhaus ellipse, which changes its length and forms intricate closed curves, tearing into parts at some point during the animation. This animation is provided within the supplementary material (PTC Community). In the shrinking ellipse, a secret meaning associated with the practical, rather than the decorative function of the plate can be seen. The vanishing ellipse is the food eaten from the plate.

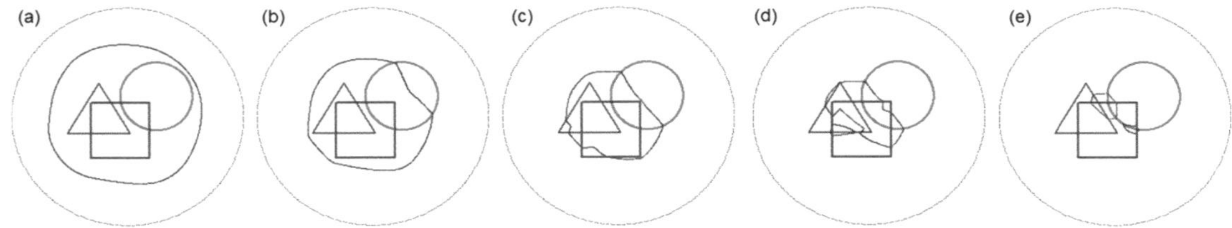

FIGURE 29.8 Frames of the drawing animation on a Tschirnhaus plate at ($L_c + L_s + L_t = S$): (a) $S = 1.920$ m; (b) $S = 1.360$ m; (c) $S = 0.910$ m; (d) $S = 0.620$ m; (e) $S = 0.330$ m (the green triangle is a first focus, the blue square is the second focus and the red circle is a third focus).

In this augmented reality, one can also create interactivity. The owner of the tablet can change the parameters of the expanded Tschirnhaus ellipse (the number and shape of the "foci", their relative position, the length of the "string", etc.) and create new sketches for the plate. If such "creativity" takes place in a museum with porcelain manufacture, then this sketch can be immediately transferred to a white plate and to get its own author's porcelain artwork.

CONCLUSIONS

A scanning method, which is used in one flat rectangular area and allows the generation of new, unusual, interesting curves with very simple properties, was introduced. This method is applicable for the generation not only of curves, but also of geometric figures, i.e. if inequalities instead of equations are considered.

The suggested method introduces a new stream into the design of porcelain dishes. The egg-shaped ellipse can help to provide new ideas for porcelain decoration and to find practical as well as decorative application. The new three-foci ellipse (elliptic-hyperbolic oval) with an equilateral triangle, a square and a circle as "foci" is introduced. The other two second-order curves, the hyperbola and the parabola, can be used to create a new porcelain design. Curves of the order of 3, 4, 5, etc. can also inspire a porcelain designer. An idea for the combination of a 3D printer and a 2D color printer in the form of 2.5D printer for porcelain production and decoration is presented and the function codes in Mathcad are provided. The superellipse or Lamé curve can provide ideas for the design of porcelain plates in the style of Lamé curves with different exponents of degree n.

The ideas proposed in this Study were presented at the porcelain factory in Meissen. Factory representatives were interested in the work and plan to use mathematical drawings on porcelain when receiving orders.

REFERENCES

1. Sahadevan, P.V. The theory of egglipse—A new curve with three focal points. *International Journal of Mathematical Education in Science and Technology*, 1987, Volume 18, pp. 29–39.
2. Williamson, B. The properties of the Cartesian oval treated geometrically. *Hermathena*, 1876, Volume 2, pp. 509–518.
3. Descartes, R. *Geometry Selected Works. P. Fermat and Descartes Correspondence*, 2nd ed. Translated from French, URSS: Moscow, Russia, 2010. (In Russian)
4. Galsworthy, J. *The Forsyte Saga*. Tauchnitz: Leipzig, Germany, 1926. p. 328.
5. Besant, W.H. *Conic Sections, Treated Geometrically*. George Bell and Sons: London, UK, 1907, pp. 50–86.
6. Levi, H.; Coxeter, H.S.M. Introduction to Geometry. *Journal of Philosophy*, 1963, Volume 60, p. 19.
7. Ochkov, V.F.; Fal'koni, A.D. Computer science, algebra, geometry: Four arithmetic curves with pokemon. *Inform. V Shkole (Computer Science School)*, 2016, Volume 9, pp. 57–61. (In Russian)
8. Ochkov, V.F.; Fal'koni, A.D. Seven computing curves or Apollonio bike. *Cloud of Science*, 2016, Volume 3, pp. 397–418. (In Russian)
9. Ochkov, V.F.; Ochkova, N.A. The project of the monument to three mathematicians or MathMetry. *Cloud of Science*, 2017, Volume 4, pp. 548–571. (In Russian)
10. Akopyan, A.V.; Zaslavskiy, A.A. *Geometric Properties of Curves of the Second Order*. MtsNMO Publishing: Moscow, Russia, 2017. (In Russian)
11. Ochkov, V.; Kalova, J.; Nikulchev, E. Optimized Fractal or FMI. *Cloud of Science*, 2015, Volume 2, pp. 544–561. (In Russian)
12. Ziatdinov, R.; Yoshida, N.; Kim, T. Analytic parametric equations of log-aesthetic curves in terms of incomplete gamma functions. *Computer Aided Geometric Design*, 2012, Volume 29, pp. 129–140.
13. Yoshida, N.; Saito, T. Interactive aesthetic curve segments. *Visual Computer*, 2006, Volume 22, pp. 896–905.

What Lies behind a Simplified Formula Or

Hybrid Calculation of the Otto Cycle

If you could take Nikanor Ivanovich's lips, and stick them on to Ivan Kuzmich's nose, and throw in some of Baltazar Baltazarovich's happy-go-lucky manner, and put all that on to Ivan Pavlovich sturdy figure—then I should have no hesitation. But as things are, I really don't know. it's such a headache!

N.V. GOGOL, *MARRIAGE*

ABSTRACT

This Study explains how it is possible to combine two mathematical programs and two kinds of mathematics (numerical and symbolic) to solve an engineering problem in the field of thermodynamics. The internal combustion engine is calculated using various mathematical programs. The problem of arguments with units in logarithmic functions is discussed.

Mathematics: The solution of an algebraic equation and of a system of equations.

Physics: Thermodynamic cycle, thermophysical properties of the working fluid, internal energy, enthalpy, entropy, efficiency of a power plant.

IT: Numerical mathematics, symbolic mathematics, hybrid mathematics.

Art: N.V. Gogol, *Marriage*

Study website: https://community.ptc.com/t5/PTC-Mathcad/Study-30/m-p/642403

At the very beginning of Study 25, which also concerned hybrid calculations, hybrid cars with internal combustion engines, electric generators, electric accumulators and electric motors were mentioned. Just a car with gasoline or diesel fuel is not very good, an electric car is also not an optimal solution, but a hybrid of these two types of vehicles is what you need!

Honestly, Study 30 would be better to be a divertissement of Study 25. However, the author of this book dedicated a separated Study, in order to emphasize the importance and relevance of hybrid calculations. By the way, many divertissements of the book deserve to be separate Studies.

The gasoline engine of a car works according to the Otto thermodynamic cycle. Previously, such engines were called carburetors,* thereby distinguishing them

* Earlier and sometimes even now we try to avoid assigning to physical processes and phenomena names of specific inventors and scientists. Here often the priority is controversial. Therefore, they try to give neutral names. Another example from Study 28: X-rays in Russian are called röntgen rays-X Another example from thermodynamics. The diagram "enthalpy-entropy" in Russia is called the h-s diagram, and, in many Western countries, the Mollier diagram (from Richard Mollier, a pioneer of experimental research in thermodynamics, particularly for water, steam, and moist air). Also, the Mendeleev-Clapeyron equation mentioned later is, in English-speaking countries, known only as the ideal gas law or the equation of state of the perfect gas. The Otto cycle is almost universally known as this, in honor of the German engineer Nikolaus August Otto, the first person to build a working four-stroke engine, a stationary engine using a coal gas-air mixture for fuel (a gas engine).

from the diesel ones. The diesel engine does not have a carburetor—a device for preparing a gasoline-air mixture. In the cylinders of the diesel engine, the fuel is injected by the nozzles into compressed air heated to a high temperature, where the fuel self-ignites. In the carburetor engine, a mixture of gasoline and air is ignited by an electric plug. But in modern gasoline engines fuel is also injected into the cylinders by injectors.

Let's do a calculation of the thermal efficiency of the Otto cycle. In its ideal version, it consists of two isochor processes: the combustion of the fuel-air mixture in the engine cylinder (vertical straight line 2–3 in the left diagram in Figure 30.1) and the exhaust gas emissions (line 4–1) and two adiabatic processes: the compression of the fuel-air mixture in the cylinder (vertical straight line 1–2 in the right diagram) and the expansion during the piston stroke (line 3–4). The isochor is a line of constant volume, while the adiabatic is a line of constant entropy. In the diesel cycle, the isobar (horizontal constant pressure line) "works" instead of one of the isochors (vertical line 23 on the left diagram in Figure 30.1). There is also the Trinkler cycle, mixed/dual cycle, when the fuel first burns along the isochor, and then burns out on the isobar.

In reference books and textbooks on thermodynamics, one can find the formula by which the thermal efficiency of the ideal Otto cycle is calculated. Figure 30.2 shows this formula, which is on the author's website for interactive use: it is possible to change the value of the compression ratio n (the ratio of the maximum volume of the cylinder to its minimum volume), set the value of c_p and the value of c_v (isobaric and isochoric heat capacities of the working cycle media tied to the universal gas

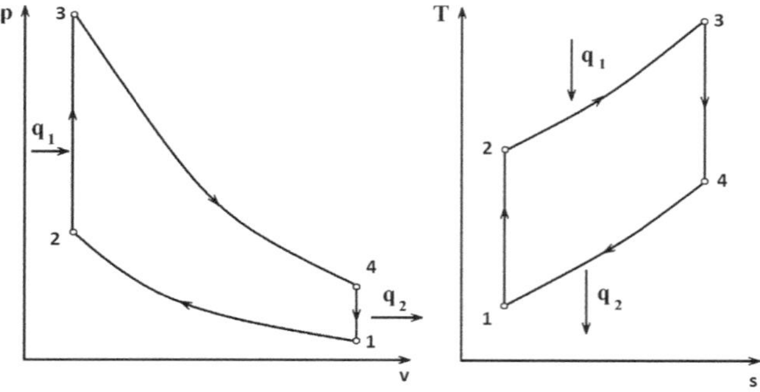

FIGURE 30.1 The "pressure-specific volume" diagram (left) and "temperature-entropy" (right) of the Otto thermodynamic cycle.

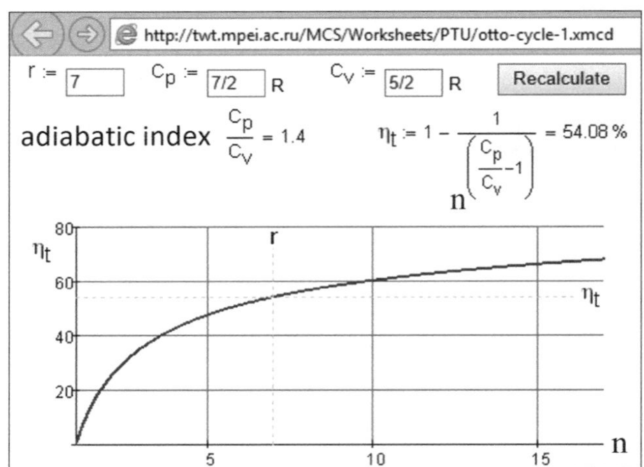

FIGURE 30.2 Online calculation of thermal efficiency of the ideal Otto cycle.

constant R) and obtain a new value of the thermal efficiency η_t shown on the graph.

Where did the formula shown in Figure 30.2 come from? The detailed point-by-point calculation of the Otto cycle will help to answer this question.

Figure 30.3 shows the calculation of the thermal efficiency of an ideal Otto cycle with the calculation of the parameters of the working fluid at four node points and with reference to the basic concepts of thermodynamics—i.e. the specific entropy s and the specific internal energy u—specific, i.e. per unit of mass (kilograms) or per unit of matter (moles).

The specific entropy is the difference of two integrals: a definite temperature integral from t_0 (it can take any value) to t (the temperature of the working fluid at the point under consideration) and the pressure integral from p_0 (here, again, any value can be taken) to p (pressure in the point under consideration).

The specific internal energy is the integral of the specific isochoric heat capacity with respect to temperature from t_0 to t.

In the calculation shown in Figure 30.3, two equations are solved at points 2 (the end of the process of compression of the fuel-gas mixture) and 4 (the end of the process of the working stroke of the piston): the algebraic equation of state of the perfect gas (i.e. Mendeleev-Clapeyron equation) and the integral equation of the specific entropy—the values of temperature and pressure are sought, that are written in the upper limits of the integrals. At the end of the calculation, shown in Figure 30.3, two answers are provided: a solution obtained through step-by-step calculation and another by using a simplified formula. Both answers coincide.

If the expression under the integral contains a constant (and for an ideal gas the values of c_p and c_v do not depend on temperature), then the constant could be taken outside the limits of the integral, the integral calculated without the constant, and then multiplied by this constant.

It is interesting to calculate the specific entropy of an ideal gas with reference to the use of units of measurement—temperature and pressure. The symbolic processor of Mathcad issued a response that contains the difference between the logarithms. To use this formula for numerical calculations, it is necessary to remember that the difference between logarithms is equal to the logarithm of the ratio (see Figure 30.4) and rewrite the formula accordingly.

If in the calculation shown in Figure 30.3, we continue conversions and simplifications of the type shown in Figures 30.4 and 30.5, then, in the end, we get the formula shown in Figure 30.2.

However, it must be remembered that, unlike ideal gases, the specific isobaric and isochoric heat

$$T_0 := 1\ K \quad p_0 := 1\ Pa \quad c_p := \frac{7}{2} R \quad c_v := \frac{5}{2} R \quad kJ := 1000\ J$$

Initial data $\quad p_1 := 1\ atm \quad T_1 := 18\ °C \quad n := 7\ \frac{m^3}{m^3} \quad T_3 := 1200\ K$

Point 1

$$v_{14} := \frac{R \cdot T_1}{p_1} = 23.891\ \frac{L}{mol}$$

$$s_{12} := \int_{T_0}^{T_1} \frac{c_p}{T}\, dT - R \cdot \int_{p_0}^{p_1} \frac{1}{p}\, dp = 69.279\ \frac{J}{mol \cdot K} \qquad u_1 := \int_{T_0}^{T_1} c_v\, dT = 6.031\ \frac{kJ}{mol}$$

Point 2

$$v_{23} := \frac{v_{14}}{n} = 3.413\ \frac{L}{mol}$$

Guess Values: $T_2 := 500\ K \quad p_2 := 20\ atm$

Constrains: $p_2 \cdot v_{23} = R \cdot T_2 \quad s_{12} = \int_{T_0}^{T_2} \frac{c_p}{T}\, dT - R \cdot \int_{p_0}^{p_2} \frac{1}{p}\, dp$

Solver: $\begin{bmatrix} T_2 \\ p_2 \end{bmatrix} := \text{Find}(T_2, p_2) = \begin{bmatrix} 634.098\ K \\ (1.545 \cdot 10^6)\ Pa \end{bmatrix}$

$$T_2 = 360.948\ °C \quad p_2 = 15.245\ atm \qquad u_2 := \int_{T_0}^{T_2} c_v\, dT = 13.16\ \frac{kJ}{mol}$$

Point 3

$$p_3 := \frac{R \cdot T_3}{v_{23}} = 28.851\ atm$$

$$s_{34} := \int_{T_0}^{T_3} \frac{c_p}{T}\, dT - R \cdot \int_{p_0}^{p_3} \frac{1}{p}\, dp = 82.538\ \frac{J}{mol \cdot K} \qquad u_3 := \int_{T_0}^{T_3} c_v\, dT = 24.923\ \frac{kJ}{mol}$$

Point 4

Guess Values: $T_4 := 500\ K \quad p_4 := 1\ atm$

Constrains: $p_4 \cdot v_{14} = R \cdot T_4 \quad s_{34} = \int_{T_0}^{T_4} \frac{c_p}{T}\, dT - R \cdot \int_{p_0}^{p_4} \frac{1}{p}\, dp$

Solver: $\begin{bmatrix} T_4 \\ p_4 \end{bmatrix} := \text{Find}(T_4, p_4) = \begin{bmatrix} 550.988\ K \\ (1.918 \cdot 10^5)\ Pa \end{bmatrix}$

$$T_4 = 277.838\ °C \quad p_4 = 1.892\ atm \qquad u_4 := \int_{T_0}^{T_4} c_v\, dT = 11.432\ \frac{kJ}{mol}$$

$$q_1 := u_3 - u_2 = 11.763\ \frac{kJ}{mol} \qquad q_2 := u_4 - u_1 = 5.401\ \frac{kJ}{mol}$$

$$\eta_t := \frac{q_1 - q_2}{q_1} = 54.084\% \qquad 1 - \frac{1}{n^{\frac{c_p}{c_v} - 1}} = 54.084\%$$

FIGURE 30.3 "Step-by-step" calculation of the thermal efficiency of the Otto cycle with calculation of the value of specific entropy and specific internal energy.

$$\int_{t_0}^{t} \frac{c_p}{T}\, dT - R \int_{p_0}^{p} \frac{1}{P}\, dP \rightarrow c_p \cdot (\ln(t) - \ln(t_0)) - R \cdot (\ln(p) - \ln(p_0))$$

$$c_p := \frac{7}{2} R = 29.101\ \frac{J}{mole \cdot K} \qquad t_0 := K \qquad p_0 := Pa$$

$$t := 18\ °C \qquad p := 1\ atm$$

$$s := c_p \cdot (\ln(t) - \ln(t_0)) - R \cdot (\ln(p) - \ln(p_0))$$

Эти единицы измерения несовместимы.

$$s := c_p \cdot \ln\left(\frac{t}{t_0}\right) - R \cdot \ln\left(\frac{p}{p_0}\right) = 69.279\ \frac{J}{mole \cdot K}$$

FIGURE 30.4 Calculation of the specific entropy of an ideal gas.

$$\int_{t_0}^{t} c_v\, dT \rightarrow c_v \cdot (t - t_0)$$

$$c_v := \frac{5}{2} R = 20.786\ \frac{J}{mole \cdot K}$$

$$t_0 := K$$

$$t := 18\ °C$$

$$u := c_v \cdot (t - t_0) = 6031.1\ \frac{J}{mole}$$

FIGURE 30.5 Calculation of the internal energy of an ideal gas.

Initial data $p_1 := 1\ atm$ $t_1 := (18 + 273.15)\ K$ $n := 7\ \frac{m^3}{m^3}$ $t_3 := 1200\ K$

Point 1 $v_{14} := \frac{R \cdot t_1}{p_1} = 23.891\ \frac{L}{mol}$

$$s_{12} := c_p \cdot (\ln(t_1) - \ln(t_0)) - R \cdot (\ln(p_1) - \ln(p_0)) \rightarrow \frac{7 \cdot R \cdot (\ln(291.15 \cdot K) - \ln(t_0))}{2} - R \cdot (\ln(atm) - \ln(p_0))$$

$$\begin{bmatrix} p_2 \cdot v_{23} = R \cdot t_2 \\ s_{12} = c_p \cdot (\ln(t_2) - \ln(t_0)) - R \cdot (\ln(p_2) - \ln(p_0)) \end{bmatrix} \xrightarrow{solve, \begin{bmatrix} p_2 \\ t_2 \end{bmatrix}} ?$$

$$s_{12} = c_p \cdot \left(\ln\left(\frac{p_2 \cdot v_{23}}{R} \right) - \ln(t_0) \right) - R \cdot (\ln(p_2) - \ln(p_0)) \xrightarrow{solve, p_2} ?$$

No find solution

FIGURE 30.6 Beginning the hybrid calculation of the Otto cycle in Mathcad.

> restart

$$> R := 8.314472 \cdot \frac{J}{mole \cdot K} : c_p := \frac{7}{2} R : c_v := \frac{5}{2} R : t_1 := (18 + 273.15)\, K : p_1 := 1\ atm : n := 7\ \frac{m^3}{m^3} :$$

$$> v_{14} := \frac{R \cdot t_1}{p_1}$$

$$v_{14} := \frac{8.314472\, J t_1}{mole\, K\, atm} \qquad (1)$$

$$> s_{12} := c_p \cdot (\ln(t_1) - \ln(t_0)) - R \cdot (\ln(p_1) - \ln(p_0))$$

$$s_{12} := \frac{29.10065200\, J (\ln(291.15\, K) - \ln(t_0))}{mole K} - \frac{8.314472\, J (\ln(atm) - \ln(p_0))}{mole K} \qquad (2)$$

$$> solve(\{p_2 \cdot v_{14} = R \cdot t_2,\ s_{12} = c_p \cdot (\ln(t_2) - \ln(t_0)) - R \cdot (\ln(p_2) - \ln(p_0))\}, \{t_2, p_2\})$$

$$\left\{ p_2 = e^{-0.4000000000 \ln(atm) + 1.400000000 \ln\left(\frac{291.1500000\, K\, atm}{t_1}\right)}, \right. \qquad (3)$$

$$\left. t_2 = 291.1500000\, e^{-0.4000000000 \ln(atm) + 0.4000000000 \ln\left(\frac{291.1500000\, K\, atm}{t_1}\right)} K \right\}$$

FIGURE 30.7 Beginning of the hybrid calculation of the Otto cycle in Maple.

capacities of real gases depend on temperature and pressure. Therefore, the calculation shown in Figure 30.3, though it is more complicated in comparison with the calculation in Figure 30.2, is more relevant due to its improvement and better approximation to the real thermodynamic process—i.e. to the cycle of the internal combustion engine, the Otto cycle. In this case, the constants c_p and c_v are replaced by complex functional dependencies on the pressure and temperature of the working fluid. The calculation shown above is a common procedure.

There is a further important aspect of the calculation, shown in Figure 30.3. It must be emphasized that the concepts of enthalpy (h) as the internal energy (u) plus the product of pressure by the specific volume (v) and of entropy (s) as the working fluid of the thermodynamic cycle, are both widely used in thermodynamics*—these are conventional values introduced into calculation practice when there are imperfections in computational technologies.

This aspect has similarity to the potential energy of the mechanical system, which we considered in Studies 21 and 26. The potential energy is evaluated from an arbitrarily given level. Analogously in Figure 30.3, the internal energy and entropy of the working fluid were calculated in terms of the two integrals with respect to temperature and pressure respectively from the arbitrarily given points t_0 and p_0. There is nothing terrible here, since we are not interested in absolute values,

* The very term "thermodynamics" is not entirely correct. Here it is necessary to talk about a kind of "thermostatic". "Thermodynamics" is what is now called heat exchange or heat transfer.

but only in their differences for different points. If the entropy does not change, this means that the processes of gas compression (line 1–2 in Figure 30.1) and its expansion during working stroke (line 3–4) are ideal.

However, problems arise when two or more mechanical or thermodynamic systems are calculated by considering different states of reference (standard) in height (mechanics) or temperature and pressure (thermodynamics). Currently, to increase energy efficiency, binary thermodynamic cycles are used—these are cycles where two working fluids simultaneously operate, for example steam in a steam turbine and air with flue gases in a gas turbine. The enthalpies and entropies of these two working bodies are calculated and tabulated from different pressure and temperature states of reference. This is inconvenient and often leads to errors in calculations.

This reminds us of the story of a bridge, which was built across the Rhine.

This bridge was built on one side by the Germans and on the other side by the Swiss. When the span of the bridge began to converge over the river in its middle, it turned out that there was an inconsistency in the structures along the vertical of a little over a meter. Where did this engineering error come from? It turned out that the Germans had a zero standard of their height as the level of the North Sea, and the Swiss used the Mediterranean! Because of the Earth's rotation, these levels differ by this exact meter. So it is with the working fluids of thermodynamic cycles: parameters of some cycles are calculated from one "level", and others from another!

However, modern hybrid calculations on the computer, which we have already examined in Study 25, allow us to abandon the numerical values of the standard states (from conditional "levels") and, consequently, from the numerical (non-hybrid) values of internal energy, enthalpy and entropy in the thermodynamic calculations.

We will explain this non-standard computational technology with the example of determining the previous thermal efficiency of the Otto cycle (Figures 30.6–30.10).

In Figure 30.6, the input data are entered in the calculation. Note that the value of t_1 is not set as $t_1{:=}\,18°C$ (see Figure 30.2), but as $t_1{:=}\,(18+273.15)K$. For a person, these two records are identical, while for the computer, as it will be shown below, they are not identical. The values of c_p and c_v remained unchanged.

The specific volume of the working fluid v_{14} on the isochore 1–4 (23.891 liters per mole) and the specific entropy s_{12} on the adiabatic 1–2 (the hybrid value consisting of symbols and numbers, with logarithms of pressure and temperature) are calculated.

Next, an attempt was made to solve the above-described system of two equations with two unknowns in the Mathcad software. There was an error—Mathcad could not do this! Failure also occurred when, by substitution, the two equations were reduced to one.

We had to resort to the Maple software, which has more powerful tools for symbolic problem solving, see Figure 30.7.

Maple solved the two equations of thermodynamics, also giving out a hybrid response, which under appropriate transformations could be turned into numbers. But this is difficult to do for various reasons. First, it is often very difficult to move Maple from "symbolic" rails to numerical ones. Secondly, in Maple, working with physical quantities is a rather complex and unusual matter. This software is focussed primarily on the theoretical analysis of a dimensional problem, and not on its specific implementation in engineering calculations. Therefore, the answer given by the Maple software (Figure 30.7) was transferred to the Mathcad software (Figure 30.8), where there are no special problems with engineering, rather than theoretical work, with units of physical quantities.

Mathcad dealt with kelvins and atmospheres and provided the numerical values of the variables t_2 and p_2, replacing them with symbolic quantities t_0 and p_0. But if, in the initial data in Figure 30.6, the operator $t_1{:=}\,(18+273.15)K$ is replaced by the operator $t_1{:=}\,18°C$, then the Mathcad software would also fail to convert the hybrid values of the variables p_2 and t_2 into numbers with pressure units (physical atmospheres) and temperatures (kelvins and degrees on a Celsius scale). By the way, we can assume that the variables Mathcad and Maple, storing dimensional values, are also some hybrid constructions, where numeric operations are performed with numbers, and symbols (units) are symbolic: simplification, substitution, replacement, etc.

The same calculations using Mathcad and Maple together are made for the final step of the hybrid calculation of the Otto cycle, see Figures 30.9 and 30.10.

The conclusion is in the style of Gogol's epigraph.

If the symbolic engine of Maple also had the simplicity and the convenience of working with Mathcad, and it took some online tools of the Mathematica software, and, perhaps, added to this the richness of the applications of the MATLAB® software and of the SMath free software—I would then immediately decide and marry, pardon, I would install this hybrid mathematical software on my computer, without regretting any money spent! And now—think about it!

And one more conclusion is also in Gogol's style, but with reference not to "Marriage" but to "Diary of a Madman".

In fact, the author of this book sent an article describing the above hybrid calculations to a leading Russian thermodynamic journal. An answer was received indicating that it was *not appropriate* to publish the article.

$$p_2 := e^{-0.4\cdot\ln(atm)+1.4\ln\left(\frac{2038.05\,K\cdot atm}{t_1}\right)} \xrightarrow{simplify \; float,5} 15.245\cdot atm$$

$$t_2 := 291.15\; e^{-0.4\cdot\ln(atm)+0.4\ln\left(\frac{2038.05\,K\cdot atm}{t_1}\right)}\cdot K \xrightarrow{simplify \; float,6} 634.097\cdot K \qquad t_2 = 360.95\ °C$$

$$v_{23} := \frac{v_{14}}{n} = 3.413\,\frac{L}{mol} \qquad p_3 := \frac{R\cdot t_3}{v_{23}} = 28.851\ atm$$

$$s_{34} := c_p\cdot(\ln(t_3)-\ln(t_0)) - R\cdot(\ln(p_3)-\ln(p_0)) \rightarrow \frac{7\cdot R\cdot(\ln(1200\cdot K)-\ln(t_0))}{2} - R\cdot(\ln(28.851107676455438\cdot atm)-\ln(p_0))$$

FIGURE 30.8 Continuation of the Otto cycle calculation in Mathcad.

$$> solve(\{p_4\cdot v_{14} = R\cdot t_4,\ s_{34} = c_p\cdot(\ln(t_4)-\ln(t_0)) - R\cdot(\ln(p_4)-\ln(p_0))\}, \{t_4, p_4\})$$

$$\left\{p_4 = e^{-0.4000000000\,\ln(28.85110768\,atm) - 1.400000000\,\ln\left(\frac{0.000002394522667\,m^3}{J}\right)},\ t_4 = 1200.\,e^{-0.4000000000\,\ln(28.85110768\,atm) - 0.4000000000\,\ln\left(\frac{0.000002394522667\,m^3}{J}\right)}\,K\right\}$$

FIGURE 30.9 Continuation of the hybrid calculation of the Otto cycle in Maple.

$$p_4 := e^{-0.4\cdot\ln(28.85110768\,atm)-1.4\ln\left(\frac{0.000002394522667\,m^3}{J}\right)} \xrightarrow{simplify} \frac{1.927651274636697618\cdot 10^7}{atm^{\frac{2}{5}}\cdot\left(\frac{m^3}{J}\right)^{\frac{7}{5}}} \qquad p_4 = 1.892\ atm$$

$$t_4 := 1200\; e^{-0.4\cdot\ln(28.85110768\,atm)-0.4\ln\left(\frac{0.000002394522667\,m^3}{J}\right)}\cdot K \xrightarrow{simplify} \frac{55389.656054268175637\cdot K}{atm^{0.4}\cdot\left(\frac{m^3}{J}\right)^{0.4}} \qquad t_4 = 277.84\ °C$$

$$h_t := \frac{q_1 - q_2}{q_1} \xrightarrow[substitute,\ q_2 = \int_{t_0}^{t_4} c_v\,dT - \int_{t_0}^{t_1} c_v\,dT]{substitute,\ q_1 = \int_{t_0}^{t_3} c_v\,dT - \int_{t_0}^{t_2} c_v\,dT} \frac{2.0\cdot 10^{-40}\cdot\left(7.572439396355366633\cdot 10^{39}\cdot atm^{\frac{2}{5}}\cdot\left(\frac{m^3}{J}\right)^{\frac{2}{5}} - 4.893921806099020 7023\cdot 10^{41}\right)}{atm^{\frac{2}{5}}\cdot\left(\frac{m^3}{J}\right)^{\frac{2}{5}}}$$

$$h_t = 54.084\%$$

FIGURE 30.10 End of hybrid calculation of the Otto cycle in Mathcad.

When the author began to ask for clarification for such a somewhat unusual response* from the editorial office of the journal, he was told that if you introduce the idea of a numerical-symbolic hybrid calculation into thermodynamic practice, then you will have to throw out all the tables with enthalpies, entropies, internal energies, Gibbs energies, Helmholtz energies, etc., that have been worked out for hundreds of years. And this can only be judged to be crazy!

It is urgent to create such hybrid mathematical software, in which you can easily carry out hybrid calculations.

* Usually the editors reject the manuscripts with such typical formulations: "it does not correspond to the journal's subject", "there is no scientific novelty", "a lot of plagiarism", "bullshit, clothed in scientific form", etc.

ASSIGNMENTS FOR READERS

1. Calculate with the hybrid method the thermal efficiency of other thermodynamic cycles—e.g. the cycles of Diesel, Trinkler, Edwards, Atkinson, Brighton/Joule, Girne, Kalina, Carnot, Lenoir, Miller, Rankin, Stirling, Humphrey, Eriksson, Brighton.

2. Try to solve the mechanical problems described in Studies 21 and 26 with the hybrid method, as well as some other problems form this book or some of your own.

3. Try to solve the problem of the efficiency of the Otto cycle with the hybrid method, but only in either Mathcad or Maple.

STUDY 31

Mad Chain Or

Physical and Mathematical Informatics

ABSTRACT

The Study describes an educational laboratory work within the framework of informatics, mathematics, and physics; specifically manifested as the sagging of a closed chain with different support points. A technology of computer processing of photos and videos of a physical experiment and the subsequent processing of the media files on the computer is illustrated. We present the π chain number (1.258—the optimal ratio of the chain length to the distance between its attachment points) and its relation to the previously unexplored problem of the shape of the sag of a closed chain hanging on two nails. A new physical and mathematical constant is found—the critical angle of the sagging of a closed chain on the "hangers". The problem of the sagging of a closed chain on a cone is detailed. The effectiveness of the computer tool "optimization with constraints" for solving theoretical mechanics problems is investigated. Three main computer tools are described: numerical mathematics, symbolic mathematics, and graphics. The importance of using units of measurement when solving physical problems on a computer is emphasized. The problem of publishing mathematical formulas in articles and books is discussed as well.

Mathematics: exponents, hyperbolic sine and cosine, catenary, closed chain, derivative, integral, parabola, curve length, differential equations

Physics: potential energy, kinetic energy, friction, suspended chain, pendulum chain, forces, moments of forces, center of mass

IT: numerical/analytical solutions of algebraic/differential equation systems, optimization, computer animation

Art: *The Big Family*, *Operation Y*, and other adventures of the character "Shurik" (Soviet movies), *Amarkord* (Italian movie by Federico Fellini), *The Surprising Adventures of Baron Munchausen* (book by German author Rudolf Erich Raspe)

Study website: https://community.ptc.com/t5/PTC-Mathcad-Questions/Round-chain-on-the-corner/td-p/553624 and https://community.ptc.com/t5/PTC-Mathcad-Questions/Ridged-chain/m-p/560385

Before *spinners* (turntables, twists) appeared, people who did not know what to do with their hands turned a coin, a box of matches or... a chain on their fingers—see Figure 31.1.

Let's also turn the chain, not on the fingers, but on two nails, on "hangers", on a cone. And we will do this not for laziness, but for scientific and cognitive purposes. In passing, we will discover a couple of new physical and mathematical constants.

THEORETICAL, IMPLEMENTATION AND EXPERIMENTAL PARTS

Usually studies of this kind are clearly divided into theoretical, implementation and experimental parts. But in this Study, it's impossible to do this—they are mixed up with each other. The author hopes that the reader will understand why this happened and will accept this quite ordinary but nonconventional technique [1–21].

THE CHAIN ON TWO NAILS

In [1] it is described how a chain is suspended at two ends and how it is all photographed and sent to the computer, where it is processed appropriately. The fingers of two schoolchildren were used as "nails" for the chain. In this school class at the junction of mathematics, physics and computer science, it was once again shown that the chain is not sagging along the parabola (and this is a very common misconception, and not only for schoolchildren and students who may be familiar with physical and computer concepts in the laboratory), but on the *catenary*, whose formula [2, 3] will be used in the calculations below.

Let us consider some previously uninvestigated modifications for a classical problem of the calculus of variations, the sagging chain problem.

Take a closed chain* and hang it not on a finger (see Figure 31.1), but on two nails (on two pins), driven into the wall,† see Figure 31.2. How will it sag? The problem is extremely idealized: the chain itself is ideal (absolutely flexible and inextensible), with the frictional forces of the chain around the nails being absent. The first condition is physically simulated with relative ease; the second, however, is almost impossible, given the fact that the chain with its joints at the attachment points clings fairly tightly to the nails. For all this construction it is necessary, for example, to tap or to shake, so that it will come to the expected equilibrium. It is possible to fasten a block with bearings on the nails and then to put a chain on it. This will complicate the geometry of the problem,

FIGURE 31.1 Character of the film,* rotating a closed chain around the finger.

* Circular in the sense that it has neither beginning nor end. Such chains without locks are hung around the neck, if the length of the chain allows it to pass over the head. In the third part of the Study, we'll talk about putting the chain on a "neck without a head"—on a cone. By the way, about the locks. They can be regarded as certain point weights on the chain, complicating its calculation and making the task more interesting (the problem of a pendant on the chain, a cabin of the cableway). By "circularity" we can also mean the absolute smoothness of the chain—the absence of friction at the points of its attachment.

† A sheet of millimeter (graph) paper is attached to the wall, with which you can measure the coordinates of individual points of the sagging chain and compare them with those that were obtained as a result of calculation. Such a work is described in [1, 2]. In our computer age millimeter paper is almost out of use. The author recently cleared through the rubbish in his office and came across rolls of millimeter paper—ordinary, semi-logarithmic and logarithmic in both coordinates. There was an idea to paste it in place of wallpaper in the room, where there is the results of the laboratory works with the chain are shown. But this idea of mathematical design of the premises is not new: the floor in the lobby of the headquarters of the Mathematical Association of America (MAA) in Washington, DC, is lined with tiles in the form of a pentagonal parquet. A search and classification of polygonal parquets is a visual and interesting task of the theory of paving modern combinatorial geometry.

* The figure shows a frame from the movie *The Big Family* (https://www.youtube.com/watch?v=941eQMgiXOA). The author of this book loves, in rare hours of leisure, to watch "for the umpteenth time" similar "good old Soviet films", something that causes surprise to his friends and relatives—how can you look at such old things!? What can I say!? "The good old Soviet film"—this is a special kind of fairy tale, epic. You can be surprised by the artificiality, and even the sheer stupidity or politicization of the script, or you can just relax and enjoy the beautiful action of brilliant actors, especially not "digging" into the plot to deeply. It's like an old-fashioned naive tale that you might have been told by your mother when you were a child. You know it by heart, but every time you listen with the same pleasure, gradually falling asleep. The fact that the "good old Soviet films" are fairy tales is something that even their creators understood perfectly, directing their attention and talent to the actors' play, and not to the plot of the film, which, we repeat, could be completely useless. Authors of modern films, having freed themselves from the bridle of censorship, try not to recount a fairy tale, but to address reality, the truth. Even so, as a rule, it turns out to be half-truth, or rather, a certain "underreporting". And this (they try to deceive you!), along with sloppy, or, frankly, mediocre actors' play, is very annoying. Gastronomic analogy. You can enjoy the taste of elite cheese, not paying attention to its appearance and/or smell. Another analogy. You can admire the well-coordinated and friendly work of the road brigade, laying asphalt, not paying attention to children and rumbling. And one more analogy, directly related to the topic of this Study. You can enjoy the solution of the physical and mathematical problem on the computer, despite the obvious lack of its practical application. After all, the friction forces of the chain about nails and the "hangers" cardinally distort the picture of the sagging of the closed chain on them.

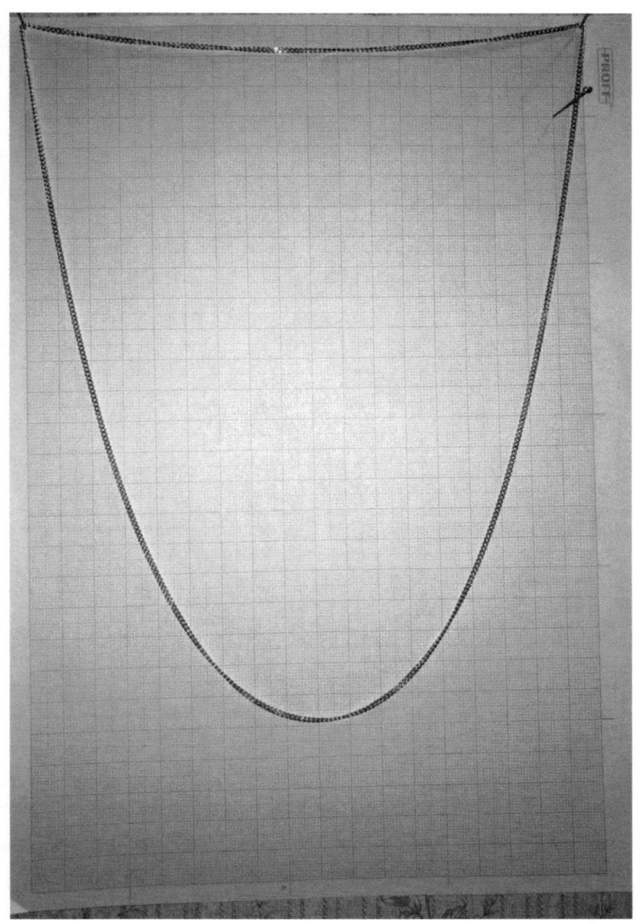

FIGURE 31.2 Real closed chain suspended on two nails.

but it will allow calculations to be compared with an application in the same way as was done in [1].

Figure 31.3 shows the beginning of the calculation of the sagging of the closed chain: input of the initial data:* the length of the chain (60 cm) and the coordinates of the attachment points of the nails, on which the closed chain is placed (0–28 and 18–15 cm).

Important note. The author does not give individual formulas in this book but places them in the context of the solutions of problems in the physical and mathematical program Mathcad. At the same time, the author perfectly understands that this prevents some readers, accustomed to traditional publications with separate numbered formulas, to understand the essence of the tasks being solved. What can we say about this!? First, all the formulas of this book are fairly clear in the texts because Mathcad uses general mathematical notation employing WYSIWIG (What You See Is What You Get) technology. Second, this style of publication meets the modern world trend, where formulas in books and articles are typed not in the text using (mathematical) editors such as LaTeX, Word (Equation), MathML, etc., but in Maple, Mathematica, Mathcad, etc. By using such "living" formulas, you can consider, it is permissible to build graphics and create animations, which significantly reduces the likelihood of errors and typos in them (see Study 1). Moreover, more and more often the chapters of books and journal articles are printouts of solutions to problems in mathematical programs. If such texts are not placed on paper but in electronic form, then they can be "live": the reader is allowed to change the original data and receive a new answer. The example in Figure 31.4: author's website for Mathcad Calculation Server technology, with which you can calculate the length of the chain by specifying the coordinates of the points of attachment of its ends h_1 and h_2, and also its clearance, the minimum distance from the "floor" h.

Figure 31.4 shows the calculation of the length of the lower part of the sagging closed chain, 51.9 cm. If in this online calculation the value of h is changed from 5.1 cm to 27.1 cm, the length of the upper short part of the chain is calculated to be equal to 18.1 cm (one mm larger than the distance between the attachment points L). In total, this gives 70 cm, which is approximately the length of the chain (see Figure 31.2) after it was removed from the nails and its length was measured directly (using a ruler), and not by an indirect (calculated) method. Thus, experiment verifies theory.

Figure 31.3 also shows user functions: the formula of the catenary† (y) and its derivative‡ (y'). Another

* As a rule, such problems are solved in dimensionless quantities. But Mathcad has a unit of measurement tool and it's a sin not to use it. This, in particular, will avoid mistakes arising from the incorrect input of formulas. The user gram-force unit (*gf*) is introduced into the calculation—one thousandth of the built-in kilogram-force unit (*kgf*). Our calculations are tuned to a system of physical quantities in cgs (centimeter-gram-second) units, so the default values of forces will be output in dynes, which we will change to gram-force.

† The canonical formula of the catenary, used in paper and electronic mathematical reference books and textbooks, has the form $a{\cdot}cosh(x/a)$. But in the calculations, we will use a non-canonical form of the catenary with two additional parameters h and x_0: $h + a \cdot cosh(((x{-}x_0)/a)\text{-}1)$. In this formula, a specific point of this curve is clearly defined—the catenary has a minimum (sagging chain, $a > 0$) or maximum (arch, $a < 0$) at the point with coordinates h—x_0, which will need to be found in the course of our calculations.

‡ The formula of the derivative of the catenary function is generated by Mathcad itself using the symbolic mathematics tool (right arrow operator). Formulas for the length of this curve and for the ordinates of its center of gravity that use a certain integral and are also involved in calculations have been found on the Internet, although they can also be easily derived using the Mathcad symbolic mathematics or "in the mind", based on our own knowledge of the basics of mathematical analysis.

$L := 60\text{cm}$ Length of the circular chain

$x_1 := 0\text{cm}$ $y_1 := 28\text{cm}$ Left chain fastening point

$x_2 := 18\text{cm}$ $y_2 := 15\text{cm}$ Right chain fastening point

$$y(x,h,a,x_0) := h + a\cdot\left(\cosh\left(\frac{x - x_0}{a}\right) - 1\right)$$ Catenary

$$y'(x,a,x_0) := \frac{d}{dx}y(x,h,a,x_0) \rightarrow \sinh\left(\frac{x - x_0}{a}\right)$$ Derivative of the catenary

$m_c := 70\frac{mg}{cm}$ Linear mass of the chain $gf := \frac{kgf}{1000}$

$$\Delta := \sqrt{(x_1 - x_2)^2 + (y_1 - y_2)^2} = 22.2\,\text{cm}$$ Distance between the points

$$PE(h_D, h_U, a_D, a_U, x_{0D}, x_{0U}) := \begin{vmatrix} L_D \leftarrow \int_{x_1}^{x_2} \sqrt{1 + (y'(x, a_D, x_{0D}))^2}\,dx \\ y_D \leftarrow \dfrac{\int_{x_1}^{x_2} y(x, h_D, a_D, x_{0D})\cdot\sqrt{1 + (y'(x, a_D, x_{0D}))^2}\,dx}{L_D} \\ y_U \leftarrow \dfrac{\int_{x_1}^{x_2} y(x, h_U, a_U, x_{0U})\cdot\sqrt{1 + (y'(x, a_U, x_{0U}))^2}\,dx}{L - L_D} \\ g\cdot m_c\cdot[y_D\cdot L_D + y_U\cdot(L - L_D)] \end{vmatrix}$$

FIGURE 31.3 Beginning of the Mathcad calculation of the sagging of a closed chain on two nails.

function called *PE* with six arguments returns the potential energy value from the sagging on the two nails of the closed chain. It depends on the parameters of the two segments of the catenary, on which the chain is divided—see Figure 31.2. We will conditionally denote these sites by indices as *D* (down, lower) and *U* (up, upper). But this division is conventional—the lower part of the chain can appear at the top, and vice versa.

The law of mechanics (the Lagrange principle [4]) states that the chain sags to minimize potential energy. Here looms a typical *optimization* problem with *constraints*, one that can be solved with the Mathcad built-in *Minimize* function, along with the *Given* keyword, see Figure 31.5.

The following calculation steps are executed in Figure 31.5:

—input of the initially assumed values for the numerical solution of the minimization problem;

—entering constraints after the word *Given*; these are length *L* of the chain (which remains constant), and the two sections (the lower *D* and the upper *U*) of the closed chain that are fixed at the ends at two given points (1 and 2);

—execution of the *Minimize* function, which returned the results: the parameters *a*, *h* and x_0 of the two parts of the catenary (*D* and *U*) such that their potential energy *PE* is minimal, and the constraints are satisfied. This result (as a vector of constants) can be transferred to the initial assumptions and assures that the answer does not change. However, we should always remember

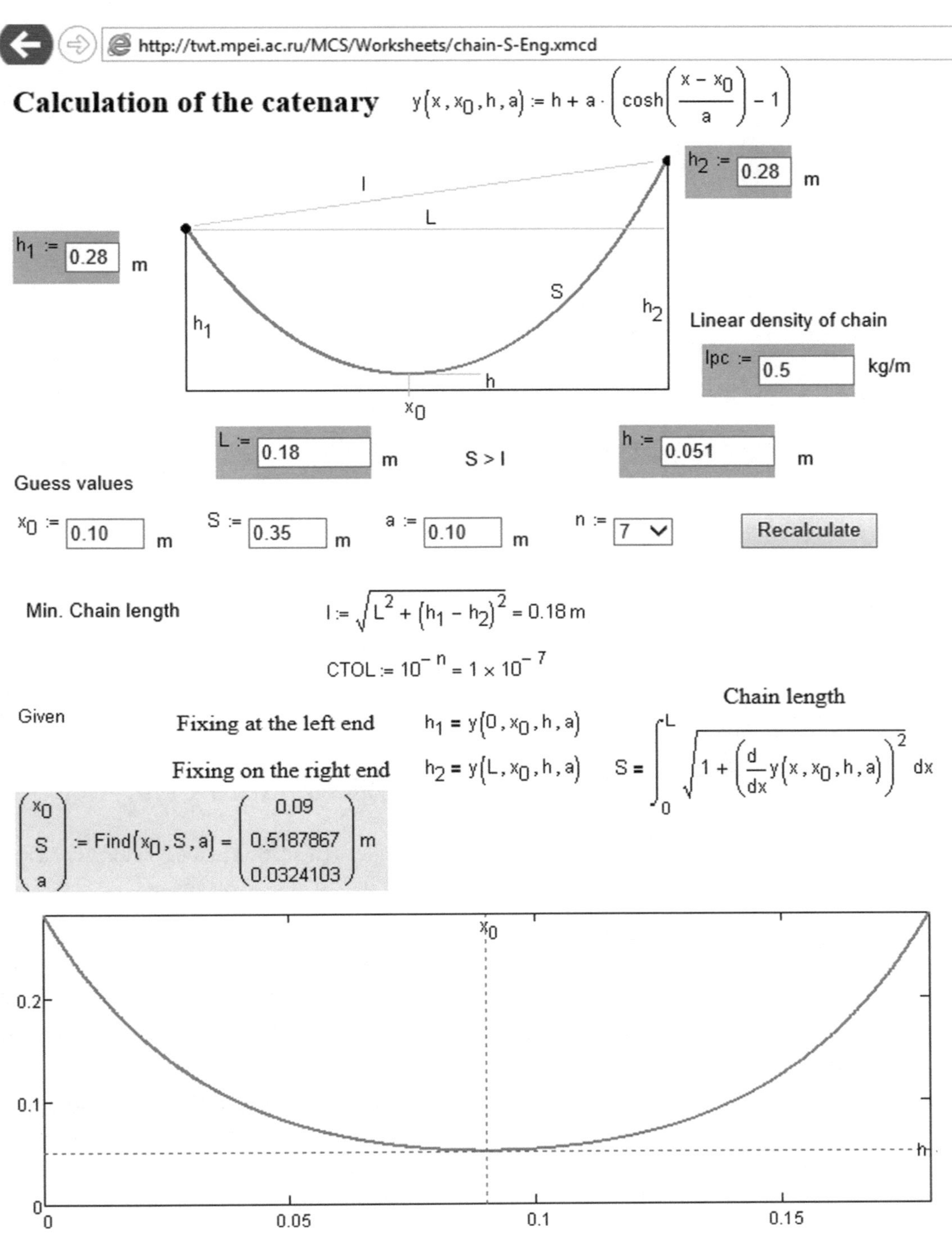

FIGURE 31.4 Online calculation of the sagging chain (see the inverse calculation in Figure 7.3).

that this answer is *approximate*, depending on the given level of accuracy in the numerical solution of the problem.

The value m_c is the linear mass of the chain, which does not affect the chain shape but will help us to check the correctness of the calculation shown in Figure 31.5, first from the point of view of the balance of forces, and second to calculate the forces that stretch the chain in correspondence of its attachment on the two nails. This new problem can be reduced to the numerical solution of an overdetermined system of eight non-linear algebraic equations with six unknowns describing the "mechanics" of a closed chain sagging on two nails

$$\begin{pmatrix} h_D \\ h_U \\ a_D \\ a_U \\ x_{0D} \\ x_{0U} \end{pmatrix} := \begin{pmatrix} 10 \\ 15 \\ 3 \\ 1 \\ 10 \\ 10 \end{pmatrix} cm \qquad \text{Guess values}$$

Given

$$\int_{x_1}^{x_2} \sqrt{1+\left(y'(x,a_D,x_{0D})\right)^2}\,dx + \int_{x_1}^{x_2} \sqrt{1+\left(y'(x,a_U,x_{0U})\right)^2}\,dx = L$$

$$y(x_1,h_D,a_D,x_{0D}) = y_1 \qquad y(x_2,h_D,a_D,x_{0D}) = y_2$$

$$y(x_1,h_U,a_U,x_{0U}) = y_1 \qquad y(x_2,h_U,a_U,x_{0U}) = y_2$$

$$\begin{pmatrix} h_D \\ h_U \\ a_D \\ a_U \\ x_{0D} \\ x_{0U} \end{pmatrix} := \text{Minimize}(PE, h_D, h_U, a_D, a_U, x_{0D}, x_{0U}) = \begin{pmatrix} 14.64 \\ 7.356 \\ 10.408 \\ 4.403 \\ 15.27 \\ 10.672 \end{pmatrix} cm$$

$$L_D := \int_{x_1}^{x_2} \sqrt{1+\left(y'(x,a_D,x_{0D})\right)^2}\,dx = 24.13\,cm$$

$$L_U := L - L_D = 35.87\,cm$$

FIGURE 31.5 Minimizing the potential energy of a hanging closed chain.

$$\begin{pmatrix} F_{y1D} \\ F_{y2D} \\ F_{y1U} \\ F_{y2U} \\ F_{xD} \\ F_{xU} \end{pmatrix} := \begin{pmatrix} 1 \\ 1 \\ 1 \\ 1 \\ 1 \\ 1 \end{pmatrix} N$$

Given

$$F_{y1D} + F_{y2D} = g \cdot m_c \cdot L_D \qquad F_{y1U} + F_{y2U} = g \cdot m_c \cdot L_U$$

$$F_{y1D} = -y'(x_1,a_D,x_{0D}) \cdot F_{xD} \qquad F_{y1U} = -y'(x_1,a_U,x_{0U}) \cdot F_{xU}$$

$$F_{y2D} = y'(x_2,a_D,x_{0D}) \cdot F_{xD} \qquad F_{y2U} = y'(x_2,a_U,x_{0U}) \cdot F_{xU}$$

$$F_{y1D}^2 + F_{xD}^2 = F_{y1U}^2 + F_{xU}^2$$

$$F_{y2D}^2 + F_{xD}^2 = F_{y2U}^2 + F_{xU}^2$$

$$\begin{pmatrix} F_{y1D} \\ F_{y2D} \\ F_{y1U} \\ F_{y2U} \\ F_{xD} \\ F_{xU} \end{pmatrix} := \text{Find}(F_{y1D}, F_{y2D}, F_{y1U}, F_{y2U}, F_{xD}, F_{xU}) = \begin{pmatrix} 1.54365 \\ 0.11098 \\ 1.74238 \\ 0.80299 \\ 0.83945 \\ 0.30399 \end{pmatrix} \cdot gf$$

$$F_1 := 2\sqrt{F_{y1D}^2 + F_{xD}^2} = 3.514 \cdot gf \qquad F_2 := 2\sqrt{F_{y2D}^2 + F_{xD}^2} = 1.694 \cdot gf$$

FIGURE 31.6 Calculation of tension forces of a chain on two nails.

with a minimum potential energy. The calculation uses the Mathcad built-in function *Find*, which returns the values of unknown equations that turns them into identities, with the given accuracy level being determined by the method and by the parameters of the numerical solution.

The physical laws shown in the calculation in Figure 31.6, are summarized below:

- the sum of the values of the vertical projections of the forces fastening the two parts of the chain on the two nails is equal to the values of the weight of the two parts of the chain; the variable notation F_{y1D} means that it is the vertical (y) projection of the force (F) with which the lower (D) part of the chain pulls the left (1) nail down;
- the values of the horizontal projections of the forces fastening the parts of the chain on the two nails (F_{xD} and F_{xU}) are equal to each other and are related to the values of the vertical projections of these forces through the values of the derivative of the catenary at these two points: $y'(x_1, a_D, x_{0D})$, $y'(x_2, a_D, x_{0D})$, $y'(x_1, a_U, x_{0U})$ and $y'(x_2, a_U, x_{0U})$;
- the tension forces of the lower and upper parts of the chain at the points of fastening are equal; the chain parts do not overstretch each other and are in equilibrium.

The calculation shown in Figure 31.6 is verified. We were convinced that the values of the catenary's parameters (variables a_D, a_U, h_D, h_U, x_{0D} and x_{0U}), found through minimization of the potential energy, would correspond to the values of the forces ensuring the equilibrium conditions of our mechanical system.

But what did the numerical computer experiment show us?

As Figures 31.7 and 31.8 show, it is a fact that the closed chain, depending on the location of the two nails, can sag in two ways!

Case 1. The chain is divided into two unequal sections (Figure 31.7).

Case 2. The chain does not bifurcate and sags in two equal halves (Figure 31.8).

At the bottom of each of Figures 31.7 and 31.8, the graphical dependences of the potential energy value of two parts of the closed chain are shown depending on the ratio of their lengths, L_U/L. The following special points are visible on the graphs:

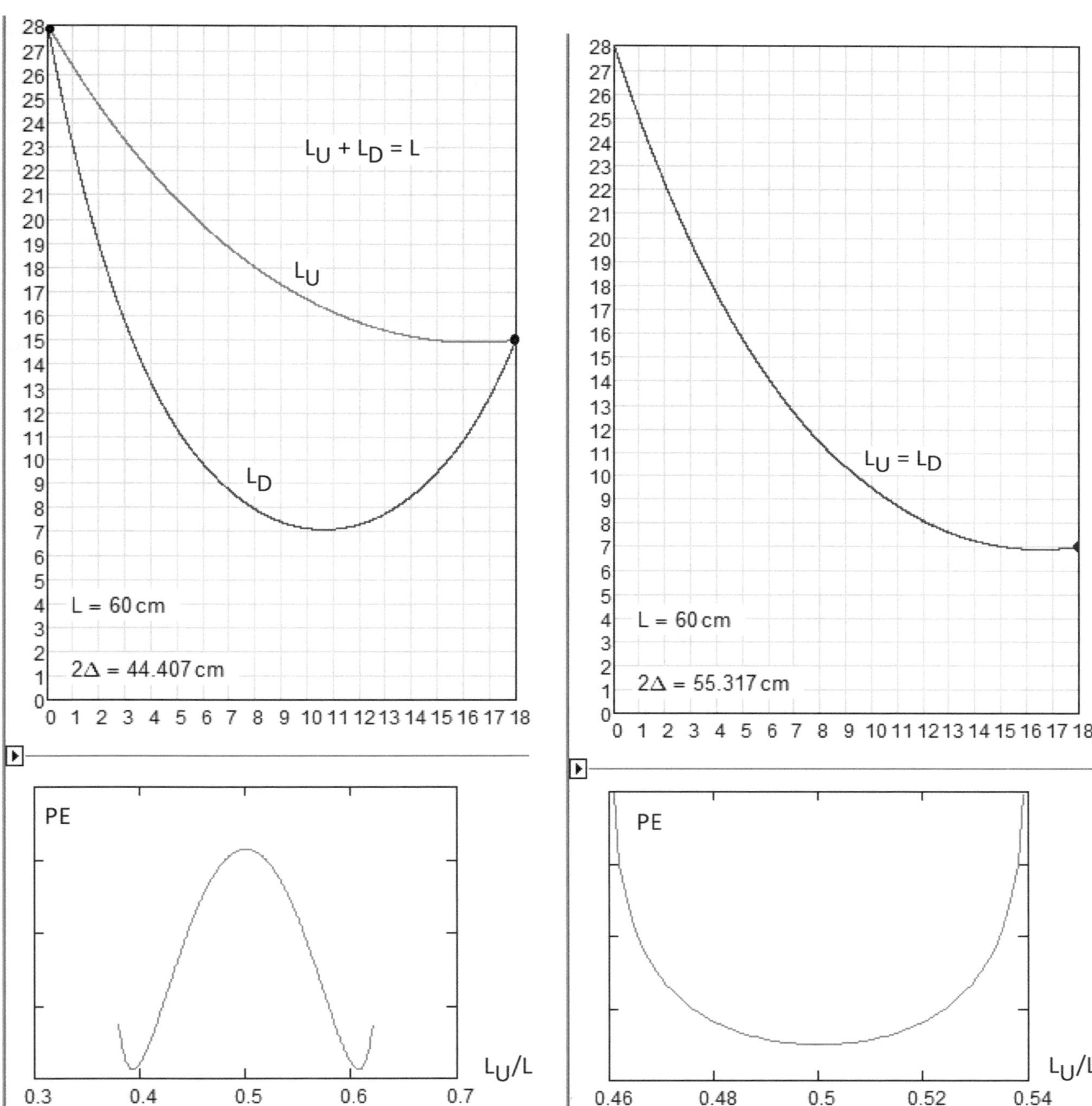

FIGURE 31.7 The closed chain hanging on two nails is divided into two unequal sections.

FIGURE 31.8 The closed chain hanging on two nails is divided into two equal sections.

- Figure 31.7, two minima (stable equilibrium) and one maximum (unstable equilibrium: in principle the chain can be of this form with two equal halves as shown in Figure 31.8, but at the slightest external disturbance it will slip into the left or right potential wells);
- Figure 31.8, one minimum (stable equilibrium): the chain sags in two equal parts).

Figure 31.9 shows a Mathcad-sheet that defines a user function that returns the value of the potential energy (PE) of the chain sagging on two nails, depending on the ratio of the length of one part to the total length of the chain (L_U / L). In this calculation case Mathcad built-in function *Find* does not return a constant vector as in Figure 31.6, but rather a vector of user-defined functions that is dependent on the parameters of the two parts of the catenary. This is a function of the variable argument L_U / L. These vector functions, named *AnsD* and *AnsU*, are decomposed into separate scalar functions with names h_D, a_D, etc. This technique allowed us to create a function called *PE* and then build its graphs, depending on the ratio of the length of the L_U to the length L for different coordinates of the points of attachment of the closed chain (see Figures 31.7 and 31.8).

To find the boundary separating the two forms of sagging of a closed chain on two nails, we will be helped by an interlude.

Interlude: The Chain Number π

Problem. How much longer should an open, conventional, two-ended chain be than the distance between the two points of its suspension (on the same level, that is) such that the tension at the ends of the chain is minimal? Let us do the calculation with using the numerical tools of Mathcad.

This is a typical problem of finding a minimum: if you increase the length of the chain, then this force will be enhanced due to the growth of its vertical projection from the increase in the weight of the chain. If the chain is shortened, the force of its tension at the attachment points will increase due to the growth of its horizontal projection. When the value of the chain length approaches the value of the distance between the points of its attachment, it ceases to be just a chain and becomes a tightly stretched string. The calculation of the string's

Given

$$y(x_1, h_D, a_D, x_{0D}) = y_1 \qquad y(x_2, h_D, a_D, x_{0D}) = y_2 \qquad \int_{x_1}^{x_2} \sqrt{1 + (y'(x, a_D, x_{0D}))^2}\, dx = L \cdot LU/L$$

$$AnsD(LU/L) := Find(h_D, a_D, x_{0D})$$

$$h_D(LU/L) := AnsD(LU/L)_0 \qquad a_D(LU/L) := AnsD(LU/L)_1 \qquad x_{0D}(LU/L) := AnsD(LU/L)_2$$

Given

$$y(x_1, h_U, a_U, x_{0U}) = y_1 \qquad y(x_2, h_U, a_U, x_{0U}) = y_2 \qquad \int_{x_1}^{x_2} \sqrt{1 + (y'(x, a_U, x_{0U}))^2}\, dx = L - L \cdot LU/L$$

$$AnsU(LU/L) := Find(h_U, a_U, x_{0U})$$

$$h_U(LU/L) := AnsU(LU/L)_0 \qquad a_U(LU/L) := AnsU(LU/L)_1 \qquad x_{0U}(LU/L) := AnsU(LU/L)_2$$

$$PE(LU/L) := \int_{x_1}^{x_2} y(x, h_D(LU/L), a_D(LU/L), x_{0D}(LU/L)) \cdot \sqrt{1 + (y'(x, a_D(LU/L), x_{0D}(LU/L)))^2}\, dx \ldots$$
$$+ \int_{x_1}^{x_2} y(x, h_U(LU/L), a_U(LU/L), x_{0U}(LU/L)) \cdot \sqrt{1 + (y'(x, a_U(LU/L), x_{0U}(LU/L)))^2}\, dx$$

FIGURE 31.9 The dependence of the potential energy of the two parts of the sagging closed chain on the ratio of their lengths*.

* In the formula for the potential energy in Figure 31.10, the constant $m_c \cdot g$ (the linear weight of the chain material, see Figure 31.5) was removed and other simplifications were made that accelerate the calculation. However, it's still quite long. The author, on his rather fast computer, drew the energy curves shown in Figures 31.8 and 31.9, in several tens of minutes. The calculation of the coordinates of each point of this curve requires the numerical solution of a rather complex system of non-linear algebraic equations with high accuracy. And this task itself is very time-consuming in addition to the other resources of the computer.

$$y(x,a,h) := h + a \cdot \left(\cosh\left(\frac{x}{a}\right) - 1\right) \quad \text{Catenary}$$

$$y'(x,a) := \frac{d}{dx} y(x,a,h) \rightarrow \sinh\left(\frac{x}{a}\right) \quad \text{Catenary derivative}$$

$$X := 0.5\ m \qquad m_c := 1\ \frac{kg}{m} \quad \text{Input data}$$

$$F_y(L) := \frac{m_c \cdot g \cdot L}{2} \quad \text{Vertical force of the suspended chain as function of its length}$$

Guess Values

$$a := 1\ m \qquad h := -0.2\ m$$

Solver Constrains

$$y(-X,a,h) = y(X,a,h) = 0\ m \qquad L = \int_{-X}^{X} \sqrt{1 + y'(x,a)^2}\ dx$$

$$Ans(L) := \text{Find}(a,h)$$

$$a(L) := Ans(L)_0 \qquad h(L) := Ans(L)_1 \qquad \alpha(L) := \text{atan}(y'(X, a(L)))$$

$$F(L) := \frac{F_y(L)}{\sin(\alpha(L))} \qquad L := 2\ X + \frac{X}{20}, 2\ X + \frac{X}{20} + \frac{X}{300} .. 4\ X$$

$$L_{opt} := \text{root}\left(\frac{d}{dL} F(L), L, 1.12\ m, 1.42\ m\right) = 1.258\ m$$

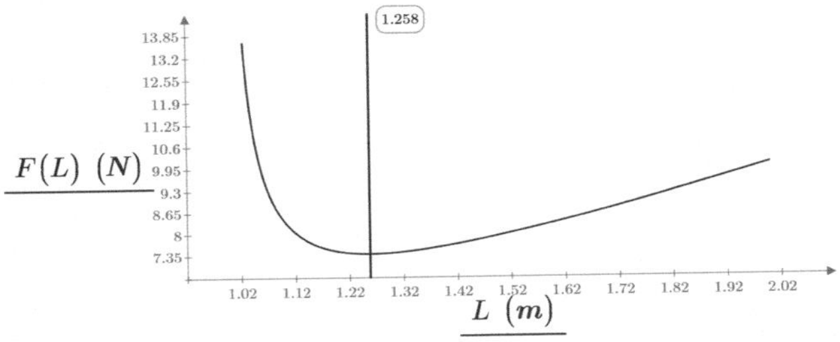

FIGURE 31.10 Determination of the optimal chain length.

vibrations is a very interesting but separate physical and mathematical problem, where the chain, now a string, cannot be considered inextensible. Here it is necessary to take into account the modulus of elasticity of the string material and other parameters.

Figure 31.10 shows the solution of this problem using Mathcad.

To solve the problem, we need to relate the force F and the length of the chain L, i.e. to define a function $F(L)$, which is called the *optimization objective function*. In the calculation in Figure 31.10, two user functions are routinely defined: a catenary whose minimum/maximum falls on the Y-axis (the X-axis passes through the two points of the chain attachments), and its derivative with respect to x. The problem is reduced to a solution of a system of two non-linear algebraic equations (where the equality of the heights of the attachment points of the chain along with the invariance of the chain length value are assumed) using the *Given-Find* block, in which the *Find* function does not return two concrete numbers (the numerical solution of the system of two equations), but defines a user function with name *Ans* (this technique is clarified in Figure 31.9). Subsequently, this function-vector (vector in a sense that the function returns two numbers) is decomposed into two separate functions, $a(L)$ and $h(L)$. Next, the optimization objective function $F(L)$ is formed, from which the graph is plotted and the minimum (where the derivative value is zero) is specified.

The number 1.258 (the ratio of the chain's optimal length to the distance between points of its attachment at the same level) can be regarded as some new physical and mathematical constant called π_c (chain number π, catenary), the ratio of the arc length of the optimal catenary to its "diameter", the distance between the attachment points. The usual number π is universally known as the ratio of the arc (the whole circle) to its diameter. It is possible to start a competition: to calculate the maximum number of digits in this constant, similarly to how it is done with respect to the "circular" number π: 3.142.

The "identify" function of Maple suggests that 1.258 might be:

$$\frac{-7}{8} + \sqrt{3} + \frac{\text{Zeta}(3)}{3} \text{ float} \rightarrow 1.2577364419554087220$$

where "Zeta(x)" is the Riemann Zeta function. A good idea would be to check whether, with sufficient numerical accuracy, the approximated π_c reaches this number. There is an inquiry into a solution by Luc Meekes from Netherlands (see https://community.ptc.com/t5/PTC-Mathcad-Questions/Symbolic-solution-of-one-optimization-problem/td-p/130281).

As a practical advice, if you want to hang a chain, rope or cable between two posts or walls at the same height, then, in the absence of any special requirements, it will be sufficient to choose a chain length that is greater by about a quarter of the distance between the attachment points, which are of course on the same level. In this case the above principle applies. If the points of attachment of the chain are at different levels, then this constant will become different (and will be a variable instead).

After the publication in the author's book [5] on this research and the emergence of the hope for the discovery of a new physical and mathematical constant, an online search on the key "1.258 catenary" was performed, which led to the publication "Optimal form of a sagging cable" [6], in which the number 1.258 already appeared. So we are a little late for the primary insight into this constant, but we consider ourselves to be the co-author.

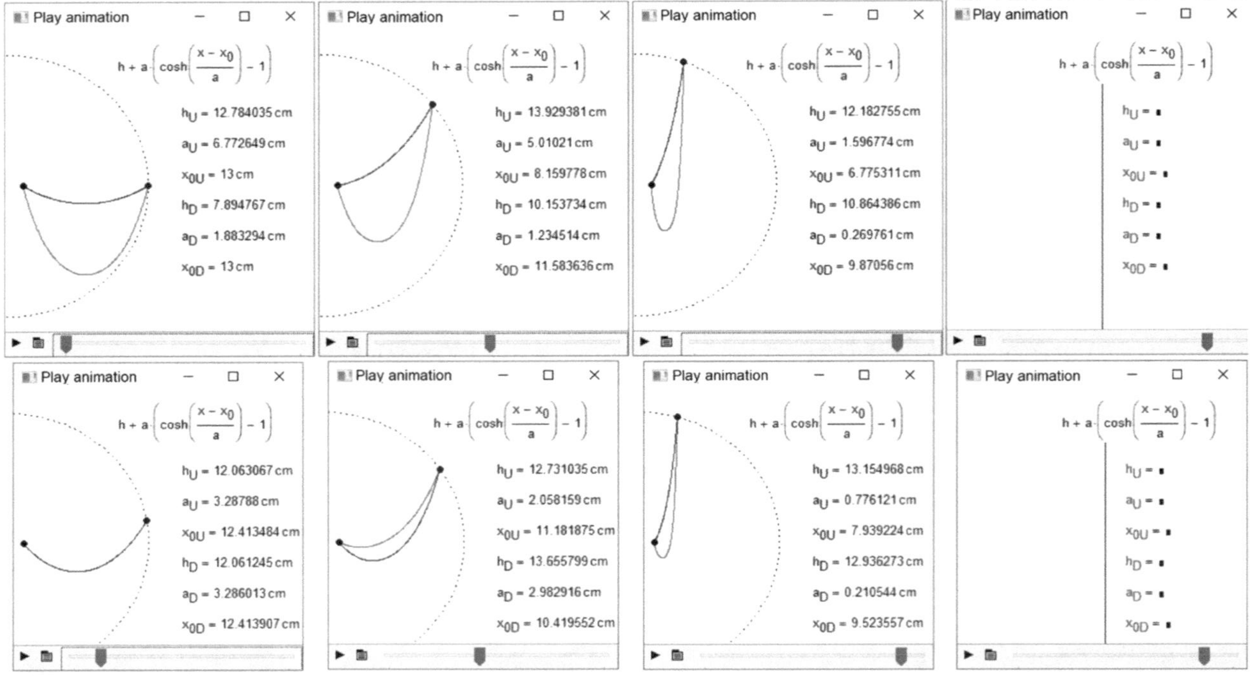

FIGURE 31.11 Animation frames of rotation of a closed chain between two nails (which replace fingers): the top row is a long chain, the bottom row is a short chain.

End of the Interlude

So that's it! Let's sketch a chain on two nails as shown in Figure 31.7, with the length gradually reduced. A numerical experiment shows that at the moment of fusion of the two sections of the chain (see Figure 31.8), the force applied to the nails acquires a minimum value. Furthermore, this force will only increase. This is a certain boundary between the two forms of sagging of the chain shown in Figures 31.7 and 31.8.

A closed chain can be rotated not only on the fingers as shown in Figure 31.1 but also in a different way: by hanging it on the left and right index fingers and rotating them about the circumference. Figure 31.11 shows the animation frames of this hand exercise in its mathematical implementation: one point of attachment of the chain is fixed, while the second is rotating in a circle around it (one-quarter of a circle to be precise, but on the other three-quarters the result will be identical or symmetrical).

What conclusions can be drawn by analyzing Figure 31.11?

First, the distance between the points of suspension of the chain is not the only criterion determining the form of its sagging. This criterion is unique to the case where the points are on the same level (see Figure 31.10).

Second, two frames of animation on the right edge of Figure 31.11 once again underline the limits of the numerical solution to the problem: under certain initial conditions, no such solution exists. This stems from the very essence of numerical methods and from the fact that a vertically sagging chain requires a different formula, rather than the one shown in the animation frames in Figure 31.11.

CHAIN ON HANGERS

A coat hanger is in the form of an obtuse-angled isosceles triangle with a hook at the top. It is hung up with clothes in the wardrobe*:

Some things (women's dresses, for example) often slip from the hanger's shoulders and fall down. In other cases,

* Conditionally all hotels in the world can be divided into two different groups. In the closets of some of them (and these, alas, are a minority) hang ordinary "home" hangers. In others, one might say, not "homemade" but "wild" hangers, which consist of two parts—the "hook", which is tightly fixed to the rod in the closet, and the hanger itself, without a hook, which must be cunningly combined with the hook. This is very inconvenient, but it is done to counter the theft of hangers by hotel guests. The author saw in one hotel also a third type. There the hangers were attached to the closet by a chain, the object of our investigation. In the old days, in public places, a chain was often attached a mug to a pot of drinking water. Now it is often possible to see in offices a chain fastened to a pen for writing so that visitors do not accidentally or intentionally take it away.

FIGURE 31.12 A hanger with close chain.

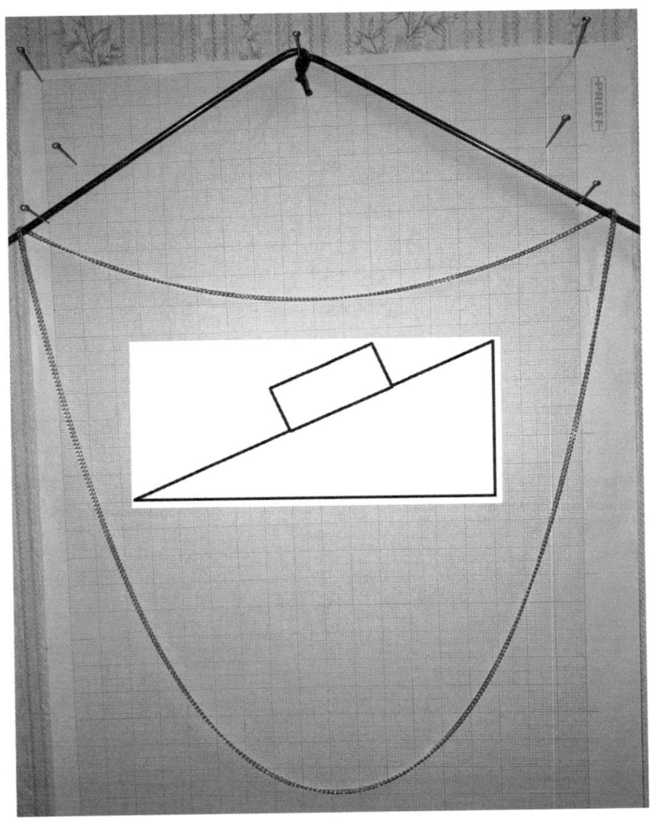

FIGURE 31.13 Closed chain on hangers.

the straps are assembled at the bottom of the hook—at the top of the triangle. Both instances are not good in terms of keeping things safe and undisturbed, so our research will have some applied value in addition to a purely scientific one!

Let's replace lingerie underwear with our closed chain and see how this hanger will hold it.

Problem. Take our familiar closed chain (Figure 31.1), which is not hung on two nails (Figure 31.2), but on two straight segments (hangers), see Figure 31.13. The forces of friction, as in the previous problem, are not taken into account, and the chain is inextensible and completely flexible. How will this chain hang?

Figure 31.14 shows the protocol for the numerical solution of this problem in Mathcad. The now-familiar *Minimize* function is used, minimizing the potential energy of the closed chain under certain constraints. The task is similar to the previous one, but the contact points of the chain are not fixed at two points, but "slide" along the segments (hangers) of the inclined lines.

It is assumed that the chain on the hangers can also hang in an asymmetrical position (obliquely, "a crooked smile"). Therefore, two unknown quantities x_L and x_R are entered into the calculation, which are the distances from the vertical axis where the hook is located to the left and right suspension points.

The chain position shown in Figure 31.13 is not an equilibrium one; rather, it is an artificial position, determined by the friction force of the chain on the hangers. As a useful analogy, the bar on the inclined plane (see the center of Figure 31.13) is stationary, though it must slide downward to reduce its potential energy. The cause of immobility is the frictional force, not the one of sliding, which is calculated by a simple formula, but the frictional force of rest, the calculation of which is much more complicated. If these forces were to not exist, the chain would either sag with the two identical halves symmetrical about the Y axis, or slip off the hangers and sag vertically, catching the base of the hook of the hanger—which condition does not depend on the chain length, but rather is only related to the angle of disclosure of the hangers. In this situation there is a qualitative difference between the hangers (Figure 31.13) and the two nails (Figure 31.2), on which two parts of the closed chain can sag either in two unequal sections (Figure 31.7) or symmetrically (Figure 31.8) to achieve the minimum potential energy. The chain on the hangers always hangs in half. It begins, however, to divide into two unequal parts at the moment when the slip begins but before slipping off the hangers.

With the help of a numerical experiment the critical opening angle of the hangers in which the chain comes off the hanger was determined. To determine this angle, an animation was created, two frames of which are shown in Figure 31.15. The animation itself can be seen on the following website, which also discusses the task: https://community.ptc.com/t5/PTC-Mathcad-Ques

$$y(x,h,a,x_0) := h + a\cdot\left(\cosh\left(\frac{x-x_0}{a}\right)-1\right)$$

$$y'(x,a,x_0) := \frac{d}{dx}y(x,h,a,x_0) \rightarrow \sinh\left(\frac{x-x_0}{a}\right)$$

$$PE(x_L,x_R,h_D,h_U,a_D,a_U,x_{0D},x_{0U}) := \int_{x_L}^{x_R} y(x,h_D,a_D,x_{0D})\cdot\sqrt{1+y'(x,a_D,x_{0D})^2}\,dx \ldots$$
$$+\int_{x_L}^{x_R} y(x,h_U,a_U,x_{0U})\cdot\sqrt{1+y'(x,a_U,x_{0U})^2}\,dx$$

$$(x_L\ x_R\ h_D\ h_U\ a_D\ a_U\ x_{0D}\ x_{0U}) := (-1\ 2\ 2.5\ 2\ 5\ 7\ -1\ 1)\,m \qquad \text{Guess values}$$

Given

$$\int_{x_L}^{x_R}\sqrt{1+y'(x,a_D,x_{0D})^2}\,dx + \int_{x_L}^{x_R}\sqrt{1+y'(x,a_U,x_{0U})^2}\,dx = L$$

$$y(x_L,h_D,a_D,x_{0D}) = x_L\cdot\tan\left(\frac{\pi}{2}-\alpha\right) \qquad y(x_R,h_D,a_D,x_{0D}) = -x_R\cdot\tan\left(\frac{\pi}{2}-\beta\right)$$

$$y(x_L,h_U,a_U,x_{0U}) = x_L\cdot\tan\left(\frac{\pi}{2}-\alpha\right) \qquad y(x_R,h_U,a_U,x_{0U}) = -x_R\cdot\tan\left(\frac{\pi}{2}-\beta\right)$$

$$\begin{pmatrix} x_L \\ x_R \\ h_D \\ h_U \\ a_D \\ a_U \\ x_{0D} \\ x_{0U} \end{pmatrix} := \text{Minimize}(PE,x_L,x_R,h_D,h_U,a_D,a_U,x_{0D},x_{0U}) = \begin{pmatrix} -1.66502 \\ 1.66501 \\ -3.35281 \\ -3.35269 \\ 3.03114 \\ 3.03188 \\ -0.00002 \\ -0.00002 \end{pmatrix} m$$

FIGURE 31.14 Calculation of the position of the closed chain on the hangers.

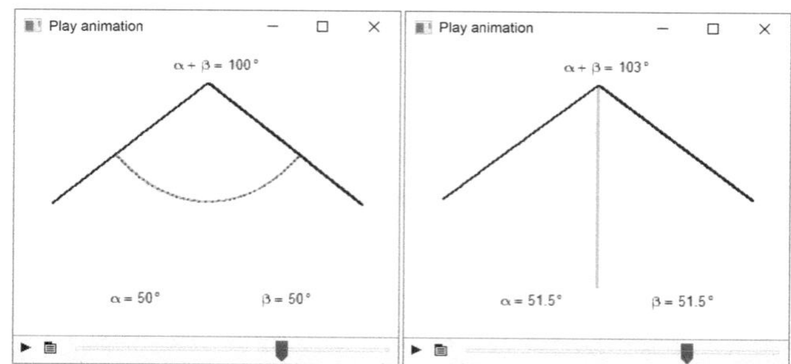

FIGURE 31.15 Two frames of the animation of closed chain jumping from the hangers.

tions/Round-chain-on-the-corner/td-p/553624. With an increase in angle α (the angle of the "opening" of the hangers) by an amount greater than 50 angular degrees, the chain comes off the hangers and hangs on the hook attachment point (at the apex of the obtuse angle).

Figure 31.16 shows the calculation of this critical angle, achieved through the numerical determination of the value of α, at which the potential energy of the chain sagging along two identical arcs of the chain line (PE_c), is equal to the potential energy of the chain hanging

Catenary $y(x,h,a) := h + a\cdot\left(\cosh\left(\frac{x}{a}\right) - 1\right)$ $y'(x,a) := \frac{d}{dx}y(x,h,a) \rightarrow \sinh\left(\frac{x}{a}\right)$

Length, linear mass and weight of chain $L := 7m$ $m_c := 70\frac{gm}{m}$ $G := g\cdot m_c\cdot L = 4.805N$

Given $X := 1m$ $Y := -1m$ $h := -1m$ $a := 1m$ Guess values

$\tan(\alpha) = -\frac{X}{Y}$ $y(X,h,a) = Y$ $\int_{-X}^{X}\sqrt{1 + y'(x,a)^2}\,dx = \frac{L}{2}$ $y'(X,a) = \tan(\alpha)$

$F(\alpha) := \text{Find}(X,Y,h,a)$ $X(\alpha) := F(\alpha)_0$ $Y(\alpha) := F(\alpha)_1$ $h(\alpha) := F(\alpha)_2$ $a(\alpha) := F(\alpha)_3$

$$PE_c(\alpha) := \left| \begin{array}{l} h_g \leftarrow \dfrac{\int_{-X(\alpha)}^{X(\alpha)} y(x,h(\alpha),a(\alpha))\cdot\sqrt{1 + y'(x,a(\alpha))^2}\,dx}{0.5\cdot L} \\ G\cdot h_g \end{array}\right. \qquad PE_v := \left| \begin{array}{l} h_g \leftarrow \dfrac{-0.5L}{2} \\ G\cdot h_g \end{array}\right.$$

$\alpha_{cr} := \text{root}\left(PE_c(\alpha) - PE_v, \alpha, 30deg, 70deg\right) = 50.34\cdot deg$ Answer

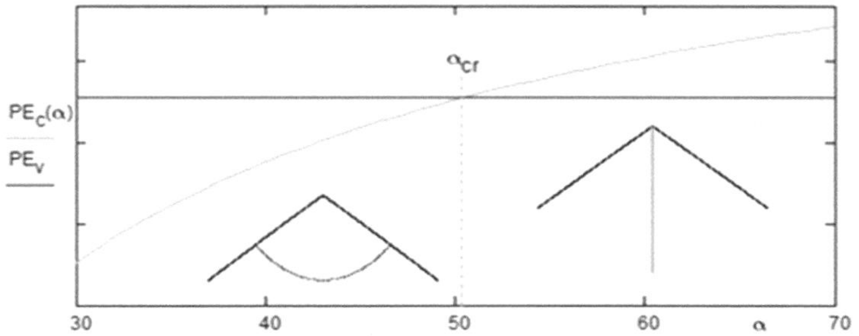

α_{cr}

$PE_c(\alpha)$

PE_v

30 40 50 60 α 70

FIGURE 31.16 Critical angle calculation of a closed chain sagging on the hangers.

vertically (PE_v) while hooked to the top of the hangers.* This angle α_{cr} was slightly more than 50 degrees, which correlates with the data obtained from the animation (Figure 31.15).

Most likely, an analytical rather than a numerical approach to solve the chain problem on the hangers will give an answer measured not in angular degrees, but in the form of the fraction $n\bullet\pi$ / m, where n and m are natural numbers greater than unity. Mathematicians are always somewhat upset when it is said that a sine of 60 angular degrees† is equal to 0.866. The orthodox mathematician here will say differently (see second line of calculation in Figure 31.17).

You can try to find analytically (symbolically) the value of α_{cr} (the values of n and m in the expression $n\bullet\pi$ / m, specifically), but you can also do it in the way shown in Figure 31.18; that is, sorting out the values of n and m in the range from 1 to 30, and remembering those ($n=7$, $m=25$) that correspond most closely to the numerical response in Figure 31.16.

* First the catenary parameters X, Y (right attachment point), h (clearance) and a (catenary scaling parameter) are determined as function of the angle α through four equations. Then the critical angle is found by solving the relationship equating the potential energy of two identical arcs of a chain line (PE_c) to the potential energy of a chain hanging vertically (PE_v). Conditionally by combining the equation of the arc length and the condition on the derivative we obtain the Whewell equation of the catenary (i.e. $L/2 = a\tan(\alpha)$).

† A hot-air balloon burst from the clouds. From the basket the flyer shouts to a man on the ground: "Excuse me, where are we?" Answer from the man: "You are in a basket of a balloon!" He was a mathematician. Only from a mathematician can we hear an absolutely accurate but completely useless answer. There is a little-known continuation of this old joke. The flyer in response shouts: "You do not understand! Our navigator's batteries are depleted, and we cannot determine our coordinates!" The mathematician looks at his smartphone and answers: "Zero point ninety-six hundredths radians of northern latitude and zero sixty-four hundredth radians of eastern longitude!" Only mathematicians measure angles in radians, and not in more familiar degrees!

$$\sin(60°) = 0.866$$

$$\sin\left(\frac{\pi}{3}\right) \rightarrow \frac{\sqrt{3}}{2}$$

FIGURE 31.17 Numeric and symbolic sine calculations in Mathcad.

```
(n)   :=  | Δmin ← 1                              = (7 )
(m)       | for  n ∈ 1.. 50                         (25)
          |   for  m ∈ 1.. 50
          |     | Δ ← |50.34deg − (n/m)·π|
          |     | if  Δ < Δmin
          |     |   | Δmin ← Δ
          |     |   | n1 ← n
          |     |   | m1 ← m
          | (n1)
          | (m1)
```

$$\frac{7}{25}\pi = 50.4\,\text{deg}$$

FIGURE 31.18 Refinement of the critical angle value of the opening of the hangers with the chain.

The calculation shown in Figure 31.18 can of course be considered as a kind of curiosity—an imitation of analytical approach to the problem solution. However, the author's emphasis of this approach is not in vain, since it is very common in our computer age. In some cases, it is very productive, while in others it is counterproductive. We entrust those readers who are strong in analytical transformations to confirm or refute the assertion that $\alpha_{cr} = 7\pi / 25$. While preparing the English version of this book a question was posed on the PTC forum in this regard: https://community.ptc.com/t5/PTC-Mathcad-Questions/Has-this-equation-one-symbolic-solution/m-p/565271#M181732. By combining "hand" and symbolic mathematics, several forum users were able to analytically solve the integral for potential

$$\alpha := \text{root}\left[1 - \frac{1}{2}\cdot\left[\frac{\text{asinh}(\tan(\alpha)) + \left(\sqrt{1+\tan(\alpha)^2}\right)\cdot\tan(\alpha)}{(\tan(\alpha))^2}\right], \alpha, 1°, 90°\right] = 50.3395°$$

FIGURE 31.19 Solution of the non-linear equation arising from the analytic calculation of the integral of the potential energy of the chain on the hangers.

energy (shown in Figure 31.16), but all had to resort to a numerical method (especially the four parameter, Mathcad *root* function, and the computation in Figure 31.19) to solve the non-linear equation arising due to the equality of the potential energy of the chain lying on the hangers with one part of the chain having come off and assumed a vertically hanging position on the hook. It was also confirmed that the critical angle is independent of the chain length.

The reader can also replace rectilinear hangers by either round ones or those in the form of a paraboloid segment, a hyperbola, or even a segment of a chain line and subsequently analyze the closed chain behavior on such "designer hangers". On the chain, one can hang a pendant (point mass) and evaluate the properties of such mechanical system.

CHAIN ON A CONE

The behavior of a closed chain on a circular, straight cone is worthy of interest. The reader can either conduct such experiment by themselves or read the report on this work in [7], the acquaintance with which moved the author to write this Study.

A few words about the chain on the cone are presented. The chain may, depending on the angle of the opening of the cone, either jump off of it or remain on the cone (the frictional forces, once again, are not taken into account). But there is also a third equilibrium—an unstable one, which is qualitatively displayed on the lower graph of Figure 31.7. At the slightest external jerk, the chain can either jump off the cone or take a stable equilibrium, thereby enveloping the cone. This was proved by the authors of [4]. It is also interesting to investigate the oblique cone along with how a closed chain will behave on it. In fact, indirect oblique hangers with a closed chain can be investigated using the program shown in Figure 31.14. The result is revealed in Figure 31.20.

At any inclination angle of the hangers, a closed chain wrapped around them does not bifurcate; rather, the angle of connection of the ends of the chain with the hangers remains straight, which eliminates the sliding of the chain along the hangers.

The hangers cannot change the angles of opening at the top (see Figure 31.20), but simply tilt them. Moreover, Figure 31.21 shows first that this slope leads to a potential energy increase of the chain along with departure from stable equilibrium, and second that there is one more critical angle when the chain slides off of the shoulders. This experiment requires more detailed analysis, which goes beyond the scope of this Study.

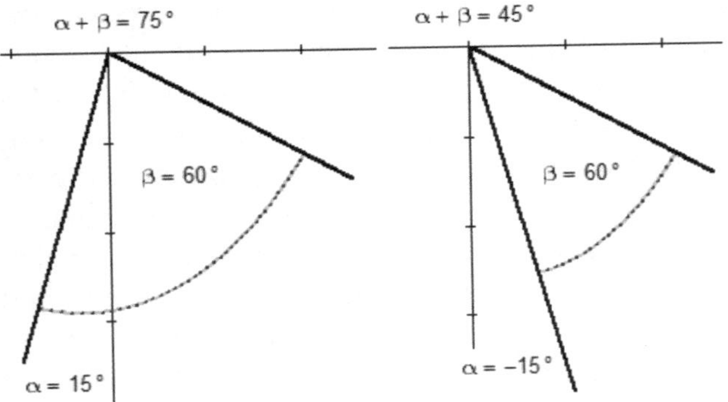

FIGURE 31.20 Chain on oblique hangers.

MAD CHAIN

The same expedient would not have answered against a mad dog, which soon after came running against me in a narrow street at St. Petersburg. Run who can, I thought; and to do this the better, I threw off my fur cloak, and was safe within doors in an instant. I sent my servant for the cloak, and he put it in the wardrobe with my other clothes. The day after I was amazed and frightened by Jack's bawling, "For God's sake, sir, your fur cloak is mad!" I hastened up to him, and found almost all my clothes tossed about and torn to pieces. The fellow was perfectly right in his apprehensions about the fur cloak's madness. I saw him myself just then falling upon a fine

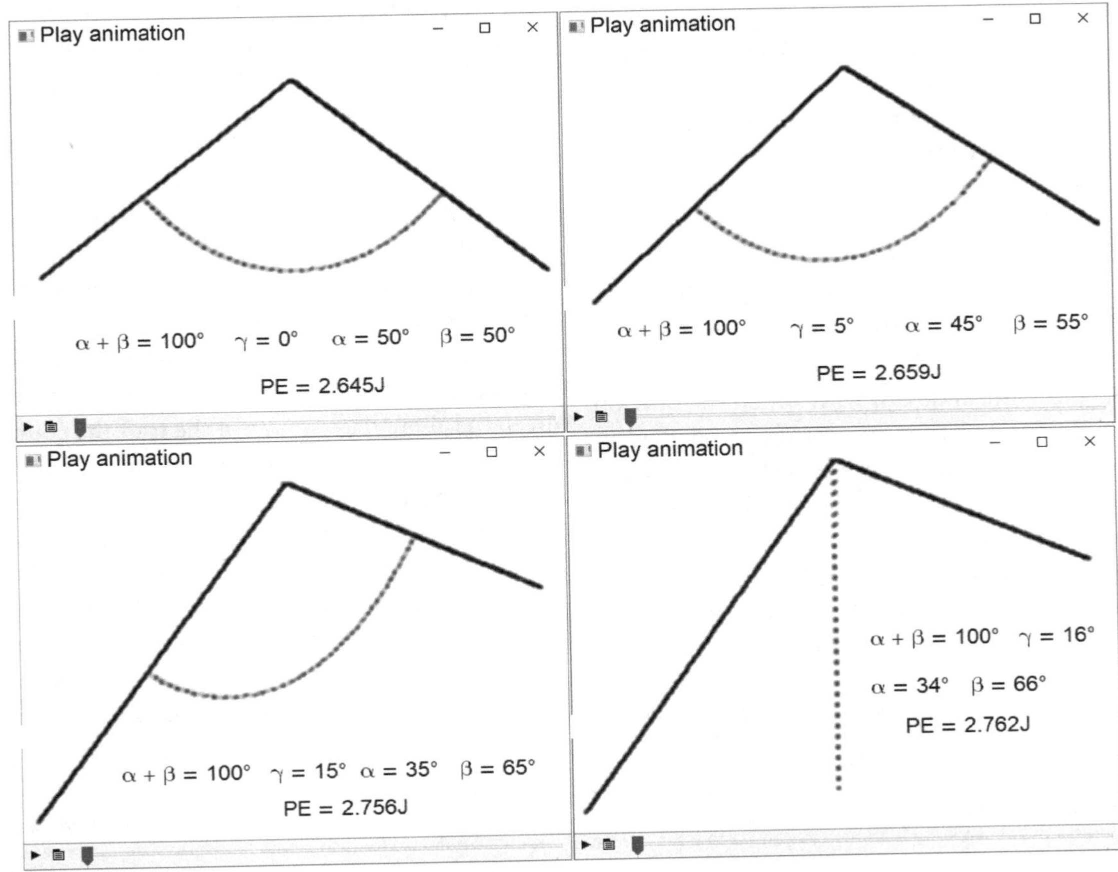

FIGURE 31.21 Critical angle of the hangers.

full-dress suit, which he shook and tossed in an unmerciful manner.

RUDOLF ERICH RASPE, *THE SURPRISING ADVENTURES OF BARON MUNCHAUSEN* (HTTPS://EN.WIKISOURCE.ORG/WIKI/THE_SURPRISING_ADVENTURES_OF_BARON_MUNCHAUSEN/CHAPTER_IV)

So far we have dealt with ***statics*** (the mechanics discipline that studies the laws of equilibrium of bodies) and partially with ***kinematics*** (the mechanics discipline that studies the motion of bodies without taking acting forces into account). To be precise, some *pseudo-kinematics* [4] (semi-kinematics and semi-dynamics) are involved in the frames of chain animation (kinematics) as shown in Figure 31.11, but the forces acting on it are implicitly taken into account through the formula of the catenary. It is important to note that theoretical mechanics has a third component (three persons in the Trinity are one God, Who loves the number three!*): dynamics, where the mechanical motion alongside the causes of its origination are studied. In dynamics, the action of forces is taken into account explicitly—the forces appear in the equations describing such motion.

Let's hang one end of our chain on a nail while pulling the other end aside as preparation for releasing the chain in free flight. If we attach a load to the chain whose weight is much greater than the weight of the chain, then we get a classical *pendulum*. Here the author recalls another "good old" (but non-Soviet) film *Amarcord*† (see note 1), where the physics teacher is shown (Figure 31.21), demonstrating a lesson to schoolchildren on this simple mechanical design with which the development of the dynamics basics often begins. Note that this teacher has a rope hanging on the catenary in his hands, while a pocket watch on a chain hangs at the waistcoat. With the help of this pocket watch, it is possible to show the oscillations of the pendulum. The demonstration is as simple as pulling the pendulum out of the pocket, hanging it on an outstretched arm at the end of the chain and slightly swinging it. Thus, this fragment of the film (produced by Fellini) has a direct relation to our study of chains. All grandfather clocks and some wall clocks have a real pendulum, thereby enabling them to simulate running time.‡

The statics-kinematics-dynamics relationship is, as already noted, akin to the Christian Trinity! Therefore, in addition to the two functions *Find* and *Minimize* that allowed us to solve the static and kinematic problems, we will also use the *Odesolve* function in our calculations, which is intended for numerical solution of ***differential equations***. Namely, these equations are obtained while modelling the oscillation of a pendulum. According to Newton's second law, the sum of the forces acting on the material point is equal to the product of the point's mass and its acceleration. Furthermore, acceleration is the second derivative (differential) of displacement with respect to the time. Hence, differential equations arise.

Comment on "diff eqns"

Here is what you can read about *diff eqns* (**диф**ференциальные **ур**авнения—**dif**ferential **equa**tio**ns**) in one online Russian dictionary of youth slang: "Definition—differential equations or system of differential equations; a training course on differential equations or in differential calculus in general; or the corresponding exam, lecture, lecture course, assignments, etc." Examples: "This problem can be solved through diff eqns", "Which of you have taken Diff Eqns?", "Diff eqns are scheduled for tomorrow to take, and I still have not written notes".

The phrase "diff eqns" appeared in student and teaching slang quite some time ago. Its meaning has always been the abbreviated name of the training course on the theory of ordinary differential equations and systems, as noted previously.

If only a formally scholastic component is seen by the teacher, then he/she quickly turns the diff eqns into a "tool of torture" for students. After all, the analytical solution of even the simplest differential equations requires knowledge of over a dozen specific techniques along with good integration skills. However, if the

* The interpretation about the Trinity is considered one of the central arguments of conflict between Orthodox and Catholic Christians (i.e. the so called the Filioque clause) https://en.wikipedia.org/wiki/Filioque.

† One of the most famous films created by the brilliant Italian director Federico Fellini. This is one of those films, the viewing of which leaves an indelible impression that can last a lifetime. You can watch it repeatedly, indefinitely, despite the simple plot (see note 1). In addition to the physics teacher, Fellini's other autobiographical film also shows other teachers: mathematics, history, fine arts, Greek, social studies and the law of God.

‡ By the way, from the pendulum clock came the unit of length meter. For a one meter pendulum, the oscillation period is 2 seconds. Each swing of such a pendulum leads to a second displacement (jerking) of gears and other parts of the watch. The *Foucault pendulum* with which help it is possible to show the rotation of the Earth and smoothly pass from physics to astronomy is also interesting.

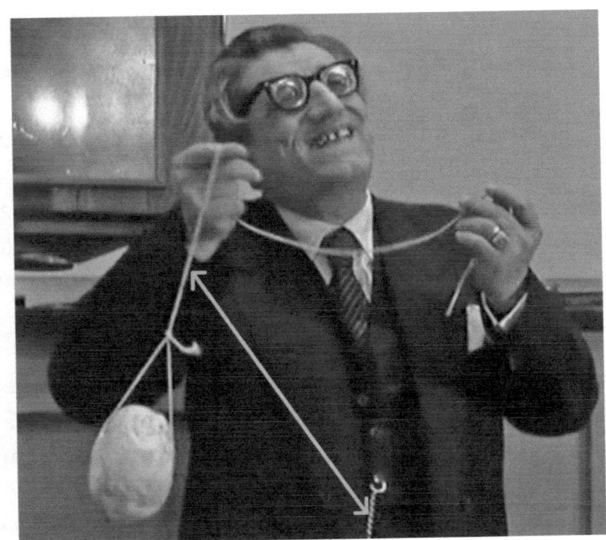

FIGURE 31.22 A still from the film *Amarkord* by Federico Fellini.

teacher deeply understands the semantic essence of differential equations, and subtly feels the balance between analytical and numerical methods of solution, he/she easily encourages students in this very "physical" realm of mathematics.

Schoolchildren face elements of differential equations theory during practically every physics lesson. However, this fact is traditionally hidden from them at school. Teachers of physics believe that it is more useful for the student to simply memorize the calculated formula (such as the one for the period of oscillation of a pendulum*) than to understand where this formula came from. Likewise, teachers of mathematics believe that it is enough to make the student learn the definition of a derivative, and thus teach derivative calculation using formal rules and tables without justifying the necessity of this difficult mathematical concept in various applications.

Previously, teachers had excuses: lack of illustrative material or unavailability/inaccessibility of computer tools for solving differential equations. The teacher in Figure 31.22, for example, could show a real physical pendulum, but without mathematically modelling it on a computer. Nowadays such justifications are untenable. Many physical and mathematical phenomena studied within the school or university program are easily modelled with computational packages that have an accessible and understandable user interface, along with good graphics and animation (of both flat and three-dimensional kinds). The appearance of these packages has enabled us to easily introduce a student into the complex world of dynamic processes. This technique should be implemented as soon as possible—at school!

The author has experience with teaching elective classes in computer science in a very advanced Moscow lyceum. In addition to a set of standard topics, the class subject focussed on the methods of computationally solving differential equations (in Mathcad)—these same *diff eqns.* The director of the lyceum, who happened to be a physicist by education and part-time professor in the department of physics at a prestigious Moscow university, said that the words "differential equations" in secondary school were tacitly forbidden; the schoolchildren would understand the algebraic equations, and you want them to solve differential equations. When the principal was shown which equations would be considered in the classroom and how they would be solved on the computer, he changed his mind and expressed his confidence that all this would be of interest to schoolchildren while simultaneously being useful, as expected. This is true for all students, not just the advanced ones in mathematics and physics.

Let's investigate on a computer the behavior of such an unusual pendulum: seven (lucky!) weights, connected by a weightless and inextensible core. We will consider weights as material points with given mass. It will be a chain-pendulum hybrid. Such a pendulum is also called a connected pendulum, i.e. a pendulum-rosary†.

There are many online sources where different approaches to solving the problem involving a multiple pendulum, double, triple, etc., are described. They basically show attempts at an analytical solution of a Lagrange equation of the second kind and pour out into a set of formulas with complex transformations. As a rule, visitors to these resources quickly cease to understand the reasoning of the website's author and lose interest in it, see Figure 31.23.

* Not just a pendulum, but the mathematical pendulum. In its differential equation the sine was replaced by the angle itself. This assumption (simplification) allowed solving it analytically and obtaining a set of formulas.

† In the alma mater of the author of this book, Moscow Power Engineering Institute, there was preserved a kind of a monument to the technical culture—a non-working elevator of continuous action. It was necessary to jump in its cars on the run. This elevator has a slang term: a lift-paternoster (from a paternoster rosary, literally "Our Father"). In this elevator the cars are connected with each other by a cable and move along a very elongated vertical oval—like a rosary in the hands of a praying monk.

FIGURE 31.23 A still from the film *Operation Y and Shurik's Other Adventures*[*].

[*] This is a frame from one more Soviet cult film "Operation "Y" and other adventures of Shurik". You can try to understand this heap of formulas (they were written on the doors and on the field of the room!) and understand that they belong to thermodynamics – see the Study 30. In this scientific discipline, derivatives and integrals are also actively used (see parts 1 and 2 of this study), as in the usual "mechanical" dynamics. The main simplification in mathematical modeling of the pendulum is the replacement of the sine of the angle by the angle itself, which can be done only at small angles. In thermodynamics there is also a fundamental simplification – this is the concept of an ideal gas, which more or less corresponds to a real gas only under certain conditions.

That's where the numerical methods will be especially useful. Of the several things that they will achieve, they will allow one to model trajectories of the movement of the pendulum's key points, which are not present in the numerous animations that are solely based on analytical solution methods and published on the Internet.

However, rather than beginning with a description of the method for solving the pendulum-chain problem, we will start with an analysis of its results.

The first frame of the animation in Figure 31.24 (created with the *Odesolve* function) shows the initial (when $t=0$) chain position before it swings.[*] The same animation can be seen here: https://community.ptc.com/t5/PTC-Mathcad-Questions/Ridged-chain/m-p/560385. Up until about 8 seconds our computer-modelled seven-substitute-link chain oscillates more or less according to commonsense belief, repeating the intricate twists of the real chain, and then a phenomenon of great importance occurs.

[*] You can specify different initial positions of the chain and get various pictures (animations) of its swing. Some interesting cases can be seen on the website for this Study.

The words "chain line" (a catenary) and "chain function", when uninitiated in mathematics, causes associations with the words "chain dog" (a watchdog). That furious, half-wild line—a dog, ready at any time to break from the chain and attack all chained and stray dogs—sometimes really raging (see the epigraph to this part of the Study). Not only a dog (or a fur coat bitten by it) can get mad, but the chain (see the lower two frames of the animation in Figure 31.24, or better, the animation itself at the above-mentioned online address) can exhibit similar behavior. After all, a normal, "healthy", oscillating chain (like any pendulum) cannot rise above its initial position without additional force!

Creating a Mathcad calculation of the pendulum's oscillation is not difficult; however, it is much more difficult to understand the results of such calculations, or to separate, in essence, the true grains from the numerical chaff.

Figure 31.25 shows a fragment of the oscillation process calculation involving a pendulum-chain with seven material points having masses m_1, m_2, …, m_7 that are connected by pairwise, rigid, weightless links of lengths L_1, L_2, …, L_7. All the points, except for the seventh one, are acted upon by three forces: the force of gravity and the two forces created by the two neighboring links. On the seventh (final) point of the material only one "pulling" force (force due to the connection) acts. This point is an ordinary pendulum in which the attachment point is in motion. There is no analytical solution of such system of differential equations.[†] Thus, it becomes

[†] There is an analytical solution to the problem of the oscillation of a double pendulum. But it is so cumbersome that it is very difficult to work with it despite the fact that the model itself is simplified—in it the sines are replaced by the angle itself (see, for example, http://vuz.exponenta.ru/PDF/book/TargKoleb.pdf). And this, as you know, can be done only for small angles. An analytical solution to the problem of seven pendulums cannot be imagined. But, "Dreaming is not harmful"—see the attempt of such a decision by M.N. Kirsanov at http://vuz.exponenta.ru/PDF/SOL/cep.html. For another example, there is also an English reference about multiple pendulum analytics: see K.L. Salisbury and D.G. Knight, "The multiple pendulum problem via Maple®", November 2010, *International Journal of Mathematical Education*, September, 1 (5): 747–755). There is an analytical solution to the problem of the motion of the planet and the satellite (ellipse!). For more than 300 years mathematicians have been looking for a similar solution for a system of three cosmic bodies, until they realized that it was impossible to do this. In addition, the development of computer solutions to problems has sharply quenched the ardor of those who are looking for analytical solutions, and the attention of mathematicians has shifted to improving numerical methods and methods for verifying solutions.

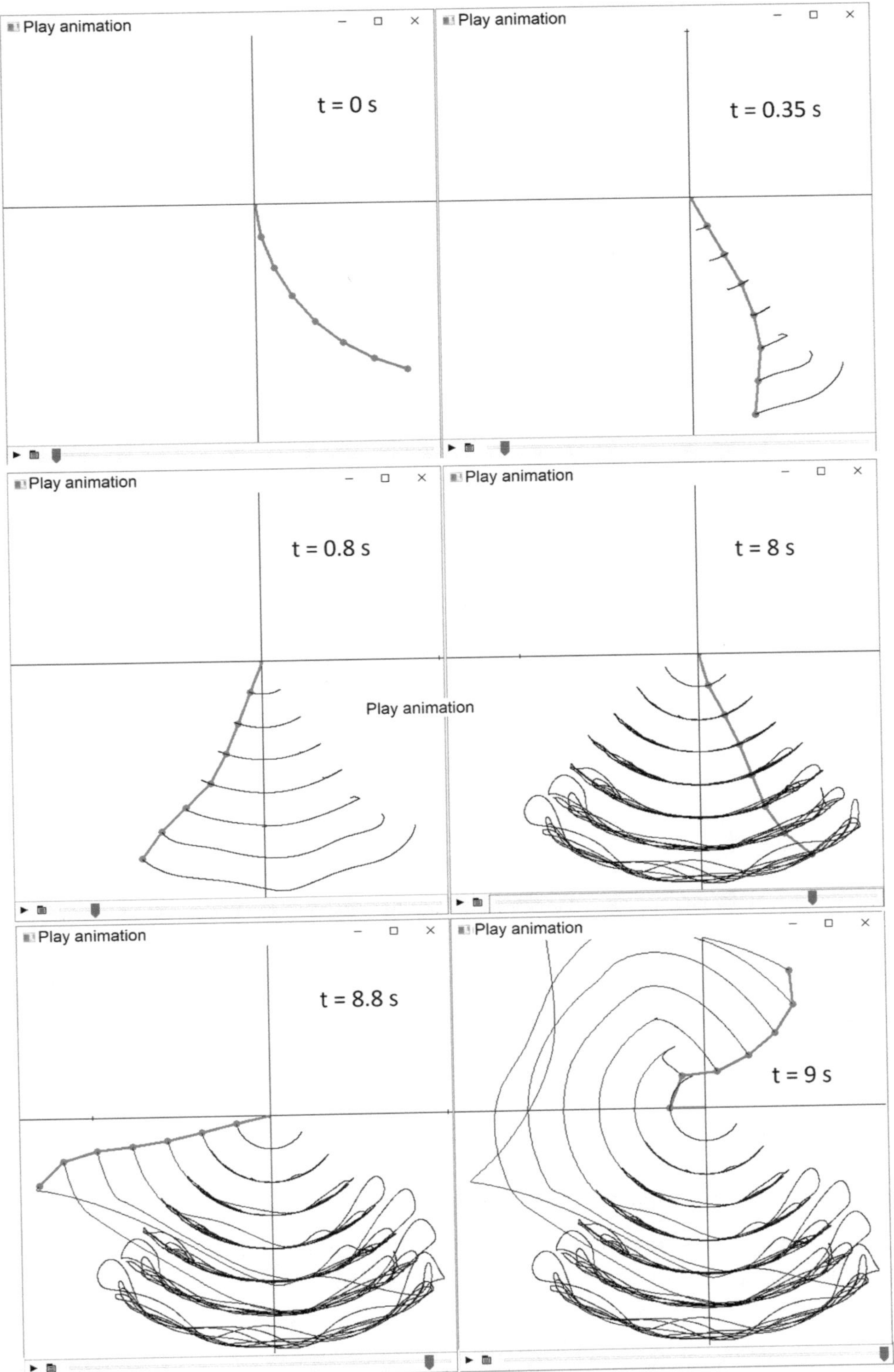

FIGURE 31.24 Animation frames of oscillations of a pendulum-chain.

6-th pendulum Initial position (x6-y6), force (F6) and velocity (x6'-y6')

$$x6(0s) = L_1 \sin(\theta_{1begin}) + L_2 \sin(\theta_{2begin}) + L_3 \sin(\theta_{3begin}) + L_4 \sin(\theta_{4begin}) + L_5 \sin(\theta_{5begin}) + L_6 \sin(\theta_{6begin})$$

$$y6(0s) = -(L_1 \cos(\theta_{1begin}) + L_2 \cos(\theta_{2begin}) + L_3 \cos(\theta_{3begin}) + L_4 \cos(\theta_{4begin}) + L_5 \cos(\theta_{5begin}) + L_6 \cos(\theta_{6begin}))$$

$F6(0s) = 0N$ $x6'(0s) = 0m/s$ $y6'(0s) = 0m/s$

System of equation $(x6(t) - x5(t))^2 + (y6(t) - y5(t))^2 = L_6^{\ 2}$

$$m_6 \cdot x6''(t) = F6(t)\frac{x6(t) - x5(t)}{L_6} - F6(t)\frac{x7(t) - x6(t)}{L_7} \qquad m_6 \cdot y6''(t) + m_6 \cdot g = F6(t)\frac{y6(t) - y5(t)}{L_6} - F7(t)\frac{y7(t) - y6(t)}{L_7}$$

7-th (last) pendulum Initial position (x7-y7), force (F7) and velocity (x7'-y7')

$$x7(0s) = L_1 \sin(\theta_{1begin}) + L_2 \sin(\theta_{2begin}) + L_3 \sin(\theta_{3begin}) + L_4 \sin(\theta_{4begin}) + L_5 \sin(\theta_{5begin}) + L_6 \sin(\theta_{6begin}) + L_7 \sin(\theta_{7begin})$$

$$y7(0s) = -(L_1 \cos(\theta_{1begin}) + L_2 \cos(\theta_{2begin}) + L_3 \cos(\theta_{3begin}) + L_4 \cos(\theta_{4begin}) + L_5 \cos(\theta_{5begin}) + L_6 \cos(\theta_{6begin}) + L_7 \cos(\theta_{7begin}))$$

$F7(0s) = 0N$ $x7'(0s) = 0m/s$ $y7'(0s) = 0m/s$

System of equation $(x7(t) - x6(t))^2 + (y7(t) - y6(t))^2 = L_7^{\ 2}$

$$m_7 \cdot x7''(t) = F7(t)\frac{x7(t) - x6(t)}{L_7} \qquad m_7 \cdot y7''(t) + m_7 \cdot g = F7(t)\frac{y7(t) - y6(t)}{L_7}$$

Solution

$$\begin{pmatrix} x1 \\ y1 \\ F1 \\ x2 \\ y2 \\ F2 \\ x3 \\ y3 \\ F3 \\ x4 \\ y4 \\ F4 \\ x5 \\ y5 \\ F5 \\ x6 \\ y6 \\ F6 \\ x7 \\ y7 \\ F7 \end{pmatrix} := \text{Odesolve}\left[\begin{pmatrix} x1 \\ y1 \\ F1 \\ x2 \\ y2 \\ F2 \\ x3 \\ y3 \\ F3 \\ x4 \\ y4 \\ F4 \\ x5 \\ y5 \\ F5 \\ x6 \\ y6 \\ F6 \\ x7 \\ y7 \\ F7 \end{pmatrix}, t, t_{end}\right] \qquad tt := 0, \frac{t_{end}}{10000} .. t_{end}$$

FIGURE 31.25 Calculation fragment in Mathcad for the oscillation of a pendulum-chain.

necessary to resort to numerical methods to generate a solution table with the subsequent interpolation of discrete values so that smooth differentiable functions can be obtained. Moreover, they need to be differentiated in order, an example of which would be the determination of the speed of motion together with the kinetic energy of the material point along it (see below).

The *Odesolve* function generates 21 user functions (3 for each of the 7 points: the abscissa x and the ordinate y of the moving material point, and the force F acting on the connection). These functions have time t as the argument.

Note One can abandon the rectangular coordinate system and work with a polar system (where the vertical, or ordinate, axis and horizontal, or abscissa, axis coordinates are replaced with angular and radial coordinates). This allows the formulation or solution of the problem to be simplified. However, the "rectangular" approach to solving the pendulum problem makes more "physical" sense—the force of gravity acts strictly vertically. The

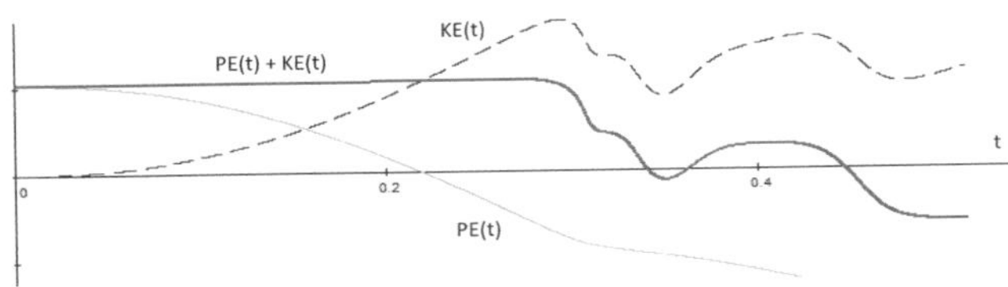

FIGURE 31.26 The graph of the time variation in the potential and kinetic energies of a seven-point pendulum-chain.

reader can therefore try to solve the pendulum-chain problem in the polar coordinate system.

It is known that the energy of the pendulum alternates between kinetic and potential forms. If we do not take friction into account, then the sum of these energies remains constant.*

An interesting point is that the Lagrangian equations obtained in the analysis of the pendulum problem (among other dynamic systems), and which have been mentioned above, follow from the account of those energies. Figure 31.26 shows the time evolution of the pendulum-chain's energy values.

What conclusions can be drawn after the analysis of Figures 31.24 and 31.26?

Up until about $t=0.3$ s, the potential energy of the seven weights (or material points) decreases smoothly while their kinetic energy increases smoothly, such that their sum remains constant. It is this that is the main indication of the solution's correct approach—its correspondence to the physics of the problem. At 0.3 $s<t<8$ s, an incubation period is observed—the accumulating error in the numerical solution of the problem disrupts the energy balance (see Figure 31.26). Nevertheless, this obviously does not affect the nature of the pendulum-chain's swing for the time being (see Figure 31.24 and the animation on the website for this Study, which should be more effective than the figure). Only when $t>8$ s is satisfied does the "frenzy" of the chain becomes obvious. The scope of the seven-point pendulum-chain problem goes beyond the *Odesolve* function. Only the first two frames of the animation in Figure 31.24 can be considered to be appropriate to the problem's physics. We had to simplify the task, specifically by reducing the number of points.

Figure 31.27 shows the animation frames of the five-point pendulum-chain oscillation with the mapping of the changes in the energy (potential, kinetic, potential-kinetic sum) of the system (the six-point pendulum also "furious"), in addition to the points' positions. The "incubation period of rabies" in the chain has sharply decreased as observed from the last animation frame with a framed section of the energy change graph (see Figure 31.27). Note that period here refers to pendulum-chain's period of oscillation, not incubation period. Figure 31.27 presents the period of the energy value oscillations, which can even be estimated numerically; however, it would be good to find an analytical formula for this period as was done for the "classical single-point" pendulum.

The solution of complex differential equations systems is a peculiar journey between Scylla and Charybdis. The analytical solution (akin to Scylla) can usually be obtained only if the original mathematical model is substantially simplified. The numerical solution (akin to Charybdis) is often distorted, usually because the accumulated computational error becomes significant, but occasionally the solution is simply interrupted halfway due to numbers that are too large or too small, and hence not valid on a digital computer.

The epigraph text of this part of the Study continues as follows:

> *After being out on the street, my servant brought not only my fur coat, but also a scrap of the chain that the mad dog lost. I decided to use this chain for experiments that I was addicted to. Specifically, I began to study this chain's oscillation, a task which included mathematical models based on the numerical solution of a system of differential equations. It turns out that the sick dog transmitted rabies not only to my fur coat, but also to my chain along with its relevant mathematical model.*

Sometimes the adjectives "mad" and "crazy" can have positive connotation and meaning. A familiar example would be a lady on a romantic date who accepts a gift (an expensive pendant on a *chain*) from her boyfriend and says to him, "You're crazy!" Another example: the best praise spoken by a lady to the cavalier at parting will be "You were mad!"

* In [44, 9], a calculation of a specific pendulum is given, in which the links are not rigid, but elastic. These are springs that can stretch and contract, obeying Hooke's law. There, the first energy (the potential energy of the cargo position and the kinetic energy of its movement) is also added to the third energy, the potential energy of the compressed/stretched spring. It is easy to take into account in such calculations the force of the medium's resistance. Corresponding animations can be seen on the website for this Study.

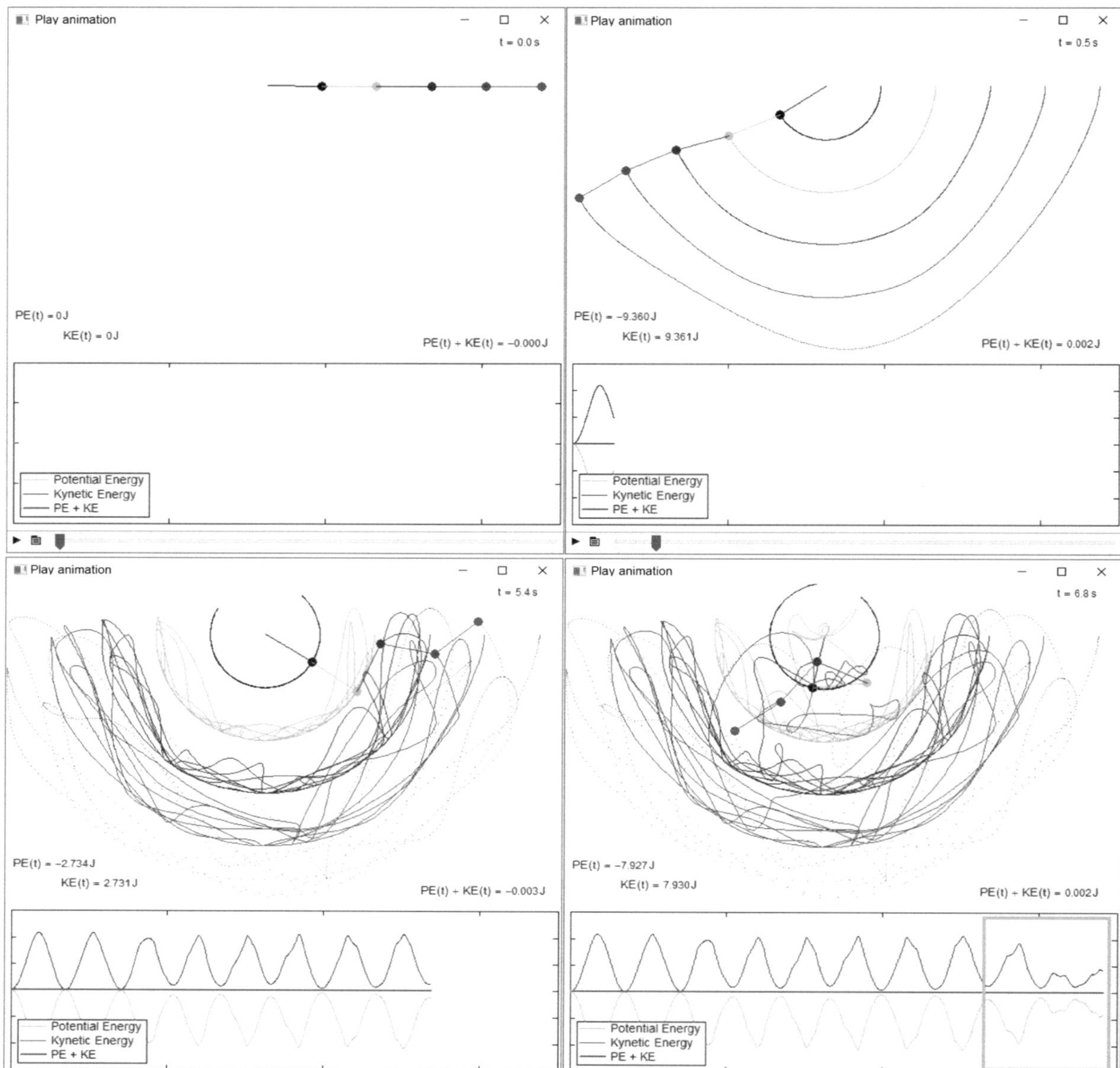

FIGURE 31.27 Animation frames of the pendulum-chain's oscillation with the display of change in potential and kinetic energies.

Putting jokes aside, let's return to the topic of discussion.

Academic websites allow us to download Mathcad files and continue performing numerical computer experiments with the pendulum-chain. In the above examples, the masses of the material points were the same. Likewise, the distances between them (bond lengths) were equal. This restriction, however, is arbitrary and can therefore be removed. In an interesting scenario, masses of loads (point loads in this discussion) decrease with increasing distance from the place of attachment in the structure. It resembles a *scourge* (shepherd's whip) whose end can reach supersonic speed and cause a characteristic sound upon a strike, frightening cattle and allowing the shepherd to manage the herd.

CONCLUSIONS

Although chains are normally thought of as something to be worn around the neck or twisted around the fingers, they also have scientific applications—conducting interesting physical and mathematical experiments, both in real life and on the computer, to be precise! This entire discussion is intended for educational and scientific purposes.

REFERENCES

1. Ochkov, V.F. *Catenary = Physics + Mathematics + Informatics*. Informatics at School. No. 3. 2018. p. 56–63. http://twt.mpei.ac.ru/ochkov/catenary.pdf.
2. Merkin, D.R. Introduction to the mechanics of a flexible thread. *Science*, 1980, p. 240. http://www.twirpx.com/file/357235.
3. Ochkov, V.F.; Ochkova, N.A. A project of a monument to three mathematicians or MathMetria. *Cloud of Science*, 2017, Volume 4, no. 4, pp. 548–571. https://cloudofscience.ru/sites/default/files/pdf/CoS_4_548.pdf.
4. Ochkov, V.F.; Leener, F.; Chudova, Yu.V.; Kapitonets, V.K.; Tarakanova, D.Yu. Physics vs informatics: rope polygon with weights in static, kinematics and dynamics or Newton vs Lagrange. *Cloud of Science*, 2017, Volume 4, no. 2, pp. 147–180. http://twt.mpei.ac.ru/ochkov/Polygon.pdf.
5. Ochkov, V.F.; Bogomolova, E.P.; Ivanov, D.A. *Physical and Mathematical Studies with Mathcad and the Internet*, 2nd ed. revised and enlarged. Publishing house "Lan": SPb, 2018. 560 p. http://twt.mpei.ac.ru/ochkov/T-2016/PhysMathStudies.pdf
6. Wang, C.Y. The optimum spanning catenary cable. *European Journal of Physics*, 2015, Volume 36, no. 2. http://iopscience.iop.org/article/10.1088/0143-0807/36/2/028001.
7. Ochkov, V.F. Mechanics in the environment of Mathcad. Collection of scientific and methodical articles. Theoretical Mechanics. Issue. 30. Edited by Professor Samsonov VA / M.: Publishing House of Moscow State University, 2018. 253 p. pp. 180–189.
8. Zubelevich, O.E.; Samsonov, V.A. Chain on the cone. Collection of scientific and methodical articles. Theoretical Mechanics. Issue 30. Publishing House of Moscow State University, 2018, pp. 131–138.
9. Ochkov, V.F.; Kolkhopp, F. Physics and computer science: The center of gravity of the black box. Informatics at school. No. 7 for 2017, pp. 65–70. http://twt.mpei.ac.ru/ochkov/Center-Gravity.pdf
10. Targ, S.M. *A Short Course of Theoretical Mechanics: Proc. for Technical Colleges*, 10th ed. Pererab. and additional.—M.: Higher education. shk., 1986, 416 p., ill. http://isopromat.ru/teormeh/literatura/targ-s-m-kratkij-kurs-teoreticeskoj-mehaniki
11. Ochkov, V.F.; Bogomolova, E.P. Teaching mathematics with mathematical software. *Journal of Humanistic Mathematics*, January 2015, Volume 5, no. 1, pp. 265–285. doi:10.5642/jhummath.201501.15. http://scholarship.claremont.edu/jhm/vol5/iss1/15
12. Solodov, Alexander; Ochkov, Valery. *Differential Models. An Introduction with Mathcad*. Springer, 2005. https://www.springer.com/gp/book/9783540208525
13. Korobov, Viktor; Ochkov, Valery. *Chemical Kinetics with Mathcad and Maple*. Springer, 2011. ISBN 978-3-7091-0531-3. https://www.springer.com/gp/book/9783709105306
14. Behroozi, F. A fresh look at the catenary. *European Journal of Physics*, 2014, Volume 35, p. 055007. http://iopscience.iop.org/article/10.1088/0143-0807/35/5/055007/pdf
15. Wang, C.Y.; Watson, L.T. The elastic catenary. *International Journal of Mechanical Sciences*, 1982, Volume 24, no. 6, pp. 349–357. https://www.sciencedirect.com/science/article/pii/0020740382900686
16. Chen, J.-S.; Li, H.-C.; Ro, W.-C. Slip-through of a heavy elastica on point supports. *International Journal of Solids and Structures*, 2010, Volume 47.
17. Boresi, A.P.; Schmidt, R.J. *Engineering Mechanics—Statics*. Baker & Taylor Books, North Carolina, 2000.
18. Christensen, H.D. Analysis of simply supported elastic beam columns with large deflections. *Journal of Aerospace Science*, 1962, Volume 29, pp. 1112–1121.
19. Lippmann, H.; Mahrenholtz, O.; Johnson, W. The heavy elastic strips at large deflections. *International Journal of Mechanical Sciences*, 1961, Volume 2, pp. 294–310.
20. Parbery, R.D. The effect of stiffness on the shape of the elastic catenary. *Civil Engineering Transaction, The Institute of Engineers, Australia*, 1976, Volume 18, pp. 98–101.
21. Wang, C.Y. A critical review of the heavy elastica. *International Journal of Mechanical Sciences*, 1986, Volume 28, pp. 549–559.

STEM Education And Or

Science Fiction

ABSTRACT

The Study shows how an engineering university can hold a lecture in the framework of the STEM-education technology. Through the example of solving problems taken from a science fiction novel, the possibilities of modern mathematical computer programs are illustrated: the utilization of built-in operators and functions of the mathematical packages Mathcad and Maple. Symbolic and numerical mathematics, 2D and 3D plots, animation, optimization, solving algebraic and differential equations and their systems, programming, working with physical quantities, links to cloud functions, etc., are described. It is shown how to use the example of solving problems in school and university, relying on the modern educational technology called STEM education (Science, Technology, Engineering and Mathematic), where in a single lecture different educational disciplines are involved—literature, visual arts, history, mathematics, computer science, physics, engineering, engineering graphics, statistics, political science and even theology (religious studies). A typical error of the symbolic mathematics of the Mathcad, Maple and Mathematica packages was found and analyzed.

Through an example of using the method of least squares, an unsolved problem in the theory of dimensional quantities is examined.

Mathematics: System of equations, optimization.

Physics: Mass, volume, surface, density.

IT: Numerical and analytical solution of algebraic and differential equations and their systems, optimization, computer animation.

Art: Jules Verne's novel, *20,000 Leagues Under the Sea.*

Study website: https://community.ptc.com/t5/PTC-Mathcad/Symbol-math-error/td-p/487278.

Jules Verne's novel *20,000 Leagues Under the Sea* encrypts many interesting physical and mathematical problems, whose solution allows us to show a wide range of tools of the mathematical packages: symbolic and numerical mathematics, flat and volumetric graphics, animation, optimization, solution of algebraic and differential equations and of their systems, programming, working with physical quantities, links to cloud functions, etc. The very process of the audience solving the problems of the novel turns out to be on trend (as they say now) modern education technology, called STEM Education (Science, Technology, Engineering and Mathematics), where topics from various academic disciplines are involved such as literature, visual arts, history, mathematics, computer science, physics, engineering, political science, "pure" art and even religion.* Along the way, the "work on the novel" revealed an interesting error in the symbolic mathematics of the Mathcad, Maple and Mathematica packages.

SIZES OF THE SUBMARINE

Here is a quotation from the original French text of the novel:†

> Voici, monsieur Aronnax, les diverses dimensions du bateau qui vous porte. C'est un cylindre très allongé, à bouts coniques. <...>. Ces deux dimensions vous permettent d'obtenir par un simple calcul la surface et le volume du *Nautilus*. Sa surface comprend mille onze mètres carrés et quarante-cinq centièmes; son volume, quinze cents mètres cubes et deux dixièmes—ce qui revient à dire qu'entièrement immergé il déplace ou pèse quinze cents mètres cubes ou tonneaux.

Captain Nemo informs his prisoner, Professor Aronnax, that the submarine on which they are located has the shape of a geometric body composed of two identical straight circular cones (the bow and the stern of the boat) and a straight circular cylinder (the hull of the boat, see Figure 32.1). The radii of the bases of the cones and the cylinder are equal. The volume of the boat V and its outer surface area S are known ("voiced" by Captain Nemo). It is necessary to determine its geometrical dimensions—the radius R of the base of two cones and of the cylinder, the height H of two cones (length of the bow and the stern) and the length L of the cylinder (length of body).

The problem is reduced to solving a system of two non-linear algebraic equations (the equation of volume $V = \ldots$ and the equation of the surface $S = \ldots$) with three unknowns R, H, and L. We have an *underdetermined* system, i.e. the number of equations is less than the number of unknowns.

In Figure 32.1 (Mathcad) it is possible to see, first, the outline of the problem and its description. Further, the symbolic mathematics tools (with the solve operator) generate two functions in the calculation—the first with the name LV and three arguments R, H and V, and the second with the name LS and three arguments R, H and S. The first function is the result of an analytical (symbolic)‡ solution of the volume equation with respect to the variable L, while the second function is the solution of the surface equation with respect the same variable L. These two functions are identical (in the submarine single body length) and form a new equation $LV(R, H, V) = LS(R, H, S)$, which is also solved analytically§ with respect to the variable H. As result, this solution produced two expressions that form a vector function with the name H and with three arguments R, V and S. By this function, a graph is built, or rather two graphs (solid line and dotted line) with fixed values of variables (parameters) $V = 1500.2\ \text{m}^3$ and $S = 1011.45\ \text{m}^2$ (see the problem conditions) and with argument R varying from 1 to 13 m in a 0.1 mm step (the left graph in Figure 32.1).¶ An enlargement of the area of the graphs of the vector function $H(R, V, S)$ in which R is changed from 2.6 to 3.3 m is also shown with the same step (the right graph in Figure 32.1), where the vertical lines are

* Rather, theology, religious studies and, if you will, atheism. In this sense, the abbreviation STEM can be expanded to the abbreviation STEAM or even STREAM, where the letter A means art (Art) and the letter R means religion (Religion). The term steam is water vapor too, which at the turn of the eighteenth and nineteenth centuries, led to the industrial revolution in Europe and North America (Industry 1.0). Stream in English: this means a flow—for example, of water, or of water vapor from the boiler to the steam engine. We can assume that the technology STEM/STEAM/STREAM at the junction of the twentieth and twenty-first centuries is making a new (fourth) industrial revolution (Industry 4.0). You can also use the abbreviation SAR: Science, Art and Religion: human civilization rests on three "whales"—on science, on art and on religion (or what religion has become in our day).

† We do not extract the quotation from the translations of the novel into English and other languages, for fear of inaccuracies.

‡ This equation is also easy to solve in the mind by transferring individual terms from right to left, etc. However, in order to avoid mistakes, it is better to do this automatically on the computer. On the other hand, it is always useful for training the mind to do such analytical calculations first "manually" and then check the answer on the computer. An example from everyday life. A modern phone keeps numbers in its memory. Nevertheless, doctors recommend that seniors and not only seniors keep their numbers in their head, dialling them manually if necessary.

§ This equation is not so easy to solve in the mind. The author asked one math teacher to do this, who began to do this work, wrote three sheets of paper and did not finish the job, however he made a well-founded assumption about the causes of symbolic mathematics errors, which we will consider below.

¶ Mathematical packages build graphs of the "explicit" function $y(x)$, tabulating the argument and function values and connecting the resulting points with straight lines or smoothing them with a certain curve. If a student in a lesson on mathematical analysis begins to build graphs, then a demanding teacher will expel such a student from his class, stomping his feet and hooting following the student. We usually build such graphs more intelligently—qualitatively, not quantitatively: we analyze the function, we look for specific points for it—zeros, extremes, inflection points, asymptotes, discontinuity points, and so on. Therefore, many teachers of mathematics in schools and in higher education institutions quite reasonably believe that machine analytics and graphics dull students, wean them to work with their head. It's more like this: it dulls the majority of students, but enriches the rest—those who are smart and conscientious.

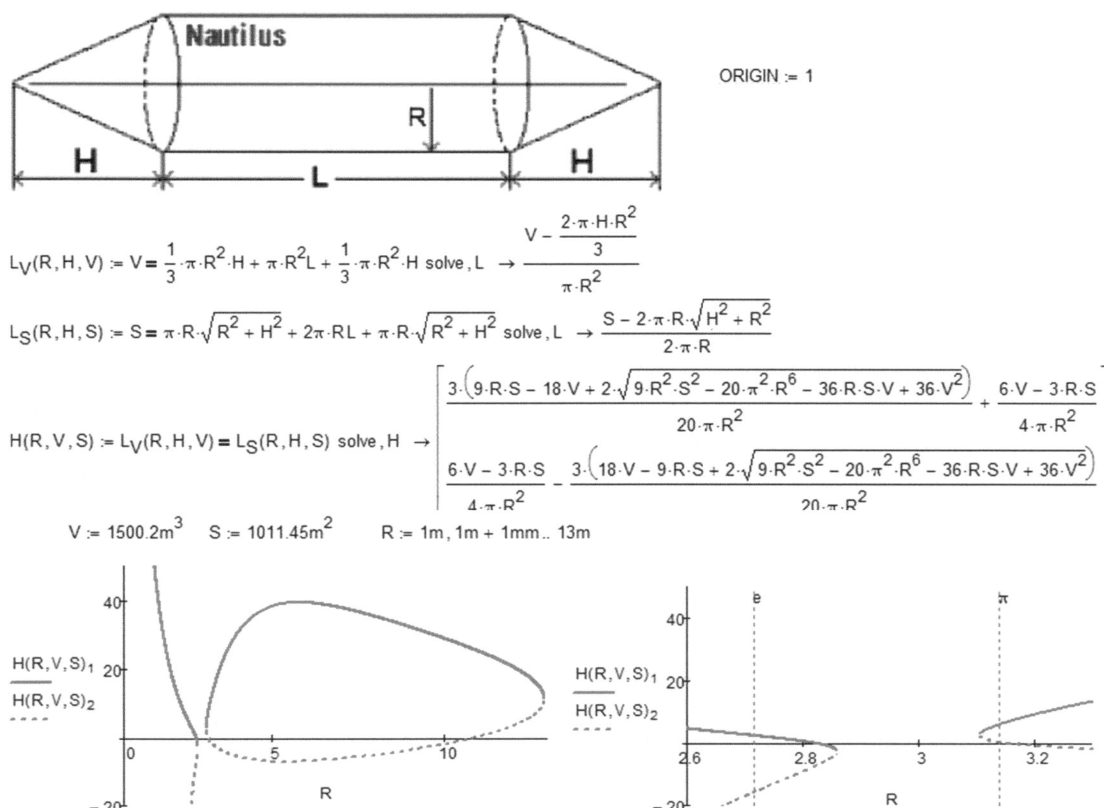

FIGURE 32.1 Analytical and graphical solution of the problem of the size of the submarine *Nautilus* (length scales for the axes R and H are different).

drawn for $R=2.71$ and 3.14 m respectively (numbers e and π as a "beautiful" assumption: two main mathematical constants could be included into the dimensions of the *Nautilus* submarine).

Note Jules Verne was a lawyer by education, not an engineer. This can explain the excessive accuracy in setting the volume and area of the submarine. One could easily set $V=1500$ m^3 and $S=1010$ m^2 or even $S=1000$ m^2. However, in science fiction novels, authors often try to operate with excessive accuracy in order to have more science and less fantasy.

GRAPH OF AN IMPLICIT FUNCTION OF TWO ARGUMENTS

Let's build the graphs obtained when solving the problem of the size of a submarine, in another way—we will not solve the equation analytically $L_V(R, H, V)=L_S(R, H, S)$ and build two separate graphs using the two expressions obtained (see the solid and dashed lines in Figure 32.1), and construct one graph of this implicit function for different R and H and fixed V and S.

In Mathcad (see Figure 32.1), there are no built-in tools for plotting a 2D function, given implicitly. But such means are in Maple—see Figure 32.2, where the *implicitplot* built-in function (implicit graph) is involved, which drew only one closed curve.* Where did the second (left) curve, shown in Figure 32.1, come from? After all, the *implicitplot* function with its arguments covered this "left" area: $H=-10, \ldots, 40$.

In Soviet times, there was a radio program *We Learn a Song*. The presenter would announce: "Now let's listen to the melody of the song on another musical instrument!" Let us also construct our plot with the help of another "tool", another mathematical package and make sure that the left curve will not be there either.

In Mathcad, it is not so easy to draw the graph of an implicit function. But if "you cannot, but really want to, then you can". Figure 32.3 shows the construction of

* A plot of an implicit 2D function can no longer be constructed by simple tabulation (see the note above). When constructing these curves, special complex numerical algorithms are used that do not use the symbolic solution of the equation.

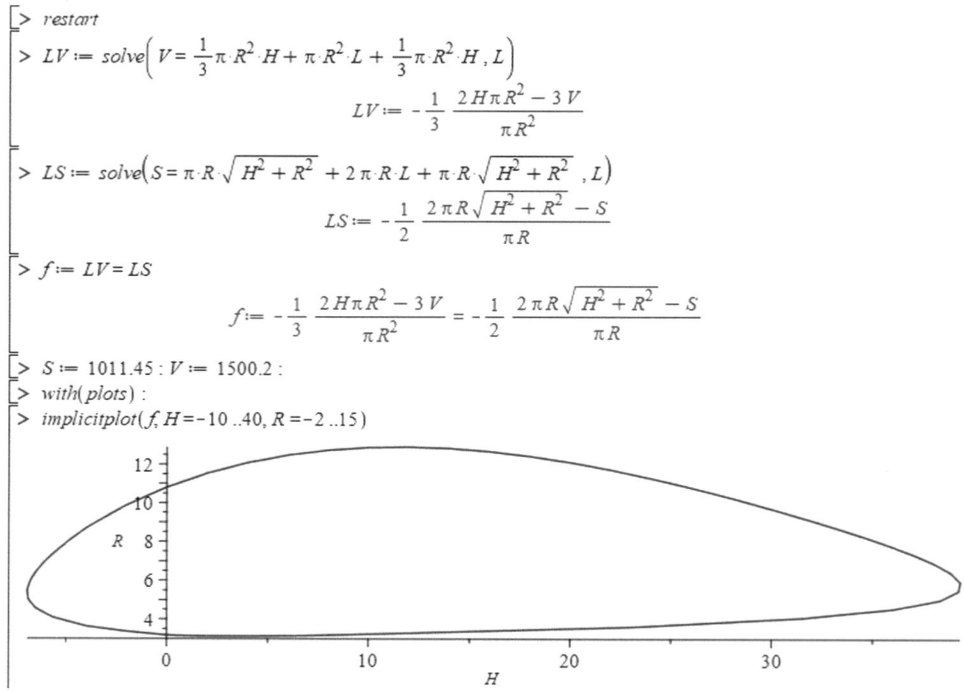

FIGURE 32.2 Plotting an implicit 2D function in Maple.

this plot by scanning a rectangular area on a Cartesian plane (see Studies 19 and 28). If at the next point the value of the function *LV* is *approximately* equal to the value of the function *LS*, then the coordinates of this point are stored ("stored") in the vectors *vR* and *vH*, which are then displayed on the Cartesian graph. So (Figure 32.2) or otherwise (Figure 32.4), but the second curve, obtained as a result of the analytical solution of the equation $LV(R, H, V) = LS(R, H, S)$, did not contain an implicit function in the graphs.

By the way, the same error with the "left" curve, obtained by solving an algebraic equation, was produced by Mathematica package, or rather, by its network version—wolframalpha.com.

SYMBOLIC MATHEMATICS ERROR

What is this error of symbolic mathematics? Is this a "pure" mistake or some feature of this computer tool that should make users alert and check their analytical solutions every time?

Figure 32.4 shows the calculation of the size of the "*Nautilus* submarine" (the values of the variables *H* and *L*) for two values of its radius *R*, recorded on the right (enlarged) graph, see Figure 32.1. The submarine volume *V* and its outer surface S are also checked. At $R = 3.14$ m (number π) the submarine has the following dimensions: $H = 6.28$ (approximately 2π) or 0.525 m and $L = 44.25$ (approximately 14π) or 48.08 m, in which both equations (the equation of volume and the equation of the surface of the boat) turn into identities. However, at the value of $R = 2.71$ m (number e), for the "left" curve, an incorrect solution is given—the equation for the surface becomes an identity, and the equation for volume does not (see the highlighted operator in Figure 32.4)! It can be said that the system of two equations is solved, but only half! What is the matter?

As suggested by the mathematician mentioned in the above note, the error appears to be as follows. The symbolic mathematics of all the listed mathematical packages (Maple, Mathcad/MuPAD and Mathematica), by substituting one expression instead of another, reduces the solution of a complex algebraic equation to the solution of a simple square one, which has two roots, one of which is correct, and the other is obtained by solving initial equations by expanding the domain of definition of unknowns. Therefore, two expressions obtained as a result of the symbolic solution of the equation LV(R, H, V) = LS(R, H, S), for some values of R, give the correct values of H, but, for others, incorrect ones. The left curve shown in Figure 32.1, turned out to be "left" not only

$$\approx(a,b) := \left|\frac{a-b}{a+b}\right| < 0.0001$$

$n := 7000$ $R_1 := -1m$ $R_2 := 15m$ $\Delta R := \frac{R_2 - R_1}{n}$ $H_1 := -10m$ $H_2 := 60m$ $\Delta H := \frac{H_2 - H_1}{n}$

```
(vR)  :=  | i ← ORIGIN
(vH)      | for R ∈ R1, R1 + ΔR.. R2
          |   for H ∈ H1, H1 + ΔH.. H2
          |     if LV(R,H,V) ≈ LS(R,H,S
          |        | vRi ← R
          |        | vHi ← H
          |        | i ← i + 1
          | (vR)
          | (vH)
```

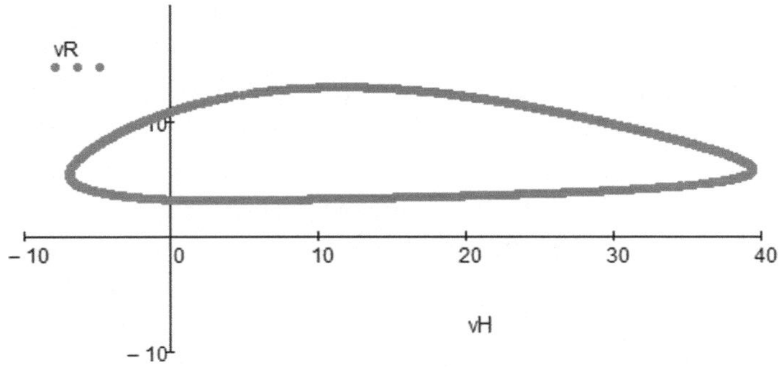

FIGURE 32.3 Plotting an implicit function in Mathcad.

$r := 3.14m$ $h := H(r,V,S) = \begin{pmatrix} 6.277 \\ 0.525 \end{pmatrix} m$ $l := LV(r,h,V) = \begin{pmatrix} 44.248 \\ 48.083 \end{pmatrix} m$ $l := LS(r,h,S) = \begin{pmatrix} 44.248 \\ 48.083 \end{pmatrix} m$

$\frac{1}{3}\cdot\pi\cdot r^2\cdot h + \pi\cdot r^2 l + \frac{1}{3}\cdot\pi\cdot r^2\cdot h = \begin{pmatrix} 1500.2 \\ 1500.2 \end{pmatrix}\cdot m^3$ $\pi\cdot r\cdot\sqrt{r^2 + h^2} + 2\pi\cdot r\cdot l + \pi\cdot r\cdot\sqrt{r^2 + h^2} = \begin{pmatrix} 1011.45 \\ 1011.45 \end{pmatrix} m^2$

$r := 2.71m$ $h := H(r,V,S) = \begin{pmatrix} 2.697 \\ -16.187 \end{pmatrix} m$ $l := LV(r,h,V) = \begin{pmatrix} 63.224 \\ 75.813 \end{pmatrix} m$ $l := LS(r,h,S) = \begin{pmatrix} 55.578 \\ 42.989 \end{pmatrix} m$

$\frac{1}{3}\cdot\pi\cdot r^2\cdot h + \pi\cdot r^2 l + \frac{1}{3}\cdot\pi\cdot r^2\cdot h = \begin{pmatrix} 1323.786 \\ 742.88 \end{pmatrix}\cdot m^3$ $\pi\cdot r\cdot\sqrt{r^2 + h^2} + 2\pi\cdot r\cdot l + \pi\cdot r\cdot\sqrt{r^2 + h^2} = \begin{pmatrix} 1011.45 \\ 1011.45 \end{pmatrix} m^2$

FIGURE 32.4 Verification of the solution obtained by analytical and numerical method (see Figure 32.1).

in its location on the chart, but in its essence! And you can say this: the second root of our system of equations turned out to be *other-worldly*. The letter M in the abbreviation STEM/STEAM can be deciphered as a *mystic* (Mystic).

It is believed that the symbolic mathematics of computer packages gives either an absolutely exact answer, or does not give any answer at all, or rather, gives a message that the answer was not found. Here we have to resort to numerical mathematics—to the mathematics of approximate calculations. But, as the above example showed, symbolic mathematics can also give incorrect answers. This is one of the reasons why, at present, engineers are more and more often abandoning analytical (symbolic) methods for solving problems and turn to numerical ones: it is better to have an inaccurate answer than "no" or "wrong"!

THE OPERATOR "APPROXIMATELY EQUAL"

In Mathcad and in other math packages, as well as in programming languages, there was not a frequently requested "approximately equal" operator.* Therefore, it had to be created—see the first operator in Figure 32.3.

* There are *six* traditional infix comparison operators in Mathcad: *equal to, not equal to, greater than, less than, greater than or equal to* and *less than or equal to*. "God loves not only the Trinity", but also the seven. Recall the seven colors of the rainbow, the seven notes, the seven wonders of the world, the seven ancient sages, etc. up to seven days a week. The STEM abbreviation, as we have already noted, lacks the letters A (Art) and R (Religion). And religion still plays a big role in our life. A specific example: if the seven colors of the rainbow are pure physics, then the seven days of the week are religion, or rather, a religion mixed with human physiology. And it is not clear whether a person was adjusted to a religion, or a religion was adjusted to a person.

The operator "approximately equal" is very much in demand in numerical mathematics. Therefore, it would be nice to make it integrated—the seventh. True, this operator depends on our understanding of what "approximately equal" means. By the way, in mathematical analysis there are the concepts of "much more >>" and "much less <<", which are also contextually dependent.

$r := 4m \quad h := H(r, V, S)_1 = 30.411\,m \quad l := LV(r, h, V) = 9.572\,m \quad l := LS(r, h, S) = 9.572\,m$

$\frac{1}{3}\cdot\pi\cdot r^2\cdot h + \pi\cdot r^2 l + \frac{1}{3}\cdot\pi\cdot r^2\cdot h = 1500.2\cdot m^3 \qquad \pi\cdot r\cdot\sqrt{r^2+h^2} + 2\pi\cdot r\, l + \pi\cdot r\cdot\sqrt{r^2+h^2} = 1011.45\,m^2$

$h + l + h = 70.393\,m$

FIGURE 32.5 Check of the calculations of the size of the submarine given in the novel.

In a program with two cycles, one of which is nested in the other, this user operator is called as an infix operator, which significantly helps the person to read and understand the calculation. And this is one of the "skates" of Mathcad, which largely determined its popularity.

AGAIN ABOUT THE SIZE OF THE *NAUTILUS*

In the Jules Verne novel, by the way, there is a direct reference to the dimensions of the *Nautilus*—the maximum diameter of the boat is 8 meters, and its length is 70 meters. The calculations shown in Figure 32.5, with a certain accuracy (0.5 of a meter) confirm this.

It is possible to check the data on the boat, given in the novel and shown in the calculation of Figure 32.5, in another way—by writing a system of three equations (the volume of the boat, its surface and total length) and two unknowns *H* (length of the bow and stern) and *L* (length of the cylindrical part of the boat). If our initial system of equations (see Figure 32.1) was *underdetermined* (two equations with three unknowns), then the new system of equations (Figure 32.6) will be *overdetermined* (three equations with two unknowns).

An attempt to solve an overdetermined system of equations (Figure 32.6) was unsuccessful. The *Find* function built into Mathcad did not find the numerical values of the unknowns *H* and *L*, which transform the three equations into three identities. Rather, almost identities, when the right and left sides of the equations differ from each other by a given small value stored in the *CTOL* system variable (0.001 by default).

Equations and systems of equations can be solved symbolically (Figure 32.1), numerically (Figure 32.8) and (God loves a trinity) graphically. Figure 32.7 shows such a graphical solution using the method of software scanning for a rectangular area already described by us. If the root of the overdetermined system of three equations with two unknowns was found, then three lines on the graph would intersect at one point. Almost at one point, if you think about the other name of numerical mathematics, the mathematics of approximate calculations.

Note In the calculation shown in Figure 32.7, not only is the user function "approximately equal" introduced, but the user function *Or* (logical addition). In Mathcad, there is the built-in infix operator with the symbol *V* (Or–Study 11), but it has only two arguments. In addition, it is often confused with the symbol Λ (logical multiplication). That is why we introduced the *Or* function, which can have more than two arguments. In addition to this, the user function *Or*, in contrast to the built-in operator *V*, also works correctly with *fuzzy logic* (see Study 12), where the arguments can be not only the two integer (Boolean) values of 0 (false) or 1 (true), but also an infinite number of non-integer ("fuzzy") values ranging from 0 to 1.

HYBRID MATHEMATICS: NUMBER AND SYMBOL

For the case noted in Figures 32.6 and 32.7, Mathcad has a built-in function called *MinErr* (Minimal Error, see Figure 32.8), which, if there is no solution to the system of equations, returns no error message (see text under function *Find* in Figure 32.6), but the values of their arguments, in which the system of equations will be closest to the system of identities. The solution found

$\Sigma L := 70m \quad R := 4m \quad S := 1011.45m^2 \quad V := 1500.2m^3$

Given

$H := 10m \qquad L := \Sigma L - H - H = 50\,m$

$\Sigma L = H + L + H$

$S = \pi\cdot R\cdot\sqrt{R^2+H^2} + 2\pi\cdot RL + \pi\cdot R\cdot\sqrt{R^2+H^2}$

$V = \frac{1}{3}\cdot\pi\cdot R^2\cdot H + \pi\cdot R^2 L + \frac{1}{3}\cdot\pi\cdot R^2\cdot H$

$\begin{pmatrix} H \\ L \end{pmatrix} := \text{Find}(H, L) =$ No solution

FIGURE 32.6 Attempt to numerically solve an overdetermined system of equations.

$$\approx(a,b) := \left|\frac{a-b}{a+b}\right| < 0.00001 \qquad Or(a) := max(a)$$

$$n := 1000 \quad H_1 := 29m \quad H_2 := 31m \quad \Delta H := \frac{H_2 - H_1}{n}$$

$$L_1 := 9m \quad L_2 := 10.5m \quad \Delta L := \frac{L_2 - L_1}{n}$$

```
(vL, vH) := | i ← ORIGIN
            | for L ∈ L1, L1 + ΔL .. L2
            |   for H ∈ H1, H1 + ΔH .. H2
            |     if Or[ ΣL ≈ (H + (S − 2·π·R·√(H² + R²))/(2·π·R) + H)
            |            V ≈ (1/3·π·R²·H + π·R²L + 1/3·π·R²·H)
            |            S ≈ (π·R·√(R² + H²) + 2π·RL + π·R·√(R² + H²)) ]
            |       | vL_i ← L
            |       | vH_i ← H
            |       | i ← i + 1
            | (vL, vH)
```

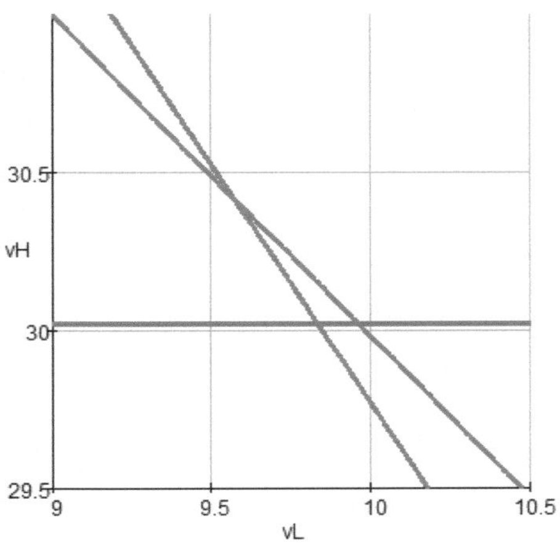

30.5, vH, 30, 29.5, 9, 9.5, 10, 10.5, vL

FIGURE 32.7 An attempt to graphically solve an overdetermined system of equations.

$$S = \pi\cdot R\cdot\sqrt{R^2+H^2} + 2\pi\cdot RL + \pi\cdot R\cdot\sqrt{R^2+H^2} \text{ solve}, L \rightarrow \frac{S - 2\cdot\pi\cdot R\cdot\sqrt{H^2+R^2}}{2\cdot\pi\cdot R}$$

$$\Sigma L := 70m \quad R := 4m \quad S := 1011.45m^2 \quad V := 1500.2m^3$$

Given

$$H := 70m$$

$$\Sigma L = H + \frac{S - 2\cdot\pi\cdot R\cdot\sqrt{H^2+R^2}}{2\cdot\pi\cdot R} + H$$

$$V = \frac{1}{3}\cdot\pi\cdot R^2\cdot H + \pi\cdot R^2\frac{S - 2\cdot\pi\cdot R\cdot\sqrt{H^2+R^2}}{2\cdot\pi\cdot R} + \frac{1}{3}\cdot\pi\cdot R^2\cdot H$$

$$H := MinErr(H) = 30.41\,m$$

$$L := \frac{S - 2\cdot\pi\cdot R\cdot\sqrt{H^2+R^2}}{2\cdot\pi\cdot R} = 9.572\,m$$

Verify

$$\frac{1}{3}\cdot\pi\cdot R^2\cdot H + \pi\cdot R^2 L + \frac{1}{3}\cdot\pi\cdot R^2\cdot H = 1500.202\cdot m^3$$

$$\pi\cdot R\cdot\sqrt{R^2+H^2} + 2\pi\cdot RL + \pi\cdot R\cdot\sqrt{R^2+H^2} = 1011.45\,m^2$$

$$H + L + H = 70.393\,m$$

FIGURE 32.8 Hybrid solution of the overdetermined system of equations.

after Verify turned out to be quite acceptable—corresponding to the figures given in the novel and falling into the triangle shown in Figure 32.7.

In the calculation in Figure 32.8, analytical (symbolic) and numerical methods for solving the problem are used in a *hybrid* way (see Study 25): first, the boat surface area equation is solved *symbolically* with respect to the variable L, and then the resulting expression is substituted into the system of equations, which is solved numerically based on the first assumption. This technique allows us to reduce the number of equations from three to two, and the number of unknowns from two to one. This speeds up the calculation, but it makes more difficult to understand in the educational process.

Therefore, it is possible to leave three equations, with which everything is more or less clear.

ANIMATION

Mathcad has *animation* tools (see Study 8). Figure 32.9 shows a frame from such an animation of the changing of the contours in the *Nautilus* submarine (two triangles respectively for the bow and stern cones and one rectangle for the body cylinder) when the radius of its cylindrical part varies. At the bottom of the animation frame is a drawing downloaded from the Internet and illustrating an attempt at a real design of a submarine. By the way, on the same website there are many drawings illustrating the novel by Jules Verne, but by which it is rather difficult to judge the size of the *Nautilus*.

This animation can be seen on the website. https://community.ptc.com/t5/PTC-Mathcad-Questions/Symbol-math-error/td-p/487278. You can also download the corresponding calculation and other files of the example.

OPTIMIZATION

Optimization tools are embedded in Mathcad, and they could also be tested on the *Nautilus* submarine problem, see Figures 32.10 and 32.11.

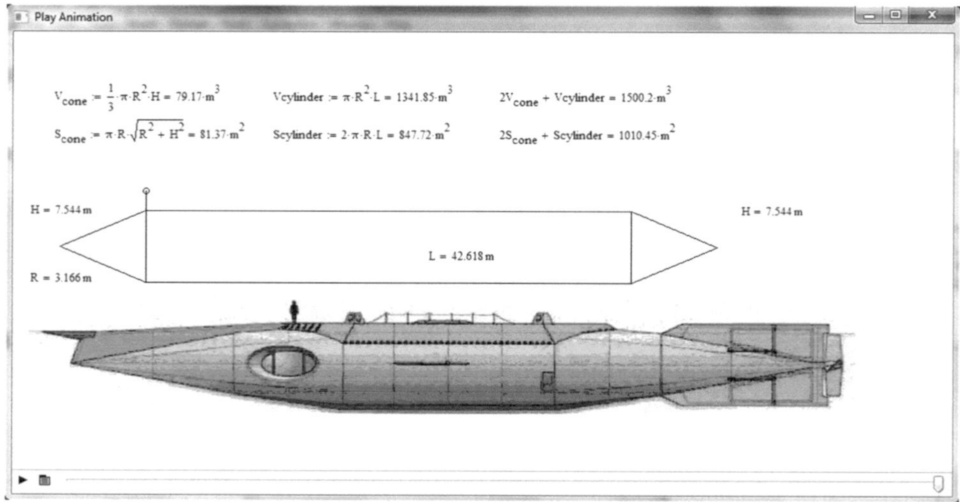

FIGURE 32.9 Animation of changing the contours of a submarine, depending on the radius of its cylindrical part.

Figure 32.10 shows how the size of the boat will change if its volume is set to $V=1500.2$ m^3, and the surface area S should be minimal. The solution involved the *Minimize* function built into Mathcad, which minimizes the given objective function S by changing the values of the arguments R, H and L from the point of the initial assumption ($R=H=L=10$ m) and fulfilling the volume constraints.

The bottom of Figure 32.10 shows a contour optimized for the outer surface of the boat, which became similar to a bathyscaphe with an internal volume of 1,500.2 m^3 and with an outer minimized area of 657.762 m^2. It is known that a sphere has the smallest surface area for a given volume. Therefore, the contour of the underwater vehicle, shown in Figure 32.10, approached a circle—the cross section of the sphere. The sphere, by the way, has the best strength properties for submersible vehicles, whose habitable compartments are made exactly in the form of a sphere (bathysphere).

In Figure 32.11, the problem of *maximizing* the volume V of a submarine with a given surface area $S=1011.45$ m^2 is solved. It also involves hybrid moments (using symbolic and numerical mathematics, see Figure 32.8), which reduce the number of optimizing parameters from three (Figure 32.10) to two (Figure 32.11). This technique made it possible to verify the correctness of the calculation of the maximum boat volume ($V=2860.642$ m^3) graphically—by the level lines of the function of two arguments by fixing the found optimal value.

The proportions of the geometric body (cone-cylinder-cone), whose contour is shown in Figure 32.10, could be found not only numerically, but symbolically by displaying in the answer not specific values in meters, but the proportions in the form of algebraic expressions. The author assigns this homework to the reader.

IMPLICIT 3D FUNCTIONS

Figure 32.12 shows the surface of an implicit function no longer of two, but of three arguments, which can also be associated with Jules Verne novel. If we assume that the lengths of the conic parts of the *Nautilus* submarine (its bow and stern) are different (we have not one variable H, but two variables H_1 and H_2), then we can construct a closed surface in Maple (Figure 32.12), which resembles a certain futuristic submarine, and a certain aerial flying device in the form of a delta wing. One of the sections of this aeronautical apparatus is shown in Figure 32.2 (Maple) and 32.3 (Mathcad). We have indeed some kind of classical section of the wing, which shows the flow of air and lifting force, which can be calculated based on the laws of gas dynamics. In this contour, you can also see the silhouette of a certain biological object—a whale, for example—a living host of the ocean depths.

Note On the closed surface shown in Figure 32.12, one can see certain flat triangles and rhombuses from which the surface is composed and which "give out" the essence of this numerical graphic operation. These flat elements will be clearly visible if the options of the *implicitplot3d* function change the value of the

$$V \equiv 1500.2m^3$$

$$S(R,H,L) := \pi \cdot R \cdot \sqrt{R^2 + H^2} + 2\pi \cdot RL + \pi \cdot R \cdot \sqrt{R^2 + H^2}$$

Given

$$\begin{pmatrix} R \\ H \\ L \end{pmatrix} := \begin{pmatrix} 10 \\ 10 \\ 10 \end{pmatrix} m$$

$$V = \frac{1}{3} \cdot \pi \cdot R^2 \cdot H + \pi \cdot R^2 L + \frac{1}{3} \cdot \pi \cdot R^2 \cdot H$$

$$\begin{pmatrix} R \\ H \\ L \end{pmatrix} := \text{Minimize}(S,R,H,L) = \begin{pmatrix} 6.842 \\ 6.126 \\ 6.117 \end{pmatrix} m$$

Verify

$$\frac{1}{3} \cdot \pi \cdot R^2 \cdot H + \pi \cdot R^2 L + \frac{1}{3} \cdot \pi \cdot R^2 \cdot H = 1500.199 \cdot m^3$$

$$\pi \cdot R \cdot \sqrt{R^2 + H^2} + 2\pi \cdot RL + \pi \cdot R \cdot \sqrt{R^2 + H^2} = 657.762 m^2$$

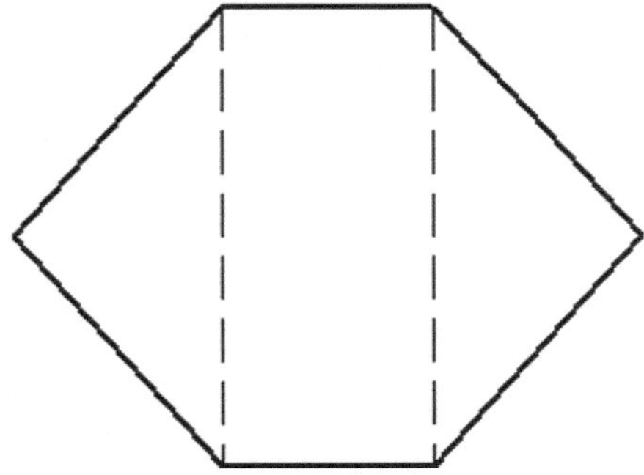

FIGURE 32.10 Minimization of the outer surface of the submarine.

$$S = \pi \cdot R \cdot \sqrt{R^2 + H^2} + 2\pi \cdot RL + \pi \cdot R \cdot \sqrt{R^2 + H^2} \text{ solve}, L \rightarrow \frac{S - 2 \cdot \pi \cdot R \cdot \sqrt{H^2 + R^2}}{2 \cdot \pi \cdot R}$$

$$S := 1011.45m^2$$

$$V(R,H) := \frac{1}{3} \cdot \pi \cdot R^2 \cdot H + \pi \cdot R^2 \cdot \frac{S - 2 \cdot \pi \cdot R \cdot \sqrt{H^2 + R^2}}{2 \cdot \pi \cdot R} + \frac{1}{3} \cdot \pi \cdot R^2 \cdot H$$

Given

$$\begin{pmatrix} R \\ H \end{pmatrix} := \begin{pmatrix} 10 \\ 10 \end{pmatrix} m$$

$$S = \pi \cdot R \cdot \sqrt{R^2 + H^2} + 2\pi \cdot R \cdot \frac{S - 2 \cdot \pi \cdot R \cdot \sqrt{H^2 + R^2}}{2 \cdot \pi \cdot R} + \pi \cdot R \cdot \sqrt{R^2 + H^2}$$

$$\begin{pmatrix} R \\ H \end{pmatrix} := \text{Maximize}(V,R,H) = \begin{pmatrix} 8.485 \\ 7.589 \end{pmatrix} m$$

$$L := \frac{S - 2 \cdot \pi \cdot R \cdot \sqrt{H^2 + R^2}}{2 \cdot \pi \cdot R} = 7.589 m$$

Verify

$$\frac{1}{3} \cdot \pi \cdot R^2 \cdot H + \pi \cdot R^2 L + \frac{1}{3} \cdot \pi \cdot R^2 \cdot H = 2860.638 \cdot m^3$$

$$\pi \cdot R \cdot \sqrt{R^2 + H^2} + 2\pi \cdot RL + \pi \cdot R \cdot \sqrt{R^2 + H^2} = 1011.45 m^2$$

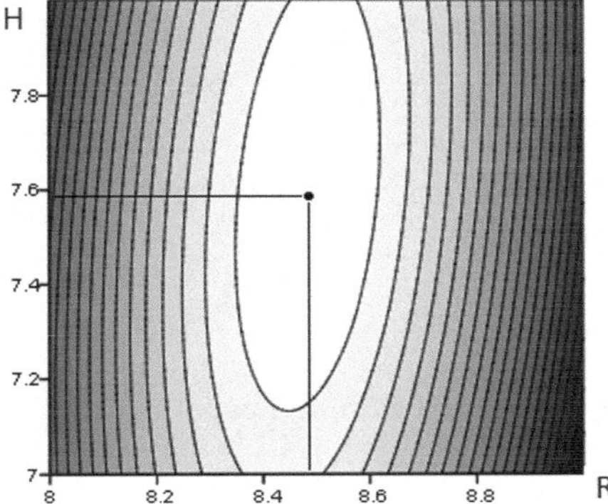

FIGURE 32.11 Maximizing the volume of the submarine.

numpoints system variable from 10,000 to 1000 or even 100. If this value is increased to 1,000,000, for example, a smooth surface will be built after a few minutes by the computer. Many users here start to think that the computer is "frozen", and overload it. And other users understand the essence of the matter and go and work on a more powerful computer—on a graphic station.

At https://community.ptc.com/t5/PTC-Mathcad-Questions/3D-body-plot/td-p/487467, it is possible to see attempts to build three-dimensional models of the *Nautilus* submarine in Mathcad using simple scanning or more complex parametric graphic tools.

TO SINK AND TO FLOAT

In the dialogues of Captain Nemo with Professor Aronnax they also touch on technical issues about submarine submersion and its ascent. Here it is an excerpt from the novel:

> The double skin of the ship is made of sheet steel, the specific weight of which is seven and eight tenths. The thickness of the outer skin is at least five centimeters, the weight is three hundred ninety-four and ninety-six hundredths of a ton.
>
> (Ces deux coques sont fabriquées en tôle d'acier dont la densité par rapport à l'eau est de sept, huit dixièmes. La première n'a pas moins de cinq centimètres d'épaisseur, et pèse trois

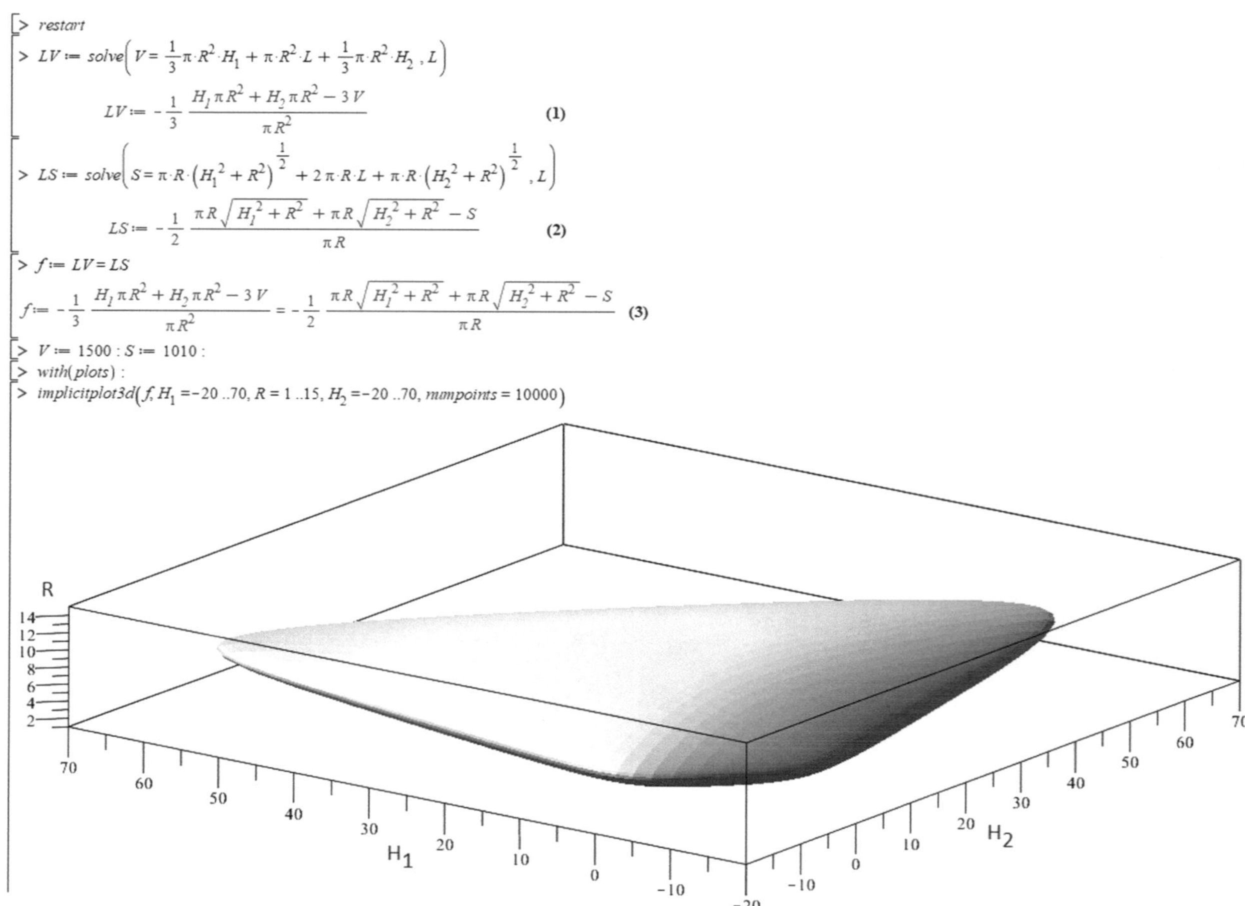

FIGURE 32.12 Plotting an implicit function of three arguments in Maple.

cent quatre-vingt-quatorze tonneaux quatre-vingt-seize centièmes.)

The inner skin, the keel—is fifty centimeters high and twenty-five centimeters wide, weighing sixty-two tons, machines, ballast and other equipment, furniture, internal bulkheads and pillars—all this weighs nine hundred sixty-one and sixty-two hundredths. Thus, the total weight of the vessel is one thousand three hundred and fifty-six and forty-eight hundredths of a ton

(La seconde enveloppe, la quille, haute de cinquante centimètres et large de vingt-cinq, pesant, à elle seule, soixante-deux tonneaux, la machine, le lest, les divers accessoires et aménagements, les cloisons et les étrésillons intérieurs, ont un poids de neuf cent soixante et un tonneaux soixante-deux centièmes, qui, ajoutés aux trois cent quatre-vingt-quatorze tonneaux et quatre-vingt-seize centièmes, forment le total exigé de treize cent cinquante-six tonneaux et quarante-huit centièmes.) …

[…] … being on the ocean surface, the *Nautilus*, under these conditions, is one-tenth above the water surface.

(lorsque le *Nautilus* se trouve à flot dans ces conditions, il émerge d'un dixième)

Therefore, in order to fully immerse the *Nautilus* in water, it is necessary to have tanks with a capacity equal to this tenth of its volume, in other words, capable of holding a hundred and fifty and seventy-two hundredths of a ton of water. In the latter case, the weight of the vessel would be one thousand five hundred and seven tons, and it would completely go under water. So it is in reality, Mr. Professor! Such tanks are available in the *Nautilus* hold. It is necessary to open the taps, as they are filled with water, and the ship plunges into the sea to the level with the surface of the water!

(Donc, reprit le capitaine, lorsque le *Nautilus* se trouve à flot dans ces conditions, il émerge d'un dixième. Or, si j'ai disposé des réservoirs d'une capacité égale à ce dixième, soit d'une contenance de cent cinquante tonneaux et soixante-douze centièmes, et si je les remplis d'eau, le bateau déplaçant alors quinze cent sept tonneaux, ou les pesant, sera complètement immergé. C'est ce qui arrive, monsieur le professeur. Ces réservoirs existent en abord dans les parties inférieures du *Nautilus*. J'ouvre des robinets, ils se remplissent, et le bateau s'enfonçant vient affleurer la surface de l'eau.)

Let's model the dynamics of immersion in water and the subsequent ascent to its surface. The submarine *Nautilus* had to do it quite often. In fact the author of the novel did not foresee the presence of air regeneration systems on board—for the removal of carbon dioxide and water vapor from the boat when a person breathes and for the oxygen enrichment of air.* Therefore, the ship had to float to the surface about every five days to ventilate the rooms. And there, judging by the text of the novel and his illustrations, Captain Nemo and Professor Aronnax also smoked expensive cigars, discussing between the puffs the design features of the boat and other engineering, as well as political issues.

Note Here it is possible to touch upon chemistry in the classroom as part of the educational technology STEAM (STREAM, SAR). In particular, describe the reactions and technologies for generating oxygen from water and from carbon dioxide in their thermodynamics and kinematics. These chemical technologies began to develop rapidly in more recent days: near-Earth stations cannot dive "every five days" into the atmosphere of the earth in order to "take a breath of air".

But back to the sink and float. Figure 32.13 shows the Mathcad calculation of such a hydrostatic and dynamic model. From a height of ten meters a diver jumps into the water. Bomb-style jumps: a diver's model is a sphere of radius $r = 30$ cm and density $\rho_{man} = 950$ kg/m^3. From these two numbers, the volume of the "diver", its mass and sectional area are calculated. A step function $\rho(h)$ is introduced—the dependence of the density of the medium on the diver's location: for $h > 0$ is air, for $h \le 0$ is water. The point of the origin is at the foot of a ten-meter tower at the water level.

Three forces act on the "diver": the force of gravity, the force of resistance of the environment and the Archimedean force. The force of resistance of the medium (the force of friction) is projected horizontally and vertically, while the other two forces act only vertically. The sums of all forces along the two axes are equal to the product of mass and acceleration, or rather, on the projection of the acceleration vertically and horizontally. In the Mathcad solver block, the initial conditions are also imputed between the keyword *Given* and the function *Odesolve* (the initial location of the diver and its initial speed also in the two directions, with the speed given by the first time derivative of the path).

Figures 32.14 and 32.15 show a graphical illustration of the numerical solution of the problem of diving into

$$r := 30\text{cm} \qquad \rho_{man} := 950\frac{\text{kg}}{\text{m}^3} \qquad \text{volume} := \frac{4}{3}\cdot\pi\cdot r^3 = 0.113\cdot\text{m}^3$$

$$\text{mass} := \text{volume}\cdot\rho_{man} = 107.442\cdot\text{kg} \qquad \text{area} := \pi\cdot r^2 = 0.283\cdot\text{m}^2$$

$$h := 10\text{m} \qquad y_0 := 0\text{m} \qquad vx_0 := 3\frac{\text{m}}{\text{s}} \qquad vy_0 := 1\frac{\text{m}}{\text{s}}$$

$$\rho(h) := \text{if}\left(h > 0, 1.225\frac{\text{kg}}{\text{m}^3}, 1000\frac{\text{kg}}{\text{m}^3}\right) \qquad k := 1 \qquad t_{end} := 30\text{s}$$

Given

$$vx(t) = x'(t) \qquad vy(t) = y'(t) \qquad vx(0\text{s}) = vx_0 \qquad vy(0\text{s}) = vy_0$$

$$x(0\text{s}) = 0 \qquad y(0\text{s}) = h$$

$$\text{mass}\cdot vx'(t) = -k\cdot\rho(y(t))\cdot\text{area}\cdot vx(t)^2\cdot\text{sign}(vx(t))$$

$$\text{mass}\cdot vy'(t) = g\cdot\text{volume}\cdot\rho(y(t)) - k\cdot\rho(y(t))\cdot\text{area}\cdot vy(t)^2\cdot\text{sign}(vy(t)) - g\cdot\text{mass}$$

$$\begin{pmatrix} x \\ vx \\ y \\ vy \end{pmatrix} := \text{Odesolve}\left[\begin{pmatrix} x \\ vx \\ y \\ vy \end{pmatrix}, t, t_{end}\right] \qquad t := 0\text{s}, \frac{t_{end}}{10000} .. t_{end}$$

FIGURE 32.13 Numerical solution of the system of differential equations of the immersion into water and the ascent to the surface.

* One of the problems faced by engineers and scientists who docked the Soviet spacecraft *Soyuz* and the American spacecraft *Apollo* was the following. The Russian habitable compartment of the spacecraft was filled with air (a mixture of nitrogen and oxygen), while the American, with pure oxygen. These two technical solutions have their pros and cons. In particular, working with pure oxygen is a greater fire hazard, but requires simpler and more compact regeneration facilities. From our days in the International Space Station (ISS), several technologies are used to guarantee the correct atmosphere for the crew. Most notably oxygen is produced by water electrolysis, with concurrent hydrogen production and disposal. Electrical energy is provided by solar panels, while water is recovered from the atmosphere and "crew liquid waste". Of course in the deep sea, it is easier to get water, while energy requires a back-up system, such as diesel, battery.

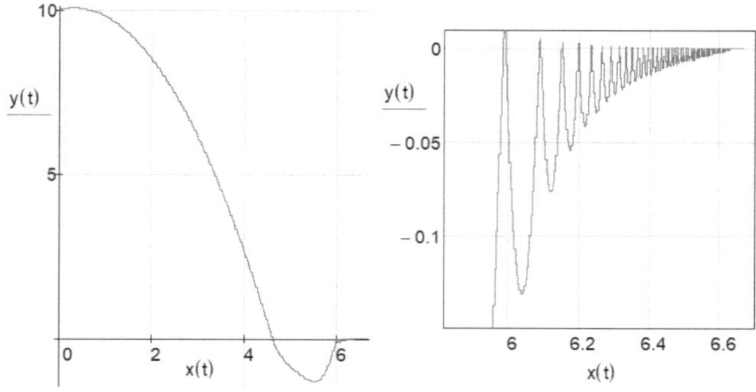

FIGURE 32.14 Diver trajectory above and below water.

water. The right graph in Figure 32.14 and its alternative display in Figure 32.15 are interesting (in form and content), the latter having the form of a certain *attractor* (a compact set of phase space): a spiral trajectory tending to the point when time tends to infinity. The animation in Figures 32.14 and 32.15 can be seen at https://community.ptc.com/t5/PTC-Mathcad-Blog/Diver/ba-p/488389. These attractors are characteristic of a dissipative system.

DENSITY OF SEAWATER

In Jules Verne's novel, there are dialogues with discussions on the change in the density of water with increasing pressure at depth, on the compressibility of water. For example:

> Or, si l'eau n'est pas absolument incompressible, elle est, du moins, très peu compressible.

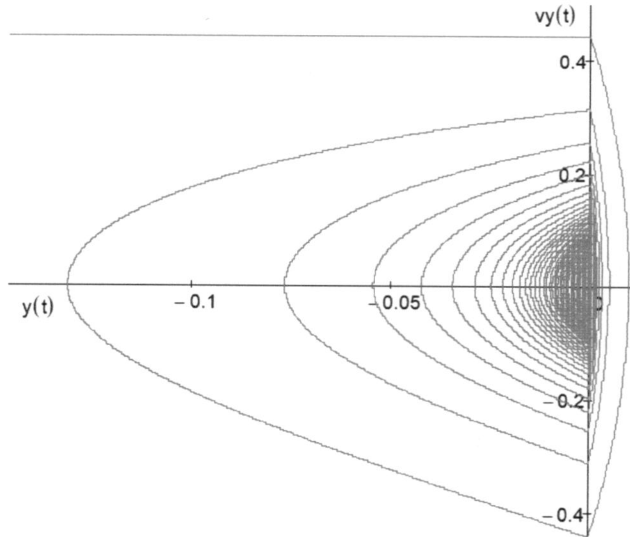

FIGURE 32.15 Diver attractor.

> En effet, d'après les calculs les plus récents, cette réduction n'est que de quatre cent trente-six dix millionièmes par atmosphère, ou par chaque trente pieds de profondeur. S'agit-il d'aller à mille mètres, je tiens compte alors de la réduction du volume sous une pression équivalente à celle d'une colonne d'eau de mille mètres, c'est-à-dire sous une pression de cent atmosphères. Cette réduction sera alors de quatre cent trente-six cent millièmes. Je devrai donc accroître le poids de façon à peser quinze cent treize tonneaux soixante-dix-sept centièmes, au lieu de quinze cent sept tonneaux deux dixièmes. L'augmentation ne sera conséquemment que de six tonneaux cinquante-sept centièmes.

In Study 23 in Figure 23.3, a website was shown, which we can calculate the density of sea water by its salinity, temperature and pressure. It is also possible to insert the corresponding cloud function into our calculation by insert a reference (a link) to the address indicated at the top of Figure 32.16. A function that returns the density of sea water depending on its salinity, temperature, and pressure will be visible in the calculation and can be used to plot a set of curves, for example.

If the user Mathcad clicks on the link shown in Figure 32.16, the corresponding calculation will be opened with the algorithm shown in Figure 23.2.

LEAGUES, MILES, KILOMETERS, METERS

Another dynamic model and its computer implementation will help us look into the very title of the novel that contains a unit of length.* In Old Russian translations of the novel, by the way, the title "80,000 kilometers (versts) under water" is found. But more "impressive" would be the name "80 million meters under water"! However, when they talk about meters, they usually mean not the distance travelled underwater, but the depth of the immersion.

Like the nautical mile, the league is tied to the length of the Earth meridian. A sea league is 5555.5 m (arc length is 1/20 degree of the Earth meridian: 3 nautical miles). And there were also land, postal and other leagues—different in various provinces of France! This

* And here is another famous science fiction novel, where there are units of measurement in the title: *Fahrenheit 451*. When translating it into Russian, there were also attempts to give to the novel a title according to the SI standard: "Celsius 233" or even "Kelvin 506".

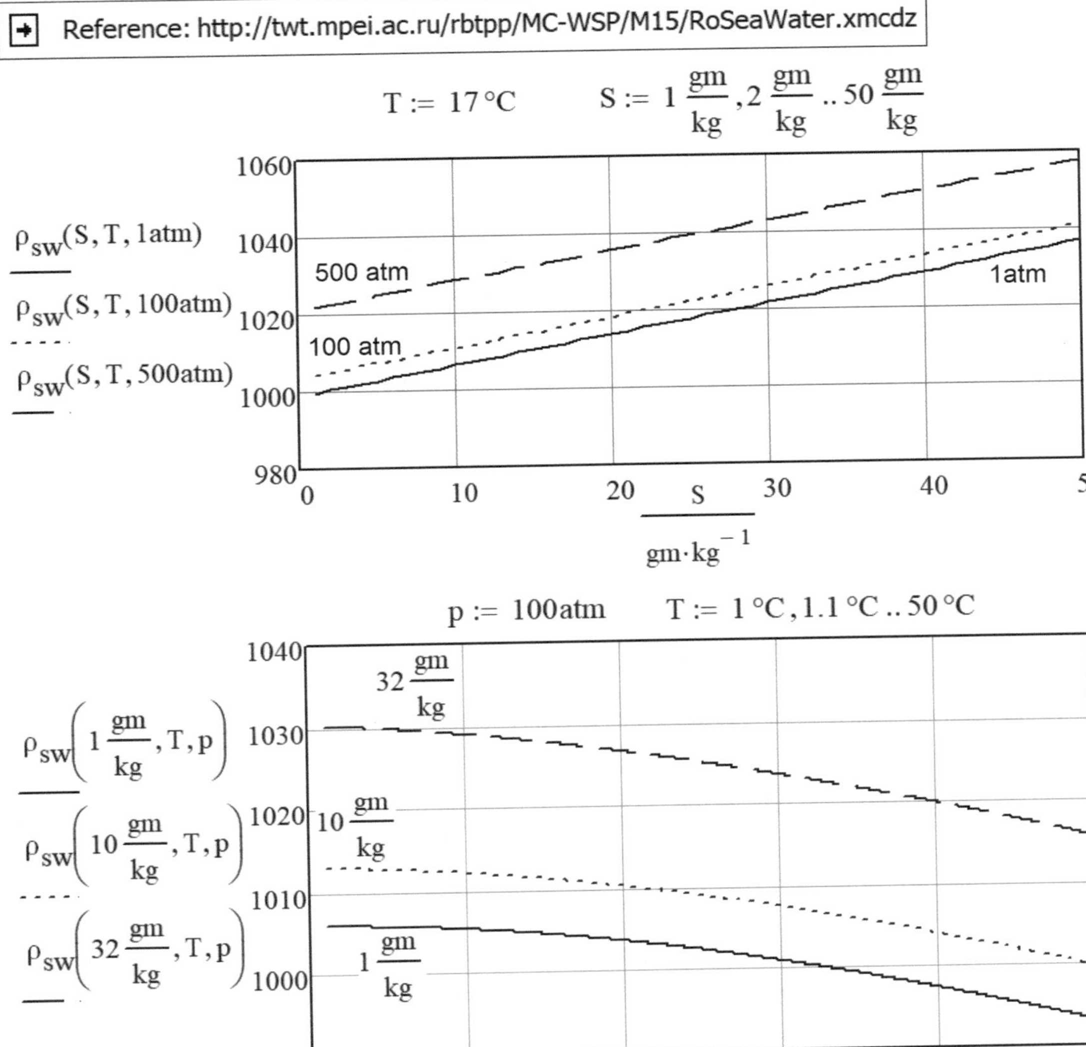

FIGURE 32.16 Using a cloud function that returns the density of sea water to build a sets of curves.

information is taken from Wikipedia. And what was meant by Jules Verne in giving the title of the novel!? They decided to eliminate all this confusion by introducing the SI system.

And where did the meter come from as the main unit of length in SI? Usually the meter is connected either with the standard stored in the International Bureau of Weights and Measures,* related to the length of the Earth meridian, or to the length of the light wave. But only a few people remember the origin of the meter: the pendulum of the floor clock. Hence the common root of these words as meter, metronome, metrology, isometry, planimetry (Figure 32.2), stereometry (Figure 32.12), etc. A pendulum with the length of one meter will make a swing in about one second. These swings through gears and other mechanical devices are translated into the motion of the clock hands. It can be assumed that the pendulum rake from the clock once attracted someone as a standard of length instead of steps, distances from different points of the human body, etc. Also, few people remember where the concept of a mathematical pendulum of one meter length and a two seconds period of oscillation of came from. Figures 32.17 and 32.18 explain this.

* An old "physical" anecdote. "What is a horsepower?" "This is the force that develops a horse one meter tall and weighing one kilogram." "Yes, where did you see such a horse!?" "And you won't see it that easily! It is kept in the Paris Chamber of measures and weights (International Bureau of Weights and Measures)!"

$$ode := m\frac{d^2}{dt^2}\alpha(t) = -m \cdot g \cdot \frac{\sin(\alpha(t))}{L}$$

$$m\left(\frac{d^2}{dt^2}\alpha(t)\right) = -\frac{m\, g \sin(\alpha(t))}{L} \quad \textbf{(1)}$$

$$ics := \alpha(0) = \alpha_0,\ D(\alpha)(0) = 0$$

$$\alpha(0) = \alpha_0,\ D(\alpha)(0) = 0 \quad \textbf{(2)}$$

$$dsolve(\{ode, ics\})$$

$$\alpha(t) = RootOf\left(\int_{\_Z}^{\alpha_0} \frac{L}{\sqrt{-2\, L \cos(\alpha_0)\, g + 2\, L \cos(\_a)\, g}}\, d\_a + t\right), \alpha(t) \quad \textbf{(3)}$$

$$= RootOf\left(\int_{\alpha_0}^{\_Z} \frac{L}{\sqrt{-2\, L \cos(\alpha_0)\, g + 2\, L \cos(\_a)\, g}}\, d\_a + t\right)$$

FIGURE 32.17 Solving the equation of the physical pendulum.

The mathematical pendulum is like the physical pendulum, but it deviates from the vertical by no more than 7 angular degrees (another "mystical", "religious" seven, see note on page 25). In the model of the physical pendulum, there is the sine of the angle in the balance equation of the acting forces. This leads to a rather complicated analytical solution of the corresponding differential equation, see Figure 32.17 (an absolutely accurate and absolutely useless answer). If the sine of the angle is replaced with the angle itself (and this, we repeat, can be done only at small angles*), then the solution will be simplified immediately and the well-known "school" formula for the period of the oscillation of the pendulum "pops up" in it, see Figure 32.18. If we substitute $L = 1$ m into this formula, or rather, into the formula for the period

* This technique is often used in geometrical optics (see Study 20).

$$ode := m\frac{d^2}{dt^2}\alpha(t) = -m \cdot g \cdot \frac{\alpha(t)}{L}$$

$$m\left(\frac{d^2}{dt^2}\alpha(t)\right) = -\frac{m\, g\, \alpha(t)}{L} \quad \textbf{(1)}$$

$$ics := \alpha(0) = \alpha_0,\ D(\alpha)(0) = 0$$

$$\alpha(0) = \alpha_0,\ D(\alpha)(0) = 0 \quad \textbf{(2)}$$

$$dsolve(\{ode, ics\})$$

$$\alpha(t) = \alpha_0 \cos\left(\frac{\sqrt{g}\, t}{\sqrt{L}}\right) \quad \textbf{(3)}$$

FIGURE 32.18 Solution of the equation of the mathematical pendulum.

of oscillation of the pendulum, then it will return those two seconds that are the basis of all the floor clocks with the pendulum. However, in this formula there is also the value of g, the acceleration of free fall, the value of which at least depends on the geographical latitude of the terrain and on the height above sea level. Therefore, over time, they abandoned the pendulum to determine the standard of length: the meter was then associated with the length of the Earth's equator, and then with a light wave of a given frequency.

A BIT OF POLITICS AND OF ECONOMICS

In Wikipedia it is possible to read the following comments about Jules Verne's *20,000 Leagues Under the Sea*.

Jules Verne, while working on the novel, was impressed by the Polish uprising of 1863. Initially, Captain Nemo was a Polish aristocrat fighting against Russian oppressors who destroyed his entire family. However, under pressure of his publisher Pierre-Jules Hetzel, who was moved not only by the political situation, but also by the common sense that a Polish aristocrat simply did not have enough money to build such an underwater ship, Verne "deprived" Nemo of any nationality, which remained a fact unknown to Aronnax and his companions. In the story only a portrait of the revolutionary Tadeusz Kościuszko, which is, among others, in the captain's cabin hints at his "Polish origin". Later, in the novel *The Mysterious Island*, Verne will reveal Captain Nemo incognito and present him as a runaway Indian prince Nana Sahib, who fled to sea after the failure of the uprising against the vengeful England, which enslaved his homeland.

The political conjuncture consisted in the fact that France, on the eve of the Franco-Prussian† war (the War of 1870), was actively seeking support from Russia. (The novel was originally serialized from March 1869 through to June 1870 in Pierre-Jules Hetzel's periodical, the *Magasin d'Éducation et de Récréation*). However, Russia was strongly offended by France after the defeat in the Crimean War (1853–1856). And this (the neutrality of Russia) largely determined the defeat of France in the war with Prussia and the loss of Alsace and Lorraine. Before the First World War, France managed to enlist the

† As we have already mentioned, the French-Prussian war has some connections also with Dostoievsky's novel *The Gambler*, which was set in (south, Roulettemberg) Germany several years before the war (see Study 6). After the war in Germany, the Gold Standard and the Gold Mark were introduced, in relation to the spoils of war and the taking of Alsace and Lorraine, while we examined the many currency conversion in the novel.

$$m_0 := 12800\text{kg} \quad F := 27000\text{kgf} \quad m_f := 8500\text{kg} \quad q := 125\frac{\text{kg}}{\text{s}} \quad t_{end} := \frac{m_f}{q} = 68\text{s}$$

$$k_f := 1.5\cdot\frac{\text{kgf}}{\text{kgf}} \quad S := \frac{(1\text{m})^2}{4}\cdot\pi = 0.785\cdot\text{m}^2 \quad \Delta t := 100\text{s}$$

Reference:http://twt.mpei.ru/tthb/g-h-psi.xmcdz   Reference:http://twt.mpei.ru/tthb/Ro-air.xmcdz

$$g(h) := g(h, 55\text{deg}) \qquad \rho_{air}(10\text{km}) = 0.374\cdot\frac{\text{kg}}{\text{m}^3}$$

Given

$$h(0\text{s}) = -1000\text{m} \quad v(0\text{s}) = 0\frac{\text{m}}{\text{s}} \quad v(t) = \frac{d}{dt}h(t)$$

$$\left(m_0 - \text{if}\left(t < t_{end}, q\cdot t, m_f\right)\right)\cdot\frac{d}{dt}v(t) \ldots = \text{if}\left(t < t_{end}, F, 0\right)$$
$$+ \left(m_0 - \text{if}\left(t < t_{end}, q\cdot t, m_f\right)\right)\cdot g(h(t)) \ldots$$
$$+ k_f\cdot S\cdot\frac{v(t)^2\cdot\text{sign}(v(t))}{2}\cdot\text{if}\left(h(t) < 0, 1000\frac{\text{kg}}{\text{m}^3}, \rho_{air}(h(t))\right)$$

$$\begin{pmatrix} h \\ v \end{pmatrix} := \text{Odesolve}\left[\begin{pmatrix} h \\ v \end{pmatrix}, t, t_{end} + \Delta t\right]$$

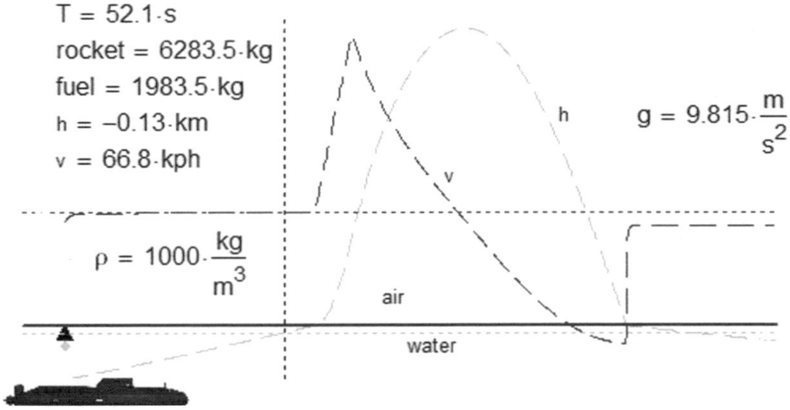

T = 52.1·s
rocket = 6283.5·kg
fuel = 1983.5·kg
h = −0.13·km
v = 66.8·kph
$g = 9.815\cdot\frac{\text{m}}{\text{s}^2}$
$\rho = 1000\cdot\frac{\text{kg}}{\text{m}^3}$
h
v
air
water

FIGURE 32.19 Flight model of a rocket launched from a submarine.

support of Russia. If it were not for Russia, the German troops would have once again marched through Paris in 1914, as they did in 1870 and at the beginning of the Second World War.

A similar ebb and flow in the relations between Russia and the West continued in the twenty-first century. In this regard, to our story in the style of STEAM, you can add history and political science. But many people rightly consider these sections of knowledge not as sciences, but as certain syntheses of science, art and even religion. So, here, let's talk not about the STEAM style, but about the STREAM style (see above).

Anyway, ending our conversation about the novel within the framework of STEM (STEAM/STREAM) educational technologies, we will consider another "dynamic" task related to the phrase the Russian Tsar,* who ruled during Jules Verne's reign, used: "Russia has only two allies—the Army and the Navy!"

Let's talk about the Navy. Currently, submarines are, as a rule, purely military ships, some of which carry nuclear and non-nuclear warheads on board. Nuclear military forces are the main tool in modern political games!† Figure 32.19 shows a model of a missile flight from a missile submarine at a depth of 1000 m. The mass

* Alexander III: by the way, according to Wiki, during Alexander's reign Russia fought no major wars, and he was therefore styled "The Peacemaker" (Wiki)

† True, some military analysts believe that in the near future the nuclear weapon awaits the fate of, for example, the cavalry, which was nullified by the use of machine guns. Anti-missile systems can greatly reduce the effectiveness of nuclear missiles.

of the rocket is 12,800 kg (variable m_0); engine fuel 27,000 kg-force (F); and fuel is consumed at a rate of 125 kg/s (q). The resistance coefficient of the environment is 1.5 (K_f) and the cross-sectional area of the rocket is 0.785 m^2 (S). On the graph in Figure 32.19, you can see the curves showing the height of the rocket and its speed.

"CLOUDY" AIR DENSITY

The calculation shown in Figure 32.19, takes into account the change in air density with the altitude above the sea level. This functional dependence is extracted from the Internet, not as a textual formula, but as a *cloud function*. The task shown in Figure 32.20 explains what it is: the flight of a parachutist is simulated from an altitude of 30 kilometers, while the parachute opens at an altitude of 1 kilometer. If you make reference to a Mathcad-sheet stored in the "cloud" (at http//:twt.mpei.ac.ru/tthb/Gas.xmcd), then the function with the name ρ_{Air} will become visible and will return the density of atmospheric air depending on the altitude above the sea level. At the beginning of Figure 32.20, this function is shown using units of length and density. However, Mathcad 15 cannot work with units when solving differential equations. This flaw has been eliminated in Mathcad Prime. In the calculation in Figure 32.15, we admit, we imitated units of measurement. How? See Figure 32.20.

Referring also to Figure 32.20, the elements of a peculiar hybridity are used—in the upper part of the calculation, the units of measurement are used, but not in the lower part: an ersatz meter, ersatz-kilogram and ersatz-second are entered there. Otherwise, we repeat, it would be impossible to solve differential equations. Note. Mathcad (not Prime) had the function "UnitsOf()" that was used to properly non-dimensionalize ordinary differential equations (ODEs) for the solve block. This eliminated the errors introduced by improperly doing it manually.

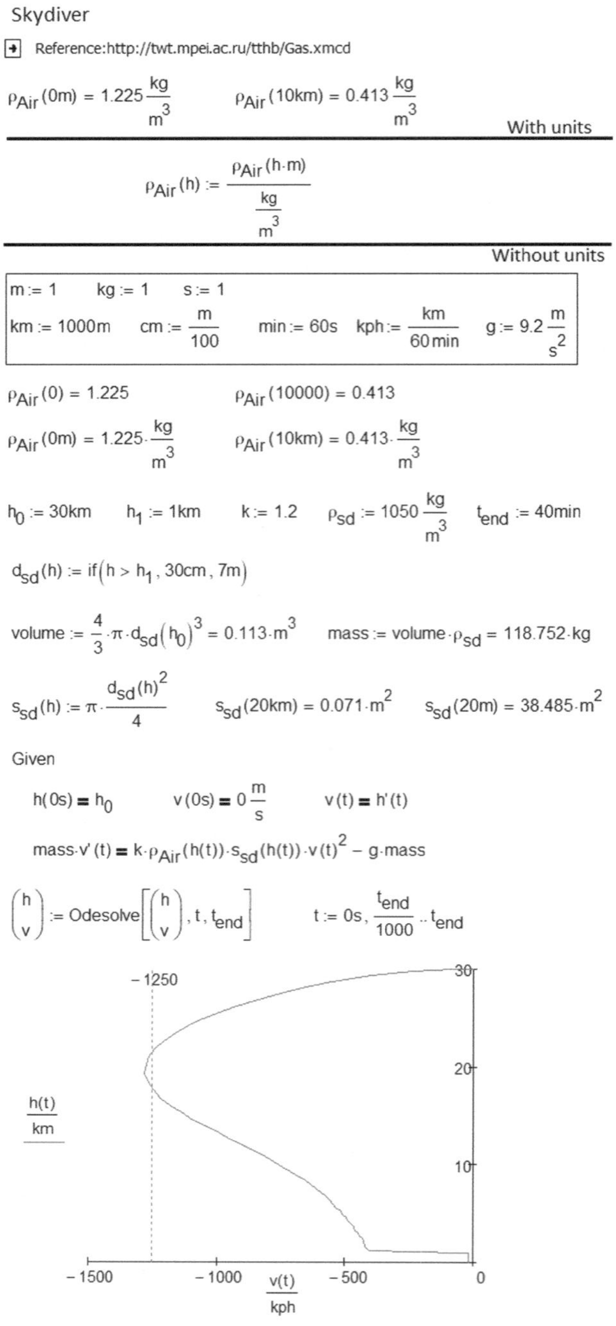

Skydiver

Reference:http://twt.mpei.ac.ru/tthb/Gas.xmcd

$\rho_{Air}(0m) = 1.225\frac{kg}{m^3}$ $\rho_{Air}(10km) = 0.413\frac{kg}{m^3}$

With units

$$\rho_{Air}(h) := \frac{\rho_{Air}(h \cdot m)}{\frac{kg}{m^3}}$$

Without units

$m := 1$ $kg := 1$ $s := 1$

$km := 1000m$ $cm := \frac{m}{100}$ $min := 60s$ $kph := \frac{km}{60min}$ $g := 9.2\frac{m}{s^2}$

$\rho_{Air}(0) = 1.225$ $\rho_{Air}(10000) = 0.413$

$\rho_{Air}(0m) = 1.225\cdot\frac{kg}{m^3}$ $\rho_{Air}(10km) = 0.413\cdot\frac{kg}{m^3}$

$h_0 := 30km$ $h_1 := 1km$ $k := 1.2$ $\rho_{sd} := 1050\frac{kg}{m^3}$ $t_{end} := 40min$

$d_{sd}(h) := if(h > h_1, 30cm, 7m)$

$volume := \frac{4}{3}\cdot\pi\cdot d_{sd}(h_0)^3 = 0.113\cdot m^3$ $mass := volume\cdot\rho_{sd} = 118.752\cdot kg$

$S_{sd}(h) := \pi\cdot\frac{d_{sd}(h)^2}{4}$ $S_{sd}(20km) = 0.071\cdot m^2$ $S_{sd}(20m) = 38.485\cdot m^2$

Given

$h(0s) = h_0$ $v(0s) = 0\frac{m}{s}$ $v(t) = h'(t)$

$mass\cdot v'(t) = k\cdot\rho_{Air}(h(t))\cdot S_{sd}(h(t))\cdot v(t)^2 - g\cdot mass$

$\begin{pmatrix} h \\ v \end{pmatrix} := Odesolve\left[\begin{pmatrix} h \\ v \end{pmatrix}, t, t_{end}\right]$ $t := 0s, \frac{t_{end}}{1000}..t_{end}$

- 1250
30
20
$\frac{h(t)}{km}$
10
- 1500
- 1000
- 500
0
$\frac{v(t)}{kph}$

FIGURE 32.20 Paratrooper jump model.

COMPETITION "MR. NAUTILUS" OR METROLOGICAL MUSK OX

Let us imagine that Captain Nemo came up with the idea to hold the contest "Mr. Nautilus" among the crew of the submarine—to determine the best sailor. And to conduct it, relying not on subjective (expert assessments), but on objective criteria.

In order to do this, you can float to the surface (to ventilate the premises, for example, see above), give the command "All hands on deck", to line up the submariners and measure their weight and growth, see Figure 32.21.

Note The vectors (transposed matrices of one column) growth and weight can be sorted so that the sailors are lined up in order of height: on the right flank, the tallest, while on the left, the shortest. However, at the present time in many armies and fleets of the world, such an order of lining up is being abandoned, considering it to be discrimination against people on a certain basis. In addition, in the marching column, the shortest soldier has to "swallow the dust". In Figure 32.21, the sailors lined up in the order in which they jumped out of the "womb" of the submarine onto its deck.

The parameters of the sailors (weight-growth) were processed by the method of the least squares (deviations of a point from a straight line) according to a linear relationship (linear regression analysis). Two deviations were taken into account—according to weight (the red line in Figure 32.21) and according to growth (the blue line). Usually they are optimized here along the "vertical" (in this case the sailor weight). Then the sailors whose parameters are closest to these two lines are selected—for the weight (the tenth sailor, red dot on the graph) and for the growth (the ninth sailor, blue dot). Then a coin was thrown, and the two sailors "Mr. Nautilus" and "Vice Mr. Nautilus" were chosen.

Our "sailor beauty contest" highlights an interesting metrological problem, whose essence is as follows. The best in terms of growth and weight of a sailor can be determined not only along the "vertical" (by the deviation of the weight from the average value) and not only along the "horizontal" (by the deviation of growth), but also along the "diagonal" according to the deviation of the point from the straight line. This deviation (this value) is easy to measure, for example, with a compass in Figure 32.21. However, what are the dimensions of this value?

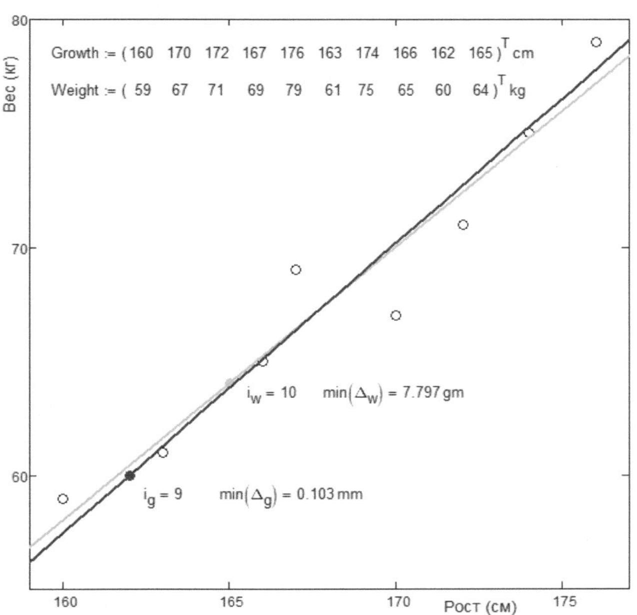

FIGURE 32.21 Parameters of the crew of the submarine.

There is a rare northern animal,* the musk ox. This is not a sheep and not a cow (or bull), but something completely different, though it has something in common with both these artiodactyla—with the sheep (thick fur) and with the cow (size). It can be assumed that our statistical task requires the use of a new unit of mass-length *kilogram-centimeter*. This is not the product of a kilogram per centimeter (as with the use of a kilowatt-hour unit for instance), but something more complicated. In the meantime, when solving the problem shown in Figure 32.22, pseudo-units of physical quantities were used: kg:= 1, m:= 1, cm:= m/100, etc. And this is not due to the flaws in Mathcad (we have already noted this, when solving the problem in Figure 32.20), but with the "flaws" of the theory of dimensional quantities: there is a purely dimensional quantity, but there is no unit of measurement! Let us recall the childish hooligan song: "**** is, but there is no word!". This approach to solve the problem of linear regression analysis of the *Nautilus* sailors will exclude a coin toss!

Figures 32.21 and 32.22 suggest an interesting idea for summarizing various beauty contests. Now there is too much subjectivity in them, which means insults, tears and even litigation. Most of these contests usually includes "beauty-women", in which it is quite difficult to choose the optimal, sorry, the most beautiful Ms. So, it is possible to measure weight and growth or other dimensions (the notorious 90–60–90, for example) of these finalists, to draw a line through the points and to choose the winner—see the filled circles in Figures 32.21 and 32.22.

The points shown in Figures 32.21 and 32.22, can be likened to reefs, through which the navigator *Nautilus* steers the course of the underwater vessel. Only the vertical axis does not indicate a length (as the horizontal axis), but a mass. This is the metrological mystery of the "map"!

However, at beauty contests there are stages where it is not the exterior data of the participants that are assessed, but their intellectual abilities, if, of course, these abilities are present.

* According to Wiki: "the modern musk ox is the last member of a line of ovibovines that first evolved in temperate regions of Asia and adapted to a cold tundra environment late in its evolutionary history. In modern times, (however) musk oxen were restricted to the Arctic areas of Northern Canada, Greenland, and Alaska. A team of Russian scientists led by Sergey Zimov aims to prove that musk oxen, along with other Pleistocene megafauna that survived into the early Holocene in northern Siberia, did not disappear from the region due to climate change, but because of human hunting."

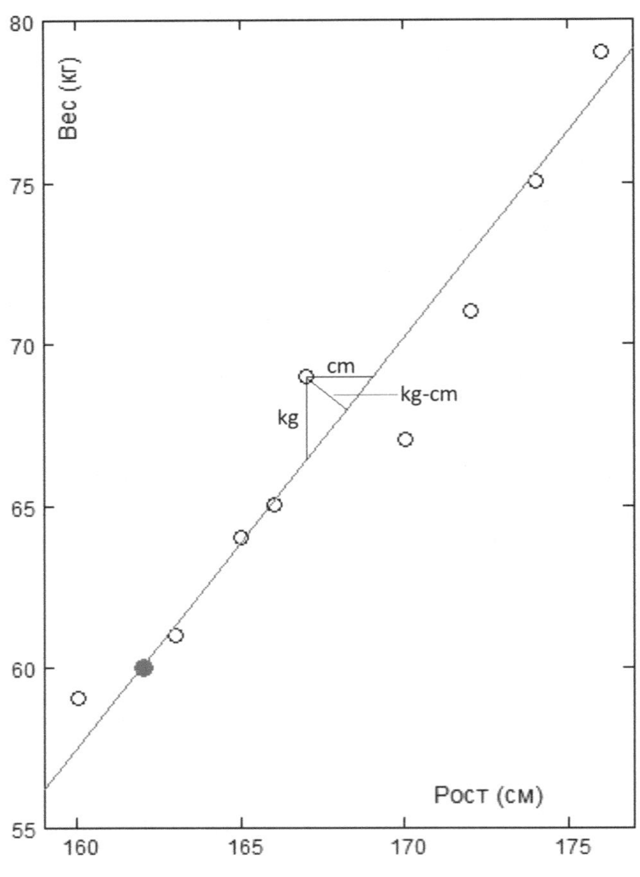

FIGURE 32.22 What is a musk ox, pardon, kilogram-centimeter.

EXIT/CONCLUSION

Currently, higher education can be obtained without attending a university. On the Internet there are many available resources of varying degrees of complexity for this purpose. This Study is, in its essence, a kind of educational material for such an independent student. And what is this occupation? In mathematics, computer science, physics, chemistry, literature, engineering graphics or history? Answer: this is a lesson in the framework of the STEM/STEAM/STREAM educational technology, not a "tearing" process of acquiring knowledge in certain traditional academic disciplines! In this case, student exams in mathematics and in other disciplines should be carried out on a computer with modern math programs and with Internet access.

Index